May 3–4, 2012
Salt Lake City, Utah, USA

I0047674

Association for
Computing Machinery

Advancing Computing as a Science & Profession

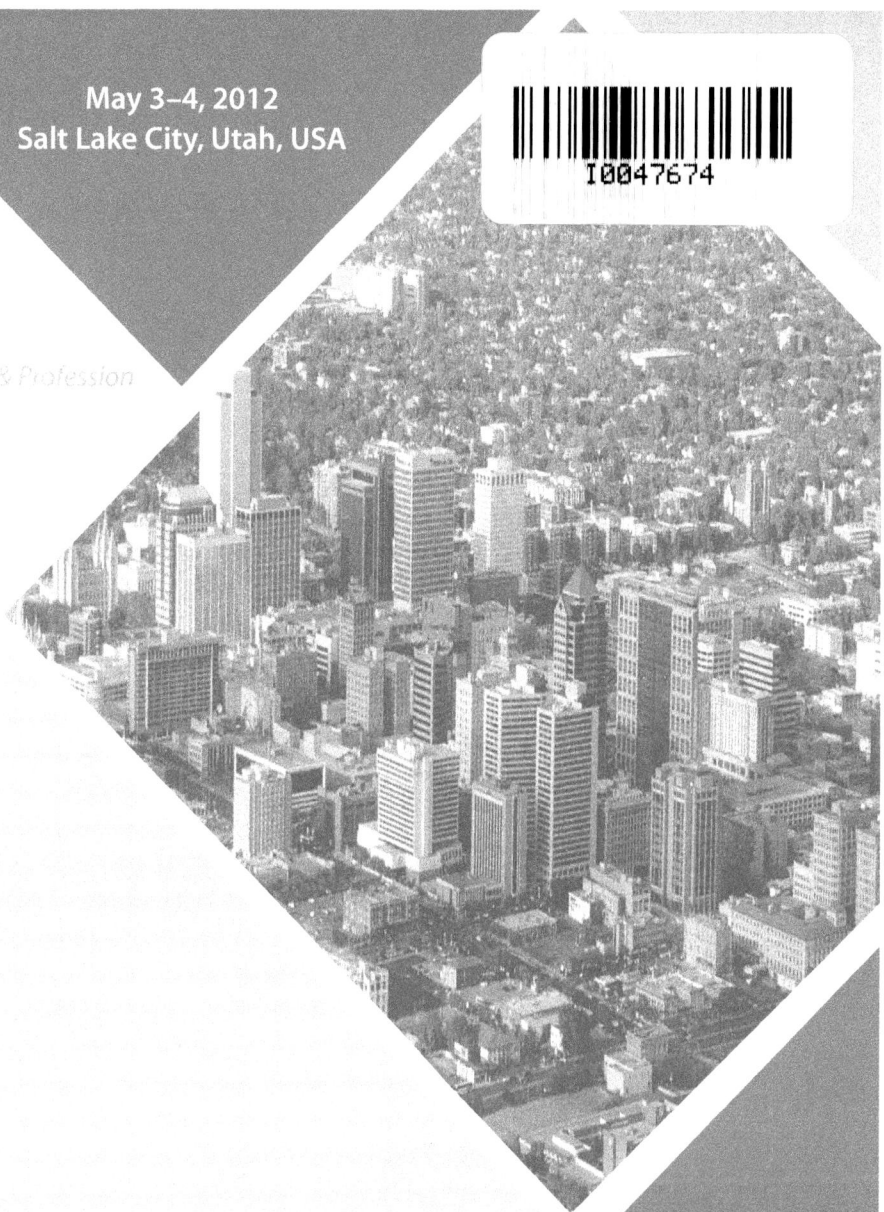

GLSVLSI'12

Proceedings of the

Great Lakes Symposium on VLSI 2012

Sponsored by:
ACM SIGDA

Technical Co-Sponsored by:
CEDA and IEEE

Supported by:
The University of Utah and Fusion-IO

**Association for
Computing Machinery**

Advancing Computing as a Science & Profession

The Association for Computing Machinery
2 Penn Plaza, Suite 701
New York, New York 10121-0701

ISBN: 978-1-4503-1244-8 (Digital)

ISBN: 978-1-4503-1726-9 (Print)

Additional copies may be ordered prepaid from:

ACM Order Department
PO Box 30777
New York, NY 10087-0777, USA

Phone: 1-800-342-6626 (USA and Canada)
+1-212-626-0500 (Global)
Fax: +1-212-944-1318
E-mail: acmhelp@acm.org
Hours of Operation: 8:30 am – 4:30 pm ET

Printed in the USA

Foreword

Welcome to the 22nd edition of the Great Lakes Symposium on VLSI (GLSVLSI) 2012 held in Salt Lake City, Utah at the Salt Lake Marriott, City Center. GLSVLSI is a premier venue for the dissemination of manuscripts of the highest quality in all areas related to VLSI, devices and system-level design. The venue of this year's GLSVLSI is Salt Lake City, which is located right next to the "Great Salt Lake." You will enjoy the beautiful Utah Mountains and surrounding scenery as well as the program over the two days of this year's GLSVLSI activity.

As for the technical meeting, GLSVLSI 2012 was a resounding success: 144 papers were submitted, including authors from 34 different countries, of which 41 papers were accepted for oral presentation at the symposium (a 28% acceptance rate). With poster papers, a total of 71 papers will be presented at the symposium and published in the conference proceedings. The final technical program consists of 23 full presentations and 18 short presentations in 10 oral sessions and 30 posters in two poster sessions.

GLSVLSI 2012 starts on Thursday, May 3rd, in the morning with an invited keynote talk followed by technical sessions on Emerging Technologies, Reliability, Circuit Design, CAD, Multi-core and NOC, Testing and Fault-Tolerance. Then, the technical program continues on Friday, May 4th with the second invited keynote talk followed by technical sessions which present the latest industrial and academic research covering topics such as Post-CMOS Circuits, Low Power, CAD, and VLSI Systems. The technical program of GLSVLSI 2012 includes two parallel tracks to allow longer presentations and discussions during the two days of the conference. Overall, there are ten regular sessions in the technical program with a poster session after lunch each day.

We are also looking forward to the two plenary speeches that will be delivered by Al Davis, University of Utah on "The Role of Photonics in Future Data Centers", and by Khaled Salama, King Abdullah University of Science and Technology on "Memristor: The Illusive Device."

Congratulations to Jia Zhao, Russell Tessier, and Wayne Burleson from University of Massachusetts, Amherst, for winning the GLSVLSI 2012 Best Student Paper Award for their paper, "Distributed Sensor Data Processing for Many-cores."

Finally, we would like to thank all the members of the Technical Program Committee and the secondary reviewers, who have done a great job in evaluating the submissions and determining the final high quality technical program of GLSVLSI 2012; the Track Chairs have done an outstanding job in ensuring that at least three reviews were provided for each submitted paper. In addition, the other members of the Organizing Committees (Jose Ayala, Theocharis Theocharides, Eby Friedman) should also be commended for providing excellent service with respect to publication, registration, organization of the event and web presence to make this new edition of GLSVLSI a great success.

We trust that you will find GLSVLSI 2012 rewarding in meeting your highest expectation of quality and technical advancement as reflected in these Proceedings and the technical presentations.

Finally, continue to watch the website (http://www.glsvlsi.org) for information regarding GLSVLSI 2013 location, committees and the call for papers.

Enjoy the symposium and looking forward to seeing you next year.

Erik Brunvand & Ken Stevens   **Joseph R. Cavallaro & Tong Zhang**
General Chairs   *Program Chairs*

Table of Contents

GLSVLSI 2012 Organizing Committee xi

GLSVLSI 2012 Organization xii

2012 22nd Great Lakes Annual Symposium
on VLSI Design (GLSVLSI) Sponsors and Supporters xvii

Keynote Address

- The Role of Photonics in Future Data Centers 1
  Al Davis *(University of Utah)*

Session 1: Emerging Technologies

- Ambipolar Double-Gate FETs for the Design of Compact Logic Structures 3
  Kotb Jabeur, Ian O'Connor, Nataliya Yakymets, Sébastien Le Beux *(University of Lyon)*

- Performance and Energy Models for Memristor-Based 1T1R RRAM Cell 9
  Mahmoud Zangeneh, Ajay Joshi *(Boston University)*

- Accelerating Thermal Simulations of 3D ICs with Liquid Cooling Using
  Neural Networks 15
  Alessandro Vincenzi, Arvind Sridhar, Martino Ruggiero,
  David Atienza *(École Polytechnique Fédérale de Lausanne)*

- Efficient CMOL Nanoscale Hybrid Circuit Cell Assignment Using Simulated
  Evolution Heuristic 21
  Sadiq M. Sait, Abdalrahman Arafeh *(King Fahd University of Petroleum and Minerals)*

Session 2: Reliability

- SNR Analysis Approach for Hardware/Software Partitioning Using Dynamically
  Adaptable Fixed Point Representation 27
  Varadaraj Kamath Nileshwar, Roman Lysecky *(University of Arizona)*

- NBTI Mitigation in Microprocessor Designs 33
  Simone Corbetta, William Fornaciari *(Politecnico di Milano)*

- A Noise-Immune Sub-Threshold Circuit Design Based on Selective Use
  of Schmitt-Trigger Logic 39
  Marco Donato *(Brown University)*, Fabio Cremona *(Sapienza University of Rome)*,
  Warren Jin, R. Iris Bahar, William Patterson, Alexander Zaslavsky, Joseph Mundy *(Brown University)*

- Efficient Selection and Analysis of Critical-Reliability Paths and Gates 45
  Jifeng Chen, Shuo Wang, Mohammad Tehranipoor *(University of Connecticut)*

Poster Session 1

- Breaking the Power Delivery Wall using Voltage Stacking 51
  Kaushik Mazumdar, Mircea R. Stan *(University of Virginia)*

- An Efficient CPI Stack Counter Architecture for Superscalar Processors 55
  Osman Allam, Stijn Eyerman, Lieven Eeckhout *(Ghent University)*

- An Optimized Multicore Cache Coherence Design for Exploiting
  Communication Locality 59
  Libo Huang, Zhiying Wang, Nong Xiao *(National University of Defense Technology)*

- Parallel Pipelined FFT Architectures with Reduced Number of Delays 63
  Manohar Ayinala, Keshab Parhi *(University of Minnesota, Twin Cities)*

- Design of an RNS Reverse Converter for a New Five-Moduli Special Set 67
  Piotr Patronik, Krzysztof Berezowski, Janusz Biernat *(Wroclaw University of Technology),*
  Stanislaw J. Piestrak *(Université de Lorraine),* Aviral Shrivastava *(Arizona State University)*

- On the Automatic Synthesis of Parallel SW from RTL Models of Hardware IPs 71
  Andrea Acquaviva *(Politecnico di Torino),* Nicola Bombieri, Franco Fummi, Sara Vinco *(Università di Verona)*

- Top-Down-Based Symmetrical Buffered Clock Routing 75
  Jin-Tai Yan, Ming-Chien Huang, Zhi-Wei Chen *(Chung Hua University)*

- Optimal Register-Type Selection during Resource Binding in Flip-Flop/Latch-Based
  High-Level Synthesis 79
  Keisuke Inoue, Mineo Kaneko *(Japan Advanced Institute of Science and Technology)*

- A Fully Integrated Switched-Capacitor DC-DC Converter with Dual Output
  for Low Power Application 83
  Heungjun Jeon, Yong-Bin Kim *(Northeastern University)*

- High-Level Modeling of Power Consumption in Active Linear Analog Circuits 87
  Laurent Bousquet, Emmanuel Simeu *(TIMA Laboratory)*

- A Novel Power-Gating Scheme Utilizing Data Retentiveness on Caches 91
  Kyundong Kim, Seidai Takeda, Shinobu Miwa, Hiroshi Nakamura *(The University of Tokyo)*

- A Zero-Overhead IC Identification Technique Using Clock Sweeping
  and Path Delay Analysis 95
  Nicholas Tuzzio, Kan Xiao, Xuehui Zhang, Mohammad Tehranipoor *(University of Connecticut)*

- RAPA: Reliability-Aware Priority Arbitration Strategy for Network on Chip 99
  Jiajia Jiao, Yuzhuo Fu *(Shanghai Jiao Tong University)*

- A High-Performance Online Assay Interpreter for Digital Microfluidic Biochips 103
  Daniel Grissom, Philip Brisk *(University of California, Riverside)*

- Reliable Logic Mapping on Nano-PLA Architectures 107
  Masoud Zamani *(Northeastern University),* Mehdi B. Tahoori *(Karlsruhe Institute of Technology)*

Session 3: Circuit Design

- Self Adaptive Body Biasing Scheme for Leakage Power Reduction in Nanoscale
  CMOS Circuit 111
  Jing Yang, Yong-Bin Kim *(Northeastern University)*

- Synchronization Scheme for Brick-Based Rotary Oscillator Arrays 117
  Ying Teng, Baris Taskin *(Drexel University)*

- A Low-Power All-Digital GFSK Demodulator with Robust Clock Data Recovery 123
  Pengpeng Chen, Bo Zhao, Rong Luo, Huazhong Yang *(Tsinghua University)*

- Link Breaking Methodology: Mitigating Noise within Power Networks 129
  Renatas Jakushokas, Eby G. Friedman *(University of Rochester)*

Session 4: CAD-I

- Unifying Functional and Parametric Timing Verification 135
  Luis Guerra e Silva *(INESC-ID / IST / TU Lisbon)*

- New & Improved Models for SAT-Based Bi-Decomposition 141
  Huan Chen, Joao Marques-Silva *(University College Dublin)*

- Lithography-Aware Layout Compaction 147
  Curtis Andrus, Matthew R. Guthaus *(University of California, Santa Cruz)*

- A Design Approach Dedicated to Network-Based and Conflict-Free
  Parallel Interleavers 153
  Aroua Briki, Cyrille Chavet *(Université de Bretagne Sud / Lab-STICC),*
  Philippe Coussy, Eric Martin *(Université de Bretagne Sud)*

Session 5: Multi-core and NOC

- **Distributed Sensor Data Processing for Many-Cores** 159
  Jia Zhao, Russell Tessier, Wayne Burleson *(University of Massachusetts)*

- **CMOS Compatible Many-Core NoC Architectures with Multi-Channel Millimeter-Wave Wireless Links** 165
  Sujay Deb, Kevin Chang, Miralem Cosic *(Washington State University)*,
  Amlan Ganguly *(Rochester Institute of Technology)*,
  Partha Pande, Deukhyoun Heo, Benjamin Belzer *(Washington State University)*

- **Voltage Island-Driven Power Optimization for Application Specific Network-on-Chip Design** 171
  Kan Wang, Sheqin Dong *(Tsinghua University)*, Satoshi Goto *(Waseda University)*

- **Design-Time Performance Evaluation of Thermal Management Policies for SRAM and RRAM Based 3d MPSoCs** 177
  David Brenner, Cory Merkel, Dhireesha Kudithipudi *(Rochester Institute of Technology)*

Session: 6 Testing and Fault-Tolerance

- **TSUNAMI: A Light-Weight On-Chip Structure for Measuring Timing Uncertainty Induced By Noise During Functional and Test Operations** 183
  Shuo Wang, Mohammad Tehranipoor *(University of Connecticut)*

- **Lazy Suspect-Set Computation: Fault Diagnosis for Deep Electrical Bugs** 189
  Dipanjan Sengupta *(University of Toronto)*, Flavio M. de Paula, Alan J. Hu *(University of British Columbia)*,
  Andreas Veneris *(University of Toronto)*, André Ivanov *(University of British Columbia)*

- **Influence of Different Layout Styles on the Performance of the Calibration of an On-Chip Programmable Voltage Reference** 195
  Dominik Gruber, Timm Ostermann *(Johannes Kepler University Linz)*

- **Input and Transistor Reordering for NBTI and HCI Reduction in Complex CMOS Gates** 201
  Saman Kiamehr, Farshad Firouzi, Mehdi B. Tahoori *(Karlsruhe Institute of Technology)*

Keynote Address

- **Memristor: The Illusive Device** 207
  Khaled Nabil Salama *(King Abdullah University of Science and Engineering)*

Session 7: Post-CMOS Circuits

- **InMnAs Magnetoresistive Spin-Diode Logic** 209
  Joseph S. Friedman, Nikhil Rangaraju, Yehea I. Ismail, Bruce W. Wessels *(Northwestern University)*

- **An Efficient Approach for Designing and Minimizing Reversible Programmable Logic Arrays** 215
  Sajib Kumar Mitra *(Samsung Bangladesh R&D Center Limited)*, Lafifa Jamal *(University of Dhaka)*,
  Mineo Kaneko *(Japan Advanced Institute of Sci. and Tech.)*, Hafiz Md. Hasan Babu *(University of Dhaka)*

- **Modeling a Single Electron Turnstile in HSPICE** 221
  Wei Wei *(Northeastern University)*, Jie Han *(University of Alberta)*, Fabrizio Lombardi *(Northeastern University)*

- **Limits of Writing Multivalued Resistances in Passive Nanoelectronic Crossbars Used in Neuromorphic Circuits** 227
  Arne Heittmann, Tobias G. Noll *(RWTH Aachen University)*

Session 8: Low Power

- **Stepwise Sleep Depth Control for Run-Time Leakage Power Saving** 233
  Seidai Takeda, Shinobu Miwa *(The University of Tokyo)*, Kimiyoshi Usami *(Shibaura Institute of Technology)*,
  Hiroshi Nakamura *(The University of Tokyo)*

- **An Efficient Power Estimation Methodology for Complex RISC Processor-Based Platforms** 239
  Santhosh Kumar Rethinagiri *(University of Valenciennes)*,
  Rabie Ben Atitallah *(University of Valenciennes)*, Jean-Luc Dekeyser *(University of Lille 1)*,
  Smail Niar *(University of Valenciennes)*, Eric Senn *(University of Bretagne South)*

- **ADAM: An Efficient Data Management Mechanism for Hybrid High and Ultra-Low Voltage Operation Caches** 245
  Bojan Maric *(Barcelona Supercomputing Center and Universitat Politecnica de Catalunya)*,
  Jaume Abella *(Barcelona Supercomputing Center)*,
  Mateo Valero *(Barcelona Supercomputing Center and Universitat Politecnica de Catalunya)*

- **A Low Stand-by Power Start-up Circuit for SMPS PWM Controller** 251
  In-Seok Jung, Yong-Bin Kim *(Northeastern University)*

Poster Session 2

- **Particle Swarm Optimization Over Non-Polynomial Metamodels for Fast Process Variation Resilient Design of NANO-CMOS PLL** 255
  Oleg Garitselov, Saraju P. Mohanty, Elias Kougianos, Geng Zheng *(University of North Texas)*

- **A Denial-of-Service Resilient Wireless NoC Architecture** 259
  Amlan Ganguly, Mohsin Yusuf Ahmed, Anuroop Vidapalapati *(Rochester Institute of Technology)*

- **Sustainable Multi-Core Architecture with On-Chip Wireless Links** 263
  Jacob Murray, Partha P. Pande, Behrooz Shirazi *(Washington State University)*,
  John Klingner *(Cornell College)*

- **SRAM Leakage in CMOS, FinFET and CNTFET Technologies** 267
  Zhe Zhang, Michael A. Turi, José G. Delgado-Frias *(Washington State University)*

- **A Novel Hybrid FIFO Asynchronous Clock Domain Crossing Interfacing Method** 271
  Zaid Al-bayati, Otmane Ait Mohamed *(Concordia University)*,
  Syed Rafay Hasan *(Tennessee Tech. University)*, Yvon Savaria *(École Polytechnique de Montréal)*

- **Density-Reduction-Oriented Layer Assignment for Rectangle Escape Routing** 275
  Jin-Tai Yan, Jun-Min chung, Zhi-Wei Chen *(Chung Hua University)*

- **NBTI Effects on Tree-Like Clock Distribution Networks** 279
  Wei Liu, Sandeep Miryala, Valerio Tenace, Andrea Calimera, Enrico Macii,
  Massimo Poncino *(Politecnico Di Torino)*

- **A Framework for High-Level Synthesis of Heterogeneous MP-SoC** 283
  Youenn Corre, Jean-Philippe Diguet, Dominique Heller, Loïc Lagadec *(University of South Brittany)*

- **Memory-Based Computing for Performance and Energy Improvement in Multicore Architectures** 287
  Kamran Rahmani, Prabhat Mishra *(University of Florida)*, Swarup Bhunia *(Case Western Reserve University)*

- **Share Memory Aware Scheduler: Balancing Performance and Fairness** 291
  Xi Li, Gangyong Jia, Yun Chen, Zongwei Zhu,
  Xuehai Zhou *(University of Science and Technology of China & Suzhou Institute for Advanced Study)*

- **Alleviating NBTI-Induced Failure in Off-Chip Output Drivers** 295
  Bhavitavya Bhadviya, Ayan Mandal, Sunil P. Khatri *(Texas A&M University)*

- **Mitigating Electromigration of Power Supply Networks Using Bidirectional Current Stress** 299
  Jing Xie, Vijaykrishnan Narayanan, Yuan Xie *(The Pennsylvania State University)*

- **Multiplexed Switch Box Architecture in Three-Dimensional FPGAs to Reduce Silicon Area and Improve TSV Usage** 303
  Marzieh Morshedzadeh Morshedzadeh, Ali Jahanian *(Shahid Beheshti University, G. C.)*

- **A Scalable Threshold Logic Synthesis Method Using ZBDDs** 307
  Ashok Kumar Palaniswamy, Spyros Tragoudas *(Southern Illinois University Carbondale)*

- **A Memristor-Based TCAM (Ternary Content Addressable Memory) Cell: Design and Evaluation** 311
  Pilin Junsangsri, Fabrizio Lombardi *(Northeastern University)*

Session 9: CAD-II

- **Extending Symmetric Variable-Pair Transitivities Using State-Space Transformations**...315
  Peter M. Maurer *(Baylor University)*

- **Crosslink Insertion for Variation-Driven Clock Network Construction**...321
  Fuqiang Qian, Haitong Tian, Evangeline Young *(The Chinese University of Hong Kong)*

- **WRIP: Logic Restructuring Techniques for Wirelength-Driven Incremental Placement**...327
  Xing Wei, Wai-Chung Tang, Yu-Liang Wu *(The Chinese University of Hong Kong)*,
  Cliff Sze, Charles Alpert *(IBM Austin Research Center)*

- **STEP: A Unified Design Methodology for Secure Test and IP Core Protection**...333
  Pranav Yeolekar, Rishad A. Shafik, Jimson Mathew, Dhiraj K. Pradhan *(University of Bristol)*,
  Saraju P. Mohanty *(University of North Texas)*

Session 10: VLSI Systems

- **Towards Systolic Hardware Acceleration for Local Complexity Analysis of Massive Genomic Data**...339
  Agathoklis Papadopoulos, Vasilis J. Promponas, Theocharis Theocharides *(University of Cyprus)*

- **A Dual-Phase Compression Mechanism for Hybrid DRAM/PCM Main Memory Architectures**...345
  Seungcheol Baek, Hyung Gyu Lee, Jongman Kim *(Georgia Institute of Technology)*,
  Chrysostomos Nicopoulos *(University of Cyprus)*

- **Verilog-AMS-PAM: Verilog-AMS Integrated with Parasitic-Aware Metamodels for Ultra-Fast and Layout-Accurate Mixed-Signal Design Exploration**...351
  Geng Zheng, Saraju P. Mohanty, Elias Kougianos, Oleg Garitselov *(University of North Texas)*

- **Efficient Folded VLSI Architectures for Linear Prediction Error Filters**...357
  Sayed Ahmad Salehi, Rasoul Amirfattahi *(Isfahan University of Technology)*,
  Keshab Parhi *(University of Minnesota)*

- **Synergistic Integration of Code Encryption and Compression in Embedded Systems**...363
  Kamran Rahmani, Hadi Hajimiri, Kartik Shrivastava, Prabhat Mishra *(University of Florida)*

Author Index...369

GLSVLSI 2012 Organizing Committee

General Chairs:

Erik Brunvard
(University of Utah, USA)

Ken Stevens
(University of Utah, USA)

Program Chairs:

Joseph R. Cavallaro
(Rice University, USA)

Tong Zhang
(Rensselaer Polytechnic Institute, USA)

Publication Chair:

Jose L. Ayala
(Complutense University, Spain)

Webmaster:

Theocharis Theocharides
(University of Cyprus, Cyprus)

Invited Speakers Chair:

Eby Friedman
(University of Rochester, USA)

GLSVLSI 2012 Organization

General Chairs: Erik Brunvand *(University of Utah, USA)*
Ken Stevens *(University of Utah, USA)*

Program Chairs: Joseph R. Cavallaro *(Rice University, USA)*
Tong Zhang *(Rensselaer Polytechnic Institute, USA)*

Publication Chair: Jose L. Ayala *(Complutense University of Madrid, Spain)*

Web Chair: Theocharis Theocharides *(University of Cyprus, Cyprus)*

Invited Speakers Chair: Eby Friedman *(University of Rochester, USA)*

GLSVLSI 2012 Program Committee

VLSI Design Track Chairs: James Stine *(Oklahoma State University, USA)*
Yang Sun *(Qualcomm, USA)*

VLSI Design Track Members: Andreas Burg *(EPFL, Switzerland)*
Yangdong Deng *(Tsinghua University, China)*
David Garrett *(Broadcom, USA)*
Manish Goel *(Texas Instruments, USA)*
Seok-Jun Lee *(Texas Instruments, USA)*
Dake Liu *(Linköping University, Sweden)*
Martin Margala *(University of Massachusetts Lowell, USA)*
Saraju Mohanty *(University of North Texas, USA)*
Rabi Mohapatra *(Texas A&M University, USA)*
Vassilis Paliouras *(University of Patras, Greece)*
Sudeep Pasricha *(Colorado State University, USA)*
Zhijie Shi *(University of Connecticut, USA)*
Gerald Sobelman *(University of Minnesota, USA)*
Miroslav Velev *(Aries Design Automation, USA)*
Xinmiao Zhang *(Case Western Reserve University, USA)*

Testing Track Chairs: Erik Larsson *(Lund University, Sweden)*
Rob Aitken *(ARM Ltd., USA)*

Testing Track Members: Yinhes Han *(Chinese Academy of Sciences, China)*
Chih-Tsun Huang *(National Tsing Hua University, Taiwan)*
Jennifer Lynn Dworak *(Southern Methodist University, USA)*
Nicola Nicolici *(McMaster University, Canada)*
Sule Ozev *(Arizona State University, USA)*
Mohammad Tehranipoor *(University of Connecticut, USA)*
Spyros Tragoudas *(Southern Illinois University, USA)*
Qiang Xu *(The Chinese University of Hong Kong, Hong Kong)*
Tomokazu Yoneda *(Nara Institute of Science and Technology, Japan)*

VLSI Circuits Track Chairs: Bozena Kaminska *(Simon Fraser University, Canada)*
Lei Wang *(University of Connecticut, USA)*

VLSI Circuits Track Members: Hyeon-Min Bae *(KAIST, Republic of Korea)*
Jing Hong Chen *(Southern Methodist University, USA)*
Jianwei Dai *(Intel Corp., USA)*
Eby G. Friedman *(University of Rochester, USA)*
Jun Ma *(Shanghai Jiaotong University, China)*
Emre Salman *(Stony Brook University, USA)*
Theocharis Theocharides *(University of Cyprus, Cyprus)*
Diego Vázquez *(University of Seville, Spain)*

Post-CMOS VLSI Chairs: Hai Li *(Polytechnic Institute of NYU, USA)*
Robinson Pino *(Air Force Research Laboratory, USA)*

Post-CMOS VLSI Track Members: Shamik Das *(MITRE, USA)*
Harika Manem *(College of Nanoscale Science and Engineering, USA)*
Kartik Mohanram *(University of Pittsburgh, USA)*
Chris Myers *(Utah University, USA)*
Guangyu Sun *(Peking University, China)*
Mehdi Tahoori *(Karlsruhe Institute of Technology, Germany)*
Wei Zhang *(Nanyang Technological University, Singapore)*

Emerging Technologies Track Chairs: Ian O'Connor *(Lyon Institute of Nanotechnology, France)*
Garrett Rose *(Air Force Research Laboratory, USA)*

Emerging Technologies Track Members: Haykel Ben Jamaa *(CEA – LETI, France)*
Jacques-Olivier Klein *(Institut d'Electronique Fondamentale, France)*
Dhireesha Kudithipudi *(Rochester Institute of Technology, USA)*
Sébastien Le Beux *(Lyon Institute of Nanotechnology, France)*
Dmitri Strukov *(UCSB, USA)*
Baris Taskin *(Drexel University, USA)*
Chris Winstead *(Utah State University, USA)*

**Secondary Reviewers
(continued):**

Mahesh Gautam	Sandeep Miryala
Mahesh Poolakkaparambil	Sandip Das
Malte Metzdorf	Sekhar Mandal
Manas Hira	Sohan Purohit
Mathias Soeken	Stefan Frehse
Matthias Beste	Suhas Satheesh
Michael Niemier	Suman Mandal
Mohamed Bawadekji	Suneil Mohan
Monica Vallejo	Surajit Kumar Roy
Mousumi Saha	Tuhina Samanta
Nicholas Roehner	Ulrich Kuehne
Oghenekarho Okobiah	Vinícius Dal Bem
Oleg Garitselov	Wei Liu
Pablo Garcia Del Valle	Wenwen Chai
Pablo Ituero	Wim Vanderbauwhede
Parthasarathi Dasgupta	Xiang Qiu
Paul Pop	Xiao Linfu
Per Karlström	Xu He
Peter Milder	Xu Yang
Ping Gao	Yoonjin Kim
Pranab Roy	Zedong Wang
Qiang Ma	Zhen Zhang
Raghupathy Ramakrishnan	Zhihua Gan
Ravi Patel	Zhixin Chen
Ridvan Umaz	Zigang Xiao
Robert Wille	
Samed Maltabas	

2012 22nd Great Lakes Annual Symposium on VLSI Design (GLSVLSI)
Sponsors and Supporters

Sponsor:

Technical Co-Sponsor:

Supporters:

The Role of Photonics in Future Data Centers

Al Davis
University of Utah
School of Computing
50 Central Campus Dr.
Salt Lake City, UT 84112
+1-801-581-3991
ald@cs.utah.edu

ABSTRACT

The most prolific information appliance today is a mobile device, typically a phone or a tablet. The near ubiquity of the internet makes it very easy to access vast amounts of information from these mobile devices. The storage and processing capability to support this infrastructure is increasingly in the datacenter and both the number of these warehouse scale computational (WSC) facilities and their size is increasing. Compound annual growth rates (CAGR) of both storage requirements and internet traffic is approximately 50%. Even more alarming is that in 2011, 2% of the total energy used in the United States was consumed by this information technology. Power and the attendant cooling requirements fundamentally affect both the cost of our IT infrastructure as well as what can be competitively designed. As VLSI technology scales, the power and speed of the transistors scale nicely but wires scale less favorably and the gap grows with every new process step. The telecom industry has long recognized the benefits of optical as opposed to electrical communication over long distances. However the definition of "long" changes with the signalling rate. Electrical signalling at high speeds consumes too much power, has significant signal integrity problems which limit bandwidth and at the WSC scale these issues border on catastrophic for future datacenters. Recent advances in silicon nanophotonics may well provide a solution to this problem. This talk will provide a brief introduction to silicon nanophotonic devices and then delve into how this technology can be used in the construction of more energy efficient datacenters of the future and discuss where this technology will be most useful and quantify the benefits for intra-datacenter networks.

Categories and Subject Descriptors

C.4. [Computer Systems Organization]: Performance of Systems

General Terms

Algorithms, Performance, Design, Economics

Keywords

Data Centers, Nanophotonics.

Bio

Alan L. Davis received the BS degree in electrical engineering from the Massachusetts Institute of Technology and the PhD degree in electrical engineering from the University of Utah. His current research interests include rapid system on chip development tools, high performance low-power embedded system architectures, embedded perception, asynchronous circuits, and VLSI. He is currently a professor of computer science at the University of Utah. He has also held industrial research positions at Burroughs, Hewlett-Packard, Fairchild Semiconductor, Schlumberger, and Intel. He is a member of the IEEE.

GLSVLSI'12, May 3–4, 2012, Salt Lake City, Utah, USA.
ACM 978-1-4503-1244-8/12/05.

Ambipolar Double-gate FETs for the Design of Compact Logic Structures

Kotb Jabeur, Ian O'Connor, Nataliya Yakymets, Sébastien Le Beux
Lyon Institute of Nanotechnology
University of Lyon, Ecole Centrale de Lyon
36 Avenue Guy de Collongue, F-69134 Ecully, France
Email: kotb.jabeur@@ec-lyon.fr

ABSTRACT
We present in this paper a circuit design approach to achieve compact logic circuits with ambipolar double-gate devices, using the in-field controllability of such devices. The approach is demonstrated for complementary static logic design style. We apply this approach in a case study focused on Double Gate Carbon Nanotube FET (DG-CNTFET) technology and show that, with respect to conventional CMOS-like static logic structures and for comparable power consumption, time delay and integration density can both be improved by a factor of 1.5x and 2x, respectively. Compared with a predictive model for 16nm CMOS technology, the gates built according to the design approach described in this work and based on DG-CNTFET offer a gain of 30% concerning Power-Delay-Product (PDP).

Categories and Subject Descriptors
B.6.1 [**Logic Design**]: Design style –*Combinational logic, logic arrays*. B.7.1. [**Integrated Circuits**]: Types and Design styles – *advanced technologies, gate arrays*.

General Terms
Performance, Design.

Keywords
Four-terminal devices, ambipolar double-gate devices, static logic, CNTFETs, emerging technologies.

1. INTRODUCTION
The "Beyond CMOS" concept has motivated much research on emerging research devices and materials, as described in the ITRS [1]. Novel materials and devices have been explored showing an ability to complement or even replace the CMOS transistor or its channel in systems on chip before silicon-based technology will reach its limits. These possible emerging technologies range from transistors made from silicon nanowires to devices made from nanoscale molecules.

Some of the most promising devices are carbon-based nanodevices such as the carbon-nanotube (CNT) field-effect transistor (FET) (CNTFET) or the graphene-nanoribbon FET. Furthermore, such a device can be built in the context of double gate devices which can be classified in terms of their gate geometry regardless of the underlying device process; Typically the front and back gates of DG devices are connected together resulting in a 3-Terminal device to improve the performance of conventional single gate bulk CMOS devices [2] [3] [4].

DG devices with independent gates are 4-Terminal devices and show the potential to provide novel logic blocks and innovative techniques for digital design thanks to their specific intrinsic properties. Many works [6] [7] have shown how such a device opens the way for fine-grain reconfigurability since DG devices can be used to realize in-field programmable ambipolar devices, i.e. devices whose p- or n-type behaviour can be programmed in-field using the fourth terminal. Also, a variety of circuits in logic and memory can benefit from independent gate operation of DG devices, such as those presented in [8] for low power and high performance.

Moreover, multiple gate structures have also shown the potential to develop new logic gates with a significant decrease in transistor count, with for example FinFETs [7] or with Double Gate CNTFET in [9], where it was demonstrated that the use of ambipolar double-gate FETs can reduce the transistor count in clocked standard cells by the use of the additional terminal of the DG-CNTFET. Independent gate control can also be used to merge parallel transistors in non-critical paths [8]. This results in a reduction of the effective switching capacitance and hence power dissipation.

In this work, we present an approach which uses double-gate ambipolar transistors in complementary static-logic style to reduce transistor count by merging transistors-in-series into a single transistor in multiple input logic circuits (such as complex gates and multiplexers), and evaluate its impact on power consumption and time delay, using a compact model, in the case of DG-CNTFETs with maximal flexibility at the layout level.

We begin by describing the technological hypotheses and modelling in section 2. We present the concept of the approach in section 3 and its application to the complementary-static logic structure with several examples, and then performance evaluation of the novel circuits obtained from this approach in the case of DG-CNTFETs is presented and compared to a 16nm CMOS technology predictive model in section 4. Section 5 is the conclusion.

2. TECHNOLOGICAL HYPOTHESES AND MODELLING

2.1 DG CNTFET fabrication

At present, various types of CNTFET devices have been produced experimentally, but there is no standard CNTFET process. In many works using CNTFET devices [10] [11], realistic and CMOS-compatible process flow steps have been suggested to manufacture such a device. A prior work, presented the structure of a DG CNTFET based on aligned semiconducting CNTs [12], applying the self-alignment technique [13] and using two top gates [14], as well as potential hybrid integration with CMOS technology and consequently with metal interconnections defined by CMOS-compatible lithography steps. In this work, the same hypotheses are kept, except that the structure used here does not have two top-gates, but a top-gate and a back-gate in order to be more faithful to the compact model used later for simulations. Figure 1(a), reported from a recent work [11] using the same compact model, shows the structure of the CNTFET with double gates. The front gate FG turns the device on or off, in the same way as the regular gate of a MOSFET; while the back-gate BG controls the device polarity setting to N- or P-type with a positive ($V_{BG}-V_S=$ +V) or negative ($V_{BG}-V_S=$ -V) voltage, respectively. The symbol for this in-field programmable CNTFET is shown in Fig 1(b).

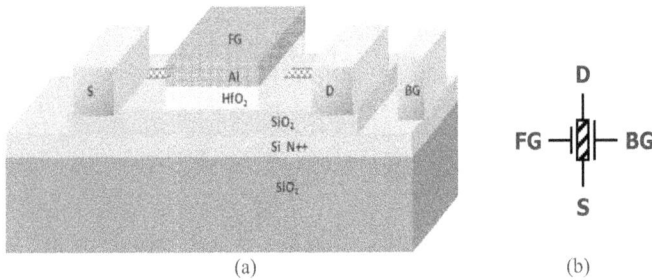

Figure 1. DG-CNTFET device (a), symbol (b)

2.2 DG-CNTFET compact model

Many compact models [15] [16] have been developed to describe CNTFET technologies such as Schottky barrier (SB) CNTFETs or MOS-like CNTFETs with a top gate or surrounding gates. However, none of these are able to model a DG CNTFET properly. In prior works [6] [9] simulations were carried out with the model presented in [17] which is limited to thermionic transport without taking into account the coupling between the FG and BG and does not include SB (sub-band) modeling or BTBT (band-to-band tunneling). Recently, a more accurate model has been presented in [10]. To the best of our knowledge; it represents the first physically accurate model of a DG CNTFET with efficient convergence and simulation speed compatible with circuit design. The compact model is detailed in [10], where it was shown to include the most significant mechanisms such as the SB at the metal–nanotube interface, charge and electrostatic modeling, BTBT effect, and quasi-ballistic transport. Furthermore, the comparison of the model with measurements from two technologies published in literature; a DG CNTFET from IBM [18] and a DG CNTFET from Stanford University (used as a MOS-like CNTFET) [19], showed the accuracy of the compact model on different technology configurations since the values of the extracted parameters are close to those in the

available technologies. We mention that although the structure of DGCNTFET suggested in this work is not exactly the same with the case modeled in [10], we suppose that the physical behaviour of the device is the same and we used the model in [10] for all simulations in this work.

3. NOVEL FAMILY OF DG-CNTFET LOGIC CELLS

In the case of the DG-CNTFET, completely new prospects for reconfigurability are possible due to its ambipolar (N- and P-type) behavior. In previous works, ambipolarity was exploited to build dynamically reconfigurable logic cells [9]. In the same perspective, a complete design methodology (using Ambipolar Binary Decision Diagrams Am-BDD) to generate reconfigurable cells based on ambipolar devices was defined in [6]. However in this work, rather than using the back-gate reconfigurability and associated states (N-type, P-type, off-state) for reconfigurability purposes, we used the back-gate signal as a free variable in order to design compact logic gates.

3.1 General concept

Equivalent logic path resistance is proportional to the number of transistors in series, and inversely proportional to the average transistor width. Hence, two n-type transistors in series (NTTS structure) or p-type transistors in series (PTTS structure) must either demonstrate a path resistance of $2R_{ch}$ with no transistor resizing or an input gate capacitance of $2C_g$ with transistor width doubling to reduce overall path resistance (where R_{ch} and C_g represent the channel resistance and gate capacitance of a single minimum width transistor, respectively). Hence two transistors in series (TTS structures) are critical for path resistance and gate capacitance optimization, with consequent impact on delay, power and area. The switching between the n- and p-states in ambipolar DG CNTFETs allows TTS structures to be substituted by a single DGFET with no loss of functionality as shown in figures 2 and 3 for the NTTS and PTTS structures respectively.

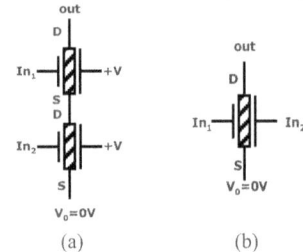

Figure 2. Direct transposition of CMOS-NTTS structure with ambipolar DGFETs (a), Single ambipolar DGFET equivalent TTS structure (b)

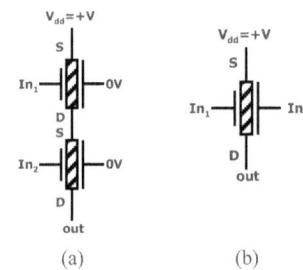

Figure 3. Direct transposition of CMOS-PTTS structure with ambipolar DGFETs (a), Single ambipolar DGFET equivalent TTS structure (b)

In the initial NTTS structure in figure 2(a), both transistors are N-type since the back gate BG is set to +V and V_0 is set to 0V (i.e. V_{BG}-V_S=+V). In this case, a path is established between "V_0" and "out" only for In_1In_2="11". In the single ambipolar DGFET structure, shown in figure 2(b), the same condition is true since for In_2="1", the back gate BG is set to +V such that the transistor is N-type and will be ON only if In_1="1" also. For other combinations In_1In_2= {"01","10","00"} the transistor will be OFF. Thus, the NTTS structure can be replaced by a single ambipolar DGFET.

By analogy, the PTTS structure in figure 3(a) obtains the same benefit since both transistors are P-type when the back gate BG is set to 0V and V_{dd} is set to +V (i.e. V_{BG}-V_S=-V). Again, In this case, a path is established between "V_{dd}" and "out" only for In_1In_2="00". In the single ambipolar DGFET structure, shown in figure 3(b), the same condition is true since for In_2="0", the back gate BG is set to 0V such that the transistor is P-type and will be ON if In_1="0". For other combinations In_1In_2= {"01","10","11"} the transistor will be OFF.

Hence, in a complementary static logic approach based on a pull-up network formed from p-type transistors and a pull-down network formed from n-type transistors, this approach can be applied in the case of many complex logic gates as shown in the next section.

3.2 Double-Gate static logic (DGSL) cells

3.2.1 Generic function

In our approach, we replace any NTTS and PTTS structures in static logic pull-down and pull-up networks respectively with equivalent ambipolar DGFETs. A generic example illustrating the transformation between a Conventional CMOS Static-Logic (CSL) structure and the Double-Gate Static Logic (DGSL) structure using the TTS approach is shown in figure 4. In the figure, both circuits implement the output function.

Out= $\neg[(i_1 \wedge i_2 \wedge i_3 \wedge ... i_n) \vee (j_1 \wedge j_2) \vee (k_1)]$.

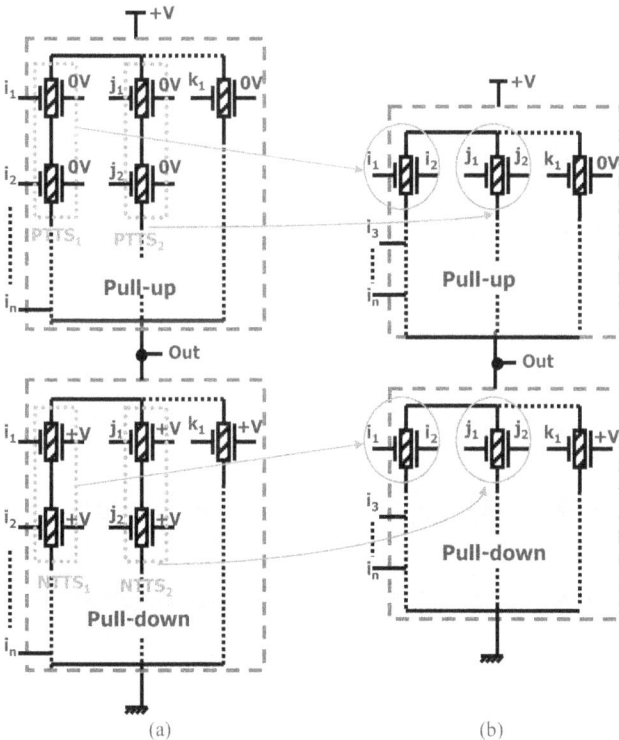

Figure 4. CSL structure (a), DGSL structure (b)

If we consider that n is the number of inputs of the function, m is the number of NTTS structures and p is the number of PTTS structures that can replace two AND-related inputs in the function path, the required number of transistors will be 2n-(m+p).

3.2.2 Examples: 2-input XOR gate and multiplexers

The approach can be applied to any complementary static logic gate containing TTS structures. In for example simple monotonic gates, such as the NAND (resp. NOR) gate, where the pull-down (resp. pull-up) network contains one NTTS (resp. PTTS) and the pull-up (resp. pull-down) network is formed from 2 transistors in parallel, only the pull-down network is substituted with a single ambipolar DGFET and we obtain a gain of a single transistor. However, in the case of more complex gates such as XOR/XNOR gates and multiplexers, the gain can be more significant and the approach can be applied to all branches of the gate.

The conventional CMOS-type 2 input XOR structure (CSL) is shown in figure 5(a). By applying the approach to this gate, figure 5(b) shows a compact XOR gate where all TTS structures were substituted with a single ambipolar DGFET, leading to a reduced transistor count with a simpler structure of 4 transistors instead of 8. Figure 5(c) shows simulation results under the conditions detailed in section IV.

We note that in figure 5(a), we choose to connect the back gates of the pull-up transistors to the ground and the back gates of the pull-down transistors to V_{dd}. But this choice is not consistent, we could connect the back gates to any different voltage with respect to the condition that the polarity of the device is set to N- or P-type with a positive (V_{BG}-V_S= +V) or negative (V_{BG}-V_S= -V) voltage, respectively.

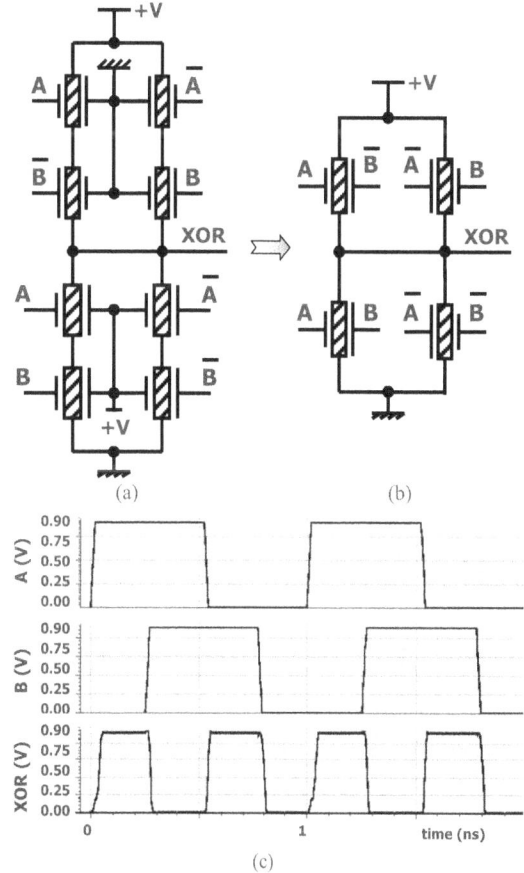

Figure 5. 2-input XOR gate: CSL structure(a) DGSL structure (b) DGSL simulated waveform (c)

To further illustrate the principle of the DGSL cells, some examples of elementary circuits are shown in figure 6. The structure of the static logic 2:1 MUX is very similar to the XOR gate so the same gain in terms of transistor count is observed. In the case of static logic 4:1 MUX, we are using 8 transistors fewer than in the conventional structure, i.e. a reduction of 40% is observed rather than 50% as in 2:1 MUX or the XOR gate .

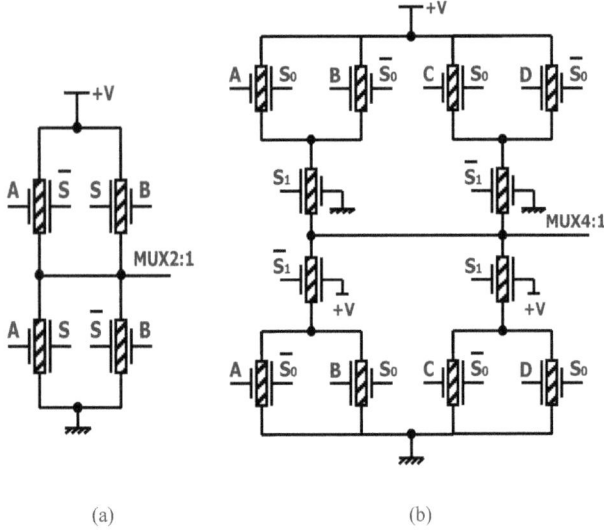

(a)   (b)

Figure 6. DGSL 2:1MUX gate (a), DGSL 4:1MUX (b)

3.2.3 Layout of DGSL structure

Figure 7(a) shows one layout of the XOR gate. This layout exploits the approach of building a complementary logic circuit along the length of a single nanotube. Here all transistors are built with the same nanotube.

In figure 7(b) we propose a second layout of the XOR gate using two nanotubes to show the possibility of direct transposition of the schematic view on a layout level.

As with any standard cell approach, the layout must be normalized to a set y-dimension, in order to place cells in rows. The inverted function (i.e. the XNOR function in this case) can be achieved by simply flipping the layout.

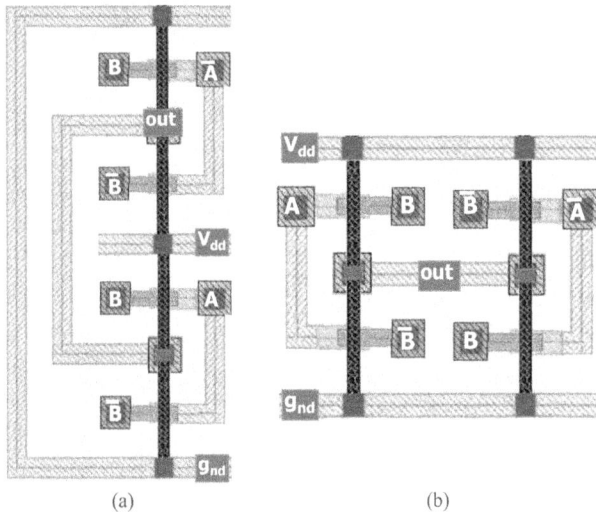

(a)   (b)

Figure 7. Layout of DGSL XOR gate: single nanotube (a) double nanotubes(b)

4. PERFORMANCE EVALUATION

While the principal goal of using the TTS association approach is to reduce transistor count in logic cells, the impact of such a reduction is expected to be extended to various performance metrics of the logic cell such as power consumption and delay.

In this section, we compare the new logic gates with their equivalent gates based on conventional CMOS-type logic. Using the same ambipolar DGFET device we extend the comparison to conventional CMOS logic built with silicon technology (SiSL) based on the Predictive Technology Model (PTM) 16nm low power (LP) transistor models [20].

4.1 Device models

While this approach is valid for any ambipolar DGFET, we evaluate performance metrics using a DG-CNTFET device model, as described in section II [10].

To obtain an objective and quantitative performance and power consumption comparison between the cells using this model with conventional cells using the CMOS PTM 16nm LP model, we fit the width of the CMOS transistors in such a way as to obtain the same I_{on} current. Table I shows the parameters used for the DG CNTFET compact model as presented in [10] and the tuned length/width of the predictive model.

In [10], the model is described in detail, with various nanotube configurations (chirality/number/diameter) for a range of values of back-gate and supply voltages. In our case, we chose configurations to enable a reasonable comparison with the PTM 16nm model in term of channel length, width, voltage and consequent I_{on}.

TABLE I Parameters of transistors used

Parameters	DG CNTFET	PTM 16nm LP
Drain access channel length	50nm	-
Source access channel length	50nm	-
Inner channel	20nm	16nm
Width	50nm	56nm(n-type)/80nm(p-type)
Chirality (n,m)	(11, 0)	-
Nb of nanotubes	12	-
Diameter of 1 nanotube	0.861176 nm	-
Supply voltage	0.9 V	0.9 V

Figures 8 and 9 show the I_{DS}/V_{GS} characteristics of the DG CNTFET model and PTM 16nm model, respectively for n-type and p-type configurations, where V_{GS} varies between 0 and 0.9 V, V_D=0.9V and V_S=0V.

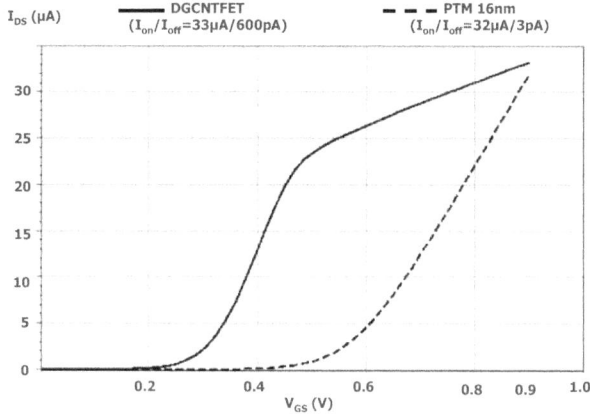

Figure 8. I_{DS}/V_{GS} characteristics of the n-branch

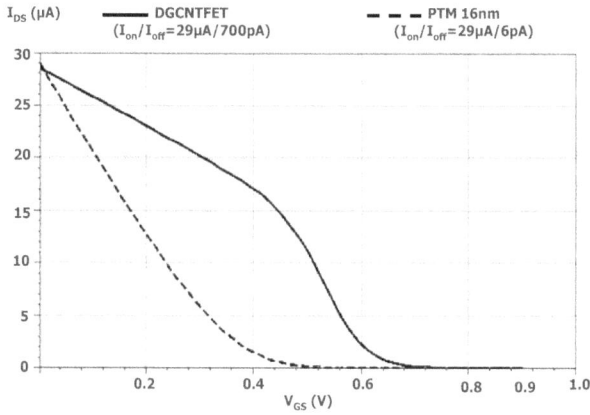

Figure 9. I_{DS}/V_{GS} characteristics of the p-branch

4.2 Program of investigation

Our comparative study was carried out between conventional static logic structures; figure 4(a), built with DG-CNTFET technology (CSL), Silicon static logic technology (SiSL) and equivalent DGSL circuits designed in this work, figure 4(b).

We suppose that there is no difference in rise and fall times between front and back gate inputs. We ran simulations with a frequency of 1GHz and equal rise and fall times (20ps) with a capacitive load of 150aF. The supply voltage was 0.9V and clock and data inputs were single rail (i.e. $+V=0.9V$, $V_0=0V$). Cyclic simulations were carried out to establish mean power consumption and worst-case time delay over all data combinations.

We ran SPECTRE simulations and used the parameters shown in table I for all transistors in the logic gates (i.e. W/L ratio of 2.5 for the DG-CNTFET, 3.5 for n-type CMOS transistors and 5.5 for p-type CMOS transistors). No resizing was carried out to balance branch resistances since this technique can be applied to all gates by using parallel transistors and has no impact from a comparative point of view.

Four different monotonic gates were simulated (XOR, XNOR, 2:1MUX, 4:1MUX) and results are shown in table II. Various performance metrics were evaluated: power consumption "P", time delay "TD", transistor count for each cell "# TRAN", the active

area "AREA" (i.e. sum of all channel areas W*L) and Power-Delay-Product "PDP".

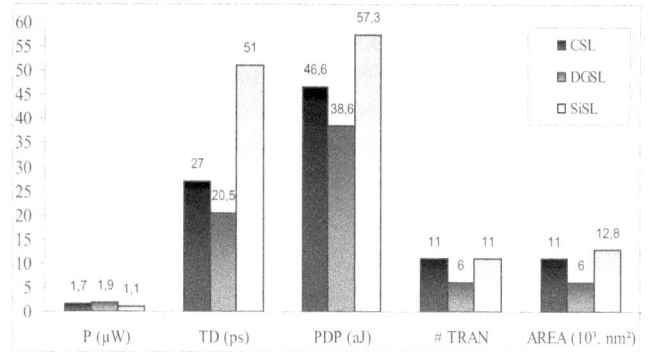

Figure 10. Comparison of average values of static logic gates

4.3 Discussion

Comparison figures show that the impact of the TTS association approach on different logic gates, proposed in this work, is almost identical.

Although fewer transistors are used, the average power consumption in this approach increases slightly (10%) compared to the conventional approach. This increase in power consumption is due to the shorter path from V_{dd} to ground that the new structures create. Since there are fewer transistors in series, the resistance per branch from (V_{dd}/G_{nd}) to output decreases to the channel resistance of a single transistor (R_{ch}) and results in a consequently higher short-circuit current during signal transition time. In the conventional structure, two transistors in series offer a path resistance of $2R_{ch}$. However this increase in current per transistor is not expected to be a reliability issue, since I_{on} in CNTFETs can attain a value 20–30X higher than that of state-of-the-art Si MOSFETs [21].

Compared with a conventional static logic structure (CSL), the new logic gates (DGSL) show an improvement in worst-case time delay of nearly 1.5X. This decrease in time delay is due to the use of fewer transistors in series, reducing the equivalent channel resistance and associated time constant with load capacitance accordingly.

One principle benefits provided by the approach is the important gain in terms of the number of transistors. While conventional static logic generally requires *2n* transistors (*n* representing fan-in), the new cells only require *2n-(m+p)* transistors (where *m* represents the number of NTTS structures and *p* represents the number of PTTS structures in the conventional gates). Table II shows clearly that for some gates only half of transistors are needed compared to conventional logic.

In this study, we also compare the power consumption and time delay of the DG-CNTFET logic gates to that of conventional Si-CMOS logic gates based on a low power (LP) 16nm predictive model. Simulations showed that CTN technology offer an improvement of 2X concerning the time delay but with an increase of power consumption of almost 2X over 16 nm silicon technology.

TABLE II COMPARISON OF CSL vs DGSL gates

	P (µW)			td(ps)			# tran			PDP (aJ)			Area ($10^3. nm^2$)		
	CSL	DGSL	SiSL	CSL	DGSL	SiSL	CSL	DGSL	SiSL	CSL	DGSL	SiSL	CSL	DGSL	SiSL
2XOR	2,3	2,5	1,3	21,0	16,0	40,0	8	4	8	48,3	40,0	52,0	8	4	9,3
2XNOR	2,3	2,5	1,4	21,0	16,0	36,7	8	4	8	48,3	40,0	51,4	8	4	9,3
MUX2:1	0,6	0,6	0,6	26,0	20,0	36,0	8	4	8	15,9	12,5	21,6	8	4	9,3
MUX4:1	1,7	1,9	1,2	40,0	30,0	91,0	20	12	20	68,0	57,0	109,2	20	12	23,4
Average	1,7	1,9	1,1	27,0	20,5	50,9	11	6	11	46,6	38,6	57,3	11	6	12,8

In fact, figures (8) and (9) illustrating the I_{DS}/V_{GS} characteristics of both model devices used for our simulations with an equal I_{on}, explain clearly the better speed and the worse power consumption of the DG CNTFET compared to the PTM 16nm. This is mainly due to the lower threshold voltage (V_{th}) of DGCNTFET compared to the PTM 16nm.

Performance is linked to I_{ds} which, in turn, is proportional to V_{th} {$I_{ds} \propto (V_{DD} - V_{th})^{1\sim2}$}. Power consumption (P) depends on static leakage current $I_{leakage}$, also dependent on V_{th} {$I_{leakage} \propto e^{-C \times Vth}$} and P depends as well on Short-circuit current { $I_{SC} \propto (\beta.\tau_{in}/12.V_{DD}).(V_{DD}- 2.V_{th})^3.f$ }. However, the PDP of gates built with DG CNT technology remains better than Si technology. By applying the approach presented in this work, we achieve an improvement of ~30% as compared to Si gates using the PTM 16nm CMOS model.

Finally, it is worth mentioning that while this work is expressed through DG CNTFETs, the main specific FET device characteristic required is that of controllable (double-gate) ambipolarity. It is expected that the circuit techniques described in this paper are also valid using other devices with this property, such as double-gate transistors based on silicon nanowires [5] or potentially in the future on graphene.

5. CONCLUSION

We have described and evaluated an approach specifically exploiting the ambipolar property of DGFETs to reduce efficiently the transistor count for logic cells in complementary static logic, leading to a greater integration density (2x) by replacing all two n-type transistors in series in the pull-down network and all two p-type transistors in series in the pull-up network with a single DGFET for equivalent functionality. When deployed onto DG-CNTFET technology and for comparable average power consumption figures, the reduction in the on-channel resistance of the function path results in a reduced delay (up to 1.5X). Compared to silicon technology based on a 16 nm predictive model, DG-CNTFET technology shows better performance but worse power consumption with a 30% of improvement concerning PDP.

6. REFERENCES

[1] EDITION International Technology Roadmap for Semiconductors: Emerging Research Devices, 2009.

[2] S. Mukhopadhyay, et. al, "Modeling and Analysis of Total Leakage Currents in Nanoscale Double Gate Devices and Circuits ISLPED, 2005

[3] L. Chang, et. al., "Direct tunneling gate leakage current in doublegate and ultrathin body MOSFETs", IEEE TED, vol. 49, Dec. 2002, pp. 2288-2295.

[4] L. Mathew, et. al, "CMOS vertical multiple independent gate field effect transistors (MIGFET)", Int. SOI Conf., 2004, pp. 187-188.

[5] J. Appenzeller et al"Dual-gate nanowire transistors with nickel silicide contacts", IEDM Technical Digest, 555-558 (2006).

[6] Kotb Jabeur, Natalya Yakymets, Ian O'Connor and Sébastien Le Beux. "Ambipolar double-gate FET binary-decision- diagram (Am-BDD) for reconfigurable logic cells", International Symposium on Nanoscale Architectures (NANOARCH), San Diego, June, 2011.

[7] Ian O'Connor et al, "Double-Gate MOSFET based Reconfigurable Cells", Electronics Letters, vol 43, Nov. 2007.

[8] K.Roy et al, "Double-Gate SOI Devices for Low-Power and High-Performance Applications", VLSID'06.

[9] Kotb. Jabeur et al, "Reducing transistor count in clocked standard cells with ambipolar double-gate FETs", IEEE/ACM international symposium on nanoscale architectures (Nanoarch'10), june 17-18 2010, Anaheim, CA,USA.

[10] S. Frégonèse et al, "A Compact Model for Dual-Gate One Dimensional FET: Application toCarbon-Nanotube FETs", IEEE Trans. on Electron Devices, 58 (1), 206 (2011).

[11] M. H. Ben Jamaa et al," FPGA Design with Double-Gate Carbon Nanotube Transistors", ECS Transactions, 34 (1) 1005-1010 (2011)

[12] L. Ding, A. Tselev, J. Wang, D. Yuan, H. Chu, T.P.McNicholas, Y. Li, J. Liu. "Selective Growth of Well-Aligned Semiconducting Single-Walled Carbon Nanotubes", Nano Letters, vol. 9, no. 2, pp. 800, 2009

[13] A. Javey et al. "Self-Aligned Ballistic Molecular Transistors and Electrically Parallel Nanotube Arrays", Nano Letters, vol. 4, no. 7, pp. 1319–1322, 2004.

[14] M. H. Ben Jamaa et al, "Novel Library of Logic Gates with Ambipolar CNTFETs: Opportunities for Multi-Level Logic Synthesis", Design, Automation and Test in Europe (DATE). March 2009.

[15] A. Raychowdhury et al, "Circuit-compatible modeling of carbon nanotube FETs in the ballistic limit of performance" in Proc. 3rd IEEE-NANO, 2003, vol. 2, pp. 343–346.

[16] J. Marulanda, et al, "Threshold and saturation voltages modeling of carbon nanotube field effect transistors (CNTFETs)", Nano, vol. 3, pp. 195–201, 2008.

[17] J. Goguet et al, "A charge approach for a compact model of dual gate CNTFET" in Proc. Int. Conf. DTISNanoscale Era, 2008, pp. 1–5.

[18] Y. Lin et al, "High-performance carbon nanotube field-effect transistor with tunable polarities", IEEE Trans.Nanotechnol., vol. 4, no. 5, pp. 481–489, Sep. 2005.

[19] A. Javey et al, "Carbon nanotube field-effect transistors with integrated ohmic contacts and high- κ gate dielectrics," Nano Lett., vol. 4,no. 3, pp. 447–450, Mar. 2004.

[20] http://ptm.asu.edu/modelcard/LP/16nm_LP.pm

[21] A. Javey, J. Guo, Q. Wang, M. Lundstrom, H. Dai, "Ballistic Carbon Nanotube Transistors". Nature, vol. 424, pp. 654–657, 2003.

Performance and Energy Models for Memristor-based 1T1R RRAM Cell

Mahmoud Zangeneh and Ajay Joshi
Electrical and Computer Engineering Department, Boston University
Boston, MA, USA
{zangeneh, joshi}@bu.edu

Abstract

Sustaining the trend of lowering energy dissipation and read /write access time and increasing density in CMOS-based memory arrays is becoming extremely challenging with each new technology generation. Hence, alternate technologies that can supplant CMOS technology need to be explored. Memristor-based resistive memory with its scaling potential and endurance is one of the viable replacements to CMOS. This paper presents accurate analytical models for the performance and the energy dissipation of a 1-transistor 1- memristor (1T1R) resistive random access memory (RRAM) cell structure. We have verified our models against detailed HSPICE simulations and our models show that the time required to write logic one into the cell is typically 30% larger than the time required to write logic zero and is in the order of ns. Unlike the access time, the energy dissipation of the cell is the same for writing logic one and logic zero (less than 350 fJ/bit). The energy dissipated while reading is roughly 120 fJ/bit which is 65% less than the energy of writing.

Categories and Subject Descriptors

B.3.1 [MEMORY STRUCTURES]: Semiconductor Memories

Keywords

Memristor, Resistive memory, Modeling

1. INTRODUCTION

Access time, density and power dissipation of memory (on-chip static random access memory (SRAM), off-chip dynamic random access memory (DRAM) or Flash Memory) have a direct impact on the overall performance of a computing system. Over the years, complementary metal oxide semiconductor (CMOS) technology scaling made it possible to continuously decrease access time and energy consumption, and increase the density of memory blocks. However, maintaining this trend has become extremely challenging due to the limitations associated with scaling CMOS technology in the nanometer regime. To maintain reliable operation under the manufacturing variations in the nanometer

regime, redundant circuits are required that increases power dissipation and decreases memory density. Moreover, any intermediate error detection or correction steps needed for reliable operation increases the memory access time.

It is therefore imperative to explore emerging devices for building memory arrays that have lower power dissipation, higher performance, higher density and reliable operation for future computing systems. Two-terminal memristor devices, with their excellent scaling potential (< 10 nm) and endurance (> 10 billion cycles) [1], can be used as storage elements and are considered viable replacements to conventional CMOS-based memory designs. The memristor can be considered as a variable resistor which can be programmed by changing the voltage drop across the memristor or changing the current injected into the memristor. Here, programming amounts to changing the value of the 'memristance' which leads to two different states for the memristor. These two states can correspond to storage of logic 0 and logic 1 in the memristor.

In this paper we model and verify the performance and energy for reading and writing a non-volatile monolithically integrated hybrid CMOS/memristor memory cell consisting of 1 CMOS transistor and 1 memristor [2]. This 1-transistor 1-memristor (1T1R) resistive random access memory (RRAM) cell structure is similar to a DRAM cell – the data is stored as the resistance of the memristor, and the transistor serves as an access switch for reading and writing data. We chose the 1T1R cell as the basic building block for a non-volatile RRAM array as it avoids sneak path problems [3] and ensures reliable operation. The two main contributions of this paper can be summarized as follows:

- We have formulated the complete end-to-end functionality (read/write/refresh) of the 1T1R RRAM cell at the circuit level.
- We have developed analytical models for both performance and energy consumption during write operation and read operation of the 1T1R RRAM cell. These models have been validated against detailed circuit simulations using HSPICE.

The rest of the paper is organized as follows. Section 2 reviews the device-level aspects of the memristors, while Section 3 summarizes the memristor models and memristor-based memory cell designs proposed by other researchers. Section 4 explains the functionality of the non-volatile 1T1R RRAM cell. Sections 5 and 6 discuss the analytical models (development and validation) for the performance and energy consumption during read/write operation of the 1T1R RRAM cell, respectively. Section 7 concludes the paper.

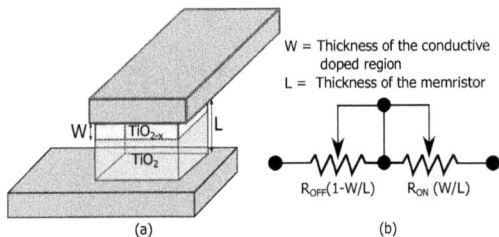

Figure 1: Variable resistor model for the memristor consisting of a highly conductive doped region and a highly resistive undoped region. (a) 3D view of the device. (b) Circuit-level model.

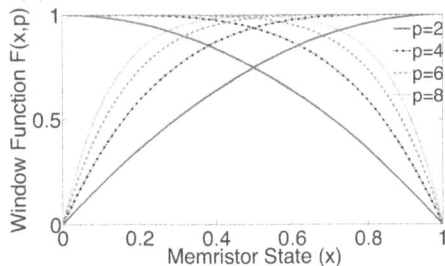

Figure 2: Dynamic window function of the memristor state showing the nonlinear behavior of the memristor for different control parameter p. The current sign function prevents the state from getting stuck at the two boundaries [6].

2. MEMRISTOR DEVICE TECHNOLOGY

Memristors provide a functional relationship between the charge and flux and was first postulated in [4]. The measurement results of a titanium dioxide nanoscale device that exhibited these memristive characteristics was first presented in [5]. The physical model for the memristor in [5] was specified by a time-dependent resistor whose value is linearly proportional to the charge q passing through it. The fabricated prototype introduced in [5] consists of a highly resistive thin layer of TiO_2 and a second conductive deoxygenized TiO_{2-x} layer (see Figure 1). The change in the oxygen vacancies due to a voltage applied across the memristor modulates the region of operation in the memristor. This results in two distinct states - a high resistance state and a low resistance state corresponding to the resistive and conductive region of operation, respectively. The effective 'memristance' of the memristor device can be calculated using Equation (1) proposed in [5].

$$M(t) = R_{ON} \frac{w(t)}{L} + R_{OFF}(1 - \frac{w(t)}{L}). \qquad (1)$$

In Equation (1), R_{ON} and R_{OFF} are the minimum and maximum memristance, respectively, $w(t)$ is the thickness of the conductive doped region as a function of time, and L is the memristor thickness. The doped region is usually considered normalized to the memristor thickness [1] and is written as $x(t) = w(t)/L$, where $x(t)$ is called the state of the memristor. The rate of change of the memristor state is a function of the memristor physical parameters and the current through the memristor. As the current itself varies with time, the change of memristor state exhibits nonlinear behavior. This nonlinear behavior can be expressed using a window function as shown in Equation (2) [1].

$$\frac{dx(t)}{dt} = \frac{\mu_v R_{ON}}{L^2} i(t) F(x(t), p) \qquad (2)$$

In Equation (2), $F(x(t), p)$ is the window function and the nonlinearity can be controlled with control parameter p. Increasing p yields a flat window function for larger memristor states. Also, μ_v is the mobility of the oxygen vacancy

dopants. Various window functions have been proposed in the literature that consider the linear ionic drift and the nonlinear behavior which appears at the boundaries of the memristor state. A linear approximation of the dynamic behavior of the state is addressed in [7]. This model, however, gets stuck at the boundaries of the state. A modified function that considers the nonlinear behavior of the memristor is proposed in [8] where it still gets stuck once the state reaches the two boundaries. A new window function that models the nonlinear behavior of the rate in the state change without getting stuck at the boundaries is proposed in [6] and given in Equation (3). We use this window function for developing the performance and energy models of the 1T1R RRAM cell.

$$F(x(t), p) = 1 - (x - sgn(-i(t))^{2p}. \qquad (3)$$

In Equation (3), $i(t)$ is the current through the memristor and sgn is a sign function that prevents the state of the cell from getting stuck at the borders. Figure 2 shows a plot of the window function for different p values.

3. RELATED WORK

There are multiple efforts in place to develop accurate models for the two-terminal memristor elements. An analytical TiO_{2-x} memristor model and the corresponding SPICE code that express both the static transport tunneling gap width and the dynamic behavior of the memristor state based on the measurement results are proposed in [9]. In [6], the authors developed a mathematical model for the prototype of memristor reported in [5] with dependent voltage and current sources as well as an auxiliary capacitor which functions as integrator to calculate the state of the memristor. The authors in [10] presented a schematic diagram of the memristor SPICE macromodel based on a simplified window function for the state change rate. However, most of these efforts use simplified analytical approaches and do not accurately consider the nonlinear behavior of the memristors. Moreover, not all of these models have been validated against measurement results. Several circuit topologies have been proposed in the literature based on the memristive structures. The authors in [11] used a Si-based memristive system to fabricate high-density crossbar arrays that can be addressed with high yield and ON/OFF ratio.

A read/write memristor based memory cell is introduced in [12] which utilizes purely system-level simulations to evaluate its functionality and lacks circuit-level verifications for the proposed closed-form expressions. An energy-efficient dual-element memristor-based memory structure is proposed in [13], in which each memory cell contains two memristors that store the complement states. Similarly, a 2-bit storage memristive cell is proposed in [14]. Both these multi-bit memory cells have large area overhead. Content addressable memory (CAM) design using memristors has been introduced in [1]. The simulation results and the measurements from the fabricated prototype of the proposed CAM cell are not yet verified. Hybrid logic circuits based on the crossbar/CMOS structures are physically viable and proposed in [15] and could offer function density of at least two orders of magnitude higher than that of their CMOS counterparts fabricated with the same design rules, at the same power density, but with higher logic delay [11]. An analysis on the peripheral circuitry of the crossbar array architecture is presented in [16]. We propose a 1T1R RRAM cell

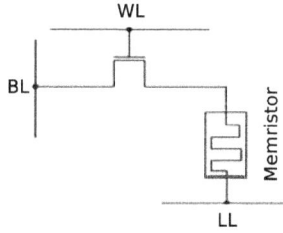

Figure 3: Schematic of the 1T1R RRAM cell. The capacitor of the conventional volatile DRAM is replaced with a memristor resulting in a nonvolatile memory cell.

based memory architecture. Unlike the proposed architecture in [16], the wordline or bitline drivers do not have direct access to the memristor. Hence, we do not have the issue of sneak paths that would be observed in the cross-point array architecture. We account for the access transistor and bitline parasitics to develop models for read/write access time and energy. We have validated our models against detailed HSPICE simulations.

4. 1T1R RESISTIVE RANDOM ACCESS MEMORY (RRAM) CELL

The difficulty associated with scaling SRAM, DRAM and Flash memory necessitates the exploration of emerging devices that can sustain the trend of lowering power, lowering access time and increasing density with each new generation. A non-volatile and low-power memory cell consisting of 1 CMOS transistor and 1 memristor is illustrated in Figure 3. This 1T1R RRAM cell stores data in the form of the resistance of the memristor element. The overall architecture of a memory array built using RRAM cells is similar to the conventional DRAM cell i.e. a wordline is used to select a row of cells, and a bitline is shared by the cells in a column for reading/writing.

During write operation, a voltage of V_{DD} is applied to the wordline, and a positive or negative voltage is applied across the memristor for writing logic 1 or logic 0, respectively. This voltage drop across the memristor is achieved by charging the bit line to V_{DD} (for logic 1) or discharging the bitline to 0 V (for logic 0) and applying a voltage of $V_{DD}/2$ at node LL. The write access time is dependent on the voltage drop across the memristor and the physical parameters of the memristor. While writing logic 1, the current flowing through the memristor increases the size of conductive region, thus reducing the 'memristance'. While writing logic 0, the current flowing through the memristor decreases the size of conductive region, thus increasing the 'memristance'. To read data out of the 1T1R cell we first discharge the bitline to 0 V, apply a voltage of V_{DD} to the wordline and apply V_{DD} to node LL. The pulse width on the wordline and loadline during reading is the same as read access time. For a fixed predefined time period, depending on the data stored in the cell (i.e. the resistance of the memristor), the bitline charges to a value that is above (for logic 1) or below (for logic 0) a threshold voltage. The bitline resistor and capacitor is calculated to be 6.8 $K\Omega$ and 200 fF for 1 mm bitline length, respectively. A differential sense-amplifier with V_{BL} as one input and threshold voltage as the other input is used to determine the value stored in the 1T1R RRAM cell. The sense amplifier has been designed to detect voltage differentials of 50 mV and larger. The read noise margin of the 1T1R RRAM cell is larger than other cell topologies,

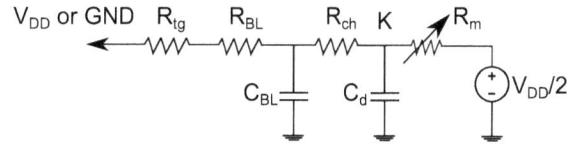

Figure 4: Equivalent circuit of the 1T1R cell in writing phase. V_{BL} is set to V_{DD} through a transmission gate switch while writing logic 1 and remains at zero while writing logic 0.

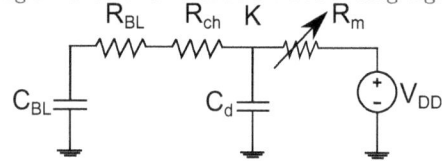

Figure 5: Equivalent circuit of the 1T1R cell in reading phase. The transmission gate switch is off while reading.

which lowers the required gain of the sense amplifier, leading to lower power consumption. Higher noise margins of the 1T1R cell particularly appear when smaller R_{ON} is chosen. This leads to larger R_{OFF}/R_{ON} ratios and larger reliabilities since the bitline voltage for reading logic 1 will increase and the following sense amplifier can distinguish between logic 1 and 0. Also increasing the memristor thickness will lead to higher noise margins since the rate of change of state will decrease according to Equation (2), leading to larger bitline voltages when reading logic 1. This will however increase the area of the 1T1R cell. An additional source of power saving during read operation is the discharging of the bitline to 0 V prior to reading as against the precharging operation performed for DRAM. It should be noted that during both read and write operation, we activate all the bitlines that share the selected wordline i.e. we read/write all the cells in the row. In case of architectures where only a part of the row is read/written, switches will need to be added onto the wordline and loadline. The unselected cells in the other rows are isolated using the CMOS access transistors.

5. PERFORMANCE MODELS

In this section, we present the models for read and write operation of a 1T1R RRAM cell. The proposed performance models can be used in the design exploration of a large memristive array architecture. The equivalent circuit model for the 1T1R RRAM cell during write operation is shown in Figure 4, while that for the read operation is shown in Figure 5. Here, R_m is the equivalent resistance of the memristor, R_{ch} is the access transistor channel resistance during writing/reading while operating in the triode region and R_{tg} is the transmission gate switch equivalent resistance which is used for pre-charging or pre-discharging the bitline capacitor. R_{BL}, C_{BL} and C_d are the bitline resistor, bitline capacitor and transistor junction capacitor, respectively.

5.1 Write Operation Model

The time required to change the state from 0 to 1 and 1 to 0 of a memristor that is directly connected to a voltage source was previously derived in [12] and is given by

$$T_w = \frac{L^2(1+\beta)}{2\mu_v V_A} \qquad (4)$$

where β is the ratio of R_{OFF}/R_{ON} and V_A is the magnitude of the applied voltage. According to equation (4), the write time of a memristor, which is a function of physical parameters of the device, increases with increase in L and β and decreases with increase in applied voltage due to the increase

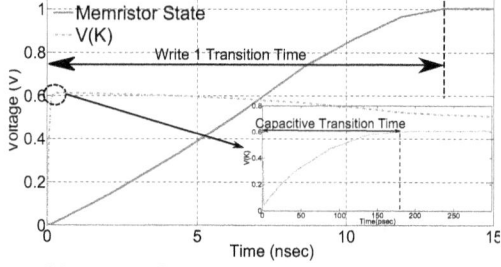

Figure 6: Variation of node "K" while writing logic 1. The write access time dominates the transient response of the bitline and parasitic cell capacitors.

in the injected flux into the device. This write model does not consider the nonlinearity of the change of state leading to an inaccurate estimation of the write time.

We use a 1T1R memory cell that avoids sneak path by isolating the memristor when not in use. The write operation model presented here for the 1T1R cell uses the window function proposed in [6]. Figure 6 shows the transient response of the node "K" (see Figure 4), where the charging time of bitline and transistor junction capacitors is less than a nanosecond whereas the write access time is several nanoseconds. Therefore, the analytical model presented in this paper neglects the transient response of the cell junction and bitline capacitors. Thus, considering Figure 4, the expression for memristor current will be

$$i_w(t) = \frac{V_{DD}}{2(R_{ch} + R_{tg} + R_{BL} + R_m(t))} \quad (5)$$

This can be used with opposite signs to determine the time for both writing logic 0 (negative) and logic 1 (positive). Considering the window function in Equation (3) and the rate of change of state in Equation (2), the write logic 1 access time can be approximated as

$$T_w = \frac{R_{TOT} I_1 - R_{OFF} I_2}{\eta} \quad (6)$$

where $R_{TOT} = R_{ch} + R_{tg} + R_{BL} + R_{OFF}$, $\eta = \frac{V_{DD} \mu_v R_{ON}}{2L^2}$, I_1 and I_2 are the solutions to the integrals $\int_{0.1}^{0.9} \frac{1}{1-x^4} dx$ and $\int_{0.1}^{0.9} \frac{x}{1-x^4} dx$, respectively. Note that the resistance of the memristor in (1) is approximated as $R_m(t) \approx R_{OFF}(1 - x(t))$ for simplicity. The integrals are determined from the window function we considered previously to model the nonlinearity of the memristor at the boundaries in equation (3) with $p = 2$. For writing logic 0 into the cell, the same Equation (6), where I_1 and I_2 are the solutions to the integrals $\int_{0.1}^{0.9} \frac{1}{1-(x-1)^4} dx$ and $\int_{0.1}^{0.9} \frac{x}{1-(x-1)^4} dx$, respectively can be used. Considering the bitline capacitor transition time will lead to a complicated differential equation which does not have a closed-form solution. However, it can be solved numerically using MATLAB *ode* solvers. Therefore, the rate in state change of the memristor is the solution to

$$\frac{C_{BL} R_{OFF}(R_{tg} + R_{BL})(\frac{dx}{dt})^2}{\eta(1-x^4)} - \frac{(R_{TOT} - x R_{OFF})(\frac{dx}{dt})}{\eta(1-x^4)} + \frac{V_{DD}}{2} = 0 \quad (7)$$

where the first term includes the bitline capacitor transition time. A comparison of the write 1 access time of the 1T1R RRAM cell for different memristor thicknesses using analytical models and HSPICE simulations is illustrated in Figure 7. For the HSPICE simulations, the 32 nm CMOS predictive technology model (PTM) [17] is used for the transistor, while memristor is modeled based on the model in [6]. Here we considered write 1(0) access time as the time required for the memristor resistance to change between 10%

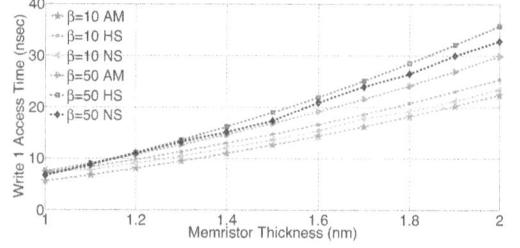

Figure 7: Analytical model vs HSPICE simulations for write 1 access time as a function of memristor thickness for different $\beta = \frac{R_{OFF}}{R_{ON}}$. AM = Analytical model ignoring bitline capacitance, HS = Hspice simulations and NS = Numerical solution accounting for bitline capacitance.

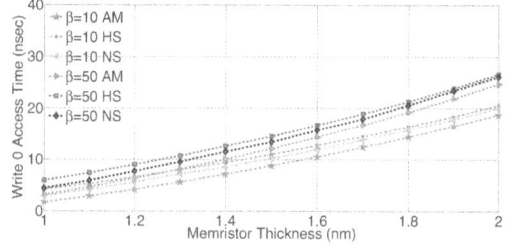

Figure 8: Analytical model vs HSPICE simulations for write 0 access time as a function of memristor thickness for different $\beta = \frac{R_{OFF}}{R_{ON}}$. AM = Analytical model ignoring bitline capacitance, HS = Hspice simulations and NS = Numerical solution accounting for bitline capacitance.

to 90% (90% to 10%) of $R_{OFF} - R_{ON}$ since the nonlinearity of the model at the borders leads to unacceptably large write access times. The proposed model underestimates the write access time by 15.1% average error as we neglected the transition times of the bitline and junction capacitors. Also it deviates from the HSPICE simulation for larger β values due to the increased nonlinearity of the device for larger R_{OFF}/R_{ON} ratios. Considering the capacitive transition times of the bitline will reduce the error to 7.3% based on the numerical solution of equation (7).

Similarly, Figure 8 shows the write 0 access time of the cell as function of the memristor thickness. The time to write logic 0 is less than the time to write 1 in the RRAM cell as the voltage drop (and hence the drive current) across memristor while writing 0 is $(V_{DD}/2)$ which is greater than that while writing logic 1 which is $(V_{DD} - V_{th} - V_{DD}/2)$ – due to the V_{th} across the pass transistor.

5.2 Read Operation Model

We define the read access time as the time required for the bitline to charge to $0.45 V_{DD}$ while reading logic 0 (See Figure 9). If a cell with logic 1 stored in it were to be read out during this time, the bitline voltage will be larger than $V_{DD}/2$ (close to $0.65 V_{DD}$ for this example). We use a differential sense amplifier with $V_{DD}/2$ and bitline voltage as two inputs to determine the data stored in the cell. As mentioned earlier, the sense amplifier has been designed to determine voltage differentials of 50 mV or larger. This approach is adopted as the memristor resistance does not change while reading 0, leading to a simplified read time model. The reading 1 process is destructive and results in the memristor state (i.e. memristance) changing during the read operation. Moreover, this change of state is nonlinear making it difficult to formulate a closed form expression. It

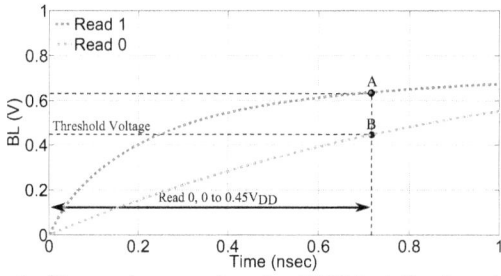

Figure 9: The read access time for 1T1R is defined as the time required for bitline voltage to increase from $0\ V$ to $0.45V_{DD}$ while reading logic 0.

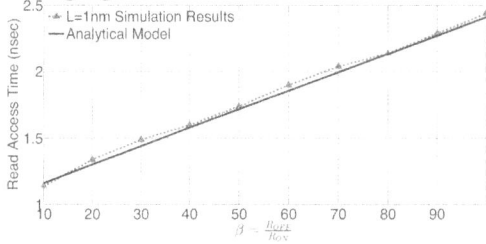

Figure 10: Analytical model versus HSPICE simulation results for read access time of the 1T1R cell as a function of β for different memristor thicknesses L. T_R is independent of memristor thickness ($R_{tg} = 582\Omega, R_{ON} = 100\Omega, C_{BL} = 200fF$).

should be noted that there is roughly 20% energy overhead with using read 0 for defining the read access time. This overhead can be decreased by using a threshold input lower than $V_{DD}/2$ to the sense amplifier.

To model the read access time, we consider the bitline voltage for the equivalent circuit in Figure 5. We need to consider the frequency response of the circuit as the read time is comparable to the charging/discharging time of C_{BL} and C_d. In order to analyze the frequency response of the bitline voltage, we use the KVL and KCL expressions which lead to the complete solution for the voltage on the bitline capacitor in Laplace domain as

$$V_{BL}(s) = \frac{1}{s(1 + 2\alpha s + \omega_0^2 s)} \quad (8)$$

where ω_0 and α are the natural frequency and the damping factor of any second order system, respectively. The time domain expression for the bitline voltage is

$$V_{BL}(t) = V_{DD} + Ae^{s_1 t} + Be^{s_2 t} \quad (9)$$

where s_1 and s_2 are the roots of the characteristic equation in (8). Considering $R_{tg} = 582\Omega, C_{BL} = 200fF, C_d = 3fF$, two poles are away by more than three orders of magnitude, and therefore the non-dominant pole can be neglected in the transient response of the cell while reading. In other words, C_{BL} is the dominant capacitor. Moreover, considering the Laplace inverse transform of the bitline voltage, we have $A \approx -1$ in Equation (9). Since the read access time of the 1T1R RRAM cell is smaller than the write access time, the transient response of the bitline capacitor cannot be neglected. In order to approximate the read access time from the time response of the cell in Equation (9), we use the method of open-circuit time constants in the high-frequency regime [18]. Neglecting (by making open-circuit) C_d, we can find the time constant associated to C_{BL}. Therefore, the final read access time can be approximated as

$$T_R = 0.69(R_{ch} + R_{BL} + R_{OFF})C_{BL}. \quad (10)$$

We show the read access time of the 1T1R cell in Figure 10 as a function of β values. It should be noted that

β	Refresh Energy	Refresh Time	Memristor Thickness	Refresh Energy	Refresh Time
10	172 fJ	3.04 ns	1.2 nm	172 fJ	3.58 ns
30	170 fJ	3.38 ns	1.4 nm	158 fJ	3.27 ns
50	170 fJ	3.74 ns	1.6 nm	132 fJ	2.52 ns
70	170 fJ	4.27 ns	1.8 nm	80 fJ	1.7 ns
90	170 fJ	4.91 ns	2 nm	24 fJ	0.48 ns

Table 1: Refresh time and energy dissipation of the 1T1R RRAM cell using HSPICE simulation for different β (left half) and memristor thickness (right half) values.

the read access time is a monotonically increasing function. The proposed model shows an acceptable matching with the HSPICE simulation results with an average error of 3.5%.

The 1T1R cell needs to be refreshed after every destructive read 1 cycle. The refresh cycle time can be modeled using the approach for write operation model. The refresh cycle time is however smaller than the write access time of the cell. The exact refresh cycle time depends on the destructiveness of the read operation and the physical properties of the memristor. Table 1 compares the refresh time and energy dissipation of the 1T1R RRAM cell using HSPICE simulation for different β and memristor thickness values.

6. ENERGY MODELS

In this section, we present the models for energy consumption during read and write operation. The proposed energy models can be used in the design exploration of a large memristive array architecture. It should be noted that the energy consumed in the word line, loadline and bitline is not explicitly included, but can be easily determined using $E = (C_{BL} + C_{LL} + C_{WL})V_{DD}^2$, where C_{BL}, C_{LL} and C_{WL} are the bitline, loadline and wordline capacitances.

6.1 Write Operation Model

As was mentioned in Section 5.1, the instantaneous current of the memristor while writing one or zero is determined by Equation (5), which depends on the memristor resistance at that instance of time. The energy dissipated in the cell can therefore be calculated as

$$E_W = \int_0^{T_W} \frac{V_{DD}}{2} i(t)\, dt = \frac{V_{DD}}{2\gamma} \int_{0.1}^{0.9} \frac{dx}{F(x)} = \frac{V_{DD}I_1}{2\gamma} \quad (11)$$

where $\gamma = \mu_v R_{ON}/L^2$ and $I_1 = \int_{0.1}^{0.9} \frac{1}{1-x^4}\, dx$ is the integral of the inverse of the window function. Also $V_{DD}/2$ is the series combination of the bitline charge voltage and the voltage at node LL (see Figure 4). According to (11), the energy dissipated in the 1T1R RRAM cell during write operation is a linear function of power supply and a quadratic function of the memristor thickness L. Moreover, the maximum memristor resistance i.e. R_{OFF} does not affect the dissipated energy. The same Equation (11) can be used for calculating the energy dissipated while writing logic 0, but with a different window function ($F(x) = 1 - (x-1)^4$) and boundary conditions (0.9 to 0.1).

Figure 11 shows the energy dissipated during write operation in a 1T1R RRAM cell as a function of memristor thickness for different values of β when calculated using analytical models and HSPICE simulations. There is an average error of 23.2% between the energy calculated using the proposed model neglecting the capacitive transition times while writing and HSPICE simulation. Using MATLAB ode solvers, we can numerically find the energy considering the capacitive transition times of the bitline which reduces the error to 11.3%. It should be noted that here dissipated energy

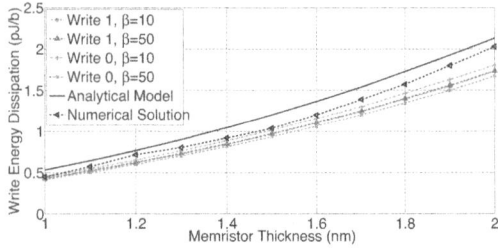

Figure 11: Analytical model versus HSPICE simulations for calculating the energy dissipated during write operation of a 1T1R RRAM cell as a function of memristor thickness for different β values ($R_{tg} = 582\Omega, R_{ON} = 100\Omega, C_{BL} = 200fF$). AM = Analytical model ignoring the bitline capacitance and NS = Numerical solution considering bitline capacitance.

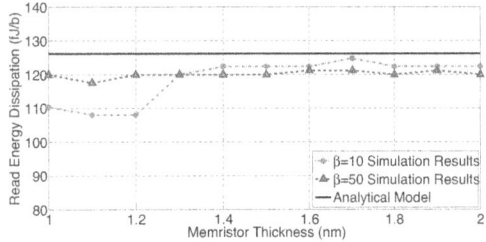

Figure 12: Read dissipated energy of the 1T1R cell as a function of memristor thickness for different β values. Read dissipated energy is independent of the memristor thickness and β ($R_{ON} = 100\Omega, C_{BL} = 200fF$).

corresponds to the required energy for the cell to change the state from 10% to 90% of its maximum value.

6.2 Read Operation Model

We propose the model for the energy dissipated while reading the 1T1R RRAM cell in this section. Using the equivalent circuit of the cell during read cycle (see Figure 5), the memristor current equation in time domain will be

$$i_R(t) = Ae^{s_1 t} + Be^{s_2 t}. \tag{12}$$

Similar to the read access time analysis, we can neglect the effects of the non-dominant pole. Therefore, the memristor current will simply be $i_R(t) \approx Ae^{s_1 t}$ where the coefficient $A = V_{DD}/(R_{ch} + R_{BL} + R_m(0))$, for reading one, $R_m(0) = R_{ON}$ and while reading zero, $R_m(0) = R_{OFF}$. The dissipated energy of the cell while reading can be expressed as

$$E_R = \int_0^{T_R} V_{DD} i_R(t)\, dt = \frac{V_{DD}^2}{(R_{ch} + R_{BL} + R_m(0))} \int_0^{T_R} e^{s_1 t}\, dt \tag{13}$$

where T_R is the read access time of the cell. Considering the dominant pole of the cell and the read access time equation in (10), the dissipated energy of the 1T1R RRAM cell is

$$E_R \approx 0.63 C_{BL} V_{DD}^2 \tag{14}$$

The read energy is therefore independent of the memristor physical parameters. A comparison of the read energy calculation using the proposed model and the HSPICE simulation is shown in Figure 12. The proposed model overestimates the dissipated energy since the open-circuit time constants method used in Equation (10) does not consider the non-dominant pole of the 1T1R cell leading to a larger cell current and energy. We can observe an average error of 6.5% between the model and the HSPICE simulation results.

7. CONCLUSION

In this paper, we derived accurate models for the performance and the energy dissipation of the 1T1R TiO_2-based RRAM cell. The models were verified against detailed HSPICE circuit simulations. The write access time is inversely proportional to minimum value of memristance and directly proportional to the square of memristor thickness. The read access time of the cell is only a function of the maximum value of memristance and does not change by the memrsitor thickness. Read operation is one order of magnitude faster than write operation in the 1T1R cell. From energy perspective, the write operation is roughly three times more energy consuming than the read operation. The write energy increases quadratically for larger memristor thicknesses while read energy dissipation depends only on the bitline capacitor and is not a function of the physical parameters of the memristor.

8. ACKNOWLEDGMENT

This work was supported in part by CELEST, a National Science Foundation Science of Learning Center (NSF SBE-0354378 and NSF OMA-0835976).

9. REFERENCES

[1] K. Eshraghian et al., "Memristor mos content addressable memory (mcam): Hybrid architecture for future high performance search engines," IEEE Trans VLSI, vol. 19, no. 8, pp. 1407 –1417, aug. 2011.

[2] S.-S. Sheu et al., "A 5ns fast write multi-level non-volatile 1 k bits rram memory with advance write scheme," in VLSI Circuits, 2009 Symposium on, 2009.

[3] J.-J. Huang et al., "One selector-one resistor (1s1r) crossbar array for high-density flexible memory applications," in IEDM, 2011.

[4] L. Chua, "Memristor-the missing circuit element," Circuit Theory, IEEE Transactions on, 1971.

[5] D. B. Strukov et al., "The missing memristor found," Nature, vol. 453, no. 7191, pp. 80–83, May 2008.

[6] Z. Biolek et al., "Spice model of memristor with nonlinear dopant drift," Radioengineering, 2009.

[7] S. Benderli and T. A. Wey, "On spice macromodelling of tio2 memristors," Electronics Letters, 2009.

[8] Y. Joglekar et al., "The elusive memristor: properties ofbasic electrical circuits," Eur. J. Phys., 2009.

[9] H. Abdalla and M. D. Pickett, "Spice modeling of memristors," in Proc. ISCAS, 2011, pp. 1832–1835.

[10] A. Rak and G. Cserey, "Macromodeling of the memristor in spice," IEEE Trans. CADICS, no. 4, pp. 632–636, 2010.

[11] S. H. Jo et al., "High-density crossbar arrays based on a si memristive system," Nano Letters, 2009.

[12] Y. Ho et al., "Dynamical properties and design analysis for nonvolatile memristor memories," IEEE Trans. CAS I, vol. 58, no. 4, pp. 724 –736, april 2011.

[13] D. Niu et al., "Low-power dual-element memristor based memory design," in Proc. ISLPED, 2010.

[14] H. Manem and G. S. Rose, "A read-monitored write circuit for 1t1m multi-level memristor memories," in Proc. ISCAS, may 2011, pp. 2938 –2941.

[15] Y.-B. Kim et al., "Bi-layered rram with unlimited endurance and extremely uniform switching," in Proc. VLSIT, june 2011, pp. 52 –53.

[16] C. Xu et al., "Design implications of memristor-based rram cross-point structures," in Proc. DATE, 2011.

[17] (2011). [Online]. Available: http://ptm.asu.edu/

[18] T. Lee, The design of CMOS radio-frequency integrated circuits. Cambridge University Press, 2004.

Accelerating Thermal Simulations of 3D ICs with Liquid Cooling using Neural Networks

Alessandro Vincenzi, Arvind Sridhar, Martino Ruggiero, David Atienza

Embedded Systems Laboratory (ESL)
École Polytechnique Fédérale de Lausanne (EPFL), Switzerland

{alessandro.vincenzi, arvind.sridhar, martino.ruggiero, david.atienza} @ epfl.ch

ABSTRACT

Vertical integration is a promising solution to further increase the performance of future ICs, but such 3D ICs present complex thermal issues that cannot be solved by conventional cooling techniques. Interlayer liquid cooling has been proposed to extract the heat accumulated within the chip. However, the development of liquid-cooled 3D ICs strongly relies on the availability of accurate and fast thermal models.

In this work, we present a novel thermal model for 3D ICs with interlayer liquid cooling that exploits the neural network theory. Neural Networks can be trained to mimic with high accuracy the thermal behavior of 3D ICs and their implementation can efficiently exploit the massive computational power of modern parallel architectures such as graphic processing units. We have designed an ad-hoc Neural Network model based on pertinent physical considerations of how heat propagates in 3D IC architectures, as well as exploring the most optimal configuration of the model to improve the simulation speed without undermining accuracy. We have assessed the accuracy and run-time speed-ups of the proposed model against a 3D IC simulator based on compact model. We show that the proposed thermal simulator achieves speed-ups up to 106x for 3D ICs with liquid cooling while preserving the maximum absolute error lower than 1.0 °C.

Categories and Subject Descriptors

B.7.2 [**Integrated Circuits**]: Design Aids—*Simulation*

Keywords

Thermal Modeling, Thermal Simulation, 3D ICs, Liquid Cooling, Neural Networks

1. INTRODUCTION

With conventional scaling of CMOS devices fast running into theoretical walls, vertical integration of IC dies seems to be the most viable solution to meet the ever rising demands for more compact and faster electronic products in the short- and medium-term [1]. However, 3D integration brings with it aggravated thermal issues due to compounded heat fluxes and larger thermal resistances. In this context, interlayer microchannel-based liquid cooling of 3D ICs has come to be accepted as the most promising solution enabling vertical integration of ICs [2].

Nevertheless, interlayer liquid cooling still remains a niche technology and has not yet been deployed for large-scale production by the electronics industry. One of the main reasons which limits interlayer liquid cooling technology is the lack of Electronic Design Automation (EDA) tools that can provide IC designers with efficient simulation of thermal behavior of ICs cooled using microchannel heat sinks. In particular, this situation becomes a major bottleneck for the development of thermal-aware design and run-time approaches for liquid cooled 3D ICs.

In the last few years, several authors have attempted to address this issue by using various modeling methodologies [3, 4, 5]. Unfortunately, most of these modeling and simulation solutions are based on either fine-grained finite-element/finite-difference methods [3] or compact modeling techniques [4, 5]. Both methods have a superlinear computational complexity with respect to size of the problem. As more and more number of IC-dies are stacked with interlayer microchannel cooling, the computational times of these solutions would become too long to be practical for effective design-space explorations. In this paper, we address the problem of computational complexity of existing thermal simulation methodologies by proposing two key innovations.

The first contribution is the use of modern Graphic Processor Units (GPUs). GPUs have become in the recent times a computing mainstay in many different applications other than graphics processing. Their massively parallel architecture makes them a much faster and cheaper alternative compared to CPUs. This has resulted in the recent years in their exploitation for a myriad of EDA tools requiring high-performance computing [6, 7]. However, it is difficult to structure algorithms in order to fully exploit the GPU architecture and any new GPU-based EDA tool must be designed keeping in mind the aspects of GPU-based programming [8].

The second contribution of this work is the exploitation of Neural Networks (NNs) to provide a well-parallelizable thermal modeling approach. Neural Network theory provides a structurally straightforward programming tool that can be trained to mimic any mathematical function. In this work, we create and train a NN using 3D-ICE [4, 5], a conventional compact model-based thermal simulator for liquid-cooled 3D ICs. Once trained, the properties of large-scale parallel operations and low data transfer overhead of this NN-based simulator make it an appropriate candidate for implementation on GPUs. Recently, a NN-based simulator for 3D ICs has been presented by the authors of [9]. However, they tackle only the thermal modeling issue of conventional air-cooled ICs, without considering liquid cooling. In this paper, we create a NN-based thermal simulator for liquid-cooled 3D ICs that efficiently runs on GPUs. In a nutshell, the main contributions of this paper are the following:

1. We present a new transient thermal simulation framework for liquid-cooled 3D ICs based on neural networks and GPUs.

2. We exploit the physical insights inspired by how heat flows propagate in microchannels heat sinks to reduce the complexity of our NN-based simulator, by introducing an innovative *proximity-based reduction*. Thus, the temperature calculation speed is significantly improved while the thermal estimation accuracy is preserved.

3. We further improve the reduction techniques to find the optimal configuration of the NN-based simulator to maximize the speed-ups. A detailed analysis on the search for the NN-model configuration is also presented.

4. The NN-model, once trained, can be reused with different floorplan configurations of the 3D ICs. This feature is crucial for the acceleration of design-space exploration of liquid-cooled 3D ICs, where a very large number of different floorplans must be comparatively evaluated for thermal and electrical characterization.

The rest of the paper is organized as follows. Section 2 briefly describes the compact modeling tool that is used to train the NN-based simulator while Section 3 reviews the basics about the thermal model of conventional air-cooled ICs based on neural networks. The description of proposed NN-based thermal simulator is introduced in Section 4 and its structure and implementation details are discussed in various subsections. Experimental results are presented in Section 5. Finally, Section 6 summarizes the work with the conclusions.

2. THERMAL MODELING FOR LIQUID-COOLED 3D ICS

Interlayer liquid cooling of 3D ICs is accomplished using microchannels that are etched behind each die that is stacked in the IC. The cross-section of a typical interlayer liquid-cooled 3D IC is shown in Figure 1. Cold fluid is injected via a reservoir from one end (the inlet) and warm fluid exits into another reservoir at the other end (the outlet).

For such a structure, a compact and transient thermal simulation methodology of liquid-cooled 3D ICs typically involves the analogy between heat transfer in materials and electric current. The equivalent RC circuit that models the liquid-cooled 3D IC is represented by the following ordinary differential equations:

$$\mathbf{G}\mathbf{X}(t) + \mathbf{C}\dot{\mathbf{X}}(t) = \mathbf{U}(t), \qquad (1)$$

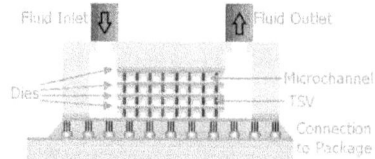

Figure 1: Cross-sectional view of a liquid-cooled 3D IC (taken from [4]).

where $\mathbf{G}$ and $\mathbf{C}$ are, respectively, the conductance and capacitance matrices. $\mathbf{X}(t)$ is the vector of temperature responses in the 3D IC, and $\mathbf{U}(t)$ is the vector of power traces. Once the system has been formulated in this manner, it is solved via numerical integration using backward Euler method. The solution at the $(n+1)^{th}$ time point is written as follows:

$$\mathbf{X}(t_{n+1}) = \mathbf{P}\mathbf{X}(t_n) + \mathbf{Q}\mathbf{U}(t_{n+1}), \qquad (2)$$

where,

$$\mathbf{P} = \left(\mathbf{G} + \frac{\mathbf{C}}{h}\right)^{-1} \cdot \frac{\mathbf{C}}{h}, \quad \text{and} \quad \mathbf{Q} = \left(\mathbf{G} + \frac{\mathbf{C}}{h}\right)^{-1}.$$

Here, h is the step size using to discretize time. In order to reduce the problem size, a modified version of the above thermal model based on porous mediums [5] was used in this work. As in [5], Eq. (2) encloses the boundary conditions of the modeled ICs with microchannels. In our work, we consider Eq. (2) as the basis for the development of the proposed NN-based thermal simulator further described in Section 4.

3. REVIEW OF NN-BASED SIMULATOR FOR CONVENTIONAL ICS

The NN-based thermal simulator proposed in this paper is based on the ability of neural networks to learn the thermal behavior of a liquid-cooled 3D IC and then, based on this training, to act as a stand-alone thermal simulator. A NN-based thermal simulator for conventional air-cooled 2D/3D ICs was proposed in [9]. The authors proposed to train a neural network to reproduce the dependence in Eq. (2): each neuron in the model represents the thermal state of a given point (node or thermal cell) in the volume of the IC. In particular, it is sufficient to define a neuron for each node in the layers of the IC where the heat dissipation occurs (active layers). Therefore, the resulting NN will have a single layer of neurons sharing the same inputs.

Each neuron in the NN expresses the output as a weighted sum of the inputs:

$$y_i = \sum_j w_{ij} x_j \qquad (3)$$

where w_{is} are coefficients, called *weights*, that represent the contribution of the input to the output. To replicate the thermal evolution in Eq. (2), a neural network performing the following operation is constructed:

$$\mathbf{X_i}(t_{n+1}) = \sum_j w_{ij}\mathbf{X_j}(t_n) + \sum_k w_{ik}\mathbf{U_k}(t_{n+1}) \qquad (4)$$

The weights in the above equation are trained to represent the coefficients of the corresponding matrices $\mathbf{P}$ and $\mathbf{Q}$ in Eq. (2). For this training, the neural network must be given training samples that describe, for each neuron, the dependencies between the inputs and the corresponding

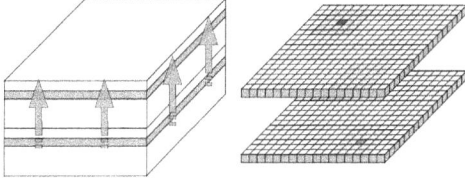

Figure 2: Heat flow and neuron connections for conventional 2D/3D ICs

output. Since the output of each neuron represents a continuous range of temperatures, all the neurons have a linear activation function. Once the neural network is trained to replicate correctly all the training samples, it can be used as stand alone thermal simulator. The accuracy of the NN-based simulator depends upon the training algorithm, the quality of inputs, the approximations involved in the construction of the NN and the error tolerance imposed on the training process.

Furthermore, the NN-based simulator proposed in [9] exploits the predominantly vertical heat flow that characterizes a conventional air-cooled IC as shown in Figure 2. This implies that there is little lateral spreading of heat, meaning temperatures of nodes in a given layer of IC that are far apart do not affect each other significantly. This model then introduces a parameter, called *proximity* that can be used to reduce the number of inputs for each neuron and hence the overall complexity of its implementation. The key idea in this process is to remove from Eq. (4) all the terms in both the summations on the RHS that do not contribute significantly to the output.

Another innovation in this simulator was that instead of using a training set made with samples representing a real scenario where a floorplan is used, power traces with random values between zero and the maximum power density of the design is fed to the neural network during training. These samples represent the worst-case scenario, and each neuron is trained for all possible heat flux distributions and heat flux levels. Hence, the run-time error of a neural network so trained will be independent of the configuration of the floorplan. This enables the NN-based simulator to be used for design space exploration (floorplanning).

4. PROPOSED NN-BASED SIMULATOR FOR LIQUID-COOLED 3D ICS

The NN-based simulator for liquid-cooled 3D ICs proposed in this work is trained in a similar manner to the model described in the preceding section. Thus, neurons represent the temperatures of nodes in the active layers. The neural network is constructed as mentioned before and trained to mimic the behavior of Eq. (2). However, the nature of heat flow in liquid-cooled 3D ICs is fundamentally different from conventional ICs and the model must be adapted to capture these different heat dissipation paths. The aspects of the proposed NN-based simulator are discussed in the ensuing subsections.

4.1 Training methodology

There are several algorithms available in the literature for training neural networks. Since the objective here is to train the NN weights in Eq. (4) to mimic the behavior of the deterministic system in Eq. (2) based on a set of inputs and outputs, a batch training algorithm like RPROP, that was

used in [9], must be used. Also, given that there are as many unknown weights to be computed per neuron as there are inputs for it, a minimum number of training samples must be provided to the training algorithm for accurate estimation of the weights. As in [9], a randomized heat flux distribution with random power traces bounded by the maximum heat flux levels expected during run-time was used to generate the training data. The use of a random training set is justified by the need of a model that is independent by the distribution of power density in the various layers of a 3D IC.

One major difference of the proposed NN-based simulator from the NN-based simulator for conventional ICs is that the flow rates of coolant affects the behavior of the system. The higher the flow rate, the lower the overall thermal resistance of the system to the heat sink, which in turn means that the coefficients of the matrices in Eq. (2) are changed. Since it is common to design a liquid-cooled 3D IC for variable flow rates, training must be performed for each flow rate value intended in the final design.

To eliminate errors introduced by the training of the network and to be able to isolate the effect of the various reduction techniques applied to the proposed model, we replace the iterative training algorithms with a direct method to compute the weights as the solution of a linear system. Each training sample provides one additional equation in the computation of the unknown weights. Since we provide a higher number of samples than unknowns, we have an overdetermined system. To solve this, we deploy the least-squares method using QR decomposition in this work.

4.2 Proximity-based reduction

The heat flow patterns in liquid-cooled 3D ICs is fundamentally different from conventional air-cooled ICs. Hence, when proximity-based reduction is applied to the proposed simulator, these patterns must be taken into account. In liquid-cooled 3D ICs, there are indeed *two* major paths of heat flow as shown in Figure 3: one vertical, from the active layers towards the microchannel heat sinks, and one along the microchannels, the path along with heat is carried by the coolant. As a consequence, the temperature of a given point on the surface of the IC will be influenced by the thermal state and the heat dissipation at nodes lying upstream or downstream along the direction of the channel, in addition to those nodes which are directly above and below it in the other active layers. Hence, *proximity regions* are defined as rectangular regions in each active layer along the channel as highlighted in Figure 3.

The width of this rectangle (the "proximity distance") defines the complexity of our model. In other words, it signifies how large an area of the IC is assumed to influence the temperatures in the node under consideration. If this proximity distance is small, then the complexity of the re-

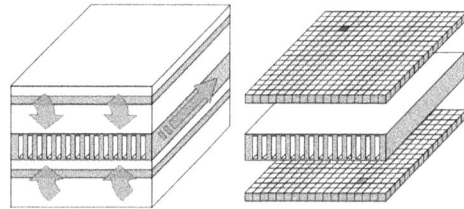

Figure 3: Heat flow and neuron connections for 2D/3D ICs with liquid cooling

Algorithm 1 Solving with constant proximity
```
1: for c_i = 1 ... N_C do
2:    I ← ExtractInput (TrainSet, Proximity, [1, 1, c_i])
3:    (Q, R) ← QRdecompose (I)
4:    for r_j = 1 ... N_R do
5:       for l_k = 1 ... N_L do
6:          O ← ExtractOutput (TrainSet, [l_k, r_j, c_i])
7:          (W[l_k, r_j, c_i], Residual) ← lssolve (O, Q, R)
8:       end for
9:    end for
10: end for
```

sulting neural network is lower, at the expense of potential loss of accuracy. Two cases are shown in Figure 3: for one neuron in the bottom-right corner of the IC, the proximity region is much smaller than the proximity of the neuron in the top-left corner.

The least squares method used for the calculation of the weights, as described in Section 4.1, gives as one of the outputs the final residual during the calculation of weights. This residual value serves as an indicator of the run-time accuracy of the NN-based simulator that has been constructed. Hence, depending upon the accuracy and performance requirements of the designer, one can fix a desired proximity distance during the training of the NN. This is illustrated in Algorithm 1, where neurons are indexed using a tuple $[layer, row, column]$ as they belong to a three dimensional structure having dimensions $[N_L, N_R, N_C]$ (see Figure 3).

Algorithm 1 extracts from each sample in the training set a matrix containing the training input for the given neuron (line 2) and executes its QR decomposition (line 3). This operation is done only for neurons that belong to the first row and to the first layer because neurons in the same row on all the layers share the same proximity region. Then, for all the remaining neurons, it extracts the target training output (line 6) and computes their weights as the unknowns of a linear system (line 7). The residual is discarded or reported as a measure of the accuracy of the NN over the training set.

4.3 Optimal proximity profile

Due to the sensible heat absorption in microchannel heat sinks, temperatures increase as the distance from the inlet increases. Hence, there is a strong thermal gradient form the inlet to the outlet. In addition to causing nodes (neurons) near the outlet to be considerably hotter than those near the inlet, this phenomenon causes unequal lateral spreading of heat in the ICs from inlet to outlet.

This means that neurons representing temperatures at nodes closer to the outlet would require a larger proximity distance than those near the inlet to attain the same level of accuracy. In other words, for the same value of proximity used for all neurons in a 3D ICs, the training residuals (or the run-time errors) for neurons near the outlet would be larger than those near the inlet.

Using a small constant proximity distances to reduce the simulator complexity would result in large errors near the outlet. On the other hand, using extremely large constant proximity distances to compensate for these errors near the outlet increases the complexity of the entire simulator, while being an overkill for neurons near the inlet. Hence, it is possible to find an *optimal* proximity distance profile for the neurons in a 3D IC as a function of the distance from the inlet, which minimizes the NN complexity, while distributing the errors fairly uniformly. This proximity distance profile would reflect the amount of lateral heat spreading that occurs at a particular point in the 3D IC along the microchannel. The residual resulting form the training algorithm can be used to estimate this optimum proximity distance for each row of neurons along the microchannel. This modified training algorithm for optimal proximity is shown in Algorithm 2.

The idea behind this algorithm is to use the given proximity to compute the weights only for the neurons in the last row and save the maximum value of the residual found in this first part (lines 1-10). Then, the proximity is set to a minimal value (line 11) and weights are computed processing one row (in all the layers) per time. If for any neuron a residual higher than the maximum found in the last row is found, then the proximity value is increased and the training of the row is restarted (line 19-21). This approach allows to identify, for each row, the minimum proximity value that will guarantee a run-time error lower than the error in the last row of neurons.

The value $MinLength$ used in line 11 and 20 to find the optimal proximity, corresponds to the minimum distance such that the operation done by ExtractInput extracts a larger set of inputs. For instance, using Figure 3 as reference, $MinLength$ is the value that turns the proximity region of the neuron in the bottom layer into a region as large as the one that belongs to the neuron in the top layer. In the setup of our model, we chose to set $MinLength$ as the length of the thermal cell in the compact model used to generate the training samples.

4.4 Running the NN-based simulator

Once the proposed NN-based simulator is trained, it can be used to simulate the temperatures of the target liquid-cooled 3D ICs using matrix vector multiplications. Essentially, the weights computed during the training are stored in the form of a sparse matrix in the GPU global memory. Then, the power traces and the thermal states are sent from the CPU memory whenever needed. These form the vector

Algorithm 2 Solving for optimal proximity
```
1: MaxRes ← 0
2: for c_i = 1 ... N_C do
3:    I ← ExtractInput (TrainSet, Proximity, 1, N_R, c_i)
4:    (Q, R) ← QRdecompose (I)
5:    for l_k = 1 ... N_L do
6:       O ← ExtractOutput (TrainSet, l_k, N_R, c_i)
7:       (W[l_k, N_R, c_i], Residual) ← lssolve (O, Q, R)
8:       MaxRes ← max (MaxRes, max(|Residual|))
9:    end for
10: end for
11: Proximity ← MinLength
12: for r_i = 1 ... N_R - 1 do
13:    for c_i = 1 ... N_C do
14:       I ← ExtractInput (TrainSet, Proximity, 1, 1, c_i)
15:       (Q, R) ← QRdecompose (I)
16:       for l_k = 1 ... N_L do
17:          O ← ExtractOutput (TrainSet, l_k, r_i, c_i)
18:          (W[l_k, r_i, c_i], Residual) ← lssolve (O, Q, R)
19:          if max(|Residual|) > MaxRes then
20:             Proximity ← Proximity + MinLength
21:             restart c_i loop
22:          end if
23:       end for
24:    end for
25: end for
```

Table 1: Structural parameters of the Test 3D IC

Parameter	Value	Parameter	Value
Layers per die	2	Channel height	$100\mu m$
Die size	14×15 mm^2	Channel width	$50\mu m$
Die height	$65\mu m$	Channel pitch	$100\mu m$

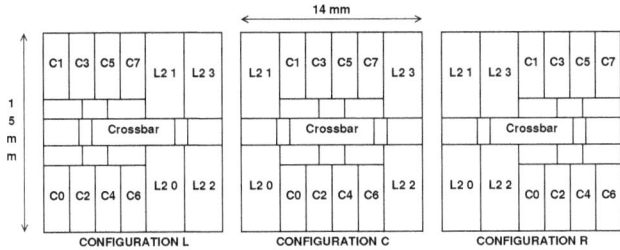

Figure 4: Niagara floorplan configurations used for testing

of inputs, which get multiplied by the weights to obtain the thermal state in the next time step. In our implementation, we used the cuSPARSE library [11] for these computations on our GPU platform.

5. EXPERIMENTAL RESULTS

To illustrate the accuracy and the performance of the proposed NN-based simulator we define a *Test 3D IC* as a stack made up of three dies interleaved by two cavities. The structural properties of the stack are shown in Table 1. To study the effect of cooling effort on the simulator, all experiments were performed using the three flow rates: 24ml/min, 36ml/min, and 48ml/min. A cell size of $500\mu m \times 500\mu m$ was used to discretize the structure and the training data for the NN-based simulator were generated by simulating it using 3D-ICE. This fixes the number of neurons in the network to 2520, as we define a neuron for each thermal cell in the three active layers.

During the training phase, data was generated using a randomized floorplan with random power traces bounded by a maximum heat flux of 90W/cm^2. The number of training samples generated was set to 20% more than the minimum number required. On the other side, during the running and measurement phase, the UltraSPARC Niagara floorplan [10] is assigned to the active layer of each die. This floorplan configuration was modified to generate three different test cases as shown in Figure 4. In the following experiments, "LLL" refers to the case when all the dies have the floorplan configuration "L" in Figure 4. Similarly, "LRL" represents a stack with the floorplans "L" in the bottom and top die, and "R" in the middle die, etc.

The NN-based simulator is run on the NVIDIA Tesla^{TM} c2070 (448 CUDA Cores running at 1.15 GHz and 6GB GDDR5) while the corresponding simulations on CPU based on 3D-ICE are run on Intel$^{®}$ Core^{TM} i7 920 (4 cores running at 2.67 GHz and 6GB of RAM).

5.1 Computation of Optimal Proximity

We tested the ability of Algorithm 2 to compute the optimal proximity by first training the Test 3D IC using Algorithm 1 with a series of different constant proximity values. During each training, the maximum value of residuals for each neuron was stored. Then, the resulting NN was run to simulate the floorplan configuration CCC and the maximum run-time error w.r.t. 3D-ICE was measured. Figure 5

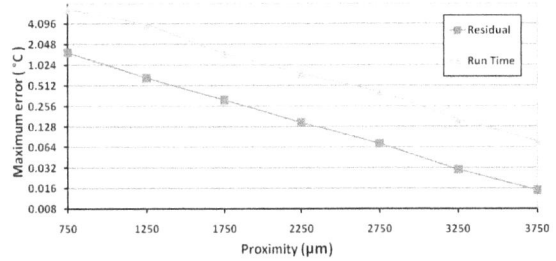

Figure 5: Max. residual and max. run-time error when training with Algorithm 1

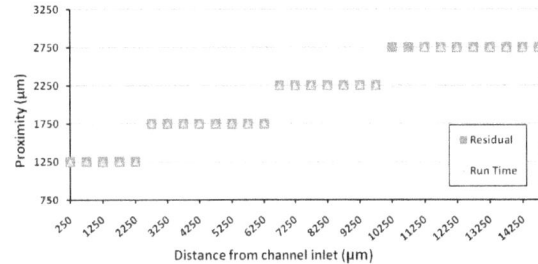

Figure 6: Profile of the optimal proximity along the channel

shows the maximum residual and corresponding the maximum run-time error for each proximity value.

As can be seen from this figure, the maximum residual from training and the maximum run-time error show the same trend. This means that the residual can be effectively used to compute the optimal profile of the proximity along the direction of the channel. Figure 6 shows the optimal proximity profile computed both using Algorithm 2 and using run-time error. To obtain the optimal proximity using the run-time error in this figure, we first train the NNs using Algorithm 1 with different proximity values and 36ml/min as flow rate. In each case, we measure the maximum run-time error as a function of distance along the channel. Then we fix, for each neuron, the minimum proximity value such that the run-time error remains lower than $0.5°C$. The minimum proximity found for the neurons representing the last row was $2750\mu m$ and therefore this value has been given as input to Algorithm 2 to verify its capability to generate the same profile.

As Figure 6 shows, the two proximity profiles show the same trend except for a few values. Even when mismatches occur, the proximity value obtained using residuals from training is higher than the one obtained using run-time measurements. Hence, using the former methods results only in reduced error at run-time at a marginal additional cost of computation. Hence, we can conclude that Algorithm 2 is a reliable method to compute the optimal proximity profile during training and the run-time errors resulting from it is less than the intended tolerance.

5.2 Error analysis

Figure 7 shows the evolution of the maximum run-time error incurred by the proposed NN-based simulator (trained using Algorithm 2), when it is used to simulate Test 3D IC with CCC configuration. Errors in this figure are reported for the three different flow rates. As can be seen from these plots, the error decreases in each case with increasing proximity values. However, given the same proximity value, simulations with different flow rates result in different er-

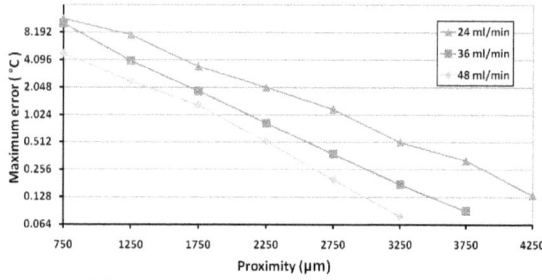

Figure 7: Maximum run-time error for the floorplan configuration CCC

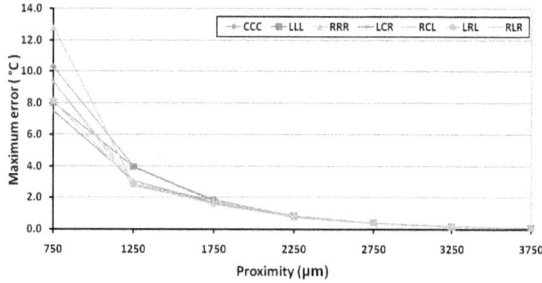

Figure 8: Maximum run-time error for different floorplan configurations

rors. Higher flow rates result in smaller errors than lower flow rates because coolant at high flow rates carries heat from the IC quicker, leading not only to lower IC temperatures, but also prevent heat spreading within the silicon by lowering the net thermal resistance to the heat sink. Hence, for a given proximity value, the NN-simulator trained for a higher flow rate better captures the heat spreading effects than an NN-simulator trained for lower flow rates.

Figure 8 reports the maximum run-time error obtained when NN-based simulator was used to simulate the Test 3D IC for the same flow rate (36ml/min) but with different floorplan configurations. As can be seen, the run-time independent of the floorplan configuration.

5.3 Performance analysis

To compare the performances of the two training techniques (Algorithms 1 and 2) simulation speed ups of NN-based simulators trained using both these techniques running on GPU, against 3D-ICE running on CPU. This experiment was repeated for different maximum proximity values and flow rates. Figure 9 shows the increase (in percentage) of speedup obtained with the introduction of the training Algorithm 2 over the Algorithm 1. As can be seen from this figure, improvements of up to 25% can be obtained by using the optimal proximity profiles.

Finally, to illustrate the scalability of our neural network-based thermal simulator, Figure 10 shows how the GPU speedup changes with increasing number of dies (in other words, the problem size), increasing flow rate and trained for various error tolerances (1.0°C, 0.5 °C and 0.1 °C). As can be seen from this figure, speedups increase with increasing problem size and also with increasing flow rate. In all these experiments, the dies are interleaved with channel cavities, similar to the Test 3D IC.

6. CONCLUSIONS

In this work we presented a Neural Network-based thermal model that can be run on massively parallel architectures

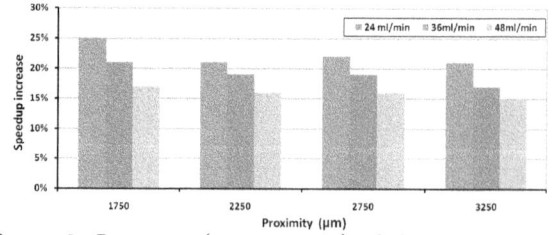

Figure 9: Increase (percentage) of the GPU speedup using Algorithm 2

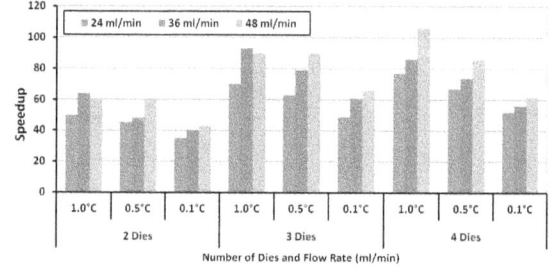

Figure 10: GPU speedup for different problem sizes, flow rates and maximum run-time error

like GPUs to accelerate thermal simulations of 3D ICs with liquid cooling. We also introduced a new training technique that exploits the horizontal heat flow due to the coolant passing through the channels to reduce the simulation time without worsening the run time error. Results show that the speedups obtained comparing the simulation times against a compact model running on CPU ranges from 35x, to limit the run-time error under 0.1°C, up to 106x if errors lower than 1.0°C are accepted.

7. ACKNOWLEDGMENTS

This work was funded in part by the Swiss NSF grant number 200021-130048 and by the Nano-Tera RTD Project CMOSAIC (ref. 123618), which is financed by the Swiss Confederation and scientifically evaluated by SNSF.

8. REFERENCES

[1] "International technology roadmap for semiconductors (ITRS)," *2009 Edition-ERD*.

[2] T. Brunschwiler *et al.*, "Interlayer cooling potential in vertical integrated packages," *Microsystem Technologies: MNSISPS*, vol. 15, no. 1, 2009.

[3] H. Mizunuma *et al.*, "Thermal modeling for 3D-ICs with integrated microchannel cooling," *Proc. ICCAD, 2009*.

[4] A. Sridhar *et al.*, "3D-ICE: Fast compact transient thermal modeling for 3D ICs with inter-tier liquid cooling", *Proc. ICCAD, 2010*.

[5] A. Sridhar *et al.*, "Compact transient thermal model for 3D ICs with liquid cooling via enhanced heat transfer cavity geometries", *Proc. THERMINIC, 2010*.

[6] Z. Feng and Z. Zeng, "Parallel Multigrid Preconditioning on Graphics Processing Units (GPUs) for Robust Power Grid Analysis", *Proc. DAC, 2010*.

[7] Y. Liu and J. Hu, "GPU-based parallelization for fast circuit optimization", *Proc. DAC, 2009*.

[8] J. Croix and S. Khatri "Introduction to GPU Programming for EDA", *Proc. ICCAD 09*.

[9] A. Sridhar *et al.*, "Neural Network-Based Thermal Simulation of Integrated Circuits on GPUs", *Transactions on Computer Aided Design of Integrated Circuits and Systems*, vol. 31, no. 1, 2012.

[10] A. Leon *et al.*, "A power-efficient high-throughput 32-thread SPARC processor", *Proc. ISSCC 2007*.

[11] CUDA Sparse Library.

Efficient CMOL Nanoscale Hybrid Circuit Cell Assignment Using Simulated Evolution Heuristic

Sadiq M. Sait
Department of Computer Engineering
Center for Communications and IT Research
Research Institute
King Fahd University of Petroleum & Minerals
Dhahran-31261, Saudi Arabia.
sadiq@kfupm.edu.sa

Abdalrahman Arafeh
Department of Computer Engineering
King Fahd University of Petroleum & Minerals
Dhahran-31261, Saudi Arabia.
arafeh@kfupm.edu.sa

ABSTRACT

Recently, many CMOS/nanodevices hybrid architectures have been proposed, the new architectures combine the flexibility and high fabrication yield advantages of CMOS technology with nanometer scale latching devices. CMOL, a novel architecture that uses two levels of perpendicular nanowires as crossbar interconnection on top of inverter-based CMOS stack, offers significant density advantages and overcomes physical barriers of lithography-based fabrication. However, the confined connectivity of CMOL nanofabric to only cells that are located within proximity square-like connectivity domain, reduces the flexibility of VLSI design automation and further complicates cells placement.

In this paper we use Simulated Evolution algorithm to solve the NP-hard problem of assigning NOR/INV gates to CMOL array. The main objective is to reduce the total number of buffers that must be inserted between cells that require long wires to connect. A novel goodness and allocation functions are introduced for efficient exploration of search space. Empirical results for ISCAS'89 benchmarks are compared with previous solutions using GA, MA, and LRMA heuristics. Our approach is able to find better solutions for all tested benchmarks and with 82% average reduction in CPU processing time.

Categories and Subject Descriptors

B.7.2 [**Hardware**]: Integrated Circuits— *Design Aids, Placement and Routing*; B.7.1 [**Hardware**]: Integrated Circuits— *Types and Design Styles, Gate Arrays*

General Terms

Algorithms, Design

Keywords

CMOL, Simulated Evolution, Combinatorial Optimization, Search Heuristics, VLSI, Nanofabric, Placement, Assignment.

1. INTRODUCTION

For the past decades, microelectronic fabrication has been the state-of-the-art technology in semiconductor industry. However, feature size scaling in CMOS has led to many implications and manufacturing difficulties. Meanwhile, A new trend is emerging for combining the flexibility and high fabrication yield advantages of CMOS technology with nanometer scale molecular devices. A self-assembly of two-terminal nanodevices with nanowire crossbar fabrics, would enable high functional density and sustain acceptable fabrication costs. Likharev and Strukov [1] introduced hybrid semiconductor/nanowire/molecular integrated circuits called CMOL, which use two levels of perpendicular nanowires as crossbar interconnection on top of inverter-based CMOS stack. Likharev and his colleagues have shown possible applications of CMOL in field programmable gate arrays (FPGA) [2], neuromophic CrossNets [3], and in memories [4].

Recently, several proposals were introduced for cell placement on FPGA-like CMOL architecture. Likharev et al utilized existing FPGA CAD tools to perform placement on 4×4 tile-based version of CMOL [5, 2] and used reserved routing cells and recursive routing algorithm for inter-tile routing. Instead of working at tiles level, Hung et al [6] encoded the CMOL cell assignment as a Satisfiability problem at cells level, where placement solution was found when all Boolean constraints are satisfied. However, when circuits sizes increased the computation time became exhibitant

Previous attempts to use sub-optimal search heuristics are reported in [7, 8, 9]. Genetic Algorithm (GA) [7] was used with two dimensional block PMX crossover operator and mutation, where the fitness function evaluates the Manhattan distance between connected cells. A more elaborate work was reported in [8]; where Memetic computing approach (MA) was used by combining Genetic Algorithm and Simulated Annealing (SA) local-based search heuristic. SA was used to enhance GA offsprings by local improvement search. Hung et al [9] extend their work on Memetic approach by integrating self-learning operators using Lagrangian Multiplier (LRMA), results reported are promising, however, more computations are needed for penalty updating mechanism. A theoretical investigation on CMOL cell assignment was reported in [10]; the authors proved mathematically that placement of 2-input NOR/INV circuits is possible and may require adding additional buffers (i.e., pair of inverters) to satisfy all connections.

The aim of this work is to investigate the cell assignment on CMOL nanofabric crossbar architecture and to overcome the restriction on nanowires length. The problem objective is to find an assignment that will result in smallest number of additional buffers. The rest of the paper is arranged as follow: in the next section we provide a background about CMOL FPGA-like architecture and Simulated Evolution algorithm (SimE), a heuristic engineered to solve our combinatorial optimization problem. Section 3 details problem formulation, and goodness function. Section 4 contains the empirical results and comparison. Finally, we conclude the paper and provide final remarks.

2. BACKGROUND

2.1 CMOL Hybrid Architecture

CMOL cell-based, field-programmable gate array (FPGA)-like architecture is based on integrating conventional inverter-based four-transistor MOSFET CMOS cell with uniform reconfigurable nanowire fabric. Two-terminal nanodevices "Latching switches", that have two metastable internal states, are self-assembled at each crosspoint in CMOL fabric and provide diode-like I-V curves for logic circuits implementation. CMOS stack is connected to nano-fabric by Metal pins that span to top and bottom nanowire levels as shown in Figure 1(a). Two CMOS inverters are connected by pin-nanowire-nanodevice-nanowire-pin connection. The electrical representation of four inverter-based CMOS cells and corresponding nanowire/nanodevices is shown in Figure 1(b). The upper right cell (Inverter A) is connected to the lower left cell (Inverter C) by activating the nanodevice 'nd1'. When two or more nanodevice on the same nanowire are activated as shown in Figure 1(b) (nd1 and nd2) the output of inverter C will be equivalent to NOR gate whose inputs are cell A and cell B. Wired-OR logic is implemented through nanowires and nanodevice.

Figure 1(c) shows the nanowire crossbar turned by angle $\alpha = \arcsin(F_{nano}/\beta F_{CMOS})$ related to the CMOS pin array, where F_{nano} is the nanowireing half-pitch, F_{CMOS} is the half-pitch of CMOS, and β is a factor larger than 1. Like other nano-fabric crossbars, CMOL's nanowires break at repeated intervals confining CMOL cells connectivity to only $M = 2r(r-1) - 1$ other cells located within its proximity "Connectivity Domain", where r is an integer value that indicates connectivity diameter and represents the constraint of CMOL placement. Reconfigurable nanodevices and nanowires can be used with inverter cells to form wired-NOR gates. If the nanowires and nanodevices shown in Figure 2(b) are activated, the CMOL circuit will be equivalent to circuit shown in Figure 2(a). The first NOR gate of the circuit can be implemented by connecting inputs 'A' and 'B' with inverter '1' to satisfy both connectivity and logic wiring for the desired gate. The abundance of available nanodevices and nanowires provide a variety of different possible configurations for the implementation of one circuitry. Among those their could be only certain configurations that satisfy connectivity domain constraint and do not require additional routing resources.

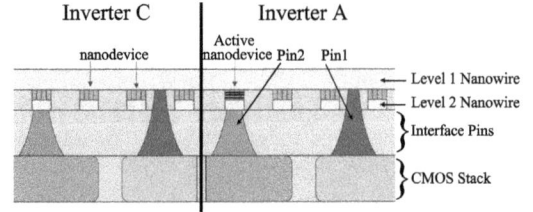

(a) Schematic side view of two CMOL cells with two levels of nanowires. Only one nanodevice is activated to connect the output of Inverter A with input of Inverter C.

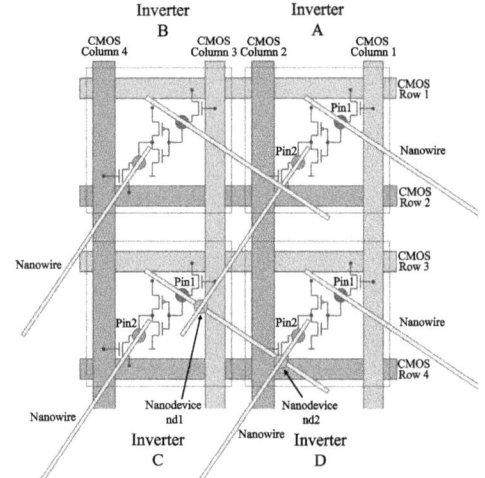

(b) Electrical representation of four CMOL cells and corresponding nanowires.

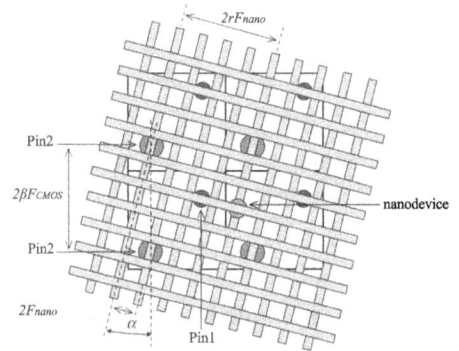

(c) Nano-fabric inclined by α on top of four CMOS cells.

Figure 1: Low-level structure of CMOL circuit: the incline angle $\alpha \ll 1$ and dimensionless parameter β satisfy two conditions, $\sin \alpha = F_{nano}/\beta F_{CMOS}$ and $\cos \alpha = r F_{nano}/\beta F_{CMOS}$ where r is an integer.

2.2 Simulated Evolution (SimE)

The SimE heuristic is similar to Simulated Annealing except that the elements that are movable have a goodness value (a number between 0 and 1). Those with goodness value close to 1 have a smaller possibility to leaving their locations, while those with smaller goodness have otherwise. The structure of the SimE algorithm is shown in

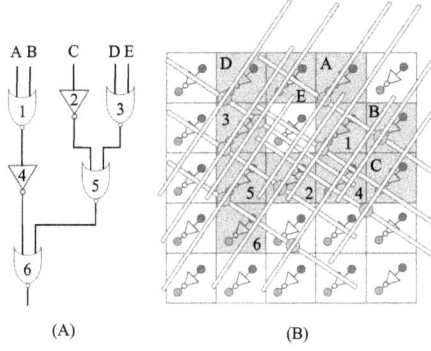

Figure 2: Example of CMOL circuit: (a) NOR/INV logical circuit; (b) CMOL implmentaion of (a), shaded cells are connected through combination of nanowires, nanodevices and CMOS pins.

Figure 3. SimE assumes that there exists a solution ϕ of a set M of n (movable) elements or modules. The algorithm starts from an initial assignment $\phi_{initial}$, and then, following an evolution-based approach, it seeks to reach better assignments from one generation to the next by perturbing some ill-suited components and retaining the near-optimal ones. A cost function *Cost* associates with each assignment of movable element m_i a cost C_i. The cost C_i is used to compute the goodness (fitness) $goodness_i$ of an element m_i, for each $m_i \in$ M. The algorithm has one main loop consisting of three basic steps, Evaluation, Selection, and Allocation. The three steps executed in sequence until the solution average goodness reaches a maximum value, or no noticeable improvement to the solution cost is observed after a number of iterations. The Evaluation step consists of evaluating the goodness $goodness_i$ of each element m_i of the solution ϕ. The goodness measure must be a single number expressible in the range [0, 1], and can be defined as

$$goodness_i = \frac{O_i}{C_i} \qquad (1)$$

Where O_i is an estimate of the optimal cost of element m_i, and C_i the actual cost of m_i in its current location. Or simply goodness can be defined as the fraction of two values related to the problem cost.

The second step of the SimE algorithm is Selection. Selection takes as input a bias value B, the solution ϕ together with the estimated goodness of each element. It partitions ϕ into two disjoint sets; a selection set S and a partial solution ϕ_p of the remaining elements of the solution ϕ. Each element in the solution is considered separately from all other elements. The decision whether to assign an element m_i to the set S is based solely on its goodness $goodness_i$. The selection operator has a non-deterministic nature, i.e, an individual with a high goodness (close to one) still has a non-zero probability of being assigned to the selection set S. It is this element of non-determinism that gives SimE the capability of escaping local minima. Allocation is the SimE operator that has the most important impact on the quality of solution. Allocation takes as input the set S and the partial solution ϕ_p and generates a new complete solution ϕ' with the elements of set S mutated according to an allocation function. The goal of Allocation is to favor im-

```
ALGORITHM Simulated_Evolution(B, Φ_initial, StoppingCond.)
NOTATION
B= Bias Value.    Φ= Complete solution.
m_i= Module i.    g_i= Goodness of m_i.
ALLOCATE(m_i, Φ_i)=Function to allocate m_i in partial solution Φ_i
Begin
Repeat
    EVALUATION:
        ForEach m_i ∈ Φ evaluate g_i;
        /*Only elements that were affected by moves of previous*/
        /*iteration get their goodnesses recalculated*/
    SELECTION:
        ForEach m_i ∈ Φ DO
            begin
                IF Random > min(g_i, 1)
                THEN
                    begin
                        S = S  ∪  m_i; Remove m_i from Φ
                    end
            end
        Sort the elements of S
    ALLOCATION:
        ForEach m_i ∈ S DO
            begin
                ALLOCATE(m_i, Φ_i)
            end
Until   Stopping Condition is satisfied
Return Best solution.
End (Simulated_Evolution)
```

Figure 3: Structure of the Simulated Evolution algorithm [11].

provements over the previous generation, without being too greedy [11].

3. PROBLEM FORMULATION

CMOL cell assignment problem can be formulated as a mapping of a given NOR/INV gate-based circuit G to a CMOL generic inverter-based cell array Ψ. Each CMOL cell can implement one inverter or one NOR gate with multiple fan-in. Each gate in G has a number of $fan-in$ and $fan-out$ gates, those comprises γ_i the netlist of gate i as described in Equation 2.

$$P : G \rightarrow \Psi \qquad (2a)$$

$$\gamma_i = \{fan-in(i) \mid fan-out(i)\} \qquad (2b)$$

Unlike conventional CMOS-based cell assignment, CMOL cell placement is constrained to *"Connectivity Domain"* of radius r. Each CMOL cell is connectable to one of its proximity cell members, any violation of this constraint would impose further processing (i.e., buffer insertion) to satisfy connectivity. However, such process would cause more congestion and could results in substantial increase of timing delay. Mathematically, the "Connectivity Domain" can be defined as follow. Given a gate and its netlist $(i, \gamma i)$ placed in location p_i, for any gate $j \subseteq G$ and j in the netlist γi the following equations should be satisfied.

$$\forall i, j \in G : (i \neq j) \Rightarrow (p_i \neq p_j) \qquad (3a)$$

$$\forall i, j \in G : dist(p_i, p_j) \leq r \qquad (3b)$$

Where p_j is the position of gate j (i.e., X and Y coordinates) in CMOL array, $dist$ is Manhattan distance and r is CMOL radius. The objective of CMOL cell mapping problem is to satisfy the constraint in Equation 3(b) for all gates of circuit G.

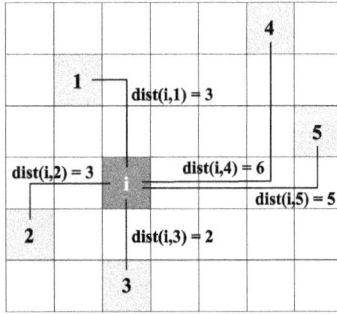

Figure 4: Evaluation of gate i's goodness; for $r = 3$ cells 1, 2 and 3 are inside i's connectivity domain (i.e., $dist \leq r$), while cells 4 and 5 are out of it (i.e., $dist > r$). $goodness_i = 3/5 = 0.6$.

3.1 Goodness Function

In Simulated Evolution, goodness function is used to evaluate individual elements in each generation, where unfit elements are selected and reassigned to other locations. In this paper each gate is considered to be as an element, while the cost of a complete solution is calculated as the number of required addition buffers to ensure connectivity between all elements. We can define the goodness function of each element as follow.

$$goodness_i = \frac{inside_i}{|\gamma_i|} \qquad (4)$$

Where $inside_i$ represents the number of gates in set γ_i that satisfy connectivity constraint (i.e., inside the connectivity domain of element i) and $|\gamma_i|$ is the cardinality of the netlist of gate i. Figure 4 shows an example on how goodness value is calculated, where two gates in γ_i are outside the connectivity domain of gate i and three otherwise. The proposed goodness function results in precise selection of those elements that violate constraint expressed in Equation 3(b), which directs the heuristic into enhancing the overall cost of the problem.

Simulated Evolution population is represented as a 2-D grid of $X \times Y$. The outer cells of the grid are reserved for I/O pins, where I/O pins moves are restricted to these reserved locations. In the initialization phase, gates are randomly assigned in the 2-D layout, while bias and stopping criteria are defined.

Selection phase uses original SimE selection function [11]; an element (i.e., gate) is selected for reallocation if its goodness score is less than a randomly generated number between 0 and 1. Selected elements are sorted in an ascending order based on their netlist size, where elements with higher cardinality of γ_i are processed first. Allocation step is based on sequential attempts to evaluate the change in cost (i.e., number of buffers) if two cells are exchange. For each element in selection set, the function starts by visiting all CMOL array locations whether empty or occupied, and evaluates the change in number of buffers if the element being processed is swapped with the element occupying the location under test. The best location is chosen and swap is performed. In some cases selected elements could be swapped or placed in already empty location. If two or more swaps have equal

cost, which also happens to be the best, the swap with lesser Manhattan distance is chosen.

4. EXPERIMENTAL RESULTS

Evaluation of search heuristic efficiency and behavior is conducted using ISCAS'89 [12] benchmarks. Benchmarks used in this work are mapping to NOR-based gates with maximum of five inputs. Simulated Evolution has been implemented using Java programming language and executed on a machine comparable to the one used by other simulations published in literature, it has 1.5 GHz Intel Pentium M processor with 512MB memory. Table 1 shows the number of cells (i.e., gates) and I/Os of benchmark circuits used; Area (*Tiles*) is the area used by CMOL FPGA CAD 1.0 tool [2], while Area (*Row $\times$ Column*) is the area used in GA [7], MA [8], LRMA [9] and SimE. We have experimented with different bias values, and a value of $B = 0.05$ has been used in all reported results to allow for non-zero selection probability even for those cells with no connectivity violations, the heuristic stops when all violations are removed or when reaching a predefined number of iterations. The median value of 20 runs is reported where each run uses different seeds for SimE random numbers. Evaluation of Simulated Evolution heuristic is shown in Figure 5. At the beginning of iterations many elements are selected for perturbation and with time the size of selection set decreases, as cells are being placed in their suitable positions. Figure 6 shows the cost (i.e., number of buffers inserted) change per iteration. It's clear that the cost and Manhattan distance are being reduced through iterations while SimE is performing hill-climbing to avoid reaching local optimum.

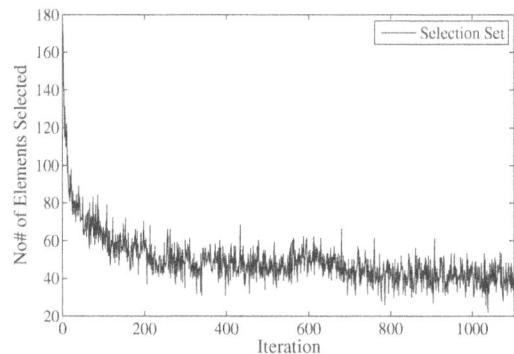

Figure 5: Change of selection set size per SimE iterations - s1238.blif

Comparison is performed with CMOL FPGA CAD 1.0, we set the connectivity radius to $r = 12$. GA, MA and LRMA use population size equals to 24 and stopping criteria when fitness score is not updated for 50 times. The crossover rate in MA and LRMA is $RC = 0.33$ and mutation rate $RM = 0.01$. Simulated Annealing used in each of GA iterations has initial temperature $T = 0.2$ and terminating temperature 0.01. Table 2 shows the final results obtained for ISCAS'89 benchmarks; (*Delay*) is the circuit's logical levels reported by SIS tool after inserting the buffers, computation time (*Time*) in seconds, (*Buf*) shows the number of inserted buffers to satisfy CMOL connectivity domain.

Simulated Evolution solutions are more effective than those of CMOL CAD 1.0 in terms of computation time, de-

Table 1: ISCAS'89 Benchmarks: *Area* is the size of CMOL 2-D grid. *AU%* is area utilization.

Circuits	Cells	Gates	Inputs	Outputs	Area (Tiles)	Area (Row × Column)	AU% (Tiles)	AU%
s27	19	8	7	4	64 (2 × 2)	25 (5 × 5)	18.75	32.00
s208	136	109	18	9	256(4 × 4)	169(13 × 13)	48.05	64.50
s298	122	85	17	20	256(4 × 4)	144(12 × 12)	48.83	59.03
s344	180	130	24	26	400(5 × 5)	196(14 × 14)	43.50	66.33
s349	184	134	24	26	400(5 × 5)	196(14 × 14)	26.50	68.37
s382	175	124	24	27	400(5 × 5)	196(14 × 14)	43.25	63.27
s386	164	138	13	13	400(5 × 5)	196(14 × 14)	54.75	70.41
s400	188	137	24	27	400(5 × 5)	196(14 × 14)	47.25	69.90
s420	299	248	34	17	400(5 × 5)	361(19 × 19)	75.00	68.70
s444	187	136	24	27	400(5 × 5)	196(14 × 14)	52.50	69.39
s510	304	266	25	13	-	361(19 × 19)	-	73.68
s526	273	222	24	27	576(6 × 6)	324(18 × 18)	57.12	68.52
s641	302	206	54	42	576(6 × 6)	676(26 × 26)	50.17	30.47
s713	321	225	54	42	-	676(26 × 26)	-	33.28
s820	447	400	23	24	-	529(23 × 23)	-	75.61
s832	454	407	23	24	-	529(23 × 23)	-	76.94
s838	606	507	66	33	-	676(26 × 26)	-	75.00
s1196	675	613	31	31	-	729(27 × 27)	-	84.09
s1238	724	662	31	31	-	784(28 × 28)	-	84.44

Table 2: ISCAS'89 Comparison With CMOL CAD, GA, MA and LRMA - ($r = 12$)

Circuits	CMOL CAD 1.0		GA [7]			MA [8]			LRMA [9]			SimE		
	Delay	Time	Delay	Time	Buf	Delay	Time	Buf	Delay	Time	Buf	Delay	Time	Buf
s27	9	1	7	0.01	0	7	0.01	0	7	0.01	0	7	0.01	0
s208	18	3	16	1.12	0	16	0.12	0	16	0.10	0	16	0.01	0
s298	13	7	11	0.17	0	11	0.11	0	11	0.09	0	11	0.01	0
s344	20	8	18	0.57	0	1	0.29	0	18	0.16	0	18	0.01	0
s349	20	7	18	0.49	0	18	0.28	0	18	0.18	0	18	0.02	0
s382	13	7	11	1.60	0	11	0.38	0	11	0.32	0	11	0.04	0
s386	16	11	10	1.05	0	10	0.33	0	10	0.34	0	10	0.02	0
s400	15	8	11	2.12	1	11	0.40	0	11	0.34	0	11	0.03	0
s420	20	8	16	8.50	1	16	3.41	0	16	1.57	0	16	0.09	0
s444	17	9	11	1.86	2	11	0.40	0	11	0.34	0	11	0.03	0
s510	-	-	18	16.56	2	18	7.56	0	18	3.42	0	18	0.16	0
s526	16	13	11	9.75	5	11	4.36	0	11	1.59	0	11	0.25	0
s641	25	8	23	82.66	15	19	39.40	4	16	22.02	0	16	2.92	0
s713	-	-	24	52.84	34	19	30.11	3	19	41.77	2	19	3.40	0
s820	-	-	15	77.52	41	12	61.71	10	12	54.09	6	12	27.72	0
s832	-	-	16	69.27	54	12	60.17	11	12	63.77	4	12	31.00	0
s838	-	-	28	201.37	50	24	85.62	7	24	100.40	4	24	2.42	0
s1196	-	-	30	234.88	84	23	208.15	19	24	179.47	9	23	23.50	0
s1238	-	-	37	268.92	121	28	267.34	31	26	353.00	9	26	53.76	0
Average	-	-	17	54.28	22	15	40.53	4	15	43.31	2	15	7.65	0

Delay: Logic Levels.
Time: Computation Time in Seconds.
Buf: Buffers Inserted.

lay and area utilization. The last two columns of Table 1 show that cell-based CMOL architecture has better area utilization $AU\%$ than that of tile-based architecture. Table 2 indicates that the tile-based approach is the most time consuming and the least effective in timing delay, it also fails to place big circuits. Results obtained from implementation of SimE for $r = 12$ are better than those obtained in GA, MA and LRMA in both computation time and Buffers count. SimE required shorter CPU processing time compared to time required by genetic crossover, mutation and Lagrangian multipliers calculation in LRMA. Table 2 shows that Simulated Evolution found the optimal solutions with zero buffers for all benchmarks, with up to 82% average computation time saving. For example, SimE required only 23.50 seconds to find the optimal solution of zero buffers and delay equals to 23 for benchmark s1196, while LRMA required 179.47 seconds and needed 9 buffers to satisfy connectivity raising timing delay to 24.

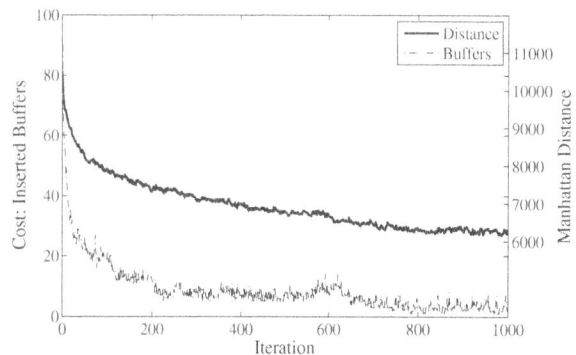

Figure 6: Correlation between Distance Minimization and Buffers Insertion. Benchmark: **s1238.blif**

25

5. CONCLUSION

In this paper we have discussed the cells placement problem in CMOL nano-hyprid circuits. We have analyzed the problem and devised a goodness measure that exploits better understanding of the limitations imposed by CMOL connectivity radius. A sub-optimal search heuristic known as Simulated Evolution (SimE) has been employed for effective and time efficient exploration of search space. Results obtained proves the robustness of our approach compared with previously implemented solutions.

6. ACKNOWLEDGMENTS

The authors acknowledge King Fahd University of Petroleum & Minerals for its support, and Dr. William N. N. Hung and Mr. Zhufei Chu for providing ISCAS'89 benchmark files. The authors would also like to thank Dr. Rajat Subhra Chakraborty for his help and support.

7. REFERENCES

[1] Dmitri B Strukov and Konstantin K. Likharev. CMOL FPGA: a reconfigurable architecture for hybrid digital circuits with two-terminal nanodevices. *Nanotechnology*, 16(6):888–900, 2005.

[2] Dmitri B. Strukov and Konstantin K. Likharev. A reconfigurable architecture for hybrid CMOS/Nanodevice circuits. In *Proceedings of the 2006 ACM/SIGDA 14th international symposium on Field programmable gate arrays*, FPGA '06, pages 131–140, New York, NY, USA, 2006. ACM.

[3] Konstantin K. Likharev. Crossnets: Neuromorphic hybrid CMOS/Nanoelectronic networks. *Science of Advanced Materials*, 3:322–332, 2011.

[4] Dmitri B Strukov and Konstantin K. Likharev. Prospects for terabit-scale nanoelectronic memories. *Nanotechnology*, 16(1):137, 2005.

[5] Dmitri B. Strukov and Konstantin K. Likharev. CMOL FPGA circuits. In *In Proc. of Int. Conf. on Computer Design, CDESŠ2006*, pages 213–219, 2006.

[6] William N.N. Hung, Changjian Gao, Xiaoyu Song, and D. Hammerstrom. Defect-tolerant CMOL cell assignment via satisfiability. *Sensors Journal, IEEE*, 8(6):823 –830, june 2008.

[7] Yinshui Xia, Zhufei Chu, William N.N. Hung, Lunyao Wang, and Xiaoyu Song. CMOL cell assignment by genetic algorithm. In *NEWCAS Conference (NEWCAS), 2010 8th IEEE International*, pages 25 –28, june 2010.

[8] Zhufei Chu, Yinshui Xia, William N.N. Hung, Lunyao Wang, and Xiaoyu Song. A memetic approach for nanoscale hybrid circuit cell mapping. In *Digital System Design: Architectures, Methods and Tools (DSD), 2010 13th Euromicro Conference on*, pages 681 –688, sept. 2010.

[9] Y. Xia, Z. Chu, W. Hung, L. Wang, and X. Song. An integrated optimization approach for nano-hybrid circuit cell mapping. *Nanotechnology, IEEE Transactions on*, PP(99):1, 2011.

[10] Gang Chen, Xiaoyu Song, and Ping Hu. A theoretical investigation on CMOL FPGA cell assignment problem. *Nanotechnology, IEEE Transactions on*, 8(3):322 –329, may 2009.

[11] Sadiq M. Sait and Habib Youssef. *Iterative Computer Algorithms with Applications in Engineering: Solving Combinatorial Optimization Problems*. IEEE Computer Society Press, California, December 1999.

[12] F. Brglez, D. Bryan, and K. Kozminski. Combinational profiles of sequential benchmark circuits. In *Circuits and Systems, 1989., IEEE International Symposium on*, pages 1929 –1934 vol.3, may 1989.

SNR Analysis Approach for Hardware/Software Partitioning using Dynamically Adaptable Fixed Point Representation

Varadaraj Kamath Nileshwar, Roman Lysecky

Department of Electrical and Computer Engineering, University of Arizona, Tucson, AZ

varad@email.arizona.edu, rlysecky@ece.arizona.edu

Abstract

During the early design phases of software development, many developers use floating point data types and libraries but often convert these applications into fixed point representations in later design phases – a time consuming process often requiring significant designer effort. While various approaches have been proposed to automate the floating to fixed point conversion process, these approaches are mainly targeted at creating optimized software implementations and do not directly support partitioning floating point implementation to hardware. We present an approach to optimize the number of bits required for a dynamically adaptable fixed-point representation using SNR analysis methods targeting computationally intensive floating-point kernels. We present a hardware/software partitioning methodology that leverages this SNR analysis to partition application kernels to custom hardware coprocessors implemented within a field-programmable gate array. Using several case study applications, we highlight the performance benefits and area requirements of the resulting hardware implementations.

Categories and Subject Descriptors

C.3 **[Computer Systems Organization]** Special-Purpose and Application-Based Systems - *Real-time and Embedded Systems.*

Keywords

Hardware/software partitioning, floating point to fixed point conversion, signal to noise ratio analysis, floating point profiling.

1. INTRODUCTION

For many embedded applications requiring support for real numbers, single and double precision floating point representations are often utilized during initial software development due to the convenience provided by built-in support within most programming languages, including C/C++/Java. However, many embedded processors do not provide hardware support for implementing floating point operations due to the high silicon area and power consumption requirements associated with implementing floating point arithmetic in hardware.

Fortunately, many applications do not need floating point representations to support real numbers. Instead, fixed point representations and arithmetic are viable alternatives for many embedded applications, often achieving greater performance with reduced area requirements. A floating point number is stored in a sign-magnitude format that allows for a floating position of the radix point specified by an exponent. The exponent provides great

flexibility in the range of numbers that can be represented. In contrast, a fixed point representation is directly stored as a two's complement binary number with a fixed radix point. For a fixed point representation, the range of numbers that can be represented is determined by the fixed radix position, and can be significantly less than a similarly sized floating point representation.

Floating point representations offer a higher dynamic range of numbers that can be represented using an equivalent number of bits. However, for those applications that do not require such dynamic range, a fixed point implementation is often the best alternative. Unlike floating point arithmetic, fixed point operations can be very efficiently performed using integer operations. Fixed point additions directly map to integer additions, and fixed point multiplication can be performed using integer multiplication followed with a shift by the radix position. However, one of greatest challenges in using fixed point arithmetic is that of software development. Due to the lack of built-in support for fixed point representations in most programming languages, software developers will often initially develop an application using floating point representation and later convert the application to a fixed point representation, which can account for 25-50% of the software development effort for some applications [3][8].

To reduce the development effort required to convert floating point application to fixed point representations, several methodologies have been proposed [1][6][12][14] that either automatically determine the necessary bit width for the fixed point representation given an initial floating point software implementation or generate the fixed point software implementation. Such automated conversion approaches are primarily targeted at the compilation phase of software development, for which the resulting output is a fixed point software implementation in which all floating point operations have been converted to a fixed point implementation.

Alternatively, instead of converting the floating point software application to utilize a fixed point software implementation, it is possible to provide significant performance improvements by using hardware/software partitioning methods to partition only the critical kernels within the target application to custom hardware coprocessors with typical performance gains of 10-100X [2][7][15]. Most hardware/software partitioning approaches are often limited to partitioning computational kernels utilizing integers or fixed point computations. When floating point representations are needed, highly pipelined hardware floating point implementations with multi-cycle latencies are required to provide acceptable performance.

By focusing on the critical kernels of the target application, an automatable hardware/software partitioning approach for floating point software applications was proposed in [13] that eliminates the need for developers to rewrite software applications utilizing fixed point implementations. Instead, this approach incorporates efficient, configurable floating point to fixed point and fixed point to floating point hardware converters at the boundary between the hardware coprocessors and memory, thereby separating the system into a floating point domain for software execution and fixed point domain for hardware coprocessor execution and eliminating the need to modify the original software application. This enables hardware coprocessor development to treat floating point computations similar to integer computations, thereby providing excellent hardware performance with minimal hardware requirements. However, ad hoc methods are utilized to determine the required fixed point representation for hardware coprocessors. Further, the initial adaptable fixed point representation utilized in this approach only considered dynamic adjustment in response to overflows, consequently requiring a fixed point representation with more bits than needed for the target application.

In this paper, we present a hardware/software partitioning approach that employs signal to noise ratio (SNR) analysis and profiling methods for determining the required size for hardware coprocessors utilizing a dynamically adaptable fixed-point representation. The dynamically adaptable fixed point implementation provides a single fixed point representation shared within each coprocessor, in which the required fixed point representation can be dynamically adjusted at runtime. By utilizing an adaptable implementation, fewer bits are required to represent real numbers. We analyze the performance benefits and requirements for several embedded applications and coprocessor alternatives, highlighting the flexibility of this approach.

2. RELATED WORK

The *fixify* environment [1] is a simulation and design space exploration tool chain developed within the Open Tool Integration Environment (OTIE) to provide support for automated floating point to fixed point conversion. OTIE uses a statistical approach to optimize fixed point representations for all data channels in the target system. This optimization process utilizes a signal to quantization noise ratio (SQNR) metric to determine the relative error with various optimization algorithms to optimize the fixed point representation with reasonable runtimes.

To leverage the efficiency provided by specialized hardware available within many digital signal processors (DSP), Menard et al. [12] propose an automated optimization framework for implementing floating point algorithms using these fixed point architectures given design specific accuracy constraints. The determination and optimization of the fixed point representation is directed by an input accuracy constraint. Initially, the data range of data processed by the target algorithm is explored to determine the radix point required to avoid overflow. Subsequently, the word length for each variable is defined in order to leverage the diversity of different fixed point data available in the target DSP. Finally, the data formats for all variables are optimized to minimize the code execution time as long as the accuracy constraint is achieved.

While floating point representations can directly provide the required dynamic range to achieve the desired SNR for most signal processing applications, the hardware requirements are prohibitive for many systems. Alternatively, block floating point representations have been proposed as a compromise between floating point and fixed point number representations [6][9]. Block floating point representation is a scalable number representation similar to floating point in that an exponent is used to adjust the radix point. However, fixed point arithmetic operations can be utilized to perform computations to provide an efficient implementation alternative. This is achieved by adjusting the floating point exponent for all variables – or a block of variables – within the target application. Given the unique aspects of block floating point representations, round off error models for block floating point formats have been

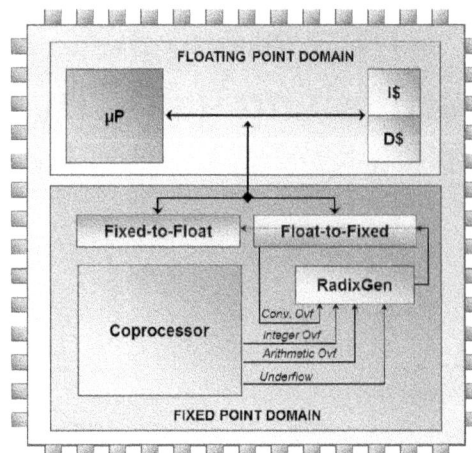

Figure 1: Microprocessor/coprocessor architecture separating computing into a floating point domain and fixed point domain.

developed to analyze the resulting error for digital filters [6]. This round off error analysis has indicated that block floating point representation can provide better SNR and range compared to standard floating point representation using the same total number of bits per sample when the block length is properly selected.

The dynamically adaptable fixed point representation is similar to block floating point implementations, but offers several advantages. Our dynamically adaptable fixed point representation has lower computational complexity and hardware requirements compared to block floating point implementation as the intermediate results for an entire *block* of computations does not need to be maintained. Additionally, the control logic required to compute and adjust the joint scale factor in block floating point representation is not required within the dynamically adaptable fixed point representation, as the scale factor for each computation can be dynamically determined.

3. DYNAMICALLY ADAPTABLE FIXED POINT REPRESENTATION

Figure 1 provides an overview of the target system architecture incorporating a main processor – with or without support for floating point operations – and loosely-coupled coprocessor architecture for accelerating floating point application kernels. At the interface between the system bus and coprocessor, configurable floating point to fixed point (*Float-to-Fixed*) and fixed point to floating point (*Fixed-to-Float*) hardware converters are utilized to separate the execution of hardware/software implementation into a floating point computing domain and a fixed point computing domain. Rather than rely on *Float-to-Fixed* and *Fixed-to-Float* converters that have been statically configured, the coprocessor architecture supports the ability to dynamically adjust the fixed point representation during execution. This allows the resulting implementation to provide greater precision and dynamic range while requiring fewer bits for the fixed point representation.

The dynamically adaptable *Float-to-Fixed* converter consists of two main components, *NormalCases* and *SpecialCases*. The *SpecialCases* component handles all the special representations in the IEEE 754 standard including positive and negative zero, positive and negative infinity, and not-a-number (NaN). The *NormalCases* component handles the normal floating to fixed point conversions.

Infinity and NaN floating point values are not representable using a fixed point representation. Thus, if a floating point input is infinity or NaN, an *Exception* output will be asserted. On the other hand, if the floating point input is the special case for representing either a positive or negative zero, the *SpecialCases* component will

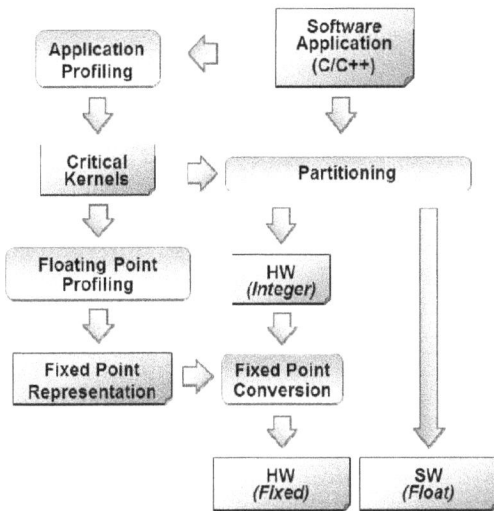

Figure 2: Hardware/software partitioning methodology targeting dynamically adaptable fixed point coprocessors.

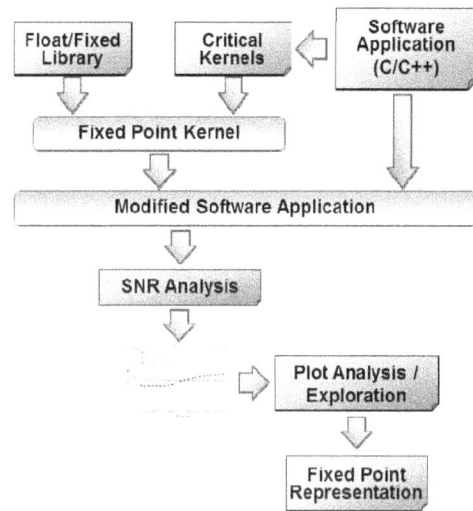

Figure 3: Signal to noise ratio based floating point profiling methodology for dynamically adaptable fixed point representation.

assert the *Zero* output signal. Finally, if the floating point input is the special case of a denormalized number, the *SpecialCases* component will deassert the *Normal* output signal. The *NormalCases* component calculates an *Overflow* output that corresponds to the number of additional bits needed within the fixed point representation to avoid an overflow for the current conversion, and computes an *Underflow* output that corresponds to the number of unutilized bits within the integer portion of the fixed point output given the current fixed point representation.

Within the dynamically adaptable fixed point implementation, the coprocessor's fixed point representation will initially provide maximum precision – i.e. a radix point initially set to the maximum number of bits within the fixed point representation. As conversion and arithmetic overflows require additional bits to represent the integer portion, the radix point can be gradually reduced to increase the fixed point representation's dynamic range. However, if the increased range is no longer needed, the radix point can be adjusted to increase precision. This process will continually adapt the fixed point representation.

To support this dynamic adaptability, the *RadixGen* component monitors the fixed point output from the *Float-to-Fixed* converter and all internal computations/registers to detect both *overflow* and *underflow* conditions. An *overflow* occurs whenever the input or intermediate results cannot be represented using the available integer bits. In response to detecting overflows, the *RadixGen* component will interrupt the coprocessor execution and shift the radix point right by one position iteratively until there are sufficient integer bits available. Conversely, an *underflow* occurs whenever the input or intermediate result requires fewer bits to represent the integer portion of the fixed point number. In response to underflows, The *RadixGen* component will interrupt the coprocessor execution and shift the radix point left by one position iteratively until the precise number of bits needed to represent the fixed point number is determined. To dynamically adapt the fixed point representation, all overflows and underflows are detected to allow the *RadixGen* component to adjust the radix point, requiring one cycle for the adjustment one cycle to re-execute the previous operation that resulted in the overflow or underflow condition.

Within this framework, overflows and underflows include floating point to fixed point conversion overflows, integer to fixed point conversion overflows, arithmetic overflows, and coprocessor underflows. Floating point to fixed point conversion overflows are

directly determined within the *Float-to-Fixed* converter. An integer to fixed point conversion overflow occurs whenever the current fixed point representation does not have enough bits allocated to the integer portion to represent an integer value needed within a fixed point calculation. Whenever an integer value is utilized within a fixed point calculation, that integer value is converted to the current fixed point representation, at which point integer to fixed point conversion overflows are detected. An arithmetic overflow occurs whenever the current fixed point representation does not have enough bits allocated to the integer portion to represent the fixed point result of an arithmetic operation, such as an addition or multiplication. All fixed point arithmetic operations are monitored to determine if an overflow occurred, outputting the appropriate overflow signal to the *RadixGen* component. An underflow condition is only possible if all current fixed point values utilized within the coprocessor do not require the current number of bits allocated to the integer portion of the fixed point representation. To detect underflows, an exclusive NOR operation (XNOR) of the two most significant bits for all fixed point registers is utilized. If the result of this operation is a logical one for all fixed point values, an underflow signal connected to the *RadixGen* component is asserted.

4. HARDWARE/SOFTWARE PARTITIONING AND FLOATING POINT PROFILING

Figure 2 presents an overview of the hardware/software partitioning methodology for floating point applications. Starting with the initial floating point software application, the application is initially profiled to determine the application's critical kernels based upon the percentage of total application execution time required by functions or loops. Utilizing an instruction accurate simulator for the target processor, detailed function and loop level profiling is used to determine the execution statistics for all loops and functions within the software application. This profiling information is utilized to determine which computational kernels will be considered for partitioning to hardware.

For each kernel within the application, an initial hardware description is created in which all floating point data types are treated as integers – single precision as a 32-bit integer, double precision as a 64-bit integer. The resulting coprocessor utilizes a loosely coupled coprocessor model in which all communication between the microprocessor and hardware coprocessors is implemented using memory mapped registers. The hardware coprocessors will execute in a mutually exclusive manner, in which

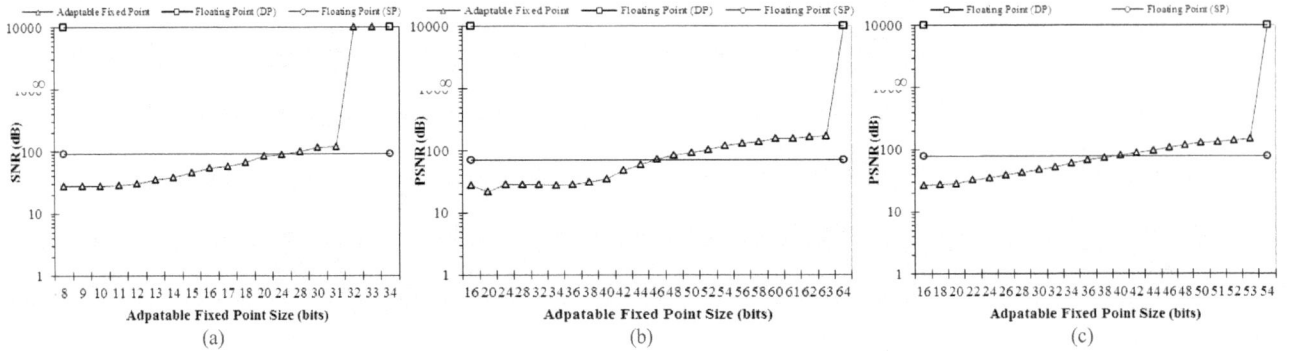

Figure 4: SNR floating point profiling results for (a) *mpeg2dec*, (b) *fft/ifft*, and (c) *epic* applications.

the microprocessor will initialize the hardware coprocessor to execute and wait for the coprocessor to complete its execution. As the required fixed point representation is not known yet – and the final hardware execution frequency depends on that fixed point representation – the partitioning process can initially select those critical kernels whose initial integer based implementation yields a net performance gain when partitioned to a hardware coprocessor.

Floating point profiling is then performed in order to determine the required width for the fixed point coprocessor implementation utilizing signal to noise ratio (SNR) analysis. Figure 3 provides an overview of the SNR floating point profiling method. For each candidate kernel within the software application, the floating point implementation is modified to utilize a float/fixed point software library that profiles the accuracy and precision of floating point operations during execution. The float/fixed library provides a seamless profiling abstraction that allows the application to be executed using either standard floating point representations, i.e. single and double precision, or utilizing a dynamically adaptable fixed point representation with varying fixed point widths.

Using the float/fixed library, the target application is executed several times to analyze the resulting SNR for various widths of the target dynamically adaptable fixed point representation. The fixed point representation's width is iteratively increased during each execution until the resulting outputs match that of the double precision floating point reference implementation. For each application kernel, a designer is provided with a plot of the measured SNR as function of increasing fixed point width, along with indicators highlighting the reference double precision floating point and single precision floating point implementations.

If the required SNR for each application kernel is known a priori, the floating point profiling can directly output the required width for the dynamically adaptable fixed point implementation. Otherwise, the application designer can analyze SNR profiling and select an appropriate fixed point width. Given the fixed point widths for each kernel, the fixed point conversion process will convert the initial integer based kernel implementation to utilize the dynamically adaptable fixed point representation.

If the resulting hardware coprocessor implementation yields a net performance gain compared to the initial software implementation and FPGA resources are available for the coprocessor, the kernel will be partitioned to hardware. This process repeats until either no additional FPGA resources are available or all candidate kernels have been evaluated.

5. EXPERIMENTAL RESULTS

We consider a target microprocessor/FPGA architecture consisting of 250 MHz MIPS processor with support for floating point calculations – provided by a floating point coprocessor – coupled with a Xilinx Virtex-6 FPGA. Using the dynamically adaptable fixed point representation and SNR floating point profiling, we

partitioned the *mpeg2dec* and *epic* applications of the MediaBench benchmark suite [10], the *fft* and *ifft* applications of the MiBench benchmark suite [5], and the *facerec* application of the ALPBench benchmark suite [11], all of which required extensive floating point calculations.

Based on the initial application profiling, the critical kernels for each application were determined. For *mpeg2dec*, the primary computational kernel performs an inverse discrete cosine transform (IDCT). Within *epic*, the critical kernel performs a two-dimensional convolution filter across nine sections – corresponding to the top, middle, and bottom sections vertically and left, center, and right section horizontally – of the input image. However, the application profiling information revealed that the majority of execution time was required to process the center of the image. For both the *fft* and *ifft* application, the predominant computational kernels perform the respective Fourier transforms. The *facerec* application performs a principle components analysis algorithm implemented within three execution stages, specifically image data preprocessing, algorithm training, and algorithm testing. Across all three stages, the primary computational kernels all perform similar matrix multiplication operations.

5.1 SNR Floating Point Analysis

For each application, SNR floating point profiling was utilized to analyze the performance of the dynamically adaptable fixed point representation with increasing widths. Figure 4 (a) presents the SNR floating point profiling results for the *mpeg2dec* application, considering fixed point widths from 8 to 34. While a fixed point width of 32 is required to match the precision of the double precision floating point implementation, only 28 bits are required to provide the same SNR compared to a single precision representation. Furthermore, the dynamically adaptable fixed presentation provides a 4dB gain in SNR compared to a statically configured fixed point representation of the same width.

For many applications, double precision floating point is not necessary, and acceptable results can be achieved with reduced SNR. For these applications, designer expertise is needed to specify an acceptable SNR or select an appropriate fixed point width given the resulting SNR profiling results. For *mpeg2dec*, several alternative implementations can be considered.

The reference IDCT operation with the *mpeg2dec* application utilizes a double precision floating point representation. However, double precision floating point is not required to perform the IDCT within the specifications for IEEE 8x8 IDCT standard [4]. As such, we utilized the floating point profiling results to determine that a fixed point width of only 24 bits is necessary to maintain compatibility with this standard. Furthermore, this bit width is less than the width required to provide the same SNR compared to a single precision floating point representation.

Table I: Application speedup and area for applications using various dynamically adaptable fixed point implementations determined using SNR profiling, floating point implementations providing equivalent SNR compared to the various dynamically adaptable alternatives, and the initial adaptable fixed point implementation presented in [1].

	Dynamically Adaptable Fixed Point						Floating Point					Initial Approach [1]		
	Speedup	MHz	Bits	Area			Speedup	MHz	Bits	Area		Speedup	MHz	Bits
				AFFHL	LUTs+FFs	DSP48				LUTs+FFs	DSP48			
mpeg2dec	1.6	77	32	493	1634	12	0.9	268	64	4067	28	1.5	67	64
	1.9	102	24	418	1282	6	1.6	380	32	1615	10			
	2.0	125	16	334	853	3	1.6	380	32	1591	10			
fft/ifft	5.9	56	64	767	8209	90	2.0	377	64	9468	101	6.0	66	64
	6.5	62	45	308	5999	72	2.4	380	32	3280	37			
	8.9	92	24	418	2243	12	2.4	380	32	3271	37			
epic	1.9	100	39	543	1204	12	1.3	293	32	652	5	1.6	61	64
	2.0	103	27	453	903	11	1.4	321	32	626	5			
facerec	1.2	84	52	870	2548	24	0.8	421	64	4203	28			
Average:	3.5				2764	27	1.6			3197	29	3.1		

Although 24 bits are required to remain standard compliant, the bit width can potentially be further decreased with some additional loss in visual quality. However, as with any lossy compression, some decrease in quality can be tolerated as long as the decrease does not overly affect the perceived quality. Hence, we further analyzed the *mpeg2dec* application to observe the visual degradation using two different video streams compared against the original output using the double precision floating point reference implementation. This subjective analysis indicated that no degradation in quality is perceivable with a SNR greater than 50dB, below which artifacts in the video sequence are clearly distinguishable. To achieve a SNR of 50dB, a 16-bit fixed point representation is required. Although a 16 bit representation may be suitable for this application, we consider all three fixed point widths, i.e. 32, 24, and 16, to evaluate the performance speedup and area requirements.

Figure 4 (b) presents the SNR floating point profiling results for the *fft* and *ifft* applications. As the intended use for the FFT and IFFT operations are unknown, we consider several different fixed point widths based upon this SNR profiling. To provide the same SNR compared to both double and single precision floating point implementations, the dynamically adaptable fixed point implementation will require 64 and 46 bits respectively. As the SNR decreases slightly with reduced fixed point width and reaches a plateau for widths between 24 and 36, we additionally consider a fixed point width of 24 bits.

For the *epic* application, whose SNR profiling results are presented in Figure 4 (c), a single precision floating point representation in the original software application is sufficient. To achieve a SNR equivalent to single precision floating point representation, a fixed point width of 39 is required. However, as with the *mpeg2dec* application, a decrease in SNR can be tolerated as long as no visual degradation is perceptible. A similar subjective visual analysis was utilized to determine that a fixed point width of 27 may be suitable for the *epic* application. Thus, we consider fixed point widths of both 39 and 27 bits.

Finally, for the facial recognition application, *facerec*, we utilized 20 images from five different individuals. The *facerec* application computes a distance metric for each pair of images indicating the difference between the detected faces in each image. A lower distance metric value corresponds to a better match, with the value of zero indicating a perfect match. To determine the required fixed point widths for this application's critical kernels, we first determined the SNR such that all actual facial matches are reported as matches by application. We note that while this requirement will lead to false positives, the occurrences of false

positives are inherent to the algorithms employed by the applications. To guarantee that all actual matches are correctly reported, a SNR of 98db is required. Using our floating point profiling, the required fixed point width of 52 was determined.

5.2 Performance and Area Requirements

We evaluated the performance improvements and area requirements using the presented dynamically adaptable hardware/software partitioning approach. For each application kernel, the dynamically adaptable fixed point coprocessors were implemented using the various identified fixed point widths. All coprocessors were implemented in Verilog and synthesized to a Xilinx Virex-6 FPGA using Xilinx ISE 11.1. Additionally, we compare both the performance and area requirements to floating point coprocessor implementations using the single or double precision representation that most closely matches the SNR achieved by each fixed point coprocessor.

Table I presents the overall application speedup compared to the original floating point software implementation and requirements for various dynamically adaptable fixed point and floating point implementations. The area required for each coprocessor is reported as the umber of look-up tables (LUTs), flip-flops (FFs), and DSP48 components utilized for each design. In addition, the LUTs and FFs required for the AFFHL converters utilized within the dynamically adaptable fixed point implementations are reported. The dynamically adaptable fixed point implementations achieve application speedups ranging from 1.2X to 8.9X. We note that the reported application speedups are compared to software execution in which a dedicated floating point processing unit is available. For systems in which a floating point processor is not available – or processors with native support for only single precision floating point computations – greater application speedups can be achieved relative to the base software.

For the *mpeg2dec* application, an application speedup of 1.6X is achieved while providing the same SNR as the reference double precision implementation. Using a fixed point width of 24 bits to meet the required SNR to remain compliant with the IEEE DCT standard, the application speedup is increased to 1.9X. This performance increase can be achieved given that the operating frequency of the hardware coprocessor can be increased from 77 MHz to 102 MHz. While further reducing the fixed point width to 16 bits – determined using the subjective visual metrics – can further increase the coprocessor frequency to 125 MHz, only a marginal performance increase is achieved. This is due to the limitations imposed by memory accesses and communication.

In contrast, the nearest equivalent floating point implementations achieve application speedups ranging from 0.9X to 1.6X. Although the coprocessor can operate at a higher frequency, this increased clock rate is achieved through pipelining within the floating point cores utilized. This ultimately leads to multi-cycle latencies because sufficient data cannot be provided to the core to keep the pipeline full. As such, lower application speedups – or even slowdowns – are achieved using the floating point coprocessor implementations across all applications. Considering the highest performing implementation for each application, the dynamically adaptable fixed point coprocessors provide an average speedup of 3.5X, compared to a speedup of 1.6X using the floating point coprocessor implementations.

For most applications, the dynamically adaptable fixed point coprocessor required fewer area resources compared to the floating point implementations. Across all applications, the fixed point coprocessors required on average 13% fewer LUTs and FFs and 7% fewer DSP48 components. However, in the case of the *fft* and *ifft* applications, the area required by the 45-bit fixed point implementation is greater than the corresponding floating point version. This area overhead is due to both larger width needed to store values within the coprocessor, i.e. 32 bits required for the single precision floating point implementation compared to 45 bits required for the adaptable fixed point implementation, and the additional logic required to detect overflow conditions. In contrast, the 64-bit and 24-bit fixed point implementations, which more closely match the width of the corresponding floating point versions, require 13% and 31% fewer logic resources compared to the floating point alternatives, respectively.

Table I further presents the overall application speedup, coprocessor frequency, and fixed point width using the initial adaptable fixed point implementation presented [13]. Because the initial adaptable fixed point implementation only adjusted the radix point due to overflows, a fixed point width of 64 bits was utilized for all applications to provide the accuracy compared to the software application. For the *mpeg2dec* and *epic* applications, this increased fixed point width results in slower coprocessor operating frequencies, and subsequently reduced application speedup, compared to our current dynamically adaptable fixed point implementation. For the *fft/ifft* applications, both approaches utilize a fixed point width of 64 bits. In this scenario, the reduced complexity of the initial adaptable implementation achieves higher performance due to a faster coprocessor frequency of 66 MHz compared to 56 MHz. Importantly, with reduced SNR requirement, the previous approach requires an increased fixed point width due to the limitation in how the fixed point representation is dynamically adapted at runtime. This increased fixed point width detrimentally impacts both the coprocessor frequency and the overall application performance.

6. CONCLUSIONS

By utilizing a dynamically adaptable fixed point representation, an average application speedup of 3.5X can be achieved over a software only implementation by partitioning floating point kernels within software applications to fixed point hardware coprocessors. Compared to coprocessor implementations utilizing standard floating point representations, the adaptable fixed point implementations achieve an additional 2X increase in performance while utilizing 13% fewer logic resources. To assist designers in selecting the required fixed point width for their applications, an efficient SNR floating point profiling method was presented to analyze the accuracy, performance, and area tradeoffs. This hardware/software partition approach provides an efficient and rapid method for partitioning floating point software applications with reduced design time, reduced designer effort, and reduced hardware coprocessor area requirements.

7. ACKNOWLEDGEMENTS

This work was supported in part by the National Science Foundation under Grant CNS-0844565.

8. REFERENCES

[1] Belanovíc, P., M. Rupp. Automated Floating-Point to Fixed-Point Conversion with the fixify Environment. Workshop on Rapid System Prototyping (RSP), 2005.

[2] Chen, W., P. Kosmas, M. Leeser, C. Rappaport. An FPGA Implementation of the Two-Dimensional Finite-Difference Time-Domain (FDTD) Algorithm. Symposium on Field-Programmable Gate Arrays (FPGA), pp. 97-105, 2004.

[3] Clark, M., M. Mulligan, D. Jackson, D. Linebarger. Accelerating Fixed-Point Design for MB-OFDM UWB Systems. Comms Design, January 26, 2005.

[4] IEEE Circuits and Systems Society Standards Committee. IEEE Standard Specifications for the Implementations of 8x8 Inverse Discrete Cosine Transform, 1991.

[5] Guthaus, M., J. Ringenberg, D. Ernst, T. Austin, T. Mudge, R. Brown. MiBench: A Free, Commercially Representative Embedded Benchmark Suite. Workshop on Workload Characterization, pp. 3-14, 2001.

[6] Kalliojarvi, K., J. Astola. Roundoff Errors in Block-Floating-Point Systems. IEEE Transactions on Signal Processing, Vol. 44, pp. 783-790, 1996.

[7] Keane, J., C. Bradley, C. Ebeling. A Compiled Accelerator for Biological Cell Signaling Simulations. Symposium on Field-Programmable Gate Arrays (FPGA), pp. 233-241, 2004.

[8] Keding, H., M. Willems, M. Coors, H. Meyr. FRIDGE: A Fixed-Point Design and Simulation Environment. Design Automation and Test in Europe Conference (DATE), 1998.

[9] Kobayashi, S., G. Fettweis. A Hierarchical Block-Floating-Point Arithmetic. Journal of VLSI Signal Processing, Vol. 24, No. 1, pp. 19-30, 2000.

[10] Lee, C., M. Potkonjak, W. Mangione-Smith. MediaBench: a Tool for Evaluating and Synthesizing Multimedia and Communications Systems. Symposium on Microarchitecture, pp. 330-335, 1997.

[11] Li, M.-L., R. Sasanka, S. Adve, C. Yen-Kuang, E. Debes. The ALPBench Benchmark Suite for Complex Multimedia Applications. Workload Characterization Symposium, pp. 34-45, 2005.

[12] Menard, D., R. Serizel, R. Rocher, O. Sentieys, O. Accuracy Constraint Determination in Fixed-Point System Design. EURASIP Journal on Embedded Systems, pp. 1-12, 2008.

[13] Saldanha, L., R. Lysecky. Float-to-Fixed and Fixed-to-Float Hardware Converters for Rapid Hardware/Software Partitioning of Floating Point Software Applications to Static and Dynamic Fixed Point Coprocessors. Journal on Design Automation of Embedded Systems, Vol. 13, No. 3, pp. 139-157, 2009.

[14] Shi, C., R. Brodersen. An Automated Floating-Point to Fixed-Point Conversion Methodology. Conference on Acoustic, Speech, and Signal Processing, Vol. 2, pp. 529-532, 2003.

[15] Stitt, G., F. Vahid, G. McGregor, B. Einloth. Hardware/Software Partitioning of Software Binaries: A Case Study of H.264 Decode. Conference on Hardware/Software Codesign and System Synthesis (CODES+ISSS), pp. 285-290, 2005.

[16] Venkataramani, G., W. Najjar, F. Kurdahi, N. Bagherzadeh, W. Bohm. A Compiler Framework for Mapping Applications to a Coarse-grained Reconfigurable Computer Architecture. Conference on Compiler, Architecture, and Synthesis for Embedded Systems (CASES), 2001.

NBTI Mitigation in Microprocessor Designs

Simone Corbetta, William Fornaciari
Politecnico di Milano – Dipartimento di Elettronica e Informazione
Via Ponzio 34/5, 20133 Milano, Italy
{scorbetta, fornacia}@elet.polimi.it

ABSTRACT

Negative-Bias Temperature Instability seriously affects nanoscale circuits reliability and performance. Continuous stress and increasing operating temperatures lead to device degradation and long-term system unavailability. The opportunity to optimize the duty-cycle of the stress/recovery phases to reduce V_{th} degradation leads to innovative research of reliability-oriented resources allocation at architectural level. This work explores the impact of different allocation strategies on the processor degradation, through a novel estimation methodology. Experimental results show that the proposed NBTI-aware allocation strategy can guarantee from 10% and up to 30% lower degradation compared to classical strategies, under different operating scenarios and under process variability.

Categories and Subject Descriptors

B.8 [**Performance and Reliability**]: Reliability, Testing, and Fault-Tolerance

General Terms

Design, Reliability

Keywords

Dynamic instruction scheduling, NBTI degradation

1. INTRODUCTION

As microelectronics scales down to 45nm and beyond, the increase in reliability concerns is becoming of paramount importance. Several physical mechanisms that are mining modern electronic devices have been spotted and studied, and are still under investigation [11]. Reliability mechanisms can be classified into two broad categories: interconnect wear-out and device wear-out. Interconnection degradation is mainly due to electromigration, caused by high density in the current flowing through resistive lines [2]; device degradation, on the other hand, are those related to transistor-level failures, such as oxide breakdown and Negative-Bias Temperature Instability (NBTI) [3]. In addition, thermal cycling and

hot-spots represent a true challenge for system reliability. In this perspective the need to optimize for reliability is turning into a hard design constraint. The birth and growth of multi-core architectures provides room to optimize system-level reliability through appropriate design techniques at the micro-architectural, architectural and Operating System levels. As an example, the presence of multiple (redundant) cores benefits from smart scheduling policies, such that thermal and reliability concerns are under direct control. In general, the presence of multiple functional units, as in superscalar processors or in MPSoC devices, allows the designers to optimize the resource allocation with emphasis on the reliability properties of the system. Run-time optimization approaches at the hardware and software level are gaining increasing attention from the scientific and engineering community. Classical power management and thermal management techniques, either DVFS-like or at micro-architecture level (e.g., clock gating), and OS-level techniques as task migration should then consider reliability as a design constraint of paramount importance. Meanwhile, the growth of density in ULSI devices makes it costly to employ early-stage evaluation of the benefits of power/thermal and reliability policies.

Purpose of the work presented in this paper is to address the problem of early-stage evaluation of NBTI mitigation techniques at architectural level, and the mitigation of NBTI-induced degradation through appropriate instruction scheduling. In this perspective, a novel methodology is proposed, and a novel policy for NBTI directed allocation strategy is compared against classical Round-Robin or Random-Scheduling strategies. The proposed strategy is meant to be used in single-core superscalar/VLIW processors or multi-core architectures, in which there generally exist different functional units accommodating the instructions execution.

1.1 Related works

Reliability-oriented management of complex systems is recently gaining interest due to the inherent unreliability of nanoscale devices and VLSI systems [5]; NBTI-aware resource usage is a relatively new concept attaining interest from the researchers, thanks to the particular feature of stress/recovery phases in the NBTI degradation process. Few works have been proposed in this context in the past. The benefits of appropriate Input Vector Control (IVC) mechanisms for NBTI mitigation have been demonstrated recently in [23]. The authors show that IVC techniques provide both mitigation of long-term degradation and leakage power control. Experimental results show that the worst-case NBTI-induced degradation is within 30% of the initial circuit delay, thus reducing the performance degradation along the critical path. For leakage power, this is in the order of 10% [23]. The authors in [18, 17] on the other hand, exploit the recovery phase typical of the NBTI process in order to mitigate degradation. The authors address emerging multi-core architectures, proposing a workload balancing scheme [17] that aims at alternating cores between full workload phases and recovery phases. The scheduling of stress and recovery phases is exploited through the concept of "workload capacity" [18], in which the impact of the degradation on the delay is computed at run-time; cores are associated to delay-driven performance metrics, used to rank the cores from their degradation status stand-point.

1.2 Novel contributions

In this work we present a novel methodology to estimate the NBTI-induced degradation at a system-level, taking into account low-level (i.e., transistor- and gate-level) analytical models to compute the NBTI degradation. We apply the proposed methodology to estimate the impact of appropriate allocation strategies at architectural level. Our work differs from the previous ones in several ways. At first we export to upper levels of abstraction the concept of IVC technique, focusing on the impact that instructions execution has on the available functional units of the processor; this is done through the definition of a novel allocation strategy for instructions selection. We then directly address the problem of workload balancing considering the cumulative effects (integrated over time) of the usage of functional units of interest. The rationale stands in the possibility to use recovery phases in a single-core processor without stopping the entire processor, but appropriately alternating the execution of instructions according to the functional units they stress. The proposed methodology is also a general approach that can be used to estimate the worst-case degradation of any hardware design, as a function of the stress duty-cycle in each RTL block of interest.

1.3 Paper structure

This paper has been organized as follows. Section 2 will introduce the reader with basic definitions of the NBTI process, and the degradation induced by threshold voltage shift, along with a brief explanation of the employed analytical models. Section 3 will discuss in details the proposed estimation methodology. Experimental results are then presented and discussed in detail in Section 4. Conclusions are drawn in Section 5.

2. NBTI OVERVIEW

Negative-Bias Temperature Instability (NBTI) is one of the most serious reliability concerns in nanoscale digital devices, gaining increasing attention beyond 45nm [11]. The degradation induced by NBTI is a complex chemical-physical process affecting both performance and availability of PMOS transistors; in these devices, NBTI degradation occurs when the gate-to-source V_{gs} control voltage is negative. This process is mainly due to bond dissociation at the silicon-oxide interface [3]: due to the high electric field, interface traps are created through a complex chemical-physical reaction, moving ions far from the interface. Traps generate charges at the interface, increasing the threshold voltage at long-term stress times. The time-varying change in V_{th} has direct impact on the performance of the transistor and on its availability; indeed, if the degraded V_{th} reaches sufficient magnitude, the PMOS device becomes unable to drive current to the load gates. However, one very important aspect of this degradation process stands in the presence of a recovery phase. Once the stress is removed, i.e. $V_{gs} > 0$, the threshold voltage degradation tends to decrease to a fraction of its original value, thus restoring part of the initial V_{th} value. Unfortunately, the cumulative stress cannot be entirely removed from the device during the recovery phase, but a fraction [3], leading to long-term permanent faults [22]. This aspect is of paramount importance in the mitigation of NBTI degradation, and it effectively acts as the driving force of our research work. As a matter of fact, we can

Table 1: Technology parameters and constant values of the analytical model in Equation 1. Values are chosen according to suggestions and technology models from ITRS [10].

Parameter	Role	Value
q	electric charge	$1.602 \times 10^{-19}\,\mathrm{C}$
t_{ox}	oxide thickness	$1.2\,\mathrm{nm}$ at $90\,\mathrm{nm}$
ϵ_{ox}	relative permittivity	3.9 for $\mathrm{SiO_2}$
K	constant	$8 \times 10^4\,\mathrm{s^{-0.25}C^{-0.5}nm^{-2}}$
C'_{ox}	oxide capacitance per unit area	$3.136 \times 10^3\,\mathrm{nFcm^{-2}}$
W_{eff}	PMOS geometry W	$1 \times 10^4\,\mathrm{nm}$
L_{eff}	PMOS geometry L	$45\,\mathrm{nm}$
E_0	reference electric field	$2\,\mathrm{MVcm^{-1}}$
E_a	activation energy	$0.49\,\mathrm{eV}$
k	Boltzmann's constant	$8.617 \times 10^{-5}\,\mathrm{eVK^{-1}}$
T_0	constant	1×10^{-8}
n	constant for H_2 diffusion	$1/6$

mitigate at some extent the degradation due to NBTI by controlling the stress and recovery phases duty cycle on the devices: lower duty cycles will result in lower stress for a PMOS transistor, and analytical models demonstrate that lower duty-cycles have a positive impact on the device lifetime [22]. At architectural level, this means that we can mitigate the degradation of a microprocessor by selectively controlling the duty-cycle usage of the functional units, preferring those that are under lower stress; in this way the device lifetime is prolonged. In more details, our work drives the scheduling of instructions in the target processor architecture according to the status of microprocessor functional units usage. The purpose of this work is to study the impact that different instructions allocation policies have on the device lifetime, as opposed to classical policies, e.g. Round-Robin, through the proposed methodology.

2.1 Analytical models

Mathematical models that can be found in literature [2, 3] explain in details the chemical-physical reaction on the basis of the NBTI degradation; however, they apply to transistor-level degradation estimation, and the algorithms for their estimation are very time consuming. Predictive analytical models have been derived for instance in [17, 4, 13] and they address gate-level computation. The degradation estimated in this work is based on analytical models shown in [22] for the long-term degradation computation as a function of the gate-level duty-cycle, and reported in Equation 1.

$$\Delta V_{th}^{t_{ref}} \approx \left(\frac{n^2 K_v^2 \alpha C t_0 t_{ref}}{\xi_1^2 t_{ox}^2 (1-\alpha)} \right)^n \tag{1}$$

where the parameters K_v and C are defined as in Equation 2a and Equation 2b, respectively.

$$K_v = \left(\frac{q t_{ox}}{\epsilon_{ox}} \right)^3 K^2 C_{ox} (V_{gs} - V_{th}) \sqrt{C} \exp\left\{ \frac{2E_{ox}}{E_0} \right\} \tag{2a}$$

$$C = T_0^{-1} \exp\left\{ -\frac{E_a}{kT} \right\} \tag{2b}$$

The definition of $\Delta V_{th}^{t_{ref}}$ approximates the degradation as a function of the reference wall-time t_{ref}, when the duty-cycle is α. The parameters in Equation 1 are explained in Table 1, whose values are derived both from ITRS technology projections and [10]. The single duty-cycle parameter α captures the effects of stress and recovery phases, to appropriately estimate the long-term degradation.

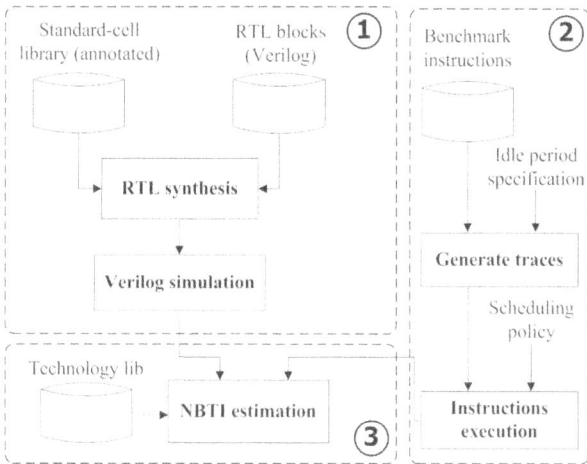

Figure 1: NBTI degradation estimation methodology overview.

3. ESTIMATION METHODOLOGY

The methodology flow is depicted in Figure 1. The proposed methodology develops a worst-case estimate of the degradation induced by the NTBI process at an architectural level, as a function of the duty-cycle usage of the different functional units of the target microprocessor. Starting from a low-level characterization of the functional units in the processor (Stage#1), our methodology emulates the execution of instructions and their impact on the duty-cycle usage of such units (Stage#2); then, the NBTI degradation is estimated in terms of threshold voltage increase ΔV_{th} (Stage#3). To this extent the proposed methodology is actually general enough to be applied to any RTL design of interest.

3.1 Stage#1: RTL blocks characterization

The purpose of this stage is to analyze the behavior of RTL blocks in the reference design (e.g., a processor), as a function of the input vectors. Starting from sythesizable RTL description (either Verilog or VHDL), the hardware design is firstly mapped to a standard-cell library. The *RTL Synthesis* block in Figure 1 computes the gate-level netlist of the design employing an annotated version of the technology library. This augmented version contains detailed information on the number of PMOS transistors that are stressed: for each cell in the standard-cell library, the number of PMOS transistors for which the V_{gs} control voltage is negative (i.e., logical 0) is statically computed. This is done by counting, for each gate in the standard-cell library, the number of PMOS transistors for which the input control voltage is a logical 0, for each input vector; this can be easily done by direct inspection of the gate circuits or through circuit-level simulation. Notice that we are not considering "weak" effects of degradation for stacked PMOS circuits as authors do in [15], since we are interested in worst-case estimation. However, as opposed to the work done in [15], we consider 3-inputs and 4-inputs gates, and not only 2-inputs gates. The mapped design collects information about the stress profile on a per-gate basis; this static information will be used later on in Stage#3 while estimating the NBTI-induced degradation in the reference design. The mapped design is simulated across several iterations, where input vectors are chosen uniformly distributed; notice that the Uniform distribution assumption of input vectors leads to a worst-case estimate, since generally input vectors follow a different pattern (e.g., Gaussian distribution). The output of Stage#1 is detailed information on the (average) number of PMOS transistors that are stressed

in *each* RTL block, such that a static characterization of the functional units has been achieved.

3.2 Stage#2: Instructions emulation

The purpose of this stage is to provide a quantitative statistical characterization of the duty-cycle of each functional unit composing the reference processor, as a function of the instruction execution trace. We select an appropriate suite of diverse benchmarks (refer to Section 4 for detailed information on the selected benchmarks), such that to cover the widest applications scenario with each benchmark bringing information regarding the statistical instructions breakdown. Using an ISS (Instruction-Set Simulator), we compute the percentage of execution of each instruction class and use it as an input to the proposed estimation flow, in the form of *Benchmark instructions* reference data. To generate traces, resembling the real execution of multiple instructions, several instructions are sampled iteratively. To be able to capture the non-continuous workload assignment of real processors (e.g., periodic as well as non periodic tasks), an idle period specification parameter is used to model the idle/active time. This is done in *Generate traces* block, that generates the emulated instructions flow. The *Instructions execution* step chooses instructions from the emulated traces, according to the desired scheduling policy: it is in this stage that the effective duty-cycles are affected by the dynamic instruction scheduling strategy. This last information will be used in Stage#3 to estimate the NBTI-induced degradation.

3.3 Stage#3: NBTI estimation

Stage#3 takes as input the static characterization of PMOS behavior from Stage#1 and the duty-cycle usage on RTL blocks from Stage#2 in order to provide an estimation of the NBTI-induced V_{th} degradation. Analytical models from Section 2 are employed in this part. In particular, the number of PMOS transistors computed from the synthesis in Stage#1 is used to represent the circuit as a bare collection of (PMOS) transistors; this matrix representation is initialized with default threshold voltage values according to the reference technology node and according to process variation parameters (see Section 5 for more details on this). The execution traces and duty-cycle usage from Stage#2 are used to compute the NBTI stress on a subset of the available PMOS transistors: at each iteration (defined by the instructions epoch) a subset of N PMOS transistors is randomly chosen and the degradation models from Section 2 are applied to that PMOS transistors. N is the average number of PMOS transistors for which the input is a logical 0, as computed by Stage#1. At each iteration, the ΔV_{th} value is then updated.

3.4 Selection policies

To compare the impact of the allocation policy on the reliability of the processor, we applied the methodology to different scheduling strategies, and compared the results (refer to Section 4 for quantitative analysis on this). We compared three different scenarios: Round-Robin (RR), Random Selection (RS) and Custom Selection (CS). In the RR policy, scheduling makes use of a circular buffer whence a new set of instructions is selected. Once the associated time quantum expires, a new set is sampled from the pool of available ones. A RR scheduling aims at fairness in the use of the computational resources, but it does not take into consideration the type of instructions and it is not suited for direct duty-cycle control on processors functional units. RS strategy, on the other hand, randomly chooses an application to be executed from the set of available ones, without any auxiliary information but the number of running applications. The proposed CS approach, last, directly takes into account the type of each instruction, according to the functional unit it requires to be executed, and tries to employ such information to drive the scheduling for reliability optimization.

More formally, we are given a set $P = \{p_1, p_2, ..., p_{NP}\}$ of NP tasks, each task p_j having M_j instructions to be executed. For simplicity, assume that M_j is known. The reference processor is composed of a set $F = \{f_1, f_1, ..., f_{NF}\}$ of NF functional units; for instance we can have integer ALUs, floating-point comparators or multipliers. In this work we assume that instructions are ready to execute as they are read from the instruction cache memory. In this way, we will focus only on balancing the use of the resources during the execution stage of the processor pipeline. Each instruction i_j generally requires a single functional unit to be executed, and this can be conveniently specified in a mapping function $E \in P \times F$. Entry e_{ij} is set to 1 if and only if functional unit $j \in F$ accommodates the execution of instruction $i \in P$. During the execution, each task is simply assumed to be a sequence of set of W instructions, representing the number of instructions that are executed (without preemption) in each available time quantum (called the "epoch"). For simplicity, we assume there exists a time quantum of exactly W instructions independently on the scheduling policy; the aim of scheduling is to decide which set should be instructions taken from. At the end of each time slot, the scheduler is invoked once again, and the next benchmark to execute is selected. Since the purpose of our policy is to balance the usage of the functional units, we continuously keep track of the resources required by each instructions window for each available tasks, and use this information to guide the scheduling decision.

The proposed policy aims at load-balancing the stress of each micro-architectural block of interest, in order to span the stress over the entire chip area, and to mitigate the appearing of hot-spots (from a reliability view-point) in a particular chip region. To this extent a suitable metric has been proposed and analyzed (refer to Section 4 for quantitative details on this). Denoting with d_j the duty-cycle usage statistics for the generic functional unit $j \in F$, and with w_i a weight associated with the generic instruction i, starting from the current Program Counter PC, the weighted value for the current epoch is computed, for each available task, as $\xi_j = \sum_{i=PC}^{PC+W} (dc_j \cdot w_i)$. The weight w_i is chosen according to the latency of the functional unit that ensures execution of the generic instruction; generally, instructions can be classified according to either integer operations or floating-point operations, as well as ALU or multiply/divide operations. The higher the latency of the corresponding functional unit, the higher the weight of the instruction. By minimizing the previous metric, we are weighting duty-cycle usage and functional unit performance. This can be easily done by sorting the ξ_j values at each scheduling period according to an ascending order.

4. EXPERIMENTAL RESULTS

We employed the proposed analysis methodology to assess the impact of different allocation strategies on the system-wide reliability of the reference single-core processor. In this section we will focus on the impact of the CS allocation strategy on different reliability-oriented metrics, when compared to classical RR or RS policies.

We selected several programs from freely available benchmark suites: WCET [7], SPLASH2 [20], MiBench [8], and FBench [21]. Each program in the different benchmark suite has been simulated using a modified version of the ESESC simulator, from the IACOMA group at the University of Illinois at Urbana Champaign [19]; during execution the instruction classes have been collected and statistically post-processed, to get the instructions breakdown

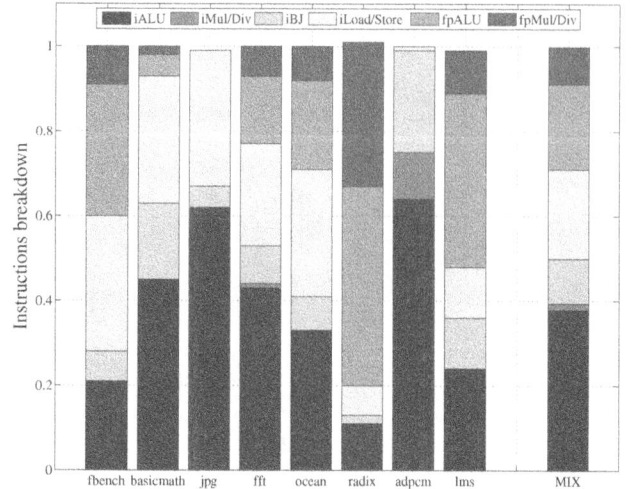

Figure 2: Instructions breakdown of the selected benchmarks, and total breakdown mix.

Table 2: Functional units characterization summary, for the simple five-stage MIPS processor.

Block	#total PMOS	#stressed PMOS (avg.)	
		absolute	*relative*
i-ALU	6653	3775	56.7%
fp-ALU	10061	6010	59.0%
fp-MULT	17083	9682	56.7%
fp-DIV	44873	22990	49.7%

for each benchmark of interest. Four different instruction classes have been selected for our experiments: integer ALU operations, floating-point ALU/multiplication/division instructions. We chose 8 benchmarks to run concurrently, such that each functional unit is used at a different mix. The instructions breakdown of the selected benchmarks is given in Figure 2. The reference processor is a simple five-stages pipeline MIPS processor, whose Verilog RTL design is freely available from OpenCores (opencores.org). The reference processor has one single integer ALU unit, one floating-point ALU, one floating point divider and a single floating-point multiplier. The RTL design has been synthesized using Cadence RTL Encounter Compiler, and a freely-available 45nm standard-cell library [12]. Verilog simulation has been conducted using NC-Verilog Simulator from Cadence as well. As specified in Section 3, the reference standard-cell library has been appropriately augmented. The results of the RTL simulation are given in Table 2: for each functional unit of interest, the number of PMOS transistors and the number of PMOS transistors that are stressed (as an average over the input vector configurations) is reported. Notice that the statistics reported in Table 2 are slightly different than the 50% assumption made in [17], providing a more precise and accurate evaluation methodology.

4.1 Impact of allocation strategy on reliability

The first analysis refers to the how long-term degradation benefits from the sole allocation strategy, while other technological and operational parameters remain constant. Figure 3 shows the trend of the V_{th} degradation process as a function of the time-wall considered (from 3 months up to 10 years), comparing RR, RS and CS scheduling for the floating-point multiplier block. The technology

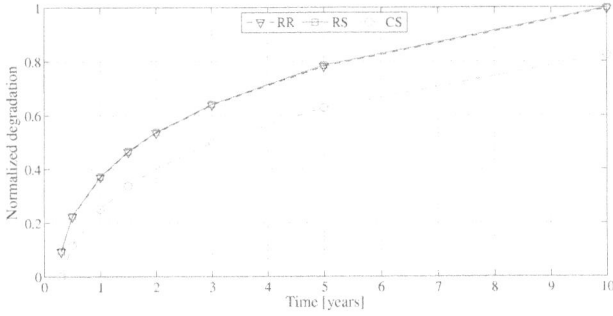

Figure 3: NBTI-induced V_{th} degradation as a function of reference lifetime.

node is set to 45nm, with supply voltage 1.1 V and oxide thickness 9Å, as suggested in [10]. Each point on the trend curve is computed as an average across every stressed PMOS, according to the results obtained from the estimation flow presented in Section 3, and reported in Table 2. As the figure shows, Round-Robin and Random-Scheduling allocation strategies have nearly the same impact on the degradation process. This is an expected behavior, since in the RS scheduling process the duty-cycle in each functional unit tends very fast to the RR one due to the huge number of simulated cycles. On the contrary, the CS allocation strategy has an average benefit of at least 10% in the degradation, having a direct impact on the system life-time: the same reference ΔV_{th}^{thr} threshold is then reached later on, providing thus a greater life-time expectation. As an example, the degradation reached by RR scheduling after 1 year of operation is reached after 2 years with CS scheduling, with a net increase in system reliability. Lower degradation is seen throughout the entire typical lifetime periods for modern electronic products: results are well suited for consumer electronics, whose typical lifetime is in the order of 2 to 3 years, as well as for industrial/military electronics, in which long-term reliability is the primary objective. Nevertheless, this result is well suited for periodic or aperiodic tasks, as those we are addressing in this work; for continuous stress workload, however, benefits are expected to decrease over time since the duty-cycle usage typical of continuous stress would sooner reach the values of RR and RS. Without loss of generality, we set the operating temperature to 423 K. In the Arrhenius expression it is generally assumed that the temperature parameter is an average value [16], thus it simply causes a shift of the final estimation. More precise models should take into account the time-varying profile of temperature [24], also considering leakage. To analyze the impact of operating temperature on the benefits due to the different allocation strategies, we computed the degradation for increasing operating temperatures. With reference wall-time of 2 years and technology parameters as in the previous case, the operating temperature has been varied in the range $[110, 210]°C$; results have shown that the benefits of the proposed methodology have actually a temperature-independent behavior when compared to both RR or RS allocation strategies. Meanwhile, the degradation gets worse with increasing operating temperatures, as expected. As compared to RS scheduling, there is a reduction of degradation in the order of 10% in either case.

4.2 The impact of technology scaling

As supply voltage scales down, it is expected that degradation increases, since as V_{DD} scales the geometry of the transistor scales as well, such that the oxide thickness t_{ox} scales, too. The results in Table 3 show this trend. The values are taken according to increasing

Table 3: Impact of technology scaling: data is given as a function of supply voltage, considering a 2 years wall-time.

V_{DD}	Degradation		
	Round-Robin	Custom	Ratio (RR/CS)
0.8V	0.461V	0.42V	1.098
0.9V	0.458V	0.418V	1.096
1.0V	0.455V	0.413V	1.101
1.1V	0.44V	0.394V	1.117
1.2V	0.413V	0.37V	1.116

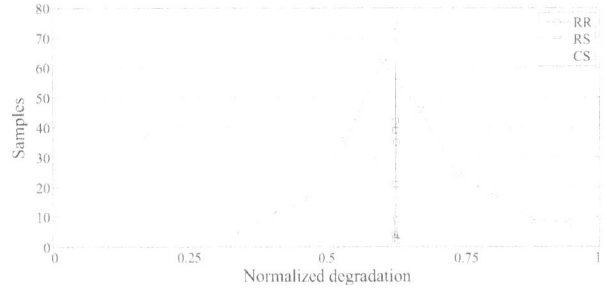

Figure 4: The impact of process variation and CS scheduling.

values of V_{DD} and t_{ox}, as frequently used in technology predictions [10]. Temperature is set to 383 K as usually done [16]. Technology values with prediction on 22 nm technology nodes have been derived from [10]. As in the previous cases, the NBTI-aware allocation strategy achieves benefits from a degradation view-point when compared to the classical RR or RS scheduling policies. In this perspective, the CS allocation is a suitable strategy to deal with NBTI degradation at architectural level, even in more aggressive technology scales. Comparing the degradation of CS and RS strategies at the same technology node, the degradation difference is in the order of 10%. Notice that similar results cannot be estimated with accuracy when considering high-κ dielectrics, since in those scenarios PBTI plays a relevant role as well [9], but in our work only NBTI is considered.

4.3 The impact of process variation

One important aspect of the proposed flow is the ability to capture the effects of process variation on a system-wide perspective. The RTL blocks characterization in Table 2 allows us to analyze the impact of V_{th} variation caused by process variability, and compute the run-time NBTI-induced degradation as a function of the allocation strategy. Process variation has negative impact on the technology parameters between devices on the same die or different die on the same wafer [1]. We are interested in particular on the first case, where different PMOS transistors can have a different (initial) threshold voltage. We assume the V_{th} distribution to be Gaussian, as shown in [1], and generate a random vector of V_{th} values according to a Gaussian distribution with mean $\mu = 0.18V$ and standard deviation $\sigma = 0.005$, such that the initial values are approximately in the range $[0.17, 0.19]V$. The shift toward lower values of the x axis in Figure 4 shows that the CS scheduling allows to have more transistors with lower (cumulative) degradation. However, it has also a grater variance, due to the increased variability of usage of the different transistors in the reference device. Indeed, RR has the higher mean degradation, but much lower variation. The mean value of the CS strategy is shifted by at least 30% toward lower values of degradation when compared to RR or RS.

This again proves the efficiency of the CS policy.

Benefits are achieved from a performance view-point also; by applying the gate delay expression as a function of supply voltage and threshold voltage [14] with $\alpha = 1.3$ for 45nm technology node, the degradation follows similar shape as in Figure 4, with an average gain in gate delay of 30% relative value. This demonstrates that the proposed strategy can be efficiently employed while ensuring low performance degradation.

5. CONCLUSIONS

This paper presents a novel methodology to estimate the NBTI degradation at architecture-level, as a function of technology parameters, RTL design, and usage (i.e., duty-cycle and allocation strategies). A novel dynamic instruction allocation strategy has been proposed, and several experiments have been conducted to show the benefits of NBTI-aware allocation strategies to prolong the device lifetime. We compared this Custom Schedule strategy with classical Round-Robin and Random-Selection policies. Results have shown a reduction of degradation in the order of 10% to 20%, in different scenarios: at different operating temperature values, as technology scales down and, above all, as process variation is accounted for. Furthermore, results show that CS is well suited for both consumer electronics, gaining benefits in the range of 2 to 3 years lifetime, and long-term reliability of industrial/military electronics. The allocation strategy is meant to be employed in the dynamic instruction scheduling portion of the microprocessor pipeline. Nevertheless, this paper focuses on the execution stage of the pipeline focusing on the functional units rather than in processor control units, since it has been shown that the processor resources that are prone to higher failure rate are ALU and control blocks [6].

6. ACKNOWLEDGMENTS

This research work is partially supported by European Community Seventh Framework Programme (FP7/2007-2013), under agreements no. 248716 (2PARMA project www.2parma.eu) and Artemis no. 100230 (SMECY project www.smecy.eu).

7. REFERENCES

[1] K. Agarwal and S. Nassif. Characterizing process variation in nanometer cmos. In *44th ACM/IEEE Design Automation Conference.*, pages 396 –399, june 2007.

[2] M. Alam, K. Kang, B. Paul, and K. Roy. Reliability- and process-variation aware design of vlsi circuits. In *14th International Symposium on the Physical and Failure Analysis of Integrated Circuits.*, pages 17 –25, july 2007.

[3] M. A. Alam and S. Mahapatra. A comprehensive model of PMOS NBTI degradation. *Microelectronics Reliability*, 45(1):71–81, 2005.

[4] S. Bhardwaj, W. Wang, R. Vattikonda, Y. Cao, and S. Vrudhula. Predictive modeling of the nbti effect for reliable design. In *IEEE Custom Integrated Circuits Conference.*, pages 189 –192, sept. 2006.

[5] S. Borkar. Designing Reliable Systems from Unreliable Components: The Challenges of Transistor Variability and Degradation. *IEEE Micro*, 25(6):10–16, Nov. 2005.

[6] T. Gupta, C. Bertolini, O. Heron, N. Ventroux, T. Zimmer, and F. Marc. Effects of various applications on relative lifetime of processor cores. In *IEEE International Integrated Reliability Workshop Final Report.*, pages 132 –135, oct. 2009.

[7] J. Gustafsson, A. Betts, A. Ermedahl, and B. Lisper. The Mälardalen WCET Benchmarks - Past, Present and Future. In *Proceedings of the 10th International Workshop on Worst-Case Execution Time Analysis*, July 2010.

[8] M. R. Guthaus, J. S. Ringenberg, D. Ernst, T. M. Austin, T. Mudge, and R. B. Brown. Mibench: A free, commercially representative embedded benchmark suite. In *Proceedings of the IEEE International Workshop on Workload Characterization.*, pages 3–14, Washington, DC, USA, 2001.

[9] K. T. Lee, C. Y. Kang, O. S. Yoo, R. Choi, B. H. Lee, J. Lee, H.-D. Lee, and Y.-H. Jeong. Pbti-associated high-temperature hot carrier degradation of nmosfets with metal-gate/high- k dielectrics. *Electron Device Letters, IEEE*, 29(4):389 –391, april 2008.

[10] MASTAR - Model for Assessment of CMOS Technology and Roadmaps http://www.itrs.net/models.html.

[11] J. W. McPherson. Reliability challenges for 45nm and beyond. In *Proceedings of the 43rd annual Design Automation Conference*, pages 176–181, New York, NY, USA, 2006.

[12] NanGate Open Cell Library http://www.si2.org/.

[13] Predictive Technology Model http://ptm.asu.edu.

[14] T. Sakurai and A. Newton. Alpha-power law mosfet model and its applications to cmos inverter delay and other formulas. *Solid-State Circuits, IEEE Journal of*, 25(2):584 –594, apr 1990.

[15] K. Saluja, S. Vijayakumar, W. Sootkaneung, and X. Yang. Nbti degradation: A problem or a scare? In *21st International Conference on VLSI Design*, pages 137 –142, jan. 2008.

[16] J. Srinivasan, S. V. Adve, P. Bose, and J. A. Rivers. The Case for Lifetime Reliability-Aware Microprocessors. In *International Symposium on Computer Architecture*, 2004.

[17] J. Sun, A. Kodi, A. Louri, and J. Wang. NBTI aware workload balancing in multi-core systems. In *International Symposium on Quality of Electronic Design*, pages 833 –838, march 2009.

[18] J. Sun, R. Lysecky, K. Shankar, A. Kodi, A. Louri, and J. Wang. Workload capacity considering nbti degradation in multi-core systems. In *15th Asia and South Pacific Design Automation Conference*, pages 450 –455, jan. 2010.

[19] Superscalar simulator http://sourceforge.net/projects/sesc/.

[20] The modified SPLASH-2 benchmark suite http://www.capsl.udel.edu/splash/.

[21] Trigonometry intense floating point benchmark http://www.fourmilab.ch/fbench/fbench.html.

[22] R. Vattikonda, W. Wang, and Y. Cao. Modeling and minimization of pmos nbti effect for robust nanometer design. In *43rd ACM/IEEE Design Automation Conference*, pages 1047 –1052, 0-0 2006.

[23] Y. Wang, X. Chen, W. Wang, V. Balakrishnan, Y. Cao, Y. Xie, and H. Yang. On the efficacy of input vector control to mitigate nbti effects and leakage power. In *International Symposium on Quality of Electronic Design*, pages 19 –26, march 2009.

[24] B. Zhang and M. Orshansky. Modeling of NBTI-Induced PMOS Degradation under Arbitrary Dynamic Temperature Variation. In *9th International Symposium on Quality of Electronic Design*, pages 774–779. IEEE, Mar. 2008.

A Noise-immune Sub-threshold Circuit Design based on Selective Use of Schmitt-trigger Logic

Marco Donato[*]
School of Engineering
Brown University
Providence, RI 02912

Fabio Cremona
Sapienza University of Rome
00185, Italy

Warren Jin
School of Engineering
Brown University
Providence, RI 02912

R. Iris Bahar
School of Engineering
Brown University
Providence, RI 02912

William Patterson
School of Engineering
Brown University
Providence, RI 02912

Alexander Zaslavsky
School of Engineering
Brown University
Providence, RI 02912

Joseph Mundy
School of Engineering
Brown University
Providence, RI 02912

ABSTRACT

Nanoscale circuits operating at sub-threshold voltages are affected by growing impact of random telegraph signal (RTS) and thermal noise. Given the low operational voltages and subsequently lower noise margins, these noise phenomena are capable of changing the value of some of the nodes in the circuit, compromising the reliability of the computation. We propose a method for improving noise-tolerance by selectively applying feed-forward reinforcement to circuits based on use of existing invariant relationships. As reinforcement mechanism, we used a modification of the standard CMOS gates based on the Schmitt trigger circuit. SPICE simulations show our solution offers better noise immunity than both standard CMOS and fully reinforced circuits, with limited area and power overhead.

Categories and Subject Descriptors

B.2.3 [**Arithmetic and Logic Structures**]: Reliability, Testing, and Fault-Tolerance—*Redundant design*

General Terms

Design, Reliability

Keywords

Schmitt Trigger, Noise immunity, Subthreshold operation.

---

[*]Should you need further information, please contact Marco Donato (email: `marco_donato@brown.edu`).

1. INTRODUCTION

Modern CMOS integrated circuits have benefited from technology scaling in the nanoscale regime, which has allowed for faster circuits and increased levels of integration. However, miniaturization has also brought many challenges. First, as node capacitances shrink with smaller device dimensions, the RMS value of thermal noise also rises. Moreover, having a lower number of electrons in the channel amplifies the effect of fluctuations of the carriers due to random telegraph signal (RTS) noise. While the effects of these phenomena could be negligible for standard designs, they become critical for ultra low-power applications (ULP). One of the most appealing approaches for ULP design is operating the circuits in the sub-threshold regime. However, given the reduced noise margins of transistors operated in the sub-threshold regime, the magnitude and the frequency of soft errors could drastically reduce the reliability of the system. Therefore, it is critical to find viable solutions to improve the noise robustness of nanoscale sub-threshold circuits.

A number of techniques to improve noise-immunity of nanoscale circuits have been proposed, many of them based on adding redundant logic to the circuit (e.g., [17], [12], [6]). However, one of the main drawbacks of these solutions is the relatively high transistor count overhead required to achieve reasonable levels of noise immunity. Increasing the number of transistors leads to higher power dissipation and virtually eliminates any benefit obtained by operating the circuit in the sub-threshold regime. Therefore, new approaches to improve noise-immunity of nanoscale circuits are required to make sub-threshold operation worthwhile.

In this paper, we propose to selectively introduce redundant logic to the circuit, based on the use of existing invariant relationships (or *logic implications*) within the circuit itself. These invariant relationships represent expected logical behavior of the circuit and must be satisfied for the circuit to be operating correctly. For instance, the invariant relationship $a = 1 \Rightarrow b = 1$ implies that b (the *implicand*) must have a logic value of 1 whenever a (the *implicant*) has a logic value of 1.

Invariant relationships may be expressed through feed-forward reinforcements in the circuit. The basic idea of using implications to reinforce correct logical behavior is itself rather general, and therefore could be implemented at the circuit level in a number of ways. One possible approach is to prevent the gate that produces the implicand from changing its output value if this contradicts the invariant relationship. One way to accomplish this is to control the switching threshold of the gate using the gate that produces the implicant. A well known circuit that shows a similar behavior is the Schmitt trigger circuit [15].

A major advantage of using the Schmitt trigger circuits is their relatively low transistor count overhead. Nevertheless, replacing every gate in the circuit with a Schmitt trigger implementation would still lead to an unacceptable transistor count. Instead, we propose selectively adding Schmitt gates to reinforce signals that are most likely to produce bit flips at primary outputs. Our SPICE simulations show that our approach has the lowest failure rate for minimum power and area overhead, proving that it is a viable solution for designing noise-immune circuits when rigid power and area constraints exist.

The rest of this paper is organized as follows. In Section 2 we will describe the motivation for our work and give some background on the building blocks of our approach. In Section 3 we first give an example explaining the importance of taking into account the effect of different noise sources in nanoscale circuits. We then present an approach for implementing the reinforcement using Schmitt-trigger based gates. In Section 4 we will describe in detail our time-domain noise simulations and give a quantitative analysis of the noise immunity of the various solutions, as well as an estimate of the introduced power and area overhead. In Section 5 we will introduce a design flow that will allow to automatically find implication paths that can be used to feed-forward reinforcement. Finally, we present conclusions and future work in Section 6.

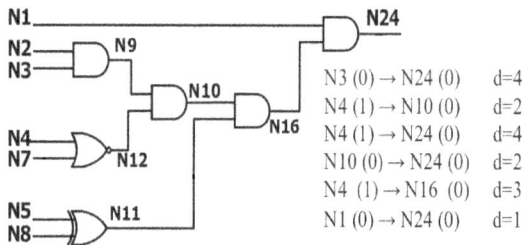

Figure 1: Example circuit with implications, taken from [1].

2. BACKGROUND

2.1 Noise-Immune Circuit Design

The basic concept of error-immune design is far from new. One of the earliest examples is the Modular Redundancy approach introduced by Von Neumann in [17] where the logic is replicated a certain number of times (e.g., three times in the case of Triple Modular Redundancy (TMR)) and all the outputs are then sent to a majority gate in order to get the final output. This technique introduces two main problems: first, the area overhead is always greater than 200% depending on how many times the main logic is replicated and the size of the majority gate; in addition, the error immunity is still determined by the majority gate that could be still affected by errors.

Other techniques use gate resizing for improving the immunity of selected gates to single event upsets (SEU) (see, for example [13]). These techniques could still be insufficient if the noise in the stages preceding the resized gate is strong enough to generate a signal flip, since the voltage at the input node of the resized gate would be already different from the correct value and there would be no means

for correcting the signal. Therefore, there is a need for reinforcing techniques that can be effective across multiple gates.

A different approach has been considered in [12], where a probabilistic framework based on Markov Random Fields (MRF) is proposed for designing noise-immune circuits. In particular, the MRF framework used feedback of the satisfiability constraints of a gate's function to reinforce correct behavior. More recently, Turtle Logic (TL) has been proposed to achieve noise immunity by exploiting port redundancy and coherence analysis of the redundant data [6]. Both these approaches show a greatly improved immunity to noise compared to standard implementations; however, they carry significant area overhead which can have a substantial negative impact on power dissipation and delay. These factors may significantly limit their suitability for sub-threshold circuit design, where performance is already traded off for reduced power dissipation.

2.2 Implications

Exploiting knowledge of invariant relationships (or implications) is a simple yet useful concept that has already found application in error detection and logic testing methodologies [1] [2] [11]. Implications are logic relationships between nodes in the circuit that hold for any input vector. An example is illustrated in Figure 1, where six different implications are enumerated.

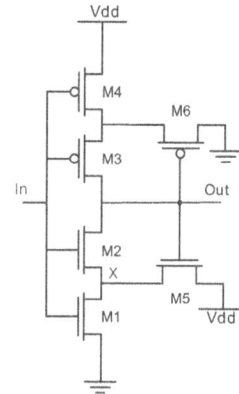

Figure 2: Schmitt trigger circuit implementing an inverter function.

If we consider for instance the first line in the implications list, it states that for any input vector for which $N3$ (the *implicant*) is equal to 0, the output $N24$ (the *implicand*) will have to be at 0 as well. The same reasoning applies to all the other implication relationships shown in Figure 1. In the works of [1] [2] [11], the authors used implications for online error detection by adding redundant circuitry (i.e., checker logic) to verify if selected implication relationships were valid. If there were an inconsistency in the checker logic, an error would be flagged.

Taking the ideas from [1], we could imagine using implication logic not just to detect errors but also to reinforce correct logic behavior. As a simple application of these implications, we could imagine using them to monitor the errors on certain paths to the primary output and, if an error is detected, flip the output signal. However, this solution would require adding not only the logic for checking the implications but also a multiplexer for selecting between the direct or inverted output. Also, the main problem would be that the multiplexing operation at the last stage would have the same issues as the voting system for the modular redundancy design. As we will show shortly, we can obtain solutions with less overhead and better noise immunity than this simple approach.

In this paper, we propose to use implicants as control signals for the gates that produce the implicands. If we consider again the circuit in Figure 1 with its implication $N3 = 0 \Rightarrow N24 = 0$, we can use the value at node $N3$ to force the value at node $N24$ through a feed-forward mechanism. Results from [1] show that thousands of these relationships exist in standard benchmark logic circuits. Therefore, we can choose a subset of these implications and use them to selectively reinforce the circuit. Depending on the architecture that is used for implementing the feed-forward system, this approach can lead to notable error-rate reduction with very modest area, delay and power overhead.

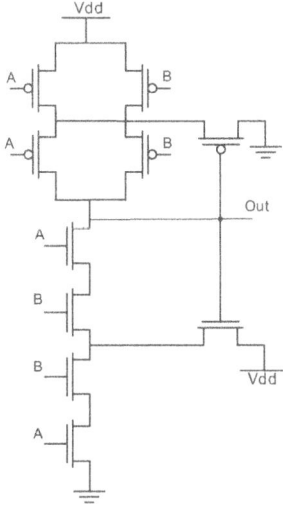

Figure 3: Schmitt trigger based NAND gate.

2.3 Schmitt Trigger Circuits

We introduced the use of implications for increasing the reliability of a circuit as a general approach. However, we need to find an architecture for implementing the feed-forwarding reinforcement. The Schmitt trigger circuit [15] is generally used to extract signals from noisy environments and can serve our purpose. In particular, we refer to the implementation shown in Figure 2. We can describe its operation as follows: let us consider the case in which the input of the trigger is at logic value 0 and the output is at 1. The NMOS pass transistor $M5$ is on, increasing the potential at node x; if the gate voltage moves from its initial value, it will have to overcome the regular switching threshold voltage of the NMOS transistors in the pull-down network (PDN) by the value at x. The same behavior can be seen for the pull-up network (PUN) and the two structures combined produce the well known hysteresis characteristic of the Schmitt trigger. This circuit has been demonstrated to be effective also in subthreshold design [9]. Moreover, Dokic has shown in [5] the possibility of extending the same design structure of the Schmitt trigger to more general logic gates such as NAND and NOR. The Schmitt implementation of a NAND function is shown in Figure 3. In all these cases the reinforcement is introduced using a feed-back mechanism. Our first modification to the circuit consists in connecting the gates of the pass transistors to a node different from the gate's output. This modification provides the means for implementing the implication reinforcement. Let us consider the example in which the implication relationship is $Ctrl = 0 \Rightarrow Out = 1$, where $Ctrl$ could be any node in the circuit and Out is the output of a NAND and an implicand related to the implicant $Ctrl$. Note that this kind of implication is unidirectional and valid only for one of the two logic levels. This requires a further modification of the

Schmitt gate structure; in this particular case we will connect the gate of the pass-transistor (PMOS) to the node $Ctrl$, and the drain of the pass-transistor will be connected to the middle node of the PDN. In this way, whenever the signal $Ctrl$ is 0, the gate will be prevented from performing the transition 1 to 0. The resulting circuit is shown in Figure 4. In the general case, the implicant will determine the type of pass-transistor (NMOS if the implicant is 1 and PMOS otherwise) while the implicand will determine to which network (pull-up or pull-down) the reinforcement will be applied, leading to four possible Schmitt gates in total.

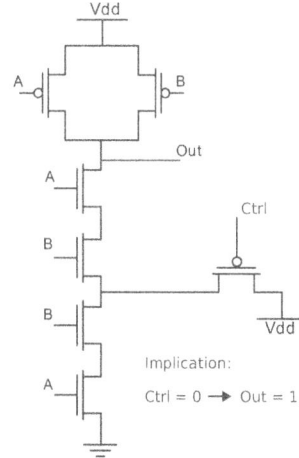

Figure 4: A modified Schmitt NAND for implication reinforcement.

3. METHODOLOGY

In this section we will show an example of how the performance of a circuit operating at sub-threshold supply voltage can be dramatically affected by noise. We choose to simulate our circuit using a 22nm FDSOI model card [3]. The importance of both thermal noise and random telegraph signal (RTS) noise for nanoscale circuits has been widely discussed [14] [8] [4] [16]. We therefore consider both sources of noise in our simulations to evaluation signal integrity.

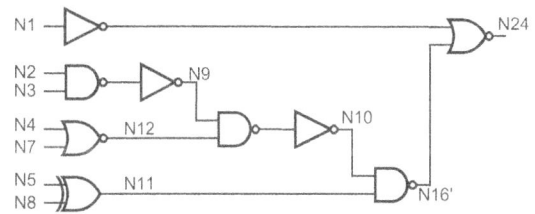

Figure 5: Implemented version of the example circuit shown in Figure 1.

3.1 Motivational Example

Let us consider again the circuit in Figure 1. The slightly modified version used for our SPICE simulations is shown in Figure 5. In order to add noise to our simulations we considered the methods described in [10] and [19]. In particular, we generated additive white Gaussian noise (AWGN) with 0 mean and $10mV$ standard deviation (a reasonable value for a 22nm technology node), RTS

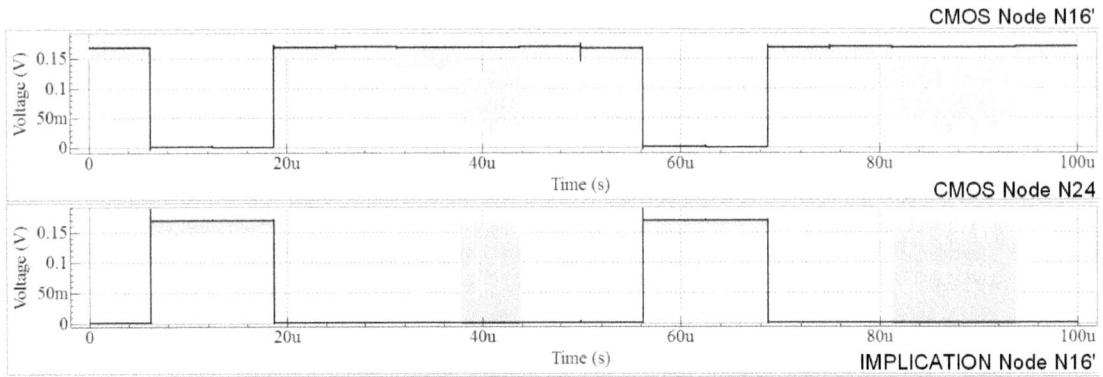

Figure 6: Example circuit's output and input to the last stage for standard CMOS implementation

noise with values of the amplitude equal to $50mV$, and we considered the Fast Slow corner library that represents the worst-case scenario. For nanoscale technologies it is very hard to control the process variations and produce a circuit following the nominal behavior, therefore we think that considering the corner libraries is the most realistic condition. The circuit was simulated using a supply voltage of $V_{DD} = 170mV$, i.e., below the threshold voltage of the devices. The signal traces from the SPICE simulations are shown in Figure 6. It is clear from this figure that the circuit is unable to perform a correct computation under the imposed noisy conditions.

In order to explain how our solution works we will empirically describe the approach. We will use the implication set listed in Figure 1 for this purpose. Since we are interested in producing the correct signal at the output, we will first consider all the implications to the output node $N24$ and then try to build a path of implications through the entire circuit. Therefore, we have the following set of implications to start with: $N3 = 0 \Rightarrow N24 = 0$, $N4 = 1 \Rightarrow N24 = 0$, $N10 = 0 \Rightarrow N24 = 0$, $N1 = 0 \Rightarrow N24 = 0$. We can immediately discard the implication with $N1$ since it is a primary input and therefore it does not provide any means for reinforcing internal nodes. The same reasoning applies to $N3$ and $N4$, therefore our choice is the implication $N10 = 0 \Rightarrow N24 = 0$. At this point we need to find an implication that has $N10$ as its implicand and a node closer to the primary inputs as implicant. In our example we will pick $N4 = 1 \Rightarrow N10 = 0$. In this way we have created a reinforcement path to the primary output where all the nodes involved in the process are reinforced along the way. The next step is modifying the circuit according to the chosen implications. The gate that produces $N24$ will have a reinforced PUN controlled by a PMOS (since the implicant is 0), while the gate that produces $N10$ will have a reinforced PUN controlled this time by an NMOS (implicant 1). The total overhead required by this solution is 6 additional transistors.

As mentioned earlier, we built a library based on a 22nm FDSOI model of standard CMOS gates and modified Schmitt gates. Considering that all devices are operating in the subthreshold regime and that the Schmitt design requires doubling the number of transistors in a stack, we decided to limit our library to only two-input gates. With regard to the implementation of a 2-input XOR gate, the authors of [18] have shown that degradation of the output signal can arise in a transmission-gate based XOR circuit due to the imbalance between the on current and the leakage current. For this reason, we decided to use the solution proposed in [7] which does not require additional inverters for generating complementary inputs and has also shown a better noise immunity when compared to other design solutions. Note however, that nothing prevents from

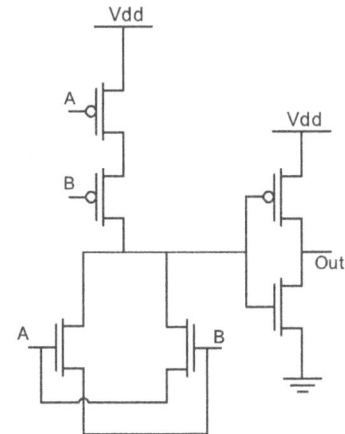

Figure 7: XOR gate adopted for the library

using any other XOR design with our approach. The adopted XOR circuit is shown in Figure 7.

4. SIMULATIONS

In this section we show the results and describe the details of the SPICE noise simulations. Let us compare the simulations for the example circuit of Figure 5 implemented in standard CMOS as well as with the implication reinforcement discussed in the previous section. The results are shown in Figure 8. The superimposed black trace is the result for the noiseless simulation of the CMOS circuit. Notice that, when simulated with additional noise, the standard CMOS implementation produces unacceptably noisy signals, whereas our implication reinforced circuit reduces noise on the output signal significantly. It is also worth noticing that in both cases the noise at node $N16'$ is the same meaning that the Schmitt gate is actually correcting the errors on the signal.

Next, we also need to compare against other reinforcement approaches. Several works on circuit reliability and noise immunity have focused their attention on strengthening just the memory elements in the circuit. One way to investigate how this approach would work would be to reinforce just the final stage of the circuit (i.e., the stage feeding directly into the memory element) with a *standard* Schmitt gate (i.e., the one shown in Figure 2). Another approach would be to fully reinforce the circuit by substituting every CMOS gate with an equivalent Schmitt gate. The results for these two solutions are shown in Figure 9. As seen from the waveforms in Figure 9, the first approach of reinforcing only the final output is not robust enough. While the second approach of fully re-

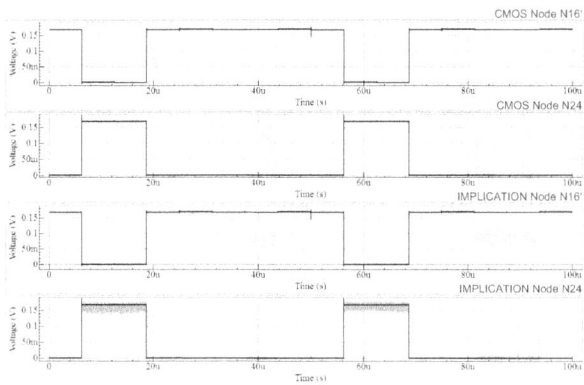

Figure 8: Comparison between standard CMOS implementation and implication/reinforced implementation.

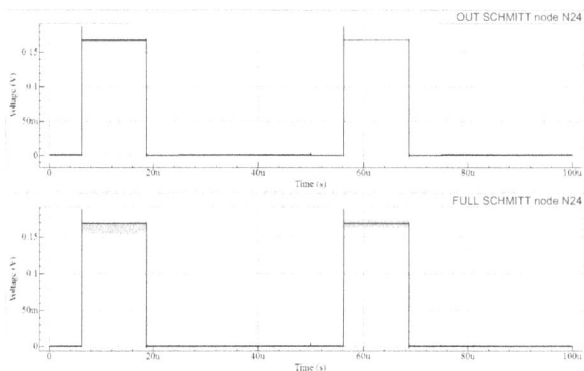

Figure 9: Example circuit's output for full Schmitt and standard CMOS with Schmitt gate on the last stage

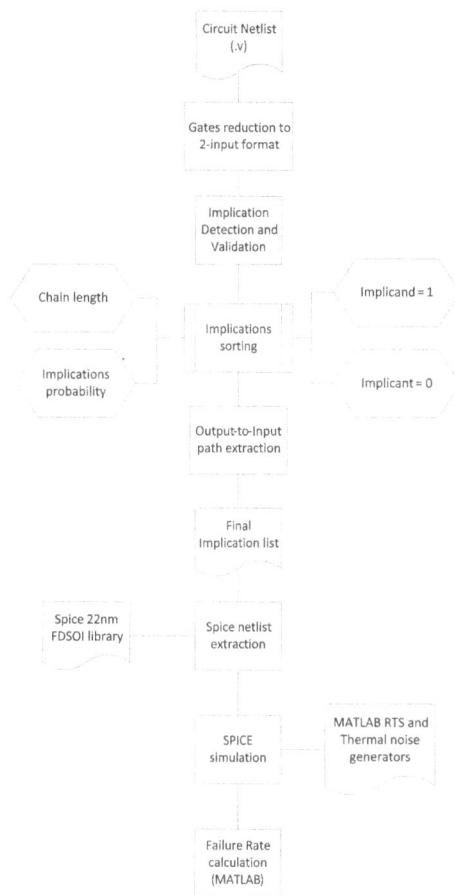

Figure 10: Detailed design flow chart

inforcing every gate in the circuit produces a clean output, its cost in terms of transistor count is more than twice that of the initial CMOS design. Hence, while our implication approach still leaves some noise on the output, it provides a much better tradeoff between noise immunity and area overhead.

The main challenge in carrying out time-domain noise simulations is the computation time. Considering the slow variability of RTS noise, with RTS glitches occuring on the ms time scale, direct time domain simulation is prohibitively long. Therefore, to better evaluate the performance of our solution, we decided to generate several samples of the injected RTS noise and perform 100 distinct simulations on a time of 100 μsec, thereby artificially increasing the frequency of the RTS noise. On each of the 100 simulations we made the inputs loop over a predefined test vector. We then considered each iteration of the loop as the time window for the error rate: if the signal crosses the error threshold for more than 10% of the time, we record an upset; this helps filtering out small glitches and transitions.

For the tested circuit we used 2 implications, producing an overhead of 18.75% in number of transistors and 24% in power consumption. With this cost we were able to reduce the failure rate from 6.5% to 0%. It is important to mention that in the case of many upsets producing failure at the primary output we count just a single error. Therefore, the failure rate does not represent the number of total errors registered in the simulations. While the CMOS implementation shows an unacceptable failure rate, the implication

design offers a faultless output, with reasonable area and power overhead compared to full logic replication.

5. DESIGN FLOW

Following the idea described in Section 3, we will now present a potential work flow that can be used for automatically inserting the implication reinforcement into a standard CMOS design. First, we need a method for identifying the implications given a circuit netlist. For this purpose we can partially re-use the flow shown in [1]. It is important to clarify that at this stage we are interested only in the implication discovery process, while the process of trimming the list trying to get rid of *weak* implications might actually hinder our purpose. In particular, the authors of [1] consider as *weak*, those implications that can be included in other implications and implications where the implicant and the implicand can be derived from a common node; while these implications can be discarded for online error detection, they might be critical for finding a path between the primary outputs and the primary inputs. Therefore, the design flow we envision is illustrated in Figure 10 and can be described as follows:

- **Implication discovery and verification:** in this initial part, a logic simulation of the circuit over a limited number of input vectors is carried out. The logic values at each node in the circuit are then used as input to a script that verifies the presence of possible implications. While the previous step produces a list of possible implications, these values have been tested only on a limited number of vectors. Therefore, we need to to check for the presence of input patterns that would violate possible implications values. This can be done implicitly using, for instance, a SAT solver.

- **Implication sorting:** once we have a verified list of implications we need to understand which implications will be more suitable for reinforcement. The best implication for our purpose is the one that holds for most of the input patterns. Once again, a logic simulation can be used for extrapolating the implications' frequency.

- **Output-to-input path extraction:** Once we have a score assigned to any implication in the list we, can build our reinforcing paths starting from those implications that have primary outputs as implicands, and then move backwards to the primary inputs considering the implications with higher score first. In this step, a threshold for the chain length could be used to set the maximum area overhead.

- **SPICE netlist extraction and simulation:** The final implication list is used in combination with the verilog netlist to extract a SPICE netlist in which the Schmitt gates are automatically inserted. This netlist is then used for circuit simulations.

6. CONCLUSIONS

In this work we have shown a very cost-effective solution for improving the reliability of sub-threshold nanoscale circuits. The implication methodology represents a general framework for selective reinforcement. We decided to use Schmitt trigger based circuits for implementing the implication reinforcement, and we have demonstrated that this solution offers good noise immunity with a limited cost in terms of area and power overhead. Future work will focus on implementing the design flow described in Section 5 for automating the selection of implications and the insertion of the reinforced gates. We will also expand the simulations to a wider range of circuits as well as to other noise-immune design solutions.

7. ACKNOWLEDGMENTS

This work was supported in part by DTRA under Grant HDTRA 1-10-1-0013.

8. REFERENCES

[1] N. Alves, A. Buben, K. Nepal, J. Dworak, and R. I. Bahar. A Cost Effective Approach for Online Error Detection Using Invariant Relationships. *IEEE Transactions on Computer-Aided Design of Integrated Circuits and Systems*, 29(5):788–801, May 2010.

[2] N. Alves, Y. Shi, J. Dworak, R. I. Bahar, and K. Nepal. Enhancing online error detection through area-efficient multi-site implications. In *29th VLSI Test Symposium*, pages 241–246. IEEE, May 2011.

[3] D. Bol, S. Bernard, and D. Flandre. Pre-silicon 22/20 nm compact MOSFET models for bulk vs. FD SOI low-power circuit benchmarks. In *IEEE 2011 International SOI Conference*, pages 1–2. IEEE, Oct. 2011.

[4] L. Brusamarello, G. I. Wirth, and R. da Silva. Statistical RTS model for digital circuits. *Microelectronics Reliability*, 49(9-11):1064–1069, Sept. 2009.

[5] B. Dokic. CMOS NAND and NOR Schmitt circuits. *Microelectronics Journal*, 27(8):757–765, Nov. 1996.

[6] L. García-Leyva, D. Andrade, S. Gómez, A. Calomarde, F. Moll, and A. Rubio. New redundant logic design concept for high noise and low voltage scenarios. *Microelectronics Journal*, 42(12):1359–1369, Dec. 2011.

[7] W. Jyh-Ming, F. Sung-Chuan, and F. Wu-Shiung. New efficient designs for XOR and XNOR functions on the transistor level. *IEEE Journal of Solid-State Circuits*, 29(7):780–786, July 1994.

[8] L. B. Kish. End of Moore's law: thermal (noise) death of integration in micro and nano electronics. *Physics Letters A*, 305(3-4):144–149, Dec. 2002.

[9] J. P. Kulkarni, K. Kim, and K. Roy. A 160 mV, fully differential, robust schmitt trigger based sub-threshold SRAM. In *Proceedings of the 2007 international symposium on Low power electronics and design - ISLPED '07*, pages 171–176, New York, New York, USA, 2007. ACM Press.

[10] T. Lee and G. Cho. Monte Carlo based time-domain Hspice noise simulation for CSA-CRRC circuit. *Nuclear Instruments and Methods in Physics Research Section A: Accelerators, Spectrometers, Detectors and Associated Equipment*, 505(1-2):328–333, June 2003.

[11] K. Nepal, N. Alves, J. Dworak, and R. Bahar. Using Implications for Online Error Detection. In *2008 IEEE International Test Conference*, pages 1–10. IEEE, Oct. 2008.

[12] K. Nepal, R. I. Bahar, J. Mundy, W. R. Patterson, and A. Zaslavsky. Designing Nanoscale Logic Circuits Based on Markov Random Fields. *Journal of Electronic Testing*, 23(2-3):255–266, Mar. 2007.

[13] R. Rao, D. Blaauw, and D. Sylvester. Soft Error Reduction in Combinational Logic Using Gate Resizing and Flipflop Selection. In *2006 IEEE/ACM International Conference on Computer Aided Design*, pages 502–509. IEEE, Nov. 2006.

[14] R. Sarpeshkar, T. Delbruck, and C. Mead. White noise in MOS transistors and resistors. *IEEE Circuits and Devices Magazine*, 9(6):23–29, 1993.

[15] O. H. Schmitt. A thermionic trigger. *Journal of Scientific Instruments*, 15(1):24–26, Jan. 1938.

[16] N. Tega, H. Miki, Z. Ren, C. P. D'Emic, Y. Zhu, D. J. Frank, J. Cai, M. A. Guillorn, D.-G. Park, W. Haensch, and K. Torii. Reduction of random telegraph noise in High-Đž / metal-gate stacks for 22 nm generation FETs. In *2009 IEEE International Electron Devices Meeting (IEDM)*, pages 1–4. IEEE, Dec. 2009.

[17] J. Von Neumann. Probabilistic logics and the synthesis of reliable organisms from unreliable components, 1956.

[18] A. Wang and A. Chandrakasan. A 180-mV subthreshold FFT processor using a minimum energy design methodology. *IEEE Journal of Solid-State Circuits*, 40(1):310–319, Jan. 2005.

[19] Y. Ye, C.-C. Wang, and Y. Cao. Simulation of random telegraph Noise with 2-stage equivalent circuit. In *2010 IEEE/ACM International Conference on Computer-Aided Design (ICCAD)*, pages 709–713. IEEE, Nov. 2010.

Efficient Selection and Analysis of Critical-Reliability Paths and Gates

Jifeng Chen, Shuo Wang, and Mohammad Tehranipoor
University of Connecticut, Storrs, CT 06269, USA
{jic09003,shuo.wang,tehrani}@engr.uconn.edu

ABSTRACT

Aging effects such as negative bias temperature instability (NBTI) and hot carrier injection (HCI) have become major concerns when designing reliable circuits at sub-45nm technologies. It is vital to efficiently identify the paths that age at a faster rate than others in the field. Moreover, gates having the most impact on the degradation of these paths must be identified for compensation purposes. In this paper, we propose (i) a new timing analysis flow, which can quickly and accurately predict path and gate degradation due to NBTI and HCI effects, and (ii) a novel algorithm that can effectively identify the smallest set of critical-reliability gates while quantitatively evaluates their relative importance to path delay degradation. This facilitates reliability-enhancement methods to efficiently mitigate reliability threats using minimum area overhead. Our simulation results on several benchmark circuits demonstrate the efficiency of the proposed technique.

Categories and Subject Descriptors

B.8.1 [**Performance and Reliability**]: Reliability, Testing, and Fault-Tolerance

Keywords

Reliability, Aging, Timing Analysis, Gate Sizing, Optimization

1. INTRODUCTION

Aggressive scaling of CMOS technology into small feature sizes ($\leq 45nm$) worsens the degradation of performance and reliability caused by aging effects, especially NBTI [1][2] and HCI [4][5]. To compensate for aging effects and meet performance requirements throughout the lifetime operation, two categories of methods may be used: (1) Guardbanding methods [6][7][14], and (2) Adaptive control methods [8][9][11]. Due to the high complexity of aging mechanisms, it is extremely difficult to perform accurate degradation prediction. As a result, unnecessary power and area overhead are often introduced. For example, aggressive guardbanding meth-

ods may introduce an over-conservative margin by oversizing a large number of gates, especially those not critical to the overall reliability of the circuit. Meanwhile, adaptive control schemes usually introduce excessive area and power overhead, as they have to keep monitoring a large number of paths and gates for performance evaluation. A gate sizing technique was proposed in [10] to identify critical gates defined as the gates that age the most in the circuit. We shift our focus toward the identification and quantification of critical-reliability gates (CRGs), defined as the minimum number of "important" gates contributing to degradation of critical-reliability paths (CRPs), i.e., paths that are sensitive to aging and could potentially become critical in the field. By identifying CRPs and CRGs, aging compensation can be achieved with a minimal area and power overhead. As will be elaborated further in later sections, CRGs can be flexibly identified according to different workload scenarios, e.g., worst-case or actual workload conditions, which makes this technique easily compatible with conventional industry design practices.

Our work makes the following major contributions: (1) We develop an efficient aging-aware timing analysis flow to identify the CRPs in the circuit given different workloads. Rather than analyzing critical path (CP) set, which is extremely large in modern designs, we focus on selection of CRPs from CPs (CRP set is substantially smaller than CP set) to considerably reduce the computational complexity of the proposed CRG selection flow. (2) We develop a technique to quantitatively evaluate the importance of the CRGs based on their contribution to CRPs' degradation. Here, we develop a new metric to evaluate *a gate's importance to circuit CRPs degradation* making our proposed technique different from the previous ones that attempt to identify the gates with maximum aging. Our results demonstrate that the most impactful gates in the circuit are not necessarily the gates that have aged the most. (3) Lastly, we develop a novel algorithm to identify the minimum number of CRGs in the circuit for sizing to ensure that performance degradation of CRPs will not cause any failure in the field.

The remainder of the paper is organized as follows. Section 2 briefly describes our proposed CRP and CRG selection flow. The details of CRP analysis and selection are presented in Section 3. Section 4 explains CRG identification based on LP optimization technique. The simulation results and analysis are presented in Section 5. Finally, the paper is concluded in Section 6.

2. PROPOSED FLOW

Figure 1 shows our proposed flow for aging-aware path delay analysis (APD) and identification of CRPs and CRGs in integrated circuits. The flow includes three major steps: (1) CP Selection, (2) CRP Selection, and (3) CRG Identification. The conventional Static Timing Analysis (STA) tool is used to extract critical paths. As investigated in [18], performance of a critical path could degrade as much as 20% over ten years operation. Thus, during the static timing analysis, if the delay of a path P_i satisfies $\Gamma_{p_{0i}} \times 120\% \geq \Gamma_{clk} - \Gamma_m$, it is selected as a potential critical path, where $\Gamma_{p_{0i}}$ is the path delay at time 0, Γ_{clk} is the clock period and Γ_m is a safe margin (i.e., guardband) added at time zero. These paths may not be critical at time zero, but may become critical at some time point in the field. CRPs and CRGs are identified in steps 2 and 3, respectively, which will be described in details in Sections 3 and 4. Gate sizing can be applied to the CRGs in order to compensate for degradation of CRPs.

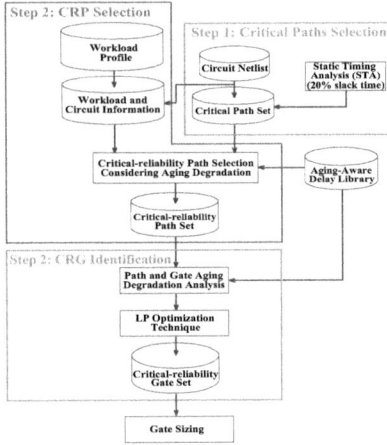

Figure 1: CRP and CRG selection flow.

3. CRP SELECTION

In this section, the conventional NBTI and HCI models are improved for CRP analysis. We then describe CRP selection and the accuracy of our CRP delay calculation flow.

3.1 Model Analysis and Improvement

The Reaction-Diffusion (R-D) model [13] is the general model used to explain and quantify the shift in threshold voltage due to NBTI and HCI effects. Also, the delay models in [16][17] are formulated to describe their dependency on threshold voltage and other parameters, such as temperature, capacitive load, stress probability, and slew rate; however, these models are incapable of incorporating complex conditions, such as multiple-input stress probabilities, which represent workload condition. First, we will briefly present and analyze these models. Then, these models will be extended for more accurate prediction based on our aging-aware delay library. Finally, the accuracy of our flow and HSpice MOSRA [12] simulation will be examined.

The R-D model formulates the increase in V_{th} for NBTI [14] and HCI [15] as:

$$\Delta V_{th\_NBTI} = A_{NBTI0} \times t_{ox} \times \sqrt{C_{ox}(V_{dd} - V_{th0})}$$
$$\times e^{\left(\frac{V_{dd} - V_{th0}}{t_{ox} E_0}\right)} \times (e^{-\frac{E_a}{k}})^{\frac{1}{T}} \times t_{stress}^{0.25}, \quad (1)$$

$$\Delta V_{th\_HCI} = A_{HCI0} \times \alpha \times f \times e^{\left(\frac{V_{dd} - V_{th0}}{t_{ox} E_1}\right)} \times t^{0.5}, \quad (2)$$

where t is the functional time of the transistor, t_{stress} is the effective stress time for NBTI, t_{ox} is the oxide thickness,

C_{ox} is the gate capacitance per unit area, α and f are the activity factor and the frequency respectively, and E_o, E_a, k and E_1 are technology-dependent constants. A_{NBTI} and A_{HCI} are constants that depend on the aging rate. Based on Alpha-power law [16] and derivation in [10], we can obtain the general form of delay estimation for gate G_j as:

$$D_j = A_j \times (1 + B_j^{\frac{1}{T}} \cdot (t_{eff})^n), \quad (3)$$

where,

$$A_j = A_1,$$
$$B_j = (A_1 B_1 A_0)^T B_0,$$

A_j and B_j are gate-dependent variables, different for NBTI and HCI. Equation (3) provides a short-range aging-degradation prediction for NBTI and HCI considering temperature T, signal probability p, time t, activity factor α and frequency f. Note that *short-range* refers to the short Euclidian distance in the parameter space between each pair of aging conditions. In [17], the authors proposed another general form for short-range gate delay calculation as a function of the input slew rate τ_j and capacitive load C_j as a two-term posynomial equation:

$$D_j = \sum_{k=1}^{2} S_k \tau_j^{m_k} C_j^{n_k}, \quad (4)$$

where τ_j is the input slew rate, C_j is the capacitive load, $S_k \geq 0$ depends on gate feature size, and m_k and n_k are real numbers and do not depend on gate feature size.

However, these short-range models are not necessarily efficient for long-range aging-degradation accurate prediction in the parameter space of aging conditions. Compared with these models, reliability analysis tools regressively approximate the result for the unknown condition on the expense of significant CPU runtime to achieve high accuracy if the Euclidian distance between the two conditions is large. In our methodology, we target a more balanced trade-off between efficiency and accuracy. An aging-aware delay library is generated to extend the models for spatially long-range aging-degradation prediction. It stores delay and output slew rate information (obtained from HSpice MOSRA simulations) for each gate according to different aging conditions including temperature T, signal probability $p_{j,m}$ (m is input pin index on a multiple-input gate), stress time t, input slew rate τ_j, and capacitive load C_j. For an aging condition not available in the library, the delay D_j is obtained by regressive approximation from the closest conditions in the library using the delay models (3) and (4). During the approximation process, once a close condition is selected as reference condition, all other available close conditions will be used as intermediate conditions during the regression process. In this way, the computational complexity will be decreased dramatically to linear level while high accuracy is maintained, which will be shown in later sections. In this paper, we improve the accuracy error to around 5% compared with direct HSpice MOSRA simulations. However, the accuracy can be further improved. For example, using smaller granularities for the parameters is one of the effective ways to improve accuracy when generating the aging-aware delay library.

We treat a path with q gates as a connection of gate primitives. By propagating the slew rate forward, the i-th CRP delay Γ_{CRP_i} under a specific aging condition can be obtained as:

$$\Gamma_{CRP_i} = \sum_{j=0}^{q} D_j(T, \tau_j, t, C_j, p_{j,1}, ..., p_{j,m}), \quad (5)$$

Note that, (5) applies to any path as long as its aging information is available. Since there is no analytical model for output slew rate estimation, we use the delay aging ratio, defined as $\lambda_j = D_j/D_k$, to approximate the output slew rate. In the path delay calculation, the output slew rate of one gate is propagated to the following gate and is regarded as its input slew rate. Thus, we calculate the corresponding output slew rate as:

$$S_j = S_k \times \lambda_j, \qquad (6)$$

where (D_k, S_k) is the delay and output slew rate stored in the aging-aware delay library corresponding to one of the closest conditions for approximation.

3.2 CRP Delay Analysis and Selection

CRPs are paths that may violate the timing constraint at some point in the field, that is, $\Gamma_{CRP_i} > \Gamma_{clk} - \Gamma_m$, where Γ_{CRP_i} is the delay of a CRP P_i in the field. For each PCP, aging information (signal stress probability, capacitive load, and switching activity) can be obtained using our in-house tools along with existing commercial tools. For example, the switching activity for gates on the PCPs can be calculated by gate-level simulation with our developed PLI routines. By collecting the switching information, the equivalent duty ratio for signal stress probability can be calculated for NBTI analysis [3]. Moreover, the switching activity can also be studied for HCI effect.

Algorithm 1 Aging-Aware Path Timing Analysis

1: Initialize delay of the i-th path $\Gamma_{pi} \leftarrow 0$
2: Specify temperature T
3: Specify stress time t
4: Specify the $1st$ gate input slew rate τ_0
5: **for** (all logic gates index j from 0 to n in i-th path) **do**
6:     Aging conditions collect for gate G_j:$(C_j, p_{j,1}, ..., p_{j,m})$
7:     Loading aging-aware delay library
8:     Find all close k conditions $(\Gamma_k, t_k, \tau_k, C_k, p_{k,1}, ..., p_{k,m})$
9:     Initialize temporary delay $D_t \leftarrow 0$
10:    Initialize temporary output slew rate $S_t \leftarrow 0$
11:    **for** each close condition $(\Gamma_k, t_k, \tau_k, C_k, p_{k,1}, ..., p_{k,m})$ **do**
12:       Data fitting for (D_t, S_t) using the math. models
13:    **end for**
14:    Finalize gate G_t approximation: $(D_j, S_j) \leftarrow (D_t, S_t)$
15:    $\Gamma_{pi} += D_j$
16:    $\tau_{j+1} = S_j$: propagate slew rate to next gate
17: **end for**

Once all the information is obtained, timing analysis is conducted to select the CRPs from PCPs using the aging-aware delay library. Generation of the aging-aware delay library is done only once for each technology library. Note that, in this work, we assume that the on-chip PLL that provides clock frequency is designed with reliable components robust to aging. However, we believe that our proposed flow can easily adopt impact of aging on clock tree as well. Algorithm 1 shows our proposed aging-aware path timing analysis procedure. Each path is regarded as a connection of gate primitives. Path delay Γ_{p_i} is initialized in Line 1; Lines 2 and 3 specify the temperature and stress time. Based on the aging condition, every closest condition stored in the aging-aware delay library is traversed from Lines 6 to 8. Lines 6 to 14 describe that, after being initialized in Lines 8 and 9, the gate delay and output slew rate can be approximated by using interpolation and extrapolation according to the delay models described in Section 3.1; Line 15 accumulates the gate delay for path delay analysis. Line 16 propagates the output slew rate forward ensuring accurate delay calcu-

lation. The process is repeated for every gate on the path to obtain the path delay as presented from Line 5 to Line 17. Considering N paths with M gates on average for each, complexity of the algorithm is $O(N \times M)$.

To verify the accuracy of our path delay calculation, we compare the delay calculated from our flow with that from HSpice MOSRA simulations for the top 100 paths extracted from the benchmark circuit s9234. The actual circuit condition when workload $WL = 0.5$ (meaning that the signal probability of being 1 at any primary input is 50%) for all parameters is considered and extracted using commercial tools. Aging degradation measured every two years for a span of ten years is shown in Figure 2. Each circle corresponds to one path at any time point, and the temperature is set to 75^oC. In Figure 3, the computational complexity between APD flow and HSpice MOSRA is compared, where the values show the ratio of HSpice MOSRA simulation time (on average for each path) over APD flow. On average, our flow shows ≥ 0.994 correlation with $\geq 244X$ speedup over the HSpice MOSRA simulation.

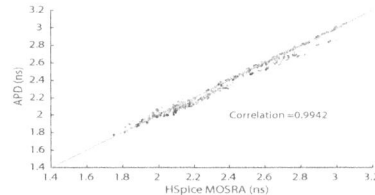

Figure 2: Accuracy analysis of our aging-aware path delay analysis flow compared with HSpice MOSRA.

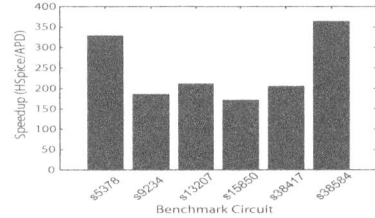

Figure 3: CPU runtime comparison between HSpice MOSRA and APD flow for different benchmark circuits.

4. CRG IDENTIFICATION TECHNIQUE

In this section, we evaluate the importance of the gates on CRPs based on their contribution to the paths' performance degradation and their relative impact on the CRPs. For the illustration purpose, Figure 4(a) shows a small circuit to demonstrate the difference between selecting gates with the largest degradation and selection of gates based on their impact on the paths. Each gate is numbered and the pair shows (d_0 and Δd_t), where d_0 is the gate delay at time zero and Δd_t shows the amount of gate degradation at time point t in the field. There are 14 paths in total in the circuit as shown in Figure 4(b). Also shown is the path delay at time zero (Γ_{t0}) and time t (Γ_{tt}). The gate and path lengths are shown using unit delay. The maximum path delay in addition to the margin (i.e. circuit timing) is set to 16 unit delays.

As seen, at time zero, all paths delay is under the timing budget; however, at time t, there are four paths (P2, P3, P11, and P12) above the timing budget, which will cause a failure in the circuit. The objective is to ensure that no path's timing become larger than the circuit timing budget.

Figure 4(c) shows the gates ordered based on their amount of aging at time t. To simplify the path delay calculation, we assume that once a gate is sized, it will not experience aging in the field. When selected based on the largest aging [10], gate G6 is chosen first. As seen in Figure 4(d), sizing this gate will impact paths P1 and P6, which are not the most critical ones. Next, gate G1 is selected, which reduces delay of paths P2 and P3. Selection of gate G4 will reduce delay of paths P11 and P12, and finally G9 will further reduce delay of paths P2, P3, P11, and P12, making all path length under the timing budget of 16 unit delay. However, using our technique, we will select only gates G9 and G7 to compensate for aging as shown in the Figure 4(d). This clearly shows that the most important gates to compensate for aging are not necessarily the gates that have aged the most.

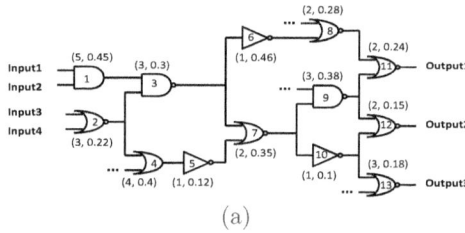

(a)

	Path	Γ_{t0}	Γ_{tt}
P_1	$G_1 \to G_3 \to G_6 \to G_8 \to G_{11} \to Output_1$	13	14.73
P_2	$G_1 \to G_3 \to G_7 \to G_9 \to G_{11} \to Output_1$	15	16.72
P_3	$G_1 \to G_3 \to G_7 \to G_9 \to G_{12} \to Output_2$	15	16.63
P_4	$G_1 \to G_3 \to G_7 \to G_{10} \to G_{12} \to Output_2$	13	14.3
P_5	$G_1 \to G_3 \to G_7 \to G_{10} \to G_{13} \to Output_3$	14	15.38
P_6	$G_2 \to G_3 \to G_6 \to G_8 \to G_{11} \to Output_1$	11	12.5
P_7	$G_2 \to G_3 \to G_7 \to G_9 \to G_{11} \to Output_1$	13	14.49
P_8	$G_2 \to G_3 \to G_7 \to G_9 \to G_{12} \to Output_2$	13	14.4
P_9	$G_2 \to G_3 \to G_7 \to G_{10} \to G_{12} \to Output_2$	11	12.12
P_{10}	$G_2 \to G_3 \to G_7 \to G_{10} \to G_{13} \to Output_3$	12	13.15
P_{11}	$G_2 \to G_4 \to G_6 \to G_7 \to G_9 \to G_{11} \to Output_1$	15	16.71
P_{12}	$G_2 \to G_4 \to G_6 \to G_7 \to G_9 \to G_{12} \to Output_2$	15	16.62
P_{13}	$G_2 \to G_4 \to G_6 \to G_7 \to G_{10} \to G_{12} \to Output_2$	13	14.34
P_{14}	$G_2 \to G_4 \to G_6 \to G_7 \to G_{10} \to G_{13} \to Output_3$	14	15.37

(b)

Gate	G_6	G_1	G_4	G_9	G_7	G_3	G_8	G_{11}	G_2	G_{13}	G_{12}	G_5	G_{10}
Δd_t	0.46	0.45	0.40	0.38	0.35	0.30	0.28	0.24	0.22	0.18	0.15	0.12	0.10

(c)

Selection Based on the Largest Aging					Selection Using Our Algorithm		
	$6 \to$	$1 \to$	$4 \to$	9		$9 \to$	7
P_2	16.72	16.27	16.27	15.89	P_2	16.34	15.99
P_3	16.63	16.18	16.18	15.80	P_3	16.25	15.90
P_{11}	16.71	16.71	16.31	15.93	P_{11}	16.33	15.98
P_{12}	16.62	16.62	16.22	15.84	P_{12}	16.24	15.89

(d)

Figure 4: Demonstrating the CRG selection using the concept of importance to CRGs: (a) an illustrative circuit, (b) paths in the circuit and their timing, (c) order of gate aging at time t, and (d) the gate selection and sizing to compensate aging.

Based on above, a fast linear programming (LP) optimization technique is developed to identify the CRGs. Aging compensation will ensure that $\Gamma_{CRP_i} \le \Gamma_{clk} - \Gamma_m$ throughout the chip's lifetime. In the LP, the predicted degradation of each gate is represented as a coefficient of the variable representing the importance of each gate. The optimal solution evaluates and quantifies the importance of each gate's contribution to the CRP degradation. Then, by ranking the importance of each gate, a minimum number of gates on each path can be selected to receive compensation for the lifetime degradation. Once the importance of each gate is identified, any pre-tapeout aging compensation techniques (e.g., gate sizing) can be implemented with a minimum area overhead. We formulate the aging compensation problem into an LP

optimization function to quantify the importance of each gate on the CRPs:

$$minmax \quad : \quad F(X) \qquad (7)$$
$$subject\ to \quad : \quad \Gamma_{CRP} - G \times X \le \Gamma_{spec}$$

where,

$$F(X) = \begin{bmatrix} f_{1,1}(x_1) & f_{1,2}(x_2) & \cdots & f_{1,m}(x_m) \\ f_{2,1}(x_1) & f_{2,2}(x_2) & \cdots & f_{2,m}(x_m) \\ & & \vdots & \\ f_{n,1}(x_1) & f_{n,2}(x_2) & \cdots & f_{n,m}(x_m) \end{bmatrix}_{n \times m},$$

$$f_{i,j}(x_j) = \begin{cases} 1, & \text{if } G_j \text{ is in path } P_i \text{ and } x_j > 0 \\ 0, & \text{otherwise} \end{cases}$$

$$\Gamma_{spec} = \begin{bmatrix} \Gamma_{clk} - \Gamma_m \\ \Gamma_{clk} - \Gamma_m \\ \vdots \\ \Gamma_{clk} - \Gamma_m \\ \Gamma_{clk} - \Gamma_m \end{bmatrix}_{n \times 1}, \quad G = \begin{bmatrix} g_{1,1} & g_{1,2} & \cdots & g_{1,m} \\ g_{2,1} & g_{2,2} & \cdots & g_{2,m} \\ & & \vdots & \\ g_{n,1} & g_{n,2} & \cdots & g_{n,m} \end{bmatrix}_{n \times m},$$

$$\Gamma_{CRP} = \begin{bmatrix} \Gamma_{CRP_1} & \Gamma_{CRP_2} & \cdots & \Gamma_{CRP_n} \end{bmatrix}_{1 \times n},$$
$$X = \begin{bmatrix} x_1 & x_2 & \cdots & x_m \end{bmatrix}_{1 \times m},$$

n is the number of CRPs and m is the number of unique gate instances on the n CRPs. Each row of matrix F and G corresponds to one CRP, while each column corresponds to one gate instance of all the non-duplicated gate instances on the CRPs. Γ_{CRP_i} is the i-th CRP's degraded delay, $g_{i,j}$ is gate degradation of the j-th gate on the i-th path with a stress time of t. Based on different lifetime timing specifications, $g_{i,j}$ is obtained from the CRP selection procedure discussed earlier. By solving the optimization problem, x_j $(j = 1, \ldots, m)$ will be obtained. The larger the x_j value is, the more important the j-th gate instance is. X is initialized to be $X_0 = [0, 0, \cdots, 0]'_{1 \times m}$ to make all gates equally important at the beginning. After solving the LP optimization problem, the gates can be ranked according to their x values rather than their degradation values. By ranking the x values from high to low, each CRP P_i can find the smallest set of gates U_i whose compensation, once is applied, can make up for the paths degradation over the lifetime. Assume that P_i has M gates, i.e., $G_1, G_2, \ldots, G_M$. Their delay degradations are $g_{i,1}, g_{i,2}, \ldots, g_{i,M}$ and corresponding x values are $x_1 > x_2 > \ldots > x_M$, respectively. Assume that the path delay Γ_{CRP_i} does not satisfy the timing requirement, i.e., $\Gamma_{CRP_i} > \Gamma_{clk} - \Gamma_m$ at some time point in the field. There are several methods [9][7], including gate sizing, that we can use to compensate for this degradation. For simplicity, we assume that once a gate is chosen for applying compensation on, its degradation is fully offset to 0 as $g_{i,j} = 0$. Then, we need to find the *critical length* l_i for the P_i to meet the specification. Note that the *critical length* l_i is the minimum number of gates necessary for aging compensation. As a result, the gate set $U_i = (G_1, G_2, \ldots, G_{l_i})$ out of $U_{P_i}(G_1, G_2, \ldots, G_M)$ is the i-th CRG set for P_i. The gate set $U = \bigcup_{i=1}^{n} U_i$ is the CRG set for the design.

Note that: (1) in the LP constraint function, $g_{i,j}$ represents the degradation for gate instance G_j. Thus, by replacing $g_{i,j}$ with the worst-case degradation, our flow is applicable to worst-case analysis; (2) our method of critical length l_i calculation for P_i is based on the assumption that $g_{i,j} = 0$ once the gate G_j is chosen for applying compensation. This is essentially assuming that the degradation of gate G_i will decrease at a scale of $g_{i,j}$ benefiting from the compensation scheme. Thus, by substituting the value $g_{i,j}$ with an actual compensation value from a different compensation scheme, the LP optimization method is scalable to any aging compensation schemes for CRG identification. For example, for compensation using gate sizing, $g_{i,j}$ is calculated in the following way:

$$g_{i,j} = D_j(t) - D_{sub}(t), \qquad (8)$$

where $D_j(t)$ represents the delay of the j-th gate at time t and $D_{sub}(t)$ represents that of the new gate. In addition, the gate sizing impact on the previous and next gates has to be considered as well. For example, when a gate is substituted for a new gate, its capacitive load has impact on the gate in the previous stage. Increasing gate size also impacts the output slew rate, which needs to be propagated forward. However this does not influence the flow of our method where gates are treated as a primitive unit and paths are treated as connection of gate primitives. Thus, the delay and degradation approximation can still be obtained from the flow with the capacitive load C_{j-1} changed in previous stage and output slew rate S_j changed in current stage.

To find the most critical gates for each CRP and obtain the entire CRG set U, [10] sorts all the gates on the CRPs according to their degradation from high to low as $G_1^*, G_2^*, \ldots, G_M^*$ with $g_{i,1}^* > g_{i,2}^* > \ldots > g_{i,M}^*$. The minimum l_i^* gates are found to meet the criterion $\Gamma_{CRP_i} - (g_{i,1}^* + \ldots + g_{i,l_i^*}^*) < \Gamma_{clk} - \Gamma_m$ and the l_i^* gates $(G_1^*, G_2^*, \ldots, G_{l_i^*}^*)$ are identified as CRGs for path P_i, and the united set of all the CRGs is the CRG set U^* for the design. In contrast, our LP technique offers several advantages over [10]: (1) Our flow is extendable to other aging effects beyond NBTI and HCI, (2) Our analysis of CRGs evaluates not only their own aging degradation, but also their contribution to CRPs degradation, and (3) Our aging analysis flow also enables the x_j to incorporate the topology-related information, such as capacitive load.

5. RESULTS

This section presents results obtained using our proposed flow on several ISCAS'89 benchmark circuits and Nangate $45nm$ standard-cell library for CRP and CRG analysis. Experiments are run on a 64-bit Windows Desktop with a 3 GHz Intel Core 2 Duo CPU and 4 GB of memory. Each benchmark circuit is processed using commercial ASIC design tools (Design Compiler, Prime Time, VCS, etc.) to extract the aging information. For the first simulation, the CRGs in the benchmark s38417 are identified for different timing margins (Γ_m) over ten years considering NBTI and HCI effects at three workload scenarios ($WL = 0.25, 0.5,$ and 0.75). The proposed flow is run to find the CRG set for the s38417 benchmark circuit; $g_{i,j} = 0$ is still held while deciding the critical length l_i for CRP P_i, for simplicity. The results are shown in Figure 5, which indicate that a more stringent timing margin will lead to a larger CRG set. For example, when the timing margin Γ_m is equivalent to 10% of the largest delay at time 0 when $WL = 0.5$, 176 gates are identified as CRG; while at the same WL value, when the timing margin Γ_m is equivalent to 5%, 493 gates are identified as CRG.

The two timing margin values 5% and 10% are also used to repeat the simulations with the stress time changing from $0.6E + 08s \; (\approx 2\,years)$ to $3.0E + 08s \; (\approx 10\,years)$. Results in Figure 6 indicate that more gates are identified as CRGs when the target lifetime is longer. Meanwhile, to satisfy the same lifetime specification, a smaller timing margin leads to

more identified CRGs. The results also demonstrate that circuit lifetime can be extended by only introducing a small area overhead as the minimum gate set is selected for aging compensation, because resources are efficiently allocated to CRGs only.

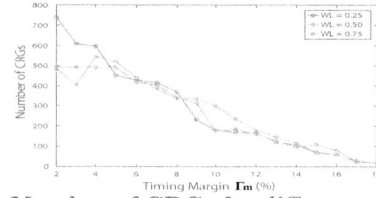

Figure 5: Number of CRGs for different timing margins (when $T = 75^oC, WL = 0.5,$ and $t = 10$ years).

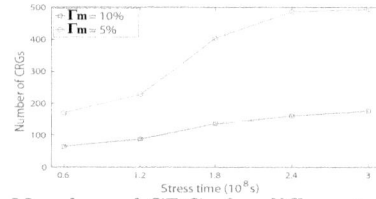

Figure 6: Number of CRGs for different stress times (when $T = 75^oC, WL = 0.5$).

Further simulations are conducted for several benchmark circuits with three timing margin values ($2\%, 7\%, 12\%$). The results are shown in Table 1. Columns 2 and 3 list the total gate and flip-flop count. Under each timing margin category, the first and the second columns list the PCPs and CRPs, respectively. The CRGs in the third column are obtained using our flow. Finally, the fourth column shows the area overhead due to sizing the CRGs. Figures 7 and 8 demonstrate the effectiveness of our CRP and CRG identification. Gate sizing is conducted on the CRGs identified by our flow. In each figure, at each time point, the number of CRPs and the number of CRGs identified are presented in the form of CRP/CRG. The horizontal line (in green) is the timing specification for the design at that time point including the timing margin. The longer the stress time is, the more CRPs and CRGs are identified, indicating that a longer lifetime requirement leads to more CRGs. In Figure 7, $\Gamma_m = 5\%$; while in Figure 8, $\Gamma_m = 10\%$. The CRP delays before gate sizing are checked and presented (in red) above the horizontal line (in green), without exception that their delays violate the timing specification at different time point. After sizing the CRGs, the CRP delays are effectively reduced and are below the horizontal line accordingly (in blue). In the figures, the CRP delays before and after gate sizing are not displayed vertically on a same line at each time point just for demonstration purpose. In Figure 9, the stress time is fixed with a varying Γ_m, which shows that a stringent timing margin leads to more CRGs. From the results, we can see that aging compensation on the CRGs can effectively reduce the aging degradation and satisfy the design with the timing specification for the lifetime operation.

These experimental results demonstrate that circuit performance degradation can be minimized if early prediction and optimization is conducted during the design stage, and that circuit aging degradation can be compensated for by focusing on a minimum set of CRGs. The experimental results also show that we can efficiently identify a small number of CRPs and CRGs. After identifying the CRGs, other aging compensation mechanisms can also be utilized more effectively with lower area and power overheads.

Table 1: CRP and CRG identification for three timing margins with 10 years NBTI and HCI degradation $(T = 75^oC, WL = 0.5)$

Bench. Circuit	# of Gates	# of FFs	$\Gamma_m = 2\%$				$\Gamma_m = 7\%$				$\Gamma_m = 12\%$			
			PCP	CRP	CRG	$Area_o(\%)$	PCP	CRP	CRG	$Area_o(\%)$	PCP	CRP	CRG	$Area_o(\%)$
s5378	1324	153	94	87	106	1.1479	83	52	41	0.5000	45	16	19	0.2592
s9234	1120	125	85	49	29	0.4934	50	26	24	0.4464	25	10	10	0.1645
s13207	1477	245	19	13	49	0.5550	7	4	21	0.1682	4	2	5	0.0168
s15850	3338	452	890	865	21	0.1936	289	273	19	0.2005	34	33	11	0.1037
s38417	10221	1523	9671	5616	493	1.1252	4927	1922	404	0.6881	1689	485	162	0.2677
s38584	12482	1246	293	267	137	0.3349	164	139	67	0.1708	55	39	15	0.0225

PCP: potential critical path; CRP: critical-reliability path; CRG: critical-reliability gate;
$Area_o$: area overhead; FF: flip-flop; Γ_m: timing margin.

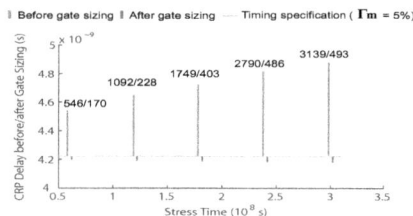

Figure 7: CRP delay comparison before/after gate sizing on the CRGs with varying stress time (when $\Gamma_m = 5\%, WL = 0.5$, and $T = 75^oC$).

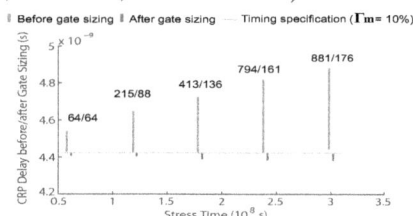

Figure 8: CRP delay comparison before/after gate sizing on the CRGs with varying stress time (when $\Gamma_m = 10\%, WL = 0.5$, and $T = 75^oC$).

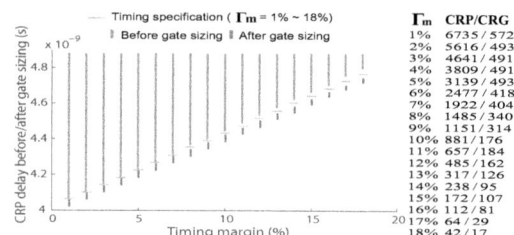

Figure 9: CRP delay comparison before/after gate sizing on the CRGs with varying Γ_m (when $WL = 0.5, T = 75^oC$, and $t = 10$ years).

6. CONCLUSIONS

In this paper, a new technique is proposed to quickly identify CRPs and CRGs. Aging compensation can therefore be focused only on the minimum number of CRGs as they contribute the most to path delay degradation. The flow is easy to be integrated into conventional industry IC design flow. Experimental results indicate that our technique is able to identify the minimum number of CRGs, minimize area overhead for design margining, and ensure performance throughout its lifetime operation.

7. ACKNOWLEDGMENTS

This work was supported in part by Semiconductor Research Corporation (SRC) under grants 2053 and 2094, and a gift from Cisco.

8. REFERENCES

[1] T. Nigam, "Impact of Transistor Level Degradation on Product Reliability," *Custom Integrated Circuits Conference*, pp. 431-438, September 2009.

[2] G. Chen, M. Li, C. Ang, J. Zheng, and D. Kwong, "Dynamic NBTI of p-MOS Transistors and Its Impact on MOSFET Scaling," *IEEE Electronic Device Letter*, vol. 23, pp. 734-736, December 2002.

[3] S. V. Kumar, C. H. Kim, and S. S. Sapatnekar, "An Analytical Model for Negative Bias Temperature Instability (NBTI)," in Proceeding of the *IEEE/ACM International Conference for Computer Aided Design*, pp. 493-496, November 2006.

[4] L. F. Wu, J. K. Fang, H. Yonezawa, Y. Kawakami, N. Iwanishi, H. Yan, P. Chen, A. Chen, N. Koike, Y. Okamoto, C. Yeh, and Z. Liu, "GLACIER: A Hot Carrier Gate Level Circuit Characterization and Simulation System for VLSI Design," *ISQED*, pp. 73-79, 2000.

[5] M. Dai, C. Gao, K. Yap, Y. Shan, Z. Cao, K. L, L. Wang, B. Cheng, S. Liu, "A Model With Temperature-Dependent Exponent for Hot-Carrier Injection in High-Voltage nMOSFETs Involving Hot-Hole Injection and Dispersion," *IEEE Transactions on Electron Devices*, pp. 1255-1258, April 2008.

[6] B. Paul, K. Kang, H. Kufluoglu, M. Alam, and K. Roy, "Temporal Performance Degradation under NBTI: Estimation and Design for Improved Reliability of Nanoscale Circuits," *Design Automation and Test in Europe*, pp. 1-6, March 2006.

[7] S. Kumar, C. Kim, S. Sapatnekar, "NBTI-Aware Synthesis of Digital Circuits," *Design Automation Conference*, pp. 370-375, June 2007.

[8] S. Kumar, C. Kim, and S. Sapatnekar, "Adaptive Techniques for Overcoming Performance Degradation due to Aging in Digital Circuits," *Design Automation Conference*, pp. 284-289, 2009.

[9] E. Mintarno, J. Skaf, R. Zhang, J. Velamala, Y. Cao, S. Boyd, R. Dutton, and S. Mitra, "Optimized Self-Tuning for Circuit Aging," *Design, Automation & Test in Europe Conference & Exhibition DATE*, pp. 586-591, April 2010.

[10] W. Wang, Z. Wei, S. Yang, and Y. Cao, "An efficient Method to Identify Critical Gates under Circuit Aging," *International Conference on Computer Aided Design*, pp. 735-740, 2007.

[11] N. Shah, R. Samanta, M. Zhang, J. Hu, and D. Walker, "Built-In Proactive Tuning System for Circuit Aging Resilience," *IEEE International Symposium on Defect and Fault Tolerance of VLSI Systems*, pp. 96-104, 2008.

[12] Synopsys, *http://www.synopsys.com/*

[13] S. Ogawa and N. Shiono, "Generalized Diffusion-Reaction Model for the Low-Field Charge Build up Instability at the $Si - SiO_2$ Interface," in *Phys. Rev. B.*, Vol. 51, Issue 7, pp. 4218-4230, February 1995.

[14] R. Vattikonda, W. Wang, and Y. Cao, "Modeling and Minimization of PMOS NBTI Effect for Robust Nanometer Design," In *Design Automation Conference*, pp. 1047, 2006.

[15] E. Takeda, C. Yang, and A. Miura-Hamada, "Hot-Carrier Effects in MOS Devices," *Academic Press*, 1995.

[16] T. Sakurai, et. al., "Alpha-Power Law MOSFET Model and Its Application to CMOS Logic," *JSSC*, pp. 548-594, 1990.

[17] H. Tennakoon, and C. Sechen, "Efficient and Accurate Gate Sizing with Piecewise Convex Delay Models," *Design Automation Conference*, pp. 807-812, June 2005.

[18] W. Wang, S. Yang, S. Bhardwaj, R. Vattikonda, S. Vrudhula, F. Liu, and Y. Cao, "The Impact of NBTI on the Performance of Combinational and Sequential Circuits," *Design Automation Conference*, pp. 364-369, Jun. 2007.

Breaking the Power Delivery Wall using Voltage Stacking

Kaushik Mazumdar
Dept. of ECE
University of Virginia
km3sj@virginia.edu

Mircea R. Stan
Dept. of ECE
University of Virginia
mircea@virginia.edu

ABSTRACT

We propose the use of *voltage stacking* for addressing some of the power delivery issues for many-core processors. To demonstrate the effectiveness of our method we first design a proxy for a many-core stacked processor in the form of a regular structure using multiple ring oscillators where we can control the voltage, frequency and switching activity for individual rings. For intermediate voltage rail regulation, we propose a push pull-based switched capacitor regulator designed specifically for balancing the stacked loads. Detailed Spice simulation results for the prototype model show a 4× reduction in supply current when using 4 layers of voltage stacking. We further validate our method by designing a voltage-stacked structure using two PIC cores.

Categories and Subject Descriptors: B.7.1 [Harware]: Integrated Circuits – *Types and Design Styles*

General Terms: Design

Keywords: voltage stacking, DC to DC conversion, charge recycling, switched capacitor, many core

1. INTRODUCTION

Technology scaling trends have brought about several *power walls* that must be overcome for Moore's Law to continue. These walls include the *power density wall* (where the power consumption density increases beyond the heat dissipation capabilities of the technology), the power and ground *power delivery pin walls* (where the chip power consumption requires increasing numbers of power and ground pins at the expense of increased cost and as a trade-off with available I/O bandwidth), the *3D IC power density wall* (where the physical stacking in the third dimension exacerbates the two-dimensional power density explosion), and the *on-chip power regulation efficiency wall* (where the relatively poor efficiencies achievable with on-chip regulators limit the effectiveness of many low power schemes that depend on on-chip regulation). Multi- and many-core solutions are becoming increasingly popular in order to improve performance, but they are particularly sensitive to the above *power delivery pin walls* due to a combination of increased power consumption, voltage scaling (hence corresponding current increases) and increased I/O bandwidth. In this paper we propose voltage stacking as a comprehensive method for addressing the power delivery pins walls for multi- and many-core processors.

Voltage stacking simply refers to the power delivery arrangement of two or more circuit blocks such that the ground of one block becomes the power connection for the next, thus the blocks being connected as a *series* stack for power delivery, with all of them sharing the same current (thus the charge being recycled in the stack), while their VDD values are added. Similar solutions have been proposed earlier for 3D ICs, for pin limited ICs, for implicit high efficient DC-DC conversion, reduction of standby power in SRAM and power supply noise suppression [1-4].

There are several attractive reasons to pursue voltage stacking of entire processor cores. The ITRS prediction for 2015, is that, for processors operating at nominal VDD, on-chip current densities will go up to 150 A/cm^2 [5]. The prediction is that the number of power and ground pins need to be about two-thirds of the total (several thousand range, see Table 1); but with more cores per chip, the number of I/O's also needs to increase which leads to one of the abovementioned power delivery walls.

Table1 ITRS predictions for high performance processors [5]

Parameters	2009	2012	2015
Avg Current density (A/cm²)	64	108	150
Power pins	1540	1783	2063
Off chip Data rate (Gb/s)	8	14	30
Avg current/pin (mA)	41	60	72

The organization of the paper is as follows. In section 2, we describe the related work and focus on the novelty of our work. In section 3, we discuss the impact that voltage stacking can have on many-core processor design through a proxy model we developed using ring oscillators. The section also contains an analysis of stacked GALS (globally asynchronous locally synchronous) many-core and how it improves upon stacked non-GALS architecture. Section 4 deals with some of the circuit design aspects of this approach and the working of the proposed push pull-based switched capacitor regulator. We also verify our claims by running simulations on real processor cores.

2. RELATED WORK

One of the earliest works on charge recycling was based on energy efficient on-chip DC-DC conversion [4]. The idea was to have multiple domains of logic, voltage stacked upon each other, to obtain what the authors called "high tension" power delivery. Charge balance was maintained between the domains through active regulation of the intermediate node using power transistor-based linear regulators. Since a close loop linear regulator was used, larger imbalances between the domains severely lowered the efficiency. To compensate for large imbalances, blocks of logic, or granules, were shifted from one logic domain to another using switching logic. This improved the efficiency of the DC-DC conversion, but at the cost of area and power overhead of granules switching logic. In [1] the authors used voltage stacking for 3D integrated circuits. In [3] the authors demonstrated the effectiveness of stacking scheme in saving standby power in SRAM by 88.6%. In [4] the authors have used voltage stacking to lower the supply current and reduce the LdI/dt and IR drop across the power grid. However, apart from [1], the focus of the other works on voltage stacking has been about maintaining the charge balance between the domains through logic/granule switching for high energy efficiency. While this regulation scheme improved the energy efficiency of low power or sub threshold circuits, it restricted the impact of voltage stacking when used for high performance power delivery.

In this work we propose voltage stacking to be used with more *relaxed* regulation of the intermediate power nodes for high performance many-core processors. Through simulation we show that decreasing the burden on regulation and allowing the cores to run asynchronously can be very beneficial from the power delivery point of view. Without the granule switching and close loop regulation, a lot of overhead can be removed. However the droops on the intermediate voltages degrade the performance of synchronous processors - hence to exploit the benefits of voltage stacking, the best option is to move from synchronous to GALS-based many core processors [6]. One contribution of our work is to show that, in addition to the intrinsic advantages, when combined with voltage stacking GALS voltage stacking has the potential to tackle the power delivery pin walls for future many-core processors. Unlike the previous works, we use multi- and many-core processors as the target application and synthesize and simulate actual cores (PIC16x from Microchip) to verify our claims. Another contribution in our power delivery scheme is the voltage regulator for the intermediate rails. In all the previous works push pull-based linear regulators have been used to compensate for charge mismatch for a short time constant. While the transient response of the linear regulator with high open loop bandwidth is fast, the close loop regulation necessitates that the activity of the different domains (i.e. charge balance) remain the same - otherwise the regulator will not be able to match up with the difference of charge. In this work, we propose the use of push pull-based *switched cap regulation* which can source or sink charge by making the fly capacitors shuttle between supply-V_{mid} and V_{mid}-ground potentials with the help of switches. Thus, instead of switching the granules, the switched capacitors will switch roles to regulate the V_{mid}, reducing the overhead.

3. VOLTAGE STACKING AND GALS

In voltage stacking, charge imbalances will inevitably come about because of variations or because of differences in circuit activities between the domains. The solution that has been used before is to actively regulate the V_{mid} node through different hierarchies of regulation including on chip decap, linear regulator and switching of granules between domains. But balancing the domains may just restrict the cores to very specific tasks and with a large overhead. So, what happens if the domains are not balanced *rigidly*, while letting V_{mid} unregulated, or with *relaxed* regulation? Having different workloads will lead to V_{mid} droop (down) or bounce (up), thus forcing one domain to have a larger supply voltage and the other a smaller one. This might cause a synchronous scheme to *break down* as the synchronous clock cannot handle the huge skew. But what if the clocks in the core are asynchronous to each other? Under DVFS the supply voltage imbalance will cause the clocks in different domains to work at different frequencies. While this defies the working of synchronous processors, it is the basic principle of GALS architecture. By modulating the working frequency in conjunction with the mid voltage, GALS many-core stacked processors can improve the throughput considerably over their synchronous counterparts.

3.1 Proxy many-core processor model

In order to demonstrate our claims, we have designed proxy models for the "GALS" and "Non-GALS" stacked cores (see fig. 1). The proxy model consists of "cores" voltage stacked one above the other, each core's logic activity being modeled using ring oscillators (RO) that can be independently turned on and off.

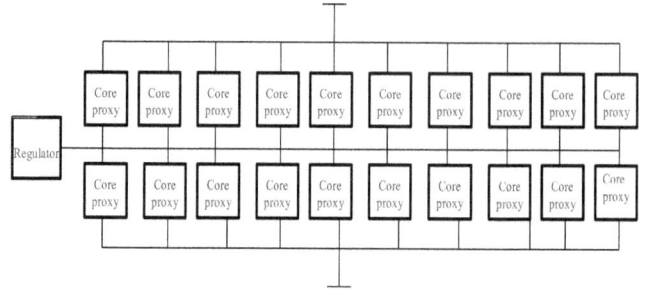

Fig 1. Many-core processor model with 2 levels of voltage stacking

Table 2. Specifications for proxy "many-core" model

Technology	Supply (V)	No of "cores"	No of sub blocks (per core)	No of inverters (per core)
FDSOI 150nm	1.2V	20	10	110

This model (details in Table 2) represents a reasonable proxy to an actual processing many-core system for several reasons: first, it can capture the transient events when switching between operating modes, or between different instructions by simply powering on/off certain oscillators; second, by running the oscillators in parallel, the peak currents drawn are sufficiently large (compared to having just one long oscillator) to actually closely mimic a real processor; third, considering the long simulation times, it is not feasible to verify the different scenarios of voltage stacking with detailed multi- and many-core structures.

3.2 GALS versus Non GALS

GALS-based stacked many-core may not need any mid voltage regulation since local clocks in each core can track the voltage droop and dynamically vary the operating frequency. This is the major advantage over non GALS synchronous cores where the operating frequency of all the cores is fixed irrespective of the voltage rail. However, the mid rail cannot be allowed to droop down completely on its own, especially when used with low voltages - otherwise transistors in worst case scenario may not even turn on due to the low rail voltage. Hence, the best tradeoff is to have a *relaxed* voltage regulation which will make sure both the domains remain active, without having the overheads of granules switching. In figures 2 and 3, we show the results of simulations that illustrate the tradeoffs between ideal GALS, non-GALS and GALS with relaxed regulation in terms of voltage droop, performance and energy/cycle. With difference in core activities, the mid voltage rail droops significantly with unregulated GALS. But with relaxed regulation, the droop is less drastic and almost comparable to that of non GALS. Figure 3 shows the improvement in power and performance of GALS with regulation over the other 2 topologies. Due to the moderate regulation, the performance does not degrade with core activity difference as in unregulated GALS, while the ability to operate at higher frequency with voltage rail gives this architecture higher performance than non-GALS.

3.3 Coarse versus Fine Grained stacking

In all the existing works so far, voltage stacking has been implemented with a finer granularity. In work [7], logic blocks

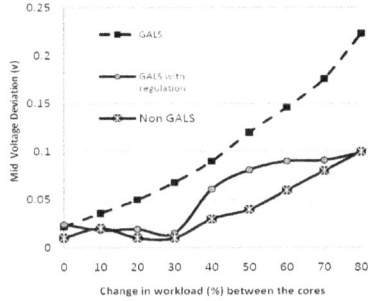

Fig 2. V_{mid} fluctuation with change in workload between levels

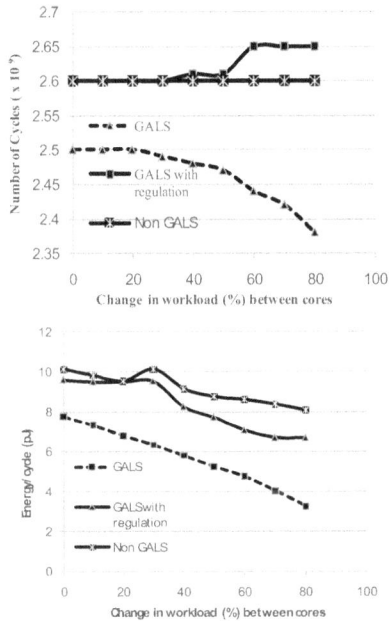

Fig 3. Degradation in performance (top) and "energy/instruction cycle" (down) with change in workload between levels

have been stacked in different domains and implicit DC-DC conversion was done by recycling the charge between the domains. However the increasing overheads of level shifter for every logic domain can nullify the gains of voltage stacking. In this work, we adopt a coarser grained approach where, instead of stacking individual logic blocks, we use stacking at the core level. Since the cores are independent of each other, the overall throughput will not be affected much by circuit issues at different levels. Level shifting is necessary only to interface between the cores (e.g. through a network-on-chip, NOC). With our proxy model, we have simulated three different scenarios. One is with 3 stacked cores (coarse), 2nd one has 3 cores, each having 2 domains of logic stacked inside (mid-granularity) and the 3rd one has 3 cores, with each having 3 levels of logic stacked inside (fine granularity). We have kept the total logic depth constant in all the three cases. The difference in performance is due to the overheads associated with the 2nd and 3rd case. As shown in Table 3,

stacking of core (coarse) instead of logic blocks (fine), leads to benefits in throughput, power and silicon area.

Table 3. Comparisons between coarse and fine grained stacking

Metrics	Core stacking	2-stage logic stacking	3-stage logic stacking
Throughput (number of cycles)	1.68x e-9	0.91x e-9	0.88x e-9
Energy/cycle (pJ)	0.23	0.66	1.19
Area (nm^2)	0.66x e-9	1.01x e-9	1.31x e-9

3.4 Multilevel Stacking

The benefits of voltage stacking improve with increasing levels of stacking. With more stacking more charge will get recycled, reducing the supply current by a larger margin. We have modified our proxy many core model and extended it to multi levels of voltage stacking. The local clock has been designed using programmable ring oscillators [6]. For inter-core communication, we have used low power level shifters [7]. Thermometric decoder and additional control signals were used to replicate changes in processor activity. Each layer has a voltage rail of 1V. Table 3 shows the 4× improvements in supply current drawn between 2 and 4 levels of stacking, coming at the cost of 16.5% reduction in performance. Even though energy/cycle increases with 2 levels of stacking, it gradually improves with more number of stacked levels. Thus increasing the number of stacked levels improves the performance of the power delivery network.

Table 3: Comparisons between multiple levels of stacking

Number of stacking layers	Supply Current (mA)	Average Number of cycles	Energy/cycle (nJ)
Parallel cores	5.235	1523x 10^2	34
2 stacks	3.769	1450x10^2	52
3 stacks	1.874	1220x 10^2	46
4 stacks	0.973	1210x10^2	32

4. Push-pull switched-capacitor regulator

The mid voltage regulation for the stacked topology has a unique feature – here the current at times needs to be both *sourced* and *sinked* unlike conventional regulation where current is only sourced by the converter. In this work, we developed a push pull-based *switched-capacitor regulator* where two *flying* capacitors change roles periodically. This switching of capacitors provides the source/sink of charges. Based on the aforementioned criteria, the regulator style chosen for this design is a modified version of a conventional switch-capacitor circuit [8]. The design is implemented with four capacitors and eight switches (see figure 4). To understand how the circuit works, consider an example with a slightly imbalanced workload, where the current offset pushes V_{mid} to droop below half-VDD. In the first phase, as the voltage droops down at the load, capacitor C_{SW1} begins charging to a voltage above half-VDD, while the voltage on C_{SW2} falls below half-VDD. In the second phase, through the on-chip switches, the capacitors C_{SW1} and C_{SW2} switch places. Since the voltage on C_{SW1} was charged to a higher voltage, it redirects this charge back onto the larger capacitor C_{D2}. This redirection of charge helps pull the load voltage back towards half-VDD. The faster the switching frequency, the closer to half-VDD the load voltage remains.

C_{D1} and C_{D2} acting as small reservoir of charge at the output, were not sufficient to reduce ripple within acceptable margins; thus we decided to apply dual phase interleaving. By having 2 converters in parallel with their switching frequency shifted over by π, we managed to reduce the ripple by 50% (see figure 5) without any additional overhead. The efficiency curve has been plotted with varying the workload (figure 6). As evidenced from the plot, the regulation efficiency is at its highest value when the loads are perfectly balanced. However after a workload difference of 60% or more, the regulation efficiency reduces rapidly.

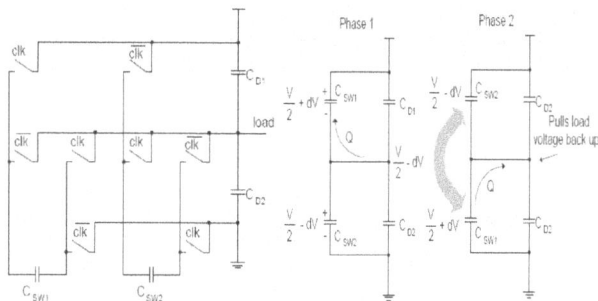

Fig 4. Switched-capacitor regulator (left). The two operational phases of the switched capacitor with $C_{D1,2}$=10pf and $C_{sw1,2}$=500pf (right). Phase one shows the voltage decreasing at the load, and during phase two this charge is redirected back to the load through the fly caps C_{sw2}.

4.1 Simulation with real Processor cores

To further validate our approach of using voltage stacking at the core level, we have fully synthesized and simulated real processor cores (8-bit PIC 16x-compatible Processor using ST65 technology) in a stacked topology. Figure 7 shows the mid voltage fluctuations with varying supply voltages for the stacked PIC cores. As the plot indicates, while the mid rail is at half VDD for lower supply voltages, it has some droop at higher supply. Figure 9 also shows the supply current with parallel cores and stacked cores for different VDD. While for lower supply voltage (1.4V), the supply current has improved by 2.7x with stacking, the improvement is less with higher voltage. This degradation in performance at higher supply voltage can be attributed to the bulk CMOS technology which has been used for synthesis and simulation of the PIC cores.

5. CONCLUSION

We have presented a novel approach of using voltage stacking for high performance many-cores (using the proxy model) with little overhead. From our model, we have seen a current decrease of 1.4× in going from non-stacking to 2 levels of stacking while an improvement of 4× in going from 2 levels to 4. When simulating real PIC cores, we found the trend to be similar with supply current getting reduced by 1.7× at 2 levels of stacking. The ITRS prediction for the number of power pins in many core processors for the year 2015 is about 2063 [5]. *Following the first order analysis of supply pins count being proportional to current, we can predict the need for supply pins to be reduced to 1213 from 2063 when using voltage stacking.* This leaves additional pins for I/O without making the chips pad- and pin-limited. Thus voltage stacking can be an effective tool in handling power pin and power

density wall for future high performance cores and will work best when there is some form of workload balancing between the domains, thus reducing the burden on regulation.

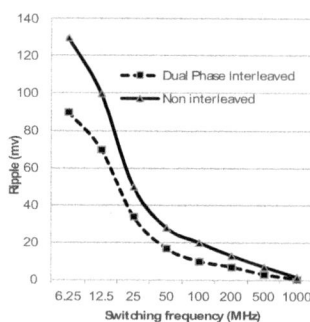

Fig 5. Suppression of ripple with increase of switching frequency

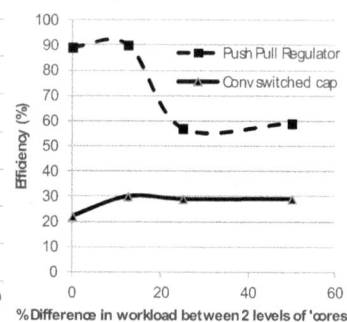

Fig 6. Efficiency degradation with change of workload between cores

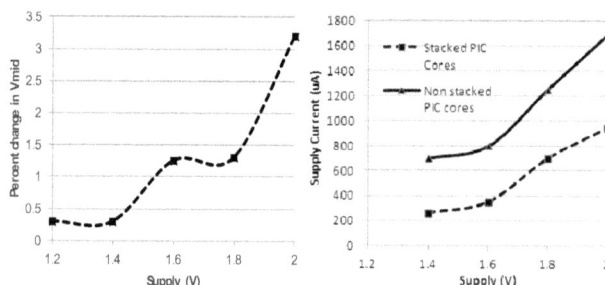

Fig 7. Mid voltage fluctuations of voltage stacked PIC cores under varying supply voltage and reduction in supply current over non stacked cores

ACKNOWLEDGMENTS

We thank AMD and Intel for partial support for this work.

REFERENCES

[1] P. Jain et al "A multi-story power delivery technique for 3D integrated circuits," *ACM International Symposium on Low Power Electronics and Design (ISLPED)*, pp. 57–62, 2008.

[2] S. Rajapandian at al "High voltage power delivery through charge recycling," *IEEE Journal of Solid State Circuits*, 41(6):1400–1410, 2006.

[3] A. Cabe and M. R. Stan "Experimental Demonstration of Standby Power Reduction using Voltage Stacking in an 8Kb Embedded FDSOI SRAM", *ACM Great Lakes Symposium on VLSI*, 2011

[4] J.Ju and C. H. Kim, "Multi-Story Power delivery technique for Supply Noise Reduction and Low Voltage Operation," *ACM International Symposium on Low Power Electronics and Design (ISLPED)*, pp. 192–197, Aug 2005.

[5] International Technology Roadmap for Semiconductors 2009

[6] Zhiyi Yu et al, "AsAP: An Asynchronous Array of Simple Processors", *IEEE Journal of Solid State Circuits*, 2007.

[7] S. Wooters, B. Calhoun and T. Blalock, "An energy efficient subthreshold level converter in 130 nm CMOS", *IEEE TCAS II*, Vol 57, issue 4, April 2010.

[8] Y. K. Ramadass and A. P. Chandrakasan. "Voltage scalable switched capacitor DC-DC converter for ultra-low power on-chip applications" in *IEEE Power Elec. Specialists Conf.*, June 2007.

An Efficient CPI Stack Counter Architecture for Superscalar Processors

Osman Allam   Stijn Eyerman   Lieven Eeckhout

Ghent University, Belgium

{osman.allam, stijn.eyerman, lieven.eeckhout}@elis.UGent.be

ABSTRACT

Cycles-Per-Instruction (CPI) stacks provide intuitive and insightful performance information to software developers. Performance bottlenecks are easily identified from CPI stacks, which hint towards software changes for improving performance.

Computing CPI stacks on contemporary superscalar processors is non-trivial though because of various overlap effects. Prior work proposed a CPI counter architecture for computing CPI stacks on out-of-order processors. The accuracy of the obtained CPI stacks was evaluated previously, however, the hardware overhead analysis was not based on a detailed hardware implementation.

In this paper, we implement the previously proposed CPI counter architecture in hardware and we find that the previous design can be further optimized. We propose a novel hardware- and power-efficient CPI counter architecture that reduces chip area by 44% and power consumption by 47% over the best possible prior design, while maintaining nearly the same level of performance and accuracy.

Categories and Subject Descriptors

C.0 [**Computer Systems Organization**]: Modeling of computer architecture; C.4 [**Computer Systems Organization**]: Performance of Systems—*Modeling Techniques*; B.8.2 [**Hardware**]: Performance and Reliability—*Performance Analysis and Design Aids*

General Terms

Design, Performance, Measurement, Experimentation

Keywords

Superscalar processor, CPI stack, Counter architecture

1. INTRODUCTION

A key role of user-visible hardware performance counters is to provide clear and accurate performance information to the software developer, who then uses the information to guide software changes

towards improved performance. An intuitive representation of performance is a cycle stack which breaks up total cycle count to execute a unit of work in its cycle components. A cycle stack consists of a base cycle component (number of cycles during which useful work is done), plus a number of 'lost' cycle components due to miss event handling such as cache misses, branch mispredictions, etc. Dividing a cycle stack by the number of dynamically executed instructions then yields a so-called Cycles-Per-Instruction (CPI) stack [3]. The power of a CPI stack is that it visually highlights the major performance bottlenecks by the large CPI components. For example, a large cache miss component implies that the workload is cache-intensive: software optimizations that improve access locality and/or reduce the working set are likely to improve performance significantly.

Constructing CPI stacks is challenging on contemporary superscalar processors. Out-of-order processors are today's prevalent superscalar processors in the server and desktop domain (e.g., Intel Xeon, AMD Opteron, IBM Power7, etc.) and they are gaining popularity in the embedded space as well (e.g., ARM Cortex A9). These processors feature superscalar out-of-order execution, speculative execution, hardware prefetching, ability to service multiple outstanding memory requests at the same time, non-blocking caches, etc. All of these features enable out-of-order processors to achieve high performance by exploiting instruction-level parallelism (ILP) and memory-level parallelism (MLP), the fundamental reason being able to hide latency through parallel execution. This ability for hiding latency complicates CPI stack construction: overlapping events should not be double-counted.

Prior work by Eyerman et al. [6] proposed a method for computing accurate and meaningful CPI stacks on superscalar out-of-order processors. The method was found to be accurate with an average error around 2.5% and a maximum error of less than 4% compared to detailed cycle-accurate simulation over a range of standardized benchmarks; this is substantially more accurate than previously proposed approaches with average errors around 20% [12]. The CPI stack construction method relies on a CPI counter architecture that is to be implemented in hardware. Eyerman et al. described the CPI counter architecture in a level of detail that is common to computer architecture papers, however, the evaluation of its hardware complexity was limited to counting the number of storage bits needed and was not based on a detailed analysis of the actual hardware implementation.

In this paper, we start off from the initial CPI counter architecture designs proposed by Eyerman et al. [6], we implement the counter architectures in hardware, and we quantify their complexity in terms of performance, area and power consumption. Our results confirm the statement made by Eyerman et al. that hardware complexity is low, however, we found several opportunities for further optimizations. In fact, we reduce the complexity of the CPI counter architecture by reducing the amount of storage needed and by removing the need for Content-Addressable Memory (CAM) lookups. This reduces hardware cost by 44% and reduces power consumption by 47%, while achieving the same performance. Overall, the counter architecture requires no more than $0.03\ mm^2$ of chip area and consumes 5.8mW at 1GHz in a 90nm CMOS standard cell chip technology. We also evaluate the impact of the optimized counter architecture on the accuracy of the resulting CPI stacks. For a set of SPEC CPU benchmarks, we found the error to be within 3% on average. The overall conclusion is that the proposed CPI counter architecture is feasible to implement in hardware at low cost, low power consumption and high performance while yielding accurate and insightful CPI stacks.

2. PRIOR WORK

Although the basic idea of a CPI stack is simple, computing accurate CPI stacks on superscalar out-of-order processors is challenging because of parallel processing of independent operations and miss events. Eyerman et al. [6] proposed a CPI counter architecture, which, in contrast to prior practice, was designed in a top-down fashion from a mechanistic analytical performance model. This counter architecture was found to be accurate within 2.5% on average (4% max error) whereas prior practice led to average errors around 20% [12]. The key difficulty in designing a CPI counter architecture relates to computing frontend miss penalty cycles: whether a speculative path resolves to a correct path (and whether the penalty cycles need to be accounted for) is not known until a branch is executed on a functional unit in the backend of the pipeline. Hence, we need to keep track of the presumed miss penalty for each in-flight branch, and only account for the penalty if the branch was correctly predicted. Eyerman et al. proposed the FMT and sFMT designs to this end; we refer to [6] for a detailed description of the (s)FMT.

3. NOVEL COUNTER ARCHITECTURE

The (s)FMT structure is fairly complex: it involves storage to keep per-entry ROB IDs, timestamps, and local I-cache/TLB counters. In addition, it also involves Content-Addressable Memory (CAM) logic, i.e., an array of comparators, to find the entry corresponding to a particular mispredicted branch as branches may be executed out-of-order. CAM logic does not scale well and consumes considerable power. We now propose the FIFO-sFMT which reduces chip area overhead and power consumption substantially. The FIFO-sFMT is a FIFO queue that contains branch dispatch timestamps only. It does not store ROB IDs, nor does it involve CAM logic. Note that the FIFO-sFMT accounts for cache misses alike the original (s)FMT designs.

The FIFO-sFMT comes with head and tail pointers. The general mechanism is that, whenever a branch is dispatched, the timestamp (current cycle count) is recorded in the FIFO-sFMT entry pointed to by the tail pointer, after which the tail pointer is incremented. When a branch commits, the head pointer is incremented. The FIFO-sFMT also comes with a novel hardware structure, called the *branch miss handler*, which consists of a timestamp register, a branch mispredict bit and a FIFO pointer, see also Figure 1. The purpose of the branch miss handler is twofold: (1) discarding false-path branches that have already dispatched, and (2) calculating the branch misprediction penalty. The branch mispredict bit is set when a mispredicted branch is resolved, i.e., the branch is executed on a functional unit and the hardware figures out that the branch was mispredicted. When committing a mispredicted branch, we compute the branch misprediction penalty, which we describe next.

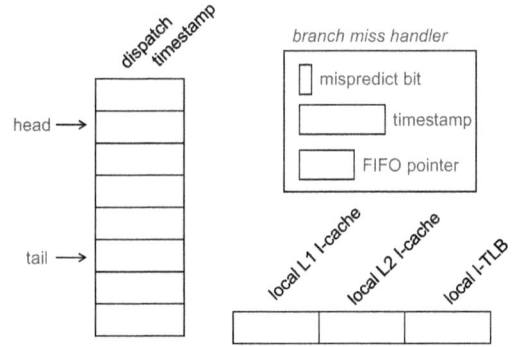

Figure 1: FIFO-sFMT with branch miss handler.

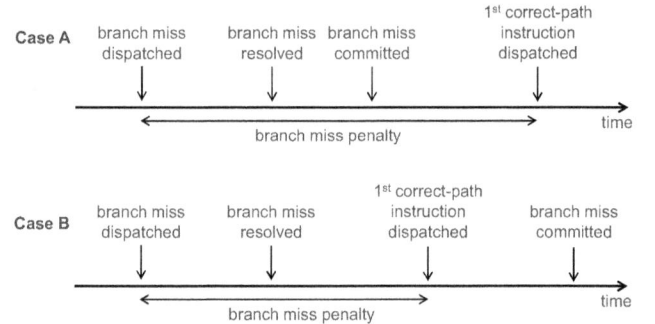

Figure 2: Two possible orderings for committing a mispredicted branch and dispatching correct-path instructions after branch resolution.

To describe the mechanism of the FIFO-sFMT, we consider two cases depending on the relative ordering of when the mispredicted branch commits versus when correct-path instructions are dispatched after the mispredicted branch was resolved.

Case A: Mispredicted branch commits before correct-path instructions are dispatched. The first case happens when the mispredicted branch is committed before new correct-path instructions are dispatched, see Figure 2, case A; this is the most frequent case according to our simulations. Upon the commit of the mispredicted branch, the FIFO-sFMT head pointer points to the dispatch time of that branch. We then store this timestamp value in the branch miss handler's timestamp register, and we set the tail pointer to the entry following the head pointer (thereby discarding the entries containing wrong-path branches). When the first correct-path instruction after the branch misprediction dispatches, the timestamp in the branch miss handler is subtracted from the current cycle count to compute the penalty of the mispredicted branch, and the mispredict bit in the branch miss handler is cleared.

Case B: Mispredicted branch commits after correct-path instructions are dispatched. The second case happens when correct-path instructions are dispatched before the mispredicted branch is committed, see Figure 2, case B. This case is detected if the timestamp in the branch miss handler is not set (because the branch miss is not yet committed) when the first correct-path instruction is dispatched and the branch mispredict bit is set. We then store the dispatch time of the first correct-path instruction in the timestamp of the branch miss handler. We also store the current FIFO-sFMT tail pointer in the branch miss handler's FIFO pointer. When the branch miss eventually commits, we compute the branch penalty by subtracting the timestamp in the FIFO-sFMT pointed to by the head pointer (i.e., dispatch time of the branch) from the timestamp

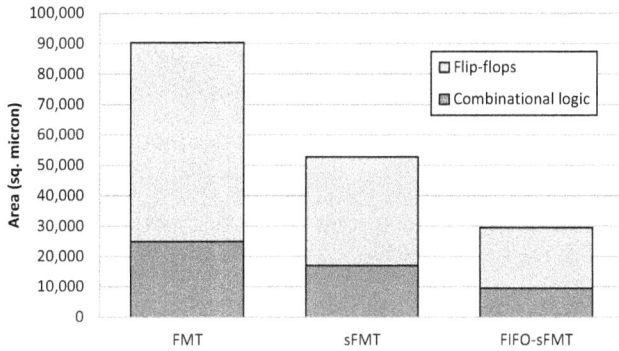

Figure 3: Area in square micron for the FMT, sFMT and FIFO-sFMT.

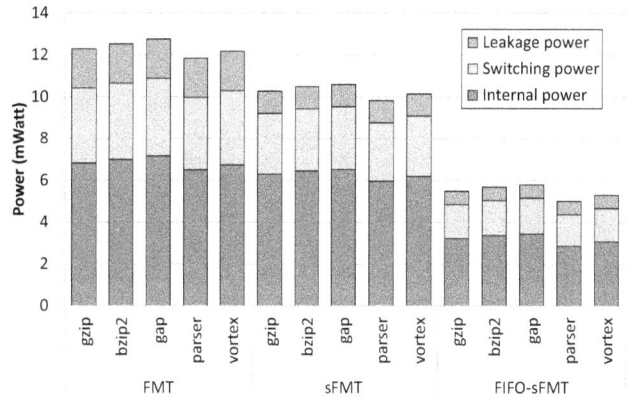

Figure 4: Estimated power consumption in mWatt for the FMT, sFMT and FIFO-sFMT.

in the branch miss handler (i.e., dispatch time of the first correct-path instruction after the mispredicted branch). After committing the branch miss, the head pointer needs to be moved to the first correct-path branch, discarding all wrong-path entries. This is done by copying the branch miss handler's FIFO pointer into the FIFO-sFMT tail pointer. Once the penalty is calculated, the mispredict bit in the branch miss handler is cleared, indicating that the branch miss penalty is accounted for.

4. EXPERIMENTAL SETUP

As part of the experimental evaluation, we implement the CPI counter architectures in hardware using VHDL. All experiments are based on standard cell logic synthesis of the counter architectures in isolation. We believe that integrating the model in a full-blown processor and performing physical synthesis would not affect our conclusions. This is due to the compactness of the design and its tolerance to global wire delays. We use a 90nm chip technology from STMicroelectronics — the most advanced technology we have access to; we assume a suppy voltage of 1V (with standard threshold voltage), and we quantify power consumption under typical operating conditions assuming a 1GHz clock frequency. Synthesis is done using Synopsys Design Compiler (version C-2009.06-SP3); RTL and GL simulation is done using Mentor Graphics ModelSim (version 6.3); and power estimation is done using Synopsys PrimeTime-PX (version D-2010.06-SP1).

As a second part of the evaluation, we also evaluate the CPI counter architecture's accuracy. This is done through microarchitecture analysis using detailed cycle-accurate simulation of a 4-wide out-of-order processor. We therefore use the SimpleScalar/Alpha v3.0 simulator [2] along with SPEC CPU2000 benchmarks.

5. RESULTS

We now evaluate the proposed CPI counter architecture through a detailed analysis of its hardware implementation. We subsequently evaluate the impact on accuracy of the obtained CPI stacks. We assume 64 entries for all three CPI counter architecture proposals.

5.1 Hardware implementation

We evaluate the CPI counter architecture in terms of area, power consumption and performance.

Area. Figure 3 quantifies chip area in square micron for the FMT, sFMT and FIFO-sFMT designs, and breaks down total chip area in two terms, namely combinatorial logic versus flip-flops. We observe that the sFMT reduces chip area by 41.6% compared to the FMT. The biggest gain comes from reducing the number of flip-flops in the design by getting rid of the local penalty I-cache/TLB

counters in the FMT. Combinational logic area savings are achieved by reducing the readout logic of the FMT as a result of reducing the bit width of its entries. The FIFO-sFMT reduces chip area by another 44% over the sFMT. Combinational logic is reduced by removing CAM logic (comparators). The number of flip-flops is reduced by getting rid of the ROB IDs. Overall, the FIFO-sFMT is slightly less than $0.03 \ mm^2$.

Note that these numbers assume a standard cell design using flip-flops. In case of a real processor design, one may use full-custom design: the FMT and sFMT would involve custom CAM-like structures, whereas the FIFO-sFMT would be designed using RAM cells. Clearly, the FIFO-sFMT would involve less design effort and less chip area than both the FMT and sFMT. Unfortunately, we could not study custom designs because we do not have access to RAM macros.

Power consumption. Figure 4 quantifies power consumption. We show a limited number of benchmarks only because of space constraints (we obtained similar results for the other benchmarks). The FIFO-sFMT reduces power consumption by 47% compared to the sFMT. The reduction comes primarily from removing power-hungry CAM logic, which results in a substantial reduction in dynamic power consumption. Leakage power is reduced due to the reduction in the amount of logic and flip-flops in the design. It is further interesting to note that the reduction in power consumption is larger for the FIFO-sFMT compared to the sFMT than for the sFMT versus the FMT. The reason is that the difference between the sFMT and FMT comes from a reduction in the amount of storage, not logic, which leads to a reduction in leakage power; dynamic power is affected less. In other words, we remove relatively inactive circuitry. The FIFO-sFMT on the other hand reduces both leakage and dynamic power over the sFMT due to, as mentioned above, reducing both the number of flip-flops and combinatorial (CAM) logic. The ROB IDs and comparators constitute active circuitry, which explains the large savings in dynamic power.

Performance. We also evaluated performance of the three CPI counter architectures. All three architectures achieved the same performance within 3.3%. The reason is that the critical path in the design is bounded by the number of entries in the structures, which is the same for all three designs.

5.2 Microarchitectural evaluation

Next to understanding the area, power and performance implications of the new CPI counter architecture, it is also important to evaluate the impact on accuracy for the obtained CPI stack. Figure 5 shows the maximum CPI component error across all of the

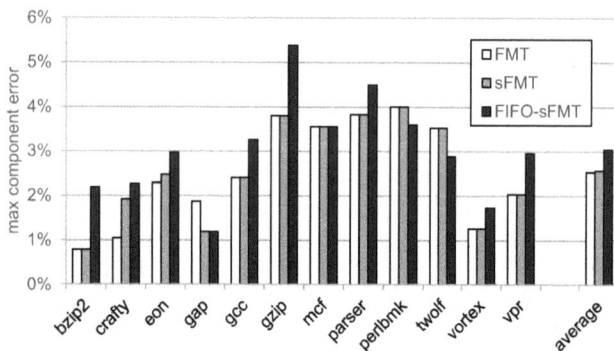

Figure 5: Maximum CPI component error for the FMT, sFMT and FIFO-sFMT.

benchmarks, for the FMT, sFMT and FIFO-sFMT designs. The errors are relative to a CPI stack obtained through detailed cycle-accurate simulation as described in [6]. The average error equals 2.5% and 2.6% for the FMT and sFMT, respectively, with a maximum error of 4%. The FIFO-sFMT has an average error of 3.0% and a maximum error of 5.4%. Although this is a slight decrease in accuracy, the error is still significantly lower than that of the other methods discussed in [6], with average errors around 20%.

There are two reasons for the slight increase in error. First, we assumed only one branch misprediction handler. In case there are multiple outstanding branch misses, we account the penalty of the last one only. Adding multiple branch misprediction handlers will reduce the error compared to the (s)FMT at the cost of involving more hardware. Second, if a branch misprediction is immediately followed by an I-cache miss (i.e., the first correct-path instruction causes an I-cache miss), the FIFO-sFMT will include the instruction cache miss penalty as part of the branch misprediction penalty. This can be solved by not incrementing the timer that generates timestamps if the first correct-path instruction after a branch misprediction results in an I-cache miss. According to our simulations, this would reduce the average error from 3.0% to 2.9%, hence, one may conclude this gain in accuracy does not justify the additional hardware needed.

6. RELATED WORK

A number of proposals have been made for computing CPI stacks for in-order architectures. For example, the Intel Itanium processor family provides a rich set of hardware performance counters for computing CPI stacks [8]. The Digital Continuous Profiling Infrastructure (DCPI) [1] is another example of a hardware performance monitoring tool for an in-order architecture. Computing CPI stacks for in-order architectures, however, is relatively simple compared to computing CPI stacks on out-of-order architectures.

The IBM POWER5 microprocessor was the first out-of-order microprocessor to implement a dedicated counter architecture for computing CPI stacks [12]. The IBM POWER5 approach, however, does not accurately compute the I-cache and I-TLB CPI components; nor does the IBM POWER5 accurately compute the branch misprediction penalty [6]. The Intel Pentium 4 [13] does not have a dedicated counter architecture for computing CPI stacks, but features a mechanism for obtaining non-speculative event counts, i.e., it does not count miss events along mispredicted control flow paths. Cycle accounting for the Intel Nehalem processor is described in [10]. Stall cycles are defined as cycles during which no instructions issue to functional units; further, accounting events to stall cycles is done using heuristics based on existing performance counters.

7. CONCLUSION

CPI stacks provide valuable and insightful performance information to software developers. Computing accurate CPI stacks on contemporary superscalar processors is non-trivial though because of various overlap effects. In this paper, we explored the hardware implementation cost of previously proposed CPI counter architectures and we proposed a new one, called the FIFO-sFMT, which reduces chip area and power consumption substantially by 44% and 47%, respectively, compared to the state of the art. The FIFO-sFMT removes the need for maintaining ROB IDs, thereby reducing chip area and leakage power, and it eliminates CAM logic, thereby reducing both chip area and dynamic power substantially. The FIFO-sFMT achieves these high savings in chip area and power consumption while maintaining high performance and accuracy.

Acknowledgements

Stijn Eyerman is supported through a postdoctoral fellowship by the Research Foundation–Flanders (FWO). Additional support is provided by the FWO projects G.0255.08 and G.0179.10, the UGent-BOF projects 01J14407 and 01Z04109, and the European Research Council under the European Community's Seventh Framework Programme (FP7/2007-2013) / ERC Grant agreement no. 259295.

8. REFERENCES

[1] J. M. Anderson, L. M. Berc, J. Dean, S. Ghemawat, M. R. Henzinger, S. A. Leung, R. L. Sites, M. T. Vandevoorde, C. A. Waldspurger, and W. E. Weihl. Continuous profiling: Where have all the cycles gone? *ACM Transactions on Computer Systems*, 15(4):357–390, Nov. 1997.

[2] T. Austin, E. Larson, and D. Ernst. SimpleScalar: An infrastructure for computer system modeling. *IEEE Computer*, 35(2):59–67, Feb. 2002.

[3] P. G. Emma. Understanding some simple processor-performance limits. *IBM Journal of Research and Development*, 41(3):215–232, May 1997.

[4] S. Eyerman and L. Eeckhout. A counter architecture for online DVFS profitability estimation. *IEEE Transactions on Computers*, 59(11):1576–1583, Dec. 2010.

[5] S. Eyerman and L. Eeckhout. Probabilistic job symbiosis modeling for SMT processor scheduling. In *ASPLOS*, pages 91–102, Mar. 2010.

[6] S. Eyerman, L. Eeckhout, T. Karkhanis, and J. E. Smith. A performance counter architecture for computing accurate CPI components. In *ASPLOS*, pages 175–184, Oct. 2006.

[7] S. Eyerman, L. Eeckhout, T. Karkhanis, and J. E. Smith. A mechanistic performance model for superscalar out-of-order processors. *ACM Transactions on Computer Systems (TOCS)*, 27(2), May 2009.

[8] Intel. *Intel Itanium 2 Processor Reference Manual for Software Development and Optimization*, May 2004. 251110-003.

[9] K. Keeton, D. A. Patterson, Y. Q. He, R. C. Raphael, and W. E. Baker. Performance characterization of a quad Pentium Pro SMP using OLTP workloads. In *ISCA*, pages 15–26, June 1998.

[10] D. Levinthal. *Performance Analysis Guide for Intel Core i7 Processor and Intel Xeon 5500 Processors*. Intel, 2009. http://software.intel.com/sites/products/collateral/hpc/vtune/performance_analysis_guide.pdf.

[11] Y. Luo, J. Rubio, L. K. John, P. Seshadri, and A. Mericas. Benchmarking internet servers on superscalar machines. *IEEE Computer*, 36(2):34–40, Feb. 2003.

[12] A. Mericas. Performance monitoring on the POWER5 microprocessor. In L. K. John and L. Eeckhout, editors, *Performance Evaluation and Benchmarking*, pages 247–266. CRC Press, 2006.

[13] B. Sprunt. Pentium 4 performance-monitoring features. *IEEE Micro*, 22(4):72–82, July 2002.

[14] M. Zagha, B. Larson, S. Turner, and M. Itzkowitz. Performance analysis using the MIPS R10000 performance counters. In *Supercomputing*, Nov. 1996.

An Optimized Multicore Cache Coherence Design for Exploiting Communication Locality

Libo Huang
School of Computer
National University of Defense
Technology
Changsha 410073, China
libohuang@nudt.edu.cn

Zhiying Wang
School of Computer
National University of Defense
Technology
Changsha 410073, China
zywang@nudt.edu.cn

Nong Xiao
School of Computer
National University of Defense
Technology
Changsha 410073, China
nongxiao@nudt.edu.cn

ABSTRACT

Supporting cache coherence in current multicore processor still faces scalability and performance problems. This paper presents an optimized cache coherence design targeting at NoC-based multicore processors. It tries to achieve the best characteristics both of the snooping and of the directory-based protocols. With the observation of network traffic locality, we design a cache coherence that aims at local and remote access separately. At the first level, snooping is achieved within a cache group and at the second level of the protocol, the coarse directories provide the caches with information about which processors must be involved in first level snooping. To support efficient coherence broadcasting, we also propose a low latency, broadcast-enabled underlying NoC design. It incorporates light weight buses into NoCs, where the snooping protocol can be performed in a broadcast fashion. Extensive experimental results demonstrate that the proposed coherence design can achieve low complexity and high performance goals.

Categories and Subject Descriptors

B.4.3 [**Input/Output and Data Communications**]: Interconnections(Subsystems)—*topology*

General Terms

Design

Keywords

network on chip, cache coherence, bus, directory

1. INTRODUCTION

One of the big challenges in multicore processors is the cache coherence problem as it was experienced in earlier multiprocessor systems [1]. Although a great deal of attention was devoted to scalable cache coherence in the contex-

t of shared-memory multiprocessors, the technological constrains entailed by multicore processors demand new solutions to the cache coherence problem [2]. A typical future multicore processor [3] could be designed as 2D array of processor cores and routers that connect all cores through a tightly integrated network-on-chip (NoC). Such architectures include two levels of on-chip caches. The L1 cache is private to its local core and the L2 cache is logically shared but physically distributed among the cores.

There are two major mechanisms used to ensure coherence among L1 caches: snooping protocol and directory protocol [4]. In NoC-based processor, hammer protocol [5] is a type of snooping. For scalability concerns, more scalable directory-based coherence protocols instead of snooping cache coherence protocols will be likely used in large-scale multicore processor [6]. Many studies have been carried out to improve directory protocol in terms of area and performance for on-chip or off-chip multiprocessors such as directory cache [7], virtual hierarchies [8], and in-network cache coherence [9]. These solutions try to optimize the protocol from either the state management or network design aspects. We think that the coherence optimization through the combination of the cache management and network design can achieve better performance-area trade-off of cache coherence. A good example guided by such concept is interconnect-aware coherence protocol [2]. It proposes an interconnect composed of wires with varying latency, bandwidth, and energy characteristics, and advocates intelligently mapping coherence operations to the appropriate wires.

The aim of this work is to address the performance-area trade-off of cache coherent access mechanism for multicore processors. We advocate an optimized coherence design exploiting data communication locality which includes first level snooping and second level directory protocols. Regarding the global coherence are rarely happen [8], it can also reduce the latency of local memory access by fast snooping. To minimize the processing delay overhead of broadcasting, we designed a specialized NoC, incorporating row/column buses into NoCs. Extensive experimental results demonstrate that the proposed coherence design can achieve low hardware complexity and high performance goals.

2. RELATED WORKS

A variety of protocols have been proposed to achieve best performance and area trade-off on both on-chip or off-chip interconnects. For the multiprocessor systems, bandwidth adaptive snooping [4] was proposed. Directory cache is a effective way to cut down directory memory overhead, that has been originally proposed in [7]. Virtual hierarchies [8] propose cache coherence that utilizes two levels of directories to provide fast local coherence and correct(and substantially slower) global coherence. This work has some similarities to our work. However, we examine a different protocol hierarchy than what is proposed in virtual hierarchies. Furthermore, the underlying NoC for hierarchical coherence is specially designed. The Group-caching approach [10] organizes on-chip L2 banks into multiple groups. Each cache group behaves like a shared L2 cache for the cores inside cache group while the cache coherence between cache groups is maintained by coherence messages.

Another research direction that related to our work is the study of in-network cache coherence [9]. It embeds tree caches within each router node that manages and steers requests towards nearby data copies. Because each intermediate node needs storage for the virtual tree cache bits, storage overhead is very high. Virtual tree coherence [11] is proposed, which keeps track of sharers of a coarse-grained region, and multicasts requests to them through a virtual tree, employing properties of the virtual tree to enforce ordering amongst coherence requests. The interconnect-aware coherence protocols [2] combines efforts on both network design and cache management to improve the memory performance.

We think that two key considerations for a coherence protocol are the storage of state information and the underlying network design. Three strategies could be used to facilitate the cache coherence design. Firstly, locality of network traffic exists for parallel applications. The work in [12] has demonstrated that the network traffic behavior conforms to the Rent's rule: a power-law relationship is clearly visible when partitioning network nodes and comparing their size versus their external bandwidth. This observation indirectly shows that the cases for global coherence are rare. So the local and global coherence can be optimized separately. Secondly, if we only track the sharers in cache groups, the state information in directory could be less precise, which can reduce the storage overhead. Finally, the underlying NoC design should provide some special support for such kind of cache protocol. Guided by these strategies, we will design a new coherence to make a good trade-off in terms of performance and area in the following section.

3. OPTIMIZED COHERENCE DESIGN

Our main idea is to employ a hybrid and hierarchical cache coherence design for exploiting communication locality. At the first level, snooping is achieved within a region and with respect to a local L2 cache level. At the second level of the protocol, the coarse directories provide the caches with information about which processors must be involved in first level snooping. Since the required information in home node become less, the implementation complexity of directories can be decreased accordingly. Furthermore, previous studies have demonstrates that most communications happen between cores in local region [12]. Using snooping to maintain the local coherence could result in high performance.

The proposed coherence design utilizes broadcasting as a local mechanism to send request to other node within local region. However, as current NoC designs do not support broadcast (due to the high hardware overhead of providing such support), the broadcast functionality has to be synthe-

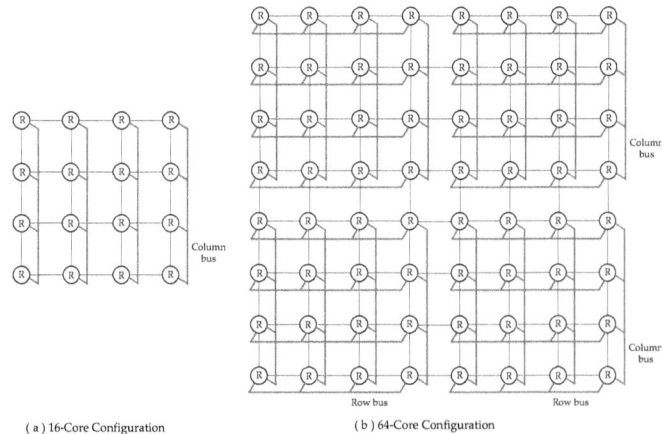

(a) 16-Core Configuration   (b) 64-Core Configuration

Figure 1: Underlying interconnect structure

sized by sending multiple point-to-point messages destined for multiple nodes. This creates high bandwidth overheads which translate to high interconnect delay and power, reducing the attractiveness of such hierarchical coherence. Existing solutions such as hybrid BENoC design [13] adds a global bus as a low latency broadcast media. But this global bus is $O(N)$ complexity and is hard to implement. So in this paper, we also present a lightweight broadcast network design, that incorporates multiple row/column buses into existing NoCs and they can be implemented within the tight delay, area budgets of NoCs. Such topology will make the proposed hierarchical coherence protocols efficient for multicore processor designs.

3.1 Underlying Network Design

The row/column buses are introduced as a simple and efficient augment for conventional NoC, offering low latency for broadcast communication services in cache coherence. Unlike NoC routers, these buses can transmit data to multiple destinations in a broadcast way. The resulting concept architecture for NoC extended with row/column buses is presented in Figure 1. The row/column bus severs as a light weight broadcast media which is inexpensive than conventional system buses in terms of area, power and system complexity.

We organize the buses in a row/column way, whose goal is to reduce the latency of physical layout for the bus organization.If all the routers are connected through one global bus, then its length is $O(N)$ complexity. To reduce this physical length of bus, *redundant bus* method is introduced, which uses several buses instead of one to constrain the longest distant of the bus in a bearable bound. Each router is connected to one row bus and one column bus. Compared with one transaction in the global bus [14], two transactions are needed for communications between different row and column buses. But its length increases at a much less rate $(O(\sqrt{N}))$. This is important for the physical layout of bus organization. Furthermore, since every row or column bus has its own access control, multiple independent bus transactions can coexist at the same time.

Figure 1(a) and (b) illustrate the topologies using buses under 16-core and 64-core configurations respectively. For 16-core configuration, only column bus is used. For 64-core configuration, the row and column buses are both used. We divide the 64-core organization into four groups. In such way, for each 16-core group, the NoC should broadcast request to all its cores. Fortunately, due to the broadcast nature of bus communication, we can implement it efficiently in each group by first broadcasting requests through one row bus and then broadcasting the received requests through four column buses.

3.2 Optimized Coherence Protocol

Based on the underlying NoC design, the proposed optimized protocol includes first level bus snooping and second level directory protocols. Bus snooping adopts conventional MESI write-back invalidation protocol [15], which utilizes column bus as the ordering point for requests. This requires additional control signals in bus for transfer the snooping results such as the information of sharers' state, correspond message and transaction termination. When a core stores data, the cache must get the exclusivity of the targeted block. This request will invalidate all other copies, retrieve dirty blocks from foreign caches and allocate the clean block if needed. These actions can take a long time on a high latency NoC, and this is the major drawback of this protocol. For the support of bus communication, this drawback can be alleviated.

The second level directory protocol requires a cache coherence protocol similar to conventional directory [16] with significant simplification. We divide the cache organization into four groups. In such way, the conventional bit-vector sharing table employed for keeping track of the possible 16 sharers can be decreased to 4 entries, thus reduce the directory complexity. The bit-vector table allows the protocol to send invalidation messages just to the caches currently sharing the data. However, for each group, the proposed protocol require broadcasting request to all its cores. Due to the broadcasting nature of row/column bus, we can implement it efficiently in proposed underlying NoC.

Some optimizations are done to combine these two level coherence protocols together more efficiently. We use the term local access to refer to the L1 cache access that belongs to group for one column bus. On the other hand, remote access imply that the home node is not in the group that L1 cache access occurs. For local access whose data are not shared by other groups, the snooping can be proceeds as usual. For remote access or local access whose data are also shared by other groups, the directory controller obtains directory information stored in cache tags, and then, the miss proceeds as usual. So in each L2 directory, we add one bit, *local*, to indicates whether there are data copies in other L1 cache groups. Each time the L1 cache read miss or L1 write happens, the L1 cache controller would first access the corresponding *local* bit and do corresponding procedure. If *local* is enabled, only bus snooping is evolved, otherwise, the directory protocol is performed.

4. EVALUATION

To evaluate the proposed coherence, we have implemented its hardware architecture and the light weight coherence protocol using a System-C based cycle-accurate Cache-NoC simulator. The simulator models a detailed pipeline structure for NoC router. We can change various network configurations, such as network size, topology, buffer size, routing algorithm, and traffic pattern. Since we are considering the performance of cache hierarchy for shared memory multicore, the memory access traces for each core are gathered

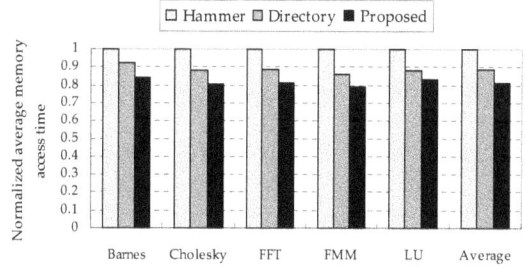

Figure 2: Average memory access time comparison results for various coherence protocols under 16-core

using the M5 full-system simulator [17] running corresponding number of thread. The benchmarks used in our simulations are chosen from the SPLASH-2 benchmark suite [18], which represent a variety of important scientific computations with different communication requirements. The input sizes are chosen to capture realistic machine behavior for these highly scalable shared memory programs.

We compare the characteristics of the proposed coherence design against conventional coherence protocol. Particularly, hammer protocol [5] avoids keeping coherence information at the cost of broadcasting requests to all cores; directory stores precise information about the private caches holding memory blocks in a directory structure. We only take the storage and network overhead into account for area comparison. CACTI [19] for 90nm technology is used to measure the coherence storage area needed in each evaluated protocol.

4.1 Impact on Memory Access Time

We use the average memory access time as the performance evaluation measure for various protocols. Figure 2 shows their comparison results for the applications evaluated in this work normalized with respect to hammer protocol. For all the applications, the hammer protocol has the lowest performance due to the generated massive network traffic. The directory protocol outperform than hammer protocol. For 4×4 mesh configuration, the directory protocol can improve the average memory access time from about 8% for *Barnes* to 14% for *FMM* (about 11% on average).

The proposed coherence achieves the best performance among protocols tested. It reduce the average memory access time when compared to hammer protocol (about 19% performance improvement on average). It also has performance advantage over directory protocol (about 7% performance improvement on average). These performance results demonstrate that the directory information for sharing group in proposed coherence design can be good enough for invalidation due to the traffic locality common in parallel applications.

4.2 Impact on Scalability

For scalability analysis, we also provide the performance results for the 64-core configuration. Figure 3 illustrates their comparison results for the applications evaluated in this work normalized with respect to hammer protocol under 64-core configuration. The performance improvement of 64-core configuration over directory protocol is more obvious than 16-core configuration, which is 9% on average. This is because when the network size increases, the network traf-

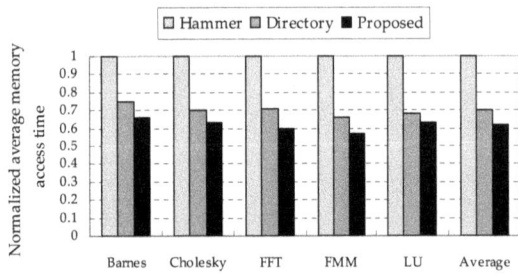

Figure 3: Average memory access time comparison results for various coherence protocols under 64-core

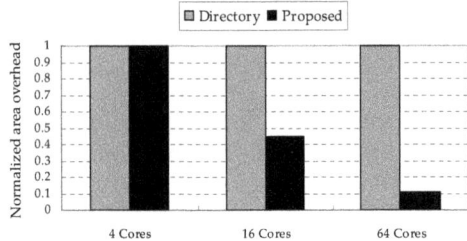

Figure 4: Area comparison of coherence storage and NoC overhead for directory and proposed protocols

fic will become more significant for hammer protocol, thus degrades the performance. Another reason is that more locality can be exploited by the proposed coherence design.

4.3 Impact on Hardware Cost

To evaluate the overhead of bus design, we implemented the conventional NoC design and row/column buses in Verilog and performed logic synthesis by using the Synopsys Design Compiler to get the area information under TSMC 90nm CMOS technology. From the implementation result, about 9.3% area overhead is observed. This is due to the added logic such as control logic and data path for row/column buses.

Figure 4 shows the area comparison results of coherence overhead between directory and proposed protocols, varying the number of cores from 4 to 64. The hammer protocol is not included in the figure because it does not theoretically require additional coherence information, while this is the focus of our comparison. The proposed protocol has group directory structure and the row/column bus overhead, which are both included in the coherence overhead in the figure. From the figure, we can see that the proposed protocol reduce the area overhead by about 45% for 16 cores and 12% for 64 cores mesh. This is significant area reduction for the proposed protocol, which demonstrates that a small overhead in the number of required bits results in a major overhead.

5. CONCLUSION

This work proposes an optimized cache coherence design for exploiting communication locality in NoC-based multicore processors. It tries to achieve the best characteristics both of the snooping and of the directory-based protocols. Our proposal is based on the special designed NoC hardware, which incorporates row/column buses into NoCs, achieving efficient data transfer for coherence broadcasting. The e-

valuation results demonstrate that the proposed design can achieve low complexity and high performance goals.

6. ACKNOWLEDGMENTS

We thank the reviewers for helpful feedbacks. This work is supported in part by 863 Program of China (2012AA010905), and NSFC (61070037, 61025009, 61103016).

7. REFERENCES

[1] M. Tomasevic and V. Milutinovic, *The Cache-Coherence Problem in Shared-Memory Multiprocessors: Hardware Solutions*, 1994.

[2] L. Cheng, N. Muralimanohar, K. Ramani, R. Balasubramonian, and J. B. Carter, "Interconnect-aware coherence protocols for chip multiprocessors," in *Proceedings of the 33rd annual international symposium on Computer Architecture*, ser. ISCA '06, 2006, pp. 339–351.

[3] D. Wentzlaff, P. Griffin, H. Hoffmann, L. Bao, B. Edwards, C. Ramey, M. Mattina, C.-C. Miao, J. F. Brown III, and A. Agarwal, "On-chip interconnection architecture of the tile processor," *IEEE Micro*, vol. 27, no. 5, pp. 15–31, 2007.

[4] M. M. K. Martin, D. J. Sorin, M. D. Hill, and D. A. Wood, "Bandwidth adaptive snooping," in *Proceedings of the 8th International Symposium on High-Performance Computer Architecture*, ser. HPCA '02, 2002, pp. 251–262.

[5] J. Owner, M. Hummel, D. Meyer, and J. Keller, "System and method of maintaining coherency in a distributed communication system," June 2006.

[6] R. T. Simoni, Jr., "Cache coherence directories for scalable multiprocessors," Ph.D. dissertation, 1992, uMI Order No. GAX93-02312.

[7] A. Gupta, W. D. Weber, and T. Mowry, "Reducing memory and traffic requirements for scalable directory-based cache coherence schemes," in *Proceedings of International Conference on Parallel Processing*, 1990, pp. 312–321.

[8] M. R. Marty and M. D. Hill, "Virtual hierarchies to support server consolidation," in *Proceedings of the 34th annual international symposium on Computer architecture*, ser. ISCA '07, 2007, pp. 46–56.

[9] N. Eisley, L.-S. Peh, and L. Shang, "In-network cache coherence," in *Proceedings of the 39th Annual IEEE/ACM International Symposium on Microarchitecture*, ser. MICRO 39, 2006, pp. 321–332.

[10] W. Zuo, S. Feng, Z. Qi, J. Weixing, L. Jiaxin, D. Ning, X. Licheng, T. Yuan, and Q. Baojun, "Group-caching for noc based multicore cache coherent systems," in *Proceedings of the Conference on Design, Automation and Test in Europe*, ser. DATE '09, 2009, pp. 755–760.

[11] N. D. Enright Jerger, L.-S. Peh, and M. H. Lipasti, "Virtual tree coherence: Leveraging regions and in-network multicast trees for scalable cache coherence," in *Proceedings of the 41st annual IEEE/ACM International Symposium on Microarchitecture*, ser. MICRO 41, 2008, pp. 35–46.

[12] W. Heirman, J. Dambre, D. Stroobandt, and J. Van Campenhout, "Rent's rule and parallel programs: characterizing network traffic behavior," in *Proceedings of the 2008 international workshop on System level interconnect prediction*, ser. SLIP '08, 2008, pp. 87–94.

[13] I. Walter, I. Cidon, and A. Kolodny, "Benoc: A bus-enhanced network on-chip for a power efficient cmp," *IEEE Comput. Archit. Lett.*, vol. 7, pp. 61–64, July 2008.

[14] S. I. Association, "International technology roadmap for semiconductors," in *http://www.itrs.net*, 2008.

[15] M. R. Marty, J. D. Bingham, M. D. Hill, A. J. Hu, M. M. K. Martin, and D. A. Wood, "Improving multiple-cmp systems using token coherence," in *Proceedings of the 11th International Symposium on High-Performance Computer Architecture*, 2005, pp. 328–339.

[16] J. Laudon and D. Lenoski, "The sgi origin: a ccnuma highly scalable server," in *Proceedings of the 24th annual international symposium on Computer architecture*, ser. ISCA '97, 1997, pp. 241–251.

[17] N. L. Binkert, R. G. Dreslinski, L. R. Hsu, K. T. Lim, A. G. Saidi, and S. K. Reinhardt, "The m5 simulator: Modeling networked systems," *IEEE Micro*, vol. 26, pp. 52–60, July 2006.

[18] SPLASH, "http://www-flash.stanford.edu/apps/."

[19] CACTI, "www.hpl.hp.com/research/cacti/."

Parallel Pipelined FFT Architectures with Reduced Number of Delays

Manohar Ayinala
University of Minnesota, Twin Cities
200, Union Street SE,
Minneapolis, MN 55455
ayina004@umn.edu

Keshab K. Parhi
University of Minnesota, Twin Cities
200, Union Street SE,
Minneapolis, MN 55455
parhi@umn.edu

ABSTRACT

This paper presents a novel approach to design four and eight parallel pipelined fast Fourier transform (FFT) architectures using folding transformation. The approach is based on use of decimation in time algorithms which reduce the number of delay elements by 33% compared to the decimation in frequency based designs. The number of delay elements required for an N-point FFT architecture is $N - 4$ which is comparable to that of delay feedback schemes. The number of complex adders required is only 50% of those in the delay feedback designs. The proposed approach can be extended to any radix-2^n based FFT algorithms. The proposed architectures are feed-forward designs and can be pipelined by more stages to increase the throughput. Further, a novel four parallel 128-point FFT architecture is derived using the proposed approach. It is shown that a radix-2^4 4-parallel 128-point design requires 124 delay elements, 28 complex adders, and four full complex multipliers.

Categories and Subject Descriptors

B.5 [**Hardware**]: Register-Transfer-level implementation; B.5.1 [**Design**]: Styles (e.g., parallel, pipeline)

General Terms

Design

Keywords

Fast Fourier transform (FFT), Decimation-in-Time (DIT), radix-2^4, folding, four parallel, eight parallel

1. INTRODUCTION

Fast Fourier transform (FFT) is one of the widely used algorithms in digital signal processing and has been of interest for many years. FFT plays an important role in different fields such as communication systems, biomedical applications, sensor and radar signal processing. FFT is a critical component in modern digital communication systems

such as digital video broadcasting (DVB), ultra wideband systems (UWB), and orthogonal frequency division multiplexing (OFDM) based systems. FFT and IFFT are core functions in such multi-carrier modulation based transmission systems [1]. Further, FFT is used for frequency domain beamforming, source tracking in the area of sensor signal processing [2] and for analyzing biomedical signals such as electrocardiogram (ECG) and electroencephalogram (EEG) in frequency domain [3].

Various algorithms have been proposed following the essential idea of decimation either in frequency or in time domain. Many parallel pipelined FFT architectures have been proposed in the literature [4] - [11] based on these algorithms. In the recent literature, four parallel designs have been proposed to increase the throughput of the FFT processors for OFDM systems [6] - [10] for a fixed 128-point FFT. These architectures are based on radix-2^3 and radix-2^4 algorithms and mixed radix (e.g., radix-2 and radix-8) algorithms. A 128-point four parallel multi-path delay feedback (MDF) architecture has been proposed in [6] based on radix-2 and radix-8 algorithms (mixed radix) combining the features of single delay feedback (SDF) and multi-path delay commutator (MDC). Another 4-parallel 128-point radix-2^4 FFT architecture based on MDF has been proposed in [7]. Further, an 8-parallel 2048-point FFT architecture has been proposed in [11] based on radix-2^3 and radix-2^4 algorithms. These architectures are based on different approaches which have their own advantages and disadvantages. The required number of delay elements in MDF architectures is less due to the delay feedback (SDF) design while the number of datapaths (consisting of butterflies) is equal to the level of parallelism. The hardware is under utilized in these architectures.

In [8], low complexity parallel architectures have been proposed and it has been shown that the hardware complexity is same in 2-parallel architectures based on decimation in frequency (DIF) and decimation in time (DIT) algorithms. Further, it was assumed that the hardware complexity will remain the same in higher parallel architectures based on DIF and DIT algorithms. Additional research reveals that this assumption is not correct when the level of parallelism is 4 or higher. In this paper, we propose novel four and eight parallel FFT architectures which can reduce the number of delay elements with 100% hardware utilization. The proposed architectures will reduce the number of delay elements compared to prior DIF based designs in [8] by 33%. Further, we present a novel four parallel 128-point FFT architecture based on the radix-2^4 algorithm.

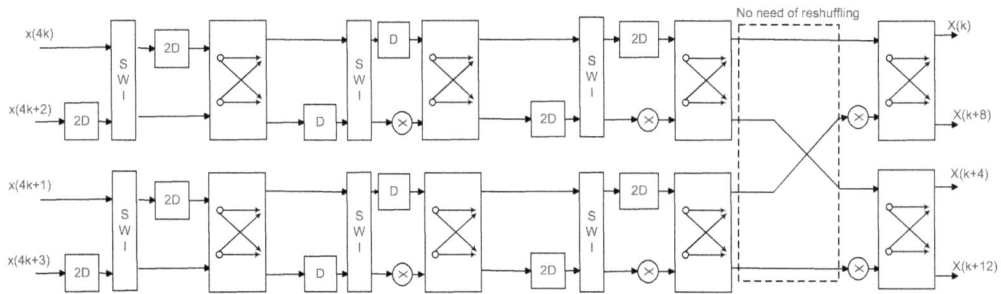

Figure 2: Proposed 4-parallel architecture for 16-point radix-2 FFT.

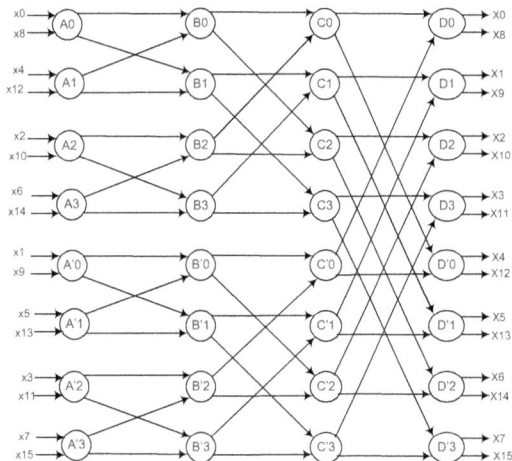

Figure 1: Data flow graph of radix-2 16-point DIT FFT

2. PROPOSED FFT ARCHITECTURE

The proposed approach to design four and eight parallel architectures is presented in this section. The approach is applicable to any radix-2^n algorithm. The proposed design is based on decimation-in-time (DIT) flow graph to reduce the number of delays in the pipelined architecture.

2.1 4-parallel design

The proposed method can be explained using two different design approaches: *adhoc* and folding approach. Further, the same method can be extended to design an eight-parallel architecture. An example of 16-point radix-2 FFT algorithm is used to describe the proposed method.

2.1.1 Ad-hoc design

Fig. 1 shows the flow graph of 16-point radix-2 DIT algorithm, where the nodes from $A0, ..., A3$ represent the top 4 butterflies (processing even samples) in the first stage of the FFT and $A'0, ..., A'3$ represent the bottom 4 butterflies (processing odd samples). Similarly, $B0, ...B4, B'0, ..., B'4$, $C0, ..., C3$, $C'0, ..., C'3$, and $D0, ..., D3, D'0, ..., D'3$ represent nodes in second, third and fourth stages, respectively. It can be observed that the even and odd samples can be processed independently until the final stage which is illustrated in Fig. 1. The outputs of the two N/2 FFTs can be combined in the final butterfly stage. We can develop a two parallel architecture for each of the N/2-point FFTs which can process two consecutive even $(4k, 4k+2)$ and odd

$(4k+1, 4k+3)$ samples, respectively. Further, the initial reordering delay elements are not required, if the input buffer is available to reorder the data before FFT processor. As the outputs of the two N/2-point FFTs arrive at the same time, the final stage does not require reshuffling circuit (delays and switches). This will reduce the required number of delays to implement the pipelined architecture.

The four parallel pipelined architecture for 16-point FFT is shown in Fig. 2. We can observe that there is no reshuffling circuit before the last stage of butterflies. The implementation uses regular radix-2 butterflies. The rest of the datapath contains switches and delay elements. The function of the switch and delay elements is to reorder the incoming samples to provide the corresponding samples at the input of each butterfly stage according to the data flow during every clock cycle. The control signal controls the multiplexers which connects the input and output of the switch in two different ways (either straight or cross paths). The control signals for switches in different stages of the architecture can be generated by using simple counter logic.

2.1.2 Folding approach

The proposed approach can also be described using folding methodology [12]. The four parallel architecture can be derived using the following folding sets.

$$A = \{A0, A1, A2, A3\} \qquad A' = \{A'0, A'1, A'2, A'3\}$$
$$B = \{B3, B0, B1, B2\} \qquad B' = \{B'3, B'0, B'1, B'2\}$$
$$C = \{C1, C2, C3, C0\} \qquad C' = \{C'1, C'2, C'3, C'0\}$$
$$D = \{D1, D2, D3, D0\} \qquad D' = \{D'1, D'2, D'3, D'1\}$$

We can derive the folded architecture by writing the folding equation [12] for the edges in the flow graph. The register minimization techniques and the forward and backward register allocation scheme [13] are applied to derive the architecture. The final architecture can be derived to be same shown as the design in Fig. 2.

In general, for an N-point FFT, the proposed four parallel architecture requires $N-4$ delay elements. An additional $N/2$ delay elements are required when the input buffer is not in place to reorder the input samples. The proposed architecture reduces the number of delay elements by 33% compared to the design in [8], which requires $3N/2-4$ delay elements. Further, the proposed architecture has only two parallel data paths (each processing two samples) compared to four data paths in [6], [7]. The proposed architecture requires $2log_2N$ butterfly units compared to $4log_2N$ in the prior designs. That is, the proposed design reduces the num-

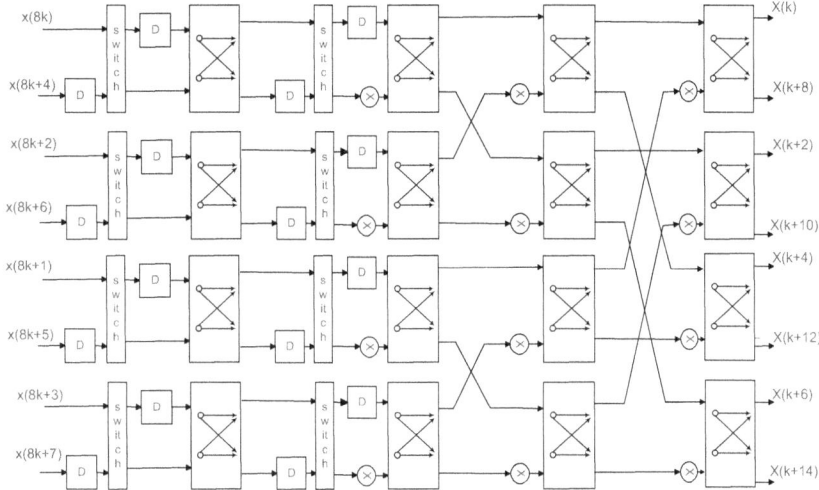

Figure 3: Proposed 8-parallel architecture for 16-point radix-2 FFT.

ber of complex adders by 50% compared to the designs in [6] and [7].

2.2 8-parallel design

In a similar fashion, an 8-parallel architecture can be derived based on the DIT flow graph. Now the input samples can be divided into four sections, $8k, 8k + 4$, $8k + 2, 8k + 6$, $8k + 1, 8k + 5$, and $8k + 3, 8k + 7$. These sample groups can be processed independently until the last two stages. The 8-parallel architecture can be derived using folding approach. Fig. 3 shows the 8-parallel architecture for a 16-point radix-2 FFT. In general, for an N-point FFT, the proposed 8-parallel architecture requires $N - 8$ delay elements excluding the input buffers and $4log_2N$ butterfly units. The 8-parallel architecture in [11] consists of 8 datapaths which leads to $8log_2N$ butterfly units.

3. PROPOSED 128-PT FFT ARCHITECTURE

The number of complex multipliers required depends on the underlying FFT algorithm. In higher radix algorithms (radix-2^3, radix-2^4), only the twiddle factors will change at each input/output stage of the butterflies. For example, in radix-2^2 algorithm twiddle factor multiplication is required in every alternate stage while it is required in every stage in radix-2 algorithm. To demonstrate that the proposed approach can be extended to any radix-2^n algorithms, we present a 4-parallel 128-point FFT architecture based on the radix-2^4 algorithm.

The 128-point FFT flow graph is based on radix-2^4 algorithm which is decimated in time. The higher radix algorithms lead to low multiplication complexity. For example, radix-2^4 algorithm leads to only one complex multiplication every 4 stages and radix-2^3 algorithm leads to one complex multiplication every 3 stages. But constant multipliers are required to implement trivial twiddle factors at the intermediate stages. The proposed 128-point FFT architecture is based on the radix-2^4 algorithm to reduce the number of constant multipliers. The radix-2^4 algorithms are described in detail in [5].

To achieve the high throughput requirement with low hardware cost, both the proposed pipelining method and radix-

2^n algorithms are exploited in this design. The proposed 4-parallel 128-point FFT architecture is shown in Fig. 4. It consists of two parallel data paths processing two input samples. Each data path consists of seven butterfly units, four constant and two full complex multipliers, delay elements and multiplexers. The function of delay elements and switches is to store and reorder the input data until the other available data is received for the butterfly operation. The four output data values generated after the first stage are multiplied by constant twiddle factors ($W_8^1 = e^{-j2\pi/8}, W_8^3 = e^{-j2\pi3/8}$). These twiddle factors can be implemented efficiently using canonic signed digit (CSD) approach. The outputs after the third stage are multiplied by the nontrivial twiddle factor. Another constant multiplier stage is required before the sixth butterfly stage. The CSD complex constant multiplier processes the multiplication of twiddle factors $W^8, W^{16}, W^{24}, W^{48}$. These twiddle factors correspond to $cos(\pi/8), sin(\pi/8)$, and $cos(\pi/4)$.

4. COMPARISONS AND ANALYSIS

The proposed 4-parallel and 8-parallel architectures require $3N/2 - 4$ and $3N/2 - 8$ delay elements, respectively for an N-point FFT. The prior 4-parallel architecture based on DIF algorithm requires $2N - 4$ delay elements [8]. This number does not depend on the radix of the algorithm, i.e., it will remain same for all radix-2^n algorithms. In general, delay feedback based architectures require $N - 4$ delay elements compared to $3N/2 - 4$ in the proposed architecture. Even though there is a difference of $N/2$ delay elements, these are required at the input stage to reorder the samples according to the flow graph. These delays are not required if we can reorder the samples before feeding them to the FFT processor (in general applications, memory will be used to store the input samples and we can read the samples in the required order instead of natural order). Further, the proposed architecture requires $4log_2N$ complex adders compared to the $8log_2N - 8$ complex adders in delay feedback designs. The proposed architectures, however, can be pipelined at any level since these do not contain feedback loops. In contrast, the delay feedback architectures cannot be pipelined at any arbitrary level. The hardware cost of

65

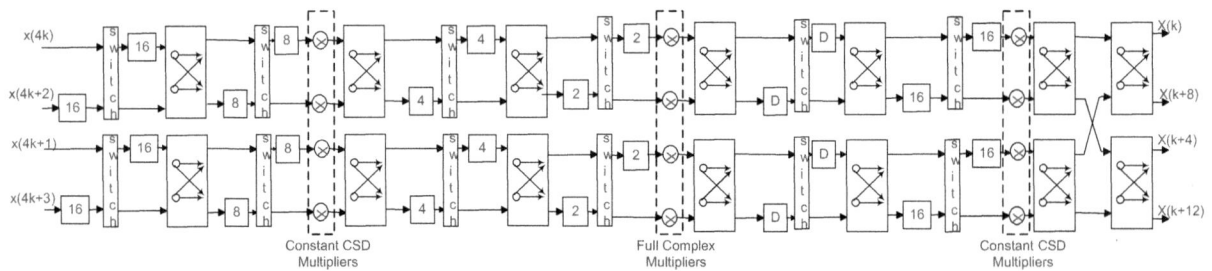

Figure 4: Block diagram of the proposed 4-parallel 128-point FFT architecture based on radix-2^4 algorithm

the proposed 128-point FFT architecture is summarized as follows:

- number of complex multipliers: 4+0.41, where the complexity of the constant multipliers is only 41% of the complex multiplier [7]

- number of complex adders: 28

- number of complex registers: 124, additional 64 registers are needed if an input buffer is needed.

Table 1: Comparison of pipelined architectures for the computation of 128-point FFT

	# C.M.	# REG	# C.A.	TP
Proposed	4+0.41	124	28	4
[8]	4+0.41	190	28	4
[6]	2+4*0.62	124	48	4
[7]	4+0.41	252	52	4
[9]	7	220	48	4
[10]	2+4*0.62	148	42	4

Table 1 compares the hardware requirement, FFT algorithm, and throughput rate of the proposed and prior architectures for computing 128-point FFT. The hardware complexity is measured in terms of required number of complex multipliers (C.M.), adders (C.A.), and delay elements (REG) and throughput (TP). We can observe that the number of adders required in the proposed design is only 50% of those in the MDF architectures [6], [7]. The number of complex multipliers is almost the same compared to the previous designs as this depends on the underlying algorithm. The number of delay elements required in the proposed design is only 65% of those DIF based designs in [8]. When compared to the design in [6] including the input buffers, the proposed design requires extra 64 delay elements while the former design [6] requires another 20 complex adders. The cost of these two additional requirements will be comparable. The throughput of proposed architecture can be increased by adding more pipelining stages to increase the frequency of operation which is not possible in the design of [6].

5. CONCLUSION

This paper has presented a novel approach to design four and eight parallel pipelined FFT architectures. The underlying algorithm can be of any radix-2^n. The proposed approach reduces the number of delay elements by 33% compared to the prior DIF based design. Further, a novel four parallel 128-point FFT architecture has been developed using proposed method. The hardware costs of delay elements

and complex adders are reduced by using proposed scheduling approach. The number of complex multipliers is reduced using higher radix FFT algorithm. The throughput can be further increased by adding more pipeline stages which is possible due to the feed-forward nature of the design.

6. REFERENCES

[1] N. Weste, D. J. Skellern, "VLSI for OFDM," *IEEE Communications Magazine*, vol.36, no.10, pp.127-131, Oct 1998.

[2] A. Wang, A. Chandrakasan, "Energy-efficient DSPs for wireless sensor networks," *IEEE Signal Processing Magazine,* vol.19, no.4, pp.68-78, Jul 2002.

[3] Y. Park, *et al.*, "Seizure Prediction with Spectral Power of EEG using Cost-Sensitive Support Vector Machines," *Epilepsia*, vol. 52, pp. 1761-1770, Oct. 2011.

[4] S. He and M. Torkelson, "A new approach to pipeline FFT processor," *Proc. of IPPS*, 1996, pp. 766 - 770.

[5] J. Lee, H. Lee, S. I. Cho and S. S. Choi, "A High-Speed two parallel radix-2^4 FFT/IFFT processor for MB-OFDM UWB systems," *IEEE Inter. Symp. on Circuits and Systems*, pp. 4719-4722, May 2006.

[6] Y. W. Lin, *et al.*, "A 1-GS/s FFT/IFFT processor for UWB applications," *IEEE Journal of Solid-state Circuits*, vol. 40, no.8 pp. 1726-1735, Aug. 2005.

[7] M. Shin and H. Lee, "A high-speed four parallel radix-2^4 FFT/IFFT processor for UWB applications", *IEEE ISCAS 2008,* pp. 960 - 963, May 2008.

[8] M. Ayinala, M. Brown, K. K. Parhi, "Pipelined parallel FFT architectures via folding transformation," *IEEE Transactions on VLSI Systems*, vol. 20, 2012.

[9] Z. Wang *et al.*, "A Novel FFT Processor for OFDM UWB Systems," *Proc. IEEE APCCAS,* pp. 374-377, Dec. 2006.

[10] S. Qiao *et al.*, "An Area and Power Efficient FFT Processor for UWB Systems," *Proc. IEEE WICOM*, Sept. 2007, pp. 582-585.

[11] S.-N Tang, J. Tsai, T.-Y. Chang, "A 2.4-GS/s FFT Processor for OFDM-Based WPAN Applications," *IEEE Transactions on Circuits and Systems II: Express Briefs*, vol.57, no.6, pp.451-455, June 2010.

[12] K .K. Parhi, *et al.*, "Synthesis of control circuits in folded pipelined DSP architectures," *IEEE Journal of Solid State Circuits*, vol. 27, no. 1, pp. 29-43, 1992.

[13] K. K. Parhi, "Systematic synthesis of DSP data format converters using lifetime analysis and forward-backward register allocation," *IEEE TCAS - II,* vol. 39, no. 7, pp. 423-440, July 1992.

Design of an RNS Reverse Converter
for a New Five-Moduli Special Set *

Piotr Patronik[1], Krzysztof Berezowski[1], Janusz Biernat[1], Stanisław J. Piestrak[2], Aviral Shrivastava[3]

[1]Inst. of Computer Engineering, Control, and Robotics, Wrocław Univ. of Technology, 50-372 Wrocław, Poland
[2]Lab. d'Instrumentation Électron. de Nancy (LIEN), Université de Lorraine, 54506 Vandœuvre-lès-Nancy, France
[3]Compiler Microarchitecture Lab, CSE Dept, Arizona State University, Tempe, AZ 85281, USA
{firstname.lastname}@pwr.wroc.pl, piestrak@univ-metz.fr, aviral.shrivastava@asu.edu

ABSTRACT

In this paper, we present a new residue number system (RNS) $\{2^n - 1, 2^n, 2^n + 1, 2^{n+1} + 1, 2^{n-1} + 1\}$ of five well-balanced moduli that are co-prime for odd n. This new RNS complements the 5-moduli RNS system proposed before for even n $\{2^n - 1, 2^n, 2^n + 1, 2^{n+1} - 1, 2^{n-1} - 1\}$. With the new set, we also present a novel approach to designing multi-moduli reverse converters that focuses strongly on moving a significant amount of computations off the critical path. The synthesis of the resulting design over the ST Microelectronics 65nm LP library demonstrates that the delay, area, and power characteristics improve the performance and power consumption of the existing complementary 5-moduli set.

Categories and Subject Descriptors

B.2.4 [**Arithmetic and Logic Structures**]: High-Speed Arithmetic — cost/performance

General Terms

Performance, Design, Theory

Keywords

Reverse Converter, 5-moduli Set, Residue Number System

1. INTRODUCTION

The residue number system (RNS) is a non-positional number system capable of improving performance and reducing power consumption of add/multiply-intensive integer datapaths by decomposing computations into r small magnitude, independent, and parallel residue channels. This results in reduced carry propagation in additions as well as size and depth of the partial product matrices in multiplications as compared to the classic 2's complement system (TCS). Consequently, the critical path delay, area, and power consumption may simultaneously reduce, leading to faster and more efficient circuits. RNS arithmetic is especially effective in digital signal and image processing applications [10, 11].

*This research was supported by Polish *Narodowe Centrum Nauki, NCN.* Grant no. DEC-2011/01/B/ST7/06120.
The research of S. J. Piestrak was done when he was with CAIRN Res. Team, IRISA/INRIA, 22305 Lannion, France, on leave from Univ. of Metz, 57070 Metz, France.

RNS datapath interacts with its TCS environment through forward and reverse conversions. Both require extra hardware, power, and time to be carried out. Especially the reverse conversion complexity grows significantly with increasing number of moduli, therefore, new RNSs are proposed along with their respective reverse converters, and each application requires careful balancing of their trade-offs [7].

Special moduli, even of the form 2^n and odd of the form $2^n - 1$ and $2^n + 1$, simplify the design of all basic arithmetic operations as well as forward and reverse conversions [7]. Therefore, many special moduli sets have been proposed. These include the classic three-moduli set $\{2^n + 1, 2^n, 2^n - 1\}$ [4,9] as well as four-moduli sets, e.g. $\{2^n + 1, 2^n - 1, 2^{n+1} \pm 1\}$ [1, 2]. However, many of those already proposed, including five-moduli sets $\{2^n, 2^{2n+1} - 1, 2^{n/2} - 1, 2^{n/2} + 1, 2^n + 1\}$ [5] and $\{2^n, 2^{n/2} - 1, 2^{n/2} + 1, 2^n + 1, 2^{2n-1} - 1\}$ [6], suffer from significant imbalance in channels' magnitude, making one channel a computational bottleneck, while leaving other with an unutilized timing slack, consequently diminishing the very advantages of the modular decomposition.

While the 5-moduli set $\{2^n + 1, 2^n, 2^n - 1, 2^{n+1} - 1, 2^{n-1} - 1\}$ [3] addresses this problem, it is co-prime only for even n, and as such offers only a coarse coverage of its $(5n - 1)$ bits of the dynamic range, stepping by 10 bits for subsequent even n. In this work, we propose a new special 5-moduli set $\{2^n + 1, 2^n, 2^n - 1, 2^{n+1} + 1, 2^{n-1} + 1\}$ that is co-prime for odd n and as such it is complementary to the one of [3]. More importantly, we propose a novel approach to designing multi-moduli reverse converters that: (i) attempts at moving non-critical computations off from the critical path, (ii) carries the constant modulo $2^n + 1$ multiplications in n-bit wide carry-save adder (CSA) trees, and (iii) employs a novel technique of subtraction-free computation of the final value.

2. RNS AND CONVERSION BASICS

An RNS is defined by a set of r pairwise co-prime integer *moduli* $\{m_1, m_2, \ldots, m_r\}$. Its *dynamic range* M equals to $M = \prod_{i=1}^{r} m_i$. An integer value $X \in [0, M - 1]$ is uniquely represented by the r-tuple $\{x_1, \ldots, x_r\}$ of residues $x_i : x_i = \frac{X}{m_i}$ calculated with respect to $\{m_1, m_2, \ldots, m_r\}$.

Given X, Y represented by $\{x_1, \ldots, x_r\}$, $\{y_1, \ldots, y_r\}$, respectively, $Z = |X \circ Y|_M$ can be computed as $z_i = |x_i \circ y_i|_{m_i}$, $1 \leq i \leq r$ for $\circ \in \{+, -, \times\}$. Each RNS digit z_i of the result depends solely on x_i and y_i and its calculation does not involve other digits z_k for any $k \neq i$. That very parallelism is the primary advantage of applying the RNS.

There are two ways of calculating $|X|_M$ from $\{x_1, \ldots, x_r\}$ given $\{m_1, m_2, \ldots, m_r\}$. The first method is the *Chinese*

Reminder Theorem (CRT) typically formulated as

$$X = \left| \sum_{i=1}^{r} \hat{m}_i \left| x_i \hat{m}_i^{-1} \right|_{m_i} \right|_M \qquad (1)$$

where $\hat{m}_i = \frac{M}{m_i}$ and $\hat{m}_i^{-1}$ is the *multiplicative inverse* of $\hat{m}_i$ [9], i.e., $|\hat{m}_i \cdot \hat{m}_i^{-1}|_{m_i} = 1$. The CRT is inherently parallel at the expense of a large multi-operand addition mod $\Pi_{i=1}^{r} m_i$, whose hardware implementation could be disadvantageous.

The second method is the *Mixed-Radix Conversion* (MRC)

$$X = x_1 + d_1 m_1 + d_2 m_1 m_2 + \ldots + d_{r-1}(m_1 m_2 \cdots m_{r-1}). \qquad (2)$$

MRC calculates the mixed-radix digits $\{x_1, d_1, d_2, \ldots, d_{r-1}\}$ from residues and represents X over the mixed-radix base $\{1, m_1, m_1 m_2, \ldots, m_1 m_2 \cdots m_{r-1}\}$. Being inherently sequential, MRC has the advantage of computing modulo operations of a magnitude of single moduli throughout the whole conversion process. A special two-moduli case of (2) (2-MRC) is often used for constructing multi-moduli reverse converters in a recursive, modulus-by-modulus fashion

$$X = x_1 + m_1 \underbrace{\left| (x_2 - x_1) m_1^{-1} \right|_{m_2}}_{d_1}. \qquad (3)$$

3. DESIGN OF THE CONVERTER

In our five-moduli RNS, we denote $m_1 = 2^n - 1$, $m_2 = 2^n$, $m_3 = 2^n + 1$, $m_4 = 2^{n+1} + 1$, and $m_5 = 2^{n-1} + 1$, while the residues corresponding to these moduli as x_1 through x_5. Similarly to [3] and many other, our converter carries out the conversion in the following three steps.

1) Convert the residues $\{x_1, x_2, x_3\}$ using an efficient CRT-based three moduli converter of [4]

$$X_1 = x_2 + 2^n \cdot \underbrace{\left| 2^n(x_3 - x_2) + 2^{n-1}(2^n + 1)(x_1 - x_3) \right|_{2^{2n}-1}}_{X_h}$$

Note that $0 \le X_1 < M_1$ and $M_1 = 2^n(2^{2n} - 1)$.

2) Apply 2-MRC to the pair of values $\{X_1, x_4\}$ over the two-moduli set $\{M_1, m_4\}$ to obtain a residue X_2, $0 \le X_2 < M_2$, and the new module $M_2 = M_1 m_4$

$$X_2 = X_1 + M_1 \underbrace{\left| \overbrace{(x_4 - X_1)}^{L_4} M_1^{-1} \right|_{m_4}}_{R_4}. \qquad (4)$$

3) Apply 2-MRC to the pair of values $\{X_2, x_5\}$ over the two-moduli set $\{M_2, m_5\}$ to obtain the final conversion result X, $0 \le X < M$, where $M = M_2 m_5$

$$X = X_2 + M_2 \underbrace{\left| \overbrace{(x_5 - X_2)}^{L_5} M_2^{-1} \right|_{m_4}}_{R_5} \qquad (5)$$

In the reminder of this paper, we develop the techniques to improve the efficiency of such conversion process.

3.1 Preliminaries

We leave without a proof the co-primality of the set $\{2^n - 1, 2^n, 2^n + 1, 2^{n+1} + 1, 2^{n-1} + 1\}$, as well as we assume the existence of X_1, i.e., the result of the conversion of residues $\{x_1, x_2, x_3\}$ with respect to moduli set $\{m_1, m_2, m_3\}$.

The inverses M_1^{-1} and M_2^{-1} have to satisfy respectively

$$\left| M_1^{-1} m_1 m_2 m_3 \right|_{m_4} = 1 \qquad (6)$$

$$\left| M_2^{-1} m_1 m_2 m_3 m_4 \right|_{m_5} = 1 \qquad (7)$$

In [1], it was observed that for $m_1 = 2^n - 1$, $m_2 = 2^n$, $m_3 = 2^n + 1$, and $m_4 = 2^{n+1} + 1$

$$M_1^{-1} = 2^n + \sum_{k=1}^{\frac{n-5}{2}} 2^{n-2k} + 2^4 - 2. \qquad (8)$$

Similarly, if $m_5 = 2^{n-1} + 1$ then

$$M_2^{-1} = \left| 18^{-1} \right|_{m_5}. \qquad (9)$$

satisfies (7) for odd n. The bit-level structure of this inverse as a function of n is given in Figure 1.

3.2 Multiplication by a Constant mod $2^\alpha + 1$

Because both 2-MRC steps require multiplication by a constant, we take the uniform approach to this computation. Let an integer $X^* = (x_\alpha, \ldots, x_0)$ and a constant $C = (c_\alpha, \ldots, c_0)$ be $(\alpha + 1)$-bit wide mod $2^\alpha + 1$. Let $X'' = (x_{\alpha-1}, \ldots, x_0)$ be a vector of α least significant bits of X^*. Then we have

$$\left| C \times X^* \right|_{2^\alpha+1} = \left| C \sum_{i=0}^{\alpha} 2^i x_i \right|_{2^\alpha+1} = \left| C 2^\alpha x_\alpha + C \sum_{i=0}^{\alpha-1} 2^i x_i \right|_{2^\alpha+1}$$

Because $|2^\alpha|_{2^\alpha+1} = -1$, therefore

$$\left| C \times X \right|_{2^\alpha+1} = \left| \sum_{i=0}^{\alpha} c_i 2^i \underbrace{\sum_{j=0}^{\alpha-1} 2^j x_j}_{2^i X''} - C x_\alpha \right|_{2^\alpha+1}$$

The value of $2^i X''$ can be easily computed by means of left circular shift with bit inversion and adding correction

$$2^i X'' = 2^i \sum_{j=0}^{\alpha-1} 2^j x_j = CLN_\alpha(X'', i) - (2^{i+1} - 1),$$

where $CLN_\alpha(U, p) = \left(u_{(\alpha-p)}, \ldots, u_0, \overline{u}_{(\alpha-1)}, \ldots, \overline{u}_{(\alpha-p+1)} \right)$ for any α-bit wide integer $U = (u_{\alpha-1}, \ldots, u_0)$ and $0 \le p < \alpha$.

Thus, the multiplication $\left| C \times X^* \right|_{2^\alpha+1}$ can be computed as

$$\left| \sum_{i=0}^{\alpha} c_i CLN_\alpha(X'', i) - c_i \sum_{i=0}^{\alpha} (2^{i+1} - 1) - C x_\alpha \right|_{2^\alpha+1}$$

and implemented by a multioperand modular adder (MOMA) [8] of α least significant bits of X^* rotated with inversion, and a conditional correction depending on the value of the most significant bit x_α of X^*.

3.3 Calculation of 2-MRC

Subsequent applications of the 2-MRC produce deep computational paths that contribute to either delay or latency. We alleviate this disadvantage by decomposing each 2-MRC step into a critical and a non-critical component, and then move the latter off the critical path and compute it in parallel.

1) Calculation of L_4: $L_4 = |x_4 - X_1|_{m_4}$ from Equation (4) is computed from the 3-moduli converter output X_1 as

$$L_4 = |x_4 - X_1|_{2^{n+1}+1} = |x_4 - (2^n X_h + x_2)|_{2^{n+1}+1}$$
$$= \left| \underbrace{x_4 - x_2}_{L_4'} - 2^n X_h \right|_{2^{n+1}+1} \qquad (11)$$

Because X_h is a $2n$-bit wide integer mod $2^{2n} - 1$, x_4 is an $(n + 2)$-bit wide integer mod $2^{n+1} + 1$, and x_2 is an n-bit wide integer mod 2^n, a direct computation of L_4 in an $(n + 1)$-bit wide CSA mod $2^{n+1} + 1$ requires six $(n + 1)$-bit wide operands. Observe that the inputs x_2 and x_4 are

$$|18^{-1}|_{2^{n-1}+1} = \begin{cases} 2^{n-2} + 2^{n-3} - \sum_{i=0}^{k-2} 2^{6i+7} + \sum_{i=0}^{k-2} 2^{6i+4} - 2^0 & \text{for } n = 6k+1, \ k \geq 1 \\ \sum_{i=0}^{k} 2^{6i} - \sum_{i=0}^{k-1} 2^{6i+3} & \text{for } n = 6k+5, \ k \geq 1 \\ 2^{n-4} + 2^{n-3} + \sum_{i=0}^{k-1} 2^{6i+2} - \sum_{i=0}^{k-2} 2^{6i+5} & \text{for } n = 6k+3, \ k \geq 1 \end{cases} \quad (10)$$

Figure 1: The distribution of significant bits in $\left|18^{-1}\right|_{2^{n-1}+1}$

available much earlier than the output X_h of the 3-moduli converter. Therefore, the residues x_2, x_4 as well as the corrective constant (available at the design time) can be added separately and scheduled in parallel to the computation of X_h, thus forming a new component L_4', so that each of additions needed to obtain L_4' and X_h requires three operands only.

Let $x_{4,i}$ and $x_{2,i}$ denote the i-th bit of vectors x_4, and x_2, respectively. Then L_4' can be computed as

$$\left| \begin{array}{l} (x_{4,n} \vee x_{4,(n+1)})x_{4,(n-1)} \ldots x_{4,0} \quad + \\ x_{4,(n+1)}\overline{x}_{2,(n-1)} \ldots \overline{x}_{2,0} - 2^{n-2} + 1 \end{array} \right|_{2^{n+1}+1} \quad (12)$$

so that L_4 can be computed from L_4' and X_h as

$$\left| \begin{array}{l} \overline{x}_{h,0}l_{4,(n+1)}'0\overline{x}_{h,(2n-1)} \ldots \overline{x}_{h,(n+2)} \quad + \\ x_{h,(n+1)} \ldots x_{h,1} \quad + \\ (l_{4,n}' \vee l_{4,(n+1)}')(l_{4,(n-1)}' \vee l_{4,(n+1)}')l_{4,(n-2)}' \ldots l_{4,0}' \end{array} \right|_{2^{n+1}+1}$$

where $l_{4,i}$, $l_{4,i}'$, $x_{h,i}$ are respective bits of L_4, L_4', X_h.

Consequently, both L_4' and L_4 are then computed in a cascade of two 3-operand MOMAs mod $2^{n+1}+1$, each composed of a single layer of CSA followed by the 2-operand adder mod $2^{n+1}+1$. Since L_4' does not depend on X_h, they can be computed in parallel, reducing the computations on the critical path to the 3-operand addition mod $2^{n+1}+1$.

2) Calculation of R_4: Applying the technique proposed in Section 3.2, we obtain the following expression for R_4

$$\left| \begin{array}{l} CLN_{(n+1)}(\overline{L_4''},1)CLN_{(n+1)}(L_4'',4) \quad + \\ \sum_{k=1}^{\frac{n-5}{2}} CLN_{(n+1)}(L_4'',n-2k) \quad + \\ CLN_{(n+1)}(L_4'',n) \quad + \\ l_{4,(n+1)} \cdot (-2M_1^{-1}) + \overline{l}_{4,(n+1)} \cdot (-M_1^{-1}) \end{array} \right|_{2^{n+1}+1}$$

Consequently, we calculate R_4 as an $\frac{n+1}{2}$-operand addition mod $2^{n+1}+1$ of terms constructed by cyclically shifting and complementing n least significant bits of L_4 and the corrective term computed from the bit $l_{4,(n+1)}$.

3) Calculation of L_5: Similarly to the calculation of L_4, we rewrite L_5 from Equation (5) as

$$L_5 = |x_5 - X_1 - 2^{3n}R_4 + 2^n R_4|_{2^{n-1}+1}$$

Because $|2^{3n}|_{2^{n-1}+1} = -8$ and $|2^n|_{2^{n-1}+1} = -2$, therefore

$$L_5 = |6R_4 + x_5 - X_1|_{2^{n-1}+1}$$
$$= |6R_4 + \underbrace{x_5 - 2^n x_2 + X_h}_{L_5'}|_{2^{n-1}+1} \quad (13)$$

Observe now that: R_4 is an $(n+2)$-bit wide integer mod $2^{n+1}+1$; X_1 is $3n$-bit wide integer mod $2^n(2^{2n}-1)$; and x_5 is an n-bit wide operand mod $2^{n-1}+1$. A direct computation of (13) in a mod $2^{n-1}+1$ CSA tree with the necessary constant correction requires chopping these vectors into 11 operands. Again, we leave on the critical path only the computations that depend on R_4, and compute L_5' separately in parallel, compressing it with some bit-level manipulations to

$$L_5' = \left| \begin{array}{l} (x_{5,(n-2)}|x_{5,(n-1)}) \ldots x_{5,0} + x_{2,(n-2)} \ldots x_{2,0} + \\ \overline{x}_{h,(n-3)} \ldots \overline{x}_{h,(0)}\overline{x}_{2,(n-1)} \quad + \\ x_{h,(2n-4)} \ldots x_{h,(n-2)} \quad + \\ x_{5,(n-1)}0 \ldots 0\overline{x}_{h,(2n-1)} \ldots \overline{x}_{h,(2n-3)} - 80 \end{array} \right|_{2^{n-1}+1}$$

where $x_{5,i}$, $x_{2,i}$, $x_{h,i}$ are respective bits of x_5, x_2, X_h.

Once L_5' is computed, we compute L_5 as

$$\left| \begin{array}{l} l_{5,(n-2)}'l_{5,(n-3)}' \ldots l_{5,0}' + \overline{r}_{4,(n-3)} \ldots \overline{r}_{4,0}r_{4,(n-2)} \quad + \\ r_{4,(n-5)} \ldots r_{4,1}(r_{4,0}|r_{4,(n+1)})\overline{r}_{4,(n-2)}\overline{r}_{4,(n-3)}\overline{r}_{4,(n-4)} + \\ 0 \ldots 0\overline{r}_{4,(n+1)}\overline{r}_{4,n}\overline{r}_{4,(n-1)}r_{4,n}r_{4,(n-1)}l_{5,(n-1)}' \end{array} \right|_{2^{n-1}+1}$$

where $r_{4,i}$ and $l_{5,i}'$ are the respective bits of R_4 and L_5'.

4) Calculation of R_5: By applying again the technique from Section 3.2 to each of the variants of the multiplicative inverse M_2^{-1} from Figure 1, we obtain three alternative equations to calculate R_5.

For $n = 6k+1$, $R_5 =$

$$\left| \begin{array}{l} CLN_{(n-1)}(L_5'',n-2) + CLN_{(n-1)}(L_5'',n-3) + \\ \sum_{i=0}^{k-2} CLN_{(n-1)}(\overline{L_5''},6i+7) \quad + \\ \sum_{i=0}^{k-2} CLN_{(n-1)}(L_5'',6i+4) \quad + \\ \overline{L_5''} + l_{5,(n+1)} - M_2^{-1}|_{2^{n-1}+1} \end{array} \right|_{2^{n-1}+1}$$

For $n = 6k+3$, $R_5 =$

$$\left| \begin{array}{l} CLN_{(n-1)}(L_5'',n-3) + CLN_{(n-1)}(L_5'',n-4) + \\ \sum_{i=0}^{k-1} CLN_{(n-1)}(L_5'',6i+2) \quad + \\ \sum_{i=0}^{k-2} CLN_{(n-1)}(\overline{L_5''},6i+5) \quad + \\ l_{5,(n+1)}| - M_2^{-1}|_{2^{n-1}+1} \end{array} \right|_{2^{n-1}+1}$$

For $n = 6k+5$, $R_5 =$

$$\left| \begin{array}{l} \sum_{i=0}^{k} CLN_{(n-1)}(L_5'',6i) \quad + \\ \sum_{i=0}^{k-1} CLN_{(n-1)}(\overline{L_5''},6i+3) + \\ l_{5,(n+1)}| - M_2^{-1}|_{2^{n-1}+1} \end{array} \right|_{2^{n-1}+1}$$

3.4   Final Multiplication and Addition

To compute the value of X, we rewrite (4) and (5) as

$$x_2 + 2^n X_h + 2^n(2^{2n}-1)R_4 + 2^n(2^{2n}-1)(2^{n+1}+1)R_5$$
$$= x_2 + 2^n \big(\underbrace{X_h + (2^{2n}-1)R_4}_{V_1} + \underbrace{(2^{2n}-1)(2^{n+1}+1)R_5}_{V_2} \big).$$
$$\underbrace{\phantom{x_2 + 2^n \big(X_h + (2^{2n}-1)R_4 + (2^{2n}-1)(2^{n+1}+1)R_5 \big)}}_{V}$$

Again, only V_2 depends on the critical path component R_5, while the calculation of V_1 from previously computed X_h and R_4 can be done separately in parallel. Both V_1 and V_2 have to be calculated through a series of constant multiplications, additions, and subtractions. However, some bit level manipulations allow to integrate constants and extract critical and non-critical components into terms V_1', V_2' that conceptually correspond to terms V_1 and V_2, yet minimize the number of vectors for the final addition.

$$V = \underbrace{X_h + (2^{2n}-1)R_4}_{V_1'} + \underbrace{(2^{2n}-1)(2^{n+1}+1)R_5}_{V_2'} - 1 + 1$$

$$V_1' = 0 \ldots 0 r_{4,(n+1)} \ldots r_{4,0}x_{h,(2n-1)} \ldots x_{h,0}$$
$$+ \underbrace{1 \ldots 1}_{2n+1} 0 \underbrace{1 \ldots 1}_{n-2} \overline{r}_{4,(n+1)} \ldots \overline{r}_{4,0} + 1.$$

$$V_2' = r_{5,(n-1)} \ldots r_{5,0}0(r_{5,(n-1)} \vee r_{5,(n-2)}) \ldots (r_{5,(n-1)} \vee r_{5,0})$$
$$\|\overline{r}_{5,(n-2)} \ldots \overline{r}_{5,0}1\overline{r}_{5,(n-1)} \ldots \overline{r}_{5,0};$$

Table 1: Left: the hardware complexity estimation of the new converter. Right: the minimum delay (M. del.), area, power and power-delay-product (PDP) at the minimum delay as a function of n

Resource	[3]	This work
FA, $n = 6k - 2$	$(5n^2 + 44n - 4)/6$	
FA, $n = 6k$	$(5n^2 + 52n - 12)/6$	-
FA, $n = 6k + 2$	$(5n^2 + 48n - 8)/6$	
FA, $n = 6k + 1$		$(5n^2 + 35n - 4)/6$
FA, $n = 6k + 3$	-	$(5n^2 + 31n + 18)/6$
FA, $n = 6k + 5$		$(5n^2 + 27n + 22)/6$
HA	-	$(5n + 23)/2$
mod $2^{n-1} - 1$	2	3
mod $2^{n+1} - 1$	2	3
mod $2^{2n} - 1$	1	1
$(3n - 1)$-bit sub	1	-
$(3n + 1)$-bit sub	1	-
$4n$-bit add	1	2
Inv., n even	$6n + 1$	-
Inv., $n = 6k + 1$		$(5n^2 + 150n + 65)/12$
Inv., $n = 6k + 3$	-	$(5n^2 + 146n - 3)/12$
Inv., $n = 6k + 5$		$(5n^2 + 130n + 65)/12$
Misc.		$(n - 5)$ OR, 1 MUX

	M. del. [ns]		Area [μm^2]		Power [mW]		PDP [mW·ns]	
n	[3]	New	[3]	New	[3]	New	[3]	New
6	5.67	-	11950	-	1.88	-	10.66	-
7	-	5.83	-	14611	-	2.48	-	14.46
8	5.83	-	16114	-	3	-	17.49	-
9	-	6.28	-	19508	-	3.53	-	22.17
10	6.6	-	21550	-	4.39	-	28.97	-
11	-	6.58	-	24671	-	4.69	-	30.86
12	6.99	-	26068	-	5.98	-	41.80	-
13	-	6.82	-	30528	-	6.25	-	42.63
14	6.99	-	31657	-	6.86	-	47.95	-
15	-	6.91	-	36533	-	7.78	-	53.76
16	7.37	-	38441	-	9.12	-	67.21	-
17	-	7.37	-	43118	-	10.32	-	76.06
18	7.92	-	47588	-	12.66	-	100.27	-
19	-	7.73	-	49882	-	12.11	-	93.61
20	8.15	-	51597	-	12.97	-	105.71	-
21	-	7.72	-	57249	-	13.74	-	106.07

As a consequence, the term V_1' can be computed using a 2-operand adder with the carry-in input set to 1, whereas the term V_2' is obtained by manipulating bits of the vector R_5. The conversion result X is obtained by concatenating x_2 with the result of 2-operand addition of V_1' and V_2' in an adder with carry-in bit set to 1, which concludes the calculation. Table 1 details the complexity of the new design as compared to the design of [3].

4. EXPERIMENTAL RESULTS

Table 1 includes also the synthesis results of our design and the complementary design of [3] over the industrial 65nm low-power cell library from STMicroelectronics using Cadence RTL Compiler v8.1. We limited our comparisons to the design of [3], wherein the authors had already thoroughly compared their design to the other 5-moduli sets, and our closely matching delay, area, and power consumption trends our design fit into that comparison.

The results summarized in Table 1 also show clearly that our design techniques have alleviated the impact of the increased complexity of the new moduli set and the resulting design is competitive to the original design of [3] w.r.t. delay, area, and power, even though it has to deal with accumulation of an increased number of partial product vectors in inner product computations.

5. CONCLUSIONS

We proposed a new balanced 5-moduli RNS and its reverse converter that complements the only existing balanced 5-moduli set available for even n introduced in [3], thus bridging the 10-bit gap between consecutive dynamic ranges available for that RNS. On the example of our converter, we demonstrated three novel design techniques that allow to reduce the critical path delay of the conversion circuit, enabling the automatic synthesis tools to produce smaller and less power consuming circuits. It is important to note that many recent results on special RNSs focus on designing simple converters with little attention being paid to the complexity of the datapath's channels: the underperformance of the unbalanced channels can quickly consume any benefits not only of a simpler converter but also of RNS in general. Both the converter of [3] as well as ours are against

this trend: they use well-balanced moduli sets for which not only a relatively simple reverse converter can be designed but also simple forward converters and most importantly balanced datapaths channels.

6. REFERENCES

[1] P. Ananda Mohan and A. Premkumar. RNS-to-binary converters for two four-moduli sets $\{2^n - 1, 2^n, 2^n + 1, 2^{n+1} - 1\}$ and $\{2^n - 1, 2^n, 2^n + 1, 2^{n+1} + 1\}$. *IEEE Trans. Circuits Syst. I*, 54(6):1245–1254, June 2007.

[2] B. Cao, C.-H. Chang, and T. Srikanthan. New efficient residue-to-binary converters for 4-moduli set $\{2^n - 1, 2^n, 2^n + 1, 2^{n+1} - 1\}$. In *Proc. ISCAS*, volume 4, pages IV536–IV539, 2003.

[3] B. Cao, C.-H. Chang, and T. Srikanthan. A residue-to-binary converter for a new five-moduli set. *IEEE Trans. Circuits Syst. I*, 54(5):1041–1049, May 2007.

[4] A. Dhurkadas. Comments on a high speed realization of a residue to binary number system converter. *IEEE Trans. Circuits Syst. II*, 45(3):446–447, Mar. 1998.

[5] M. Esmaeildoust, K. Navi, and M. Taheri. High speed reverse converter for new five-moduli set $\{2^n, 2^{2n+1} - 1, 2^{n/2} - 1, 2^{n/2} + 1, 2^n + 1\}$. *IEICE Electron. Expr.*, 7(3):118–125, 2010.

[6] A. Molahosseini, C. Dadkhah, and K. Navi. A new five-moduli set for efficient hardware implementation of the reverse converter. *IEICE Electron. Expr.*, 6(14):1006–1012, 2009.

[7] A. Omondi and B. Premkumar. *Residue Number Systems: Theory and Implementation*. Imperial College Press, London, UK, 2007.

[8] S. Piestrak. Design of residue generators and multioperand modular adders using carry-save adders. *IEEE Trans. Comput.*, 43(1):68–77, Jan. 1994.

[9] S. J. Piestrak. High-speed realization of a residue to binary number system converter. *IEEE Trans. Circuits Syst. II*, 42(10):661–663, Oct. 1995.

[10] T. Shahana, R. James, B. Jose, K. Jacob, and S. Sasi. Performance analysis of FIR digital filter design: RNS versus traditional. In *Proc. Int. Symp. Commun. & Inf. Techn.*, pages 1–5, 17–19 Oct. 2007.

[11] T. Toivonen and J. Heikkila. Video filtering with Fermat number theoretic transforms using residue number system. *IEEE Trans. Circuits Syst. Video Technol.*, 16(1):92–101, Jan. 2006.

On the Automatic Synthesis of Parallel SW from RTL Models of Hardware IPs *

A. Acquaviva
Dip. Informatica e Automazione
Politecnico di Torino, Italy
andrea.acquaviva@polito.it

N. Bombieri, F. Fummi, S. Vinco
Dip. Informatica
Universita' di Verona, Italy
name.surname@univr.it

ABSTRACT

Heterogeneous multicore system-on-chips (MPSoCs) provide many degrees of freedom to map functionalities on either SW and HW components. In this scenario, enabling the remapping of HW IPs as SW routines allows to fully exploit the computation power and flexibility provided by heterogeneous MPSoCs. On the other hand, reuse of existent IP cores is the key strategy to explore this large design space in a reasonable amount of time and to reduce the error risk during the MPSoC design flow. A methodology for automatic generation of parallel SW code taking into account these aspects is currently missing. This paper aims at overcoming this limitation, by presenting a methodology to automatically generate parallel SW IPs starting from existent RTL IP models.

Categories and Subject Descriptors

C.3 [**Computer Systems Organization**]: SPECIAL PURPOSE AND APPLICATION-BASED SYSTEMS

Keywords

RTL IP reuse, Parallel Embedded Software Generation

1. INTRODUCTION

To handle the heterogeneous and intensive workload imposed by multimedia and interactive applications, MPSoCs are getting more and more complex, incorporating multicore processors, digital signal processors (DSPs), application specific instruction set processors (ASIPs), configurable logic and hardware accelerators [10]. In this context, reuse of pre-designed and pre-verified IP blocks is the key strategy to explore this large design space in a very short time, reduce the error risk during the MPSoC design flow and increase the flexibility of the platform.

*This work has been partially supported by the European project TOUCHMORE FP7-ICT-2011-7-288166.

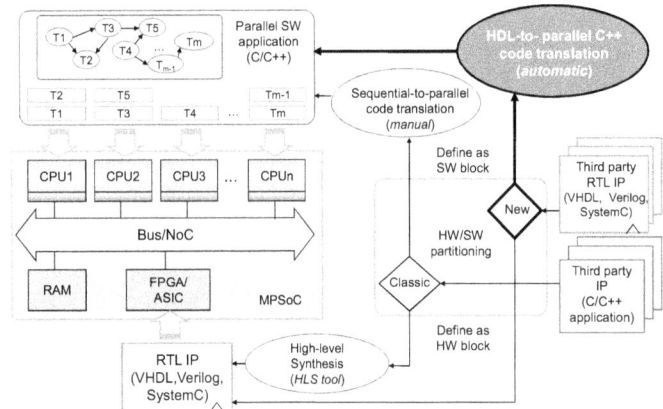

Figure 1: HW/SW partitioning in MPSoC design space exploration

Designers may reuse both third party IPs implemented as C/C++ applications, and RTL IPs implemented in HDL (i.e., VHDL, Verilog, SystemC) (see Fig. 1). In the first case, IP reuse can be considered efficient when the IP is defined to become a HW block, since several high-level synthesis (HLS) methodologies and tools are available [5]. In contrast, when the IP code is defined to become SW, it often undergoes a sequential-to-parallel translation to guarantee CPU load balancing and system performance. Such a translation relies on designer experience and application knowledge for manual SW code parallelization: Thus, in many cases, it leads to re-implement the parallel IP functionality from scratch rather than reuse the sequential code.

On the other hand, many IPs are also available to designers as RTL descriptions. Nevertheless, they mainly fit to be reused as HW blocks, as any modification to the RTL model for design space exploration (e.g., re-timing) is manual, time consuming and error prone.

HDL to C/C++ model translation has been investigated in the last decades aiming to speed up the HDL model verification (i.e., VHDL, Verilog) by using C/C++ simulators [1, 11]. In these works, all the implementation details included in the RTL descriptions strictly related to HW models are uselessly maintained during the translation to the SW domain, for generating sequential rather parallel code. Different approaches are presented in [4, 6]. [6] presents a methodology for generating C++ code starting from SystemC descriptions, by overloading of a subset of SystemC constructs. [4] proposes a methodology to abstract RTL IPs implemented in any HDL into embedded SW (i.e., C++) based on an abstraction algorithm that eliminates

```
1: while (simulation_time < simulation_end_time) do
2:     T_c := T_n;
3:     Update_signals();
4:     for each process p_i ∈ P do
5:         if p_i is currently sensitive to a signal s and an event e_s
             occurred on s in this simulation cycle then
6:             process_queue := {process_queue, p_i};
7:         end if
8:     end for
9:     for each p_i ∈ process_queue do
10:        p_i executes until it suspends;
11:        process_queue := {process_queue \ p_i};
12:    end for
13:    T_n :=     earliest_of{simulation_end_time, next_event
           _time, next_process_time};
14: end while
```

Figure 2: HDL scheduling algorithm

many HDL details (e.g., the clock). Nevertheless, both approaches generate sequential C++ code. Different methodologies to allow parallel execution of SystemC descriptions are proposed in [7,8]. Nevertheless, such methodologies aim at speeding up the hardware simulation but they do not deal with automatic generation of SW applications.

In order to effectively exploit the parallelism made available by multicore platforms, a methodology for the automatic remapping of HW blocks to parallel SW would be needed to achieve a fast and error-prone reuse of the HDL code and to reduce the effort of parallel translation from sequential IP code.

This paper presents a strategy allowing designers to reuse RTL IP models for automatically generating SW applications for MPSoC platforms. The degree of parallelism of such parallel SW applications can be settle by designers along the design space exploration. The experimental results show that the greater the complexity of the reused IP, the higher the speed up achieved by augmenting the degree of parallelism. Furthermore, the trade-off between the RTL IP size and the synchronization/communication overhead introduced by the SW support determines the most efficient granularity of parallelism.

2. CONCURRENCY IN HDL MODELS

A HDL model behavior description consists of *sequential* and *concurrent* statements. Concurrent statements are used to define interconnected blocks and *user-defined processes*, which represents single independent sequential processes and that jointly describe the overall behavior or structure of a design. Concurrent statements are sequentially executed (in any standard HDL simulator) so that the execution order does not affect the result. Synchronization among concurrent statements is handled by the *HDL dynamic scheduler* which is included into the kernel of any HDL simulator.

Fig. 2 summarizes the HDL scheduling algorithm. The simulation of a HDL model consists of an iterative run of the concurrent statements (i.e., processes) (row 1). Each such repetition is defined *simulation cycle*. In each cycle, the current time, T_c, is set equal to the time of the next simulation cycle, T_n (row 2). Then, all signal values in the description are updated (row 3). If, as a result of this computation, an event occurs on a given signal, concurrent statements that are sensitive to that signal will resume and will be executed as part of the simulation cycle (rows 4-12). The time of the next simulation cycle, T_n, is determined (row 13) by setting it to the earliest of (i) the time at at which simulation ends,

(ii) the next time at which an event occurs, or (iii) the next time at which a process resumes. If $T_n = T_c$, the next simulation cycle (if any) will be a delta cycle. Simulation is finished when there are no more timed notifications.

3. PARALLEL SOFTWARE GENERATION

The main structure of the SW application generated by the proposed methodology is shown in Fig. 3 and detailed in the next subsections.

3.1 Mapping of HDL statements into SW

Each HDL statement of the RTL IP description is first translated into a corresponding C++ statement.

Given an HDL model $M_{HDL} =< I_{HDL}, DS_{HDL}, SS_{HDL}, CS >$, in which: I_{HDL} is the model interface, DS_{HDL} is the set of declaration statements, (including declarations of constants, types, and objects, i.e., variables, signals, components), SS_{HDL} is the set of sequential statements (which includes the statements contained into processes or subprograms) and CS is the set of concurrent statements(which includes processes and signal assignment statements), the equivalent SW model is defined as $M_{SW} =< I_{SW}, DS_{SW}, SS_{SW}, T, SW\_P, gs >$, in which:

- The application interface I_{SW} is a function $F (io\_struct*)$. The C++ data structure *io_struct* is generated by mapping all ports of I_{HDL} into C++ structure fields $el\_type_x in\_i\ el\_type_z\ out\_j$ (with the exception of clock).

- The set of declaration statements DS_{SW} is generated by translating all components of DS_{HDL} into C++ statements. Each HDL signal s_i of M_{HDL} is translated into a two variable structure $sig_i = \{sig.old, sig.new\}$, representing the old and the current values of the signal. An extended datatype library is adopted to support HDL datatypes operators [3].

- The set of sequential statements SS_{SW} is generated from SS_{HDL}.

- $T = \{t_0, .., t_i, t_0^s, .., t_i^s\}$ is the set of tasks, i.e., the concurrent pieces of work in parallel computation, each one composed of sequential statements. They are divided into *user-defined tasks* (t) and *scheduling tasks* (t^s), as explained in Section 3.2.

- $SW\_P = \{sw\_p_0, .., sw\_p_j\}$ is the set of SW processes (threads), each one composed of a group of tasks (see Fig. 3), as explained in Section 3.3.

- gs is the global scheduler, which handles task communication and synchronization between SW processes, as explained in Section 3.4.

3.2 Definition of tasks

In this methodology, *tasks* represent unit of computation, divided into:

- *user-defined task*: these tasks represent the execution of the starting concurrent HDL processes. Each concurrent statement is CS is intuitively mapped into a user-defined task t. Such tasks can be executed in parallel without altering the model functionality (rows 9-12 of Fig. 2), as explained in Section 2;

- *scheduling task*: the scheduler step that detects any events on signals for each process (rows 4-8 of Fig. 2),

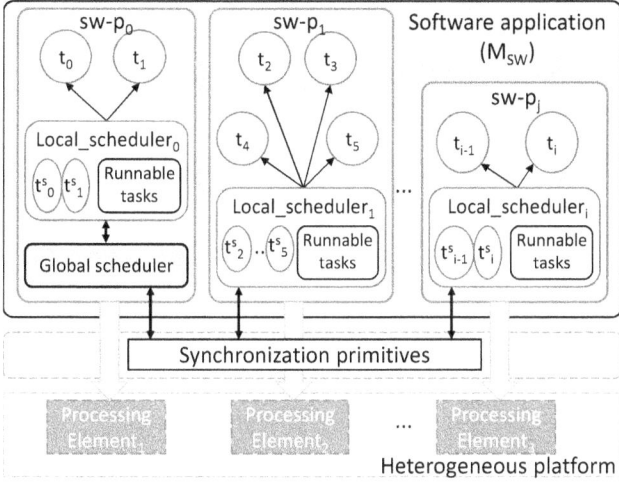

Figure 3: Main structure of the generated SW

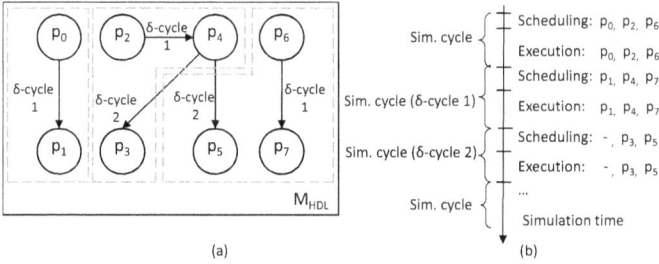

Figure 4: HPC-DAG process communication graph (a), and corresponding process execution order (b)

in order to handle the process queue, can be performed in parallel. As a consequence, for each user-defined process, a scheduling task t^s is added in T.

3.3 Assignment of tasks to SW processes

In the assignment step, all tasks of T are grouped into SW processes (i.e., threads) of M_{SW}. The assignment strategy aims both at keeping the load balanced and at reducing communication among tasks of different processes. This work proposes a *static assignment* algorithm, in order to reduce the management costs of dynamic load balancing during simulation. An extended analysis on dynamic task assignment will be considered in future work.

The assignment strategy relies on a *HDL process communication graph (HPC-DAG)* that is automatically extracted from the RTL description of M_{HDL}. Such graph represents the synchronization and communication net among processes of the starting HDL model. The HPC-DAG consists of a directed acyclic graph in which each vertex represents a user-defined process and each edge represents communication between processes. Fig. 4(a) shows an example. Each vertex with no incoming edges (e.g., p_0, p_2, p_6) represents a user-defined process sensitive to one or more input ports (e.g., the clock port). Such processes are run for first in any simulation cycle. In contrast, each vertex with incoming edges represents a user-defined process that is made runnable by a write on a signal and will be run in the next delta cycle.

Fig. 4(b) shows the process execution order along the simulation time, which has been extracted from the HPC-

```
1: Local_scheduling_step_i() {
2:   for each task t ∈ sw_p_i do
3:       if t is currently sensitive to a signal sig_w and an event e_sig_w
             occurred on sig_w in the last simulation cycle (sig_w.old ≠
             sig_w.new) then
4:           sig_w.old := sig_w.new
5:           runnable_tasks_i := {runnable_tasks_i, t};
6:       end if
7:   end for
8: }
9: Local_execution_step()_i {
10:   for each t ∈ runnable_tasks_i do
11:       t executes;
12:       runnable_tasks_i := {runnable_tasks_i \ t};
13:   end for
14: }
```

Figure 5: Local scheduler ls_i pseudocode

```
1: while (application_return = false) do
2:   while (all runnable_tasks are not empty) do
3:       SYNCH(num_sw_p);
4:       for each SW process sw_p_j ∈ SW_P do
5:           Local_scheduling_step_j();
6:       end for
7:       SYNCH(num_sw_p);
8:       for each SW process sw_p_j ∈ SW_P do
9:           Local_execution_step_j();
10:      end for
11:      SYNCH(num_sw_p);
12:   end while
13: end while
```

Figure 6: Global scheduler pseudocode

DAG of Fig. 4(a). At each simulation cycle, p_0, p_2, and p_6 are first executed. Then, p_1, p_4, and p_7 execute after the first delta cycle, while p_3 and p_5 execute after the second delta cycle. The longest path in the HPC-DAG represents the number of delta cycles involved in a simulation cycle. The user-defined processes of M_{HDL} that are scheduled to be run in the same simulation (delta) cycle identify the tasks of M_{SW} that can be run in parallel.

The HPC-DAG and the corresponding process execution order are exploited for assigning tasks to parallel SW processes as follows. The set of user-defined processes of M_{HDL} represented by vertex with no incoming edges identify the tasks to be uniformly distributed to the parallel SW processes of M_{SW}. The set of user-defined processes that can be scheduled after a delta cycle are distributed and added to the SW processes according to a greedy algorithm that aims at minimizing communication between tasks associated to different SW processes of M_{SW}. Finally, each SW process is completed by assigning the scheduling task t^s of each user-defined task of the set. Considering the example of Fig. 4: $\{t_0, t_1, t_0^s, t_1^s\}$ are assigned to sw_p_0, $\{t_2, t_3, t_4, t_2^s, t_3^s, t_4^s\}$ are assigned to sw_p_1, and $\{t_5, t_6, t_7, t_5^s, t_6^s, t_7^s\}$ are assigned to sw_p_2.

3.4 Communication and synchronization

A *local scheduler* (ls_i) for each SW process sw_p_i and a *global scheduler* (gs) are used to handle synchronization between user-defined tasks of M_{SW} (see Fig. 3). The local scheduler uses a shared data structure, containing the two variable representation corresponding to each signal M_{HDL}, to determine if its tasks must be activated. If so, tasks are made runnable by setting of a local array(i.e., *runnable_tasks*), as shown in Fig. 5.

IP	PIs (#)	POs (#)	Gates (#)	FF (#)	RTL IP (loc)	Processes (#)
ECC	25	32	993	79	174	4
CRC	56	34	9,213	385	492	9
FFT	92	114	87,397	1,359	3,335	38

Table 1: Benchmark characteristics

The global scheduler (Fig. 6) drives the M_{SW} execution. For each computational cycle, the global scheduler reproduces the same number of simulation cycles as in the M_{HDL}, by alternating a scheduling step (rows 4-6) and an execution step (rows 8-10) until all *runnable_tasks* sets are empty (rows 2-12). The global scheduler also synchronizes the scheduling and execution steps of all parallel SW processes (*num_sw_p*) by means of the appropriate synchronization structures provided by the runtime environment.

4. EXPERIMENTAL RESULTS

The proposed methodology has been implemented with different synchronization mechanisms (i.e., *HW synchronization primitives*, *pthreads barriers*, and *spinlock*) and has been applied to three RTL IPs having different architectural characteristics: an error correction code (ECC) model, a cyclic-redundancy checking (CRC) model, and a Fast Fourier Transform (FFT) model. Table 1 summarizes their characteristics, by reporting the number of primary inputs (*PIs*), primary outputs (*POs*), gates, flip flops (*FFs*), lines of code, and processes. The generated code has been executed by using as stimula generator the original RTL testbench, abstracted to activate the parallel execution instead of the RTL synchronization points (e.g., `wait` invocations) and to set the data structure fields instead of device ports. Experiments have been conducted by compiling the generated SW with gcc 4.4.4 and running on a SMP machine 8x2.8GHz Intel Xeon with Debian 5.0 x86 O.S..

Table 2 reports the experimental results. Column *Tasks* reports the number of tasks in which the SW applications has been decomposed (which corresponds to the number of processes of the starting RTL models). Column *Thr. or Pr.* reports the number of SW processes in which the tasks are grouped, each one mapped to a different CPU. Columns *Time* and *Speedup* report, respectively, the wallclock time of execution and the speedup with respect to the reference implementation (i.e., the sequential code).

The results show that the inherent overhead of synchronization is not negligible and mainly depends on two factors. The first is the number of processes of the starting RTL IP that communicate each other (i.e., how many edges exist in the HPC-DAC of the IP model). The results related to the FFT module are the worst as it consists of many asynchronous and communicating processes. The second factor is the number of sequential statements in the body of the synchronous processes. Since the generated tasks synchronize at list at each clock cycle, this factor mainly determines the computation over communication ratio. For an RTL IP designed for a short clock period, the number of sequential statements will be low and the synchronization frequency will be dominant.

We found that the average overhead for each synchronization step of the three evaluated mechanisms are the following: 49.2K, 36.6K, and 1.8K clock cycles for HW primitives, pthread barriers and spinlocks, respectively. As a consequence, to take advantage of parallelism, the synchronous

IP	Lines of code	Tasks (#)	Thr. or Pr.(#)	Time (s)	Speedup
ECC seq. code	102	1	1	4.02	1.00x
ECC-HW prim.	244	4	4	15.11	0.27x
ECC-barriers	250	4	4	12.35	0.33x
ECC-spinlocks	256	4	4	9.52	0.42x
CRC seq. code	247	1	1	9.91	1.00x
CRC-HW prim.	581	9	4	24.24	0.41x
CRC-barriers	589	9	4	17.11	0.58x
CRC-spinlocks	597	9	4	14.42	0.69x
FFT seq. code	876	1	1	5.67	1.0x
FFT-HW prim.	3,412	38	4	171.61	0.03x
FFT-barriers	3,427	38	4	98.12	0.06x
FFT-spinlocks	3,442	38	4	9.44	0.6x

Table 2: Experimental results: time comparison and speed-up

processes should consume at least 7.2K floating point operations between two synchronization points if using HW primitives, 5.5K if pthread barriers, and 196 floating point operations if spinlocks. The evaluation of synchronization mechanisms provided by lighter operating systems (e.g., RTEM O.S. [9]) and those provided by the AMD Fusion Family of APUs [2](in which the tasks would be assigned to GPUs) is part of our current and future work.

5. CONCLUDING REMARKS

This paper presented a methodology to automatically generate parallel SW IPs from RTL models for heterogeneous multicore architectures. The methodology has been tested on RTL IPs with different architectural characteristics and by using three different synchronization mechanisms.Future work will be focused on the exploitation of optimization techniques for aggregating tasks in order to better balancing the computational effort and in producing ad-hoc code for more efficient parallel architectures such-as GP-GPUs.

6. REFERENCES
[1] Aldec DVM. http://www.aldec.com.
[2] AMD. The AMD Fusion Family of APUs. http://fusion.amd.com.
[3] N. Bombieri, F. Fummi, V. Guarnieri, F. Stefanni, and S. Vinco. Efficient implementation and abstraction of systemc data types for fast simulation. In *Proc. of IEEE FDL*, pages 125–131, 2011.
[4] N. Bombieri, F. Fummi, and G. Pravadelli. Abstraction of RTL IPs into embedded software. In *Proc. of ACM/IEEE DAC*, pages 24–29, 2010.
[5] P. Coussy, D. D. Gajski, M. Meredith, and A. Takach. An introduction to high-level synthesis. *IEEE Design and Test of Computer*, 13(24):8–17, 2009.
[6] F. Herrera, H. Posadas, P. Sanchez, and E. Villar. Systematic embedded software generation from SystemC. In *Proc. of ACM/IEEE DATE*, pages 142–147, 2003.
[7] K. Huang, I. Bacivarov, F. Hugelshofer, and L. Thiele. Scalably distributed SystemC simulation for embedded applications. In *Proc. of IEEE SIES*, pages 271–274, 2008.
[8] A. Mello, I. Maia, A. Greiner, and F. Pecheux. Parallel simulation of SystemC TLM 2.0 compliant MPSoC on SMP workstations. In *Proc. of ACM/IEEE DATE*, pages 606–609, 2010.
[9] RTEMS Operating System. http://www.rtems.com.
[10] STMicroelectronics. SPEAr embedded microprocessors, 2010.
[11] W. Stoye, D. Greaves, N. Richards, and J. Green. Using RTL-to-C++ translation for large SoC concurrent engineering: A case study. *IEEE Electronics Systems and Software*, 1(1):20Ü–25, 2003.

Top-Down-Based Symmetrical Buffered Clock Routing

Jin-Tai Yan and Ming-Chien Huang

Department of Computer Science and Information
Engineering, Chung-Hua University,
Hsinchu, Taiwan, R. O. C.

Zhi-Wei Chen

College of Engineering,
Chung-Hua University,
Hsinchu, Taiwan, R. O. C.

ABSTRACT

It is important for a synchronous design to minimize the clock skew in a clock tree. In this paper, based on the length-matching benefit in exact routing, an efficient four-stage algorithm is further proposed to generate a symmetrical buffered clock tree with smaller clock skew under a given slew-rate constraint. For symmetrical buffered clock routing, compared with Shih's approach, the experimental results show that our proposed approach can use extra 2.54% of total resource to reduce 85.78% of clock skew in a symmetrical buffered clock tree with satisfying the slew-rate constraint for tested benchmarks in less CPU time on the average.

Categories and Subject Descriptors

B.7.2 [**Integrated Circuits**]: Design Aids – *Placement and routing*

General Terms: Algorithms, Design

Keywords: Clock routing, Buffer insertion, Skew

1. INTRODUCTION

For earlier clock trees, the works[1-3] focused on skew minimization on length or delay. Clearly, the MMM approach[1] used a hierarchical partitioning method for all the sinks and a bottom-up-based method to construct a clock tree with length skew minimization. However, the bottom-up-based MMM approach only obtained a length-matching result from clock source to all the sinks. In considering the timing effect, the concept of zero-skew design[3] in Elmore delay model was widely accepted in a complicated clock tree. To satisfy zero-skew requirement in a clock tree, the deferred-merge embedding(DME) algorithm[4-6] was proposed to reduce total wirelength in a zero-skew clock tree. In modern chip designs, slew-rate in a circuit is the maximum changing rate of signals. It is known that buffer insertion can be used to improve the clock speed or satisfy the slew-rate constraint in a clock tree. Hence, buffered clock tree[7-8] was usually applied to reduce the signal-phase latency and satisfy the slew-rate constraint.

Recently, clock skew in a buffered clock tree could be further improved by aligning a well-designed wiring structure and assigning the locations feasible buffers. Shih *et al.*[9] introduced the concept of a *symmetrical structure* for a buffered clock tree and

propose a fast heuristic approach to construct a symmetrical buffered clock tree. In the construction of a symmetrical buffered tree, the lengths of the connecting segments and the locations of inserted buffers of all paths from the clock source to all the sinks are almost the same. By using the concept of exact length-matching wiring and location-matching buffering, the latencies from the clock source to all the sinks would be near equal if the number of branches at each level in a clock tree is the same. As a result, the clock skew in a symmetrical buffer clock tree would be further minimized. Based on the symmetrical wiring and buffering feature, the skew in a symmetrical buffered clock tree could be minimized without considering any timing model. However, Shih's heuristic approach permits the phenomenon of multiple branches in a symmetrical buffered clock tree. If the number of branches is greater than 3, it is known that the symmetrical clock result may increase the routing difficulty in a Manhattan routing model. Although the phenomenon of multiple branches reduces the number of levels in a symmetrical clock tree, it may lead to use more wiring capacitance at any level and increase the difficulty of the balanced partition for all the sinks. As a result, the unbalanced partition at any level will yield more pseudo sinks to reduce the benefit of the symmetrical structure in skew minimization. Besides that, Shih's approach did not consider the effect of the attached capacitances on all the sinks and the feasible locations of inserted buffers for skew minimization in a symmetrical buffered clock tree. Due to the unbalanced distribution of the sinks at the lowest level in a symmetrical clock tree, the infeasible distribution of inserted buffers may increase the clock skew.

In this paper, based on the length-matching benefit[10] in exact routing, an efficient four-stage algorithm is further proposed to generate a symmetrical buffered clock tree with smaller clock skew under a given slew-rate constraint. In the construction of a symmetrical buffered clock tree, firstly, a balanced trunk is constructed. Next, based on the construction of exact routing, level-by-level symmetrical branches are iteratively constructed. Furthermore, timing-matching isolated leaves are constructed to balance the timing delays of all the isolated leaves. Finally, feasible buffers are inserted to satisfy the given slew-rate constraint. For symmetrical buffered clock routing, compared with Shih's approach[9], the experimental results show that our proposed approach can use extra 2.54% of total resource to reduce 85.78% of clock skew in a symmetrical buffered clock tree with satisfying the slew-rate constraint for tested benchmarks in less CPU time on the average.

2. PROBLEM FORMULATION

Given a set of n clock sinks, $\{s_1, s_2, ..., s_n\}$, with their coordinates, $(x_1, y_1), (x_2, y_2),..., (x_{n-1}, y_{n-1})$ and (x_n, y_n), inside a routing plane, it is assumed that the n clock sinks can be sorted as a horizontal sink list, $s_{\alpha(1)}\text{->}s_{\alpha(2)}\text{-> }...\text{->}s_{\alpha(n)}$, in a x-directional increasing order and the n clock sinks can be sorted as a vertical sink list, $s_{\beta(1)}\text{->}s_{\beta(2)}\text{-> }...\text{->}s_{\beta(n)}$, in a y-directional increasing order, where $\alpha()$ and $\beta()$ are a permutation function of $(1, 2,..., n)$, respectively. To obtain a more balanced partition of the n clock sinks, the *horizontal balanced region*(HBR) of the n clock sinks inside the routing plane can be defined as a x-directional region, $[x_{\alpha(n/2)}, x_{\alpha(n/2+1)}]$, if n is even or $[x_{\alpha((n-1)/2)}, x_{\alpha((n+1)/2+1)}]$, if n is odd, where $(x_{\alpha(i)}, y_{\alpha(i)})$ is the coordinate of the sink, $s_{\alpha(i)}$, $1 \leq i \leq n$. Similarly, the *vertical balanced region*(VBR) of the n clock sinks inside the routing plane can be defined as a y-directional region, $[y_{\beta(n/2)}, y_{\beta(n/2+1)}]$, if n is even or $[y_{\beta((n-1)/2)}, y_{\beta((n+1)/2+1)}]$, if n is odd, where $(x_{\beta(i)}, y_{\beta(i)})$ is the coordinate of the sink, $s_{\beta(i)}$, $1 \leq i \leq n$.

For a synthesized clock tree, if the lengths of all the branches at the same level are the same, the clock tree will be defined as a *symmetrical clock tree*. For a synthesized symmetrical clock tree, if the locations of inserted buffers on all the branches at the same level are the same, the symmetrical clock tree will be defined as a *symmetrical buffered clock tree*. It is known that the symmetrical buffering and wiring structure in a synthesized clock tree can be timing-model independent. For symmetrical buffered clock-tree synthesis(SBCTS), given a clock source, s_0, a set of n clock sinks, $\{s_1, s_2, ..., s_n\}$, inside a routing plane, a slew-rate constraint, t, and a library of buffers, $\{b_1, b_2, ..., b_m\}$, the SBCTS problem is to construct a symmetrical buffered clock tree to minimize its clock skew with satisfying the given slew-rate constraint. Given a clock source, s_0, a set of 14 clock sinks, $\{s_1, s_2, ..., s_{14}\}$ inside a routing plane with an array of 72x72 grids, it is assumed that a single wire width is used and a slew-rate constraint based on a single type of buffers is modeled as an upper bound of the driving length. If the upper bound of the driving length for a given slew-rate constraint is set as 12, a symmetrical buffered clock tree with 47 buffers can be constructed for skew minimization and shown in Figure 1.

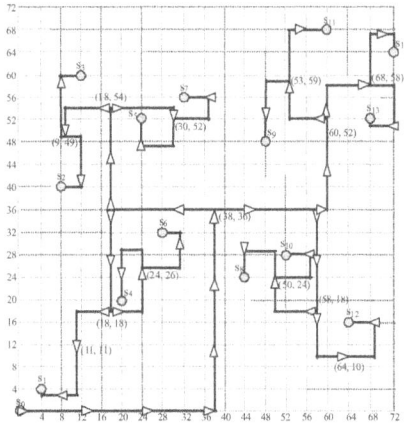

Figure 1 A symmetrical buffered clock tree of 14 given sinks

3. SYMMETRICAL BUFFERED CLOCK ROUTING

To minimize the clock skew, the trunk and branches of a symmetrical clock tree can be constructed by using exact length-matching routing and the remaining isolated leaves of a symmetrical clock tree can be constructed by using timing-matching connections. Furthermore, the feasible buffers can be inserted to satisfy the given slew constraint. For symmetrical buffered clock routing, the routing process can be divided into four main stages: *Balanced trunk construction, Iterative construction of symmetrical branches, Timing-matching connections of isolated leaves for skew minimization* and *Slew-constrained buffer insertion*.

3.1 Balanced Trunk Construction

Given a clock source, s_0, and a set of n clock sinks, $\{s_1, s_2, ..., s_n\}$ inside a routing plane, the HBR and VBR of the n clock sinks

inside the routing plane can be found. Based on the exact length-matching routing, the horizontal and vertical geometrical centers, x_c and y_c, of the n clock sinks can be further obtained by using the coordinates, (x_1, y_1), (x_2, y_2),..., (x_{n-1}, y_{n-1}) and (x_n, y_n), of the n clock sinks. In the destination assignment of a balanced trunk, as the geometrical center at (x_c, y_c) is located inside HBR or VBR, the center at (x_c, y_c), will be assigned as the destination of the balanced trunk. In contrast, as the geometrical center at (x_c, y_c), is not located inside HBR or VBR, the center at (x_c, y_c) will be replaced with the nearest coordinate inside HBR or VBR from the center at (x_c, y_c) to maintain the balanced result and the replaced center will be assigned as the destination of a balanced trunk. If the destination of a balanced trunk is located inside the HBR, it means that all the sinks will be firstly partitioned into two horizontal subsets. In contrast, if the destination of a balanced trunk is located inside the VBR, it means that all the sinks will be firstly partitioned into two vertical subsets. Clearly, the balanced trunk in the synthesized clock tree can be constructed by using a routing path from the clock source, s_0, to the trunk destination in (x_c, y_c). Refer to the 14 given sinks in a clock tree in Figure 1, by using the exact routing for the first sub-segment, the positions of the horizontal and vertical geometrical centers can be obtained as 38 and 36. Since the coordinate of the geometrical center, (38, 36), is located inside the HBR, the geometrical center can be assigned as the trunk destination in the symmetric clock tree. Since the coordinate of the clock source is (0, 0), the balanced trunk of the clock tree can be obtained from the clock source to the trunk destination.

3.2 Iterative Construction of Symmetrical Branches

After the balanced trunk from the clock source, s_0, to the trunk destination, (x_c, y_c), in a symmetrical clock tree is constructed, the routing plane can be partitioned into four routing sub-planes by using a "+"-type wiring structure at the destination, (x_c, y_c). In iterative construction of symmetrical branches, based on the exact length-matching routing, the level-by-level symmetrical branches can be constructed as follows: Firstly, the destination, (x_c, y_c), is assigned as the branch node at the first level. Based on the coordinate of the branch node, any non-zero horizontal coordinate, x_i, $1 \leq i \leq n$, must be modified as the absolute value, $\|x_i\|-\|x_c\|$, and any non-zero vertical coordinate, y_i, $1 \leq i \leq n$, must be modified as the absolute value, $\|y_i\|-\|y_c\|$, in non-zero length modification. According to the result of the modified coordinates, (x_i, y_i), $1 \leq i \leq n$, for n clock sinks, the global geometrical center at (x'_c, y'_c) of the modified coordinates for n clock sinks can be obtained and the initial coordinates of four global branch nodes at the second level can be obtained as $(x_c+x'_c, y_c+y'_c)$, $(x_c-x'_c, y_c+y'_c)$, $(x_c-x'_c, y_c-y'_c)$ and $(x_c+x'_c, y_c-y'_c)$, inside four partitioned routing planes, respectively. On the other hand, the local geometrical center at (x^i_c, y^i_c) of the modified coordinates for all the sinks inside the i-th partitioned routing sub-plane, $1 \leq i \leq 4$, can be obtained and the coordinates of four local branch nodes at the second level can be obtained as $(x_c+x^1_c, y_c+y^1_c)$, $(x_c-x^2_c, y_c+y^2_c)$, $(x_c-x^3_c, y_c-y^3_c)$ and $(x_c+x^4_c, y_c-y^4_c)$, inside four partitioned routing planes, respectively. For all the sinks inside the i-th partitioned routing sub-plane, $1 \leq i \leq 4$, its corresponding HBR(VBR) can be assigned as the balanced region of the sub-plane if the center at (x_c, y_c) is located inside its corresponding HBR(VBR). By assigning the sum of two values, x'_c and y'_c, as the length of the symmetrical branch from the only branch node at the first level to any branch node at the second level, a *Manhattan circular boundary*(MCB) of available locations with the assigned length, $x'_c+y'_c$, from the center at $(x_c,$

y_c), can be constructed. If the initial coordinate of any global branch node in the second level is not located inside its balanced region, the branch node will be reassigned on the location onto the MCB and inside its balanced region that is the nearest to the corresponding local branch node. Based on the final coordinates of four global branch nodes, two "Y"-type branches with two introduced internal nodes from the only branch node at the first level to four global branch nodes at the second level can be constructed according to the location of the geometrical center at (x_c, y_c).

By iteratively using the exact routing to assign the global branch nodes at the next levels, the level-by-level symmetrical "Y"-type branches can be sequentially obtained in the iterative construction of symmetrical branches. Until the number of the clock sinks inside any partitioned routing sub-plane is fewer than or equal to 4, the level-by-level construction of symmetrical "Y"-type branches will stop. Refer to the position of the geometrical center of the 14 sinks, the routing plane is partitioned into four routing sub-planes. By using the exact length-matching routing, the global geometrical center of the modified coordinates for the 14 clock sinks is at (20, 18) and the length of the next symmetric branch is 38. Hence, the coordinates of four global branch nodes in the second level can be obtained as (58, 54), (18, 54), (18, 18) and (58, 18), and the coordinates of four local branch nodes in the second level can be obtained as (60, 58), (20, 50), (16, 18) and (54, 22). Because the coordinate of the global branch node, (58, 54), is not located inside its corresponding balanced region, the coordinate of the branch node will be moved to the coordinate, (60, 52), under the length constraint onto the MCB. Finally, two "Y"-type branches with two introduced internal nodes at (18, 36) and (58, 36), from the center at (38, 36), to four branch nodes at (60, 52), (18, 54), (18, 18) and (58, 18) can be assigned as shown in Figure 2.

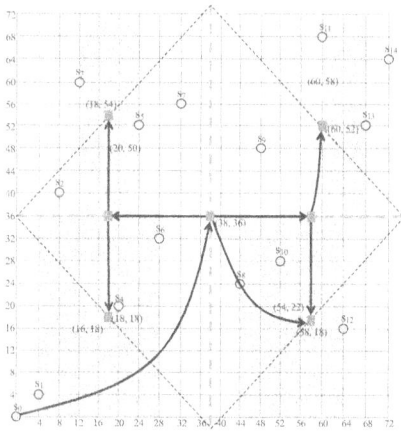

Figure 2 Iterative construction of symmetrical "Y"-type branches

As the number of the clock sinks inside any partitioned routing sub-plane is fewer than or equal to 4, the sub-plane can be only partitioned into two routing sub-planes by using a horizontal or vertical line through the corresponding global branch node. By using the exact routing to assign two global branch nodes at the final level in the partitioned sub-planes, two symmetrical branches at the final level can be further constructed inside any partitioned routing sub-plane and the iterative construction of symmetrical branches can be completed. Refer to the construction of symmetrical "Y"-type branches in Figure 2, any routing sub-plane only contains 3 or 4 clock sinks. By using the exact routing, the global geometrical center of the modified coordinates for the 14 clock sinks is at (7, 7) and the length of the next symmetric branch

is 14. Finally, the coordinates of 8 global branch nodes at the final level can be obtained as (65, 58), (53, 59), (30, 52), (9, 49), (11, 11), (24, 26), (50, 24) and (64, 10), and 8 symmetrical branches can be constructed as shown in Figure 3.

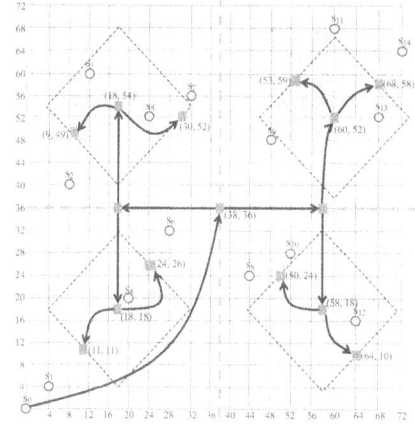

Figure 3 Construction of symmetrical branches at the final level

3.3 Timing-Matching Connections of Isolated Leaves for Skew Minimization

After level-by-level symmetrical branches are constructed, all the branch nodes at the final level will be directly connected to the corresponding sinks to construct isolated leaves in the clock tree. To minimize the clock skew in the clock tree, it is important to assign the timing-matching lengths of all the leaf links between the branch nodes and the corresponding sinks. For any sink, s_i, in n clock sinks, its leaf link, L_i, has the connected minimum Manhattan length, l_i, and the attached capacitance, C_i. Based on the assumption of RC delay computation, any leaf link can be modeled as a π–type RC connection. For any leaf link, L_i, with its length, l_i, and its capacitance, C_i, the addition(subtraction) of an extra capacitance, ΔC_i, on the attached capacitance, C_i, can be delay-equivalent to the addition(subtraction) of an extra length, Δl_i, on the length, l_i, in RC delay computation. By using RC delay computation, if a given extra capacitance, ΔC_i, is added on the attached capacitance, C_i, in the leaf link, L_i, the extra delay-equivalent length, Δl_i, can be subtracted on the length, l_i, as

$$\Delta l_i = \sqrt{(l_i + \frac{C_i}{c_0})^2 + \frac{2l_i \Delta C_i}{c_0}} - (l_i + \frac{C_i}{c_0}),$$

where c_0 is the wiring capacitance per unit length. Similarly, if a given extra capacitance, ΔC_i, is subtracted on the capacitance, C_i, in the leaf link, L_i, the extra delay-equivalent length, Δl_i, can be subtracted on the length, l_i, as

$$\Delta l_i = (l_i + \frac{C_i}{c_0}) - \sqrt{(l_i + \frac{C_i}{c_0})^2 - \frac{2l_i \Delta C_i}{c_0}}.$$

In the assignment of timing-matching lengths for all the n leaf links, L_1, L_2,... and L_n, firstly, the sink, s_j, $1 \leq j \leq n$, with minimum attached capacitance, C_j, can be found. For any sink, s_i, $1 \leq i \leq n$, $i \neq j$, with its attached capacitance, C_i, the extra capacitance, ΔC_i, can be obtained as $(C_j - C_i)$ and the extra delay-equivalent length, Δl_i, can be obtained. Next, the sink, s_k, $1 \leq k \leq n$, with the maximum total length, $(l_k + \Delta l_k)$, can be treated as the sink with the maximum latency. Furthermore, for any sink, s_i, $1 \leq i \leq n$, with its attached capacitance, C_i, the extra capacitance, ΔC_i, can be obtained as $|C_i - C_k|$ and the extra delay-equivalent length, Δl_i, can be obtained. Finally, the timing-matching length on the leaf link, L_i, can be assigned as $(l_k - \Delta l_i)$ if $C_i \geq C_k$ or $(l_k + \Delta l_i)$ if $C_i < C_k$. Refer to the construction of level-by-level symmetrical branches in Figure 3, 14 branch nodes at the final level must be connected to14 corresponding sinks to form 14 leaf links. If it is assumed that the attached capacitances on 14 clock sinks are the same, the timing-matching length on any leaf link will be assigned as 16 and 14 timing-matching isolated leaves will be constructed as shown in Figure 4.

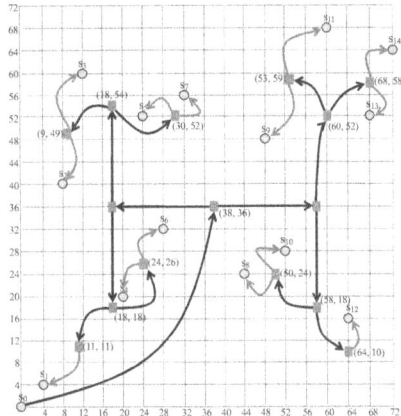

Figure 4 Timing-matching routing of isolated leaves

3.4 Slew-Constrained Buffer Insertion

To satisfy the given slew-rate constraint, feasible buffers must inserted to reduce the driving capacitance. In the construction of a symmetrical clock tree, the binary-tree structure from the clock source to all the branch nodes at the final level is guaranteed to be symmetrical. To guarantee the zero-skew design from the clock source to all the branch nodes at the final level, an identical buffer must be inserted on any branch node at the final level. For slew-constrained buffer insertion, a two-way approach is proposed to insert feasible slew-constrained buffers onto a balanced trunk, some hierarchical branches and some leaf links. It is assumed that a given slew-rate constraint based on some types of buffers is modeled as a set of upper bounds of the driving capacitances and the slew rate is approximated by adding the wiring capacitance starting from the latest inserted buffer.

For the binary-tree structure from the clock source to all the branch nodes at the final level, the buffer insertion can be done level-by-level from the branch nodes at the final level to the clock source in a bottom-up manner. By tracing along a tree edge or combing two tree edges, based on different resource-constrained factors, identical buffers from the buffer library are inserted for all branches at the same level if the slew rate is equal to the given constraint. On the other hand, for the link structure from any branch node to its corresponding sink, the buffer insertion is done in a top-down manner. By tracing along any leaf link, based on different resource-constrained factors, identical buffers from the buffer library are inserted for all leaf links if the slew rate is equal to the given constraint. Clearly, symmetrical buffering from the branch nodes at the final level can be performed in a two-way style. Refer to the symmetrical clock tree in Figure 4, it is assumed that only one type of buffers is used for slew-constrained buffer insertion and the upper bound of the driving length for a given slew-rate constraint is set as 12. By running two-way slew-constrained buffer insertion, 47 buffers can be inserted in the clock tree and shown in Figure 1.

4. EXPERIMENTAL RESULTS

For symmetrical buffered clock routing, our proposed approach has been implemented by using standard C++ language and run on a Pentium QuadCore CPU 2.66 GHz machine with 2GB memory.

Four benchmarks, *ispd09f11*, *ispd09f12*, *ispd09f21*, and *ispd09f22*, in ISPD'09 clock network synthesis contest[11] and five IBM benchmarks[3], *r1*, *r2*, *r3*, *r4* and *r5*, are used to test the clock skew in a symmetrical buffered clock tree. For fair comparison, ngspice[12] simulation based on 45nm process technology is used to evaluate the clock skew. In the experiment, clock skew from ngspice simulation, total resource usage and running time are reported in Table I and II. Compared with Shih's approach[9], for ISPD'09 benchmarks, the experimental results in Table I show that our proposed approach uses extra 2.19% of total resource to reduce 86.88% of clock skew in less CPU time on the average. On the other hand, for IBM benchmarks, the experimental results in Table II show that our proposed approach uses extra 2.88% of total resource to reduce 84.67% of clock skew in less CPU time on the average. As a result, our proposed approach only uses extra 2.54% of total resource to reduce 85.78% of clock skew in a symmetrical buffered clock tree for tested benchmarks in less CPU time on the average.

Table I Experimental results for ISPD'09 benchmarks

	#sinks	Shih's Approach[9]			Our Approach		
		Skew(ps)	Cap(fF)	Time(s)	Skew(ps)	Cap(fF)	Time(s)
ispd09f11	121	0.110	95749	0.05	0.000	98627	0.02
ispd09f12	117	0.051	96609	0.05	0.000	98974	0.02
ispd09f21	117	2.321	108755	0.06	0.246	109938	0.03
ispd09f22	91	1.160	69696	0.02	0.232	71374	0.02

Table II Experimental results for IBM benchmarks

	#sinks	Shih's Approach[9]			Our Approach		
		Skew(ps)	Cap(fF)	Time(s)	Skew(ps)	Cap(fF)	Time(s)
r1	267	1.510	13829	0.07	0.318	14072	0.05
r2	598	1.770	31056	0.28	0.307	31873	0.15
r3	862	2.310	44188	1.05	0.324	45116	0.27
r4	1903	2.540	98450	3.35	0.336	102392	0.61
r5	3101	3.010	171228	5.56	0.423	175626	0.93

5. CONCLUSIONS

Based on the length-matching benefits in exact routing and the timing-matching connections, an efficient algorithm is proposed to generate a symmetrical buffered clock tree with smaller clock skew under a given slew-rate constraint.

6. REFERENCES

[1] M. Jackson, A. Srinivasan, and E. S. Kuh, "Clock routing for high performance ICs," *Design Automation Conference*, pp.573–579, 1990.

[2] A. Kahng, J. Cong, and G. Robins, "High-performance clock routing based on recursive geometric matching," *Design Automation Conference*, pp.322–327, 1991.

[3] R. S. Tsay, "Exact zero skew," *Design Automation Conference*, pp.336–339, 1991.

[4] K. D. Boese and A. B. Kahng, "Zero-skew clock routing trees with minimum wirelength," *International Conference on ASICON*, pp.17–21, 1992.

[5] T. H. Chao, Y. C. H. Hsu, and J. M. Ho, "Zero skew clock net routing," *Design Automation Conference*, pp.518–523, 1992.

[6] M. Edahiro, "A clustering-based optimization algorithm in zero-skew routings," *Design Automation Conference*, pp.612–616, 1993.

[7] R. Chaturvedi and J. Hu, "Buffered clock tree for high quality IC design," *International Symposium on Quality Electronic Design*, pp.381–386, 2004.

[8] X. W. Shih, C. C. Cheng, Y. K. Ho, and Y. W. Chang, "Blockage-avoiding buffered clock-tree synthesis for clock latency-range and skew minimization," *Asia South-Pacific Design Automation Conference*, pp.395-400, 2010.

[9] X. W. Shih and Y. W. Chang, "Fast timing-model independent buffered clock-tree synthesis," *Design Automation Conference*, pp.80–85, 2010.

[10] M. M. Ozdal and R. F. Hentschke, "Exact rout matching algorithms for analog and mixed signal integrated circuits," *International Conference on Computer-Aided Design*, pp.231-238, 2009.

[11] C. N. Sze, P. Restle, G. J. Nam, C. Alpert, "ISPD2009 clock network synthesis contest," *International Symposium on Physical Design*, pp.149–150, 2009.

[12] http://ngspice.sourceforge.net/

Optimal Register-Type Selection during Resource Binding in Flip-Flop/Latch-Based High-Level Synthesis

Keisuke Inoue   and   Mineo Kaneko
Japan Advanced Institute of Science and Technology (JAIST)
1-1 Asahidai, Nomi, Ishikawa, 923-1292 JAPAN
{k-inoue, mkaneko} @jaist.ac.jp

ABSTRACT

Flip-flop (FF)/latch-based design has advantages on such as area and power compared to single register-type design (only FFs or latches). Considering FF/latch-based design at high-level synthesis is necessary, because resource binding process significantly affects the quality of resulting circuits. A major downside of FF/latch-based design is the increase in resources (functional units and registers) due to the modification of the lifetimes of operations and data. Therefore, as a first step, this paper addresses the datapath design problem in which resource binding and register-type selection are simultaneously optimized for resource optimization. An efficient comprehensive framework is presented, which has flexibility to incorporate other design objectives. Experiments show that the proposed approach can generate resource-efficient FF/latch-based datapaths.

Categories and Subject Descriptors

B.5.1 [Register-Transfer-Level Implementation]: Design—Data-path Design

General Terms

Design

Keywords

Register-type selection, FU binding, register binding

1. INTRODUCTION

In traditional datapath design, flip-flops (FFs) are predominantly used as the elements of registers. A major reason is that each function block between FF-based registers can be separated in view of timing, which makes timing analysis simple. Instead of FFs, using transparent latches (latches for short) has a large potential to improve die-area, power dissipation, and propagation delay, mainly due to the

GLSVLSI'12, May 3–4, 2012, Salt Lake City, Utah, USA.
Copyright 2012 ACM 978-1-4503-1244-8/12/05 ...$10.00.

fact that latch-based registers have around half of the number of transistors, compared to FF-based registers. A disadvantage of latches, however, is that they are subject to races that violate hold time constraints (hold violation), because transparency allows short-path delay to pass through adjacent function blocks, different from timing analysis in FF-based design. Two-phase non-overlapping clocking scheme was used to fix the hold violation, which in turn increases the design cost to manage two clocks [1].

It is a promising way to combine use of FF-based and latch-based registers (FF/latch-based design) to maximize advantages from each other, rather than a single register-type design. In logic-level, FF/latch-based design with single clocking scheme has been studied to overcome the design overhead of using two clocks (e.g., [2]). These approaches tried to find out the FFs that do not change the behavior of a circuit when replaced with latches. However, sometimes it is difficult to fully extract circuit potential, since the number of replaceable FFs is affected by resource (functional unit (FU) and register) binding in high-level synthesis (HLS).

In this paper, we propose a resource binding technique in FF/latch-based HLS, which considers the resource cost as the objective along with the hold constraint. We introduce a novel HLS task, namely *register-type selection*, which is a decision of the register-type (FF-based or latch-based) of registers. Register-type selection is also optimized during resource binding. Our technique generates a resource binding solution that validates many opportunities for FF/latch-based design, leading to a highly-optimized datapath.

2. RESOURCE SHARING CONDITIONS

2.1 FF-Based Design

Figure 1(a) shows an example scheduled data flow graph (DFG) where '+' and '*' mean addition and multiplication, respectively. The names of data are annotated next to the corresponding arcs ($\mathbf{a} \sim \mathbf{j}$). In FF-based design, operations and data are bound to FUs and registers, so that two operations and data having lifetime overlapping are not bound to a same FU and a same register, respectively. The lifetime of an operation is the time-interval (a set of continuous clock-cycles (CCs)) between the birth and death CCs of the operation, during which the operation executes. The lifetime of a data is the time-interval between the birth and death CCs of the data, where the former is the CC at which the data is generated and the latter is the latest CC at which the data is referenced as an input to another operation. Figure 1(b) shows a resource binding solution with the minimum re-

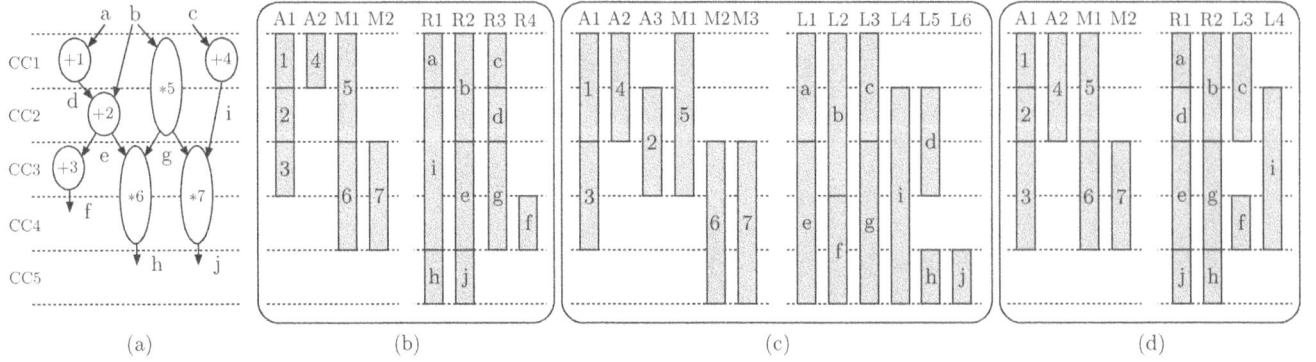

Figure 1: Resource binding solutions with the minimum resources in three different types of datapath designs. (a) An example scheduled DFG. (b) FF-based design. (c) Latch-based design. (d) FF/latch-based design.

sources, where lifetimes of operations and data are drawn by rectangles, $\mathbf{A}i$ and $\mathbf{M}i$ denote i^{th} adder and i^{th} multiplier, respectively, $\mathbf{R}i$ denote i^{th} FF-based register, and the operations and the data drawn in a same column are bound to a same FU and a same register, respectively.

2.2 Latch-Based Design

Given a resource binding solution in FF-based design, replacing FF-based registers with latch-based registers could cause timing violation. With a simple modification in the lifetimes, it is possible to synthesize datapaths with no hold violation due to register sharing. Each data should not be changed until one more CC after the death CC of the data, which can be implemented with a modification in the lifetime of data with the following equation. $t_d(a_i)$ and $t'_d(a_i)$ mean the death CC and the modified death CC of the lifetime of a data a_i, respectively.

$$t'_d(a_i) = t_d(a_i) + 1 \tag{1}$$

There is another type of hold violation in latch-based design. In general, an FU is shared by several operations in similar to registers. Then, switching of a multiplexer in front of an FU could cause hold violation, even if there is no hold violation due to register sharing. We can avoid this type of hold violation by modifying the lifetime of each operation for a similar reason to the sharing of data. $t_o(o_i)$ and $t'_o(o_i)$ mean the death CC and the modified death CC of the lifetime of an operation o_i, respectively.

$$t'_o(o_i) = t_o(o_i) + 1 \tag{2}$$

Figure 1(c) shows a resource binding solution in latch-based design with the minimum number of resources, where $\mathbf{L}i$ denote i^{th} latch-based register.

However, it requires extra resources: one adder, one multiplier, and two registers, which will negate the benefit of using latches, and increase the overall design cost.

2.3 FF/Latch-Based Design

Hold violations due to both FU and register sharing occur only if the output register is latch-based. In other words, if the output register is FF-based, the lifetimes of the relevant data and operations do not need to be extended. Therefore, using both FF-based and latch-based registers appropriately, it can be possible to synthesize resource-efficient

and hold violation-free datapaths, which is one motivation of using FF/latch-based design. Formally, (1) and (2) are modified in FF/latch-based design as follows.

(1') For each data a_i, if all the output data of operations o_j that reference a_i lastly (i.e., $t_o(o_j) = t_d(a_i)$) are assigned to FF-based registers, $t'_d(a_i) = t_d(a_i)$; otherwise, $t'_d(a_i) = t_d(a_i) + 1$.

(2') For each operation o_i, if its output data is assigned to an FF-based register, $t'_o(o_i) = t_o(o_i)$; otherwise, $t'_o(o_i) = t_o(o_i) + 1$.

Figure 1(d) shows a resource binding solution in FF/latch-based design, where two FF-based registers are replaced with latch-based registers $\mathbf{L3}$ and $\mathbf{L4}$, without increasing resources. We can see that Fig. 1(d) has the minimum resource cost among Fig. 1(a)–(d).

2.4 Problem Formulation

As a first step in the FF/latch-based HLS, we tackle the following design problem.

Problem. Given a scheduled DFG and resource constraint, find a resource binding solution and a register-type selection, so that (1') and (2') are met for all the data and operations, resource constraints are met as much as possible, and FF-based registers are replaced with latch-based registers as much as possible. □

3. HEURISTIC ALGORITHM

Figure 2 summarizes the proposed algorithm. Our algorithm first performs FU binding with FF-based registers, and completely performs register binding on the FF-based registers for a sub set of data (Steps 2–5). After that, resource binding on latch-based registers is performed for the operations and the remaining data with the modified lifetimes (Steps 6 and 7). The intuition behind our algorithm is to select and apply the minimum number of FF-based registers to the 'resource critical' parts of a given scheduled DFG, and latch-based registers to the other non-critical parts. Note that, due to the heuristic aspect, resource constraint cannot be always guaranteed. However, our algorithm tries to bind resources to meet a given resource constraint as much as possible. If the current resources are not enough, we increase the number of resources (Step 1).

Input: A scheduled DFG, and resources.
Output: A resource binding solution and a register-type selection.
Step 1: If the resources are not enough to bind the remaining unbound operations and data, then compensate for the shortfall.
Step 2: Construct the WOCG G_{cg} for unbound data.
Step 3: If there is a vertex with a positive weight, find and remove a maximum weighted path P from G_{cg}; otherwise, go to **Step 6**.
Step 4: Bind the data in P to FF-based registers.
Step 5: If all data are bound, go to **Step 7**; otherwise, go to **Step 1**.
Step 6: Modify the lifetimes of unbound data, and bind them to latch-based registers.
Step 7: Modify the lifetimes of operations according to the register binding solution, and bind them to FUs.

Figure 2: Summary of the proposed algorithm.

3.1 Preliminaries — Path-Based Binding

We construct the compatibility graph for data, which is a graph model of the compatibility of data (i.e., the possibility of register sharing of data), and commonly used to solve resource binding problem (e.g., [5]). The compatibility graph for our problem at hand is defined as follows.

Definition 1. The *weighted and ordered compatibility graph (WOCG)*, G_{cg}, is a directed acyclic graph. The vertex set is composed of vertices each of which represents a data (let a_i be the data corresponding to vertex u_i). Arcs represent the compatibility between data. That is, for each vertex-pair, the arc (u_i, u_j) is added to G_{cg} if a_i and a_j do not have lifetime overlapping even before modification, and a_i is generated earlier than a_j. $w_v(u_i)$ and $w_e(u_i, u_j)$ represent weights on u_i and (u_i, u_j), respectively. □

Register binding on FF-based registers is equal to decompose G_{cg} into a set of directed paths so that the data corresponding to one path are assigned to a same register.

3.2 Preliminaries — Terms Definition

The following are notations used in the definition of weighting functions. Let a_i be the output data of operation o_i.

- $\mathcal{M}_i$: Set of operations o_j, such that o_i and o_j are of the same operation type, the lifetime of o_j in FF/latch-based design is undecided, and $t_o(o_i) = t_o(o_j)$. Note that o_i itself can be an element of $\mathcal{M}_i$.
- m_i : The minimum number such that the lifetimes of m_i operations in $\mathcal{M}_i$ cannot be extended due to the current resource (FU) constraint.
- $\mathcal{N}_i$: Set of data a_j, such that the lifetime of a_j in FF/latch-based design is undecided, and $t_o(o_i) = t_d(a_j)$.
- n_i : The minimum number such that the lifetimes of n_i data in $\mathcal{N}_i$ cannot be extended due to the current resource (register) constraint.
- $\mathcal{Q}_i$: Sub set of data in $\mathcal{N}_i$, that are referenced by o_i.
- s_j : The number of operations that reference data a_j lastly.
- C_i : The cost of an FU of the same type with o_i.
- x_i : 0-1 variable that takes 1 if a_i must be assigned to an FF-based register.
- $t_{i,j}$: The number of registers that become FF-based if data a_i and a_j share a same register.

3.3 Vertex Weighting Function

We define the vertex weighting function so that a maximum weighted path includes the data relating to critical CCs as much as possible, leading to resource reduction. The basic idea is to heuristically quantify the impact of assigning a data to an FF-based register on resource reduction.

Reducing FUs: For each data a_i, the lifetimes of at least m_i operations in $\mathcal{M}_i$ cannot be extended due to the resource constraint. This is equal to choose at least m_i operations from $\mathcal{M}_i$ and assign their output data to FF-based registers. However, since the possible number of such combinations is exponentially increased, enumerating all the combinations explicitly is not practical. Instead, we add the weight $m_i/|\mathcal{M}_i|$ to the vertex u_i in G_{cg}, which represents the degree of priority to assign a_i to an FF-based register. We multiply $1/C_i$ to this weight so that an FU with a lower cost has a priority to be chosen.

Reducing total registers: For each data a_i, the lifetimes of at least n_i data in $\mathcal{N}_i$ cannot be extended due to the resource constraint. For the same reason with reducing FUs, we add the weight $n_i/|\mathcal{N}_i|$ to u_i in G_{cg}. The presence or absence of data-dependency, and the degree of data-dependency between a_i and $\mathcal{N}_i$ should be considered. We represent it by $\sum_{a_j \in \mathcal{Q}_i}(1/s_j)$, which is multiplied to this weight.

Reducing FF-based registers: As will be shown in the next section, the register-type for some data a_i are determined to be FF-based during the procedure. We represent it by x_i.

Based on the above observations, we define the vertex weight of u_i in G_{cg} as the weighted sum of the metrics for FUs and registers, as follows.

$$w_v(u_i) = \alpha * \frac{m_i}{|\mathcal{M}_i|} * \frac{1}{C_i} + \beta * \frac{n_i}{|\mathcal{N}_i|} * \sum_{a_j \in \mathcal{Q}_i} \frac{1}{s_j} + \gamma * x_i \quad (3)$$

where α, β, and γ are real values $(0 \leq \alpha, \beta, \gamma \leq 1)$ that represent the importance of each product term. Note that if $\mathcal{M}_i = \emptyset$ and $\mathcal{N}_i = \emptyset$ hold, we assume $m_i/|\mathcal{M}_i| = 0$ and $n_i/|\mathcal{N}_i| = 0$ to avoid zero-divide in (3), respectively.

3.4 Arc Weighting Function

It is possible that register sharing of a particular data-pair determines the register-type of some registers. To reduce the number of FF-based registers, it would be better to assign such data-pairs to different registers as much as possible. We reflect this heuristic by giving a negative weight to each arc (u_i, u_j) in G_{cg} as follows.

$$w_e(u_i, u_j) = -\delta * t_{i,j}, \quad (4)$$

where δ is a real value $(0 \leq \delta \leq 1)$ that represents the importance of this parameter.

4. EXPERIMENTAL RESULTS

We have developed a computer program based on our heuristic algorithm, implemented in C language. For comparative experiments, we have tested the following methods.

Full FF: All the registers are FF-based registers. Given resource constraints, resource binding is performed by integer linear programming (ILP) to bind operations and data using given resources as much as possible.

Full latch: All the registers are latch-based registers. The minimum resources are computed by using the left-edge algorithm.

Table 1: Results of resource binding and latch replacement.

Bench. (#data)	Full FF		Full latch		FF/latch — Conv.			FF/latch — ILP				FF/latch — Our heuristic			
	FU	FF	FU	latch	FU	FF	latch	FU	FF	latch	time[s]	FU	FF	latch	time[s]
JWF (22)	(2,2)	8	(4,3)	10	(2,2)	8	0 (0%)	(2,2)	5	3 (38%)	0.32	(2,2)	5	3 (38%)	0.02
	(3,2)	8	(4,3)	10	(3,2)	7	1 (13%)	(3,2)	2	6 (75%)	0.65	(3,2)	2	6 (75%)	0.05
	(3,3)	9	(6,3)	12	(3,3)	7	2 (22%)	(3,3)	5	4 (44%)	0.54	(3,3)	5	4 (44%)	0.02
EWF (42)	(2,3)	13	(4,4)	16	(2,3)	12	1 (8%)	(2,3)	8	5 (38%)	81.54	(2,3)	8	5 (38%)	0.04
	(3,4)	11	(6,4)	14	(3,4)	10	1 (9%)	(3,4)	6	5 (45%)	N/A	(3,4)	6	5 (45%)	0.03
	(4,4)	11	(6,4)	14	(4,4)	10	1 (9%)	(4,4)	6	5 (45%)	1833.65	(4,4)	6	5 (45%)	0.05
FDCT (50)	(5,7)	13	(8,11)	16	(5,7)	13	0 (0%)	(5,7)	9	4 (31%)	N/A	(5,7)	9	4 (31%)	0.07
	(7,9)	12	(12,8)	17	(7,9)	12	0 (0%)	(7,9)	9	3 (25%)	N/A	(7,9)	9	3 (25%)	0.05
	(8,10)	12	(12,8)	18	(8,10)	11	1 (8%)	(8,10)	10	2 (17%)	N/A	(8,10)	10	2 (17%)	0.06
FFT (97)	(10,5)	18	(18,7)	27	(10,5)	18	0 (0%)	(10,5)	—	—	N/A	(10,5)	16	2 (11%)	0.18
	(11,6)	17	(19,7)	28	(11,6)	17	0 (0%)	(11,6)	—	—	N/A	(11,6)	16	1 (6%)	0.21
	(12,7)	16	(22,8)	28	(12,7)	16	0 (0%)	(12,7)	—	—	N/A	(12,7)	16	0 (0%)	0.17
Average							5%							37%	

FF/latch — Conv.: For the resource binding solutions in "Full FF," FF-based registers are replaced with latch-based registers as much as possible, using ILP. This design corresponds to conventional latch replacement approaches in logic-level synthesis.

FF/latch — ILP: Our problem is solved by ILP. This design is used to empirically evaluate the quality-of-result and the efficiency of our heuristic algorithm.

FF/latch — Our heuristic: Our problem is solved by the heuristic algorithm proposed in Section 3.

All the ILPs were solved with the commercial ILP solver IBM ILOG CPLEX 11.0.0 [7] (the details of the above ILP formulations are omitted due to space limitation.) These five methods have been tested using 2.40 GHz AMD^R Dual Opteron^{TM} processor, and gcc4 as compiler. We used two types of FUs: ALU (addition/subtraction) and MUL (multiplication). Scheduling was performed using list scheduling algorithm [8] under the assumption that addition and subtraction are single-cycle operations, and a multiplication is a two-cycle operation. As benchmarks, we used the Jaumann Wave digital Filter (JWF), the fifth-order Elliptic Wave digital Filter (EWF), the Fast Discrete Cosine Transform (FDCT), and the 16-point Fast Fourier Transform (FFT). For the parameters α, β, γ, and δ in our heuristic algorithm, α, β, and γ are set as 1, and δ is set as 0.2. The cost of an adder and a multiplier are set as 1 and 3, respectively.

Table 1 shows the results of latch replacement, where the column "FU" denotes (#ALU, #MUL), the column "FF" and "latch" denote the number of FF-based and latch-based registers in the total registers, respectively. In the column "time[s]," 'N/A' means that the CPLEX cannot terminate within one hour, and the solution is a temporary best solution if obtained. On average, our heuristic algorithm was able to replace 37% of the total registers with latch-based registers, while "FF/latch — Conv." was able to achieve only 5%. Therefore, register-type selection should be optimized in HLS. Our heuristic algorithm outputs solutions that are the same with the ILP-based exact solutions for all the cases except FFT, and works efficiently.

5. CONCLUDING REMARKS

In this paper, we have discussed a resource binding problem in flip-flop/latch-based high-level synthesis. Considering the register-type at an early design stage, the resulting datapaths can be further optimized, compared to logic-level design. We propose a compatibility path-based heuristic resource binding algorithm to simultaneously optimize resource binding and register-type selection for total resource optimization. Experimental results showed the effectiveness of our algorithm. As a future work, we have to develop a design methodology considering the total resource cost (e.g., area and power) along with interconnection cost (i.e., wiring and multiplexers), which requires further investigation.

Acknowledgment

This work is partly supported by Grant-in-Aid of Scientific Research (C), 22560326, 2010-2012, Japan Society for the Promotion of Science.

6. REFERENCES

[1] D. Chinnery et al., "Automatic replacement of flip-flops by latches in ASICs," *Closing the Gap Between ASIC and Custom*, Springer, 2007.

[2] T.-Y. Wu and Y.-L. Lin, "Storage optimization by replacing some flip-flops with latches," *Proc. EURO-DAC*, pp. 296–301, Sep. 1996.

[3] W. Yang et al, "Low-power high-level synthesis using latches," *Proc. ASP-DAC*, pp. 462–465, Jan. 2001.

[4] Y. Chen and Y. Xie, "Tolerating process variations in high-level synthesis using transparent latches," *Proc. ASP-DAC*, pp. 73–78, Jan. 2009.

[5] T. Kim and X. Liu, "Compatibility path based binding algorithm for interconnect reduction in high-level synthesis," *Proc. ICCAD*, pp. 435–441, Nov. 2007.

[6] F.J. Kurdahi and A.C. Parker, "REAL:A program for register allocation," *Proc. DAC*, pp. 210–215, Jun. 1987.

[7] IBM ILOG CPLEX, http://www.ilog.com/

[8] G. De Micheli, *Synthesis and Optimization of Digital Circuits*, New York: McGraw Hill, 1994.

A Fully Integrated Switched-Capacitor DC-DC Converter with Dual Output for Low Power Application

Heungjun Jeon and Yong-Bin Kim
Department of Electrical and Computer Engineering
Northeastern University
Boston, MA, USA

hjeon@ece.neu.edu and ybk@ece.neu.edu

ABSTRACT

This paper presents a fully integrated on-chip switched-capacitor (SC) DC-DC converter that supports two regulated power supply voltages of 2.2V and 3.2V from 5V input supply and delivers the maximum load currents up to 8mA at both of the outputs. The entire converter system uses two 2-to-1 converter blocks. The upper output voltage (3.2V) is generated from the 2-to-1_up converter by means of averaging the 5V input and the generated lower output voltage (2.2V), which is generated from 2-to-1_dw converter. Since 2-to-1_up converter is less sensitive to the bottom-plate parasitic capacitance loss, they are implemented with MOS capacitors, which show higher capacitance density ($2.7fF/\mu m^2$, $\alpha=6.5\%$) than MIM capacitors ($1fF/\mu m^2$, $\alpha=2.5\%$) while they have bigger bottom-plate parasitic capacitance ratio (α). The proposed implementation saves the area and quiescent currents for the control blocks since each block shares required analog and digital control blocks. The proposed converter is designed using high-voltage 0.35μm BCDMOS technology. Both output voltages are regulated by means of pulse frequency modulation (PFM) technique using 18-bit shift registers and digitally controlled oscillators (DCOs). Over the wide output power ranges from 5.4mW to 43.2mW, the converter achieves the average efficiency of 70.0% and the peak efficiency of 71.4%. 10-phase interleaving technique enables the output voltage ripples of the both outputs less than 1% (<40mV) of the output voltages when 400pF of output buffer capacitors are used for both outputs.

Categories and Subject Descriptors

B.7.0 [Integrated Circuits]: General

Keywords

Dual Output, Switched-Capacitor, DC-DC Converter

1. INTRODUCTION

The use of multiple supply voltages on a single chip has become very common due to the coexistence of low/high power digital circuits and analog/RF circuits in recent integrated systems. It is not desirable approach to add multiple high-efficiency off-chip DC-DC converters, which are mostly implemented with off-chip inductors, for generating multiple output voltages due to the increased cost/size, the degraded of supply impedance, and the

limited allowance for power pins. Since the integrated voltage regulators are cost and size effective and show fast load-transient response, integrating voltage conversion blocks on the silicon chip is a very attractive approach. Linear regulators have been widely used for on-chip DC-DC converters. However, the most significant drawback of linear regulators is their linear efficiency drop with increasing dropout voltage. Therefore, the alternatives are required to achieve high efficiency across a broad range of output voltages. Since the on-chip capacitors have significantly higher quality factor, higher energy density, and lower cost than on-chip inductors in standard CMOS process, SC based on-chip converter have been receiving increased attention from both academia and industry [1-5].

In this paper, a novel design technique, supporting two regulated output voltages (2.2V and 3.2V) out of 5V input at the maximum load currents up to 8mA, is proposed and designed using high-voltage 0.35μm BCDMOS technology. Two types of flying capacitors (MIM and MOS Capacitors) are used to maximize the power density and efficiency at the limited area. The proposed architecture uses closed-loop feedback control scheme by means of digitally controlled pulse frequency modulation (PFM) to regulate output voltage in the wide range of load current levels. Two sets of four dynamic comparators are used compare four reference voltage levels with the scaled output voltages to determine the mode of control. Top and bottom voltage levels are used for fast startup and fast transient response of the varying output load current while two intermediate levels are used for stably locking the output voltage. Section 2 presents the core design of proposed dual output converter in terms of operating principle and charge transfer and loss mechanisms. System architecture and simulation results are presented in Section 3 and Section 4, respectively. The paper is finally concluded in Section 5.

2. PROPOSED CORE DESIGN
2.1 Operating Principle

In general, a SC DC-DC converter consists of capacitors and switches driven by two non-overlapped clock signals. Each of signals is set as close as 50% duty cycle with a minimal dead-time (different phased ϕa and ϕb switches, as shown in Figure 2(c), are never closed at the same time) for the maximum efficiency and the maximum charge transfer to the load.

As shown in Figure 1, the proposed dual output topology is made in combinations of the two conventional 2-to-1 topologies. Conventional 2-to-1 topology in Figure 1 (a) can be symbolized as shown in Figure 1 (b), which has two input terminals and one output terminal. To present the loss due to bottom-plate parasitic capacitors, the bottom-plate parasitic capacitance C_{BP} is modeled as $C_{BP}=\alpha C$, where C is the actual capacitance and α is the process and layout dependent parameter. For an ideal operation, the output

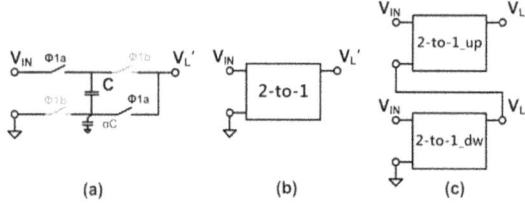

Figure 1 (a) Conventional 2-to-1 step-down topology (b) Simplified block diagram (c) Proposed 4-to-3 step-down topology

terminal produces the average voltage between the two input voltages. In the same way, new 4-to-3 topology is formed. That is, one input terminal of the 2-to-1_up is fed directly from the input (V_{IN}) and the other terminal is fed out of the output (V_L') of the 2-to-1_dw block. Therefore, the generated output voltage V_L ($=(V_{IN}+V_L')/2-\Delta V_L$) is the average value of V_{IN} and V_L' ($= 1/2V_{IN} - \Delta V_L'$). ΔV_L and $\Delta V_L'$ represent the voltage difference of the delivered output voltages when there is load and there is no load. ΔV_L and $\Delta V_L'$ arise from the fundamental conduction loss and they limit the maximum attainable efficiency to $\eta_{lin}=V_L'/(1/2V_{IN})$ for 2-to-1_dw and $\eta_{lin}=V_L/\{(V_{IN}+V_L')/2\}$ for 2-to-1_up. Figure 2 shows the transistor level implementations of the 2-to-1_dw(up) blocks and their gate driving signals. Since the breakdown voltage of MOS transistors is 5.5V, all switches can withstand any voltage levels between 0V and input (5V). All the gate driving signals in Figure 2 are generated from level shifter and non-overlap clock generator blocks in Figure 5 to minimize the switching loss and shoot-through current loss. Upper NMOS transistors are implemented by means of a triple-well device to isolate the voltage from the substrate.

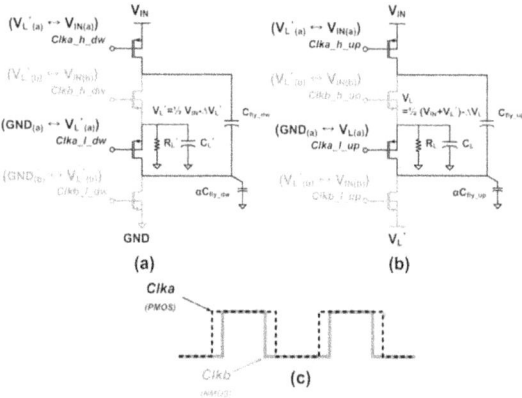

Figure 2 Transistor level implementation of the converter core; (a) 2-to-1_dw (b) 2-to-1_up (c) Non-overlap clock signals

2.2 Charge Transfer and Loss Mechanisms

Figure 3(a) shows 2-way interleaved structure, where $\phi 1a$ ($\phi 1b$) and $\phi 2a$ ($\phi 2b$) are 180° out of phase signals while $\phi 1a$ ($\phi 2a$) and $\phi 1b$ ($\phi 2b$) represent non-overlapping clock signals as shown in Figure 2(c). Figure 3(b) represents equivalent circuit during each half cycle of the clock. Assuming that those blocks deliver charges to the load capacitors at DC voltage V_L' and V_L, the charge extracted from the input voltage source (Q_{EXT}) during each half cycle of the clock ($\phi 1a$ and $\phi 2b$ are on) can be derived by

$$Q_{EXT} \approx \frac{C_{up}}{2}\{(V_{IN}-V_L)-(V_L-V_L')\} + \frac{C_{dw}}{2}\{(V_{IN}-V_L')-V_L'\}$$

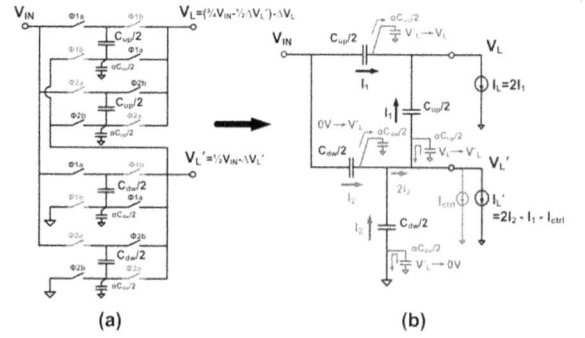

Figure 3 Proposed dual output topology with 2-way interleaved structure to provide voltages of V_L' (=2.2V) and V_L (=3.2V) out of V_{IN} (=5V) input.

$$Q_{EXT} \approx Q_{EXT(up)} + Q_{EXT(dw)} = C_{up}(\Delta V_L) + C_{dw}(\Delta V_L') \quad (1)$$

where $V_L=(3/4V_{IN}-\Delta 1/2V_L')-\Delta V_L$ and $V_L'=1/2V_{IN}-\Delta V_L'$.

Since the total charge delivered to the each load is the sum of the charge transferred from both top and bottom $C_{up}/2(C_{dw}/2)$ capacitors as shown in Figure 3(b), the total charge transferred to the load can be derived by

$$Q_L + Q_L' \approx 2Q_{EXT(VIN)}$$

$$\approx 2C_{up}\left\{\left(2.5V + \frac{V_L'}{2}\right) - V_L\right\} + 2C_{dw}(2.5V - V_L') \quad (2)$$

Since the efficiency for each output can be defined as $Q_L V_L/Q_{EXT(up)}V_{IN}$ and $Q_L'V_L'/Q_{EXT(dw)}V_{IN}$, this again sets the fundamental efficiency limitations of $V_L'/2.5V$ for the 2-to-1_dw and $V_L/(2.5V+V_L'/2)$ for the 2-to-1_up.

Besides of the conduction loss, the loss due to the bottom-plate parasitic capacitor is significant and has to be considered. In our design, MIM capacitors ($1fF/\mu m^2$) and MOS capacitors ($2.7fF/\mu m^2$) are used, and the bottom plate capacitance ratios (α) were assumed 2.5% for MIM capacitors and 6.5% of MOS capacitors. During every half cycle, each top bottom-plate capacitor $\alpha C_{up}/2$ ($\alpha C_{dw}/2$) is charged to V_L (V_L'), while each bottom bottom-plate capacitor $\alpha C_{up}/2$ ($\alpha C_{dw}/2$) is discharged to V_L (GND). The charged electrons in the bottom-plate capacitors of the 2-to-1_dw block are discharged to ground; all stored charge is dumped into ground, but for the charged electrons in the bottom-plate capacitor of the 2-to-1_up block are discharged to the load V_L'. As a result, the energy lost every cycle due to those bottom plate capacitors can be given by

$$E_{BP} \approx \alpha C_{up}(V_L - V_L')^2 + \alpha C_{dw}V_L'^2 \quad (3)$$

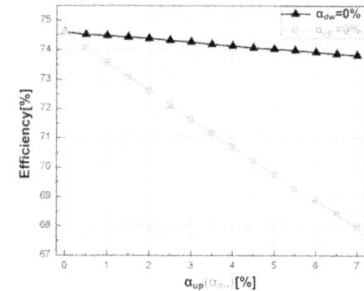

Figure 4 Efficiency drop dependencies with respect to increasing bottom-plate parasitic capacitance ratio (α=0% to 7%); Black (Grey) represents the efficiency drop with increasing α_{up} (α_{dw}) while α_{dw} (α_{up}) is kept constant at 0%.

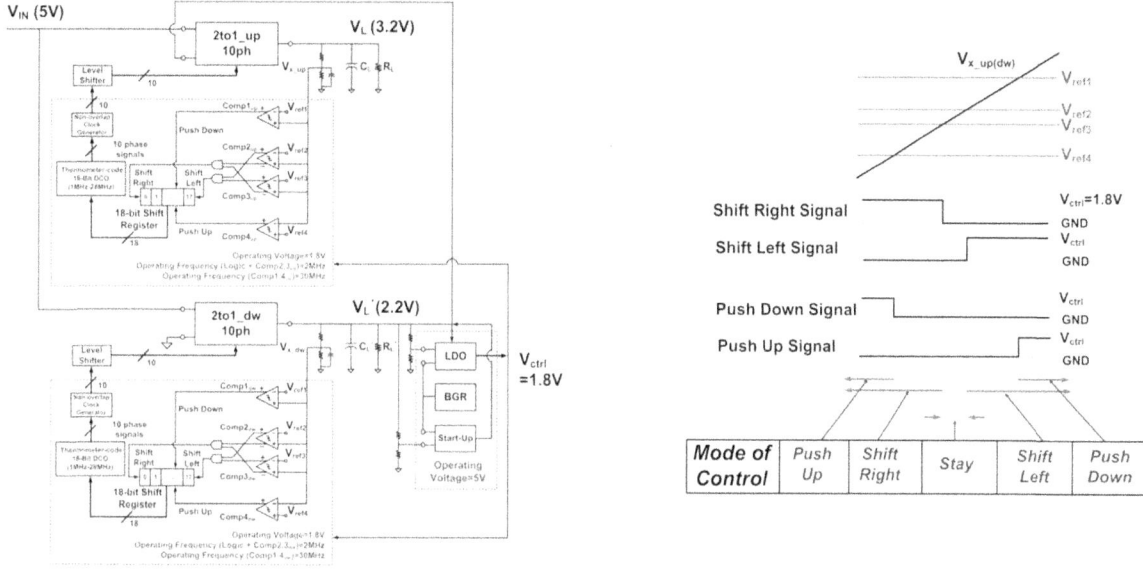

Figure 5 Architecture of dual output switched capacitor DC-DC converter system

Figure 4 shows the efficiency drop dependencies due to the increasing bottom–plate parasitic capacitance of flying capacitors used in 2-to-1_up and 2-to-1_dw while either α_{up} (when α_{dw} is swept) or α_{dw} (when α_{up} is swept) is set to 0%. Simulation results are obtained at its maximum load condition (delivering 8mA of load currents to both outputs) while the average output voltages are being maintained at $V_L' \approx 2.2V$ and $V_L \approx 3.2V$. With increasing α_{up} (0% to 7%), the total efficiency drops less than 1%, which is six times less than the efficiency drop with increasing α_{dw} (0% to 7%). 1% of efficiency drop arises from the loss during the charging phase (V_L' to V_L) of either of $\alpha_{up}C/2$ capacitor and from the increased V_L' due to the transferred charge from $\alpha_{up}C/2$ capacitors as shown in Figure 3(b) and Equation (3).

Since the overall efficiency is less sensitive to the increasing α_{up} and their larger capacitance density (2.7fF/μm^2) in 0.35μm BCDMOS Technology, MOS capacitors are used for implementing the flying capacitors of the 2-to-1_up while MIM capacitors are used for implementing 2-to-1_dw since they have less bottom-plate capacitance ratio (α_{dw}) than MOS capacitors. This trades off with bigger area since MIM capacitors have smaller capacitance density (1fF/μm^2).

The minimum required capacitances for each flying capacitor that satisfies the design requirements (the maximum load currents (I_L, I_L') of 8mA) are determined based on the load current handing capabilities of the proposed converter. From equations (1), (2) and Figure 3(b), the load current handing capabilities for both output loads at a fixed frequency and ΔV_L ($\Delta V_L'$) can be obtained by

$$I_L = 2I_1 \approx 2Q_Lf_{sw} = 4C_{up}\Delta V_Lf_{sw} \qquad (4)$$

$$I_L' + I_{ctrl} + 0.5I_L = 2I_2 \approx 2Q_L'f_{sw} = 4C_{dw}\Delta V_L'f_{sw} \qquad (5)$$

$$I_L' \approx 4C_{dw}\Delta V_L'f_{sw} - 4C_{up}\Delta V_Lf_{sw} - I_{ctrl} \qquad (6)$$

where ΔV_L and $\Delta V_L'$ represent the voltage difference of output voltages when there is load and there is no load as descried earlier in this paper. Since our target voltages are 2.2V and 3.2V, from Figure 3(a), $\Delta V_L'$ and ΔV_L are determined to be 0.3V and 0.4V, respectively. Since the maximum control current (I_{ctrl}) required by the control block is 300μA, the required control current is chosen to be 0.5mA with the margin of 200μA. Therefore, the required

$2I_2$ in Eq. (5) is 12.5mA (=8mA+0.5mA+4mA) because both I_L and I_L' are maximum output load current in this case and they are predetermined as design goals. For the given specifications (ΔV_L is 0.4V, $\Delta V_L'$ is 0.3V, and the maximum switching frequency is 28MHz), the minimum required C_{up} and C_{dw} are estimated as 178.57pF and 372pF, respectively. In our design, C_{up} of 200pF and C_{dw} of 400pF are chosen.

As can be observed from equations (4) and (6), with the fixed values of ΔV_L ($\Delta V_L'$) and C_{up} (C_{dw}), I_L (I_L') can be controlled by changing switching frequency (f_{sw}). With changing load current, therefore, the output voltage can be regulated by mean of pulse frequency modulation (PFM). In this design, PFM control scheme is used with 18bit shift register and 18bit DCO which are designed to be operating in the range of 1MHz to 28MHz.

3. ARCHITECTURE

Figure 5 shows the overall architecture of the proposed dual output DC-DC converter. The complete system consists of two 10 phase 2-to-1 blocks, two 18-bit shift registers with push-up(down) function, two 18-bit thermometer code digitally controlled oscillators (DCOs), non-overlap clock generators, level-shifters, 8 dynamic comparators [7], a low-drop output (LDO) voltage regulator and a start-up circuit. The DCO is controlled by an 18-bit thermometer code produced by the shift resister. As shown in Figure 5, the load voltages are scaled to $V_{x\_up(dw)}$ with feedback resistors, and four reference voltages (V_{ref1-4}) are generated from 5V input with resistor ladders and capacitors. Four dynamic comparators (Comp1-4$_{up(dw)}$) compare $V_{x\_up(dw)}$ to the four different reference voltages to determine the mode of control. For fast start-up and fast transient response with a large load current transition, Comp1$_{up(dw)}$ and Comp4$_{up(dw)}$ are operated with 30MHz of clock frequency while Comp2$_{up(dw)}$ and Comp3$_{up(dw)}$ are operated with 2MHz of clock frequency for stable voltage locking between V_{ref2} and V_{ref3}. If $V_{x\_up(dw)}$ is less than V_{ref4}, Push-Up mode is enabled and thermometer code transits to its maximum value, which generates the maximum switching frequency. In the similar way, if $V_{x\_up(dw)}$ is bigger than V_{ref1}, thermometer code drops to the pre-programmed value. If $V_{x\_up(dw)}$ enters between V_{ref1} and V_{ref4}, the switching frequency increases or decreases

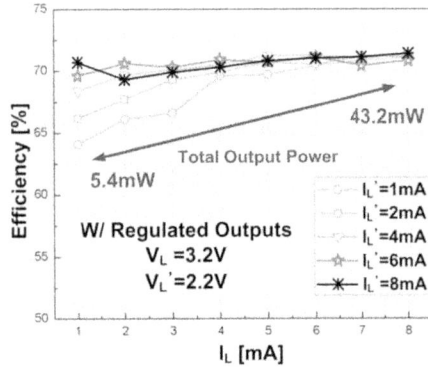

Figure 6 Efficiency versus I_L while I_L' is varying between 1mA and 8mA

linearly with the step period of 0.5μS ((2MHz)$^{-1}$) until $V_{x\_up(dw)}$ is locked between V_{ref2} and V_{ref3}.

4. SIMULATION RESULTS

The proposed SC power converter is designed using high-voltage 0.35μm BCDMOS technology and simulated with HSPICE. Two types of flying capacitors (MIM and MOS Capacitors) are used to maximize the power density and efficiency at the limited area. Each 2-to-1_up block uses 20pF of MOS capacitor (the total capacitance of 200pF for 10 phase) for its flying capacitor to maximize the power density, while each 2-to-1_dw block uses 42pF of MIM capacitor (the total flying capacitance of 420pF for 10 phase) to minimize the loss due to bottom-plate parasitic capacitance. MOS capacitors are used for the output buffer capacitors (400pF for each) to reduce the output ripple voltages and to maintain the moderate level of transient response for varying load currents. Figure 6 shows the simulated efficiency with different load current levels between 1mA and 8mA, while the output voltages are regulated at DC level of 2.2V (3.2V) for V_L' (V_L). The proposed converter achieves 70.0% of the average efficiency in the output power ranges between 5.4mW and 43.2mW and the maximum efficiency (71.4%) is achieved when it delivers the maximum power (43.2mW). The control logic blocks including BGR (Bandgap Reference) circuit and bias circuits consume the power between 0.46mW and 1mW over the operating power transfer ranges. Figure 7 shows simulated transient response with load current (I_L) transition from 1mA to 8mA, and vice versa. With Push-up and Push-down modes, the converter settles within 450ns (1μs) for 1mA (8mA) to 8mA

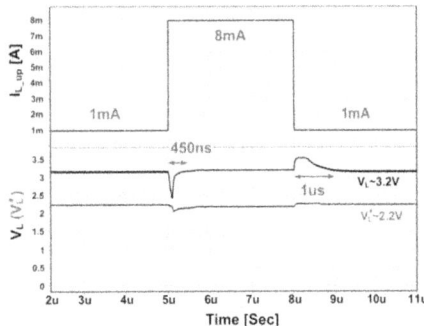

Figure 7 Transient response of V_L (V_L') with varying load current I_L (1mA to 8mA and vice versa) while I_L'=8mA

(1mA) transition while 2-to-1_dw delivers 8mA of the load current (I_L'). The interference between two outputs are inevitable, but this can be minimized with increasing clock frequency of Comp1,4$_{up(dw)}$ or increasing the load capacitance. Comparison with recently published SC converters designed using 0.35μm CMOS technology is listed in Table 1. While other SC converters are able to support only one step-down output voltage at a time, proposed converter can support two different voltages at the same time. In addition, since the switching frequency of the proposed converter is regulated digitally over wide frequency ranges between 1 and 28MHz with different load conditions, it maintains higher peak and average efficiency even with less flying and output buffer capacitance, hence less area.

Table 1 Comparison with Recently Published SC Converters

	[4]	[5]	[6]	This work
Process (CMOS)	0.35μm	0.35μm	0.35μm	0.35μm BCDMOS
V_{in}	2.5V	5V	3.4~5V	5V
V_{out} Regulated	0.9~1.5V	1V	3.3V	3.2V and 2.2V
C_{fly}	6.72nF (on-chip)	1.2nF (on-chip)	1uF (off-chip)	C_{flyup}=200pF MOS-cap C_{flydw}=420pF MIM-cap
C_{out}	470nF (off-chip)	N/A	1uF (off-chip)	400pF (x2) MOS-cap
f_{sw}	0.2~1MHz	15MHz	100kHz	1~28MHz
η	≤66.7%	31%	≤65%	≤71.4% (η_{Ave}=70.0%)
P_{out}	0.4~7.5mW	10mW	3.3~24.7mW	5.4~43.2mW
I_L(I_L')	0.5~5mA	N/A	1~7.5mA	1~8mA

5. CONCLUSION

This paper presents a fully integrated on-chip SC DC-DC converter with dual outputs (2.2V and 3.2V). The proposed converter is designed using high-voltage 0.35μm BCDMOS technology. The converter achieves the average efficiency 70.0% and the peak efficiency 71.4%. Using 10-phase interleaving technique, the output voltage ripples of the both outputs are maintained less than 1% (<4mV) of the output voltages when 400pF of output buffer capacitors are used for both outputs.

6. REFERENCES

[1] Y. Ramadass, A. Fayed, and A. Chandrakasan, "A fully-integrated switched-capacitor step-down DC-DC converter with digital capacitance modulation in 45 nm CMOS," *IEEE J. Solid-State Circuits*, vol. 45, no. 12, pp. 2557–2565, Dec. 2010.

[2] H.-P. Le, M. D. Seeman, S. R. Sanders, V. Sathe, and E. Alon, "Design Techniques for Fully Integrated Switched-Capacitor DC-DC Converter," *IEEE J. Solid-State Circuits*, vol. 46, no. 9, pp. 2120–2131, Sep. 2011.

[3] T. Van Breussegem and M. Steyaert, "Monolitic Capacitive DC-DC Converter With Single Boudary-Multiphase Cotrol and Voltage Domain Stacking in 90 nm CMOS," *IEEE J. Solid-State Circuits*, vol. 46, no. 7, pp. 1715–1727, July 2011

[4] L. Su, D.Ma, and A. Brokaw, "Design and analysis of monolithic stepdown SC power converter with subthreshold DPWM control for selfpowered wireless sensors," *IEEE Trans. Circuits Syst. I*, vol. 57, no. 1, pp. 280–290, Jan. 2010.

[5] K. P. Viraj and G. A. J. Amaratunga, "A monolithic CMOS 5V/1V switched capacitor DC–DC step-down converter," in *Proc. IEEE Power Electron. Spec. Conf.*, pp. 2510–2514, Jun. 2007.

[6] C. L.Wei and H. H. Yang, "Analysis and design of a step-down switchedcapacitor-based converter for low-power application," in *Proc. ISCAS*, 2010, pp. 3184–3187.

[7] Heungjun Jeon and Yong-Bin Kim, "A Novel Low-Power, Low-Offset, and High-Speed CMOS Dynamic Latched Comparator," *Anal. Integr. Circuits and Signal Processing*, vol. 70, no. 3, pp. 337-346, July 2011

High-level Modeling of Power Consumption in Active Linear Analog Circuits

Laurent Bousquet
TIMA Laboratory
46, Av. Félix Viallet
F-38031, Grenoble
laurent.bousquet@imag.fr

Emmanuel Simeu
TIMA Laboratory
46, Av. Félix Viallet
F-38031, Grenoble
emmanuel.simeu@imag.fr

ABSTRACT

This work presents an approach for including energy consumption information in high-level modeling of active linear electrical circuits. The method introduced here proposes to start from a description of the system at a high-level of abstraction (transfer function or state space model) and refine it in order to generate the electrical circuit corresponding, in the form of a SPICE netlist, for example. The following steps of the proposed approach automatically extract the state space representation corresponding to this circuit. During the state space representation extraction, the information needed to find the power consumption is regarded as an output of the state space model. These outputs allow an instantaneous monitoring of the power consumption of the system at a high-level of abstraction. SystemC AMS has been used to model and simulate the examples presented in this paper.

Categories and Subject Descriptors: H.1.0 [Information Systems]: Models and Principles - General; J.2 [Computer Applications]: Physical Science and Engineering – Electronic

Keywords: High-level modeling, power consumption analysis, circuit analysis, state space model, SystemC AMS

1. INTRODUCTION

The design and simulation of System-on-Chips becomes more and more challenging because of their growing heterogeneity. There is a need to have global simulations of these systems in order to validate them. In this purpose, an extension of SystemC: SystemC AMS has been developed to avoid the using of several CAD tools [1]. The advantages of this platform comparing to VHDL/VHDL-AMS are its C++ basis (SystemC and SystemC AMS are both C++ libraries), its simulation speed and the fact that an analog module is always encapsulated in a digital one, facilitating the modeling of mixed-systems. Several examples proving the legitimacy of this platform in the mixed-systems modeling are available in the literature. These works deal with chemistry [2], digital, analog and RF [3][4] or mechanical systems [5].

Three models of computation or modeling formalisms are available in SystemC AMS. The first one is called Timed Data Flow (TDF). It regards signals as uniformly sampled directed signals. The second MoC, Linear Signal Flow (LSF) regards signals as continuous-time directed signals.

The Electrical Linear Network (ELN) MoC allows the modeling of conservative and continuous-time behaviors. The TDF and LSF MoCs allow a description at higher levels of abstraction than the one offered by the ELN MoC. The benefit of using high-level models is reduced simulation time. Moreover, because of the complexity, designers need a global view of a system and it is impossible with a high-level of detail.

Nowadays, the interest for low power systems is increasing. System designers work with behavioral models (based on the specifications) that do not include the power consumption. A method has been proposed for including this information in a high-level model of a passive linear circuit [6]. In the same way, it could be interesting to have, early in the design flow, an idea of the power consumption of an active linear analog system in order to choose the one that fits the best to the specified behavior and to the power consumption specifications. So there is a need to enrich the high-level description of a system with its power consumption. The problem is that it is impossible to determine the consumption of a circuit without any information on its implementation, component values, etc.

The idea proposed in this paper is to start from a behavioral model (transfer function, state space model) which is a high-level model and refine it until the obtaining of the circuit (low-level). Then, it is possible to automatically extract relevant power consumption information from this low-level description and to reassemble them in a new high-level model so that they are propagated during the simulation. Several candidates can be implemented so that there will be several power consumptions and the user will be able to choose the one that best suits the power specifications. Figure 1 presents a global view of this work.

The paper is structured as follows: in Section II an example of a transfer function refinement process is given (Sallen-key topology filter). Section III describes the extraction of the state space equation system over a state variable filter. In Section IV the power consumption estimation is analyzed and the simulation results are shown. Finally, Section V gives the conclusions and introduces the future works.

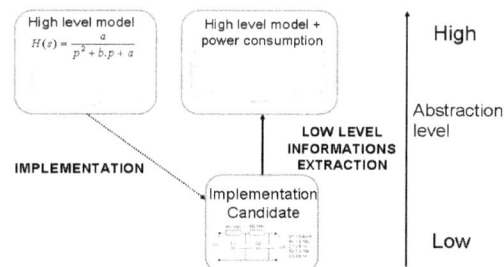

Figure 1. Global view of the method presented.

2. REFINEMENT OF A TRANSFER FUNCTION

Starting from a transfer function, we generate automatically a corresponding circuit. The basic structures of the filter can be a must, for example: using Rauch or Sallen-Key cells. There will be little differences between the two structures refinement process, but the global idea remains the same. Supposing that we want to refine a 6th order transfer function using the Sallen-Key basic cell, the global filter will be an association in series of three second order Sallen-Key cells. Figure 2 shows a Sallen-Key low pass cell and equation (1) its transfer function under the standard form.

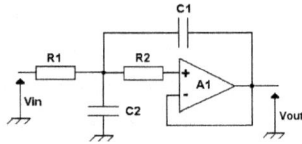

Figure 2. Sallen-Key low pass cell.

$$\frac{Vout}{Vin} = \frac{\dfrac{1}{R_1 R_2 C_1 C_2}}{p^2 + p(R_1 C_2 + R_2 C_2) + \dfrac{1}{R_1 R_2 C_1 C_2}} \quad (1)$$

Starting from a given filter specification, it is possible to generate automatically a transfer function using Matlab. The specifications of the filter are the following: the maximum attenuation under 50 KHz is 6 dB and the minimum attenuation over 100 KHz is 40 dB. The denominator coefficients of the 6th order low pass transfer function are given in table 1. The numerator is equal to the constant term of the denominator. Then, it is required to factorize the transfer function to obtain a zero-pole decomposition. In this case, there is no zero, but six poles, three pairs of conjugate complex. In order to form three 2nd order transfer functions, two conjugate poles are multiplied, the three transfer function denominators obtained are given in table 2. Each numerator is equal to the constant term.

Table 1. 6th order low pass transfer function denominator coefficients.

Term	Coefficient value
constant	$613,4506.10^{30}$
p	$8,1312.10^{27}$
p^2	$53,8886.10^{21}$
p^3	$226,4191.10^{15}$
p^4	$634,2163.10^9$
p^5	$1,1262.10^6$
p^6	1

Table 2. 2nd order low pass transfer functions denominator coefficients.

Term	1st TF coeff.	2nd TF coeff	3rd TF coeff
constant	$8,5.10^{10}$	$8,5.10^{10}$	$8,5.10^{10}$
p	$1,51.10^5$	$5,63.10^5$	$4,12.10^5$
p^2	1	1	1

After that, we can fix the resistor values (1KΩ) and by identification between each column of the table 2 and the equation (1), we are able to find the capacitor values. Table 3 shows the component values of each cell. If the Rauch cells have

been imposed for the refinement, the identification would have been applied with the Rauch cell transfer function standard form.

To make the simulations, two SystemC AMS models have been created and compared. The first one is a LSF model. It consists of a transfer function block having the coefficients given in table 1. The second is an ELN model: three cells (figure 2) associated in series, each one having the components values given in table 3.

Table 3. Passive component values of the filter

Cell	Component name	Component value
1st cell	R_{11}, R_{12}	1KΩ
	C_{11}	13,3nF
	C_{12}	0,888nF
2nd cell	R_{21}, R_{22}	1KΩ
	C_{21}	3,55nF
	C_{22}	3,31nF
3rd cell	R_{31}, R_{32}	1KΩ
	C_{31}	4,85nF
	C_{32}	2,43nF

Figure 3 shows a Bode diagram for the two models. The curves are superimposed, showing that the low-level model corresponds to the transfer function and correspond to the specifications. The simulation time cost is about 10% superior using the ELN model with respect to the LSF model, according to [7].

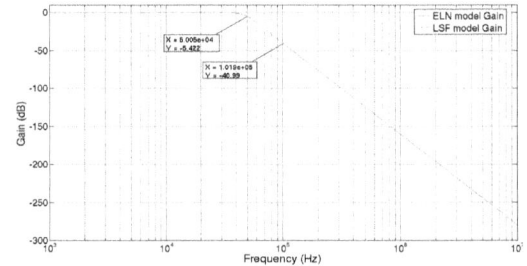

Figure 3. Bode diagram of the two compared models.

3. HIGH-LEVEL MODEL AUTOMATIC GENERATION

The type of systems we are focusing on are dynamic linear systems The state space representation allows to model these systems, its equations are shown below, equation (2) is the state equation and equation (3) is the output equation. u is the input vector, y is the output vector, x is the state vector and A, B, C and D are the time invariant matrices of the system.

$$\dot{x} = Ax(t) + Bu(t) \quad (2)$$

$$y(t) = Cx(t) + Du(t) \quad (3)$$

We will present how to generate automatically a high-level model starting from a low-level description, such as a SPICE netlist. The method will be presented onto a 2nd order state variable filter. This circuit and its netlist are shown in figure 4. The idea is to analyze the electrical circuit topology in order to extract an extended state space representation that contains the power consumption information. In the following sub-sections, it will be explained how the matrices A, B, C and D of the equations (2) and (3) are generated from a SystemC AMS circuit description or SPICE netlist. The main steps are: the generation of a matrix representing the circuit (circuital matrix), the arrangement of this matrix and the extraction of the state space matrices. This method is similar to the one presented in [6], but we propose to enrich the component set managed in order to deal with active linear circuits.

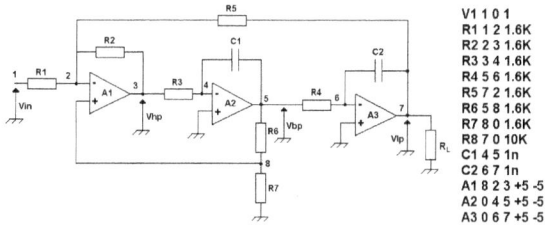

Figure 4. A 2nd order state variable filter and its netlist.

The circuital matrix is obtained from the analysis of a standard SPICE netlist. Each component introduces at least one equation. Inductances and capacitances define state equations. Resistances define a topology equation. In the following table 4, the behavioral equation and the KCL defined for each component managed by our method is shown. With respect to [6], the ideal transformer, the gyrator and the op-amp in negative reaction have been added.

Table 4. Components equations.

Component	Equation	KCL
Resistor	$-i_R.R+(u_{n1}-u_{n2})=0$	$n_1:+i_R$ $n_2:-i_R$
Capacitor	$-C\overset{\bullet}{u}_{n2n3}+i_C=0$	$n_2:+i_C$ $n_3:-i_C$
Inductor	$-L.\overset{\bullet}{i}_L+(u_{n1}-u_{n2})=0$	$n_1:+i_L$ $n_2:-i_L$
Voltage Source	$-V+(u_{n1}-u_{n2})=0$	
Current Source		$n_1:+I$ $n_2:-I$
VCVS	$-E.(u_{n1}-u_{n2})+(u_{n3}-u_{n4})=0$	
CCVS	$-H\cdot i_{n1n2}+(u_{n3}-u_{n4})=0$	
CCCS		$n_3:+F.i_{n1n2}$ $n_4:-F.i_{n1n2}$
VCCS		$n_3:+G(u_{n1}-u_{n2})$ $n_4:-G(u_{n1}-u_{n2})$
Reference Node	$u_0=0$	
Ideal transformer	$V_2.X-V_1.Y=0$ $I_1.X-I_2.Y=0$	
Gyrator	$F-\Psi.i=0$ $-V-\Psi.u=0$	
Op-amp	$-V^-+V^+=0$	

Once the equations are defined, a matrix representing the circuit can be constructed. This matrix, called the circuital matrix is shown in figure 5.The rows of the matrix correspond to the equations and the columns correspond to the variables. The matrix is filled with the coefficients of the variables. A zero in position (i,j) indicates that equation i is not dependent on the variable j. It is needed to precise that a node potential (u_n) is equal to the potential difference between this node and the reference node $(u_n = u_{n0})$.

The arrangement of the circuital matrix consists of performing some permutations between the rows and between the columns in order to obtain a decomposition of figure 6. In the decomposition, the state equations have been put in the upper rows, the desirable variables (inputs, state variables and their derivatives) in the left columns and sub-matrix X4 has been made upper triangular. The circuital matrix owns 2 state variables: u_{45} and u_{67}, thus 2 state equations on the rows of $X1$ and $X2$. The columns of $X1$ and $X3$ contain the desirable variables. Columns and rows of $X4$ have to be arbitrary swapped in order to obtain an upper triangular sub-matrix. Figure 7 shows sub matrix $X4$ after the arrangement.

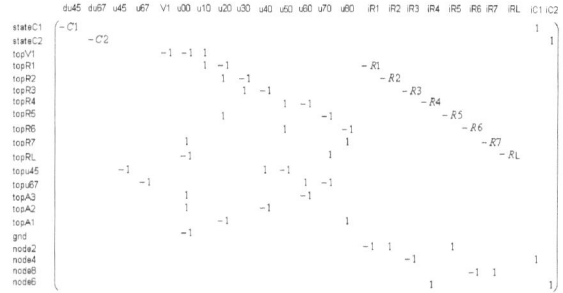

Figure 5. State variable filter circuital matrix.

The last row (n) of the arranged matrix allows to express the variable u_{00} as a function of the desirable variables. The row above allows to express u_{60} as a function of the desirable variables and u_{00}, and so on, the results are written in a relation matrix (rows: undesirable variables, columns: desirable variables). Once this relation matrix is known, the non-null terms of $X2$ are replaced by a zero and the dependency coefficients of the undesirable variable are used to adjust the corresponding row of $X1$. The matrices A and B of the state space representation are automatically extracted from $X1$, as shows the equation (4). The 2x2 matrix of the equation (4) corresponds to the matrix A of the equation (2) and the vector 2x1 corresponds to the matrix B of the equation (2).

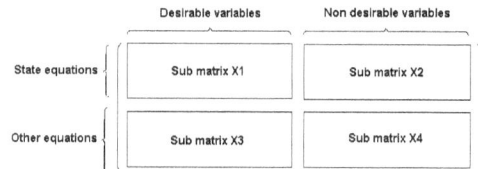

Figure 6. Decomposition of the circuital matrix after the arrangement.

The steps performed so far do not depend on what variables have been defined as the outputs of the state space representation. It is now necessary to define what the wanted outputs are in order to extract the matrices C and D.

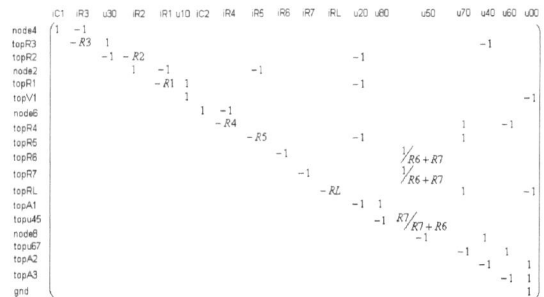

Figure 7.Submatrix X4 after the arrangement.

In our case-study we have considered that the output is the node voltage u_{70}. Find its expression consists of find its row in the relation matrix.. Equation (5) shows the resulting matrix C. The vector 1x2 of the equation (5) corresponds to the matrix C of the equation (3). The matrix D is equal to 0.

$$\begin{pmatrix} \dot{u}_{45} \\ \dot{u}_{67} \end{pmatrix} = \begin{pmatrix} -9,37.10^5 & 6,25.10^5 \\ -6,25.10^5 & 0 \end{pmatrix} \begin{pmatrix} u_{45} \\ u_{67} \end{pmatrix} + \begin{pmatrix} 6,25.10^5 \\ 0 \end{pmatrix} Vin \quad (4)$$

$$u_{70} = \begin{pmatrix} 0 & -1 \end{pmatrix} \begin{pmatrix} u_{45} \\ u_{67} \end{pmatrix} \quad (5)$$

4. HIGH-LEVEL MODEL INCLUDING THE POWER CONSUMPTION

Using this method under passive circuit allows to find easily the power consumption by adding one component in the output vector of the state space model. This component is the input current. The power consumption of the circuit is given by the product of the input current and the input voltage. For active circuits using op-amps, it is different because the current consumption depends on the op-amps supply.

Firstly, a part of the power consumption is due to the quiescent power consumption of the op-amps. This consumption is given by the product of the quiescent current (given by the op-amp datasheet) and the supply voltage. The state variable filter owning three op-amps, the op-amps power consumption is given by equation (6).

$$P_Q = 3.i_Q.(V_{DD} - V_{SS}) \quad (6)$$

An interesting idea for the dynamic power consumption computation consists of add the currents flowing to the ground and multiply them by the voltage supply. It is possible to insert these currents into the state space model output vector found in the previous section and given by equation (5). The expression of the currents i_{R7} and i_{RL} are extracted from the relation matrix using the same method as u_{70} in the previous section. Equation (7) represents a new high-level model of the state variable filter including the currents allowing to compute its power consumption.

$$\begin{pmatrix} u_{70} \\ i_{RL} \\ i_{R8} \end{pmatrix} = \begin{pmatrix} 0 & -1 \\ -3,13.10^{-4} & 0 \\ 0 & -1.10^{-5} \end{pmatrix} \begin{pmatrix} u_{45} \\ u_{67} \end{pmatrix} \quad (7)$$

The dynamic part of the power consumption is now given by equation (8).

$$P_D = (i_{R7} + i_{RL}).(V_{DD} - V_{SS}) \quad (8)$$

Finally, the global power consumption of the state variable filter is given by equation (9).

$$P_C = (V_{DD} - V_{SS}).[(3.i_Q + (i_{R7} + i_{RL})] \quad (9)$$

The simulations have been done with a sinusoidal input (amplitude 1V) and an op-amp quiescent current of 1,4 mA (TL081). Figure 8 shows the voltage u_{70} in function of the frequency and figure 9 shows the power consumption. 10^5 Hz is the resonance frequency for the output of the second op-amp (u_{50} maximum value). At this frequency, u_{50} amplitude is equal to 1 and u_{70} amplitude is equal to 0,707, the result of this combination is a power consumption peak at 10^5 Hz.

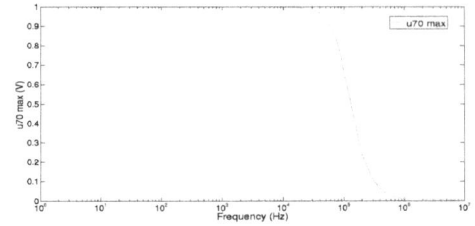

Figure 8. Output voltage (u_{70}) of the state variable filter.

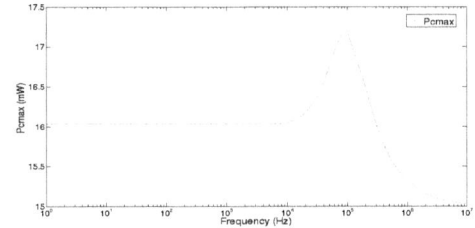

Figure 9. Power consumption of the state variable filter.

5. CONCLUSION

In this paper, we have shown that it is possible to obtain a high-level model of the power consumption of analog blocks. It is needed to start from a high-level model and refine it in order to obtain a circuit description. At this point, the inverse method is done: the abstraction. Low-level specific information concerning the power consumption are extracted and used to define a new high-level model including the power consumption information. The added value of this work is that we provide a methodology that can be applied to active-RC amplifier circuits. Future works will concern multi-physical systems such as micro-electromechanical systems (MEMS).

REFERENCES
[1] "Standard SystemC AMS extensions language reference manual," Open SystemC Initiative, March 8 2010.
[2] F. Cenni, E. Simeu, S. Mir, "Macro-modeling of analog blocks for SystemC AMS simulation: A chemical sensor case-study," *17th International Conference on Very Large Scale Integration, VLSI-SoC*, 2009, Florianópolis, Brazil, Oct 2009.
[3] S. Adhikari, C. Grimm, "Modeling switched capacitor sigma delta modulator nonidealities in SystemC AMS", *13th International Forum on specification & Design Languages (FDL)*, Southampton, UK, Sept 2010.
[4] M. Vasilevski F. Pecheux, N. Beilleau, H. Aboushady, K. Einwich, "Modeling and refining heterogeneous systems with SystemCAMS: Application to WSN", *Design Automation & Test in Europe (DATE) Conference*, Munich, Germany, Mar 2008.
[5] E. Markert, M. Dienel, G. Herrmann, U. Heinkel, "SystemC AMS Assisted Design of an Inertial Navigation System*," Sensors Journal*, IEEE, Vol. 7 Issue. 5, pp 770-777, May 2007.
[6] L.Bousquet, F.Cenni, E.Simeu, "Inclusion of power consumption information in high-level modelling of linear analog blocks", *Journal of Low Power Electronics (JOLPE)*, vol 7, pp 541-551, Dec 2011.
[7] F. Paugnat, L. Bousquet, K. Morin-Allory, L. Fesquet, "A performance comparison between the SystemC AMS models of computation", *edaWorkshop*, May 2011.

A Novel Power-Gating Scheme Utilizing Data Retentiveness on Caches

Kyundong Kim [†], Seidai Takeda [†], Shinobu Miwa [‡], Hiroshi Nakamura [†,‡]

[†]Graduate School of Engineering, The University of Tokyo
[‡]Graduate School of Informaiton Science and Technology, The University of Tokyo
Tokyo, Japan
{kim, takeda, miwa, nakamura}@hal.ipc.i.u-tokyo.ac.jp

ABSTRACT

Caches are one of the most leakage consuming components in modern processor because of massive amount of transistors. To reduce leakage power of caches, several techniques using power-gating(PG) were proposed. Despite of its high leakage saving, a side effect of PG for caches is the loss of data during a sleep. If useful data is lost in sleep mode, it should be fetched again from a lower level memory. This consumes a considerable amount of energy, which very unfortunately mitigates the leakage saving. This paper proposes a new PG scheme considering data retentiveness of SRAM. After entering the sleep mode, data of an SRAM cell is not lost immediately and is usable by checking the validity of the data. Therefore, we utilize data retentiveness of SRAM to avoid energy overhead for data recovery, which results in further chance of leakage saving. To check availability, we introduce a simple hardware whose overhead is ignorable. We also examined leakage saving potential of our approach. For both L1 data and instruction caches, our scheme results in more than 2 times of smaller leakage energy compared to conventional PG scheme.

Categories and Subject Descriptors
B.3.2[Memory Structures]:Design Styles-*Cache memories*

Keywords
Low-power, Cache, Leakage

1. INTRODUCTION

As technology scales down, leakage energy becomes one of the most important issues in VLSI design. Allowing large numbers of transistors on-die and decreasing threshold voltage cause significant amount of leakage power in latest chips. On-chip caches are one of the main candidates for leakage reduction. Since caches occupy large area in the die and are implemented with high density, they consume significant fraction of leakage power.

Power-gating(PG) is a promising technique to reduce leakage power because of its high leakage saving capability. In PG scheme, a high threshold transistor, which is called sleep transistor, is inserted between the circuit and power supply (or ground). By turning off the sleep transistor, circuitry enters sleep mode, and then the effective supply voltage of the circuit *gradually* decreases to near zero. After that, leakage power is almost completely shut off. Turning-off should be done when the circuit is in idle state because the circuitry is non-operational in sleep mode. Fortunately,

cache has lower per-area activity than logic component like execution units. Applying PG to caches provides a better of chance in leakage saving.

Several PG schemes for caches have been developed in recent years. DRI cache[1] is a scheme for leakage saving on an instruction cache. The scheme resizes the cache to fit just the working set of the code, and turning off the rest of cache. Cache decay[2] is a finer grain approach than DRI cache. This approach shuts down individual cache line when the line is not accessed for a predetermined amount of time.

However, utilizing PG on caches has a critical drawback in terms of extra energy consumption for data recovery. When a SRAM cell is transferred into sleep mode, *its data is usually regarded as being invalid at that time*. If this lost data is referenced later, extra energy which may not be spent in non-PG case is induced by accessing to a lower level memory. This overhead decreases net leakage saving during a sleep interval. Furthermore, when energy overhead overwhelms leakage saving, it consumes more energy than non-PG scheme. Transition to sleep mode is not desirable at such a time. In summary, data recovery energy decreases not only amount of leakage saving interval but also chances for leakage saving.

In previous studies., PG was considered as a non-state-preserving technique. However, the effective supply voltage of SRAM cell does not become near zero immediately when the sleep transistor is turned off. That is, the effective supply voltage *gradually* decreases. If the power gated SRAM cell wakes up before its effective supply voltage reaches a threshold, the states of SRAM cells automatically recover. If the power gated SRAM cell wakes up before its effective supply voltage reaches a threshold, the states of SRAM cells automatically recover. In this case, additional energy for data recovery is not required.

In this paper, we propose a novel PG scheme utilizing data retentiveness on caches. By utilizing data retentiveness, accessing lower hierarchy memory for data recovery is not required in short sleep. This results in larger overall leakage saving. We also address a method to check the availability of data with ignorable overhead. We examined leakage saving potential of our proposal compared to that of conventional PG scheme on L1 instruction and data caches.

The contributions of our work are summarized below.
- We propose a new PG scheme to utilize data retentiveness on caches. It provides additional chances for leakage saving.
- We propose a method to check the availability of data in cache.
- We observe that leakage power saving of conventional PG is highly improved after considering data retentiveness.

In section 2, we presents the deprived chance of leakage saving in conventional PG scheme. Section 3 presents the concept of data retentiveness and its utilization on caches. In section 4, we address experimental results. Section 5 presents related work. Finally, we summarize this paper in section 6.

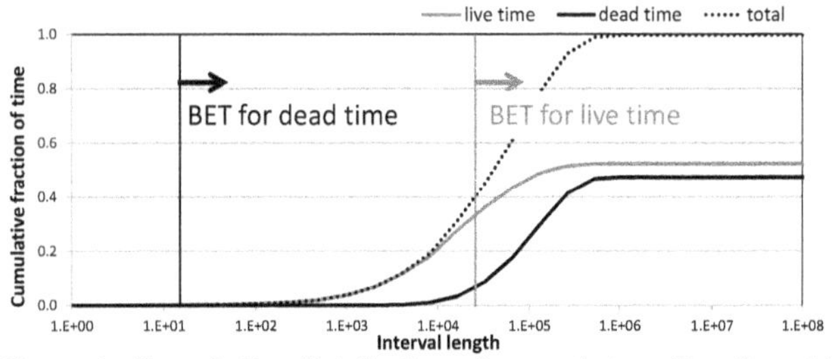

Figure 1: Cumulative distribution of access interval(perlbench)

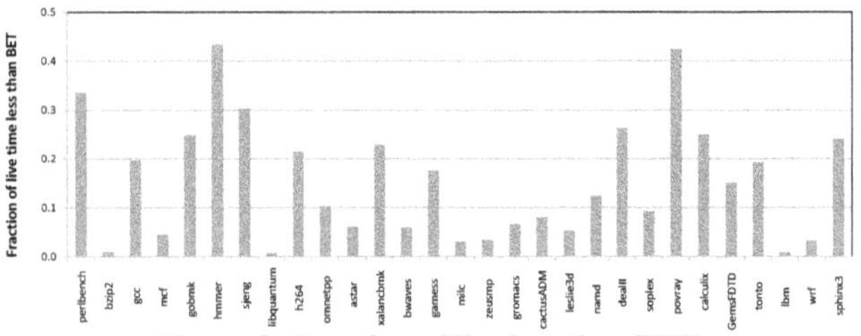

Figure 2: Fraction of live less than BET

2. LEAKAGE SAVING LIMITATION OF CONVENTIONAL PG

In conventional PG, caches still have a large room to improve leakage saving. For some idle time intervals, leakage saving is limited because entering sleep mode consumes more energy. In this section, we address the reason why such idle time intervals exist. And then we specify the fraction of these idle time intervals in conventional PG.

2.1 Break-even Time Depending on Cache Behavior

Break-even time(BET) provides a concept which directly indicates that whether sleep is a proper choice or not with an idle time. When an SRAM cell goes into sleep mode, extra energy is consumed which is not incurred in active mode. If this extra energy overwhelms saved leakage energy, entering sleep mode wastes more energy. BET is the minimum sleep time to pay back the energy overhead. Therefore, only the sleep time longer than BET guarantees energy saving.

In applying PG to caches, there are two energy overhead components. One is for turning on/off sleep transistor. The other is for accessing lower level cache to recover data. The former is significantly small compared to the latter one. The energy for data recovery is much bigger and requires long time of sleep to pay back. Our experimental evaluation with hspice and CACTI[3] shows that the energy overhead by data recovery is paid back with $13\mu s$ of sleep for 32KB L1 cache in 45nm technology. Therefore energy overhead for data recovery heavily lengthens BET. Here, we clarify whether data recovery energy is caused or not in the perspective of cache behavior.

In order to classify whether data recovery energy is incurred or not, cache decay uses terminology *live time* and *dead time* which represent two states of cache line. A line is live if the following request to the line results in hit. Otherwise, a line is classified into dead. In other words, a dead line is no longer accessed and is waiting for eviction in a replacement manner. Sleeping of dead lines does not cause additional energy overhead for data recovery, while sleeping live cache lines causes it. Both sleeping live lines and dead lines cause data lost. Therefore, both requires an access to a lower cache when the lost data is referenced after lines wake up. However, this access is not *extra* for dead lines because it is also required in no sleep case. Thus, extra energy is consumed *only* for live line sleep. Consequently, BET for live time is longer than that for dead time.

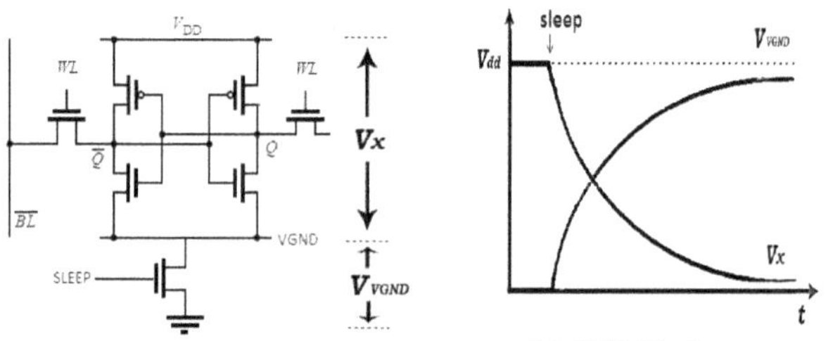

(a) Voltage division of SRAM   (b) VGND change

Figure 3: Voltage variation in sleep mode

2.2 Distribution of Live Time on Caches

Figure1 shows cumulative distribution of access interval for perlbench. We simulated one billion cycles with a processor simulator and then sum all access intervals of each cache line for L1 data cache. Red and black vertical lines show BET of dead time and live time respectively. Since leakage saving is achievable only with the idle time whose interval length is longer than BET, the idle times up to these lines do not have opportunity for leakage saving.

The BET of dead time is very short and almost all of dead time have leakage saving chance. On the contrary, the BET of live time is very long. The fraction of live time under its BET is 0.33. For these live time intervals, leakage saving is impossible because energy overhead overwhelms leakage saving.

Figure 2 shows the fraction of live time whose interval length is less than BET for SPEC CPU2006[4] benchmark. Each value is normalized to total idle time for L1 data cache. In hmmer and povray, the fraction of live time less than BET is up to 0.4. If we can reduce energy overhead, more leakage saving can be achieved with sleeping these live times. To reduce overhead energy, we propose a new PG scheme to utilize data retentiveness. Next section introduces our proposal in detail.

3. PG SCHEME UTILIZING DATA RETENTIVENESS ON CACHES

One important observation of our work is that when an SRAM cell is power gated, it does not lose its data immediately. Although the state of the SRAM cell is unstable, it is possible to check whether data of SRAM cell is lost or not. If data of SRAM cell is available, data can be used after waking up. Therefore, even in sleep mode, SRAM cell still stores usable data in some case. In this paper, we refer to this as *data retentiveness*.

We propose a new PG scheme utilizing data retentiveness on caches. Our novel PG scheme works in simple manner with caches whose PG is controlled by each line. When a data of cache line in sleep mode is requested, availability of its data is checked. If data is valid, the data is used just after waking up the line. The rest of this section addresses how to know the availability of data. Its hardware implementation is also specified.

3.1 Relationship between Data Retentiveness and Voltage of VGND

VGND voltage makes important role in PG caches in terms of data retentiveness in sleep mode. Figure 3 shows the variation of V_x and V_{VGND} after entering sleep mode. Here V_x refers to the effective supply voltage which is the voltage between Vdd and VGND.

Assuming that initially '0' is written at node Q and '1' at $\overline{Q}$ in non-sleep mode. When a cache line goes into sleep mode, the leakage path between VGND and real GND is turned off. Since the equivalent resistance of sleep transistor is larger than that of SRAM cell, more current flows from Vdd to VGND rather than that from VGND to GND. Hence, charge from Vdd is accumu-

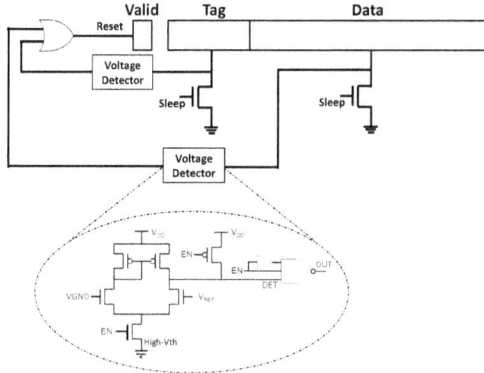

Figure 4: Availability check scheme

lated to VGND node and then V_{VGND} gradually increases. When the current through SRAM cell and sleep transistor become equal, V_{VGND} is saturated. In this paper, we refer to this voltage as V_{sat}.

In this sequence, the voltage of node Q, whose voltage is '0', also increases until it equals the V_{VGND}. When the voltage of node Q increases to the voltage where node Q can not go back to '0' after waking up, the data of SRAM is lost. Therefore, V_{VGND} is the determining factor for data availability.

[5] reported that the lowest data retention voltage for SRAM cell was 190mV in 90nm technology. Drowsy[6] cache, a state-preserving leakage saving technique, sets 30% of Vdd voltage for data retention. DRG-cache[7] reported that data is lost when V_{VGND} increases to over 80% of Vdd. In this paper, we conservatively assume that SRAM cell's data is preserved until V_x is over 30% of Vdd voltage. This assumption well follows previous works. We represent time interval when data is available by T_{avail}. Our experimental result shows that T_{avail} ranges from 6.19μs to 9.72μs depends on temperature and sleep transistor width. Details are addressed in section 4.

3.2 Hardware Implementation

Figure 4 shows hardware implementation to check the availability of data. Cache decay reset valid bit immediately after going into sleep mode. On the other hands, our scheme control valid bit in different manner. The main features of our design is below.

First, we mount a voltage detector on each virtual ground line. The voltage detector checks the voltage of VGND to guarantee the data validity. We use an on-chip voltage detector based on clocked sense amplifier presented in [8] with a little modification. The voltage detector compares V_{VGND} to V_{ref}. When V_{VGND} is higher than V_{ref}, voltage detector outputs '1' which indicates that data is invalid. We set V_{ref} as 0.77V(=70% of Vdd).

Second, the output of voltage detector is connected to valid bit of each cache line. When voltage detector outputs '1', the valid bit is reset. If valid bit is not reset after a cache line switches a sleep state, the cache data is available and can be recovered by just waking it up. Enable signal is activated only when cache is accessed, therefore the power consumption of voltage detector is small enough.

Our experimental evaluation with hspice shows that the energy consumption of voltage detector is under 50fJ for one activation. In standby mode, under 10nW of leakage power is consumed. PMOS holder and high threshold NMOS help to reduce leakage power in standby mode. Area overhead is less than 0.7%. Both overheads are very small and ignorable.

4. EXPERIMENTAL RESULTS

To evaluate the potential of leakage saving with utilizing data retentiveness, we make a comparison with conventional PG scheme for L1 instruction and data caches. We use oracle knowledge policy. In oracle policy, a cache line never goes into sleep mode when energy overhead overwhelms leakage saving.

Table 1: Processor configuration

Functional Units	2 Integer and 1 FP ALUs
	1 Integer multiplier/divider
	1 FP multiplier/divider
L1 I-cache	32KB, 64B line, 4-way
	3 cycles latency
L1 D-cache	32KB, 64B line, 4-way
	3 cycles latency
L2 unified cache	1MB, 64B line, 8-way
	15 cycle latency
Memory latency	200 cycles

For dead time, a cache line sleeps if an idle time interval is longer than the BET. In this case, proposed PG scheme saves same leakage energy as conventional PG scheme. For live time, when an idle time interval is longer than the BET which includes data recovery energy, both of the conventional and proposal PG go into sleep mode. When an idle time interval is shorter than T_{avail}. proposal scheme has additional leakage saving chance comparing to conventional PG scheme. This is because our PG scheme does not suffer from data recovery energy in such a interval.

For the evaluation of T_{avail} and data recovery energy, we use CACTI and hspice with 1.1V Vdd Nangate 45nm[9]. Evaluation result of CACTI for data recovery energy(which is also L2 cache access energy) is 0.63nJ for 32KB cache size and 64B line size in 45nm. The width of sleep transistor is designed as 5% and 10% of total width of SRAM cell transistors. To obtain the information of idle time for L1 instruction and data caches, we use a cycle accurate processor simulator Onikiri2[10]. We assume a 2GHz out-of-order processor whose configuration resembles as much as possible that of a Alpha processor. The configuration is shown in Table 1. In our simulation, we use 29 applications from the SPEC CPU2006[4]. We simulated 1 billion cycles with ref data sets.

4.1 Interval of Data Availability

Table 2 shows experimental result of T_{avail} and V_{sat} under different temperature and sleep transistor sizes. T_{avail} varies from 4.87μs to 9.72μs. Depending on temperatures, T_{avail} heavily varies. As temperature increases, leakage current exponentially increases which means that the speed of charge accumulation to V_{VGND} increases. This result in shorter T_{avail}. Temperature also affects to V_{sat}. Result shows that V_{sat} is lower with high temperature. V_{sat} is determined by IV-curve characteristic of transistor because V_{VGND} is saturated when leakage current of SRAM cell and sleep transistor become same.

Table 2: Interval of data availability

Sleep transistor size	Temperature ($^\circ$C)	V_{sat} (V)	T_{avail} (μs)
5%	25	1.06	8.87
5%	40	1.00	7.60
5%	60	0.98	6.19
5%	80	0.97	4.87
10%	25	0.91	9.72
10%	40	0.87	8.70
10%	60	0.86	7.44
10%	80	0.79	6.39

4.2 Leakage Saving on Caches

4.2.1 Effect of T_{avail}, BETs and Temperature

T_{avail} and the BET of live time are important factors to determine leakage saving improvement of our proposal. As we mentioned in section 2, the BET of live time determines the fraction of leakage energy whose saving is limited in conventional PG. T_{avail} determines leakage saving improvement of our proposal. In our experimental evaluation, T_{avail} is shorter than the BET of live time in all configurations.

For live time, effectiveness of leakage saving is different depending on the interval length of idle time. Considering BETs and T_{avail}, interval length of idle times are classified 4 types. Below

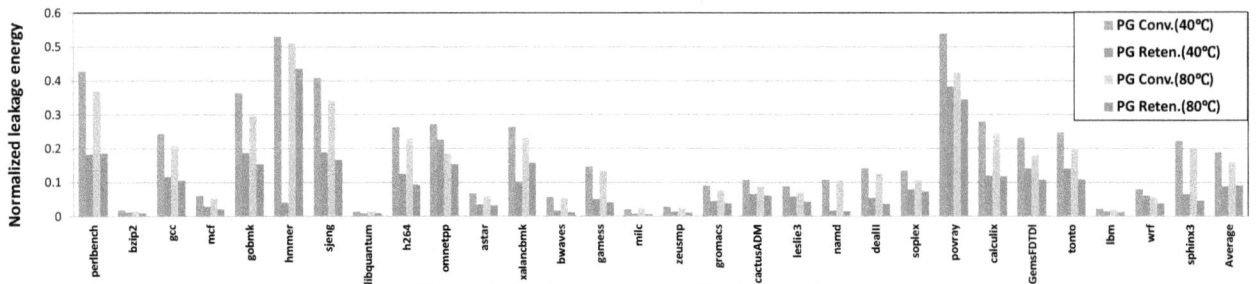

Figure 5: Leakage energy of L1 data cache

BET of dead, they occupy under 1% for total idle time and their effect to total leakage energy is ignorable. For intervals whose length are between BET of dead and T_{avail}, our proposal achieves leakage saving improvement. For intervals whose length are between T_{avail} and BET of live, there is no leakage saving because they do not meet BET. When an interval is longer than BET of live, leakage energy is saved in both conventional PG and our proposal. For dead time, almost all of leakage energy is saved in both conventional PG and our proposal because more than 99% of dead idle time has longer interval length than BET.

T_{avail} and BET are effected by temperature. In higher temperature, more leakage flows in active mode which means that a sleep saves more leakage. Hence, saved leakages energy pay back overhead energy earlier which result in shorter BET. In conventional PG scheme, the leakage energy consumption is lower with higher temperature. This is because shorter BET increases the opportunity for leakage saving. On the other hands, in our proposal, higher temperature does not always results in smaller leakage energy because higher temperature also shorten data available interval.

In the perspective of leakage saving improvement from conventional PG, lower temperature provides better saving chance because of longer data available interval. Comparing to non-PG, higher temperature achieves higher leakage saving because of shorter BET both conventional PG and our proposal.

4.2.2 Effect of Idle Time Distribution

We evaluate our proposal in both L1 data and instruction cache. Figure5 shows leakage energy for L1 data caches when the width of sleep transistor is 5% of total width of SRAM cell transistors. The leakage energy is normalized to non-PG scheme. For L1 data cache, average leakage energy for conventional PG varies from 15% to 18% depending on the temperature. Average leakage energy for our proposal varies from 8% to 9%.

For L1 instruction cache, average leakage energy for conventional PG ranges from 17% to 19%. With our proposal, average leakage energy varies from 7% to 8%. Average leakage energy of our proposal is 2 times smaller than that of conventional PG in both L1 data and instruction caches.

When a idle time distribution of application has large fraction of live time interval below T_{avail}, our proposal achieves a lot of saving improvement. For data cache, perlbench, gobmk and sjeng are this kind of applications. Short(<10K cycles) live time intervals occupy more than 30% of total idle time distribution.

On the other hands, when a application has dead time dominant contribution, there is slice room for leakage saving improvement because such applications do not suffer from energy overhead for data recovery. Libquantum and bzip2 are such applications and the leakage saving in both conventional PG and our proposal are under 2%.

Hmmer for L1 data cache shows interesting result. There are 40% of difference in leakage saving improvement for proposal between 40 °C and 80 °C condition. This is because more than 90% of live time exists between T_{avail} of 40 °C and 80 °C condition.

5. RELATED WORK

Drowsy Cache[6] is one of the state-preserving approaches for leakage saving on caches. It saves leakage power by putting cache lines into low-power mode which keeps enough effective voltage to a cache line for state-preserving. By keeping the state even in low-power mode, extra cache miss does not incurs. However, drowsy cache should pay a cost of lower leakage saving than cache decay in low-power mode because of incomplete shutdown. The requirement of two power supply is another shortcoming of drowsy cache in terms of higher area overhead and design complexity.

DRG-cache[7] is another approach to achieve both of leakage saving and data retention. In DRG-cache design, the voltage of VGND is saturated to the voltage where data is never lost. However, as well as drowsy cache, less leakage saving capability for data retention is a shortcoming of this work. Additionally, the saturated VGND voltage should be designed even lower because temperature and process variation increase risk of data lost.

6. CONCLUSION

In this paper, we propose a new PG scheme for cache which outperforms conventional PG by having more leakage saving chances for live time. The main idea of our work is to utilize data retentiveness of SRAM when using PG. Depending on the voltage of VGND, even in sleep mode, data of SRAM can still available which mitigates energy overhead for data recovery.

In conventional PG, the leakage saving was limited for the live times whose interval length does not meet its BET. Our proposal reduces energy overhead for recovery which leads to further leakage saving chance for those live times. Experimental results for L1 instruction and data caches show that our proposal help saving up to 49% more leakage than conventional PG. On average, normalized leakage energy consumption varies from 7% to 9% which is more than 2 times smaller than that of conventional PG. Evaluation with practical control policy is our future work.

Acknowledgments

This research was partially supported by a New Energy and Industrial Technology Development Organization (NEDO) research project entitled "Development of normally-off computing platform technology", VLSI Design and Education Center(VDEC), and Synopsys, Inc.

7. REFERENCES

[1] M. Powell, et al., Gated-Vdd: a circuit technique to reduce leakage in deep-submicron cache memories. In *Proc. of ISLPED*, pp.90–95, 2000.

[2] S. Kaxiras, et al., Cache decay: exploiting generational behavior to reduce cache leakage power. In *Proc. of ISCA*, pp.240–251, 2001.

[3] N. Kim, et al., Drowsy instruction caches: leakage power reduction using dynamic voltage scaling and cache sub-bank prediction. In *Proc. of MICRO*, pp.219–230, 2002.

[4] A. Agarwal, et al., DRG cache: a data retention gated-ground cache for low power. In *Proc. of DAC*, pp.473–478, 2002.

[5] CACTI. http://www.hpl.hp.com/research/cacti/.

[6] SPEC CPU2006 suite, The Standard Performance Evaluation Corporation. http://www.spec.org/cpu2006/.

[7] A. Kumar, et al., Fundamental data retention limits in sram standby experimental results. In *Proc. of ISQED*, pp.92–97, 2008.

[8] K. Usami, et al., On-chip detection methodology for break-even time of power gated function units. In *Proc. of ISLPED*, pp.241–246, 2011.

[9] Nangate open cell library. http://www.nangate.com/.

[10] Processor Simulator Onikiri2. http://www.mtl.t.u-tokyo.ac.jp/~onikiri2/.

A Zero-Overhead IC Identification Technique Using Clock Sweeping and Path Delay Analysis

Nicholas Tuzzio, Kan Xiao, Xuehui Zhang, and Mohammad Tehranipoor
Dept. of Electrical & Computer Engineering, University of Connecticut
Storrs, CT, USA
npt05001,kanxiao,xhzhang,tehrani@engr.uconn.edu

Abstract

The counterfeiting of integrated circuits (ICs) has become a major issue for the electronics industry. Counterfeit ICs that find their way into the supply chains of critical applications can have a major impact on the security and reliability of those systems. This paper presents a new method for uniquely identifying ICs through path delay analysis. There is no overhead in terms of area, timing, or power for this method, since it extracts the intrinsic path delay variation information of the IC. Simulation results from 90nm technology and experimental results from 90nm FPGAs demonstrate the effectiveness of our technique.

Categories and Subject Descriptors

B.7 [**Integrated Circuits**]

Keywords

IC identification, counterfeit ICs, process variations, clock sweeping, path delay analysis.

1. INTRODUCTION

A counterfeit IC is an electronic component whose material, performance, or characteristics are knowingly misrepresented by the vendor, supplier, distributor or manufacturer [1] [2]. They can be parts which have been remarked to resemble different parts, defective parts diverted from disposal and sold, or previously used parts salvaged from scrap electronics [1]. We will focus on a subset of counterfeit ICs that physically and functionally appear to be the correct IC, which are extremely difficult to detect.

According to one estimate, the United States Department of Defense may have purchased between $15-100M USD worth of counterfeit ICs in 2005 alone [3]. If counterfeit ICs were to end up in the supply chain for mission-critical or life-saving applications, the results of the failure of an unreliable or insecure counterfeit part could be catastrophic. To prevent these issues, we propose a novel technique for uniquely identifying ICs without the use of specialized hardware structures. Unique identification helps prevent these issues because it allows us to differentiate between counterfeit ICs and authentic ICs with a high degree of success. Because this method does not require any modifications to the hardware it will be used on,

it can be used on both new IC designs and legacy designs already in production or in use.

Several techniques have been developed to combat IC counterfeiting. A technique for creating unique identifiers by exploiting an IC's internal process variation was described in [4]. Silicon Physical Unclonable Functions (PUFs) were developed in [5], and many types of PUF have been proposed for ASICs and FPGAs [6] [7]. Another approach to combating IC counterfeiting are metering schemes. Passive metering involves identifying and tracking ICs throughout their lifetime [8]. Active metering prevents an IC from functioning properly until certain activation criteria are met [9]. Work has been done to create unique IC identifiers without any additional hardware. In [10], the authors describe a technique for uniquely identifying ICs by analyzing the unique leakage currents of a series of gates in a circuit.

None of these techniques represent a comprehensive solution to the IC counterfeiting problem. PUF-based identification has yet to be widely accepted due to their reliability issues and their area and logic overhead. They also require high levels of process variation to be present in the circuit and cannot always be inserted in legacy designs. Metering schemes are subject to these problems as well. The method for identifying ICs without additional hardware presented in [10] depends on complex and time-consuming computations and current-measuring techniques which are not often used in industrial tests.

In this paper, we describe a technique for uniquely identifying ICs without any area overhead, and without using uncommon steps in the design or test processes. This technique uses path delay information to create unique binary identifiers. We use a method called "clock sweeping" to obtain this path delay information during testing. We can analyze our ability to accurately identify ICs under measurement and environmental noise. This technique represents a novel improvement on existing ideas for several reasons. First, our technique can be applied to ICs already in production, including legacy designs. Second, it uses data that can be obtained through use of existing pattern sets and testing hardware capabilities. Third, no additional hardware is necessary- there is no area, power, or timing overhead to the technique. Finally, because the circuit is not added to or modified in any way, no attacks on the circuit are possible.

This paper will be organized as follows: Section 2 will present background theory for our methods. Section 3 will describe the ID creation and IC identification methodologies. Section 4 will present simulation results and experimental results from an FPGA implementation. We will conclude with final remarks in Section 5.

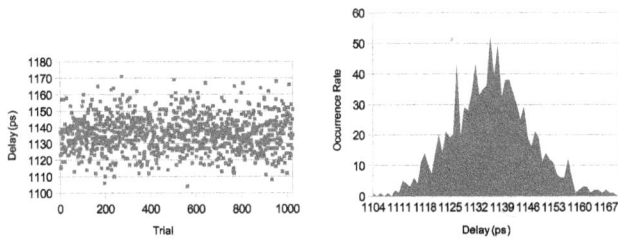

(a) Scatter plot   (b) Distribution plot

Figure 1: Path delays for 1,024 simulations of the most critical path in s38417.

2. THEORY AND BACKGROUND

2.1 Process Variations and Path Delay

In theory, the specifications and functionality of different ICs of the same design should be identical. In practice, this is not the case. This happens because of our inability to accurately create IC structures at smaller technology nodes, due to process variations. Thus, values such as the threshold voltage of a transistor (V_{th}), the length of the gate of the transistor (L), or the oxide thickness of the transistor (T_{ox}) can only be guaranteed to be within some range of values. Variation on these and other parameters is important to take into consideration, because these parameters directly affect device performance. For example, the delay of a traditional CMOS inverter can be expressed in two equations, where the high-to-low and low-to-high propagation delays t_{PHL} and t_{PLH} are effected by variation on the three previously mentioned parameters.

Figures 1a and 1b show how process variations can affect the delay of a path in a set of ICs. We measured the delay of the most critical path in the ISCAS'89 benchmark s38417 1,024 times at 25°C using HSPICE. Differing levels of process variations between the Monte Carlo simulations resulted in the path having a different propagation delay in each simulation. Figure 1a shows the path delay for each simulation, and Figure 1b shows the distribution of these path delays. These delays are centered around an average value because process variation results in small changes to the parameters which control the path delay.

2.2 Clock Sweeping

"Clock sweeping" is the process of applying patterns to a path multiple times with different frequencies to find a frequency at which the path cannot propagate its signal, often for purposes of speed binning. By observing the frequencies at which the path can and cannot propagate its signal, we can measure the delay of the path with some degree of precision. Our ability to perform clock sweeping on a path is limited by the degree of control we have on the clock that controls the capturing memory elements (i.e., the flip-flops), the degree to which we can excite paths in the circuit, and the lengths of the paths in the IC.

Figure 2 shows a visual example of clock sweeping being performed on several paths. Assume that paths P1 through P8 are paths in the circuit which end with a capturing flip-flop, and have some delay in nanoseconds. Each of the eight paths can be swept, or tested, at the frequencies f_1 through f_5. All paths are able to propagate their signal at f_1, as this is the rated frequency of the IC design. However, at f_2, the path P3 will usually fail to propagate its signal. At frequency f_3, path P3 will always fail to propagate its signal. Path P8 will succeed in propagating its signal at all five clock frequencies-in this example, it is too short to test. All of the paths have some number of frequencies they will pass at, some they may fail at, and some they are guaranteed to fail at. Process variations change which frequency each path will fail at between different ICs.

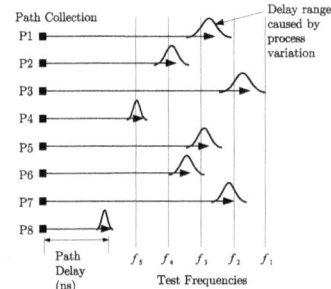

Figure 2: A visual example of clock sweeping being performed on several paths.

3. METHODOLOGY

Our method for uniquely identifying ICs can be broken down into three main parts: (a) Test preparation, (b) ID generation and optimization, and (c) IC identification.

3.1 Test Preparation

The data that we will use to create unique IDs for every IC will be the path delay information collected from the ICs. This data can be collected during the manufacturing test process, but this requires some additional preparation. We can collect this information through application of transition-delay fault (TDF) patterns. TDF patterns seek to identify slow-to-fall or slow-to-rise patterns and as such must be run with a specific clock frequency in mind; by applying the same TDF patterns at different frequencies (f_2, f_3, etc.), we perform the clock sweeping process described above.

Figure 3a shows the pattern generation procedure. We will obtain path delay information using the TDF patterns that would be generated anyways during for manufacturing tests. Some of the TDF patterns will be applied multiple times at different frequencies. However, not all patterns need to be tested at all test frequencies. Consider path P2 in Figure 2. If the circuit is fault-free, path P2 will pass tests at the frequencies f_1 through f_3. It will either fail the test at f_4 or f_5. Thus, if P2 passes tests at f_1, we can call the circuit fault-free, and we only need to perform the additional tests for f_4 and f_5 for identification purposes. We can save time by not testing the circuit at f_2 or f_3. By carefully selecting the set of frequencies that each pattern needs to be applied at, we can significantly reduce the pattern application time and thus the indentification time. Once the test preparation step has been completed, and TDF patterns for the different sets of frequency tests have been created, ICs can use these tests to generate unique IDs.

3.2 ID Generation and Optimization

Once the tests have been applied in the manufacturing test stage, we possess a large amount of path delay information from each manufactured ICs. We will use this data to generate a binary ID for each IC. Three steps are performed to generate the IDs for each IC: (1) Stability checking, a path selection procedure that selects the most stable measured paths from each IC. (2) ID generation, a path delay comparison procedure which generates an unoptimized ID for each IC. (3) ID optimization, an analysis and optimization procedure which increases the randomness and decreases the size of the IDs.

Stability Checking: We are using path delay information to generate IDs. Path delay measurements using clock sweeping do have the potential to be affected by measurement and environmental noise. To reduce the effect of this issue, we will attempt to discard unstable paths early on. Measuring an IC multiple times gives us multiple measurements of each path in that IC. Analysis of multiple measurements of a path can tell us whether that path is stable- it reported the same delay during each measurement- or unstable- reporting different delays on different occasions. To select stable paths, we will measure some subset of the IC set multiple times and select the m most stable paths to be used for further analysis on all n ICs.

ID Generation: Once the m most stable paths have been selected, we begin to generate IDs for each IC. To generate an ID, we will perform comparisons between the delays of different paths from the ICs. The goal is to perform a comparison that is somewhat uncertain, so that the result the comparison generates is random. We obtain a list of similar paths by sorting a list of m paths from an IC in ascending or descending order. We will call this sorting the "golden ranking", and apply this ordering to the m paths of all n ICs. Once the golden ranking has been obtained, the path delay data for every IC is sorted by this ordering. For each IC, we will traverse this ordered list of delays one element at a time. Whenever a delay in the list is greater than the next delay in the list, we will append a "1" onto the ID for that IC. Whenever a delay in the list is less than the next delay, we will append a "0" onto the ID for that IC. If m paths have been analyzed for each of the n chips, we will end up with n IDs of $m - 1$ bits each. The first/golden IC, which was used to determine the ordering, will have an ID of all ones or zeros, depending on whether the sort was ascending or descending.

ID Optimization: There are several optimization techniques we can use on these ICs. Some paths that we are comparing in the ID Generation step may produce the same value for all ICs because the path delays are very different. This will reduce the average Hamming distance between IDs as this bit will be the same in all ICs. If enough paths have been measured, it may be possible for an IC designer to pick and choose which comparisons they are going to perform to create their IDs. The comparisons can be chosen by analyzing each bit of the $m - 1$ bit "unoptimized IDs", and deciding which bits have the least bias across all IDs. A bit should be included in the "optimized" ID if the number of zeros in that bit position is close to 50% of the number of IDs- if the number of zeros in that bit position is in the range of $\frac{n}{2} \pm \varepsilon$, where ε is some small constant. This process selects the most unpredictable comparisons across the IC data set. By performing this optimization technique, we obtain with shorter, more random IDs, at the cost of discarding some of the data obtained during the measurement process.

3.3 IC Identification

This process will be run on every IC when they are first manufactured and going through manufacturing tests. These tests will produce an ID for every manufactured IC, will will be stored in a database for later access. Parts of this process will be run again later when we wish to identify an IC. Figure 3b shows these two scenarios. In order to identify an IC from the market, referred to as an IC under authentication (IUA) in Figure 3b, we will need: (1) the ID that the IC is presenting, and (2) the database of IDs from when the ICs were manufactured. Once we have the ID for the IUA, we can check to see if it is in the database using Hamming distance analysis. Hamming distance analysis is done by comparing the ID from the IUA to every ID in the database. If the ID from the IUA is an exact match to one of the IDs in the database, then we have identified the CUT.

Noise in the measurement process might make it so that there are a small number of bits in the ID which are different between the IC's ID in the database and its ID in the field. We addressed this issue through the stability checking technique described in Section 3.2, but differences may still be present. Therefore, we define a *Hamming distance threshold*, based on analysis of the IDs during manufacturing tests. This threshold could be based off the minimum Hamming distance between two different ICs in the manufactured set. The threshold could also be based off of the maximum Hamming distance between IDs from the same IC, which would be cre-

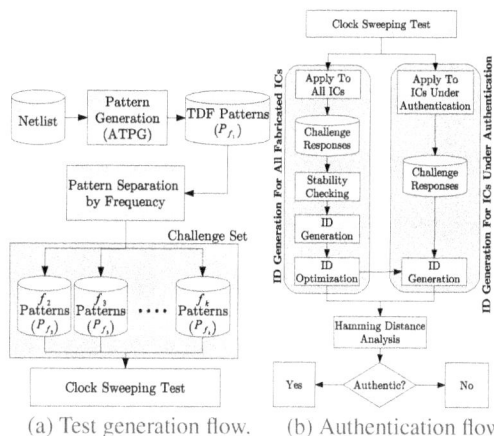

(a) Test generation flow.   (b) Authentication flow.

Figure 3: Test generation and authentication flow.

ated during the stability checking process. Exact thresholds would be derived from analysis of the ID set and measurement noise rates.

3.4 Overhead Analysis

This IC identification procedure incurs no area, power, or timing overhead. There is an overhead in testing time and computational time. Each IC must undergo an additional k rounds of TDF testing at k different frequencies. If we have n ICs with P_{f_1} TDF patterns and N_{sff} flip-flops that we would like to test at k frequencies, the worst case increase in testing time would be $n \cdot (k-1) \cdot P_{f_1} \cdot N_{sff} + N_{sff}$ clock cycles. This is a worst-case scenario as not all P_{f_1} TDF patterns will need to be tested at all k frequencies, as shown in Figure 3a. The stability checking procedure requires us to measure a subset of $n_r \ll n$ ICs j times each, so there is a testing overhead of $n_r \cdot j \cdot (k-1) \cdot P_{f_1} \cdot N_{sff} + N_{sff}$ clock cycles in the worst-case scenario. Judging the relative stabilities of each path in the ICs requires us to individually analyze each of the m paths that were measured j times each in n_r ICs, resulting in a complexity of $O(n_r m j)$ time; however, n_r is much less than n, and j is going to be on the order of dozens to hundreds. The ID generation procedure requires us to individually analyze each of the m paths in each of the n ICs, resulting in an $O(nm)$ runtime. The ID optimization technique analyzes each of the $m-1$ bits across all of the n IDs for a general runtime of $O(nm)$ as well. In general, m will be much smaller than n or n_r.

4. RESULTS AND ANALYSIS

4.1 Simulation Results

To evaluate this methodology, we performed a series of simulations using HSPICE. The simulations were performed on an implementation of the ISCAS'89 benchmark s38417. This circuit was simulated at the 90nm technology node. To measure the path delay, we identified the 256 top critical paths from the circuit, and performed 128 Monte Carlo simulations. This provides us with 128 simulated ICs worth of data, with 256 data points for each IC. During the Monte Carlo simulations, the threshold voltage V_{th} had at most 5% inter-die variation and 5% intra-die variation, the oxide thickness T_{ox} had at most 2% inter-die variation and 1% intra-die variation, and the transistor gate length L had at most 5% inter-die variation and 5% intra-die variation, with these values representing 3-sigma deviations from specified values. All paths for all ICs were simulated over a range of temperatures from 23°C to 27°C to represent small deviations from room temperature.

We used the simulation data to create 128 255-bit IDs using the methodology described in Figures 3a and 3b. Figure 4a shows the

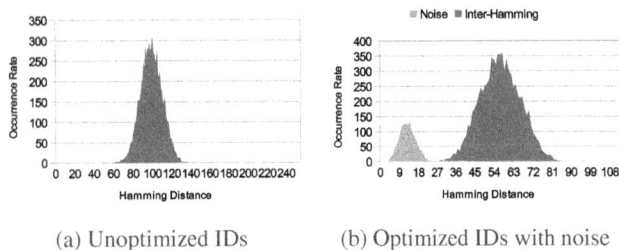

(a) Unoptimized IDs   (b) Optimized IDs with noise

Figure 4: Hamming distance analysis on 128 simulated s38417 circuits.

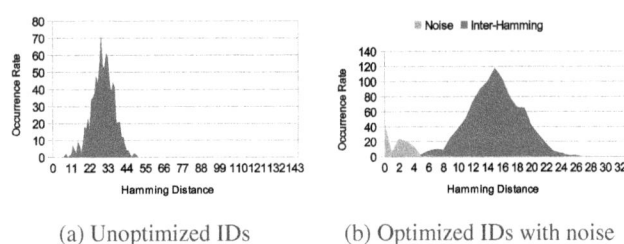

(a) Unoptimized IDs   (b) Optimized IDs with noise

Figure 5: Hamming distance analysis on 44 FPGA s38417 circuits.

results of Hamming distance analysis on these unoptimized IDs. The 255-bit IDs have, on average, a 99-bit or 39% inter-Hamming difference. The fact that this average is less than 50% means that some bit-positions in the IDs have a bias towards zero or one. By optimizing the IDs to remove these bits, using the optimization procedure described in Section 3.2, we improve the average Hamming distance of the ID set and reduce the size of the IDs, as shown in Figure 4b. The optimization process reduced the size of the IDs from 255 bits to 114 bits, and increased the average inter-Hamming distance from 39% to 50%.

Figure 4b also shows the noise rates of the simulated IDs. Each IC was simulated at temperatures from 23°C to 27°C at 1°C increments to represent small variations around room temperature. We performed Hamming distance analysis on each set of IDs coming from the same IC to see how many bits of the ID would change over the ID range. The average number of bits changing as a result of temperature was 13 bits, or about 11%.

The Hamming distance distribution and noise rate distribution are both roughly normal distributions. By treating them both as normal distributions with their own averages and standard deviations, we can come up with probability density functions (PDF) for each set. We can use these two PDFs to find the Hamming distance at which it is equally likely that the two IDs are from different ICs or from the same IC. In the sets shown in Figure 4b, this Hamming distance is 27 bits. If two IDs are compared and have less than 27 different bits between them, they are most likely from the same IC. If two IDs are compared and have more than 27 different bits between them, they are most likely from different ICs.

4.2 FPGA Implementation Results

We implemented this methdology on 44 90nm Xilinx Spartan-3E FPGAs. We implemented the ISCAS'89 benchmark s9234 on our FPGAs, along with RAM for TDF pattern storage and structures for clock control. The IDs were sent from the FPGA boards to a computer for analysis by using an additional microcontroller. The paths in the FPGA were swept at the frequencies f_{1-16}, with $f_1 = \frac{1}{6.4ns}$ and $f_{16} = \frac{1}{3.4ns}$. The frequency step size was 200ps. To obtain noise information, we measured 4 of the implementations 8 times each. After selecting stable paths, the unoptimized IDs were 145-bit strings, with a Hamming distance distribution as shwon in Figure 5a. The average inter-Hamming distance was 30 bits, or about 21%. This is lower than in the simulation example. The optimization process reduces the size of the ID from 145 to 33 bits long, and the distribution of the inter-Hamming distances between these 33-bit IDs are shown in Figure 5b. By optimizing the IDs, we change the average inter-Hamming distance from 30 out of 145 bits to 15 out of 33 bits, or about 45%. In practice, a 33-bit ID would be short, but creating IDs of a greater size only requires measuring more paths.

Figure 5b also shows the noise rates for these IDs. We measured 4 different implementations of s9234 8 times each and compared the IDs from each implementation to each other. On average, there was a 2 out of 33 bit difference between IDs from the same implementation, or about a 6% difference.

Again, we can perform further analysis to compare the noise and inter-Hamming rates. At a difference of 5.75 bits, it is equally likely that any two IDs being compared are from the same IC or from different ICs. IDs with a difference of less than 5.75 bits are probably from the same IC and IDs with a difference of more than 5.75 bits are probably from different ICs.

5. CONCLUSION

In this paper, we proposed a technique that allows us to uniquely identify ICs without any area, power, or timing overhead. In both the simulation and implementation results, we were able to create unique IDs with no collisions for a set of different implementations of the same circuit. In both cases, the average Hamming distance between the IDs was nearly 50%, and the levels of noise were sufficiently low that we could distinguish between IDs from the same IC and IDs from different ICs. These results would improve with a larger data set, a more accurate clock with a smaller frequency step size, and larger levels of process variations.

6. ACKNOWLEDGEMENTS

This work is supported in part by grants from National Science Foundation (NSF) under CNS 0844995 and Army Research Office (ARO).

7. REFERENCES

[1] H. Livingston, "Avoiding counterfeit electronic components," *Components and Packaging Technologies, IEEE Transactions on*, vol. 30, pp. 187–189, march 2007.

[2] M. Tehranipoor and C. Wang, *Introduction to Hardware Security and Trust*. Springer, 2012.

[3] J. Stradley and D. Karraker, "The electronic part supply chain and risks of counterfeit parts in defense applications," *Components and Packaging Technologies, IEEE Transactions on*, vol. 29, pp. 703–705, sept. 2006.

[4] K. Lofstrom, W. Daasch, and D. Taylor, "Ic identification circuit using device mismatch," in *Solid-State Circuits Conference, 2000. Digest of Technical Papers. ISSCC. 2000 IEEE International*, pp. 372–373, 2000.

[5] B. Gassend, D. Clarke, M. van Dijk, and S. Devadas, "Silicon physical random functions," in *Proceedings of the 9th ACM conference on Computer and communications security*, CCS '02, (New York, NY, USA), pp. 148–160, ACM, 2002.

[6] G. Suh and S. Devadas, "Physical unclonable functions for device authentication and secret key generation," in *Design Automation Conference, 2007. DAC '07. 44th ACM/IEEE*, pp. 9–14, june 2007.

[7] J. Guajardo, S. S. Kumar, G.-J. Schrijen, and P. Tuyls, "Fpga intrinsic pufs and their use for ip protection," in *Proceedings of the 9th international workshop on Cryptographic Hardware and Embedded Systems*, CHES '07, (Berlin, Heidelberg), pp. 63–80, Springer-Verlag, 2007.

[8] F. Koushanfar and G. Qu, "Hardware metering," in *Design Automation Conference, 2001. Proceedings*, pp. 490–493, 2001.

[9] Y. M. Alkabani and F. Koushanfar, "Active hardware metering for intellectual property protection and security," in *Proceedings of 16th USENIX Security Symposium on USENIX Security Symposium*, (Berkeley, CA, USA), pp. 20:1–20:16, USENIX Association, 2007.

[10] Y. Alkabani, F. Koushanfar, N. Kiyavash, and M. Potkonjak, "Information hiding," ch. Trusted Integrated Circuits: A Nondestructive Hidden Characteristics Extraction Approach, pp. 102–117, Berlin, Heidelberg: Springer-Verlag, 2008.

RAPA: Reliability-Aware Priority Arbitration strategy for Network on Chip*

Jiajia Jiao
School of Micro Electronics
Shanghai Jiao Tong University
Shanghai,China

jiaojiajia@ic.sjtu.edu.cn

Yuzhuo Fu
School of Micro Electronics
Shanghai Jiao Tong University
Shanghai,China

fuyuzhuo@ic.sjtu.edu.cn

ABSTRACT

Reliability issue, especially from transient errors due to scaling IC technology, low voltage supply, high frequency and heavy thermal effects, particles emission etc, has become a challenge for NoC design. Focus on this problem, an effective Reliability-Aware Arbitration Strategy simplified as RAPA, is proposed in this paper to decide which flits should be prioritized in the network transmission for higher application-level reliability. Different from pervious performance-oriented arbitration strategies, it includes the application-level reliability requirement to determine the reliability priority ranking. Flits patching mechanism is also used for avoiding starvation. The evaluation metric is redefined to emphasizing application-level reliability. Finally, we verify the reliability based prioritization policy on cycle accurate platform. And the simulation results show that the averaged successful delivery rate is upgraded from three nine of round robin (RR), old age based arbitration(OA) to five nine of our method RAPA. Especially, 67.15%, 41.83% reliability improvement in rest unreliable space on average are obtained over typical RR policy and OA based arbitration policy respectively with guaranteed performance.

Categories and Subject Descriptors

C.1.2 [Computer Systems Organization]: Multiprocessors; Interconnection architectures; C.1.4 [Parallel Architectures]: Distributed architectures

General Terms

Design, Algorithms, Reliability

Keywords

NoC, reliability, arbitration, application-level, transient error

1. INTRODUCTION

With scaling IC technology development, multi-core system becomes the mainstream like Intel's SCC and Tilera's TILE64/100 in the industry, which needs high communication bandwidth and lower latency. NoC is the most promising solution to meet such communication requirement. But the shrinking feature size makes IC more likely to be effect by factors like

* This work was supported a University Project Grant from Cisco Research Center in the USA .

crosstalk, electronic migration, electromagnetic interference, alpha particle hits, and cosmic radiation. And the induced transient error errors will degrade the NoC performance and reliability dramatically, or even result in system crash. That's why reliability issue becomes critical for NoC.

Traditionally, NoCs have been designed with various fault tolerant mechanisms [1-5] for guaranteed reliability. These effective methods mostly focus on ECC (Error Correct/Check coding), retransmission, multi-path routing for correct link transmission and redundant computation for harden router inner logic (RC, VCA, SA etc). Different from the previous works, we try an enhanced prioritization mechanism for NoC reliability.

The rest of this paper is organized as follows. The next section describes related works. Section 3 describes the basic background, which includes basic NoC architecture and fault model. The effective scheme for reliability-aware arbitration strategy RAPA is proposed in Section 4. Simulation and analysis is given in Section 5 while conclusion is drawn in Section 6.

2. RELATED WORK AND OUR CONTRIBUTIONS

To our knowledge, this is the first to improve NoC reliability from the view of arbitration strategy. Here, we describe the most closely related works as follows.

Fault tolerant methods in NoC. Rich methods have been proposed to guarantee NoC communication reliability, including retransmission mixed with error detection, spare link or router, fault tolerant routing algorithm. Kim, et al.[1] explore error detection and correction schemes based on the router architecture and use the model to evaluate the on-chip network in the presence of these error protection schemes. Murali, et al. [2] characterize efficient error recovery mechanisms for the NoC design environment based on area, power, and performance design constraints. Yung-Chang Chang etc [3] propose an innovative router-level fault tolerance scheme with spare router to provide redundancies and diversify connection paths between adjacent routers. S.D. Mediratta etc [4] explores and caches alternative path between source-destination pairs for data transmission after fault detection. In the paper [5], novel fault tolerant routing scheme for NoC, is shown to have a low implementation overhead and adapt to design time and runtime faults better than existing turn model, stochastic random walk, and dual virtual channel based routing schemes.

Prioritization and Fairness in NoC. Existing local arbitration (Local Age, Local Round Robin) and state-of-the-art QoS-oriented prioritization [6] policies have been designed in NoC. Similarly, a large number of arbitration policies [7-10] have been proposed in multi-chip multiprocessor networks and long-haul networks. The goal of these mechanisms is to provide fairness or performance improvement. Especially, Reetuparna Das etc [10] exploit the slack of application-level latency through application-aware prioritization mechanism, and a serial of experimental results verify this method effective in performance optimization.

Compared with the prior works, our proposed method has the following main contributions as a complementary:

1) It is the first to use a light weight prioritization mechanism in router for high application-level reliability (successful delivery rate), rather than high performance (throughput, latency etc).

2) The redefinition of comprehensive metric is given for application-level reliability evaluation.

Besides above description, we also do a lot of simulation and analysis based on cycle accurate platform for verification.

3. BACKGROUND

3.1 NoC Architecture
Considering regularity and scalability, typical Mesh topology and XY distributed routing algorithm is selected as the basic NoC architecture. The typical five ports router with wormhole flow control mechanism is as Fig1 shows. Note that each dash-dotted box is duplicated by five times in each router. Each port in the router consists of VCs, a RC unit, a VCA, and Output register and controllers. Each VC has input status registers VC.RC, VC.VCA and VC.Utilization for storing its RC, VCA, and VC utilization state. OC integrates the function of switch allocation (SA). The five ports are connected with a 5x5 crossbar. From pipeline view, each hop has three pipelines. The first stage is data receiving and storage in VC. RC and VCA are only designed for head flit while input VC. In the second stage, the RC extracts the type of the coming flit and determines the output port from the given destination address. The VCA grants the available VC in next hop from given CREDIT information and SA in output channel grants the input port to output port. In the third stage, the flit in output is transmitted through link. In this work, we concentrate the the VCA and SA in color region.

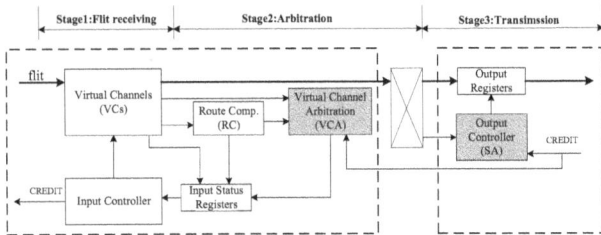

Fig1. Router architecture

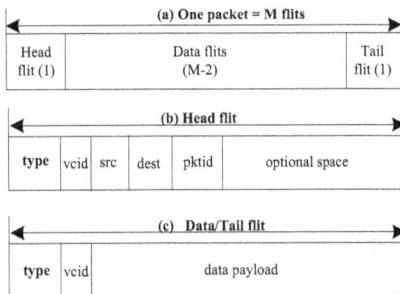

Fig2. Transmission format

Each incoming flit is buffered in a virtual channel (VC) of the router, until it wins in VCA stage and is allocated an output VC in the next router. Following this, the flit arbitrates for the output crossbar port in the SA stage. Thus, there are two arbitration stages where the router must choose one flit among several flits competing. Current routers use simple, local arbitration policies to decide which flit should be scheduled next (i.e., which flit wins arbitration). A typical policy, referred to as RR (round robin) in this paper, is to choose flits in different VCs in a round-robin order such that a flit from the next non-empty VC is selected every cycle.

3.2 Fault Model
In this section, we also present our fault model to evaluate the impacts of transient errors and the efficiency of proposed fault tolerant NoC design. According to the error style, the possible faults are divided into two kinds:

- Flit error, occurs in VCs or inter-router links. Namely, some bits upset in flit due to SEU(Single Event Upset) or crosstalk are injected at arbitrary user-defined location and time. From the Fig2, it is clear that there are invalid error like optional space and valid error, including the upsets in type, vcid, src, dest, pktid, data payload.

- Control error, is produced by transient faults in RC, VCA or SA logic. Such error behaves wrong computation result. For example, a fault in RC leads to incorrect next-hop RC information, which may cause misrouting.

Our fault model supports above two kinds of errors and has been integrated into a system-level network simulator for verification. Considering SRAM and Flip-Flop sensitive to transient errors, in this paper we inject flit error in SRAM based VCs and Flip-Flop based registers and set the same error injection rate for the two types of components.

4. PROPOSED NOC ARBITRATION STRATEGY: RAPA

4.1 Motivation
NoC reliability due to transient errors is affected by both the hops of routing path and inner router latency. In addition to the router pipeline stages, a flit can cost several cycles waiting in a router until it wins a VC slot and gets scheduled to traverse the switch. The less the cycles are spent on transmission, the higher communication reliability is obtained. Therefore, application level arbitration strategy in NoC router has a direct effect on NoC reliability. To analyze arbitration strategy impacts on reliability, we formulate this problem as follows. The possibility of correct flit transmission is denoted as P and the k th flit corresponds to Pk. Pij represents the error injection rate per cycle while Lij gives the spent cycles in the i th stage in router and the j th hops. S denotes the number of router pipeline. H is the hops of the flit routing path. And N represents the total number of delivered flits while $P_{success}$* means the general successful delivery rate. Then, we can clearly get the equation (1) and (2). It is obvious Lij has an important effect on the value of $P_{success}$* while Lij is determined by VCA and SA arbitration policy. Therefore, we launch the reliable-aware prioritization research for NoC.

Furthermore, different types of data in applications have different importance degree. Eg: the A type information in H.264 is the basic and most important for graph construction while B or C type is less important and even be ignored. Flits with application-level reliable requirement have more weight in final reliability evaluation. So the equitation (2) is modified as the new metric in (3), where q flits require application level reliability and α is the reliability coefficient.

$$P = \prod_{j=1}^{H} \prod_{i=1}^{S} (1 - P_{ij})^{L_{ij}} \qquad (1)$$

$$P_{success}{}^{*} \approx \frac{\sum_{k=1}^{N} P_k}{N} \qquad (2)$$

$$P_{success} = \alpha * (P_{success}{}^{*}(1..q))$$
$$+ (1 - \alpha) * (P_{success}{}^{*}(q+1..N))$$
$$\approx \frac{\alpha * (\sum_{k=1}^{q} P_k) + (1 - \alpha) * (\sum_{k=q+1}^{N} P_k)}{N}$$
$$(q < N, 1 \geq \alpha > 0.5) \qquad (3)$$

4.2 Framework of RAPA

Our goal is to improve the application-level reliability by prioritizing flits belonging to high reliability requirement over others in NoC. To enable the prioritization, we combine two principles: reliability ranking and fairness ranking as following description.

Reliability ranking is used to prioritize reliability-aware flits transmission. Application-level intrinsic requirement is considered to determine a good rank. flits born with critical information require high reliability and will be transmitted with high ranked priority. Firstly application-level intrinsic requirement can be obtained with the help of compiler-directed data store and access. Then the application reliable requirement is exposed to NoC by packets compression and decompression in network interface. Finally, the requirement information is contained in each generated flit. This reliable requirement of a flit is unchanged during its whole transmission in NoC. Namely, it is not necessary to re-compute the application-level reliable ranking. Ideally, many ranking priority levels will describe the reliable requirement accurately in a fine-grained manner. In practice, however, the number of ranking levels has to be constrained in NoC, because it negatively affects flit and flit's buffer sizes, arbitration logic complexity and latency in router. For these reasons, we have a compromise between the reliable benefits and overhead to set proper ranking levels.

Fairness ranking is essential to avoid low-ranked flits starvation, which may cause potential network congestion. To overcome this problem, we adopt the approach of flit patching. Several flits through the same VC are grouped into a batch dynamically. A flit belonging to an older batch is always prioritized over a flit in a younger batch. Due to the accepted wormhole mechanism in NoC, the ordering of batches is inherent in router without extra overhead. This is different from packet batching in [10], and behaves: 1) the granularity is flit rather than packet, 2) the batching rule is not time-based batching or packet-based in static manner, but arbitration-based. When the top flit in VC fifos sends the VCA requests, the following flits in the same VC belong to a packet and grouped into a batch. In the SA case, only a flit in the register is a batch. And our arbitration-based flit patching starts a new batch based on arbitration information. If a request fails, the corresponding batch number of VC or SA register increase by 1 from initialized 0. Until the upper bound of ranking levels, a batch number wrap-around occurs and the batch number restarts from 0. The larger the batch number is, the higher the priority is.

Rule in Fig4 summarizes the prioritization order applied by each router in order to enable reliable-aware prioritization. Each router prioritizes flits in the specified order when making flit scheduling, arbitration, and buffer allocation decisions. To trade off the overhead and reliability benefits, we set the ranking levels of flit patching, application requirement to 4, 2 respectively. Namely, the patching number is 2 bits while application requirement flag is 1 bit in each flit. And the 2 bits size counters are configured for each VC and SA register. Then, the centralized arbiter logic should collect these PARA's arbitration information for high ranked flits.

Rule: flit prioritization rules in each router

Step 1: Oldest Batch First: Flits belonging to older batches are prioritized over flits belonging to younger batches (the goal is to avoid starvation).

Step 2: Application requirement Highest Rank First: Flits with higher rank values (belonging to higher-ranked applications) are prioritized over packets with lower rank values (the goal is to maximize NoC reliability)

Step 3: Basic Router Rule: If the above two rules do not distinguish the flits, then use the simple round robin policy at the router

---

Fig4. Flit prioritization rule of PARA

5. SIMULATION AND EVALUATION

5.1 Simulation Configuration

We extend cycle accurate NoC simulator Nirgam[11] to support our RAPA mechanism and transient error injection. Parameter configuration is listed in Table 1. Each flit has certain possibility with application level reliable requirement and its app_flag is marked with 1. Otherwise the flag is set to 0. Here, we set the value of possibility to 0.01($\approx$ q/N), namely 1% packets have high reliability requirement. All components in the router micro architecture are all direct error sources. But to simplify the error occurrence, we only inject the flit error (single bit upset) as Fig2 (b) and Fig2 (c) described in input VC and output SA registers per cycle. The counter for flit patching number will increase in wrap around way, and be reset to 0 when the corresponding patch in the VC or flit in SA register is transmitted out.

We do comparative evaluation RAPA with RR (local Round Robin strategy) and OA (local Old Age based strategy) through the two metrics: the new defined successful with reliability-criticality ($P_{success}$), and the general throughput (Tp, represents the total number of delivered flits). The latter represents the performance while the former is used to evaluate reliability. The reliability coefficient α is set to 0.9 for the results Psuccess statistic.

Table1. Parameters configuration

Parameters	Value
Basic topology	4x4 Mesh
Routing algorithm	XY
Flit size payload	4Bytes
Packet size payload	32Bytes(a cache line)
Packet accruing strategy	CBR(Constant Bit Rate)
Traffic trace	Transpose
VC Buffer size (depth)	4
VC number	2
Injection Rate(flit /node/cycle)	1
(packet /node/cycle)	{0.0055,0.01,0.02,0.04,0.05, 0.1,0.2,0.33,0.5,1}
Error rate(flit/cycle)	{1E-5, 1E-6,1E-7,1E-8}
Random characteristic	averaged 5 samples
Simulation time	100000 cycles

5.2 Results and Analysis

The traffic load rate and injection rate both determine the performance and reliability, so we do the 40 mixed configurations. Fig5 shows the overall results of RAPA comparison with typical performance-oriented RR (local Round Robin strategy) and OA (local Old Age based strategy).

Firstly, it is obvious the three arbitration strategies bring different performance and reliability. RR behaves the worst (compared with OA and RAPA, averaged throughput of RR is decreased by 11.35%, 9.29% over them respectively) due to its only fairness consideration while OA and RAPA both takes advantages of history information to improve arbitration efficiency. Therefore, it is available to enhance both performance and reliability through optimized arbitration strategy.

Secondly, RAPA achieves better reliability than OA and RR. The most important reason its reliability-awareness, which prioritizes the more critical flits to transfer first. After all, RAPA breaks the original fairness and blocks the normal transmission. The reliability improvement is at the expense of throughput dropping by 2.278% over OA. More attractive results are that our RAPA can upgrade the redefined successful delivery rate from 0.999 to 0.99999 on average of 40 configurations.

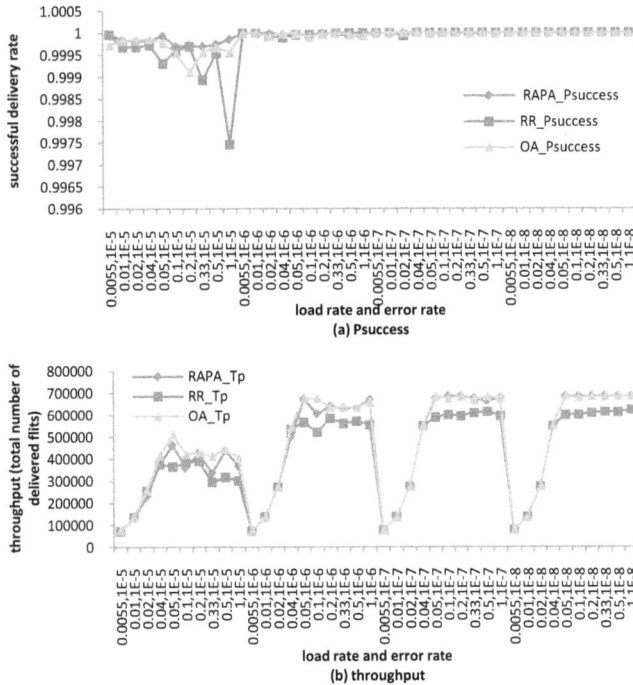

Fig5. Overall RAPA comparison with RR and OA

Then, to evaluate reliability improvement in a fine-grain way, we also bring a new definition relative successful delivery rate, which is given in unreliable space for application level reliability in eq (4), and denoted as $RP_{success}$(RAPA/RR). The RR can be replaced by OA to get $RP_{success}$(RAPA/OA). Comprehensive $RP_{success}$ is averaged by statistic 40 available configurations (different error rates and load rates). The value of $RP_{success}$(RAPA/RR) is 67.15% while this of $RP_{success}$(RAPA/OA) is 41.83%.

$$RP_{success}(RAPA/RR) = \frac{P_{success}(RAPA) - P_{success}(RR)}{1 - P_{success}(RR)} \quad (4)$$

Finally, to further analyse the error rate and traffic load effects, we need to observe the curves in Fig4 from different perspectives. On one hand, higher error rate leads to small successful delivery rate and low throughput no matter what RAPA or OA, RR under certain load rate. From the curves, our method at the error rate 1E-5 performs the worst and low error rate such as 1E-7 or 1E-8 can achieve five nine (0.99999) or higher successful delivery rate in Table2 and near to ideal throughput. The trend of throughput change is similar to successful delivery rate variation. On the other hand, the load rate seems to have little effect on $P_{success}$ but only influence the throughput at the specified error rate as Fig5 shows. We also observe that high error rates as well as high load rates

result in successful delivery rate and throughput curves ups and downs. It owes to random characteristic of transient error impacts.

6. CONCLUSION
We introduce a novel, comprehensive reliable-aware prioritization framework to improve the application-level reliability in NoCs. The high reliability requirements of critical flits are transmitted in high priority, while using error status to feedback the arbitration and guaranteeing starvation freedom. The randomly-generated transpose workload on a 4x4 Mesh, RAPA behaves 67.15%, 41.83% reliability improvement in rest unreliable space on average better than typical round robin policy and old age arbitration policy respectively. And the performance loss is only 2.28% compared with old age based strategy. We conclude that the proposed arbitration strategy RAPA provides a promising way to enhance reliability with little degraded performance.

7. REFERENCES
[1] A. Kahng, B. Li, L.-S. Peh, and K. Samadi, ORION 2.0: A fast and accurate NoC power and area model for early-stage design space exploration, in Proceedings of the IEEE DATE, pp. 423–428,(2009)

[2] J. Kim, D. Park, C. Nicopoulos, N. Vijaykrishnan, and C.R. Das, "Design and analysis of an NoC architecture from performance, reliability and energy perspective", in ACM Proceedings of the symposium on Architecture for networking and communications systems, pp. 173–182 (2005)

[3] Srinivasan Murali, Luca Benini, Theocharis Theocharides, N. Vijaykrishnan, Mary Jane Irwin, Giovanni De Micheli. "Analysis of error recovery schemes for networks on chips", Design & Test of Computers, vol.22, no.5, pp. 434- 442, (2005)

[4] Yung-Chang Chang, Ching-Te Chiu, Shih-Yin Lin, Chung-Kai Liu, "On the Design and Analysis of Fault Tolerant NoC Architecture Using Spare Routers", Asia and South Pacific Design Automation Conference, pp. 431 - 436, (2011)

[5] S.D. Mediratta, J. Draper, "Characterization of a Fault-tolerant NoC Router", IEEE International Symposium on Circuits and Systems, pp. 381-384, (2007)

[6] Sudeep Pasricha, Yong Zou , "NS-FTR: A Fault Tolerant Routing Scheme for Networks on Chip with Permanent and Runtime Intermittent Faults". 16th ASP-DAC, pp. 443 – 448,(2011)

[7] E. Rijpkema, K. Goossens, A. Radulescu, J. Dielissen, J. van Meerbergen, P. Wielage, and E. Waterlander. "Trade-offs in the design of a router with both guaranteed and best-effort services for networks on chip". DATE,(2003).

[8] E. Bolotin, Z. Guz, I. Cidon, R. Ginosar, and A. Kolodny. "The Power of Priority: NoC Based Distributed Cache Coherency". In NOCS'07, (2007)

[9] J. W. Lee, M. C. Ng, and K. Asanovic. "Globally-Synchronized Frames for Guaranteed Quality-of-Service in On-Chip Networks". In ISCA-35th, (2008)

[10] Jin Ouyang, Yuan Xie. "LOFT: A High Performance Network-on-Chip Providing Quality-of-Service Support", MICRO '43 Proceedings of the (2010)

[11] Reetuparna Das, Onur Mutlu, Thomas Moscibroda, Chita R. Das. "Aérgia: exploiting packet latency slack in on-chip networks". ISCA '10 ,Volume 38 Issue 3, June(2010)

[12] Nirgam: http://nirgam.ecs.soton.ac.uk/

A High-Performance Online Assay Interpreter for Digital Microfluidic Biochips

Daniel Grissom, Philip Brisk
Department of Computer Science and Engineering
University of California, Riverside
{grissomd, philip}@cs.ucr.edu

ABSTRACT

We introduce an online interpreter to execute biochemical assays on droplet-based digital microfluidic biochips (DMFBs). Online interpretation enables adaptivity, e.g., response to faults during assay execution, variable-latency assay operations, and concurrent workloads whose composition is not known statically. Our online method routes droplets dynamically, making decisions in milliseconds while running on a low-cost Intel Atom™ processor.

Categories and Subject Descriptors

B.7.2 [**Integrated Circuits**]: Design Aids; B.8.2 [**Performance and Reliability**]: Performance Analysis and Design Aids; J.3 [**Life and Medical Sciences**]: Biology and Genetics, Health

General Terms

Algorithms, Design, Performance

Keywords

Digital Microfluidic Biochip (DMFB), Electrowetting-on-Dielectric (EWoD), Virtual Architecture, Routing, Deadlock.

1. INTRODUCTION

Digital microfluidic biochips (DMFBs) are cyber-physical MEMS devices that manipulate droplets of liquid on a 2-dimensional grid (Fig. 1). A DMFB is a planar array of indistinguishable *cells*; a cell is the abstraction of a square region on top of each electrode. Cells can perform basic operations—e.g., droplet movement, merging, mixing, splitting, storage—that form building blocks for larger chemical reactions called *assays*.

DMFBs will pave the way for programmable chemistry: the chemist of the future will specify assays using domain-specific languages. The assay representation is compiled into a sequence of *droplet actuation cycles*. Each cycle specifies a set of signals to be sent to the DMFB to actuate droplet movement. A program running on a computing device connected to the DMFB traverses the sequence of cycles, sends the appropriate signals for each cycle, and holds the signals for an appropriate period of time to ensure that all droplets complete their movements.

Historically, DMFB compilation has been performed offline. This works fine under ideal circumstances in which the behavior of the system can be characterized statically, and execution proceeds without error; however, it cannot handle any form of variability.

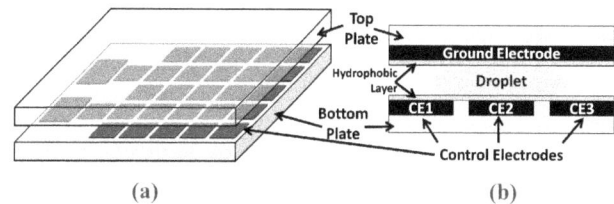

Figure 1. (a) DMFB with a 2D array of control electrodes; (b) DMFB cross-section: a droplet is centered on top of electrode CE2 and overlaps adjacent electrodes, CE1 and CE3. A voltage applied to CE1 or CE3 induces motion to the left or right. Other feasible operations include droplet splitting, merging, mixing, and storage.

This paper introduces online assay interpretation for DMFBs, as an alternative to static compilation. Online interpretation will enable interesting new capabilities in the areas of control flow and dynamic scheduling. For instance, a particular assay could be executed based on the results of a previous assay or environmental condition. Other potential benefits include the ability to detect and respond dynamically to faults or to dynamically adjust a schedule to account for variable-latency operations [2].

Static compilation methods employ long-running algorithms that produce highly optimized results, effectively minimizing assay completion time. Online interpretation, in contrast, must overlap algorithmic decision-making steps with the execution of each cycle. A DMFB typically runs at around 100 Hz, meaning that each cycle lasts for around 10ms. If the online decision-making algorithm runs for longer than 10ms, then the length of the cycle must be extended, thereby increasing assay execution time.

A secondary consideration is the cost of the computing device that is connected to the DMFB, especially when integrated into products that perform low-cost portable point-of-care diagnostics. Ideally, 10ms runtimes could be achieved on low-cost battery-operated embedded computers, as opposed to higher-performance power-hungry desktop PCs. The interpreter described here meets these constraints on a single-threaded Intel Atom™ processor.

2. RELATED WORK

Static DMFB compilers must solve three NP-complete problems: scheduling [7], placement [8], and routing [1, 3, 5, 9]. High quality solutions are achieved using long-running iterative improvement algorithms [7, 8] or optimally via Integer Linear Programming [7] or A* Search [1].

Obviously, these methods do not meet the real-time constraints imposed by the interpretation framework. For example, the BioRoute router reports droplet routing times as low as 40ms on a 1.2 GHz Sun Blade-2000 machine with 8GB of memory. Our interpreter, in contrast, is able to meet the timing constraints while running on a low-cost Intel Atom™ processor.

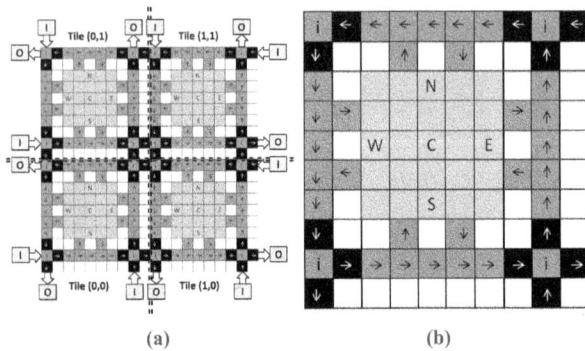

Figure 2. (a) A 2x2 virtual architecture represented as an array of tiles (separated by double, dashed, black lines) showing potential I/O locations; (b) A tile, comprised of a work chamber ('C'), 4 streets, and 4 intersections ('i').

Griffith and Akella [5] impose a *virtual architecture* on the DMFB that restricts the functions that different cells can perform. Similar to the layout of a city, the DMFB is organized into streets, intersections, *rotaries* (e.g., traffic circles), and city blocks (where all assay operations happen). This limits the flexibility of the DMFB, but facilitates the adaptation of network routing algorithms for droplet transport. To prevent deadlock, they limit the injection rate of droplets into the system. Our interpreter takes a similar approach, but adopts mesh network deadlock-free routing algorithms [4], rather than limiting the injection rate.

3. VIRTUAL ARCHITECTURE

As shown in Fig. 2, our interpreter imposes a virtual architecture onto the DMFB, and exploits its restrictive structure to achieve fast algorithmic runtimes. Input and output reservoirs are placed on the perimeter of the DMFB.

The city blocks are referred to as (reaction) *chambers*, because all assay operations occur there. Each chamber can perform one operation (e.g., merging, mixing or splitting) or can store up to four droplets. External devices such as heaters or optical detectors can be affixed to the DMFB above or below a chamber.

All streets are 1-way; eight 1-one streets come together at rotaries (Fig. 2(a)), which offer an abstraction like a network router. Droplets travel clockwise through the rotary. Without loss of generality, a droplet traveling north that enters a rotary could continue straight, or turn left (west) or right (east); our routing algorithms do not allow droplets to reverse directions, so a droplet would not enter a rotary traveling north and then exit traveling south.

Each chamber and the four adjacent streets surrounding it is called a *tile*, as shown in Fig. 2(b). For a 5x5 chamber size, each tile is a 10x10 array of cells. Tiles are repeated to form a virtual architecture, e.g., Fig. 2(a) shows a 2x2 array of tiles.

3.1 Simplifying Placement and Routing

The virtual architecture simplifies the placement and routing steps of an assay as follows:

Placement: we sacrifice the ability to place an assay operation at any DMFB location. Instead, the interpreter dynamically *binds* operations to chambers.

Routing: traditional routers move droplets across the DMFB in a chaotic and unorderly fashion while ensuring separation between all droplets at all times. The DMFB imposes a city-like network of streets that all droplets must follow. This limits the number of routes between each source-destination, which simplifies routing.

3.2 Intermediate Bytecode Language

Conceptually, the set of signals sent to a DMFB during each cycle can be treated like a machine language. If the DMFB is comprised of N cells, then N binary signals are sent to the device (e.g., a '1' activates an electrode, and a '0' leaves it off).

The virtual architecture is a virtual machine with its own *intermediate bytecode language* that operates at a higher-level of abstraction than the DMFB machine language described above. There are two types of bytecode instructions: operations (*O-type*) and transport (*T-type*):

Each O-type instruction has the form *(opcode, chamber-id)*, where the opcode specifies the operation to perform, and the chamber-id specifies which chamber to perform the operation. All chambers support four basic opcodes: *{start-mix, stop-mix, split}*. If a chamber has an external device affixed to the outside of the chip, such as a heater or detector, then it may support additional opcodes such as *{heater-on, heater-off, detector-on, detector-off}*.

Each T-Type instruction has the form *(droplet-id, src, dst)*; the droplet-id specifies a droplet originating at the source (src), and the instruction is to transport the droplet to the destination (dst). The source may be a chamber or an input reservoir, and the destination may be a chamber or an output or waste reservoir. The droplet-id field is necessary to handle the situation where a chamber is storing multiple droplets, but can be dropped when convenient: an input reservoir generates a new droplet, so no droplet-id field is necessary; similarly, if the chamber only contains one droplet, then its id is implicitly known. If a T-type instruction transports a droplet to a chamber, it is stored implicitly, until an O-type instruction initiates an operation.

4. ONLINE INTERPRETER

Assays are specified as directed acyclic precedence graphs (DAG) [7], which are input to the interpreter. The interpreter is decomposed into two phases: a *virtual machine layer (VML)*, which schedules the DAG and binds its operations to chambers, converting the assay to an intermediate bytecode representation. The *droplet transportation protocol (DTP)* converts each T-type instruction into a path from source-to-destination, and moves all droplets along their respective paths one cell at a time.

The DTP converts the intermediate byte code representation of the assay into machine language: the cells that are activated during each cycle can be derived from the transport information, coupled with the state of each chamber (as set by prior O-type instructions). For example, if a chamber is performing a mixing operation, the cells to activate (at each cycle) are known.

The interpreter can run in either online or offline mode. In offline mode, all scheduling, binding, and routing decisions are made up-front, and the output is a statically compiled sequence of cycles. In online mode, the VML and DTP collaborate to interpret the assay in real-time. The VML schedules and binds assay nodes dynamically, generating O-type instructions (for the operations it wants to perform) and T-type instructions (to transport the droplets to their appropriate destinations before the operations can commence). In real-time, the DTP executes T-type instructions one cell at a time, and informs the VML when each droplet arrives at its destination. The VML executes each O-type instruction when all of the droplets on which the operation depends (as specified by the DAG), arrive at their destinations. For example, when two droplets are set to be mixed, the DTP must route both droplets to the chamber before the VML can execute the O-type instruction that initiates the mixing operation.

4.1 Virtual Machine Layer (VML)

The VML uses *modified list scheduling (MLS)* [7] coupled with a fast and simple binder based on the *left-edge algorithm* [6]. MLS was chosen because of its speed and simplicity. The main goal of the left edge binder is to minimize storage overhead. For example, if two droplets are to be stored, it is better to store them together in one chamber, rather than separately in two chambers. Storing them together maximizes the number of free chambers that become available to perform other assay operations concurrently.

4.2 Droplet Transportation Protocol (DTP)

The primary job of the DTP is to select a path for each droplet and then to route all of the droplets along their respective paths while preventing interference among droplets. If two droplets occupy adjacent cells, they will mix. To prevent this, an *interference region* is defined to be the cells directly adjacent to a droplet (Fig. 3(a)). When a droplet moves, its interference region expands to include the union of the source and destination cells (Fig. 3(b)). As long as no droplet enters the interference region of another, undesired mixing is prevented [1, 3, 5, 9].

The DTP must prevent deadlocks from occurring when multiple droplets are in transport at the same time. To accomplish this goal, we adapted deadlock-free routing algorithms from mesh networks. Our implementation uses a variant of XY routing, but other deadlock-free routing algorithms can also work. Conceptually, XY routing moves each droplet from its source position (x_1, y_1) to its destination position (x_2, y_2), by first traveling along the x-axis to (x_2, y_1) and then traveling along the y-axis to complete the route. XY routing is deterministic and non-adaptive, but worked well enough for our purposes. XY routing prevents specific turns from occurring during the route, as shown in Fig. 3(c).

Several modifications to XY routing are required to account for rotaries (whose internal structure is quite different from mesh routers), the I/O reservoirs on the perimeter of the chip, and the process by which droplets enter and exit chambers. First, we need to introduce some terminology: the four streets and intersections surrounding a chamber form a counter-clockwise traffic circle called a *chamber rotary* (Fig. 4(a)). In Fig. 4(b), *exchange rotaries*, which allow droplets to move from one tile to its neighbors, are formed between tiles (earlier, we referred to them simply as "rotaries"). Larger counter-clockwise cycles can also be formed by combining multiple chamber rotaries (Fig. 4(b)).

We now add four additional rules to the basic XY routing algorithm to prevent deadlock in the DMFB's virtual architecture:

Chamber Entries and Exits: Droplets may not make prohibited turns when leaving source and entering destination chambers. To ensure routability in light of prohibited turns, entries and exits are placed on all four sides of the chamber.

Droplet I/O: To prevent forbidden turns, input, output, and waste reservoirs are placed on the DMFB perimeter and the allowable turns that a droplet may make at an entry point are limited.

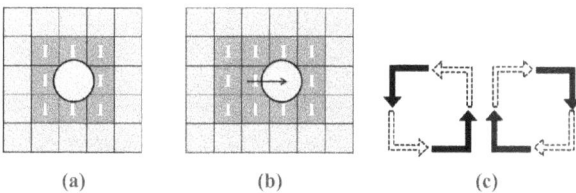

Figure 4. (a) A chamber rotary (the cycle formed by four streets and intersections) (b) An exchange rotary (the clockwise inner loop) and counter-clockwise turns (outer loop) of a tiled design.

Exchange Rotaries: In Fig. 5(a), a droplet *clips* an exchange rotary if it touches one intersection before leaving. In Fig. 5(b) and (c), a droplet *passes through* an exchange rotary if it touches at least two intersections. As droplets move clockwise within an exchange rotary, a clip implies a left turn, and passing through implies that the droplet continues traveling straight or turns right. Fig. 5(d) depicts exchange rotary deadlock when four droplets attempt to pass through; no droplet can progress without violating spacing constraints (Figs. 3(a) and (b)). In Fig. 5(e), deadlock is eliminated if at least one droplet clips the exchange rotary. *To prevent deadlock in an exchange rotary, at most three droplets that wish to pass through may enter concurrently.*

Chamber Rotaries: Fig. 6(a) illustrates chamber rotary deadlock. Droplet 16 creates a dependency chain which causes deadlock; however, if it does not enter the chamber rotary, then a bubble is created which ensures that the sequence of droplets can proceed, starting with Droplet 1. *To prevent deadlock in a chamber rotary, no droplet may enter an exchange rotary unless the system can guarantee that there is space for it to exit into the next street; if the street is full, then the droplet must wait for space to become available prior to entering the exchange rotary.* Droplets attempting to enter a street from an adjacent chamber or input reservoir must also wait until that street has room; in Fig. 6(b), Droplets 1, 2, 3, 5 and 6 must wait for this reason.

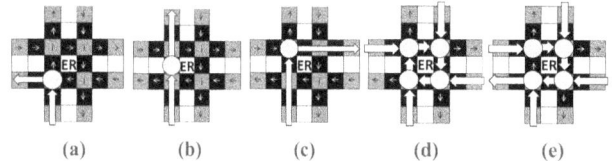

Figure 5. (a) Clipping an exchange rotary ('ER'); (b) Passing through an ER while traveling straight; (c) Passing through an ER while turning right; (d) Deadlocked ER; (e) Non-deadlocked ER.

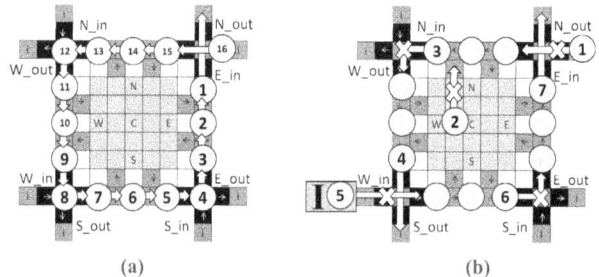

Figure 3. (a) The interference region ('I') of a droplet at the beginning of a cycle; (b) the interference region at the end of a cycle; (c) prohibited turns (white) in XY routing.

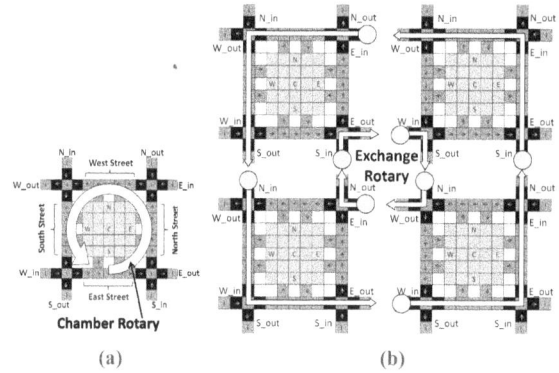

Figure 6. (a) Deadlock in a chamber rotary; (b) Chamber rotary with street capacity rules being enforced to prevent deadlock.

5. EXPERIMENTAL RESULTS

5.1 Experimental Setup

We implemented our interpreter in C++. We compare against a C++ implementation of a static compilation framework that uses modified list scheduling [7], Su and Chakrabarty's placer [8], and Cho and Pan's high-performance droplet router [3]. We assume a droplet actuation frequency of 100 Hz, as in other works [9].

We evaluated the system on an Inforce SYS9402-01 development board, with a 1GHz Intel Atom™ E638 processor and 512MB RAM, running TimeSys 11 Linux. We chose a low-end processor to show that our interpreter performs well on cheap hardware.

5.2 Experiments Performed

We used the PCR sequencing graph, shown in Fig. 7, for our experiments. Droplet input times are assumed to be 0s. We executed PCR using the two synthesis methods described above. Before obtaining results, we optimized the I/O locations for both methods to ensure that neither method had an unfair advantage. Our online interpreter was run on a 20x20 DMFB (2x2 tile array) and the traditional method on a 17x17 DMFB.

We also ran a stress test on the DTP under a heavy transport load. We generated random traffic on 2x2, 3x3, 4x4, and 8x8 arrays of tiles. For each array, we inject 5 droplets at each input reservoir as quickly as possible. Each droplet stops at two random chambers before exiting at a random output reservoir. These tests required the DTP to route up to 160 droplets concurrently.

5.3 Results and Discussion

Table 1 compares the online interpreter (ON) with the offline compiler (OFF). A routing sub-problem is defined as the time when at least one droplet is routed between assay operations. ON successfully routed the PCR assay in 14ms, while OFF required 10.66 seconds, 769 times longer. The lengths of the computed routes were 670ms and 520ms, respectively; when accounting for scheduling, the total assay completion times were 19.67s and 19.52s respectively. Thus, the total slowdown experienced by our online interpreter was less than 1% of the total execution time. Furthermore, the maximum time ON spent computing routes during any cycle was 2.33ms, well within the 10ms droplet-actuation cycle of a 100Hz DMFB.

Table 2 shows the results of the random traffic stress test. For the largest test, the DTP required 2.7s to compute 480 routes (3 per droplet). The offline method took 10.6s to compute 15 routes for a smaller example. Clearly, the offline router would be a poor choice for use as an interpreter, despite the fact that it produces higher quality routes than the DTP.

6. CONCLUSION

This paper introduced an online assay interpreter for DMFBs and compared it with a static compiler. The router (DTP) can compute routes within milliseconds on low-powered commodity hardware, and the overall impact on assay completion time is negligible. The virtual architecture greatly simplifies placement and routing, but it sacrifices route optimality. A secondary drawback of the virtual architecture is that it is not area efficient, although this is offset by its ability to guarantee deadlock-free routing.

Here, we have established the feasibility of online interpretation. In the future, we plan to exploit the interpreter to introduce new capabilities to the DMFB, such as the ability to execute assays with control flow, to handle variable-latency operations, and to dynamically detect faulty cells and reconfigure the virtual architecture around them. Lastly, we plan to formally verify the

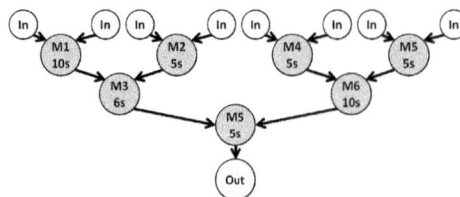

Figure 7. PCR sequencing graph annotated with mixing times.

Table 1. PCR droplet routing results for online vs. offline router.

PCR Routing Results						
Routing Sub-Problem	# Routes		Route Comp. (ms)		Sub-Prob Length (# cycles)	
	ON	OFF	ON	OFF	ON	OFF
1	8	8	5	5294	7	13
2	3	2	6	1626	30	17
3	1	2	2	1620	12	12
4	1	2	1	1469	12	6
5	1	1	0	655	6	4
SUMS:	14	15	14	10664	67	52

Table 2. Random traffic results for our online droplet router.

Random Traffic Stress Test - Online Routing			
DMFB Size (# Chambers)	# Droplets/ #Routes	Completion Time (s)	Total Computation Time (ms)
2x2	40/120	2.19	235.77
3x3	60/180	2.33	514.88
4x4	80/240	2.26	809.57
8x8	160/480	3.69	2780.24

correctness of our criteria for deadlock free routing, as proofs were omitted from this paper due to space limitations.

7. ACKNOWLEDGMENTS

This work was supported in part by NSF Grant CNS-1035603. Daniel Grissom was supported by an NSF Graduate Research Fellowship.

8. REFERENCES

[1] K. F. Böhringer. Modeling and controlling parallel tasks in droplet-based microfluidic systems. IEEE Transactions on Computer-Aided Design of Integrated Circuits and Systems, 25(2): 334-344, February, 2006.

[2] K. Chakrabarty, P. Pop, and T-Y. Ho. Digital microfluidic biochips: recent research and emerging challenges. In Proc. 9th International Conference on Hardware-Software Codesign and System Synthesis (CODES-ISSS), pp. 335-344, Taipei, Taiwan, October 9-14, 2011.

[3] M. Cho and D. Z. Pan. A high-performance droplet router for digital microfluidic biochips. IEEE Transactions on Computer-Aided Design of Integrated Circuits and Systems, 27(10): 1714-1724 November, (2008).

[4] W. J. Dally and B. P. Towles. Principles and practices of interconnection networks. Morgan Kaufammn, 2004.

[5] E. J. Griffith, S. Akella, and M. K. Goldberg. Performance characterization of a reconfigurable planar-array digital microfluidic system. IEEE Transactions on Computer-Aided Design of Integrated Circuits and Systems, 25(2): 345-357, February, 2006.

[6] F. J. Kurdahi and A. C. Parker. REAL: a program for Register Allocation. In Proc. ACM/IEEE Design Automation Conference (DAC), pp. 210-215, Miami, FL, USA, June 28-July 1, 1987.

[7] F. Su and K. Chakrabarty. High-level synthesis of digital microfluidic biochips. ACM Journal on Emerging Technologies in Computing Systems, 3(4): article #16, January, 2008.

[8] F. Su and K. Chakrabarty. Module placement for fault-tolerant microfluidics-based biochips. ACM Transactions on Design Automation of Electronic Systems, 11(3):682-710, July, 2006.

[9] P-H. Yuh, C-L. Yang, and Y-W. Chang. BioRoute: a network-flow-based routing algorithm for the synthesis of digital microfluidic biochips. IEEE Transactions on Computer-Aided Design of Integrated Circuits and Systems, 27(11): 1928-1941, November, 2008.

Reliable Logic Mapping on Nano-PLA Architectures

Masoud Zamani
Northeastern University
Boston, United States
mzamani@ece.neu.edu

Mehdi B. Tahoori
Karlsruhe Institute of Technology
Karlsruhe, Germany
mehdi.tahoori@kit.edu

ABSTRACT

Programmable nano-architectures fabricated using bottom up self assembly are promising alternatives to CMOS circuits. However, extreme process variation and high defect rate are major challenges in this nanotechnology. In this paper, we present variation and defect aware logic mapping algorithms for these nano-architectures. Simulation results show that the proposed logic transformations and the mapping algorithms based on them can improve manufacturing yield by 65% with only 12.5% area overhead.

Categories and Subject Descriptors

B.8.1 [**Performance and Reliability**]: Metrics—*Reliability, Testing, and Fault-Tolerance*

General Terms

Reliability

Keywords

Crossbar array, reliable mapping, nano-PLA

1. INTRODUCTION

Emerging nanoscale devices such as *Carbon NanoTubes* (CNTs) and semiconductor *NanoWires* (NWs) are promising alternatives to traditional CMOS devices [1]. Regular structures such as crossbar arrays are easier to fabricate in self-assembly nano fabrication. Crossbars are formed from two sets of perpendicular parallel nanowires and active elements at the intersections (crosspoints). *Nano Programmable Logic Array* (nano-PLA) is a crossbar architecture based on programmable diodes and non-programmable *Field Effect Transistors* (FETs) as the building blocks [2].

However, process variation and high defect rate are major challenges in the nano fabrication of these devices and architectures. Simulated Annealing algorithm is used for variation tolerant logic mapping into a crossbar [3]. In this method, mappings on vertical and horizontal lines are swapped to reduce variation cost of a crossbar. A set of inte-

ger linear programming formulations has been introduced to provide defect-aware logic mapping into a single crossbar [4]. Since the search space of these methods is limited to a single crossbar with only swapping the rows and columns, the efficiency would be low. Furthermore, by using various algorithms, the rows and the columns of a mapping are swapped, in order to provide defect tolerant logic mapping(e.g. [5, 6, 7, 9]). Although, these methods result in high yield for a single crossbar, they are not efficient in yield improvement of the entire circuit mapped into a crossbar array. It is due to the fact that, the modifications on the mapping (applied by these methods) are limited to the inside of a single crossbar.

In this paper, we present a combination of intra block (local) and inter block (global) logic mapping algorithms for nano-PLAs in order to tolerate variation and defects. This method is based on a set of logic transformations to modify the implementation and mapping (while preserving the logic functionality), to bypass defective devices and meet timing constraints. As simulation results confirm, this approach allows us to achieve better yield compared to previous techniques.

The rest of the paper is organized as follows. Section 2 presents preliminaries on nano-PLA. The logic transformations are described in Section 3. Section 4 presents the logic mapping algorithm. Section 5 provides simulation results. Finally, section 6 concludes the paper.

2. PRELIMINARIES

CNTs and NWs with appropriate doping can be crossed to fabricate diode, FET, and bipolar junction in the implementation of programmable interconnect, logic blocks, and memory elements [1]. Self-assembly fabrication process restricts aligned lines of CNTs and NWs to parallel lines in one direction superimposed with another set of parallel wires in the orthogonal direction.

In nano-PLA architecture, the logic operation is implemented on programmable diodes [2]. Using diode in logic implementation necessitates the use of restoration stages between subsequent logic blocks. Each block of nano-PLA consists of three stages: input stage (to selectively invert or buffer the inputs), programmable logic block, and programmable interconnect switch (to route nets between logic blocks). The operation of the nano-PLA consists of two phases: a *precharge* phase followed by an *evaluation* phase.

The correct operation of nano-PLA highly depends on timing of precharge and evaluation phases. The inputs of each logic block must be ready, before its evaluation phase starts. Therefore, the propagation delay of a logic block and its variation must be considered for appropriate adjustment of the evaluation and the precharge signals. In other words, the computation delay of each crossbar block must be considered in the logic mapping in order to tolerate delay variations. A delay model for diode-based crossbars (including nano-PLA) has been introduced in [8]. In this model, the delay at the output of a logic block is determined by the maximum switching delays of its inputs. When an input is replicated, the output delay depends on the minimum delays of replicated inputs.

3. LOGIC TRANSFORMATIONS

The mapping of a logic function into a crossbar array (aka configuration), i.e. which crosspoints are used (activated) and which are deactivated, has a considerable impact on the delay as well as fault-free functionality (i.e. not using defective devices) of the mapped circuit. Here, we present a set of *logic transformations*, which preserves the logic functionality, while changing the way crossbar resources are used for the implementation of that function. Some of these transformations are local (intra block), meaning that they preserve the functionality of the portion of the logic function mapped into a block, while modifying the configuration of that block. On the other hand, global transformations (inter block) may modify the portion of the logic function mapped into different logic blocks while preserving the entire logic functionality of the circuit.

3.1 Intra Block Transformations

3.1.1 Swapping

In the swapping transformation, the configurations of two rows or two columns of a crossbar are swapped ($A + B = B + A$). This transformation is used as the basic operation in variety of algorithms (e.g [3, 4, 5, 6, 7, 9]) . Swapping results in cascading changes, i.e. it introduces constraints outside of the logic block, by forcing the switching crossbars, both at the input and the output of this logic block, to route in a specific order. Since swapping changes the order of the inputs, such routing order imposed by swapping may be in contradiction with delay improvements achieved by routing algorithms.

3.1.2 Input Duplication

The crossbar array is a regular architecture, and the crossbar itself is a complete structure (there is a crosspoint at every intersection, in contrast to FPGA). Therefore, during logic mapping there are always unused input rows. These unused input rows can be used to duplicate some of the inputs [9]. Input duplication preserve the functionality, because $A + B = A + A + B$. Similar to swapping, duplication results in cascading changes (to the preceding switch blocks). Any duplication on input lines must be done by switching crossbars. This increases the number of used paths in switching crossbar, adding to the routing complexity. In other words, the duplication method introduces extra constraints on routing in the form of additional paths.

3.1.3 Output Decomposition

In wired-OR logic, an output can be decomposed into two or more parts, each mapped into a separate vertical line. However, all these lines must be wired to construct the output. This transformation does not change the functionality, because $(A+B+C+D) = (A+B)+(C+D)$ [8]. Unlike the other two intra block transformations, all changes done by output decomposition are non-cascading, i.e. all modifications introduced by decomposition are done inside the logic block. Since all decomposed vertical lines are wired at the logic block, output decomposition does not introduce any extra routing constraints. Therefore, unlike the other two intra block transformations, output decomposition can be done after any optimizations applied at routing crossbars.

3.2 Inter Block Transformations

In an inter block transformation, the configuration of a block is modified to be distributed between two blocks. In this transformation, the entire logic function mapped into a block, or a portion of that is decomposed into two sets of functions each mapped into a separate crossbar. In the follows we propose two inter block transformations: 1) function decomposition, and 2) block decomposition.

3.2.1 Function Decomposition

A function decomposition transformation decomposes the function of some of the outputs into sub-functions, where the sub-functions together form the original function of the decomposed outputs. If $O_i = i_1 + i_2 + ... + i_m$ is an output of a block, then it can be decomposed into two sub-functions O_i^1 and O_i^2, where $O_i^1 = i_1 + i_2 + ... + i_t$ and $O_i^2 = O_i^1 + i_t + i_{t+1} + ... + i_m$ (for any $t < m$). Since the blocks of nano-PLA produce OR function, function decomposition does not affect the functionality of O_i ($O_i = O_i^2$).

The function decomposition transformation reduces the complexity of logic mapping into a block (the block which produces O_i^1), by reducing the complexity of O_i to O_i^1. However, it adds to the complexity of the block which produces O_i^2, in terms of an extra output and a set of extra inputs. It also introduces extra complexity to routing crossbar to route extra lines for the decomposed outputs.

3.2.2 Block Decomposition

In block decomposition transformation, a subset of the output functions generated by a block B_i is mapped into a new block B_j. So if B_i originally implements n outputs, it will implement m ($m < n$) outputs and the remaining $n - m$ outputs are implemented by another block.

Block decomposition introduces area overhead in terms of an extra logic block and the corresponding routing block. However, since block decomposition reduces the number of the outputs of each block, it reduces the mapping complexity of that block.

Figure 1.a shows the configurations of two logic blocks of a nano-PLA architecture. For the sake of simplicity, the restoration units as well as routing crossbars are not shown in this figure. These configurations correspond to the mapping of two logic functions ($O_3 = I_1 + I_3$ and $O_4 = I_1 + I_2 + I_3 + I_4$). Using a set of intra and inter block logic transformations, the configurations are modified in Figure 1.b. The intra block transformations include the swapping of I_1 and I_2 (block 1), the duplication of I_3 over two rows (block 1), and the output decomposition of O_4 (block 2). Using the function decomposition transformation O_2 is decomposed into two sub-functions, one of them ($O_2^1 = I_2 + I_3$) is implemented on block 1, and another sub-function ($O_4 = O_2^1 + I_4$) is implemented on block 2.

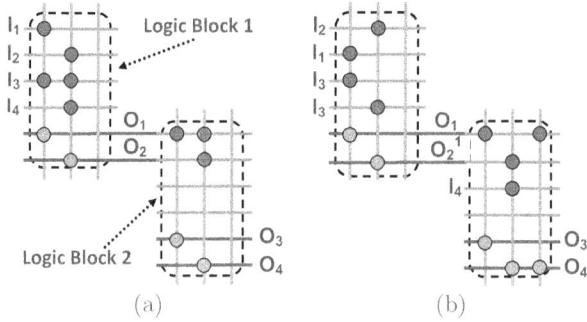

Figure 1: Mapping of two logic functions (O_3 and O_4) into two logic blocks of a nano-PLA a) The configurations of the logic blocks before applying any logic transformation b) The configurations of the logic blocks after applying a set of intra and inter block logic transformations.

4. MAPPING ALGORITHMS

The timing requirements of nano-PLA impose an upper bound on the delay between two consecutive nano-PLA blocks. This bound is expressed as the maximum acceptable delay of the path from the inputs to the successor block. A logic mapping is called **Reliable Logic Mapping** if 1) no defective crosspoint is used, and 2) the delays of all paths satisfy the timing constraints.

The proposed reliable mapping algorithm consists of two phases: 1) *local mapping* (an intra block mapping algorithm) and 2) *global mapping* (two inter block mapping algorithms). The local mapping uses intra block transformations to provide a reliable mapping for each nano-PLA block. If the local mapping fails in providing reliable mapping for a block, then that block is decomposed into two nano-PLA blocks using inter block transformations (global mapping).

4.1 Intra Block Mapping Algorithm

A heuristic algorithm is developed to provide reliable mappings for nano-PLA blocks. The algorithm applies the intra block transformations on each block. We represent the configuration of a block by a binary matrix called *Configuration Matrix* (CM), where each entry of this matrix corresponds to a crosspoint. Furthermore, each row (column) of this matrix corresponds to an input (output) of the block. $CM[i][j]$ is 1 if the corresponding crosspoint is activated; otherwise, it is 0. Figure 2 shows an example of a configuration matrix.

The proposed intra block mapping algorithm converts a configuration (called CM_1) to a reliable configuration (called CM_2). In the beginning, it applies a set of random swapping transformations on the CM_1 to avoid local optima. In each step, the algorithm maps a non-zero entry of the CM_1 ($CM_1[i][j]$) into the CM_2 ($CM_2[t][k]$), then it changes the $CM_1[i][j]$ to zero. Therefore, the algorithm terminates successfully, when all entries of the CM_1 are 0. After mapping the $CM_1[i][j]$ into the $CM_2[t][k]$, the row t and the column k of the CM_2 are assigned to the input i and the output j, respectively. Using the swapping transformation, the algorithm first tries to map a $CM_1[i][j]$ into any $CM_2[t][k]$, where the row t (column k) has been assigned to the input i (the output j). If using such a row (column) is not possible (due to reliability constraints), the algorithm assigns a new row (column) for that input (output), using input

duplication (output decomposition) transformation. If the local mapping algorithm fails, then the last updated CM_1 and CM_2 are saved to be used in the global mapping (the inter block mapping algorithm); otherwise, only the CM_2 is saved.

Figure 2: The configuration matrix of logic block 1 in Figure 1.a.

4.2 Inter Block Mapping Algorithms

If the local mapping (intra block mapping algorithm) fails in providing reliable mapping for a block, then the global mapping is applied to that block. Two global mapping algorithms are developed. The first algorithm (function decomposition algorithm) uses function decomposition transformation to map the remaining entries of the failed block into one of the existing blocks (without adding area overhead in terms of using an extra block). However, if in this way a reliable mapping is not found, the second algorithm (block decomposition algorithm) is executed to decompose the block by exploiting an extra block.

4.2.1 Function Decomposition Algorithm

The function decomposition algorithm is executed on all of the CM_1 matrices which have at least one non-zero entry. The algorithm tries to map the unmapped entries of each CM_1 into the existing CM_2 matrices. For example, if the function of an output is $i_1 + i_2 + i_3 + i_4$, but a part of its functionality (e.g. $i_1 + i_2$) is already mapped on an output (O_1). The algorithm tries to map $O_1 + i_3 + i_4$ into one of the existing CM_2 matrices.

At first, all of the unmapped or partially mapped outputs are inserted to a list (called *unmapped list*). An output is added to this list if it has at least one non-zero entry in a CM_1. Furthermore, the input set of each CM_2 matrix is determined. This set includes all of the inputs which are used in the configuration of the corresponding block. Finally, for each output in the unmapped list a set (called *function input set*) is determined. The function input set of an output includes all of the required inputs in the mapping of that output. An input i is added to the function input set of an output j if $CM_1[i][j] = 1$.

The algorithm selects the outputs from the unmapped list, one by one. It tries to map the selected output into one of the existing CM_2 matrices. However, a CM_2 matrix can be used as a candidate for mapping an output if: 1) the "function input set" of that output is a subset of the input set of the CM_2 matrix, and 2) the matrix has at least one unused row and one unused column (if needed). The first condition eliminates the extra complexity in the local mapping of the candidate matrix, in terms of adding extra inputs. The unused column is required to map the output (if needed), while the unused row is used to include the already mapped part of the function. If using the local mapping algorithm the output can be mapped into the candidate CM_2, the output is removed from the unmapped list. The algorithm repeats this procedure until all entries of the unmapped list are removed.

4.2.2 Block Decomposition Algorithm

If the function decomposition algorithm cannot successfully find a reliable mapping for a block, a block decomposition transformation is used. The block decomposition partitions the block to two separate blocks. However, the efficiency of the block decomposition transformation depends on the partitioning algorithm. Since increasing the number of unused rows and columns in a block increases the flexibility of the local mapping, the objective of the block decomposition transformation is to maximize the number of unused rows and columns in both the blocks.

In order to maximize the number of unused columns, half of the outputs are mapped into one block and the rest are mapped into another block. Therefore, in each block at least half of the columns will be unused. On the other hand, each shared input between the blocks reduces the number of unused rows by two (one from each block), while each unshared input reduces the number of unused rows only by one. Therefore, the total number of unused rows in the blocks increases by reducing the number of the shared inputs between the blocks. An input is shared between two blocks, if at least one of the outputs in each block is a function of that input. In order to reduce the number of the shared inputs between the blocks, partitioning algorithm selects the outputs with the maximum number of shared inputs to be mapped into the same block. These outputs must also have the maximum number of unshared inputs with the outputs of another block. Therefore, the partitioning is done based on the joint differential input dependency of the outputs. Joint differential input dependency of two outputs is defined as the difference between the number of shared and unshared inputs of those outputs. Furthermore, joint differential input dependency of an output and a set of outputs is defined as the difference between the number of shared and unshared inputs of that output with all of the outputs of the set. The outputs with highest joint differential input dependency are mapped into the same block.

5. SIMULATION RESULTS

The proposed algorithm has been compared to *Second-order Cardinality Algorithm* (SCA) (a heuristic algorithm which uses intra block transformations) [8]. A set of MCNC benchmark circuits is synthesized to PLA blocks, by using RASP synthesis tool [10]. The AND-plane and OR-plane of each PLA block are separated to be mapped into two 16×16 crossbars. For each circuit, 1000 different crossbar arrays with defect rate = 20% are generated. Furthermore, a Gaussian distribution ($\mu = 50$, $\sigma = 15$) is used to generate the switching delay of each crosspoint.

Two kinds of yield are calculated: 1) *Block Yield* (B_Y), and 2) *Circuit Yield* (C_Y). B_Y is defined as the percentage of the reliable block mappings. If a circuit has been synthesized to n blocks, then each block must be mapped into N different crossbars. If the mapping algorithm is able to provide m_i reliable mappings for block i, then $B_Y = \sum_{i=1}^{n} m_i/(n \times N)$. C_Y for a circuit is defined as the percentage of the cases, where algorithm can provide reliable mapping for all of the blocks of a circuit. For example, if in t cases out of N cases, all n blocks of the circuit are reliable mapped, then $C_Y = t/N$.

In these experiments, the timing constraint (threshold delay) of nano-PLA is set to 80 ($\mu + 2\sigma$). The simulation results

Table 1: Percentage of successful reliable mapping of blocks (B_Y) and the circuits (C_Y)

Circuit	SCA [8]		Proposed Method		
	B_Y	C_Y	B_Y	C_Y	Overhead
alu4	93.9%	0.0%	99.9%	99%	12.0%
apex4	96.5%	0.0%	99.9%	99%	6.9%
b9	100%	100%	100%	100%	0.0%
C880	100%	100%	100%	100%	0.0%
C1355	99.6%	91.9%	100%	100%	0.7%
C3540	98.7%	3.0%	100%	100%	2.61%
duke2	98.4%	1.0%	100%	100%	3.2%
ex5p	79.4%	0.0%	100%	100%	41.3%
k2	97.7%	0.0%	100%	100%	4.6%
rd84	66.8%	0.0%	100%	100%	66.3%
t481	99.8%	84.8%	100%	100%	0.4%
Average	93.6%	34.5%	99.9%	99.9%	12.5%

for each of the algorithms have been shown in Table 1. As seen in the table, the proposed algorithm can achieve more than 99.9% block and circuit yields, while SCA results in 96.6% block yield and 34.5% circuit yield. It must be noted, although for some cases the block yield is high (more than 97%) but the circuit yield is very low. This is due to the fact that, if at least one of the blocks of a circuit cannot be mapped reliably, the mapping of that circuit will not be reliable.

6. CONCLUSIONS

Major challenges for emerging nano-architectures are extreme variations as well as high defect rate. In this paper, we have proposed a set of logic mapping transformations used in reliable implementation of circuits on nano-PLA architecture. The transformations are embedded in a mapping flow. Simulation results on a set of benchmark circuits show the efficiency of the proposed method with more than 65% improvement in the circuit yield, compared to previous work.

7. REFERENCES

[1] W. Lu and C. M. Lieber, "Nanoelectronics from the bottom up," *Nature Materials*, vol. 6, pp. 841-850, 2007.

[2] A. Dehon, "Nanowire-Based Programmable Architectures," *ACM Journal on Emerging Technologies in Computing Systems*, vol. 1, pp. 109-162, 2005,

[3] C. Tunc and M. B. Tahoori, "Variation tolerant logic mapping for crossbar array nano architectures," *ASP-DAC*, pp. 855-860, 2010.

[4] J. S. Yang and R. Datta "Efficient Function Mapping in Nanoscale Crossbar Architecture," *DFT*, pp. 190-196, 2011.

[5] Y. Zheng and C. Huang "Defect-aware logic mapping for nanowire-based programmable logic arrays via satisfiability," *DATE*, pp. 1279-1283, 2009.

[6] W. Rao, A. Orailoglu, and R. Karri "Logic Mapping in Crossbar-Based Nanoarchitectures," *IEEE Design Test of Computers*, vol. 26(1), pp. 68-77, 2009.

[7] S. Goren, H. Ugurdag, and O. Palaz "Defect-Aware Nanocrossbar Logic Mapping through Matrix Canonization Using Two-Dimensional Radix Sort," *ACM Journal on Emerging Technologies in Computing Systems*, vol. 7(3), pp. 12:1-12:16, 2011.

[8] M. Zamani and M. B. Tahoori, "Variation-aware Logic Mapping for Crossbar Nano-architectures," *ASP-DAC*, pp, 317-322, 2011.

[9] Y. Yellambalase, M. Choi and Y. Kim "Inherited Redundancy and Configurability Utilization for Repairing Nanowire Crossbars with Clustered Defects," *DFT*, pp, 98-106, 2006,

[10] "cadlab.cs.ucla.edu/software_release/rasp/htdocs"

Self Adaptive Body Biasing Scheme for Leakage Power Reduction in Nanoscale CMOS Circuit

Jing Yang
Dept. of ECE
Northeastern University
617-373-7780, Boston, MA, 02115

jyang@ece.neu.edu

Yong-Bin Kim
Dept. of ECE
Northeastern University
617-373-2919, Boston, MA, 02115

ybk@ece.neu.edu

ABSTRACT

This paper presents techniques to determine the optimal reverse body bias (RBB) voltage to minimize leakage currents in modern nanoscale CMOS technology. The proposed self-adaptive RBB system finds the optimum reverse body bias voltage for minimal leakage power adaptively by comparing subthreshold leakage current (I_{SUBTH}), gate tunneling leakage (I_{GATE}), and band-to-band tunneling leakage currents (I_{BTBT}) in standby mode. The proposed circuit has been designed and tested using 65nm bulk CMOS technology at 25°C under a supply voltage of less than 1V. The optimal RBB was achieved at -0.372V with 1.2% error in the test case of the paper, and the simulation result shows that it is possible to reduce the total leakage current significantly as much as 86% of the total leakage using the proposed circuit techniques.

Categories and Subject Descriptors

B.7.1 [types and design styles] advanced technologies

General Terms

Design, Performance

Keywords

Self-adaptive body bias voltage, Subthreshold leakage, Gate tunneling leakage, Band-to-band tunneling, Power consumption

1. INTRODUCTION

To achieve higher density, higher performance, and lower power consumption, CMOS transistors have been scaled aggressively for decades and the supply voltage has been reduced along with the shrinking of device dimensions. Therefore, the threshold voltage (V_{TH}) of a transistor has to be commensurately scaled to achieve high performance. Unfortunately, due to the exponential relationship between leakage current and threshold voltage, decreasing of V_{TH} leads to a dramatic increasing of subthreshold leakage. Reverse body biasing (RBB) is one of the most widely used methods to reduce subthreshold leakage, which has already been introduced by [1]. However, the increasing RBB increases drain-induced barrier lowering (DIBL) and substrate doping

density, which leads to an increasing of band-to-band tunneling leakage. At the same time, increasing of RBB will increase V_{TH}, causing short channel effect in ultra-small technology, which results in an increase of gate tunneling leakage.

Previous research [2] [3]shows, in the standby mode, a certain fixed RBB can reduce I_{SUBTH} while maintains a relatively small I_{GATE} and I_{BTBT}. And it has been shown in [4] that correctly applying body bias reduces the impact of die-to-die and within die parameter variations. Therefore, the implement of optimal RBB can lead to both reduction of leakage power and yield improvement.

In this paper, a novel leakage monitoring schematic is proposed, and the principle of the scheme is illustrated in Figure 1. To monitor the leakage current components, I_{SUBTH}, I_{BTBT}, and gate induced drain leakage (I_{GIDL}) currents, are generated from a test device and convert the difference of I_{SUBTH} and I_{BTBT} current into voltage signals assuming the gate tunneling leakage is not as large as the other two leakage components. The converted voltage called "Vc" goes to the charge pump shown in the Figure 1 to charge and discharge the capacitor in the charge pump. The voltage in the capacitor is fed back to the leakage monitoring circuit through the output stage in the Figure 1 to control the body bias voltage of the test device to reduce the total leakage and to form a closed loop. The closed loop is necessary to make the scheme adaptive and to maintain the optimal body bias status continuously. An optimal RBB is achieved and maintained in the steady state if the two leakage currents become equal.

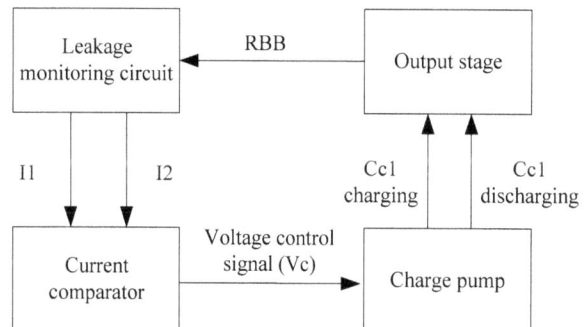

Figure 1 block diagram of optimal RBB generating system

This paper is organized as follows: Section 2 presents the three major leakage current components in ultra-small CMOS technology, and Section 3 explains the novel schematic of our optimal RBB system. Section 4 presents the results of the achieved optimal RBB and its VT variations, followed by the conclusion of the paper in Section 5.

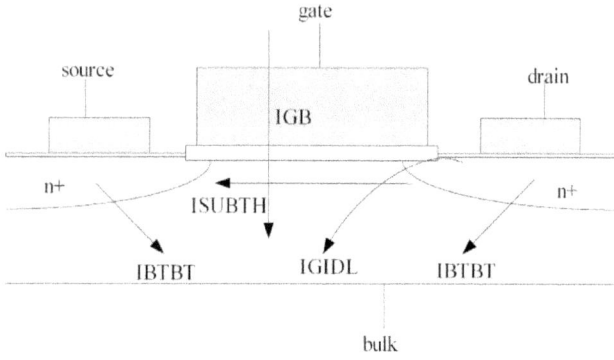

Figure 2 Leakage components in 65nm bulk CMOS transistor in the standby mode

2. Leakage current components

There are several leakage components in scaled CMOS transistors as illustrated in Figure 2. The major leakage components in nanoscale CMOS technology are I_{SUBTH}, I_{GATE} and I_{BTBT}. Thus, the total leakage current can be expressed as

$$I_{LEAKAGE} = I_{SUBTH} + I_{GATE} + I_{BTBT} \quad (1)$$

Those three major leakage currents components need to be understood for leakage reduction efforts and they are explained in this section.

2.1 Subthreshold leakage

When the gate-to-source voltage in a MOS transistor is below V_{TH}, the transistor is not completely turned off, instead, there is weak inversion region having some minority carriers along the length of the channel. This makes a small current flow from drain to source in NMOS case, which is called the subthreshold leakage and is expressed as:

$$I_{SUBTH} = I_0 e^{\frac{V_{GS}-V_{TH}}{nkT/q}} \cdot \left(1 - e^{-\frac{V_{DS}}{kT/q}}\right) \quad (2)$$

Where

$$I_0 = \mu_0 \cdot C_{ox} \cdot (W/L) \cdot \left(\frac{KT}{q}\right)^2 \cdot \left(1 - e^{1.8}\right) \quad (3)$$

V_{GS} is the gate to source voltage, V_{TH} is the threshold voltage, kT/q is the thermal voltage, n is the body effect coefficient [5], and V_{DS} is the drain to source voltage.

2.2 Gate tunneling leakage

In order to keep short-channel effect under control and to maintain a good subthreshold turn-off slope, gate oxide thickness is reduced in proportion to channel length. This results in an exponential increase in the gate leakage due to the direct tunneling of electrons through gate oxide into the substrate. This current is referred to as gate leakage and is expressed as below:

$$I_{GATE} = B \cdot \left(\frac{V_{GB}}{T_{ox}}\right)^2 \cdot (W \cdot L) \cdot e^{\frac{T_{ox}}{V_{GB}} \cdot \alpha} \quad (4)$$

Where B and α are the fitting parameters, V_{GB} is the gate to substrate voltage and T_{OX} is the gate oxide thickness.

Considering I_{GATE} has an inverse exponential relationship with V_{GB}, it is less sensitive to RBB. What is more, gate leakage current issue can be overcome while maintaining excellent gate control by using high-k dielectric material. Therefore, when

considering RBB adjustment to reduce leakage power, I_{GATE} can be ignored as in [2].

2.3 Band-to-band tunneling (BTBT)

In a high electric field ($>10^6$V/cm), there are electrons tunneling across the reverse biased PN junction of drain-substrate and source-substrate. If both n and p regions are heavily doped, the BTBT significantly increases and becomes a major contributor to the total leakage current, and the current is expressed as below:

$$I_{BTBT} = A \cdot J_S \cdot (1 - e^{-\frac{V_{RB}}{nkT/q}}) \quad (5)$$

Where A is the junction area, J_S is the maximum reverse saturation current density, n is the emission coefficient (usually set to 1), and V_{RB} is the reverse bias voltage across the junction. Compare with the high drain-substrate leakage, source-substrate leakage can be ignored.

The gate induced drain leakage (I_{GIDL}) is the current from drain to substrate caused by high electric field between gate and drain. This leakage mechanism becomes worse by high drain-to-body voltage and high drain-to-gate voltage. In this paper, I_{GIDL} is taken into account with I_{BTBT}. Since both currents are from drain to substrate, they are not easy to measure from the simulation results and they are all related to drain-to-body voltage. In a CMOS transistor, there also exists gate induced source leakage, but it is relatively small compare with I_{GIDL}, thus can be ignored in leakage analysis.

3. Optimal body biasing technique

From equations (2) (3) and (5), it can be seen that there exists an optimum RBB voltage to minimize the total standby leakage. I_{SUBTH} and I_{BTBT} have opposite dependence on body bias. As I_{GATE} is insensitive to body bias, the standby leakage is minimized at the RBB at which I_{SUBTH} and I_{BTBT} are approximately equal.

3.1 Prior Arts of leakage monitor

The ratio of I_{SUBTH} and I_{BTBT} is not easy to determine due to process variations and the complexity of calculating the electric field across the junction. In addition, the ratio varies with technology and doping profile [2]. Therefore, a leakage monitoring circuit to compare I_{SUBTH} and I_{BTBT} is the prerequisite to find the optimal RBB voltage. I_{SUBTH} and I_{BTBT} comparison circuits have been designed in [2][3] [9] and [10].

C.Neau [2] introduced a model to calculate $I_{SUBTH}/2 - I'_{SUBTH} - I_{DG}/2 - I_{GIDL}/2$. While ignorance of I_{GB} causes a minor error in leakage calculation, the drain-to-source voltage drop in PMOS current mirror also causes errors in mirroring leakage components in the approach. M.Nomura [3] developed a more precise model with consideration of I_{GB} and I_{DG} and maintaining the drain voltages of current mirrors with op-amps. However, the use of op-amp causes a voltage drop issue between the connected PMOS and NMOS on the current subtraction, which affects the mirror of the leakage currents as well. K.K.Kim [9] and H.Jeon [10] provided improved model, which separate I_{SUBTH} and I_{BTBT} by using current mirrors. However, the circuits are complicated and delicate to design, and the mismatch among current mirrors leads to erroneous results in leakage current calculation.

3.2 Proposed leakage comparator circuit

In this paper, a novel leakage current monitoring scheme is proposed. The proposed scheme compares the subthreshold and the junction leakage by calculating the subtraction of I_{SUBTH} and I_{BTBT} to generate an optimum RBB at which the two leakage current components become equal. Figure 3 shows the circuit that generates leakage currents, compares them, and converts the subtraction of the currents into voltage signal.

The source and gate tied NMOS transistor N1 is in the "off" state and the leakage current through N_1 is leakage I1 and is given by $I_1 = (I_{SUBTH} + I_{DG} - I_{GB})$. Current through the drain of a gate-grounded "off" NMOS transistor N4 stacked on top of another NMOS transistor N5 is given by $I_2 = (I'_{SUBTH} + I_{DG} + I_{GIDL}) \approx (I_{DG} + I_{GIDL})$. Considering stack effect of N4 and N5, the subthreshold leakage I'_{SUBTH} through N4 is tremendously reduced by a factor of 10 compared with I_{SUBTH} of a single transistor and it can be even smaller in 65nm NMOS technology[6][7] with lower supply voltage. Therefore, ignoring the subthreshold current of N4 and N5, $I_1 = I_{SUBTH} + I_{DG} - I_{GB}$ is mirrored by N2, N3 and N6. An accurate current mirror is required as mentioned in [8] since the current level is quite low. In a similar way, $I_2 = I_{DG} + I_{GIDL}$ is mirrored by transistors P1, P2 and P3. And the difference between the current I1 and I2 provide a means to compare I_{SUBTH} vs. $I_{BTBT} = I_{GIDL} + I_{GB}$. The op-amp E2 with a feedback resistor R_f works converts $I_A = I_{BTBT} - I_{SUBTH}$ current to voltage signal V_A.

Level shifter using op-amp is used to adjust the voltage level of V_A and it generates V_A'. The use of level shifter makes the output voltage swing from zero to supply voltage.

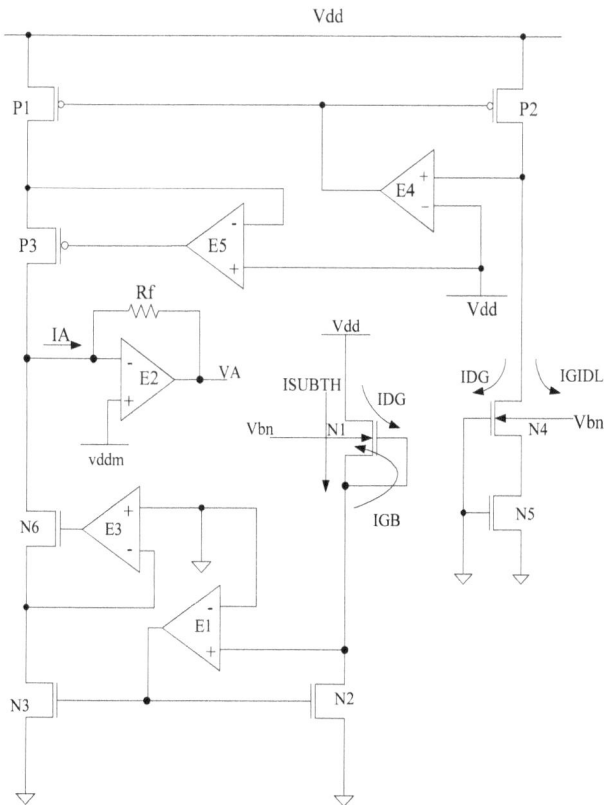

Figure 3 proposed leakage monitoring circuit

3.3 RBB regulator analysis

V_A drives gates of the two CMOS transistors P20 and N25. Figure 4 shows the circuit schematic of the charge pump and output stage of RBB. There are three operation modes for the gate inputs of P20 and N25:

(1) If $I_{SUBTH} > I_{BTBT}$, $I_A < 0$, $V_A > 0$, and N25 is turned on and capacitor Cc1 is discharging. Gate voltage of P22 decreases, drain current of P22 increases, and the absolute value of V_{bn} (|RBB|) increases. Increasing of |RBB| leads to decreasing of I_{SUBTH} and increasing of I_{BTBT}, which moves whole circuit toward mode (2).

(2) If $I_{SUBTH} = I_{BTBT}$, $I_A = 0$, $V_A = 0$, and P20 is slightly on, Cc1 will be charging slowly towards Vdd. Gate voltage of P_{22} increases slowly along with decreasing of drain current through P_{22}, and when voltage on Cc1 is greater than the threshold voltage of P22, P22will turn off and the absolute value of RBB voltage will decrease until a certain value. Then the voltage change through Cc1 is fixed and |RBB| is hold.

(3) If $I_{BTBT} > I_{SUBTH}$, $I_A > 0$, $V_A < 0$, and $P_{20}$is in the "on" state, Cc1 is charging towards Vdd. The gate voltage of P_{22} increases with its drain current decreasing, which drives |RBB| to decrease. And the decreasing of |RBB| leads to an increasing of I_{SUBTH} and decreasing of I_{BTBT}. The whole circuit will go back to mode (2)

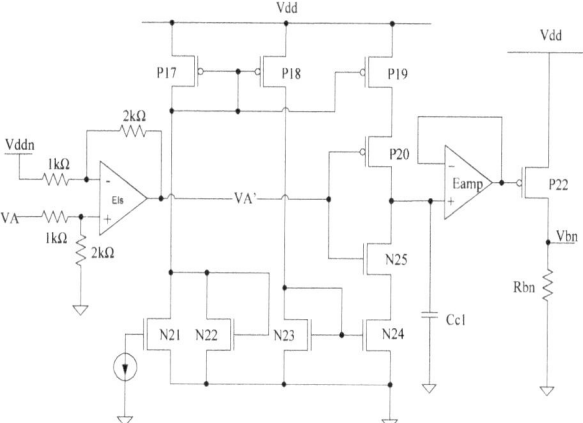

Figure 4 charge pump and output stage to generate RBB

From the above three modes, we can see that under the control of V_A, the output of V_{bn}(RBB) is continuously regulated through charging and discharging capacitor Cc1 in the charge pump until the optimal RBB is reached.

The implemented optimal RBB circuit accomplishes optimal RBB of NMOS transistors. In the exactly same way, an optimal RBB of PMOS transistor can be achieved as well.

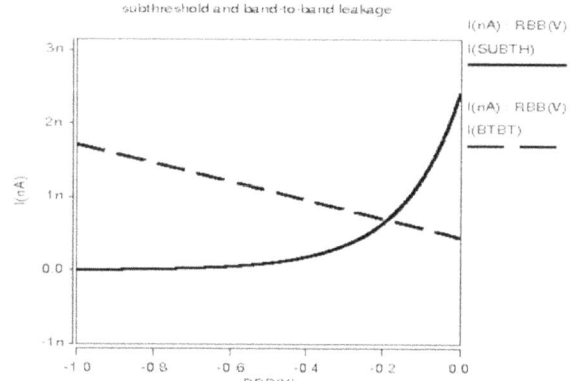

Figure 5 I_{SUBTH} and I_{BTBT} for single NMOS transistor in the standby mode with W/L=200nm/70nm at supply voltage of 1V and temperature of 25 °C

4. Experiments and results

4.1 Experimental results

Figure 5 shows an optimal RBB when leakage of I_{SUBTH} and I_{BTBT} equals for single NMOS transistors in the standby mode. For NMOS transistor that has W/L=200nm/70nm, optimal RBB voltage is achieved at around -0.2V minimizing the total leakage currents.

Figure 6 shows simulation results of the proposed current monitoring circuit using 65nm bulk CMOS technology at the supply voltage of 0.8V and temperature of 25°C. The W/L ratio of the NMOS transistors is 1μm/70nm. The standby mode NMOS transistors N1 and N4 have different leakage components as shown in Figure 3, and they are denoted as I1 and I2 as explained in Section 3.2. The subtraction of I1 from I2 is given by $I_{SUBTH} - I_{BTBT}$, and the zero value of $I_{SUBTH} - I_{BTBT}$ indicates the optimum RBB voltage point to minimize the total leakage current. As shown in Figure 6, the optimum RBB voltage turns out to be 0.37182V in the case.

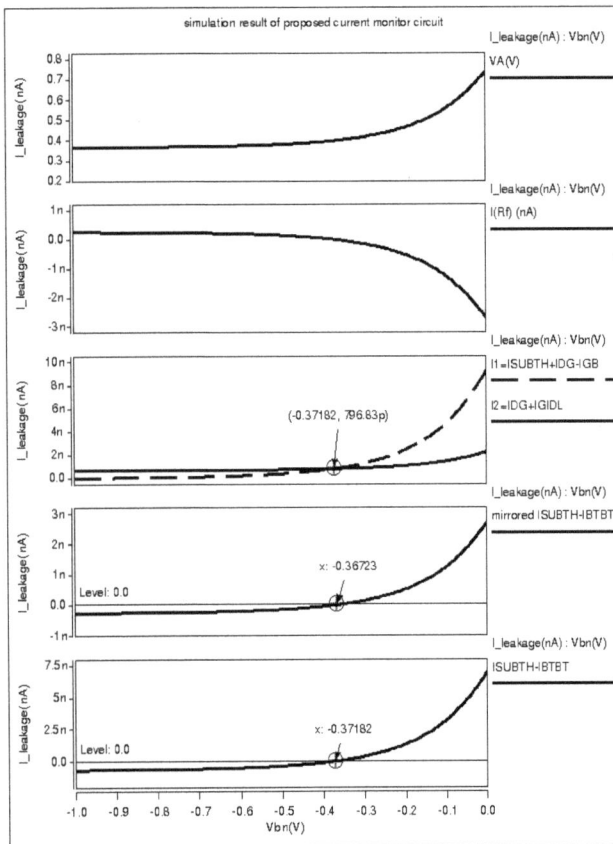

Figure 6 Simulation results of the proposed current monitoring circuit

The mirrored current through N_6 and P_3 form the current monitor and the subtraction of the two current (almost same as $I_{SUBTH} - I_{BTBT}$) reaches its zero value at the optimum RBB of 0.36723V, which is 1.2% error compared to $I_{SUBTH} - I_{BTBT}$ from N1 and N4. This result proved that the proposed current monitor circuit is precise enough to mirror the leakage components and perform the subtraction accurately. The total leakage current is 1.5937nA at the achieved optimal RBB at -0.372V compared with 11.46nA at zero body bias voltage, which demonstrates a significant reduction as much as 86% of the total leakage.

$I_{SUBTH} - I_{BTBT}$ flows through Rf and is converted to voltage signal. The top two curves in Figure 6 show the trend of the relationship between $I_{SUBTH} - I_{BTBT}$ and the corresponding voltage output signal VA.

4.2 VT variations of leakage current

Systematic and random variations in supply voltage and temperature caused leakage of low power circuits to vary significantly under design values and these variations are posing increasing challenge to nano-scale circuit design. The demand for low power causes supply voltage scaling, making voltage variations a significant design challenge. What is more, the quest for growth in operational frequency occurs at high junction temperatures and with die temperature variations. Sections below show how temperature and supply voltage changes affect optimal RBB in the proposed circuit.

4.2.1 Temperature variations with RBB

Temperature variations change the mobility of electrons and holes. An increase in operating temperature causes the mobility to decrease. From equation (2) and (3), with the increasing of T and decreasing of μ_0, I_{SUBTH} will decrease and the curve will be more flat. One the other hand, under high temperatures, electrons have more energy and the tunneling across the reverse biased pn junction of drain-substrate and source-substrate tunnels will be more easily, which results in an increasing of band tunneling current. Equation (5) well explained that, increasing of T leads to an increasing of I_{BTBT}, which makes the curve sharper. Therefore, the optimal RBB will decrease.

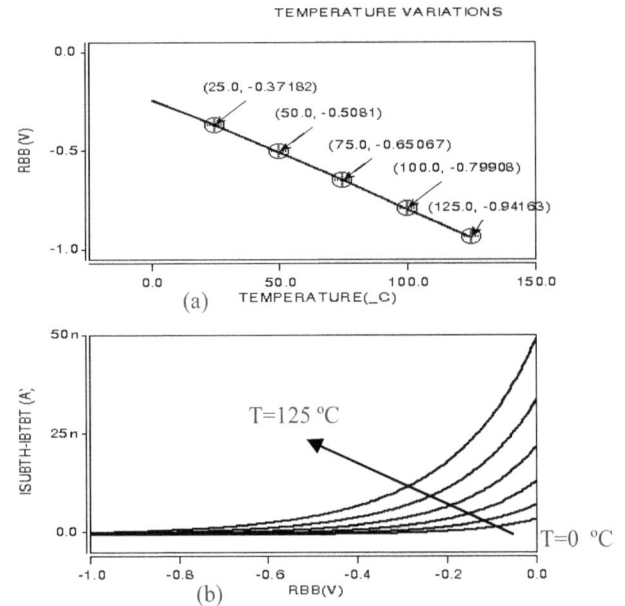

Figure 7 effects on temperature variations on leakage currents of proposed circuit

The above Figure 7 (b) shows the $I_{SUBTH} - I_{BTBT}$ curve at T ranging from 0 °C to 125 °C. Figure 7 (a) shows the corresponding RBB under temperature variations. We can tell that with T increasing, we need the substrate to be more biased in order to achieve less leakage.

4.2.2 Voltage variations with RBB

Large current consumption across long power lines with parasitic impedance causes varied, low and unstable voltage levels on the power line. This variation results in uneven supply voltage distribution and temperature hot spots, across a die, causing transistor subthreshold leakage variation across the die. From Equation (2), increasing of VDD will result in increasing of I_{SUBTH}, making the curve more "curvy", and Equation (5) shows that I_{BTBT} will increase with increasing of VDD also, since the maximum reverse saturation current Js is a function of both the Electric field at the junction and the applied reverse bias. Therefore, the optimal RBB will decrease.

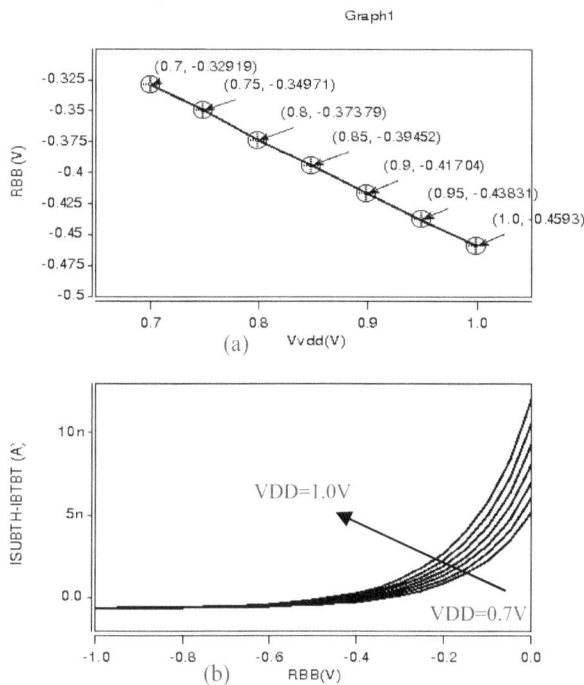

Figure 8 effects on supply voltage variations on leakage currents of proposed circuit

Figure 8 (b) above shows the $I_{SUBTH} - I_{BTBT}$ curve at VDD ranging from 0.7V to 1.0V. Figure 8 (a) shows the corresponding RBB under supply voltage variations. We can tell that with VDD increasing, we need the substrate to be more biased in order to achieve less leakage also.

5. Conclusion

As technology scaling goes down below 90nm, the standby leakage increase significantly. This makes the circuit design technique for minimizing leakage power more important. In this paper, a leakage monitoring circuit is proposed and designed to determine the optimal RBB by comparing leakage components of I_{SUBTH}, I_{BTBT} and I_{GATE}. It is possible to accomplish the minimum total leakage power during standby mode by monitoring the leakage current components. Using the information found in the monitoring circuits, an optimal body bias (RBB) voltage is found and applied to the substrate to accomplish the minimal leakage during standby state. The RBB is generated by the converted voltage from leakage current subtraction using charge pump. The voltage of the capacitor in the charge pump is held once the optimal RBB is reached. Otherwise, it keeps charging and discharging the capacitor to adaptively adjust RBB. The optimal RBB was achieved at -0.372V with a small in the test case in this paper. RBB also varies with supply voltage and temperature, with the increasing of supply voltage and/or temperature; we need more negative biased RBB to achieve

leakage reduction. The result shows that it is possible to reduce total leakage significantly as much as 86% of the total leakage using the proposed algorithm. This research can be a good reference for the future research in leakage minimization.

6. REFERENCES

[1] T.Kuroda, Testuya Fujita, et. Al., "A 0.9-V, 150-MHz, 10-mW, 4mm2, 2-D discrete Cosine Transform Processor with Variable Threshold-Voltage (VT) scheme," IEEE JSSC, vol. 31, No.11, Nov. 1996, pp.1770-1779

[2] C.Neau and K.Roy, "Optimal Body Bias Selection for Leakage Improvement and Process Compensation Over Different Technology Generations, " in Proc. IEEE ISLPED, Aug. 2003, pp.116-121

[3] M.Nomura, et. Al, "Delay and Power Monitoring Schemes for Minimizing Power Consumption by Means of Supply and Threshold Voltage Control in Active and Standby Modes", IEEE J.Solid State Circuits, Vol.41, No.4, Apr.2006

[4] J.Tschanz, J.Kao, et. Al, "Adaptive Body Bias for Reducing Impacts of Die-to-Die and Within-Die Parameter Variations on Microprocessor Frequency and Leakage",2002 ISSCC Digest of Technical Papers, pp 477-479

[5] Kaushik Roy, "Leakage Current Mechanisms and Leakage reduction Techniques in Deep-Submicrometer CMOS Circuits", proceedings of the IEEE, Vol. 91, No.2, Feb.2003

[6] S.Narendra, et. Al., "Scaling of Stack Effect and its Application for Leakage Reduction", proc. ISLPED, 2001, pp.195-200

[7] K.Sathyaki and P.Paily, "Leakage Reduction by Modified Stacking and Optimum ISO Input Loading in CMOS Devices", proc.IEEE ICACC, 2007, pp220-225

[8] J.Ramirez-Angulo, R.Carvajal, et.al., "Low Supply Voltage High Performance CMOS Current Mirror with Low Input and Output Voltage Requirements" IEEE Trans.Circuits Syst.II, Exp.Briefs, Vol.51, No.3, Mar.2004, pp.124-129

[9] K.K.Kim and Y. Kim, "A Novel Adaptive Design Methodology for Minimum Leakage Power Considering PVT Variatons on Nanoscale VLSI Systems," IEEE Trans. VLSI Syst., Vol.17, No.4, pp.517-528, Apr.2009

[10] H.Jeon, Y.Kim, "Standby Leakage Power Reduction Technique for Nanoscale CMOS VLSI Systems," IEEE Trans.Instrum Meas., Vol.59, No.5, pp.1127-1133, May.2010

Synchronization Scheme for Brick-based Rotary Oscillator Arrays

Ying Teng
Department of Electrical and Computer
Engineering
Drexel University
Philadelphia, PA 19104
yt74@drexel.edu

Baris Taskin
Department of Electrical and Computer
Engineering
Drexel University
Philadelphia, PA 19104
taskin@coe.drexel.edu

ABSTRACT

In this paper, a brick-based rotary oscillator array (ROA) synchronization scheme is proposed, which directs all the rotary traveling wave oscillators (RTWOs) in the ROA to rotate in a pre-determined direction. This synchronization scheme increases the speed of the ROA synchronization process by eliminating the repetitive start-up trials due to start-ups from incorrect points on the oscillatory array. Simulation results confirm the effectiveness of the ROA synchronization scheme. Furthermore, the synchronization scheme is applied to an ROA-based clock generation and distribution network designed for an ISPD 10 clock benchmark in order to demonstrate its application at a larger scale.

Categories and Subject Descriptors

B.7.2 [**Hardware**]: INTEGRATED CIRCUITS—*Design Aids*

General Terms

Design

Keywords

VLSI, clocking, network, resonant, synchronization

1. INTRODUCTION

Resonant clocking has become an appealing alternative to the traditional clock tree network design thanks to providing a low power, low skew solution for clock generation and distribution [1–5, 8, 13]. Rotary traveling wave oscillator (RTWO) [13] based resonant clocking technology is a particularly promising approach, which provides rail-to-rail signal amplitude and multiple clock phases compared to its competitors (such as standing wave oscillator [1] and distributed LC-based oscillators [2, 7]). However, the uncertainty of the oscillation signal rotation direction still remains an issue, which impedes the utilization of the RTWO as a

multi-phase clock signals generator. Earlier studies have detected this problem [10, 13, 14] and proposed that a single RTWO will oscillate in the direction with less capacitive loading [13]. Several solutions have been proposed in the previous studies to dedicate a direction with less capacitive loading: The insertion of varactors in order to selectively scale the capacitance loading, proposed as early as in the pioneering paper [13]. This solution increases the power consumption of the oscillation system and requires post-silicon tuning. The study in [14] advocates a physical structure modification of the RTWO in order to direct the signal rotation direction. The RTWO physical modifications in [14] require a more detailed analysis in order to accurately estimate its influence on the operating frequency. The study in [11] proposes the structure of an ROA-brick, which can effectively maintain a uniform signal rotation direction on the RTWOs in the ROA-brick. Despite the uniformity of oscillation on an ROA-brick, there still remains two signal rotation direction possibilities: Clockwise or counter clockwise. For a zero skew operation, a uniform oscillation direction, i.e. ROA bricks, is sufficient. However, the knowledge of oscillation direction—clockwise or counter clockwise—is necessary in order to benefit from the multi-phase property of rotary clocking, e.g. for useful skew or multiple clock domains. Thus, a more comprehensive and adaptive design method which enables the control of directionality is desirable.

In this paper, a unique synchronization scheme is proposed which can effectively control the signal rotation direction of the RTWOs on an ROA constructed by ROA-bricks. The ROA-bricks, proposed in [11], are used as the building blocks of the rotary clock network so as to provide a uniform, albeit unknown, oscillation direction. Instead of repetitive start-up trials and detecting a desired oscillation direction out of random start ups, the proposed synchronization scheme strategically directs the ROA into a pre-determined oscillation direction.

The paper is organized as follows: The structure and operations of the RTWO and ROA are reviewed in Section 2. The proposed synchronization scheme is presented in Section 3. The experiment setup and test results are presented in Section 4. The conclusion is presented in Section 5.

2. BACKGROUND AND MOTIVATION

An ROA is proposed in [13] as an array of checkerboard connected identical RTWOs, which is a simple and effective design structure when small RTWOs are used to deliver the

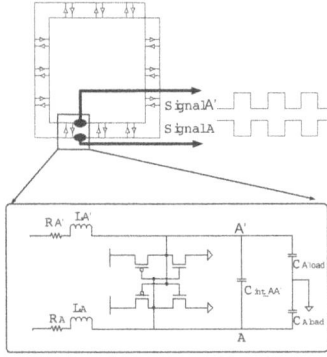

Figure 1: The structure of the RTWO.

clock signal to the whole circuit design area. Brick-based ROA, proposed in [11], is a variant of a checkerboard topology where the redundant RTWOs are eliminated and the uniformity of the signal rotation direction on all the RTWOs in the ROA is guaranteed.

2.1 Rotary Traveling Wave Oscillator

As shown in Fig. 1, the RTWO (ring) is composed of two differential transmission lines cross connected with each other. The anti-parallel inverter pairs are connected between these two transmission lines. The inverter pairs are used to provide rotation lock and overcome the energy loss during the signal propagation. Any point on the RTWO can provide a traveling square wave with a 50% duty cycle. The oscillation signals on the outer ring and inner ring at the same point, such as A and A', are of π phase difference.

The traveling signals along the transmission line of the RTWO provide multiple phase signals for sampling.

2.2 Brick-based ROA

The ROA-brick, proposed in [11], is the smallest ROA structure which can maintain the uniformity of the signal rotation direction of all the RTWOs in the ROA. Fig. 2 shows the topology of an ROA-brick {R1, R2, R4, R3}. As discussed in Section 2.1, points A and A' should maintain a phase difference of π. In order to maintain the π phase difference between A and A', the four edges along the highlighted path of propagation contribute either all $\pi/4$ or $-\pi/4$ phase difference ($| \frac{\pi}{4} \times 4 |=| \frac{-\pi}{4} \times 4 |= \pi$). The same sign of phase contributions indicate that each ring R1, R2, R4, R3 is rotating in the same direction; clockwise or counter clockwise.

A brick-based ROA theoretically guarantees a uniform signal rotation direction on all the rings as long as it is composed of a group of connected ROA-bricks [11].

2.3 Oscillation and Synchronization

When the brick-based ROA starts to oscillate, all the rings of the ROA need to coordinate their individual oscillating frequencies and signal rotation directions. This process is called synchronization.

The *frequency synchronization* of an ROA is essential to provide consistent clocking between the self-oscillation frequencies of each RTWO ring connected in locked-in-phase pattern in the ROA. For instance, consider the ROA-brick illustrated in Fig. 2. The mismatch of natural oscillation fre-

quencies of each RTWO is caused by the mismatch of total capacitance loading of each RTWO. The total capacitance loading of each RTWO is contributed by two parts: (1) Ring loading: The capacitance loading of the transmission lines and inverter pairs of the adjacent rings. (2) Circuit loading: The capacitance of the clock networks loaded on each RTWO. Due to the mismatch of the ring loading (the RTWOs in ROA-brick do not have ring loading mismatch) and circuit loading, each RTWO in the brick-based ROA needs to coordinate its own natural frequency with that of the adjacent RTWOs and finally reach a uniform oscillation frequency for the entire ROA system.

The *signal rotation direction synchronization* is essential to provide multi-phase operation capability. The study in [13] reveals that the signal on a single RTWO tends to propagates in the direction of less capacitance. This dominant direction on an individual RTWO can be different when the RTWO is not connected to the ROA versus when it is connected, as necessitated in brick-based ROAs. Thus, after the brick-based ROA is started up, some RTWOs need to change its natural signal rotation direction in order to adapt to the uniform signal rotation direction of the ROA system.

The brick-based ROA synchronization process contains both frequency and signal rotation direction synchronization. In the proposed synchronization scheme, the two are mutually coupled. This is because the uniform oscillation frequency, which is the average oscillation frequency of all the RTWOs, can be easily obtained through phase averaging at the junctions of adjacent RTWOs as long as all the rings are rotating in the same direction.

3. SYNCHRONIZATION SCHEME

The proposed synchronization process divides the system startup process into two periods: Pre-synchronization period and synchronization-period. In the pre-synchronization period, only one ring is powered up and allowed to propagate its oscillation signal to the other rings in the ROA. In the synchronization-period, all the rings are powered up and reach synchronization with certainty. In the proposed synchronization scheme, the "master" ring to start-up initially during the pre-synchronization period is selected method-

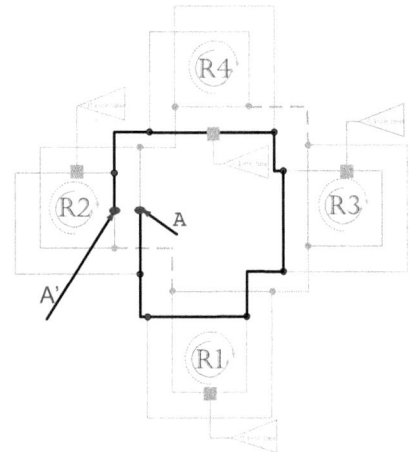

Figure 2: ROA-brick driving local clock networks.

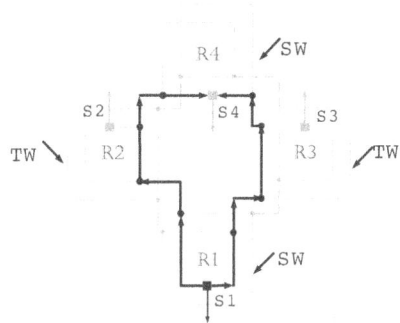

(a) R1 is not a master ring candidate.

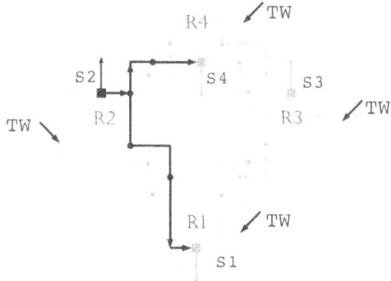

(b) R2 is a master ring candidate.

Figure 3: Master ring selection based on traveling wave (TW) and standing wave (SW) resonance implications.

ologically as it has direct implications on the signal rotation direction synchronization.

3.1 Master Ring Selection

The RTWO in an ROA which possesses the highest driving capability determines the uniform signal rotation direction and direct all the other RTWOs to rotate in its direction [13]. In previous applications, since all the RTWOs in the ROA are loaded with clock networks with relatively balanced capacitive loads [8], the difference of driving capability among RTWOs are not significant. In order to start up the ROA circuit efficiently and guarantee the uniformity of the signal rotation direction of the RTWOs in the ROA, one RTWO is selected as the master ring, which is assigned zero circuit loading. In other words, no registers are connected to the tapping points on the master ring, whereas the remaining rings are used to synchronize the circuit. The master ring starts to oscillate in the pre-synchronization period. The remaining rings, which are called slave rings, are delayed in the power up until the synchronization-period. Before the synchronization period, the master ring forms a steady oscillation signal, which is propagated to the other rings through the junctions between adjacent rings. The master ring is constructed based on the following rules:

1. The master ring is in only one ROA-brick.

2. The master ring is only loaded with ring capacitance (i.e. zero circuit load capacitance).

3. The tapping point of the master ring has asymmetrical adjacent tapping points.

Rule 1 guarantees that the master ring has the smallest ring capacitance loading. The RTWO selected by Rule 1 only has two adjacent RTWOs, which is the smallest ring capacitance loading that could be obtained in the ROA structure. Rule 2 requires that the master ring is not assigned with circuit loading which further guarantees the master ring to maintain the highest driving capability. Rule 3 is related to the oscillation starting point on the master ring through the startup circuit. In the proposed synchronization scheme, the oscillation on the master ring starts to propagate from its tapping point position. For instance in Fig. 3(b), R2 and R3 are the candidates of the master rings. R2 is a candidate that satisfies Rule 3 because the adjacent tapping points of S2—S1 and S4—are asymmetrical. S4 is 1 edge away from S2 and S1 is 3 edges away from S2. The asymmetrical tapping point location leads to asymmetrical circuit loading distribution among rings, which is leveraged to define the "known" dominant direction for oscillation during start-up.

3.2 Signal Rotation Direction Control

By picking R2 or R3 as the master ring, the uniform signal rotation direction is known through the synchronization process. As shown in Fig. 4(a), if R2 is selected as the master ring, the oscillation begins to propagate from S2 seeing imbalanced capacitance loadings between the two directions. Thus, a counter clockwise traveling wave is formed on R2 during the pre-synchronization period. When the signal propagates to R1 and R4, these two rings form traveling waves due to the capacitance imbalance between the two directions seen from their junctions with R2. The traveling wave propagation directions on R1 and R4 are opposite. Usually, the circuit loading is less than the ring loading. Thus, R1 forms a counter clockwise traveling wave, the signal rotation direction of which is the same as that on R2 so as to maintain stronger oscillation signals. Referring to R4, both the circuit loading and ring loading are on the right hand side seen from the junction between R2 and R4. Thus, R4 is driven to form a clockwise traveling wave, the signal rotation direction of which is conflicting with that on R2. The signal amplitude on R4 is attenuated. When the signals of R1 and R4 converge on R3, both of these oscillation signals try to direct R3 to follow their signal rotation directions. Due to the higher power of the oscillation signal on R1 and the clock network loading position on R3, the oscillation signal on R3 will follow the signal rotation direction of R1, which is counter clockwise. Thus, during the pre-synchronization period, R2, R1 and R3 have formed a uniform signal rotation direction and the attenuated oscillation signal on R4 will be easily overwritten when all the rings are powered up.

Alternatively, as shown in Fig. 4(b), if R3 is selected as the master ring and the oscillation signal begins to propagate from S3, a traveling wave with clockwise signal rotation direction forms during the pre-synchronization period. Through similar analysis, it is derived that R3, R1 and R2 will have a uniform signal rotation direction and the signal rotation direction on R4 will be conflicting. The signal amplitude is also attenuated on R4 due to the conflict. Thus, when all the rings are powered up, the signal rotation direction on R4 is easy to be overwritten so as to adapt to the uniform clockwise signal rotation direction on the ROA-brick.

Consider the following analysis for the case where Rule 3 is not included in the master ring construction rules. In this undesirable case, all the rings on the periphery of the

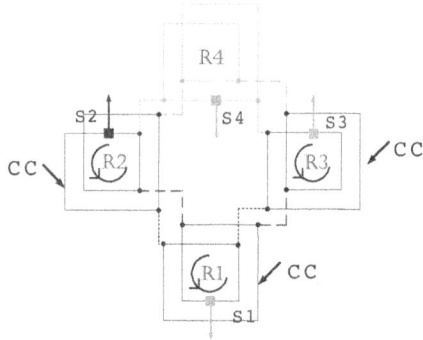

(a) Counter clockwise (CCW) signal rotation direction.

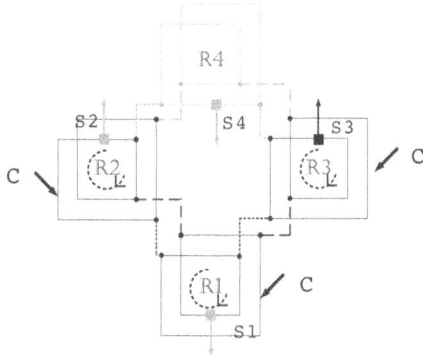

(b) Clockwise (CW) signal rotation direction.

Figure 4: Master ring affecting rotation direction.

brick-based ROA can be randomly picked as the master ring. Note that, this case illustrates the current state-of-the-art, as start-up sequencing of resonant rotary clocking has not been analyzed critically. Suppose R1 in Fig. 3(a) is selected as the master ring and the oscillation signals start to propagate from S1. The oscillation signals propagating in both directions see almost equal capacitance due to the symmetrical structure of R2 and R3 and the symmetrical tapping point locations of S2 and S3. Thus, a standing wave (SW) is more likely to be established on R1 rather than a traveling wave (TW) during the pre-synchronization period. Traveling waves are formed on R2 and R3 due to the asymmetrical capacitance loading on both sides seen from the their junctions with R1 and the signal rotation directions are opposite. When oscillation signals on R2 and R3 converge on R4, the equal power, opposite direction traveling waves form a standing wave on R4 similar as that on R1 and the tapping point S4 becomes a standing point. Thus, during the pre-synchronization period, standing waves are formed on two rings and the traveling waves on R2 and R3 choose their signal rotation direction freely. When all the rings are powered up, the signals on R1 and R4 starts to form their traveling waves, yet their rotation directions are unknown. Thus, selecting R1 as the master ring does not effectively assists the synchronization process.

One assumption for the proposed synchronization scheme is that all the tapping points have relatively balanced capacitance loading. For instance, in Fig. 4(a), when R2 is the master ring and all the capacitance are loaded on R1, R1 may not form a counter clockwise traveling wave as R2. If such

an extremely unbalanced circuit loading was present, signal rotation direction on R3 would remain undetermined. Fortunately, such a capacitive imbalance is not likely as the previous literature shows the need for capacitance balancing [8, 12] among all the tapping points and design techniques provide capacitive balancing on RTWO tapping points. Note that extreme capacitive imbalance is not only detrimental to the synchronization process but to resonant oscillation in general: Traveling waves are terminated in presence of extremely imbalanced capacitive load [12].

4. EXPERIMENTS

The simulations are performed based on a $90nm$ technology process. The operating frequency of the brick-based ROA is arbitrarily picked between $6 - 7GHz$. The simulations are performed to serve two primary goals: (1) The verification simulations of the effectiveness of the synchronization scheme. (2) The simulations of the synchronization scheme applied on a large scale ROA-based clock generation and distribution network.

4.1 Signal Rotation Direction Synchronization

In order to verify the synchronization scheme presented in Section 3.1, three sets of simulations are performed by applying the synchronization scheme to an ROA-brick. In these three simulations, R2, R3 and R1 are selected as the master ring, respectively. The simulation results shown in Fig. 5 show the uniform signal rotation direction when either R2 or R3 is selected as the master ring. The waveform on top shows that when R3 is selected as the master ring, all the RTWOs in the ROA-brick are rotating in the clockwise direction. The second waveform shows that when R2 is selected as the master ring, all the RTWOs in the ROA-brick are rotating in the counter clockwise direction. To verify the signal rotation direction controllability of the master ring, this experiment is performed repeatedly by continuously changing the capacitance loading on the other three rings from $10fF$—$2pF$. The waveforms shown in Fig. 6 is one of the simulations when the capacitance loading on the other three rings is $1pF$. All the simulation results confirm the robustness of the signal rotation direction controllability of the synchronization scheme.

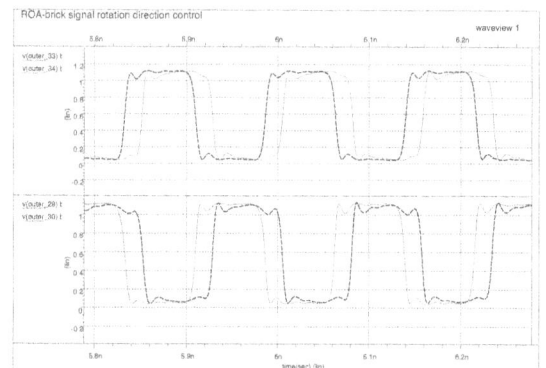

Figure 5: ROA-brick signal rotation direction.

When R1 is used as the master ring for synchronization, the uniform signal rotation direction remains undetermined as postulated in the theoretical analysis based around the illustration in Fig. 3(a). As shown in Fig. 6, the ROA-brick is restarted three times and the signal rotation directions after each time the circuit reach synchronization is inconsistent: Clockwise, clockwise and counter clockwise. If the synchronization scheme proposed in this paper is not applied, a phase detector circuit is necessary to observe the oscillation direction and restart synchronization as many times as necessary—elongating the start-up time—or trigger circuit functionality when correct directionality is detected.

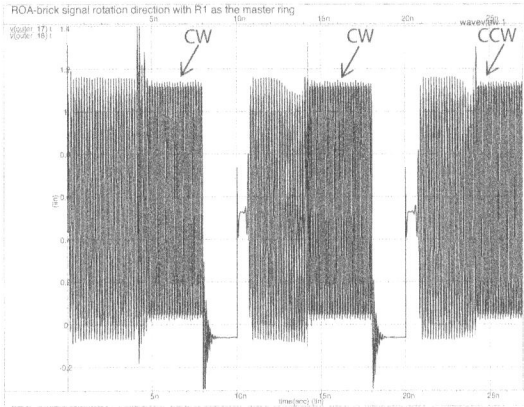

Figure 6: Random ROA-brick rotation directions.

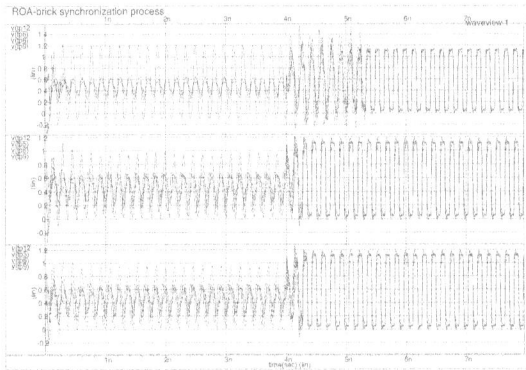

Figure 7: Synchronization process by different master ring selection.

Besides the signal rotation direction controllability, the synchronization scheme present in Section 3 also exhibits high efficiency. The simulation results are shown in Fig. 7, from top to bottom showing the synchronization results when R1, R2 or R3 are used as the master ring, respectively. As shown in the plot, the length of pre-synchronization period is $4ns$, during when only the master is powered up. When all the RTWOs are powered up, different master rings selection costs different times to reach synchronization. When R1 is selected as the master ring, it takes about $1.5ns$ after all the RTWOs are powered up to reach synchronization. This is due to standing waves are formed on R1 and R4 during the pre-synchronization period. On the other hand, when R2 or R3 are used as the master ring, it takes less than $0.5ns$ to reach synchronization. This is because three out of four RTWOs in the brick have formed a uniform signal

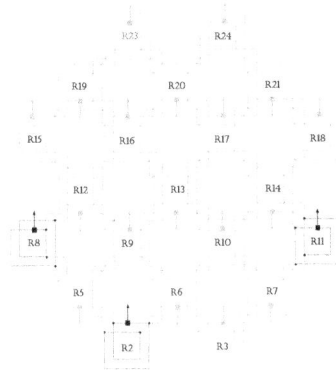

(a) Brick-based ROA topology designed for an ISPD10 benchmark.

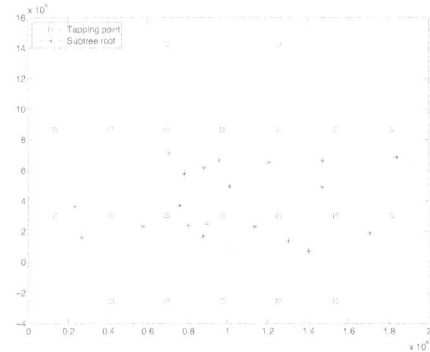

(b) Subtree roots and tapping points locations.

Figure 8: Brick-based ROA for an ISPD10 benchmark.

rotation direction, which significantly accelerates the synchronization speed. More importantly, the pre-determined oscillation direction eliminates the need for repetitive startups and control circuitry, which substantially improves the synchronization speed (e.g. start-up time).

4.2 ROA-based Clock-tree Circuit Design

In this section, the application of the synchronization scheme on the ROA-based clock network design is presented. The experiment is performed on the circuit #06 ISPD10 benchmark. First, a brick-based ROA composed of 21 RTWOs is generated for the benchmark. Second, all the benchmark registers are clustered using a modified deferred merge embedding (DME) [6, 9] solution to generate unbuffered steiner tree based networks with relatively balanced capacitance loading, as explained in [8]. Third, one of the RTWOs is selected as the master ring. The clock networks are connected to the slave rings aiming for minimum total wirelength.

The topology shown in Fig. 8(a) is the brick-based ROA for the ISPD10 benchmark. The ROA contains 21 RTWOs, 20 of which are used for clock distribution. R2, R8 and R11 are selected to be the master ring candidates and one of them is used to be the master ring in each simulation. The register, subtree root of each cluster and tapping point locations are shown in Fig. 8(b).

The simulation results are shown in Fig. 9. The simulation waveforms from top to bottom correspond to the simulations when R8, R11 or R2 are used as the master ring, respectively.

121

The simulation results show that the brick-based ROA takes less than $3ns$ to reach the synchronization state in the experiments when R8 or R11 is selected as the master ring. When R2 is selected as the master ring, the ROA is not guaranteed to reach synchronization because the case depicted in Fig. 3(a) occurs. The standing wave in this case causes the ROA to take longer time to reach synchronization.

The simulation results shown in Fig. 10 show that by using either R8 or R11 as the master ring, the uniform signal rotation direction of the ROA can be controlled. The simulation result on top corresponds to when R11 is selected as the master ring. The oscillation signals on all the RTWOs in the ROA are rotating in the clockwise direction. The bottom simulation result shows a counter clockwise rotation direction which corresponds to R8 being selected as the master ring.

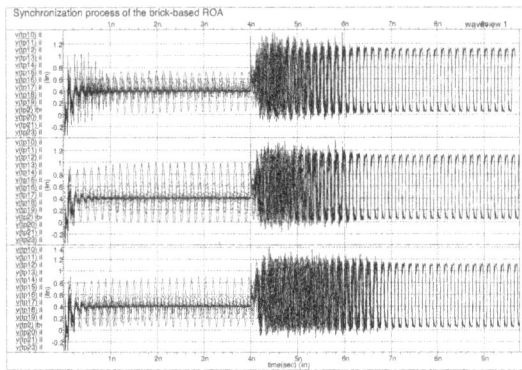

Figure 9: Brick-based ROA synchronization simulation results using three different master rings.

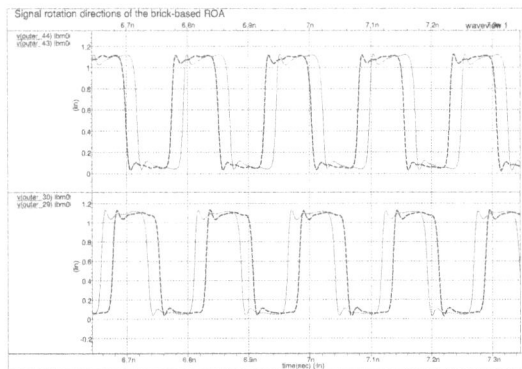

Figure 10: Brick-based ROA rotation direction.

5. CONCLUSIONS

A brick-based ROA synchronization scheme is proposed in this paper, which directs all the RTWOs in the ROA to rotate in a determined signal rotation direction. Furthermore, this synchronization scheme effectively increases the speed of the ROA synchronization. Simulation results confirm the effectiveness of the ROA synchronization scheme on a large scale ROA clock network.

6. REFERENCES

[1] W. Andress and D. Ham. Standing wave oscillators utilizing wave-adaptive tapered transmission lines. *IEEE Journal of Solid-State Circuits*, 40(3):638–651, March 2005.

[2] S. Chan, P. Restle, K. Shepard, N. James, and R. Franch. A 4.6GHz resonant global clock distribution network. In *IEEE International Solid-State Circuits Conference (ISSCC) Proceedings*, pages 342–343, Febrary 2004.

[3] V. Chi. Salphasic distribution of clock signals for synchronous systems. *IEEE Trans. Comput.*, 43(5):597–602, May 1994.

[4] V. Cordero and S. Khatri. Clock distribution scheme using coplanar transmission lines. In *Design, Automation and Test in Europe (DATE) Proceedings*, pages 985–990, March 2008.

[5] A. Drake, K. Nowka, T. Nguyen, J. Burns, and R. Brown. Resonant clocking using distributed parasitic capacitance. *IEEE Journal of Solid-State Circuits*, 39(9):1520–1528, September 2004.

[6] M. Edahiro. A clustering-based optimization algorithm in zero-skew routings. In *ACM/IEEE Design Automation Conference (DAC) Proceedings*, pages 612–616, June 1993.

[7] X. Hu and M. Guthaus. Distributed resonant clock grid synthesis (ROCKS). In *ACM/IEEE Design Automation Conference (DAC) Proceedings*, pages 516–521, June 2011.

[8] J. Lu, V. Honkote, X. Chen, and B. Taskin. Steiner tree based rotary clock routing with bounded skew and capacitive load balancing. In *Design, Automation and Test in Europe (DATE) Proceedings*, pages 455–460, March 2011.

[9] R.-S.Tsay. Exact zero skew. In *IEEE International Conference on Computer-Aided Design (ICCAD) Proceedings*, pages 336–339, November 1991.

[10] K. Takinami, R. Walsworth, S. Osman, and S. Beccue. Phase-noise analysis in rotary traveling-wave oscillators using simple physical model. *IEEE Trans. Microwave Theory Tech.*, 58(6):1465–1474, June 2010.

[11] Y. Teng, J. Lu, and B. Taskin. Roa-brick topology for rotary resonant clocks. In *IEEE International Conference on Computer Design (ICCD) Proceedings*, pages 273–278, October 2011.

[12] V.Honkote and B.Taskin. Skew-aware capacitive load balancing for low-power zero clock skew rotary oscillatory array. In *IEEE International Conference on Computer Design (ICCD) Proceedings*, pages 209–214, October 2010.

[13] J. Wood, T. Edwards, and S. Lipa. Rotary traveling-wave oscillator arrays: A new clock technology. *IEEE Journal of Solid-State Circuits*, 36(11):1654–1665, November 2001.

[14] Y. Zhang, J. Buckwalter, and C. Cheng. On-chip global clock distribution using directional rotary traveling-wave oscillator. In *IEEE Conference on Electrical Performance of Electronic Packaging and Systems (EPEPS) Proceedings*, pages 251–254, October 2009.

A Low-Power All-Digital GFSK Demodulator with Robust Clock Data Recovery *

Pengpeng Chen
Department of Electronic Engineering
Tsinghua National Laboratory for Information
Science and Technology
Tsinghua University, Beijing, 100084, China
cpp05@mails.tsinghua.edu.cn

Rong Luo
Department of Electronic Engineering
Tsinghua National Laboratory for Information
Science and Technology
Tsinghua University, Beijing, 100084, China
luorong@mail.tsinghua.edu.cn

Bo Zhao[†]
Department of Electronic Engineering
Tsinghua National Laboratory for Information
Science and Technology
Tsinghua University, Beijing, 100084, China
zhao_bo@mail.tsinghua.edu.cn

Huazhong Yang
Department of Electronic Engineering
Tsinghua National Laboratory for Information
Science and Technology
Tsinghua University, Beijing, 100084, China
yanghz@tsinghua.edu.cn

ABSTRACT

This paper presents an all-digital Gaussian frequency shift keying (GFSK) demodulator with robust clock data recovery (CDR) for low-intermediate-frequency (low-IF) receivers in wireless sensor networks (WSN). The proposed demodulator can detect and adapt to the intermediate frequency of the received signal automatically. In addition, the CDR can tolerate the frequency deviation of the input clock. An implementation of the demodulator with CDR is realized with HJTC 0.18 μm CMOS technology. The chip is designed for GFSK signals with a center frequency of 200 kHz, a modulation index of 1 and a data rate of 100 kbps. Experimental results show that the chip consumes 0.53 mA from a 1.8 V power supply, and only a 11 dB input signal to noise ratio (SNR) is required for 10^{-3} bit error rate (BER). The tolerance range for IF offset is ±12.5% at 11 dB input SNR, and the CDR can tolerate frequency deviation of the input clock of ±0.1%.

Categories and Subject Descriptors

B.7.1 [**Hardware**]: INTEGRATED CIRCUITS—*Types and Design Styles*

General Terms

Design

---

*This work was supported by China Postdoctoral Science Foundation(2011M500308).

†Bo Zhao is the corresponding author.

Keywords

GFSK demodulator, low-IF receiver, IF offset tolerance

1. INTRODUCTION

Wireless sensor networks consist of a large number of low-cost micro-sensor nodes. WSN nodes are battery-powered, so low power consumption is very important to ensure a long life time of the sensor nodes. In addition, the power consumption of the transceiver is the largest in a WSN node. Therefore simple modulation schemes such as GFSK are often adopted.

There has been a lot of research about receivers. Low-IF receivers are widely used in low power applications because of their high integration level, simpleness and low power consumption [12], [8], [6]. Because there is no information in the amplitude for GFSK signals, the low-IF receiver can use limiting amplifiers instead of voltage controlled amplifiers (VGAs) and analog-to-digital converters (ADCs) to save power [2], as shown in Fig.1. The limiting amplifiers keep only zero-crossing information, which can be used to realize the demodulation with a digital, analog or mixed-signal demodulator.

There are large IF offsets in low-IF receivers caused by the leakage of RF signals or carrier signals. For demodulators in low-IF receivers, the tolerance of IF offsets is an important factor that affects the performance of the whole transceiver.

In recent years, mainly five methods are proposed to realize GFSK demodulation: 1) IF differentiator based method: Analog time-domain IF differentiator and mixer can realize frequency discrimination [5]. However, to determine the IF differentiator's center frequency, resisters and capacities which have large process induced variations are used. These variations result in DC components in the baseband which seriously depress the demodulator's performance. 2) Zero-crossing detection method: Generating pulses at zero-crossing points can demodulate the GFSK signal [13], [10]. A capacitance is used to decide the width of the pulses [13], while the capacitance has large process induced variations which make the pulse width inaccurate [14]. 3) Quadrature Dis-

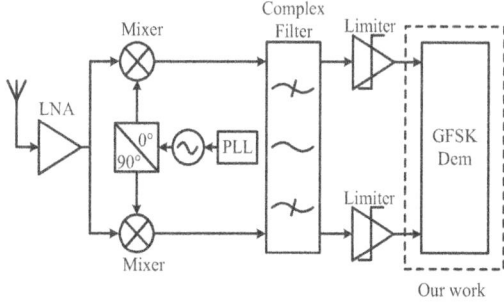

Figure 1: Block diagram of a low-IF Receiver.

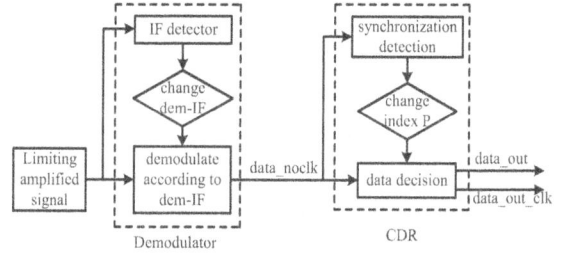

Figure 2: Block diagram of the proposed all-digital GFSK demodulator with CDR.

criminator Based Method: Double the frequency of the input GFSK signal is used as a clock of a delay circuit to sample the GFSK signal [11]. Unfortunately, the Sallen-Key filter, which is used to turn a square wave into a sine wave, consumes a large amount of power. The work in [4] uses a Bessel low-pass filter to realize $90°$ delay and frequency-to-voltage transformation, but a PLL which increases power consumption severely is needed to tune the time constant of the filter to predefined values. 4) DLL based method: A delay-locked loop (DLL) and a replica delay line are adopted to realize GFSK demodulation and IF offset cancelation [1], but it can not be used in situations where the GFSK signal's frequency deviation is larger than half of the intermediate frequency. 5) TDC based method: The time-to-digital-converter (TDC) based method has no need for a DLL so as to save power and chip area [2]. Nevertheless, it needs a high intermediate frequency, 5 MHz in [9] and 6 MHz in [2], which means a large gain-bandwidth product for amplifiers in the complex filter and more power consumption. Moreover, because of the thermometer-coded output, a large number of delay cells and D flip-flops, which consume a lot of power, are needed to achieve a good accuracy for the GFSK cycle period.

In this paper, we present a low-power all-digital GFSK demodulator with robust CDR. The contributions of this paper can be summarized into the following aspects:

• A low-power all-digital GFSK demodulator with CDR.
• An all-digital demodulator with automatic IF correction.
• An all-digital CDR with clock deviation tolerance.

The rest of our paper is organized as follows. Section II describes the digital demodulator with automatic IF correction and the robust CDR in detail. Section III gives an implementation and the experimental results. Section IV concludes the paper.

2. ALL-DIGITAL DEMODULATOR WITH ROBUST CDR

The flow of our proposed digital demodulator with robust CDR is shown in Fig.2. The demodulator realizes frequency-to-voltage conversion and automatic IF correction. The CDR outputs the digital data from the baseband wave $data\_out\_clk$. When there is a rising edge in $data\_out\_clk$, data is outputted in $data\_out$.

2.1 All-Digital Demodulator

Fig.3 shows the flow of the proposed digital demodula-

tor. The demodulator is designed for GFSK signals with intermediate frequency F_c, frequency deviation F_s and data rate F_d. The input is a digital clock with a frequency of f_o and the limiting amplified GFSK signal. The output is the demodulated signal $data\_noclk$.

There are two functions that the demodulator realizes, which are described as follows:

1) Frequency to voltage conversion. The modulated signal is sampled by the input clock and accumulated. The space between two adjacent edges is shown by the number of clock periods. When the intermediate frequency is F_c, the space between two adjacent edges is $f_o/2F_c$. So if the current space between two adjacent edges is smaller than $f_o/2F_c$, the current frequency is higher than F_c, then the output signal $data\_noclk$ is high. Or else, the $data\_noclk$ is low. In fact, the current space between two adjacent edges is inversely proportional to the current frequency, so we can determine by how much the current frequency varies from the intermediate frequency. It can be seen that the signal is demodulated according to the signal's intermediate frequency. We call the intermediate frequency adopted by the demodulator dem-IF. If dem-IF and the input signal's intermediate frequency is not the same, there is IF offset, and the performance of the demodulation above will decrease.

2) Automatic IF correction. In our demodulator, the IF offset tolerance is realized by choosing a different dem-IF. When the demodulator is receiving a signal, the demodulator can output the baseband signal, and detect the widest spaces and the narrowest spaces at the same time. If the input signal's intermediate frequency is F_c and frequency deviation is F_s, then the smallest space between two adjacent edges is $f_o/2(F_c + F_s)$ clock periods, and the largest space is $f_o/2(F_c - F_s)$ clock periods. Firstly, we set the dem-IF F_c for example. If the detected space is smaller than $f_o/2(F_c + F_s)$, then the intermediate frequency of the input signal is larger than dem-IF, the dem-IF should be increased. If the detected space is larger than $f_o/2(F_c - F_s)$, the dem-IF should be reduced. Considering the affect of noise, a threshold M_{thrsd} is defined. If the number of spaces larger than $f_o/2(F_c - F_s)$ reaches the threshold before the number of spaces smaller than $f_o/2(F_c + F_s)$, the dem-IF should be reduced and the number of wider and narrower spaces are counted from zero again. If the number of spaces larger than $f_o/2(F_c - F_s)$ reaches the threshold after the number of spaces smaller than $f_o/2(F_c + F_s)$, the dem-IF should be increased and the number of wider and narrower spaces are counted from zero again.

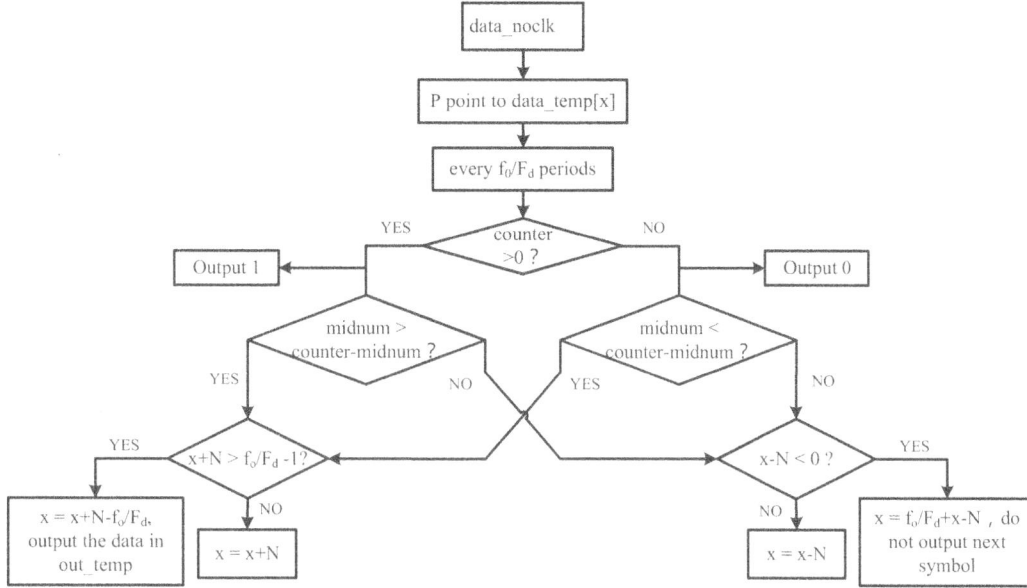

Figure 4: Flow of CDR.

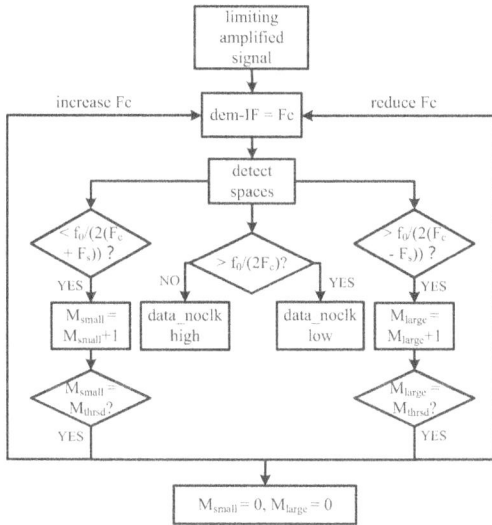

Figure 3: Flow of the proposed demodulator.

It can be seen that the signal's intermediate frequency is tracked automatically and the tolerance of IF offset can be very large as long as the range of the dem-IF for the demodulator is increased.

2.2 Robust CDR

Due to the affect of noise and the resolution of the digital clock, the duty cycle of *data_noclk* is not accurate, so a CDR is needed to get the digital information from *data_noclk*.

The flow of the proposed CDR is shown in Fig.4. To store one bit of data (one symbol) in the sampled *data_noclk*, a f_o/F_d bits shift register named *out_temp* is used. The data in the register moves 1 bit left every clock and *data_noclk* is inputted into the right bit. An index P is used to point to a position x, i.e. *out_temp*[x] in the register where to start counting the symbol. The moving step of the index P is set N. For every f_o/F_d clock periods, the data in *out_temp*[x] is cumulated (plus 1 when *out_temp*[x] is high and minus 1 when *out_temp*[x] is low.). The sum of *out_temp*[x] in the first $f_o/2F_d$ clock periods is got and named *midnum*, and the sum of *out_temp*[x] in the f_o/F_d clock periods is got and named *counter*. The *midnum* and *counter* are initially set to 0 before the start of every f_o/F_d periods.

The CDR realizes symbol synchronization and tolerance of clock deviation:

1) Symbol synchronization. At the end of every f_o/F_d clock periods, if *counter* is larger than 0, the previous one symbol is 1. Then, if *midnum* is smaller than *counter* − *midnum*, the number of zeros in the first $f_o/2F_d$ periods is larger than the zeros in the second $f_o/2F_d$ periods. That means x should be reduced to $x - N$ to make the start-cumulate point coincide with the start of symbols. If *midnum* is larger than *counter* − *midnum*, x should be increased to $x + N$. If *counter* is smaller than 0, the previous symbol is

125

Table 1: Parameter setting of the digital demodulator.

dem-IF (kHz)	$f_o/2(F_c + F_s)$	$f_o/2F_c$	$f_o/2(F_c - F_s)$
180	43	56	77
190	41	53	72
200	40	51	67
210	38	48	63
220	37	46	59

0. For the situation that the previous symbol is 0, the flow is shown in Fig.4.

2) Tolerance of clock deviation. There may be frequency deviation for the clock signal. In our CDR, the frequency deviation of the clock is tolerated by the circulating movement of index P. Assuming that x is the start of the current symbol and the frequency of the clock is higher than the ideal clock, then the start of next symbol will be smaller than x and x should be reduced. So x will keep getting smaller as long as the frequency of the clock is higher than the ideal clock. If $x - N$ is smaller than 0, the next symbol will be counted at $out\_temp[f_o/F_d + x - N]$. Because the symbol in the register out_temp is already outputted, there will be no output data for next f_o/F_d periods. If the frequency of the clock is lower than the ideal clock, x will keep getting larger. If $x + N$ is larger than $f_o/F_d - 1$, the next symbol is counted at $out\_temp[x + N - f_o/F_d]$. Because the symbol in out_temp is still not outputted, the data in out_temp will be outputted first before the data in the next f_o/F_d periods.

The tolerance range of clock frequency deviation depends on the moving step N of P. We know that x will keep getting smaller (larger) as long as the frequency of the clock is higher (lower) than the ideal clock. Additionally, larger frequency deviation of the clock means that larger changing speed of x is needed. So large N can realize a good CDR frequency tracking capability, but makes the resolution of the start-cumulate point poor. If the frequency of the provided clock is relatively stable and accurate, then N should be set to a small number to get a good resolution of the start-cumulate point. N can be made programmable in the design. In fact, the clock is usually provided by a crystal oscillator, and the frequency of the crystal oscillator usually has high accuracy. So we prefer to set N to a small number. The latency of the digital CDR loop is f_o/F_d clock periods, i.e. a bit of data. Large latency ensures good CDR jitter/noise tolerance performance.

3. EXPERIMENTAL RESULTS

The demodulator with CDR is realized using the verilog-HDL language and is implemented with HJTC 0.18um CMOS technology. The chip is designed for GFSK signals with an intermediate frequency of 200 kHz, frequency deviation of 50 kHz and a data rate of 100 kbps. The input clock is 20 MHz. 5 dem-IFs are implemented in the demodulator. The corresponding $f_o/2F_c$, $f_o/2(F_c + F_s)$ and $f_o/2(F_c - F_s)$ are shown in Table 1.

The relation between bit error rate (BER) and SNR is measured and shown in Fig.5. For 0.1% BER, only 11 dB SNR is required. Fig.6 compares the measure result of the demodulator with automatic IF correction and without automatic IF correction. Because the dem-IF range of the designed demodulator is 180 kHz to 220 kHz, the IF cover-

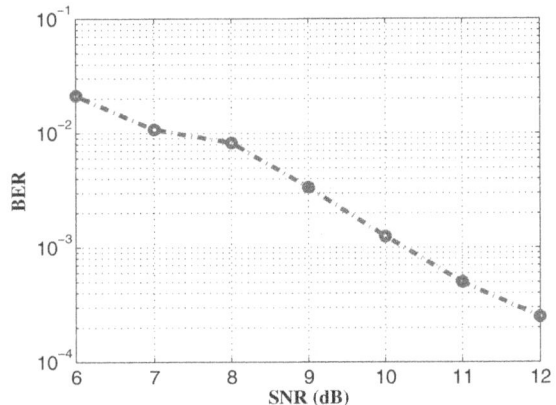

Figure 5: Measured BER vs. SNR.

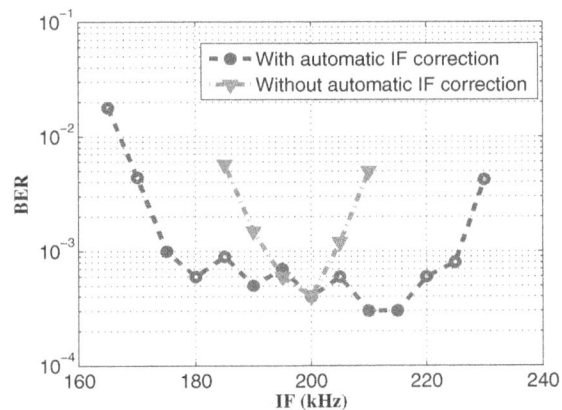

Figure 6: Measured BER vs. IF offset at 11dB SNR.

Table 2: Performance Comparison.

References	CMOS Technology (μm)	Frequency (Hz)	Data Rate (b/s)	SNR (dB)	IF Offset Tolerance	Power Consumption (mW)
[11], 2006	0.25	2 M	1 M	16.5	10%	6
[9], 2007	0.18	5 M	1 M/250 k	14.9/7.4	7%/-12%~+9%	3.6[a]
[3], 2007	0.18	3 M	N/A	19.5	3.3%	2.64
[7], 2009	0.18	2 M	0.1 to 2.0 M	16.0	N/A	0.81[a]
This work	0.18	200 k	100 k	11.0	12.5%[b]	0.95[a]

[a]The power consumption includes CDR.

[b]The demodulator is measured at 11 dB input SNR to get the IF offset tolerance.

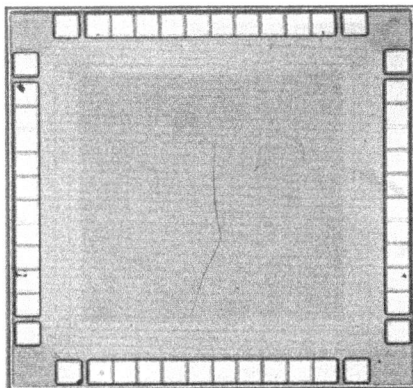

Figure 7: Die Photo.

ing range of the demodulator is -12.5%~+12.5% (180 kHz to 220 kHz). It shows that our method can tolerate IF offset effectively. The moving step N is set to 10 and the tolerance range of the clock frequency is -0.1%~+0.1% (19.98 MHz~20.02 MHz).The die photo of the chip is shown in Fig.7.

Table 2 summarizes the demodulator's performance and lists the comparison of different IF circuits. It shows that the proposed demodulator has excellent IF offset tolerance and relatively low power consumption. In our design, a large part of power is consumed by the CDR circuit due to the use of the large register out_temp. In fact, out_temp can be made much smaller considering the effect of N, and thus the power consumption can be reduced further more. Whereas, a complete f_o/F_d bits shift register out_temp ensures a more stable CDR performance. In our design, we adopted the complete f_o/F_d bits shift register out_temp.

4. CONCLUSION

A new kind of GFSK demodulator with CDR is presented in this paper. The proposed automatic IF correction method can realize ultra large IF offset tolerance. The CDR can tolerate frequency deviation of the clock. An implementation is realized with HJTC 0.18um CMOS technology. The chip consumes 0.53 mA from a 1.8 V power supply. Only 11 dB SNR is required for 0.1% BER. The IF offset tolerance is ±12.5% at 11 dB SNR, and the CDR can tolerate ±0.1% clock frequency.

5. REFERENCES

[1] S. Byun, C.-H. Park, Y. Song, S. Wang, C. Conroy, and B. Kim. A low-power CMOS Bluetooth RF transceiver with a digital offset canceling DLL-based GFSK demodulator. *IEEE Journal of Solid-State Circuits*, 38(10):1609 – 1618, 2003.

[2] C.-P. Chen, M.-J. Yang, H.-H. Huang, T.-Y. Chiang, J.-L. Chen, M.-C. Chen, and K.-A. Wen. A Low-Power 2.4-GHz CMOS GFSK Transceiver With a Digital Demodulator Using Time-to-Digital Conversion. *IEEE Transactions on Circuits and Systems I: Regular Papers*, 56(12):2738 – 2748, 2009.

[3] Y.-C. Chen, Y.-C. Wu, and P.-C. Huang. A 1.2-V CMOS Limiter / RSSI / Demodulator for Low-IF FSK Receiver. In *IEEE Custom Integrated Circuits Conference*, pages 217 –220, 2007.

[4] B. Chi, J. Yao, P. Chiang, and Z. Wang. A 0.18-μm CMOS GFSK Analog Front End Using a Bessel-Based Quadrature Discriminator With On-Chip Automatic Tuning. *IEEE Transactions on Circuits and Systems I: Regular Papers*, 56(11):2498 – 2510, 2009.

[5] H. Darabi, S. Khorram, B. Ibrahim, M. Rofougaran, and A. Rofougaran. An IF FSK demodulator for Bluetooth in 0.35 μm CMOS. In *IEEE Conference on Custom Integrated Circuits*, pages 523 –526, 2001.

[6] B. Guthrie, J. Hughes, T. Sayers, and A. Spencer. A CMOS gyrator low-IF filter for a dual-mode Bluetooth/ZigBee transceiver. *IEEE Journal of Solid-State Circuits*, 40(9):1872 – 1879, 2005.

[7] D. Han and Y. Zheng. A Mixed-Signal GFSK Demodulator Based on Multithreshold Linear Phase Quantization. *IEEE Transactions on Circuits and Systems II: Express Briefs*, 56(9):719 –723, 2009.

[8] H. Jeong, B.-J. Yoo, C. Han, S.-Y. Lee, K.-Y. Lee, S. Kim, D.-K. Jeong, and W. Kim. A 0.25-μm CMOS 1.9-GHz PHS RF Transceiver With a 150-kHz Low-IF Architecture. *IEEE Journal of Solid-State Circuits*, 42(6):1318 – 1327, 2007.

[9] H.-S. Kao, M.-J. Yang, and T.-C. Lee. A Delay-Line-Based GFSK Demodulator for Low-IF Receivers. In *IEEE International Solid-State Circuits Conference. Digest of Technical Papers*, pages 88 –589, 2007.

[10] E. Lee. Zero-crossing baseband demodulator. In *Sixth IEEE International Symposium on Personal, Indoor and Mobile Radio Communications*, volume 2, pages 466 – 470, 1995.

[11] T.-C. Lee and C.-C. Chen. A mixed-signal GFSK demodulator for Bluetooth. *IEEE Transactions on Circuits and Systems II: Express Briefs*, 53(3):197 – 201, 2006.

[12] A. Maxim, R. Poorfard, R. Johnson, P. Crawley,

J. Kao, Z. Dong, M. Chennam, T. Nutt, and
D. Trager. A Fully-Integrated 0.13 μm CMOS Low-IF
DBS Satellite Tuner. In *Symposium on VLSI Circuits,
Digest of Technical Papers*, pages 37 – 38, 2006.

[13] W. Sheng, B. Xia, A. Emira, C. Xin, A. Valero-Lopez,
S. T. Moon, and E. Sanchez-Sinencio. A 3 V, 0.35 μm
CMOS Bluetooth receiver IC. In *IEEE Radio
Frequency Integrated Circuits (RFIC) Symposium*,
pages 107 –110, 2002.

[14] V. Torre, M. Conta, R. Chokkalingam, G. Cusmai,
P. Rossi, and F. Svelto. A 20 mW 3.24 mm^2 Fully
Integrated GPS Radio for Location Based Services.
IEEE Journal of Solid-State Circuits, 42(3):602 – 612,
2007.

Link Breaking Methodology:
Mitigating Noise within Power Networks

Renatas Jakushokas
jakushok@ece.rocheste.edu

Eby G. Friedman
friedman@ece.rochester.edu

Department of Electrical and Computer Engineering
University of Rochester
Rochester, New York 14627

ABSTRACT

A link breaking methodology is introduced to reduce voltage degradation within mesh structured power distribution networks. The resulting power distribution network combines a single power distribution network to lower the network impedance, and multiple networks to reduce noise coupling among the circuits. Since the sensitivity to supply voltage variations within a power distribution network can vary among different circuits, the proposed methodology reduces the voltage drop at the more sensitive circuits, while penalizes the less sensitive circuits. The proposed methodology is evaluated for two case studies, demonstrating a reduction in the voltage drop in sensitive circuits. Based on these case studies, the voltage is improved by, on average, 4% at those nodes with the highest sensitivity. The voltage after application of the link breaking methodology is, on average, 96% of the ideal power supply voltage. Lowering the noise on the power network enhances, on average, the maximum operating frequency by 11% by utilizing the proposed link breaking methodology.

Categories and Subject Descriptors

B.7.1 [**Hardware**]: Integrated Circuits—*Types and Design Styles*

Keywords

Power Distribution Networks, Power Integrity, Power Noise

1. INTRODUCTION

The increasing density and performance of integrated circuits (IC) requires advancements in design methodologies for the global interconnects, particularly the on-chip power networks, clock networks, and long distance on-chip signals. The on-chip power distribution network typically provides hundreds of amperes to the load circuits while utilizing up to 40% of the overall metal resources [1, 2]. With advancements in technology, higher current is required;

This research is supported in part by the National Science Foundation under Grant Nos. CCF-0811317 and CCF-0829915, grants from the New York State Office of Science, Technology and Academic Research to the Center for Advanced Technology in Electronic Imaging Systems, and by grants from Intel Corporation and Qualcomm Corporation.

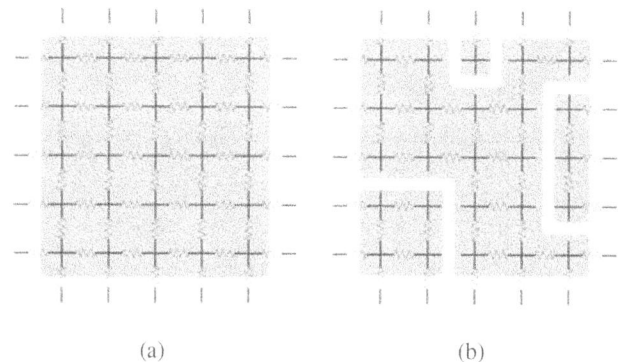

Figure 1: Mesh structured power distribution network. (a) single power distribution network focused on reducing the network impedance. (b) multiple power distribution networks lower the noise at the expense of increasing the network impedance.

therefore, more efficient on-chip power distribution networks have become an essential element of modern IC design flows.

A change in voltage at the power node of a gate can significantly increase the delay of a logic gate [3, 4], degrading the overall performance of a system [5]. Since different circuits are affected differently by a drop in the power supply voltage, the power distribution network should be designed to satisfy multiple constraints. The voltage level for those gates along the critical path can tolerate the least voltage degradation, while the gates along a non-critical path may satisfy speed constraints despite a higher voltage drop [6]. Circuits such as PLLs (phase lock loops) and VCOs (voltage controlled oscillators) are highly sensitive to changes in the power supply [7], while digital logic circuits can tolerate much higher variations in the power supply voltage.

Separate power networks can be designed to independently supply current to different parts of a circuit; thereby shielding different parts of an IC from each other. Separate power networks are widely used in mixed-signal circuits, where the current is supplied by different power networks to the analog and digital circuits [8]. For systems requiring the same voltage level, this approach however may inefficiently utilize metal resources due to the additional area and routing constraints [6]. The number of I/O pads is also a limiting resource, preventing the use of an excessive number of separate power networks [9]. In Fig. 1, a single and multiple separate power networks are illustrated. With a single network, as shown in Fig. 1(a), the sensitive circuit (for example, a PLL) and aggressor circuit (exemplified by a large digital logic circuit) share the same power network, thereby lowering the power network impedance. A sensitive circuit can however be highly affected by the noise gen-

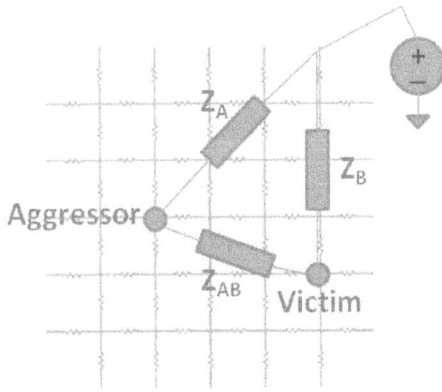

Figure 2: Aggressor and victim circuits sharing a mesh structured power distribution network. The objective is to increase Z_{AB}, while insignificantly increasing Z_A, resulting in shielding the victim from an aggressor.

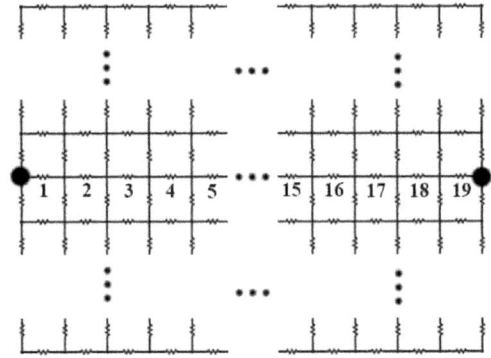

Figure 3: 20×20 node mesh structured network. The effective resistance is between the two bold nodes. The links are numbered based on the location along the x-axis.

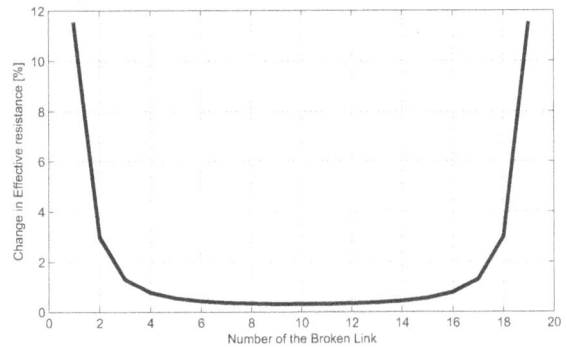

Figure 4: A change in the effective resistance between the left and right nodes within a 20×20 mesh structured power distribution network (see Fig. 3) as a function of a specific location of a disconnected link between two nodes.

erated from an aggressor circuit. With multiple power networks, as shown in Fig. 1(b), one network can be dedicated to an aggressor circuit while another network can be dedicated to the sensitive circuits, minimizing noise coupling between the aggressor and sensitive circuits. This approach however results in an increase in the power network impedance and routability constraints. The methodology proposed in this paper utilizes a single power network to provide a low network impedance and reduced routability constraints, while disconnecting (or breaking) links within the on-chip power network between an aggressor and sensitive circuit, thereby reducing noise coupling to the sensitive circuits.

The paper is organized as follows. The primary design objective for reducing voltage variations is formulated in Section 2. The sensitivity of the victim circuits to variations in the voltage within the power network is characterized in Section 3. In Section 4, the proposed link breaking methodology is described. An algorithm for breaking links for a large number of aggressor and victim circuits connected to a common on-chip power distribution network is also described in this section. In Section 5, two design cases are evaluated. The degradation in the supply voltage and propagation delay before and after applying the proposed link breaking methodology is summarized. Additional discussion related to enhancing the voltage levels within an on-chip power distribution network and the computational runtime of the algorithm is presented in Section 6. Finally, the conclusions are summarized in Section 7.

2. REDUCTION IN VOLTAGE VARIATIONS

Every circuit connected to the power distribution network can be considered as an aggressor and a victim. Two parameters are therefore assigned to each circuit, one characterizing the aggressiveness and the second the sensitivity of a circuit. The aggressor parameter is directly related to the current sunk by a circuit. Simultaneously, every circuit exhibits a different sensitivity to variations in the power network voltage. For example, a PLL is highly sensitive to voltage variations as compared to digital logic. Two circuits with a different critical path may also exhibit a different sensitivity to voltage variations: a slower critical path requires a smallest power drop, while a fast critical path can better tolerate a large voltage drop on the power network. A sensitivity factor is therefore assigned to each circuit connected to the power network. A more detailed discussion of the sensitivity factor is presented in Section 3.

In a system with multiple aggressors and victims, the objective

LINK-BREAKING
1. Determine voltage drops at all k nodes
2. Calculate initial $delay_{ini}$ function based on (2)
3. Generate x randomly perturbed systems
4. Determine voltage drops at k nodes for x systems
5. Calculate $delay$ function based on (2) for x systems
6. For every x systems
7.   Generate six different networks,
        where a link is broken at every direction
8.   Determine new $delay$ values, maintaining network
        with lowest $delay$
9.   Goto 7, if enchantment is achieved
10. Select system with lowest $delay$
11. If $delay_{ini} > delay$, $delay_{ini} \leftarrow delay$ and goto 3

Figure 5: Pseudocode for link breaking algorithm.

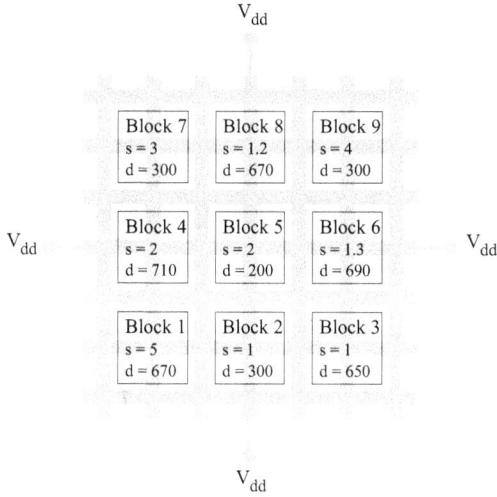

Figure 6: Nine circuit blocks are connected to a mesh structured power distribution network. Four power supplies provide the current. The numbers indicated within the blocks represent the sensitivity factor (s) and propagation delay in ps (d) when applying one volt to the block. Note that the minimum propagation delay is achieved when applying a full power supply.

is to minimize the effect of the voltage drop over the entire system. To improve the performance of an IC, the voltage drop is reduced in those circuits with high sensitivity at the expense of increasing the voltage drop in the less sensitive circuits. Three specific nodes, the victim, aggressor, and power supply, within a mesh structured power distribution network, are illustrated in Fig. 2. The objective is to increase the network impedance between the victim and the aggressor nodes (Z_{AB}), thereby reducing the influence of the aggressor on the victim node, while only minimally increasing the effective impedance between the aggressor and the power supply (Z_A).

A 20×20 node mesh structured network is illustrated in Fig 3. The normalized effective resistance between the left and right nodes as a function of a specific disconnected link at a particular location (along the x-axis) is depicted in Fig. 4. The x-axis describes the location (or link number depicted in Fig. 3) of the disconnected link between two nodes. The largest increase in the effective resistance is achieved when breaking the link closest to either node. An 11% increase in the resistance is caused by breaking a single link. This change confirms that breaking links within a mesh structured power distribution networks may result in a large change in the effective impedance; effectively shielding the victim from the aggressor.

3. SENSITIVITY FACTOR

The sensitivity factor describes the relative importance of a change in voltage on the performance of a circuit. A method to describe the sensitivity factor is to investigate the sensitivity of the supplied voltage on the performance (for example, the propagation delay) of a particular circuit. The sensitivity factor is [10]

$$s = \left. \frac{\frac{\Delta delay}{delay(x)}}{\frac{\Delta V}{V(x)}} \right|_{x=V_{dd}} = \frac{\Delta delay}{\Delta V} \cdot \frac{V_{dd}}{delay_{min}}, \qquad (1)$$

where $\Delta delay$ and $delay_{min}$ are, respectively, the change in the delay and the minimum delay of a circuit. The minimum delay is

achieved assuming a full V_{dd} at the power rail of the circuit. ΔV is the change in the supply voltage at the node supplied to the circuit. The sensitivity factor is dependant on the type of circuit.

4. LINK BREAKING METHODOLOGY

An algorithm for determining which links should be removed, thereby shielding the sensitive circuits, is described in this section. Since each circuit within a network can be characterized as both an aggressor and a victim, each node of interest is associated with a matrix composed of two parameters $[i, s]$. Parameter i is an aggressor related parameter, and is equal to the current sunk from the network. Parameter s is related to the victim, expressing the sensitivity of the circuit connected to the node. The objective is to enhance overall performance, such as minimize the worst case delay.

$$delay_{worst} = max(delay_1, delay_2, ..., delay_k), \qquad (2)$$

where

$$delay_j = delay_{min-j} \cdot \left[\frac{s_j}{V_{dd}} \cdot \Delta V_j + 1 \right]. \qquad (3)$$

ΔV_j is a change in the voltage at node j due to load currents and the impedance of the mesh structured power distribution network. $delay_{min-j}$ is the minimum propagation delay of circuit j achieved by applying the maximum supply voltage V_{dd}. s_j is the sensitivity factor of circuit j.

Pseudocode of the LINK-BREAKING algorithm for the proposed methodology is provided in Fig. 5, with the objective of minimizing the worst case propagation delay.

In line 1, the voltage drop at k nodes (all aggressor/sensitive nodes) is determined. Based on the voltage and sensitivity of the circuits, the initial value of the delay function $delay_{ini}$ is determined, as listed in line 2. The revised number of power networks x is generated, where each network is perturbed by removing a random link. In lines 4 and 5, the voltage drop and delay function are determined for each of the perturbed networks. A search for a local minimum is evaluated for each perturbed system in lines 6 to 9. The network with the lowest $delay$ value is selected in line 10. The process is repeated until the value of the delay function cannot be further reduced.

Since k nodes of interest are typically lower than the overall number of nodes in a system, a random walk procedure can be used to efficiently determine the voltages [11], trading off accuracy with runtime. The number of parallel random walk procedures is based on the target accuracy.

5. CASE STUDIES

Two study cases are presented in this section. In both cases, the circuit is composed of nine blocks. For the first case, one block sinks significantly greater current, representing the case of a single dominant aggressor. In the second case, the current and delay of the nine blocks are varied, representing general circuits. The design objective is to minimize the worst case propagation delay.

A mesh structured power distribution network with 20×20 number of nodes is considered. A block diagram of the circuit is schematically illustrated in Fig. 6. Four one volt power supplies are connected at the center of the four edges (left, right, top, and bottom). The maximum permitted degradation in supply voltage is 0.3 V.

The current is different among the circuit blocks (see Table 1) for both study cases. The supply voltage map before and after application of the link breaking methodology, as well as the resulting power network, is illustrated in Fig. 7. The current sunk for each

Table 1: Sensitivity factor, sunk current, minimum delay, supply voltage, and propagation delay before and after application of the link breaking methodology for the nine circuit blocks. The improvement or degradation in the supply voltage, propagation delay, and maximum operating frequency are also listed. Case 1 represents the case where a single block sinks significantly higher current as compared to the other blocks. In Case 2, the sunk current, sensitivity factor, and delay are different for different blocks, representing a general design case.

Block number	1	2	3	4	5	6	7	8	9	$f_{max} = \frac{1}{delay_{worst}}$
Sensitivity factor (s)	5	1	1	2	2	1.3	3	1.2	4	——
Delay [ps] @ $V_{dd} = 1V$	670	300	650	710	200	690	300	300	300	——
Case 1 (see Figs. 7(a) and 7(b))										
Sunk current	1	1	1	1	1	1	1	10	1	——
Voltage before methodology [mV]	945	944	944	937	922	934	925	850	924	——
Voltage after methodology [mV]	966	934	924	947	936	906	939	700	909	——
Voltage improvement [%]	2.1	-1.1	-2.0	-1.0	1.6	-3.0	1.5	-17.6	-1.6	
Delay before methodology [ps]	904	336	727	847	245	794	390	375	413	1.11 GHz
Delay after methodology [ps]	855	349	762	857	243	844	387	445	446	1.17 GHz
Delay improvement [%]	5.4	-3.9	-4.8	-1.2	0.8	-6.3	0.8	-18.7	-8.0	5.4
Case 2 (see Figs. 7(d) and 7(e))										
Sunk current	1	5	5	2	2	3	1.3	4	1.2	——
Voltage before methodology [mV]	907	861	850	901	874	875	907	876	901	——
Voltage after methodology [mV]	958	825	781	928	838	864	852	716	890	——
Voltage improvement [%]	5.5	-4.3	-8.1	2.9	-4.1	-1.2	-6.1	-18.2	-1.3	——
Delay before methodology [ps]	1050	366	800	910	268	860	411	369	448	0.95 GHz
Delay after methodology [ps]	870	378	847	870	283	869	463	430	463	1.15 GHz
Delay improvement [%]	17.1	-3.2	-5.8	4.5	-6.0	-1.1	-12.7	-16.5	-3.3	17.1

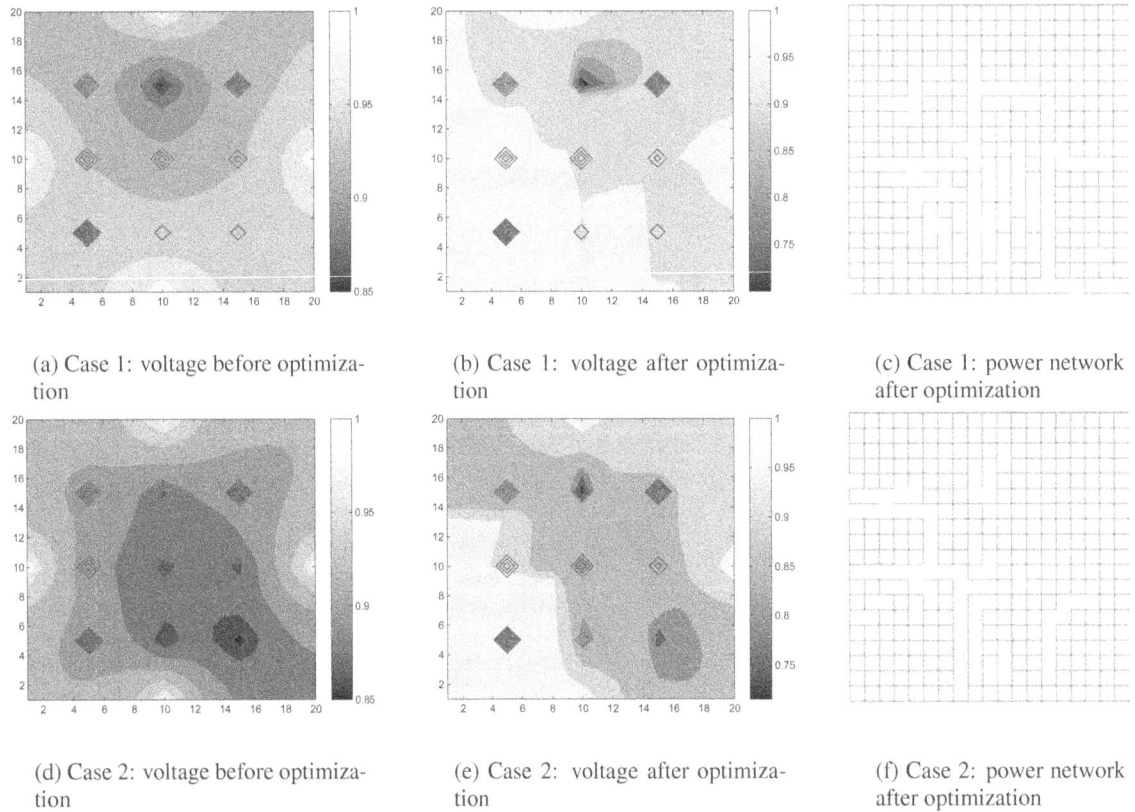

(a) Case 1: voltage before optimization

(b) Case 1: voltage after optimization

(c) Case 1: power network after optimization

(d) Case 2: voltage before optimization

(e) Case 2: voltage after optimization

(f) Case 2: power network after optimization

Figure 7: Map of voltage variations before and after application of the link breaking methodology for two case studies. The diamond shapes represent the location of the aggressor/victim circuit blocks. The size of the diamond represents the relative sensitivity factor of a particular block. The resulting power network after the link breaking methodology is also illustrated. The first case represents the case where a single block sinks significantly higher current as compared to the other blocks. In the second case, the sunk current, sensitivity factor, and delay are different for different blocks, representing a general design case.

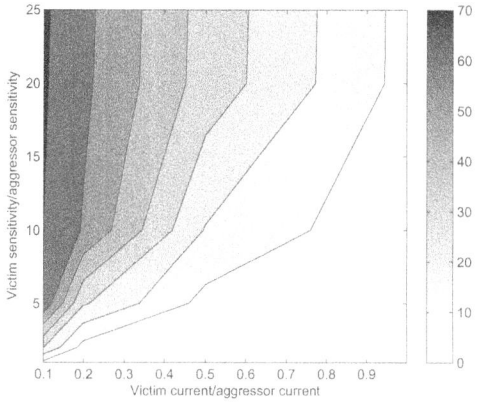

(a)

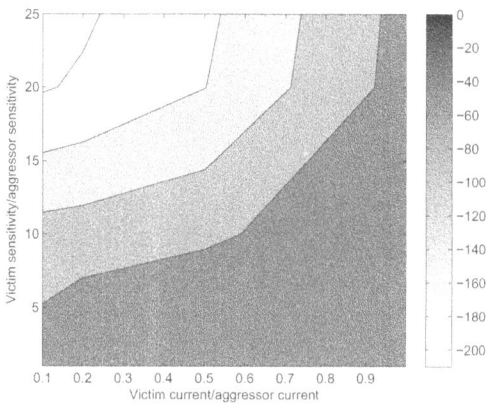

(b)

Figure 8: Change in voltage drop at the (a) victim and (b) aggressor circuit. The darker shade represents a greater reduction in the voltage drop at the victim and a lower increase in the voltage drop at the aggressor.

of the cases, voltage before and after application of the methodology, sensitivity, propagation delay, and improvement in the supply voltage and propagation delay are listed in Table 1.

Case 1 (Figs. 7(a) and 7(b)) illustrates the case where the current sunk by block 8 (the aggressor) is significantly higher as compared to the other circuit blocks. The highest degradation in the supply voltage is within the aggressor circuit; however, the voltage drop is reduced in those circuit blocks with a higher sensitivity and minimum delay, resulting in a reduction in the worst case delay and a higher maximum operating frequency. The increase in the supply voltage at block 1 is 2%, achieving 97% of the ideal power supply voltage and resulting in an improvement in the propagation delay of 5%. Note that the improvement in the propagation delay is greater than the supply voltage due to the high sensitivity factor. In Case 2 (Figs. 7(d) and 7(e)), the voltage at block 1 is increased by 5%, achieving 96% of the ideal power supply voltage and resulting in an improvement in the propagation delay of 17%.

6. DISCUSSION

The voltage drop within a power distribution network is investigated for circuit blocks with different current levels and sensitivities. The minimum propagation delay ($delay_{min}$) is maintained

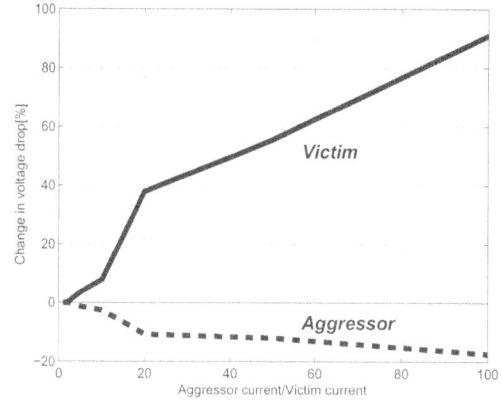

Figure 9: Change in voltage within the power distribution network for the victim and aggressor circuits as a function of the ratio of the current sunk by the aggressor and victim circuits. The sensitivity factor is assumed equal for both circuits.

the same. A 20×20 mesh structured power distribution network with two power supplies and two current sources (one aggressor and one victim) is considered. The voltage improvement at the victim and degradation at the aggressor are illustrated, respectively, in Figs. 8(a) and 8(b). Note that by assigning a higher sensitivity to the victim circuit, the voltage drop on the power network at the victim is reduced. Simultaneously, the voltage drop at the aggressor is increased, while the aggressor is less sensitive to voltage variations. The tradeoff between reducing the voltage drop at the victim while increasing the voltage drop at the aggressor is an important aspect of the proposed link breaking methodology.

The improvement and degradation of the voltage drop at, respectively, the victim and aggressor are depicted in Fig. 9 for different ratios of the current sunk by the victim and aggressor, assuming the two circuits have equal sensitivity. Note that a higher change in voltage is achieved at the victim when the current sunk by the aggressor is greater. This effect is due to the dominance of the aggressor on the victim circuit before applying the link breaking methodology.

The computational runtime of the algorithm, depicted in Fig. 5, is evaluated for differently sized power distribution networks. The algorithm has been executed on a Linux eight-core with 8 GB RAM system. The runtime as a function of the number of nodes in the power network is depicted in Fig. 10. The runtime of the link breaking methodology can also be accelerated by utilizing multigrid-like techniques [12] and ignoring those current sources located far from the target nodes. The number of aggressor and/or victim circuits is not a dominant factor affecting the runtime of the algorithm, as illustrated in Fig. 11. Initially, the runtime increases exponentially with the number of aggressor and victim circuits. With a further increase in the number of circuits, the computational runtime decreases due to the smaller number of links that can be disconnected. For those cases where only a small number of circuit are evaluated within a large power distribution network, the random walk method [11] can be used to estimate the voltage variations, significantly accelerating the link breaking methodology.

The worst case voltage drop (located at the aggressor) cannot be reduced by utilizing the link breaking methodology, since the methodology always increases the worst case power network impedance. However, the effect of the aggressor on other circuits with a higher sensitivity and propagation delay can be reduced, resulting in enhanced overall system performance.

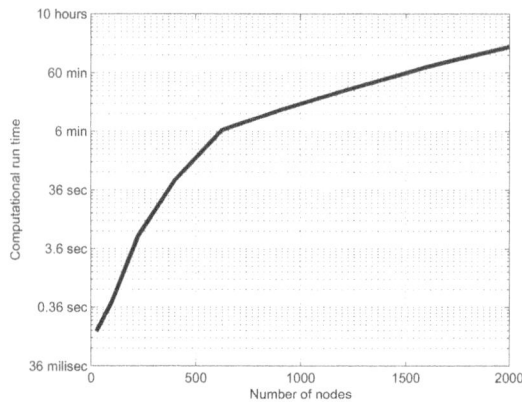

Figure 10: Computational runtime of the link breaking methodology as a function of the number of nodes within the power distribution network.

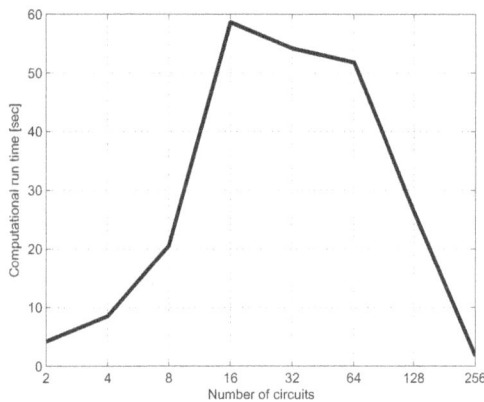

Figure 11: Computational runtime of the link breaking methodology as a function of the number of victim and aggressor circuits. The runtime initially increases with a higher number of circuits. After reaching a peak, the runtime decreases due to the smaller number of links that can be removed.

7. CONCLUSIONS

The design of the power distribution network is an essential part of an IC design flow. The network is typically designed as a single network or multiple separate networks. The advantages of a single network are a reduced network impedance and fewer routability constraints, while multiple separate networks have the advantage of lower noise coupling. The proposed link breaking methodology utilizes a single network, disconnecting links between the aggressive and sensitive circuits; thereby, isolating the victim from the aggressor. This approach reduces the noise, while maintaining a low network impedance.

Sensitivity to changes in the supply voltage will vary for different circuits. Voltage variations at the more sensitive circuits need to be reduced at the expense of increased voltage variations at the less sensitive circuits. A smaller voltage drop is also important in long critical paths as compared to shorter less critical logic paths. The aggressiveness and sensitivity of circuits are considered during the link breaking process. The methodology is evaluated for two cases. The objective for these case studies is reduced worst case propagation delay by increasing the supply voltage at blocks with high propagation delay. An average enhancement of 4% in power supply

voltage at nodes with high sensitivity and high propagation delay is achieved, resulting in, on average, 96% of the ideal power supply voltage at these nodes. As a result, an average improvement of 11% in the maximum operating frequency is achieved when utilizing the proposed link breaking methodology.

8. REFERENCES

[1] D. Blaauw, R. Panda, and R. Chaudhry, "Design and Analysis of Power Distribution Networks," *Design of High-Performance Microprocessor Circuits*, Chandrakasan, Bowhill, and Fox, Eds., Chapter 24, pp. 499–522, IEEE Press, 2001.

[2] R. Jakushokas, M. Popovich, A. V. Mezhiba, S. Kose, and E. G. Friedman, *Power Distribution Networks with On-Chip Decoupling Capacitors, Second Edition*, Springer, 2011.

[3] C. Tirumurti, S. Kundu, S. Sur-Kolay, and Y.-S. Chang, "A Modeling Approach for Addressing Power Supply Switching Noise Related Failures of Integrated Circuits," *Proceedings of the IEEE Design, Automation and Test in Europe Conference and Exhibition*, Vol. 2, pp. 1078 – 1083, February 2004.

[4] L. H. Chen, M. Marek-Sadowska, and F. Brewer, "Buffer Delay Change in the Presence of Power and Ground Noise," *IEEE Transactions on Very Large Scale Integration (VLSI) Systems*, Vol. 11, No. 3, pp. 461–473, June 2003.

[5] M. Saint-Laurent and M. Swaminathan, "Impact of Power-Supply Noise on Timing in High-Frequency Microprocessors," *IEEE Transactions on Advanced Packaging*, Vol. 27, No. 1, pp. 135–144, February 2004.

[6] V. Kursun and E. G. Friedman, *Multi-Voltage CMOS Circuit Design*, John Wiley & Sons, Inc., 2006.

[7] X. Lai and J. Roychowdhury, "Fast, Accurate Prediction of PLL Jitter Induced by Power Grid Noise," *Proceedings of the IEEE Custom Integrated Circuits Conference*, pp. 121 – 124, October 2004.

[8] P. Larsson, "Measurements and Analysis of PLL Jitter Caused by Digital Switching Noise," *IEEE Journal of Solid-State Circuits*, Vol. 36, No. 7, pp. 1113 –1119, July 2001.

[9] L. A. Arledge Jr. and W. T. Lynch, "Scaling and Performance Implications for Power Supply and Other Signal Routing Constraints Imposed by I/O Pad Limitations," *Proceedings of the IEEE Symposium on IC/Package Design Integration*, pp. 45 –50, February 1998.

[10] M. Alioto and G. Palumbo, "Impact of Supply Voltage Variations on Full Adder Delay: Analysis and Comparison," *IEEE Transactions on Very Large Scale Integration (VLSI) Systems*, Vol. 14, No. 12, pp. 1322 –1335, December 2006.

[11] P. G. Doyle and J. L. Snell, *Random Walks and Electric Networks*, Mathematical Association of America, 1984.

[12] J. N. Kozhaya, S. R. Nassif, and F. N. Najm, "Multigrid-Like Technique for Power Grid Analysis," *Proceedings of the IEEE/ACM International Conference on Computer-Aided Design*, pp. 480–487, November 2001.

Unifying Functional and Parametric Timing Verification

Luis Guerra e Silva
INESC-ID / IST / TU Lisbon
Lisbon, Portugal
lgs@inesc-id.pt

ABSTRACT

This paper proposes a unified modeling framework for timing verification of IC designs that, through an elegant SMT-based formulation, seamlessly integrates functional timing analysis and parametric delay modeling. Such framework enables accurate timing verification by simultaneously ignoring false paths and accounting for process variability. By casting the timing verification problem as a general SMT instance it is possible to benefit from the continuous advances in performance and robustness of modern SMT engines. The proposed framework is validated for a representative set of benchmarks, using Microsoft's Z3 SMT solver.

Categories and Subject Descriptors

B.7.2 [**Hardware**]: Integrated Circuits—*Design Aids*

General Terms

Algorithms, Design, Theory, Verification

Keywords

Timing Verification, False Paths

1. INTRODUCTION

Timing verification is concerned with predicting, prior to fabrication, the timing-critical paths of an IC design, and assessing whether it will be able to operate at its target clock frequency. Such task is becoming increasingly challenging, given the sheer size and complexity of modern IC designs and the uncertainty introduced by nanometric fabrication technologies, extremely sensitive to process variations.

The need to adequately model process variability has motivated the introduction of parametric delay models, where cell and interconnect delays are no longer given by fixed real numbers, but instead by affine functions of the parameters of the fabrication process. Several compatible parametric static timing analysis (PSTA) techniques that make use of such models have likewise been proposed [20, 18, 10].

While modern timing verification tools incorporate sophisticated, variability-aware, parametric delay modeling techniques, they only consider the topology of the circuit, still lacking the ability to account for its logic behaviour. Unfortunately, for reasons hard to assess, high-level synthesis systems are prone to generate circuits with many *false paths* [3], i.e. paths that cannot be exercised by any input pattern [13]. Topology-based timing verification tools, the standard in industrial design flows, can often produce conservative timing estimates by considering such paths as critical, even though they cannot be exercised in real circuit operation. This can lead to wasteful overdesign, which constitutes a serious problem in the competitive market of microelectronics, where continuously increasing performance and complexity must be packed into the same die area.

The inability of industrial timing verification tools to identify false paths was slightly mitigated by allowing the designer to tag known false paths, to be ignored during timing verification. However, it is impossible to manually identify and tag all the false paths of any useful design block which, most often, may contain thousands of cells. Therefore, the integration of automated and systematic false path identification capabilities into modern timing verification tools is badly needed for enabling performance optimization to be guided in a computationally efficient manner.

Due to the existence of false paths, the problem of computing the delay of a circuit can no longer be solved in linear time, being instead an NP-complete [11] problem. During the 90s, the research work on false paths was extensive and, among others, several promising modeling and algorithmic approaches have been proposed [1, 4, 7, 16, 15, 22]. The added complexity that entails solving the false path problem and the existence of comfortable design margins, that could accomodate conservative timing estimates, were the main reasons why, despite all the research work conducted on the topic, automated false path detection techniques never found their way into industrial timing verification tools.

Recent years have seen a renewed interest on false path identification techniques, with most contributions [19, 23, 5] targeting practical aspects of their integration into deterministic timing verification. Variability-aware timing verification has also been addressed [14, 12, 21]. However, the difficulty of combining, in a single formulation, logic constraints and parametric delay models, led to *ad hoc* solutions. False paths were ignored either by performing Monte Carlo electrical simulation, for several parameter settings, or through a pre-processing step, that would perform functional analysis assuming worst-case parameter settings.

Indeed, the integration of functional analysis and parametric delay modeling is a complex task, since it requires a computational framework capable of simultaneously manipulating Boolean and linear numeric constraints which, until recently, was not readily available. However, the advent of Satisfiability Modulo Theories (SMT) [2], which generalizes Boolean Satisfiability (SAT) by adding several first-order theories such as equality, arithmetic, quantifiers, etc, has opened the possibility of formulating problems that combine Boolean constraints with other types of constraints, namely linear constraints on real values.

Leveraging on the contemporary SMT technology, we propose a unified modeling framework for timing verification of IC designs that, through an elegant SMT-based formulation, seamlessly integrates functional timing analysis and parametric delay modeling. Such framework achieves accurate timing verification by simultaneously ignoring false paths and accounting for process variability.

To the best of our knowledge, this is the first work to cast the timing verification problem as an SMT instance. This approach has several advantages. Firstly, the expressive power of SMT formulas enables the representation of complex relations between timing quantities and/or process parameters, which can be used to model a wealth of timing verification problems. Secondly, the increased support of SMT solvers for non-linear arithmetic enables easy extension to support future non-linear delay models. Lastly, by casting the timing verification problem as an SMT instance, it is possible to benefit from the continuous advances in performance and robustness of modern SMT engines.

The remainder of this paper is organized as follows. Section 2 introduces the parametric timing modeling, underlying the work presented in this paper, as well as the essential aspects of SMT, necessary to understand the formulations discussed in upcoming sections. Section 3 presents an overview of the proposed modeling framework. Sections 4 and 5 detail functional and timing modeling for cells and interconnect, respectively. The experimental results are presented and discussed in Section 6. Finally, Section 7 presents brief concluding remarks and foresees future work.

2. BACKGROUND

2.1 Parametric Timing Modeling

The timing information of a circuit is modeled by a *timing graph* $G = (V, E)$, where *vertices*, $v \in V$, correspond to pins in the circuit, and directed *edges*, $e \in E$, correspond to pin-to-pin delays in cells or interconnect. The *primary inputs*, $u \in PI(G)$, are vertices with no incoming edges. All vertices with no outgoing edges are *primary outputs*, $w \in PO(G)$, but there may also be primary outputs with outgoing edges. A *complete path* is a sequence of edges, connecting a primary input to a primary output. A *path* is a sequence of edges connecting any two vertices

Edges are annotated with the corresponding *delays*. At most four delays can be annotated on each edge, depending on the input and output rise/fall transitions: d^{FF}, d^{FR}, d^{RF} and d^{RR}. Since it is out of the scope of this paper to discuss the delay computation procedure, in the following, we will assume that the timing information of any circuit is already made available in the form of an annotated timing graph. Output pins of cells are also annotated with the corresponding logic function.

This work assumes a PSTA model [20, 18, 10], where delays are described by affine functions of process and operational parameter variations, corresponding to a first-order linearization of every delay, d, around a nominal point, λ_0, in the parameter space. Considering the parameter space to have size p, and representing d as a function of the incremental parameter variation vector, $\Delta\lambda = \lambda - \lambda_0$, around a nominal value λ_0, we obtain

$$d(\Delta\lambda) = d_0 + \sum_{i=1}^{p} d_i \Delta\lambda_i \qquad (1)$$

where $d_0 = d(\lambda_0)$ is the nominal value of d and d_i is the sensitivity of d to parameter λ_i, $i = 1, 2, \ldots, p$, computed at the nominal point λ_0. Parameter variations are assumed to lie within a given range, $\Delta\lambda_i \in \left[\Delta\lambda_i^{min}, \Delta\lambda_i^{max}\right]$,

2.2 Modes of Operation

The primary task in any timing verification run is arrival time computation. The *arrival time*, represented by at, is a conservative estimate of the earliest or the latest time instant that a signal transition can reach a given circuit point, when traveling from an input (or the output of a sequential element). The meaning of the arrival time values depends on whether we assume the early or the late mode of operation. In *early mode*, we are concerned with computing the earliest time instant that a signal transition can reach a given circuit point. Conversely, in *late mode* we are concerned with computing the latest time instant that a signal transition can reach a given circuit point. Both early and late mode analyses are of practical interest. Setup constraints are verified through late mode analysis, while hold constraints are verified through early mode analysis. In the following, and without loss of generality, we assume the late mode.

Even though functional analysis in static timing verification does not require any specific input excitation, some assumptions on the variation of logic values of circuit nodes must be made. Two possibilities have been considered and extensively studied in the literature: transition mode and floating mode. In the *transition mode* of operation [8], circuit nodes are assumed to switch from a known initial logic value to a known final logic value (e.g. $1 \rightarrow 0$). In the *floating mode* of operation [4], circuit nodes are assumed to switch from an unkown initial logic value to a known final logic value (e.g. $? \rightarrow 0$). In this work we assume the floating mode of operation. Even though the transition mode provides more accurate timing estimates, it has a significantly higher computational cost.

2.3 Satisfiability Modulo Theories

This subsection is not meant to be a comprehensive introduction to SMT, as it would probably exceed the length of the entire paper (see [2] for that). It is solely intended to informally introduce a few basic concepts necessary for the reader to understand the use of SMT throughout the paper.

Satisfiability is one of the quintessential problems in computer science, which consists of determining whether a given formula, enconding one or more constraints, has a solution. SAT is the best known of the constraint satisfaction problems, where the formula is built using logical connectives, over the Boolean variables.

An SMT instance is a formula in first-order logic [17] where function and predicate symbols can be interpreted resorting to a variety of underlying theories. SMT can be seen as a generalization of SAT, but where some of the Boolean variables are replaced by predicates. A *predicate* is a Boolean-valued function, $P : X \rightarrow \{0, 1\}$, designated by predicate on X. A predicate can be seen as a condition that evaluates to either 1 (*true*) or 0 (*false*), depending on the values of its variables. For instance, $a > 3.14$ is a predicate that evaluates to 1 if the real-valued variable a assumes a value larger than 3.14 and evaluates to 0 otherwise. Predicates are classified according to the theory they belong to. For example, linear inequalities over real variables are evaluated using the rules of the theory of linear real arithmetic.

The SMT problem is concerned with determining a set of variable assignments that make the corresponding formula *true*, or prove that no such assignments exist and the formula is always *false*. An example SMT formula, on a real-valued variable a and two Boolean-valued variables b and c is,

$$(a > 3.14) \wedge (\neg b \vee c)$$

A *model* for this problem, i.e. a set of satisfying variable assignments, is $a = 3.15$, $b = 0$ and $c = 1$.

SMT formulas are built by combining logical connectives, such as $\neg$ (negation), $\wedge$ (conjunction), $\vee$ (disjunction), $\oplus$ (exclusive disjunction), $\Rightarrow$ (implication) and $\Leftrightarrow$ (equivalence), with Boolean variables and predicates. The structure of the predicates depends on their underlying theory. For example, predicates under the theory of linear real arithmetic, can be built by combining real-valued variables and constants with comparison operators ($<, =, >, \leq, \geq, \neq$) and arithmetic operators ($+, -, *, /$), among others.

3. MODELING FRAMEWORK

First and foremost, we should clearly state the general problem to be addressed by our modeling framework. *Given the timing graph for a combinational circuit, as described in Section 2, we want to determine the largest arrival time at its primary output vertices.* This problem, which is often designated by *circuit delay computation*, constitutes the cornerstone of any timing verification procedure.

The problem variables are, in first instance, the values that need to be computed, i.e. the arrival times at the primary output vertices. However, such arrival times are dependent on the arrival times at the intermediate vertices, as well as on the delays between them. Delays, on the other hand, are dependent on the Boolean values assumed by the vertices and on the values of the process parameter variations. Through this simple dependency analysis we are able to conclude that the variables of our problem should be:

- a Boolean-valued variable b_v, for every vertex v, corresponding to the final logic (Boolean) value assumed by vertex v after the transition;

- a real-valued variable at_v, for every vertex v, corresponding to the arrival time at vertex v;

- a real-valued variable $\Delta\lambda_i$, for every parameter of order i, corresponding to the parameter variation.

Formulating the timing verification problem as an SMT instance, requires capturing both functional and timing constraints, for all circuit elements, into an SMT formula, φ. While the expressive power of SMT admits a wealth of different formulations, in the following we shall assume that

such SMT formula is a conjunction of other partial formulas, capturing functional (φ^b) and timing (φ^t) constraints:

$$\varphi = \left(\bigwedge_i \varphi_i^b \right) \wedge \left(\bigwedge_j \varphi_j^t \right) \tag{2}$$

Therefore, for the whole formula to be satisfied, all the partial formulas must also be satisfied. φ can be progressively built by traversing the timing graph in a levelized breadth-first fashion, and augmenting it with the partial formulas that capture the relations between boolean values and arrival times of input and output vertices of cells and interconnect, as detailed in Sections 4 and 5.

The SMT solver is not an optimization engine, since it is only able to find *a* solution that satisfies the formula, not the *best* solution according to some given criteria. However, we need to determine the *largest* arrival time at the primary output vertices, not just some valid arrival time value. This is the typical case where an optimization problem must be cast into a sequence of decision problems.

The topological arrival time is cheap to compute (linear time), yet it assumes that all paths can be exercised. Since some paths may be false, the topological arrival time is actually an upper bound to the true arrival time (assuming late mode). Therefore, we can start by checking whether the topological arrival time is the true arrival time or not. If it is not, then we can check that for a slightly smaller value, and continue iterating until we reach the true arrival time. On each iteration, we must check whether the arrival time at some primary output can be not smaller than the given *required arrival time*, *rat*. This "question" can be asserted into the SMT formula by adding,

$$\varphi_{PO}^t = \bigvee_{w \in PO(G)} (at_w \geq rat) \tag{3}$$

Once all the functional and timing constraints are built into the SMT formula, as detailed in Sections 4 and 5, other "questions" may be asked, by adding proper assertions and subsequently running the SMT solver. For example, this same formulation can be used to check setup times, or arrival times for specific parameter variation settings (corners). Moreover, it can also be used for generating input test patterns capable of producing specific arrival time values at particular points in the circuit (ATPG). For any problem instance, the answer will always be a set of assignments for the logic values, arrival times and process parameter variations, that satisfy the problem constraints, or the proof that no such assignments exist.

4. CELL MODELING

This section details how digital cells are modeled in our SMT-based timing verification framework, both at functional and timing level.

4.1 Functional Constraints

The logic function of a given output pin, for any given combinational cell in the library, is described in a particular field of the Liberty (`.lib`) file, in terms of one- or two-operand elementary logic operations: `&` (AND), `|` (OR), `!` (NOT) and `^` (XOR). Exemplifying, for the single output pin of a 3-input `AOI21` cell we would have the following formula: "`!(A | (B1 & B2))`". This representation enables arbitrarily complex functions to be described in terms of elementary

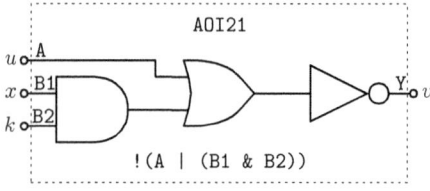

Figure 1: Parsing tree for logic function of AOI21 cell.

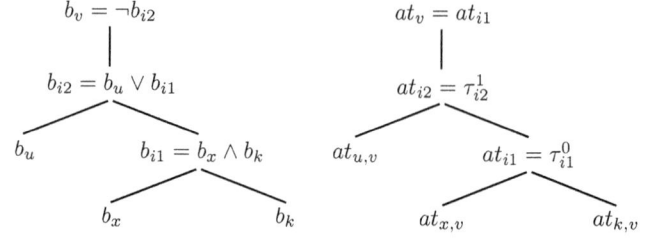

Figure 2: Functional and timing constraints.

logic operations. A trivial parser for this type of logic expressions was developed. Such parser maps each expression into a parsing tree, as illustrated in Figure 1, where internal nodes are logic operations and leaf nodes represent input pins of the cell. As we will see, the parsing tree is a convenient representation that can be easily traversed for generating all the required functional and timing constraints.

Modeling the functional behaviour of a combinational cell amounts to adding to the SMT formula, φ, all the constraints that assign to the logic value variables of every output pin their corresponding value, in terms of the logic value variables of the input pins. Since the parsing tree of each output pin already contains its logic function factored into elementary logic operations, we can trivially generate all the necessary constraints by traversing the tree in a bottom-up fashion and generating the constraints for each operation, as summarized below:

$$\varphi^b_{AND} = \{b_v = b_u \wedge b_x\} \tag{4}$$

$$\varphi^b_{OR} = \{b_v = b_u \vee b_x\} \tag{5}$$

$$\varphi^b_{NOT} = \{b_v = \neg b_u\} \tag{6}$$

$$\varphi^b_{XOR} = \{b_v = b_u \oplus b_x\} \tag{7}$$

where v is assumed to be the output pin and u and x the input pins. This procedure is illustrated, for the AOI21 cell, in the left tree of Figure 2. Since only the root and the leaves of the parsing tree actually correspond to vertices in the circuit, with associated logic value variables, we must add variables to represent the logic values in the intermediate nodes of the parsing tree, that we designate by $b_{i1,2,...}$.

4.2 Timing Constraints

The timing constraints for any given cell must enable the computation of the arrival times at its output pins from the logic values and arrival times at its input pins, and from the input/output pin delays.

We start by adding an artificial arrival time variable, for each cell input/output combination, representing what would be the arrival time at the output if its transition was triggered by the corresponding input. The value of this variable will be computed by adding the arrival time value at the input pin to the proper delay value, chosen according to the logic value variables of the input and output pins. Assuming the input and output pin vertices to be u and v, respectively,

$$at_{u,v} = \begin{cases} at_u + d^{FF}_{u,v} & \text{if } \neg b_u \wedge \neg b_v \\ at_u + d^{FR}_{u,v} & \text{if } \neg b_u \wedge b_v \\ at_u + d^{RF}_{u,v} & \text{if } b_u \wedge \neg b_v \\ at_u + d^{RR}_{u,v} & \text{if } b_u \wedge b_v \end{cases} \tag{8}$$

Eqn. (8) can be added to the SMT formula, φ, by nested application of the $ite(t_1, t_2, t_3)$ (if-then-else) operator, sup-

ported by most SMT engines, whose result is t_2, when t_1 is true, and t_3 otherwise. Therefore, we obtain,

$$\varphi^t_{ATD} = \{ at_{u,v} = at_u + ite(\neg b_u \wedge \neg b_v, d^{FF}_{u,v}, \\ ite(\neg b_u \wedge b_v, d^{FR}_{u,v}, \\ ite(b_u \wedge \neg b_v, d^{RF}_{u,v}, \\ d^{RR}_{u,v}))) \} \tag{9}$$

Computing the true arrival time at any given cell output pin involves considering its logic function as well as the logic values, arrival times and delays at the relevant input pins. For the two-input AND and OR cells, the arrival time at the output pin is computed considering whether each input pin assumes a controlling value or not. For the AND cell the controlling value is $c = 0$ and for the OR cell the controlling value is $c = 1$. The arrival time for the output pin vertex v of an AND/OR cell, assuming the controlling value to be c and the input pin vertices to be u and x, is given by

$$\tau^c_v = \begin{cases} \min(at_{u,v}, at_{x,v}) & \text{if } b_u = c \wedge b_x = c \\ at_{u,v} & \text{if } b_u = c \wedge b_x = \neg c \\ at_{x,v} & \text{if } b_u = \neg c \wedge b_x = c \\ \max(at_{u,v}, at_{x,v}) & \text{if } b_u = \neg c \wedge b_x = \neg c \end{cases} \tag{10}$$

Eqn. (10) can be added to the SMT formula using nested ite operations, as in the case of Eqn. (8). However, a more efficient encoding would be to use implications, resulting in

$$\varphi^t_{AND/OR} = \{b_u = c \Rightarrow at_v \leq at_{u,v}\} \wedge \\ \{b_x = c \Rightarrow at_v \leq at_{x,v}\} \wedge \\ \{b_u = \neg c \wedge b_x = \neg c \Rightarrow at_v \leq at_{u,v} \vee at_v \leq at_{x,v}\} \tag{11}$$

For the XOR cell, no single input pin can independently determine the logic value at the output pin, therefore,

$$\varphi^t_{XOR} = \{at_v \leq at_{u,v} \vee at_v \leq at_{x,v}\} \tag{12}$$

For the NOT cell, assuming input pin vertex u and output pin vertex v, we obtain,

$$\varphi^t_{NOT} = \{at_v = at_{u,v}\} \tag{13}$$

The timing constraints for complex cells can now be easily generated. Since cell delays have already been incorporated into the artificial arrival time variables given by Eqn. (8), we can plug such variables into the corresponding input pin leaves of the parsing tree and, by traversing it in a bottom-up fashion and applying Eqns. (11), (12) or (13), according to the logic function of each internal node, all the required timing constraints are generated. New arrival time variables $at_{i1,2,...}$ must be added for the internal nodes of the parsing tree. This procedure is illustrated, for the AOI21 cell, in the right tree of Figure 2.

5. WIRE MODELING

This section details how the interconnect (wires) between digital cells are modeled in our SMT-based timing verification framework, both at functional and timing level.

5.1 Functional Constraints

In a digital circuit, the role of signal wires is to carry digital signals with minimal voltage degradation, such that the voltage on each endpoint corresponds to the same logic value. The logic function of wires can thus be thought as of identity. Therefore, for every wire edge $\langle u, v \rangle$, a constraint enforcing the equality between the logic values at vertices u and v must be added to the SMT formula, φ:

$$\varphi_{WIRE}^{b} = \{ b_v = b_u \} \qquad (14)$$

Most often this equality does not need to be explicitly stated in the form of a constraint. It is sufficient to define a single logic variable b_{uv} that will be used in place of both b_u and b_v. This approach reduces the number of variables, thus contributing to improve the performance of the SMT engine.

5.2 Timing Constraints

Timing constraints for wires are quite easy to generate. Since both ends of a wire must assume the same logic value, to compute the arrival time at the output pin of the wire we only have to select the rise/rise or fall/fall delay corresponding to that logic value and add it to the arrival time of the input pin. Assuming the input pin vertex to be u and the output pin vertex to be v, we obtain,

$$\varphi_{WIRE}^{t} = \{ at_v = at_u + ite(b_v, d_{u,v}^{RR}, d_{u,v}^{FF}) \} \qquad (15)$$

6. EXPERIMENTAL RESULTS

The modeling framework described in the previous sections was coded in C++. We have used Microsoft's Z3 Theorem Prover v3.2 [6] as our SMT solver, given its performance, robustness, and the availability of a C++ API. Z3 is one of the most reputed SMT solvers, and is the workhorse behind several of Microsoft's software verification tools. For benchmark circuits we have used the traditional ISCAS'85 combinational suite. All the results presented in this section were obtianed on a machine with an Intel Core i7 @ 3.07GHz and 12GB of available RAM. For all the runs a single processor was used.

6.1 Validation

The first stage of our experimental procedure was dedicated to validate that the proposed framework was actually ignoring the false paths, and therefore was able to compute the true arrival time values of any given circuit. Therefore, we evaluated our implementation with a few benchmark circuits, for which correct results are published in [16, 9]. Such benchmarks assume the unit delay model. The results are reported in Table 1, where column "TD/RD" presents the topological vs. the real (true) delay computed by the proposed framework, column "SMT" presents the CPU time in seconds taken by such computation and column "SAT" presents the CPU time in seconds taken by the SAT approach described in [9]. For all approaches, the computed delays were the same, which validates our proposed framework. CPU times in our case are larger, particularly because the ones reported in column "SAT" were obtained for a significantly slower machine. Nevertheless, this was to be ex-

Design	TD/RD	SMT	SAT
c432	17/17	0.08	0.03
c499	11/11	0.22	0.02
c880	24/24	0.09	0.04
c1355	24/24	0.75	0.12
c1908	40/37	1.34	0.26
c2670	32/30	0.74	2.83
c3540	47/46	1.30	0.54
c5315	49/47	1.63	1.27
c6288	124/123	853.02	11.19
c7552	43/42	1.23	0.17
cbp.12.2	40/23	0.67	1.53
cbp.16.4	44/27	1.01	1.03
cla.16	34/34	0.06	0.04
tau92ex1	27/24	0.79	0.63
mult-csa	78/78	299.60	5.90

Table 1: Results for unit delays.

pected, since our formulation is significantly more complex, as necessary for handling process variability.

6.2 Evaluation

For evaluating the proposed framework in the context of process variability, we have synthesized and mapped the benchmark circuits to the Nangate OCL $45nm$ technology. As process parameters, we have considered the widths and thicknesses of the 10 metal routing layers, resulting in a total of 20 parameters. Variational delay computation was subsequently performed, and the resulting affine delay formulas were annotated into the corresponding timing graphs. Table 2 presents a brief characterization of the resulting benchmark circuits, where "#PI" and "#PO" columns report the number of primary inputs and outputs, "#C" and "#N" columns report the number of combinational cells and nets, and "#V" and "#E" columns report the number of vertices and edges in the corresponding timing graph.

Table 3 reports the experimental results for true arrival time computation, using the proposed timing verification framework. Column "%OPT" reports the percentual reduction of the true arrival time, computed with the aid of functional analysis, over the topological arrival time. When they are the same, a value of 0 is reported. Columns "Formula", "Solve" and "Total" report CPU times in seconds for formula generation, SMT solve and total, respectively.

Analyzing the results presented in Table 3 we conclude that for some benchmarks it is possible to obtain non-negligible improvements on arrival time estimation, with a fair computational cost, given the complexity of the problem. The cost of formula generation seems to be related to circuit size and the improvement on arrival time estimation. The latter implies, in general, more iterations (see Eqn. 3), which can impact the cost of both formula generation and SMT solve. We believe that the results presented in Table 3 compare favorably to the results presented in Table 1 since, in a variability context, the problem is much harder, and the increase in CPU time is nonetheless limited.

While the performance of the proposed modeling framework does not yet make it adequate for the characterization of large digital blocks, it can be used for characterizing small critical blocks. Nevertheless, we believe that this approach has enormous potential, since our implementation is rather simplistic and can be much improved, which should enable great savings in terms of CPU time. Moreover, the problem can be easily partitioned for parallelization in modern multicore machines.

Design	#PI	#PO	#C	#N	#V	#E
c432	37	7	88	124	356	457
c499	41	32	170	211	595	736
c880	60	26	169	232	697	910
c1355	41	32	170	211	595	736
c1908	33	25	202	235	708	921
c2670	157	63	278	511	1204	1474
c3540	50	22	469	520	1841	2622
c5315	178	123	597	781	2551	3429
c6288	32	32	1005	1470	3556	5006
c7552	208	107	764	986	2820	3593

Table 2: Benchmark characterization.

Design	%OPT	Formula	Solve	Total
c432	4.25%	0.73	7.94	8.67
c499	0.03%	1.15	0.23	1.38
c880	0%	0.84	<0.01	0.84
c1355	0.24%	1.22	0.19	1.41
c1908	5.49%	1.70	8.49	10.19
c2670	0%	1.16	<0.01	1.16
c3540	3.19%	13.21	162.07	175.28
c5315	0%	8.54	<0.01	8.54
c6288	0.88%	302.33	1717.48	2019.81
c7552	0%	9.67	<0.01	9.67

Table 3: Results for variability-aware delays.

7. CONCLUSIONS AND FUTURE WORK

This paper proposes an SMT-based timing verification framework that enables accurate computation of timing estimates by integrating functional and variation-aware timing constraints, that characterize modern digital IC designs and associated fabrication technologies. While experimental evidence shows that the performance of the proposed framework is not yet adequate for application to large digital blocks, we still believe that it constitutes a significant advancement of the state-of-the-art, as it enables better accuracy in timing estimates, and provides a general variability-aware SMT formulation, that can benefit from the continuous advances in SMT engines.

Due to space restrictions, several relevant implementation details were left out of this paper, as well as the application of the proposed framework to sequential circuits. We intend to publish them in a more comprehensive journal paper, together with several strategies for significantly improving the performance of formula generation and SMT solve.

8. ACKNOWLEDGMENTS

I would like to thank João Marques-Silva for introducing me to SMT and for preliminary discussions on efficiency issues. This work was supported by FCT (INESC-ID multiannual funding) through the PIDDAC Program funds.

9. REFERENCES
[1] P. Ashar, S. Malik, and S. Rothweiler. Functional Timing Analysis using ATPG. In *Proceeding of The European Design Automation Conference*, 1993.

[2] C. Barrett, R. Sebastiani, S. A. Seshia, and C. Tinelli. Satisfiability Modulo Theories. In A. Biere, M. J. H. Heule, H. van Maaren, and T. Walshy, editors, *Handbook of Satisfiability*, volume 185 of *Frontiers in Artificial Intelligence and Applications*, chapter 26, pages 825–885. IOS Press, February 2009.

[3] R. Bergamaschi. The Effects of False Paths in High-Level Synthesis. In *Proceedings of ICCAD*, November 1991.

[4] H.-C. Chen and D. H. C. Chu. Path Sensitization in Critical Path Problems. *IEEE Transactions on CAD*, 12(2):196–207, February 1993.

[5] O. Coudert. An Efficient Algorithm to Verify Generalized False Paths. In *Proceedings of DAC*, pages 188–193, June 2010.

[6] L. de Moura and N. Bjorner. Z3: An Efficient SMT Solver. In *Proceedings of TACAS*, pages 337–340, Budapest, Hungary, March-April 2008.

[7] S. Devadas, K. Keutzer, and S. Malik. Computation of Floating-Mode Delay in Combinational Circuits: Practice and Implementation. *IEEE Transactions on CAD*, 12(12):1924–1936, December 1993.

[8] S. Devadas, K. Keutzer, S. Malik, and A. Wang. Certified Timing Verification and the Transition Delay of a Logic Circuit. In *Proceedings of DAC*, pages 549–555, June 1992.

[9] L. G. e Silva, J. Marques-Silva, L. M. Silveira, and K. Sakallah. Realistic Delay Modeling in Satisfiability-Based Timing Analysis. In *Proceedings of ISCAS*, Monterrey, CA, USA, May-June 1998.

[10] L. G. e Silva, J. Phillips, and L. M. Silveira. Effective Corner-Based Techniques for Variation-Aware IC Timing Verification. *IEEE Transactions on CAD*, 29(1):157–162, 2010.

[11] M. R. Garey and D. S. Johnson. *Computers and Intractability: A Guide to the Theory of NP-completeness*. W. H. Freeman and Company, 1979.

[12] R. Garg, N. Jayakumar, and S. Khatri. On the Improvement of Statistical Timing Analysis. In *Proceedings of ICCD*, pages 37–42, 2006.

[13] V. Hrapčenko. Depth and Delay in a Network. *Soviet Math. Dokl.*, 19(4):1006–1009, 1978.

[14] J.-J. Liou, A. Krstic, L.-C. Wang, and K.-T. Cheng. False-Path-Aware Statistical Timing Analysis and Efficient Path Selection for Delay Testing and Timing Validation. In *Proceedings of DAC*, pages 566–569, New Orleans, LA, June 2002.

[15] J. Marques-Silva and K. A. Sakallah. Efficient and Robust Test-Generation Based Timing Analysis. In *Proceedings of ISCAS*, pages 303–306, 1994.

[16] P. McGeer, A. Saldanha, P. Stephan, R. Brayton, and A. Sangiovanni-Vincentelli. Timing Analysis and Delay-Fault Test Generation Using Path Recursive Functions. In *Proceedings of ICCAD*, November 1991.

[17] E. Mendelson. *Introduction to Mathematical Logic*. Chapman & Hall / CRC, 1997.

[18] S. Onaissi, K. Heloue, and F. Najm. A Linear-Time Approach for Static Timing Analysis Covering All Process Corners. *IEEE Transactions on CAD*, 27(7):1291–1304, 2008.

[19] D. Tadesse, D. Sheffield, E. Lenge, R. I. Bahar, and J. Grodstein. Accurate Timing Analysis using SAT and Pattern-Dependent Delay Models. In *Proceedings of DATE*, pages 1–6, 2007.

[20] C. Visweswariah, K. Ravindran, K. Kalafala, S. Walker, S. Narayan, D. Beece, J. Piaget, N. Venkateswaran, and J. Hemmett. First-Order Incremental Block-Based Statistical Timing Analysis. *IEEE Transactions on CAD*, 25(10):2170–2180, 2006.

[21] L. Xie, A. Davoodi, K. Saluja, and A. Sinkar. False Path Aware Timing Yield Estimation under Variability. In *Proceedings of the IEEE VLSI Test Symposium (VTS)*, pages 161–166, 2009.

[22] H. Yalcin and J. P. Hayes. Hierarchical Timing Analysis using Conditional Delays. In *Proceedings of ICCAD*, November 1995.

[23] J. Zeng, M. S. Abadir, J. Bhadra, and J. A. Abraham. Full Chip False Timing Path Identification: Applications to the PowerPC Microprocessors. In *Proceedings of DATE*, pages 1–5, Apr 2010.

New & Improved Models for SAT-Based Bi-Decomposition

Huan Chen
Complex & Adaptive Systems Laboratory
School of Computer Science and Informatics
University College Dublin, Ireland
huan.chen@ucd.ie

Joao Marques-Silva
Complex & Adaptive Systems Laboratory
School of Computer Science and Informatics
University College Dublin, Ireland
jpms@ucd.ie

ABSTRACT

Boolean function bi-decomposition is pervasive in logic synthesis. Bi-decomposition entails the decomposition of a Boolean function into two other simpler functions connected by a simple two-input gate. Existing solutions are based either on Binary Decision Diagrams (BDDs) or Boolean Satisfiability (SAT). Furthermore, the partition of the input set of variables is either assumed, or an automatic derivation is required. Most recent work on bi-decomposition proposed the use of Minimally Unsatisfiable Subformulas (MUSes) or Quantified Boolean Formulas (QBF) for computing, respectively, variable partitions of either approximate or optimum quality. This paper develops new group-oriented MUS-based models for addressing both the performance and the quality of bi-decompositions. The paper shows that approximate MUS search can be guided by the quality of well-known metrics. In addition, the paper improves on recent high-performance approximate models and versatile exact models, to address the practical requirements of bi-decomposition in logic synthesis. Experimental results obtained on representative benchmarks demonstrate significant improvement in performance as well as in the quality of decompositions.

Categories and Subject Descriptors

B6.3 [**Logic Design**]: Design Aids—*automatic synthesis*

General Terms

Algorithms, Design

Keywords

bi-decomposition, logic synthesis, satisfiability

1. INTRODUCTION

Boolean function bi-decomposition is ubiquitous in logic synthesis. It is arguably the most widely used form of func-

tional decomposition. Bi-decomposition [5–7,12,14,15] consists of decomposing Boolean function $f(X)$ into the form of $f(X) = h(f_A(X_A, X_C), f_B(X_B, X_C))$, under variable partition $X = \{X_A | X_B | X_C\}$, where $f_A(X_A, X_C)$ and $f_B(X_B, X_C)$ are functions simpler than $f(X)$. In practice variable partitions are often not available, and so automatic derivation of variable partitions is required.

The quality of bi-decomposition is mainly determined by the quality of variable partitions, as an optimal solution results in simpler sub-functions f_A and f_B. Typically, two *relative* quality metrics [5,6,12,13], namely *disjointness* and *balancedness*, are used to evaluate the resulting variable partitions, for which smaller values represent preferred bi-decompositions. In practice, disjointness is in general preferred [5, 12], since it represents the reduction of common variables to f_A and f_B, which in turn often simplifies the resulting Boolean function. Similar to recent work on functional decomposition [5,6,12,13], this paper addresses these two relative metrics, namely disjointness and balancedness. *Absolute* quality metrics are an alternative to relative quality metrics, and include total variable count (Σ) and maximum partition size (Δ) [7]. Nevertheless, absolute quality metrics scale worse with the number of inputs [7].

Research on decomposition of Boolean functions can be traced back to 1950s [2, 8]. Traditional approaches [4, 11, 15, 16] are based on BDDs for bi-decomposing Boolean networks. However, BDDs impose severe constraints on the number of input variables functions can have. Also, approaches based on BDD-based bi-decomposition lack mechanisms for deriving variable partitions; hence the partition of the input set of variables needs to be assumed. Moreover, it is also generally accepted that BDDs do not scale for large Boolean functions. As a result, recent work [5,6,12,13] proposed SAT-based models for addressing performance and quality in decompositions.

SAT-based bi-decomposition has a number of advantages and also disadvantages. In terms of the quality of computed partitions, SAT-based models can either be *approximate* or *exact*. Approximate solutions [5, 12] proposed Boolean Satisfiability (SAT) and Minimally Unsatisfiable Subformulas (MUS) for manipulating Boolean functions. These approaches achieve significant performance gains. However, the approximate models find solutions with brute-force search [12], heuristic shuffling of CNF IDs [12], heuristic searching of seed variable partitions [5] and interfacing of standalone MUS solvers [5]. These solutions prevent *controllable* quality in practice. In contrast, exact models [6] use QBF for manipulating Boolean functions and constraints. This results in guaranteed quality of bi-decompositions. However, solving the QBF formulas representing the exact models can have significant computational cost.

This paper develops extensions to earlier work on approximate [5, 12] and exact [6] SAT-based bi-decomposition, and proposes a new heuristic model. The paper has three main contributions. First, the paper develops a new approximate Group-oriented MUS-based model that in practice yields good quality metrics. This new model offers significant performance improvement over recently proposed exact solutions [6] as well as to the majority of approximate solutions [5, 12]. Moreover, this new model yields bi-decompositions of quality better than existing approximate models [5, 12]. The second contribution addresses techniques for achieving target quality metrics with exact bi-decomposition models [6]. The third contribution proposes a flow for aggregating all existing bi-decomposition models [5, 6, 12], as well as the new models, with the purpose of achieving a wide range of trade-offs between quality of decomposition metrics and runtime performance.

The paper is organized as follows. Section 2 provides the preliminaries. Section 3 reviews Satisfiability-based models for bi-decomposition. Section 4 proposes the new models and the practical extensions. Section 5 presents the experimental results. Finally, section 6 concludes the paper and outlines future work.

2. PRELIMINARIES

Variables are represented by set $X = \{x_1, x_2, \ldots, x_n\}$. The cardinality of X is denoted as $||X||$. A partition of a set X into $X_i \subseteq X$ for $i = 1, \ldots, k$ (with $X_i \bigcap X_j = \emptyset, i \neq j$ and $\bigcup_i X_i = X$) is denoted by $\{X_1|X_2|\ldots|X_k\}$. A Completely Specified Function (CSF) is denoted by $f : \mathcal{B}^n \to \mathcal{B}$. Similar to the recent work [5, 6, 12], this paper assumes CSFs.

2.1 Boolean Function Bi-Decomposition

DEFINITION 1. *Bi-decomposition [15] for Completely Specified Function (CSF) $f(X)$ consists of decomposing $f(X)$ under variable partition $X = \{X_A|X_B|X_C\}$, into the form of $f(X) = f_A(X_A, X_C) <OP> f_B(X_B, X_C)$, where $<OP>$ is a binary operator, typically OR, AND or XOR.*

This paper addresses OR, AND and XOR bi-decomposition because these three types form other types of bi-decomposition [12]. Bi-decomposition is termed *disjoint* if $||X_C|| = 0$. A partition of X is trivial if $X = X_A \bigcup X_C$ or $X = X_B \bigcup X_C$ holds. Similar to earlier work [5, 6, 12, 13], this paper addresses *non-trivial* bi-decompositions.

2.2 Boolean Satisfiability

A formula in Conjunctive Normal Form (CNF) $\mathcal{F}$ is defined as a set of sets of literals defined on X, representing a conjunction of disjunctions of literals. A literal is either a variable or its complement. Each set of literals is referred to as a clause c. Moreover, it is assumed that each clause is non-tautological. Additional SAT definitions can be found in standard references (e.g. [3]).

DEFINITION 2. (MUS). *[5] $\mathcal{M} \subseteq \mathcal{F}$ is a Minimally Unsatisfiable Subformula (MUS) iff $\mathcal{M}$ is unsatisfiable and $\forall_{c \in \mathcal{M}}, \mathcal{M} \setminus \{c\}$ is satisfiable.*

DEFINITION 3. (GROUP-ORIENTED MUS). *(E.g. [5]) Given an unsatisfiable CNF formula $\mathcal{C} = \mathcal{D} \cup \bigcup_{G \in \mathcal{G}} G$, where*

$\mathcal{G} = \{\mathcal{G}_1, \ldots, \mathcal{G}_k\}$, *and $\mathcal{D}$ and each $\mathcal{G}_i$ are disjoint sets of clauses, a group-oriented MUS of $\mathcal{C}$ is a subset $\mathcal{G}'$ of $\mathcal{G}$ such that $\mathcal{D} \cup \bigcup_{G \in \mathcal{G}'} G$ is unsatisfiable and, for every $\mathcal{G}'' \subset \mathcal{G}'$, we have that $\mathcal{D} \cup \bigcup_{G \in \mathcal{G}''} G$ is satisfiable.*

2.3 Quality Metrics

The quality of variable partitions mainly impacts the quality of bi-decomposition [5–7, 12], and indirectly impacts the decomposed network, e.g. delay, area and power consumption [7]. Similar to [5, 6, 12, 13], this paper measures the quality of variable partitions through two *relative* quality metrics, namely *disjointness* and *balancedness*. Assume a variable partition $\{X_A|X_B|X_C\}$ for $f(X)$, where X_A, X_B and X_C are the sets of the input variables to decomposition functions f_A, f_B and common to f_A and f_B, respectively.

DEFINITION 4. (DISJOINTNESS). $\epsilon_D = \frac{||X_C||}{||X||}$ *denotes the ratio of the number of common variables to inputs. A value of ϵ_D close to 0 is preferred, as $\epsilon_D = 0$ represents a disjoint bi-decomposition.*

DEFINITION 5. (BALANCEDNESS). $\epsilon_B = \frac{\big|||X_A||-||X_B||\big|}{||X||}$ *denotes the absolute size difference between X_A and X_B. $\epsilon_B = 0$ represents a balanced variable partition.*

In practice, disjointness is preferred since a lower value represents a smaller number of shared input variables of the resulting decomposed circuit that typically has smaller area and power footprint. A lower balancedness typically corresponds to smaller delay of the decomposed network.

3. SATISFIABILITY-BASED MODELS

Traditional bi-decomposition algorithms [4, 11, 15, 16] are based on BDDs. This can have significant impact on different aspects of logic synthesis (see [5] for a review). Recent work on bi-decomposition mainly focuses on the performance of models when decomposing large functions [5, 12] as well as the quality of the bi-decompositions [6]. This section briefly reviews Satisfiability-based models in accordance with the precision of targeted quality, namely approximate and exact.

3.1 Approximate Models

The primary objective of using approximate models is to achieve good performance with approximate quality.

SAT-Based and MUS-Based Models. The approximate models are either SAT-based [12] or MUS-based [5]. These models provide a collection of solutions to the OR, AND and XOR bi-decompositions under known and unknown partition of variables. In practice, the partitions of variables are often unknown, and therefore, an automatic derivation of partitions is required.

A distinct feature provided by SAT-based models is that they are capable of automatically deriving variable partitions. The main difference between these SAT-based approaches is in the underlying representation of the constraints for modeling the bi-decomposition, including SAT [12], plain-MUS [5] and group-oriented-MUS [5]. The performance of these models is determined by (1) the modeling of constraints, and (2) how the constraints are solved. For example, the group-oriented MUS model [5] exhibits remarkably good performance when partitioning input variables for bi-decompositions. Consider the explicit groups of clauses for the modeling of bi-decomposition [5]:

$$\mathcal{D} = \{f(X) \wedge \neg f(X') \wedge \neg f(X'')\}$$
$$\mathcal{G}_{i_a} = \{(x_i \equiv x_i')\}, \mathcal{G}_{i_b} = \{(x_i \equiv x_i'')\} \quad (1)$$

PROPOSITION 1. [5] A completely specified function $f(X)$ can be decomposed into $f_A(X_A, X_C) \vee f_B(X_B, X_C)$ for some functions f_A and f_B if and only if the Boolean formula of the set of clauses $\mathcal{C}$, with

$$\mathcal{C} = \mathcal{D} \cup \mathcal{G}_A \cup \mathcal{G}_B \quad (2)$$

is *unsatisfiable* under a non-trivial partition, where the sub-set $\mathcal{G}_A \subset \{\bigcup_i \mathcal{G}_{i_a}\}$, the sub-set $\mathcal{G}_B \subset \{\bigcup_i \mathcal{G}_{i_b}\}$.

Notice that $\mathcal{D}$ contains clauses for representing the target Boolean function and its two complements. However, $\mathcal{D}$ does not contribute to the size of a group-oriented MUS, and can hence be viewed as *don't care* conditions w.r.t the size of the modeling. This in part motivates the gain of performance and this concept is adapted for the new models proposed in this paper.

The partitioning of variables occupies most of the run time in bi-decomposition [5–7, 12]. Essentially, the derivation of partitions is the process of switching the input variables between the two partitions. Switching of variables can be captured by selecting the groups of the input variables.

3.2 Exact Models

Exact models [6] provide controllable quality of bi-decomposition for various optimizations in logic synthesis. Existing exact solutions are based on QBF [6], and address the problem of computing bi-decompositions with optimum variable partitions. The optimality of achieved variable partitions is measured in terms of the existing metrics, namely disjointness and balancedness. Besides the novel QBF formulation, [6] shows how bi-decompositions can be computed with *optimum* values for target metrics. These QBF-based models guarantee the optimum quality of variable partitions. Nevertheless, exact models incur a performance penalty when compared with approximate models, in particular, with group-oriented MUS-based models [5].

4. NEW MODELS AND EXTENSIONS

Algorithm 1 proposes Quality-Guided-Group-MUS that targets high quality and efficient bi-decomposition.

4.1 OR Bi-Decomposition

Group-oriented MUS model [5] shows remarkably good performance, and it is used as the underlying construction in Algorithm 1. Similar to [5], during MUS search, the group $\mathcal{D}$ of clauses remains unchanged and only the groups $\bigcup_i \mathcal{G}_{i_A}$ and $\bigcup_i \mathcal{G}_{i_B}$ of clauses for variable partitions are considered:

$$\mathcal{F} = \mathcal{D} \bigcup_i \{\overbrace{((x_i \equiv x_i')}^{\mathcal{G}_{i_A}} \wedge \overbrace{(x_i \equiv x_i''))}^{\mathcal{G}_{i_B}}\} \quad (3)$$
$$\underbrace{}_{\mathcal{G}_{i_C}}$$

Observe that the subset $\mathcal{C}' = \mathcal{D} \cup \bigcup_i \mathcal{G}_{i_a}' \cup \bigcup_i \mathcal{G}_{i_b}'$ computed from (3) indicates the variable partitions. $\mathcal{G}_{i_a}'$ and $\mathcal{G}_{i_b}'$ are defined as follows: $((\mathcal{G}_{i_a}' \equiv \mathcal{G}_{i_a}), (\mathcal{G}_{i_b}' \equiv \mathcal{G}_{i_b})) = (1,1), (1,0), (0,1),$ and $(0,0)$ denote $x_i \in X_C$, $x_i \in X_B$, $x_i \in X_A$ and x_i can either be in X_A or X_B, respectively.

Algorithm 1: Quality-Guided-Group-MUS ($\mathcal{F}$,X)

Input: $\mathcal{F}$ — UNSAT formula of a seed variable partition
Input: X — input variables of Boolean function $f(X)$
Output: $\mathcal{M}$ — MUS with approximate good quality
Data: $i \leftarrow 2$ — do not involve the two seed variables

```
1  for i ≤ ||X|| do
2      if DIFF(F) ≥ 0 then
3          F' ← INCREASE_X_B_DECREASE_X_C(&F, i)
4          if SAT_SOLVE(F') == UNSAT then
5              F ← F'
6          else
7              F' ← INCREASE_X_A_DECREASE_X_C(&F, i)
8              if SAT_SOLVE(F') == UNSAT then
9                  F ← F'
10             end
11         end
12     else
13         | ...
14     end
15     i ← i + 1
16  end
17  M ← F
```

4.1.1 Approximating Optimum Balancedness

The refinement of a quality metric initially starts from an UNSAT formula of a seed variable partition [12], where the set X_A and the set X_B each takes at least one variable. The improvement of balancedness essentially corresponds to reducing the size difference between X_A and X_B:

$$DIFF_{AB} = ||X_A|| - ||X_B|| \quad (4)$$

If (4) results in a non-negative value (line 2), then the size (or cardinality) of set X_B is increased whereas the size of X_C is thereby reduced (line 3). This behaviour is captured by deleting the clauses of one group $\mathcal{G}_{i_B}$ of (3), which produces a formula $\mathcal{F}'$ with reduced size (line 3). The resulting $\mathcal{F}'$ can be UNSAT (line 4) and then be substituted for the original $\mathcal{F}$ (line 5). As a result, the balancedness is improved by the guide of the cost function (4). In contrast, formula $\mathcal{F}'$ can be SAT which implies an unchanged $\mathcal{F}$ at the current step, since $\mathcal{F}$ is only used as a constant reference in function INCREASE_X_B_DECREASE_X_C(...) (line 3).

4.1.2 Approximating Optimum Disjointness

Observe that achieving optimum Disjointness consists of reducing the size of X_C. Such a reduction can either be done by removing the group $\mathcal{G}_{i_A}$ or the group $\mathcal{G}_{i_B}$ from formula (3). As a result, if the previous steps (line 3 to 5) failed to produce a sub-formula $\mathcal{F}'$ with the deletion of group $\mathcal{G}_{i_B}$ (line 3), the deletion of group $\mathcal{G}_{i_A}$ is thus performed (line 7) and, followed by a SAT checking (line 8). This greedy scheme of removing either group $\mathcal{G}_{i_A}$ or group $\mathcal{G}_{i_B}$ (line 2 to 14) insists on an improvement of disjointness by looking for an UNSAT subformula $\mathcal{F}'$ with reduced size. Algorithm 1 also considers the case when $DIFF_{AB}$ has negative value. This situation is similar to previous ones, and can be implemented by replicating lines 2-11 to lines 12-14, but swapping lines 3 and 7. In addition, functionally non-support inputs are identified by preprocessing and are shifted to either X_A or X_B. Thus, simultaneously removing $\mathcal{G}_{i_A}$ and $\mathcal{G}_{i_B}$ is not required.

It it important to point out that this way of greedy MUS search is novel compared to [12, 14] mainly in that the used model (3) is different from the ones with *control variables* [12] and the ones with BDD-based *variable grouping* [14]. In addition, an implicit cost function (4) is introduced to guide the greedy search. Moreover, the techniques proposed in [5] target the efficient computation of MUSes, but are unable to take quality into account during either *plain* or *group* MUS search.

4.2 AND/XOR Bi-Decomposition

AND bi-decomposition is the dual of OR bi-decomposition and can be converted from the construction of OR models [5, 12, 14]. The proposed models can decompose $\neg f$ into $f_A \vee f_B$. By negating both sides, f is decomposed into $\neg f_A \wedge \neg f_B$ [12]. The XOR bi-decomposition is similar to OR bi-decomposition [5,12] and can be explained with an analogous derivation of the OR model. The derivation of AND/XOR bi-decomposition is omitted due to lack of space.

4.3 Application-Oriented Optimization

4.3.1 Targeting Desired Quality

Practical uses of bi-decomposition [4, 11, 16] require the ability to control the quality metrics of computed variable partitions. As noted earlier, approximate models, including *LJH* [12] and *STEP-M* [5], address the quality of variable partitions, by enumeration of control variable assignments [12] or unguided searching of MUSes [5], but offer no guarantees that the results respect any quality criterion. Moreover, the enumeration of control variables grows exponentially for searching the *optimum* variable partition. Hence, approximate models are unable to guarantee quality criteria of the computed bi-decompositions. This section proposes the use of cardinality constraints in the QBF-based Models *STEP-Q* [6] for optimizing and controlling the quality of variable partitions.

This paper focus on two quality metrics, namely disjointness and balancedness. Different applications require distinct combinations of disjointness and balancedness. The QBF models *STEP-Q* allow a wide range of possible practical uses. As shown in [6], cardinality constraints defined over the control variables α_x and β_x serve to constrain the computed solutions (see [6] for the definitions of α_x and β_x). Consider small natural numbers p_D, p_B, q_A, $q_B \in \mathbb{N}$. As a result, a number of possible sets of constraints can be envisioned:

- Completely disjoint and completely balanced:
$$[\sum_{x \in X}(\overline{\alpha_x} \cdot \overline{\beta_x}) = 0] \wedge [\sum_{x \in X}(\alpha_x \cdot \overline{\beta_x} - \overline{\alpha_x} \cdot \beta_x) = 0] \quad (5)$$

- Completely disjoint and approximately balanced:
$$[\sum_{x \in X}(\overline{\alpha_x} \cdot \overline{\beta_x}) = 0] \wedge [\sum_{x \in X}(\alpha_x \cdot \overline{\beta_x} - \overline{\alpha_x} \cdot \beta_x) = p_B] \quad (6)$$

- Approximately disjoint and completely balanced:
$$[\sum_{x \in X}(\overline{\alpha_x} \cdot \overline{\beta_x}) = p_D] \wedge [\sum_{x \in X}(\alpha_x \cdot \overline{\beta_x} - \overline{\alpha_x} \cdot \beta_x) = 0] \quad (7)$$

- Approximately disjoint and customized balancedness:
$$[\sum_{x \in X}(\overline{\alpha_x} \cdot \overline{\beta_x}) = p_D] \wedge [\sum_{x \in X}(\alpha_x \cdot \overline{\beta_x}) = q_A] \wedge [\sum_{x \in X}(\overline{\alpha_x} \cdot \beta_x) = q_B] \quad (8)$$

The values of p_D and p_B range from completely disjoint and balanced case to a fully customized case. Figure 1 illustrates the concepts for case (8). The additional constraints, coupled with the basic QBF model *STEP-Q* [6], allow fairly flexible modeling of constraints on variable partitions.

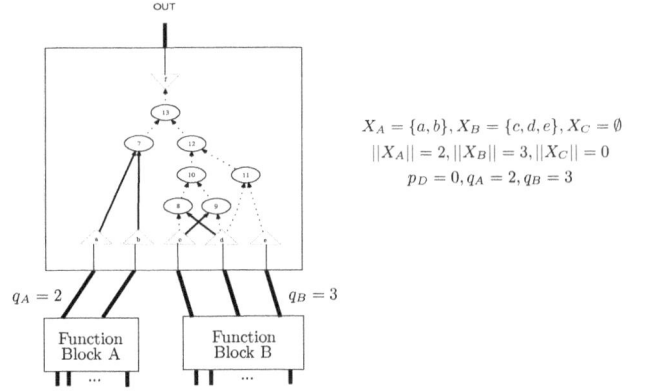

$X_A = \{a, b\}, X_B = \{c, d, e\}, X_C = \emptyset$
$\|X_A\| = 2, \|X_B\| = 3, \|X_C\| = 0$
$p_D = 0, q_A = 2, q_B = 3$

Figure 1: Example AIG (And-Inverter Graph) of the application-oriented optimization technique

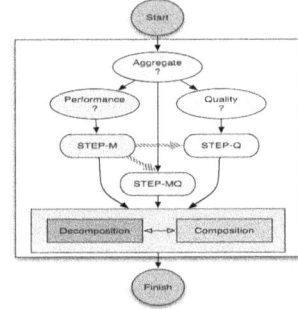

Figure 2: Optimization flow for bi-decomposition

4.3.2 Optimization Flow for Bi-Decomposition

In practice, bi-decomposition is recursively applied in logic synthesis. As stated before, different applications may require different targets for the quality metrics. This applies to both approximate and exact bi-decomposition solutions. In addition, there is often a bound on the allowed runtime. Therefore, it is reasonable to develop different models for tackling different objectives when using bi-decompositions. Figure 2 outlines the flow of bi-decomposing a Boolean function for different targets. Performance sensitive application finds *STEP-M* model to compute an approximate solution. Model *STEP-Q* directs the quality of bi-decomposition to fulfill the quality requirement of the quality sensitive applications. The new model *STEP-MQ* aggregates these two aspects to compute an approximate quality with a good performance. The precedent uses of *STEP-M*, as shown by the dashed arrows in Figure 2, offer *STEP-Q* and *STEP-MQ* the preprocessing of upper bounds for QBF searching [6] and candidates for variable partitions, respectively. After obtaining of variable partitions, the original Boolean function is decomposed into decomposition functions, and which are connected by a composition function [16], by using Craig Interpolation [10, 12, 13]. The decomposition functions are required to be *functionally* shared. Moreover, the interactive search of variable partitions should be *incremental*.

5. EXPERIMENTAL RESULTS

The models from Section 4 are implemented in the Boolean function bi-decomposition tool *STEP* — *S*atisfiability-based *func*T*ion d*E*com*P*osition. *STEP* is written in C++ (and compiled with G++ 4.4.3), it is implemented as an add-on

Table 1: Performance comparison

Circuit	Circuit Statistics			LJH [12]				STEP									
				LJH-P		LJH-Q		STEP-MP [5]		STEP-MG [5]		STEP-QD [6]		STEP-QB [6]		STEP-MQ	
	#In	#InM	#Out	#Dec	Time (s)	#Dec	Time (s)	#Dec	Time (s)	#Dec	Time (s)	#Dec	Time (s)	#Dec	Time (s)	#Dec	Time (s)
C7552	207	194	108	10	536.48	10	625.13	17	71.79	17	16.56	17	50.72	17	25.64	17	22.48
s15850.1	611	183	684	-	TO	-	TO	294	306.35	294	42.83	294	152.53	294	90.58	296	55.97
s38584.1	1464	147	1730	1065	446.23	1065	1912.06	1055	53.04	1055	23.12	1055	572.78	1055	117.25	1056	28.86
C2670	233	119	140	40	26.17	40	258.68	40	13.58	40	3.86	40	39.89	40	16.83	40	6.97
i10	257	108	224	131	407.22	131	2582.97	150	130.30	150	17.18	150	299.46	150	54.37	153	49.22
s38417	1664	99	1742	1202	5321.27	-	TO	1203	2944.47	1203	2658.25	1203	4718.92	1203	3487.92	1203	2739.31
s9234.1	247	83	250	102	55.79	102	130.43	114	18.32	114	12.23	114	100.10	114	27.50	115	8.39
rot	135	63	107	49	29.25	49	28.53	62	2.03	62	0.81	62	17.88	62	4.42	62	1.43
s5378	199	60	213	107	7.16	107	47.19	111	5.22	111	3.31	111	82.88	111	11.24	111	2.66
s1423	91	59	79	26	57.06	26	53.45	40	6.50	40	1.63	40	22.14	40	5.13	40	1.93
pair	173	53	137	117	12.84	117	84.42	114	13.89	114	10.50	114	202.11	114	33.00	114	6.74
C880	60	45	26	16	6.61	16	64.72	16	2.84	16	2.03	16	6.65	16	7.44	16	2.29
clma	415	42	115	39	1356.93	-	TO	34	1206.85	39	40.90	39	106.27	39	48.01	39	178.80
ITC_b07	49	42	57	14	18.08	14	16.38	18	3.70	18	1.47	18	2.44	18	2.07	18	1.10
ITC_b12	125	37	127	80	18.18	80	17.80	79	2.21	79	0.44	79	13.14	79	1.97	79	1.11
sbc	68	35	84	51	4.43	51	8.80	59	1.00	62	0.57	62	10.28	62	2.80	59	0.81
mm9a	39	31	36	22	10.07	22	103.38	28	8.17	28	4.16	28	28.29	28	10.20	28	4.17
mm9b	38	31	35	20	18.97	20	95.90	26	13.00	26	7.57	26	34.50	26	13.30	26	7.12

to **ABC** [1], and it uses **ABC**'s circuit representation and manipulation. **MiniSAT** [9] is used for SAT solving.

This section evaluates the performance and quality of bi-decomposition between SAT-based models. Given a circuit, each Boolean function of Primary Output (PO) is decomposed into simpler sub-functions using the proposed models. Sequential circuits are converted into combinational circuits. In this section, **LJH-P** (the fastest performance mode of **LJH** model [12]), **LJH-Q** (the best quality mode of **LJH** model), **STEP-MP** (plain-MUS mode of **STEP** [5,6]), **STEP-MG** (group-oriented-MUS mode of **STEP**), **STEP-QD** (QBF model of **STEP** for optimum Disjointness) and **STEP-QB** (QBF model of **STEP** for optimum Balancedness) are compared with the new model **STEP-MQ**. The model **STEP-MQ** implements Algorithm 1.

The experiments were performed on a Linux server with an Intel Xeon X3470 2.93-GHz processor and 6GB RAM. Experimental results were obtained on industrial benchmarks ISCAS'85, ISCAS'89, ITC'99 and LGSYNTH. Circuits with *zero* decomposable PO functions were removed from the tables of results. For each circuit, the total timeout was set to 6000 seconds. Due to space restrictions, only representative experimental results (for the *large* circuits, with #InM > 30) are shown. This section mainly presents the experimental results for OR bi-decomposition [1].

5.1 Performance of Models

Performance of models significantly affects the practical uses of bi-decomposition in logic synthesis, partly because bi-decomposition is recursively exploited in a number of internal loops of logic synthesis. This section evaluates the performance of the techniques proposed in this paper. Two performance metrics, CPU time and the number of decomposable functions, were used for assessing performance. Smaller CPU times indicate that decomposing a complete circuit will be faster and a larger number of decomposable functions represents an enhanced decomposability of the tool, indicating the tool is able to decompose more functions in the allowed CPU time, assuming more decomposable functions do exist.

Table 1 presents the performance data for OR bi-decomposition of the *large* circuits. The experimental data is sorted by decreasing number of maximum support variables

---

[1] AND and XOR bi-decomposition using the **LJH** model is unavailable in the *Bi-dec* tool [12].

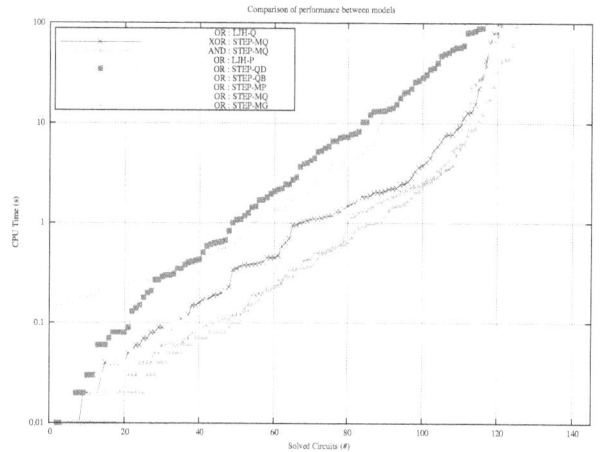

Figure 3: Performance comparison

(#InM). Columns #In, #InM, #Out, #Dec and Time(s) denote the number of primary inputs, maximum number of support variables in POs, PO functions (to be decomposed), decomposed POs and total CPU time, respectively. The results clearly show that **STEP-MQ** significantly outperforms **LJH** models and the other three **STEP** models, while achieving similar decomposibility.

Figure 3 shows the performance data for bi-decomposition of *small* and *medium-size* circuits. The number of solved circuits are plotted in the figure. Moreover, the *total* number of solved circuits out of *all* 145 circuits are sorted and it is showed descendingly by the legends. It is clearly that **STEP-MQ** solved more instances than the **LJH** models and the majority of **STEP** models. **STEP-MQ** performs slightly worse than **STEP-MG** but it is important to emphasize that **STEP-MQ** produces much better quality of bi-decompositions than **STEP-MG**. In addition, **STEP-MQ** produced similar quality to **LJH-Q**, but with much better performance.

5.2 Quality of Models

The quality of variable partitions is tightly related with the quality of decompositions [5–7,12,13]. This paper evaluates the quality of bi-decompositions using the same quality metrics as in [5,6,12,13], namely disjointness and balancedness.

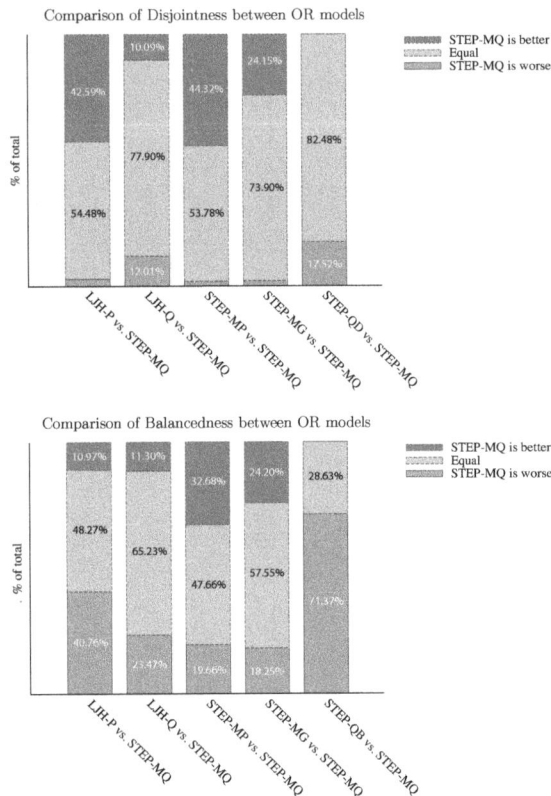

Figure 4: Quality metrics comparison

Following [5, 6, 12], disjointness is preferably considered in part because a smaller disjointness typically corresponds to optimally decomposed circuit during logic circuits [5,12]. Figure 4 summarizes the quality of variable partitions achieved in the decomposed circuits. To guarantee a fair comparison, only the functions that can be decomposed by the both two approaches are calculated. Due to space restrictions, only OR models are shown[1]. **STEP-MQ** computes approximate solutions. Unlike the *exact* models [6] **STEP-{QD,QB}**, the disjointness and balancedness cannot be guaranteed by **STEP-MQ**. **STEP-MQ** produces inferior balancedness compared to the SAT-based models [12], this phenomenon is because low disjointness and low balancedness are sometimes mutually exclusive [5,12]. As can be observed, **STEP-MQ** achieved significantly better disjointness than the major *approximate* models, including **LJH-P** and **STEP-{MP,MG}** models.

6. CONCLUSION

This paper extends SAT-based models for bi-decomposition, and develops a new model. The relative inefficiency of the *exact* models limits their use on *very large* Boolean functions. The nature of unguided MUS and SAT searching limits the use of existing *approximate* models for *controllable* qualities. This paper develops new approximate group-oriented MUS-based models for bi-decomposition, and extends existing SAT-based models for addressing practical requirements of bi-decomposition. A key feature is that new models allow MUS searching to be guided by an implicit cost function. Comprehensive experiments with the existing and the new

bi-decomposition models demonstrate that the new models achieve significant performance gains and, more importantly, also achieve visible improvement in the quality of bi-decompositions. Future work will address a tighter integration between tools, including **STEP** [5,6] and **ABC** [1], in logic synthesis, targeting area, delay and power reduction.

Acknowledgment

The authors would like to thank Prof. Jie-Hong Roland Jiang for kindly providing the SAT-based Boolean Function bi-decomposition tool *Bi-dec*. This work is partially supported by SFI PI grant BEACON (09/IN.1/I2618).

7. REFERENCES

[1] Berkeley Logic Synthesis and Verification Group. ABC: A System for Sequential Synthesis and Verification, Release 70930. In *http://www.eecs.berkeley.edu/~alanmi/abc/*.

[2] R. Ashenhurst. The decomposition of switching functions. In *Proceedings of an International Symposium on the Theory of Switching*, pages 74–116, 1957.

[3] H. K. Buning and T. Lettman. *Propositional Logic: Deduction and Algorithms*. Cambridge University Press, 1999.

[4] S. Chang, M. Marek-Sadowska, and T. Hwang. Technology mapping for TLU FPGA's based on decomposition of binary decision diagrams. *IEEE Trans. on CAD*, 15(10):1226–1236, 1996.

[5] H. Chen and J. Marques-Silva. Improvements to satisfiability-based boolean function bi-decomposition. In *VLSI-SoC*, pages 142–147, 2011.

[6] H. Chen and J. Marques-Silva. QBF-Based Boolean Function Bi-Decomposition. In *DATE*, 2012.

[7] M. Choudhury and K. Mohanram. Bi-decomposition of large boolean functions using blocking edge graphs. In *ICCAD*, pages 586–591, 2010.

[8] H. Curtis. *A new approach to the design of switching circuits*. Van Nostrand, Princeton, NJ, 1962.

[9] N. Eén and N. Sörensson. An extensible SAT-solver. In *SAT*, pages 502–518, 2003.

[10] J.-H. Jiang, C.-C. Lee, A. Mishchenko, and C.-Y. Huang. To SAT or Not to SAT: Scalable Exploration of Functional Dependency. *IEEE Trans. Comp.*, 59(4):457–467, Apr. 2010.

[11] Y. Lai, M. Pedram, and S. Vrudhula. BDD based decomposition of logic functions with application to FPGA synthesis. In *DAC*, pages 642–647, 1993.

[12] R.-R. Lee, J.-H. Jiang, and W.-L. Hung. Bi-decomposing large boolean functions via interpolation and satisfiability solving. In *DAC*, pages 636–641, 2008.

[13] H.-P. Lin, J.-H. Jiang, and R.-R. Lee. To sat or not to sat: Ashenhurst decomposition in a large scale. In *ICCAD*, pages 32–37, 2008.

[14] A. Mishchenko, B. Steinbach, and M. Perkowski. An algorithm for bi-decomposition of logic functions. In *DAC*, pages 103–108, 2001.

[15] T. Sasao and J. T. Butler. on bi-decomposition of logic functions. In *IWLS*, pages 1–6, 1997.

[16] C. Scholl. *Functional decomposition with application to FPGA synthesis*. Springer Netherlands, 2001.

Lithography-Aware Layout Compaction

Curtis Andrus
Department of Computer Engineering
University of California, Santa Cruz
1156 High St. MS SOE3
Santa Cruz, CA 95064
candrus@gmail.com

Matthew R. Guthaus
Department of Computer Engineering
University of California, Santa Cruz
1156 High St. MS SOE3
Santa Cruz, CA 95064
mrg@soe.ucsc.edu

ABSTRACT

Optical Proximity Correction (OPC) tools can suffer if the original layout is inherently difficult to print. Most routing techniques are unaware of the lithographic process, but several have been proposed to make the layout easier for the OPC tool to correct. This paper proposes a generalized preprocess step for OPC that uses a modified 1D layout compactor to expand or shrink geometry to decrease the amount of OPC work needed. This lithography-aware compactor estimates the OPC effort required and then uses a non-linear formulation to adjust the layout geometry. We describe a method for performing this compaction on small layouts and then extend it to handle larger hierarchical standard cell layouts.

Categories and Subject Descriptors

B.7.2 [**Integrated Circuits**]: Design Aids

Keywords

Design for Manufacturing (DFM), Optical Proximity Correction (OPC)

1. INTRODUCTION

Resolution Enhancement Techniques are methods of enhancing the mask to improve the transfer to the wafer. The goal is to generate a mask such that the printed geometry is as close as possible to the desired layout. Methods for doing this include *Optical Proximity Correction* (OPC) [8] and more recently inverse lithography techniques [14]. OPC introduces small corrections to the original mask until the printed shape is as desired. Rule-based OPC improves the mask by adding additional features such as serifs and scattering bars [6]. Another method breaks the edges of shapes in the layout into fragments, which are then moved in order to optimize the layout's printability. In this case, the goal is to minimize the total *Edge Placement Error* (EPE), the distance between the printed contour location and the desired location in order to minimize systematic variation [11].

These techniques are generally effective, but they are limited by the original mask geometry. Unfortunately, certain shape configurations can be difficult for OPC to correct. For example, constraints imposed by the mask manufacturer may make it impossible to correct the shape as much as the OPC tool desires, reducing its effectiveness. These lithographic "hotspots" must be corrected manually. In order to automate this process, lithographic effects cannot be ignored in the design stage. There are already examples of tools in this area, such as a router that has been modified to avoid areas of high EPE [13].

There has been a significant amount of past work on layout compaction [2, 5]. At the lowest level, a *Leaf Cell Compactor* operates on circuit elements and not hierarchical designs. Two common methods of leaf cell compaction are *Constraint Graph Compaction* and *Virtual Grid Compaction*. These two types of compactors only perform the compaction in one-dimension at a time, so they require to passes to fully compact the layout. Other compactors consider the two-dimensional compaction problem, which attempts to find the optimal packing of geometry. The problem is NP-complete [15], but algorithms exist for finding good solutions.

Extensions to the basic compaction methods include hierarchical compaction which compacts hierarchical layouts instead of just leaf cells. Compactors have also been modified to consider more than just area [4, 12]. Yield, crosstalk and power are examples of other costs to take into account when compacting a layout. The Enhanced Network Flow algorithm extends constraint graph compaction to support convex cost functions of the edge lengths [1]. They use this method to compact the layout while minimizing probability of spot defects.

As the purpose of a layout compactor is to generate the mask geometry, it is an ideal target for litho-aware design. There exist yield optimization techniques for compaction, but little work has been done to optimize compaction for lithographic effects. In this paper, we present a modified compaction method that takes these effects into account to decrease the effort required of the OPC program. In particular, this work contributes:

- A layout compaction technique that allocates more space to shapes of high OPC demand.

- A generic non-linear optimization approach to layout compaction.

- A simple hierarchical method that uses whitespace between cells to improve printability.

Our method differs from the one described by Cobb [8] in that an approximation of initial aerial image intensity is used to guide the movements instead of EPE measurements. Additionally, Cobb's method tries to match the contours to the target mask, while our proposed method makes changes to the actual mask.

Section 2 describes how the basic compaction process was modified to consider lithographic effects. Section 3 explains the non-linear optimization used to improve printability. Section 5 provides experimental results using a commercial OPC and Verification tool [9] and Section 6 concludes the paper.

2. OPC AWARE COMPACTION

We enhance a typical compactor to account for lithographic effects by:

- Approximating OPC demand at each point in the layout.

- Adding a weight for each edge in the constraint graph based on the approximate OPC (higher weights → more OPC required).

- Modifying the optimization to allocate more space to edges with higher weights.

In addition, we also implemented a basic hierarchical extension of this technique.

2.1 Approximating OPC Demand

The amount of OPC required is approximated on a 2D grid of regularly spaced sample locations. The value is computed at each pixel to form the *demand image*. We utilize an *Insufficient Intensity Method* from [7] that predicts OPC demand based the aerial image of the layout. The layout is pixelized to form mask image M. Light simulation is approximated by convolution with a Gaussian function G in the compactor, but this method could be easily extended to handle a more accurate lighting system. The insufficient intensity I_s is then:

$$I = G * M \qquad I_s = \max(tM - I, 0) \qquad (1)$$

where $*$ denotes 2D convolution. t is a user-defined threshold value below which light is considered insufficient. The final demand image D is computed as:

$$D = I_s * Q \qquad (2)$$

where Q is the quasi-inverse kernel as described in [7]. This image then represents the amount of OPC that will be required to fix the areas with insufficient intensity. Points with higher OPC demand have a higher probability of being modified in the pattern of the OPCed layout. Convolutions are performed efficiently using the FFTW library and can be easily parallelized. Fig. 1(b) shows the results of this method on a test layout.

Once the demand image has been computed, this information is encoded in the constraint graph by adding weights to each edge. Each edge in the constraint graph corresponds to two segments in the layout. Some edges, such as those necessary for connectivity or via overlap, ensure that the layout remains logically the same so do not require a weight. Wire spacing and width edges are the focus of this weighting step. The weight of this edge is computed by integrating the demand image over the region between the two segments

$$w = \int_{y_0}^{y_1} \int_{x_0}^{x_1} D(x, y) \, dx \, dy \qquad (3)$$

where the region between the two segments is the rectangle $[x_0, x_1] \times [y_0, y_1]$.

2.2 Optimization Changes

Now that the constraint graph has a concept of OPC demand, the optimization step must be changed. The goal is no longer simply to minimize the area, but to expand/shrink the edges in the constraint graph based on their weight. There are several possible ways to do this based on the constraint graph. One would be to modify the DRC rules, increasing the rule for edges with high weights.

(a) Mask Geometry

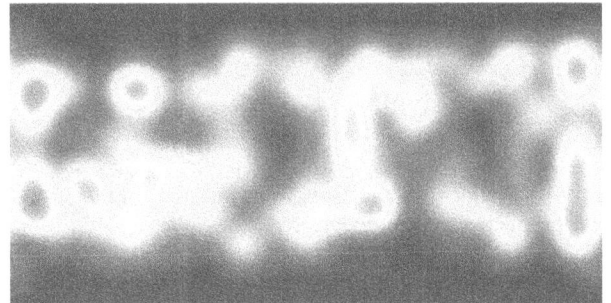

(b) Insufficient Intensity Map

Figure 1: OPC demand of a layout.

This is similar to the OPC-aware routing performed in [13], where blockages were added in critical areas. This approach is simple, and requires no modification to the actual optimization algorithms. Unfortunately doing this could introduce constraints that are impossible to satisfy. Another option is to modify the cost function to put pressure on heavy edges to increase their length. This is the approach we take and will be explained in detail in the next section.

Because the compactor is no longer optimizing for area, it could generate arbitrarily large layouts. To get around this, the modifications introduce a maximum area constraint, requiring the program to find a minima of the cost function in a limited amount of space. Since OPC demand typically decreases as shape width and spacing increases, allocating a larger amount of area should allow the compactor to find better solutions. This idea is important for the hierarchical compaction technique.

2.3 Hierarchical Compaction

In a standard cell layout, it is common to have whitespace between each cell. The hierarchical compactor uses this whitespace to give cells more room to expand, which should help decrease the OPC demand as explained in the previous section. For simplicity, our implementation uses a 2-level hierarchy. The bottom level is the standard cells and the top level contains the power rails and intracell routing. It is assumed that metal1 is used for inter-cell and not intra-cell routing which is typically the case. This way, each cell can be compacted independently, working only on metal1 and keeping the other layers fixed. Slight modifications to this methodology can easily incorporate the compaction of other layers.

At the beginning of the compaction process, the amount of whitespace in between cells is calculated. The whitespace around each cell is then included in the maximum area constraint. Each cell is compacted independently, but the neighboring geometry is considered when computing OPC demand. The hierarchical compactor computes the OPC demand image for the entire layout at once. Each leaf cell uses this same image when generating edge weights.

3. NONLINEAR OPTIMIZATION

3.1 Cost Function

This section discusses the cost function and optimization technique used in minimization. The notation x_k is used to denote the kth rectangle edge position, and $\vec{x} = (x_1, x_2, \ldots, x_n)$ is the vector containing all positions for a graph. The length of the ith edge is $w_i = x_j - x_k$ for some positions x_j and x_k, and α_i is the associated weight computed using the methods of the previous section. The α_i are determined from the OPC demand image and remain fixed throughout the optimization process. The goal is to define a cost function $c(\vec{x})$ that tries to minimize the total area while maximizing the spacing between individual edges, giving priority to edges with higher weights. This cost function should have the following properties:

- Edges with large weights are expanded.

- Edges are expanded in a balanced way (lower-weight edges are still expanded, but more priority is given to higher-weight edges).

- Higher area is penalized.

With these properties in mind, a suitable cost function is:

$$c(\vec{x}) = \frac{area(\vec{x})}{w_1^{\alpha_1} w_2^{\alpha_2} \ldots w_m^{\alpha_m}} \qquad (4)$$

Algorithm 1 Piecewise Linear Optimization

Require: Constraint graph with initial position $\vec{x^0}$, with max move amount Δx
1: $count \leftarrow 0$
2: $c_{prev} \leftarrow c(\vec{x^0})$
3: **for** $i = 0$ to max_iterations **do**
4:    $dc \leftarrow$ linear approximation of c near $\vec{x^i}$
5:    Constrain each movement to $(-\Delta x, \Delta x)$
6:    $\vec{x^{i+1}} \leftarrow$ location that minimizes dc using LP_SOLVE
7:    $c_{curr} \leftarrow c(\vec{x^{i+1}})$
8:    **if** $c_{curr} > c_{prev}$ **then**
9:       $\Delta x \leftarrow \max(1, \frac{\Delta x}{2})$
10:    **end if**
11:    **if** $|c_{curr} - c_{prev}| < 0.01$ **then**
12:       $count \leftarrow count + 1$
13:       **if** $count > 2$ **then**
14:          Successive iterations have converged on a solution
15:          **break**
16:       **end if**
17:    **else**
18:       $count \leftarrow 0$
19:    **end if**
20:    $c_{prev} \leftarrow c_{curr}$
21: **end for**
22: $\vec{x} \leftarrow$ last value of $\vec{x^i}$

3.2 Piecewise Linear Approximation

As Equation 4 is not a linear function, a different optimization approach is needed. In [10], the authors optimize a cost function by approximating it by a series of linear optimization problems. Our compactor takes a similar approach. From an implementation perspective this method is ideal because the algorithm can use the linear programming algorithms already implented by LP_SOLVE [3]. Given initial edge positions $\vec{x^0}$, the nonlinear cost function can

be locally approximated with a linear function, computed using the first terms in the Taylor series

$$dc(\vec{x}) = x_1 \frac{\partial c}{\partial x_1}\big|_{x_1 = x_1^0} + \cdots + x_n \frac{\partial c}{\partial x_n}\big|_{x_n = x_n^0} \qquad (5)$$

where $\frac{\partial c}{\partial x_i}$ is inversely proportional to the length of the edges that start or end at x_i. For example, if x_1 is on the positive end of the 1st edge and negative end of the 2nd edge, the partial derivative is

$$\frac{\partial c}{\partial x_1} = c(\vec{x^0})(\frac{\alpha_2}{w_2} - \frac{\alpha_1}{w_1}) \qquad (6)$$

Linear programming can be used to find the minimum of dc, but if the solution is too far away from $\vec{x^0}$ then the error between the cost function and the linear approximation will be large. By adding additional constraints to the problem, we can limit the variables to lie within a fixed distance from the starting position:

$$x_i - x_i^0 \quad < \Delta x \qquad (7)$$
$$x_i - x_i^0 \quad > -\Delta x \qquad (8)$$

As the constraints are linear, this restricted problem can still be solved via linear programming. Its minimum location will move the positions in a direction that decreases the original cost function while satisfying the DRC constraints. To continue the minimization, the cost function can be reapproximated from the new location, followed by another pass of linear programming.

3.3 Algorithm Details

Pseudocode of the optimization algorithm is shown in Algorithm 1. Starting at Line 3, the algorithm performs a fixed number of iterations to find the optimal positions. Once successive iterations no longer change the cost, it is assumed to have converged (Lines 11-16). If an iteration ever increases the cost, then the step has crossed over a local minimum. To ensure that the algorithm can converge to this minimum, the allowed movement amount is halved (Lines 8-9).

4. EXPERIMENTAL SETUP

This section evaluates the performance of the compactor on various 45nm layouts. Performance is measured using the internal OPC demand measure of the compactor program and using an external tool for final accurate evaluation. Accurate lithography simulations are done using the OPC tool developed by Gauda Inc. [9] described in [16], and the performance is measured using the Edge Placement Error (EPE) statistics reported by the program.

For testing purposes the compactor generates a single number that represents the total OPC demand of the layout. This is the average value of each pixel in the OPC Insufficient Intensity demand image. The optimization is performed on the metal1 layer, and all other layers are fixed. For the OPC, fragmentation size was 50nm with a 20nm constraint imposed on width and spacing of layout geometry.

5. EXPERIMENTAL RESULTS

5.1 Example Test Case

The compactor was evaluated on several layouts that are typically hard for OPC to correct. In these layouts, several shapes are very close to each other, making it difficult for the OPC tool to generate a layout that is correct and satisfies the constraints of the mask manufacturer. The two original layouts are shown in Fig. 2(a) and Fig. 3(a). The OPC demand estimations are shown in Table 1 and the results of Gauda's tool are shown in Table 2. Images of the compacted layouts and contours are shown in Fig. 2 and Fig. 3.

	OPC Demand	
Layout	Original	Compacted
pattern1	0.036256	0.018286
pattern2	0.035174	0.040524

Table 1: Compactor Results for Test Pattern Layouts

		pattern1	
		Original	Compacted
Pre-OPC	Mean	18.0956	19.0472
	Max	31.3892	12.0157
	Min	-50	-50
	σ	21.8335	22.0650
Post-OPC	Mean	1.1415	1.0544
	Max	4.4766	2.2577
	Min	-4.4762	-6.9113
	σ	1.5166	1.6477
		pattern2	
		Original	Compacted
Pre-OPC	Mean	27.7789	12.4581
	Max	50	33.2761
	Min	-50	-50
	σ	31.5891	16.8256
Post-OPC	Mean	1.0763	0.8840
	Max	4.2153	4.0209
	Min	-6.0541	-4.4938
	σ	1.6660	1.2740

Table 2: EPE Results for Test Pattern Layouts (nm)

Before OPC, the compacted layouts have significantly better EPE statistics than the original, despite the fact that the OPC demand estimation increased on pattern2 in Table 1. Visually, the compacted contours have no bridging errors (i.e. contours of separate shapes merging), but there are several bridging errors in the original contours. After correction the bridging errors are gone, but the original contours still contain some ringing effects which are not as visible in the compacted contours. The EPE statistics of the compacted layouts are also slightly better. The compactor efficiently allocates space among geometry in the layout. The additional space in the critical locations gives the OPC tool more freedom to correct the layouts.

5.2 Leaf-Cell Evaluation

To further evaluate the lithograph-aware compaction on actual layouts, 10 small leaf-cell layouts were processed using the compactor. Table 6 contains information about the test layouts. The number of rectangles reported is the number of rectangles generated by the program and used in the compaction process. The height of each layout is 1570nm.

We first evaluate the compaction results with the OPC demand. In all cases the area was not allowed to increase from the original layout. Table 6 shows the results. In 5/10 cases the program succeeded in decreasing the OPC demand. These layouts are highly constrained through adjacent layer constraints, which in many cases this prevents them from being optimized effectively.

For a more accurate evaluation, we measure the EPE results of the original and compacted layouts in Gauda's program. Table 3 shows the EPEs before OPC, and Table 4 shows how the EPEs changed after OPC.

	Original	Compacted	Runtime (s)
OPC Demand	0.025859	0.021062	376.976

Table 5: Hierarchical compaction results for the multiplyadd layout

Layout	Rect. (#)	Width (nm)	Pre	Post	Time (s)
BFX1	25	570	0.0228	0.0276	0.35
AND2	34	760	0.0317	0.0287	0.44
NOR3	34	760	0.0213	0.0234	0.42
MUX2	61	1330	0.0276	0.0268	0.94
NND4	65	1710	0.0258	0.0254	1.19
HA	70	1900	0.0270	0.0257	1.18
OAI22	73	1710	0.0255	0.0266	1.18
BFX32	77	1710	0.0244	0.0257	1.14
DLH	91	2090	0.0248	0.0239	1.16
DFF	134	3230	0.0196	0.0246	1.84
Average	66.4	1577	0.0250	0.0258	0.93

Table 6: Comparison of pre- and post-OPC demand before and after compaction.

5.3 Hierarchical Evaluation

In addition to leaf cells, the basic hierarchical compaction was tested on a standard cell layout. A 16x16 multiply add module using the previous 45nm library of standard cell layouts was synthesized with Synopsys Design Compiler and placed and routed using Cadence Encounter. The layout is approximately 27000nm x 25000nm area and contains 501 standard cells. The compactor moves shapes into the previously blank regions and manages to successfully decrease the OPC demand of the layout, as shown in Table 5.

A runtime of 377 seconds means an average of 0.752 seconds per cell. Once the demand image has been computed, each cell can be compacted independently of the other, meaning this process can be easily parallelized and extended to run on much larger layouts.

6. CONCLUSION AND FUTURE WORK

In this paper we extended the capabilities of a traditional layout compactor to account for lithographic effects. The piecewise linear optimization method is capable of reallocating space between geometry to give more room to parts of the layout that are difficult to OPC. This produces a new, equivalent layout which, in several cases, gives the OPC tool more freedom to perform the necessary corrections. In several benchmarks, the compactor was shown aid the OPC tool in creating a mask that can be successfully printed.

Giving more space between shapes is not always the best solution for improving OPC. Due to the behavior of the lithography system, increasing the width and spacing does not always mean that printability will improve. "Forbidden pitches" are pitches between shapes that cause problems for manufacturing. In the future, we plan to experiment with more advanced cost functions that discourage these pitches. The optimization method described in this paper applies to a large variety of cost functions. Our method can easily handle more complicated interactions between shapes (instead of just functions between two edges).

Acknowledgments

The authors wish to thank Luigi Capodieci of Global Foundries for initial discussions on the topic and Ahmet Karakas and Ilhami

	Original EPE (nm)				Compacted EPE (nm)			
	Mean	Max	Min	σ	Mean	Max	Min	σ
BFX1	20.4	50.0	-29	19.9	10.9	22.7	-50	15.8
AND2	12.6	37.2	-50	15.9	23.2	36.2	-50	27.0
NOR3	11.6	36.3	-50	15.0	16.2	29.5	-50	21.2
MUX2	21.3	50.0	-50	25.1	23.1	50.0	-50	29.5
NND4	13.6	33.4	-50	15.0	14.9	37.0	-50	19.8
HA	15.4	50.0	-50	19.2	23.4	35.1	-50	28.1
OAI22	17.0	50.0	-50	21.9	17.3	50.0	-50	23.5
BFX32	18.4	50.0	-50	22.6	19.9	33.9	-50	25.0
DLH	18.1	50.0	-50	22.8	21.7	50.0	-50	27.8
DFF	20.4	50.0	-50	19.4	19.6	50.0	-50	25.0
Avg.	16.88	45.69	-48	19.68	19.02	39.44	-50	24.27

Table 3: Compaction EPE results (simulated using Gauda's OPC tool). EPE values of +/- 50nm indicate that the shape is not printing.

	Original EPE (nm)				Compacted EPE (nm)				
	Mean	Max	Min	σ	Mean	Max	Min	σ	Time (s)
BFX1	0.46	2.87	-3.14	0.78	0.61	3.04	-4.44	1.01	5.0
AND2	0.59	2.89	-5.25	0.99	0.57	2.80	-2.16	0.84	5.3
NOR3	0.71	2.74	-5.13	1.08	0.99	4.01	-6.64	1.48	5.3
MUX2	0.55	2.89	-5.81	0.93	0.56	2.50	-9.13	0.99	3.9
NND4	0.55	3.08	-6.51	0.97	0.57	7.08	-5.90	0.98	5.2
HAX1	0.65	4.01	-3.84	1.01	0.52	2.42	-2.95	0.77	4.9
OAI22	0.47	3.01	-4.42	0.80	0.87	4.92	-31.8	2.76	2.5
BFX32	0.63	3.41	-4.81	0.97	0.51	4.87	-5.54	0.88	4.9
DLH	0.47	2.54	-3.28	0.81	0.51	5.31	-3.82	0.90	4.0
DFF	0.40	2.46	-2.58	0.70	1.08	3.88	-50.0	2.67	2.0
Avg.	0.548	2.99	-4.477	0.904	0.679	4.083	-12.238	1.328	4.3

Table 4: Compaction EPE results after correction using Gauda's OPC tool.

Torunoglu of Gauda Inc. for the use of their software. This work was supported in part by the National Science Foundation under grant CCF-1053838.

7. REFERENCES

[1] C. Bamji and E. Malavasi. Enhanced network flow algorithm for yield optimization. In *DAC*, pages 746–751, 1996.

[2] C. Bamji and R. Varadarajan. *Leaf cell and hierarchical compaction techniques.* Kluwer Academic Publishers, Norwell, MA, USA, 1997.

[3] M. Berkelaar. lp_solve. http://lpsolve.sourceforge.net.

[4] Y. Bourai and C.-J. R. Shi. Layout compaction for yield optimization via critical area minimization. In *DATE*, pages 122–127, 2000.

[5] D. G. Boyer. Symbolic layout compaction review. In *DAC*, pages 383–389, 1988.

[6] J. F. Chen, T. L. Laidig, K. E. Wampler, and R. F. Caldwell. Practical method for full-chip optical proximity correction. volume 3051, pages 790–803. SPIE, 1997.

[7] T. Chen, G. Liao, and Y. Chang. Predictive formulae for OPC with applications to lithography-friendly routing. In *DAC*, pages 510–515, 2008.

[8] N. Cobb. *Fast optical and process proximity correction algorithms for integrated circuit manufacturing.* PhD thesis, UC Berkeley, 1998.

[9] Gauda. http://www.gauda.com/, 2011.

[10] R. E. Griffith and R. A. Stewart. A Nonlinear Programming Technique for the Optimization of Continuous Processing Systems. *MANAGEMENT SCIENCE*, 7(4):379–392, 1961.

[11] M. Guthaus, N. Venkateswaran, V. Zolotov, D. Sylvester, and R. Brown. Optimization objectives and models of variation for statistical gate sizing. In *GLSVLSI*, pages 312–316, 2005.

[12] T. Iizuka, M. Ikeda, and K. Asada. Timing-driven cell layout de-compaction for yield optimization by critical area minimization. In *DATE*, pages 884–889, 2006.

[13] J. Mitra, P. Yu, and D. Z. Pan. Radar: Ret-aware detailed routing using fast lithography simulations. In *DAC*, pages 369–372, 2005.

[14] A. Poonawala. *Mask Design for Single and Double Exposure Optical Microlithography: An Inverse Imaging Approach.* PhD thesis, UC Santa Cruz, 2007.

[15] M. Schlag, Y.-Z. Liao, and C. Wong. An algorithm for optimal two-dimensional compaction of VLSI layouts. *Integration, the VLSI Journal*, 1(2-3):179 – 209, 1983.

[16] I. Torunoglu, A. Karakas, E. Elsen, C. Andrus, B. Bremen, and P. Thoutireddy. OPC on a single desktop: a GPU-based OPC and verification tool for fabs and designers. Number 1, page 764114. SPIE, 2010.

(a) Layout   (b) Pre-OPC Contours

(c) Post-OPC Contours

(a) Layout   (b) Pre-OPC Contours

(c) Post-OPC Contours

(d)   Compacted (e) Compacted Pre-OPC Con-
Layout   tours

(d) Compacted Layout

(f)   Compacted   Post-OPC
Contours

(e) Compacted Pre-OPC Con- (f) Compacted Post-OPC Contours   tours

Figure 3: Test Layout pattern2 (images approximately same scale)

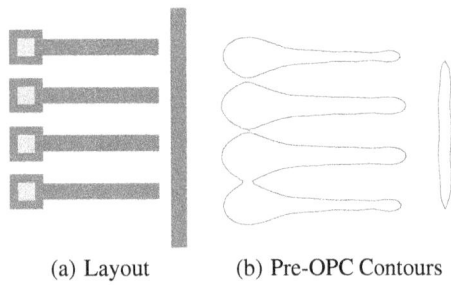

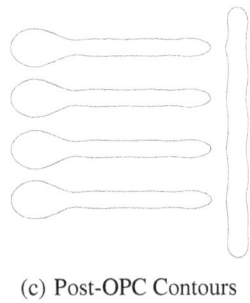

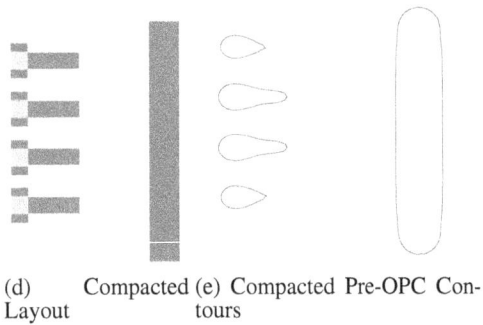

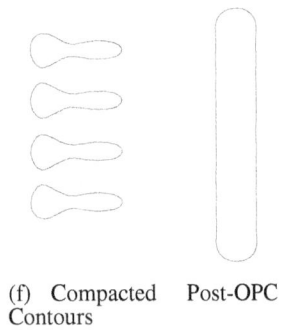

Figure 2: Test Layout pattern1 (images approximately same scale)

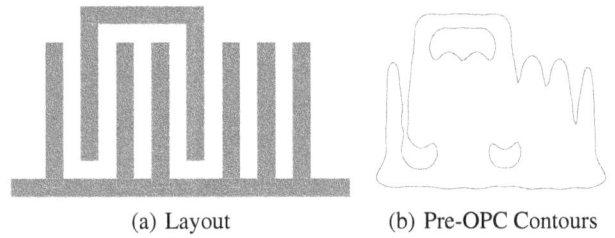

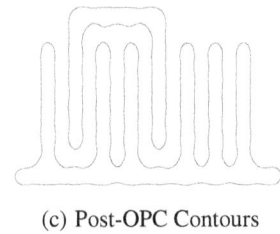

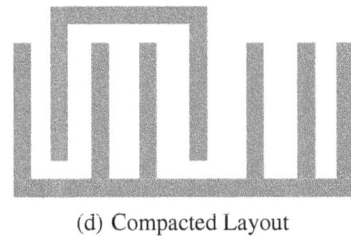

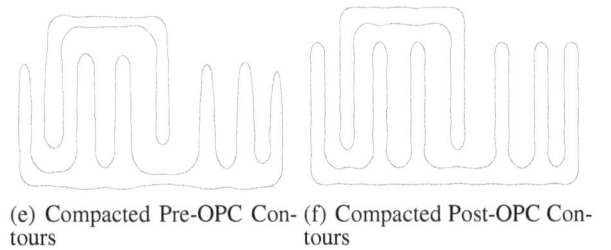

been found. Otherwise, postponed data are processed by the second step of the algorithm. Initial mapping is performed as follows, until all the columns of the mapping matrix have been processed:

1-Column selection: This step selects the most constrained and not yet explored column in the mapping matrix. The most constrained column (read or write column for a given time instance) is the column in which there remains the fewest number data to be mapped in memory banks.

2-Assigning and reporting: For the selected column, our algorithm searches for a valid memory mapping according to memory and network constraints. Then the algorithm updates the corresponding cells with respect to memory constraint #2.

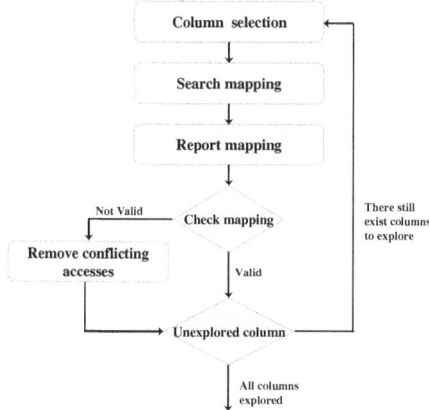

Figure 4. Initial mapping

3-Verification: This algorithm checks if the updated cells respect the constraints, otherwise these data are not assigned, and the related mapping (regarding the memory constraint #2) are removed from the matrix (see Figure 6) and postponed.

```
// M: Mapping matrix
// TA: Targeted network constraints.
// TS: Targeted memory constraints.
// Ci: Column i of the memory mapping M which can corresponds
//      to write or read access for a given cycle column
// Lci : Mapping solution for the column Ci (read or write access).
// RC : list of the column of the mapping matrix from which mapping
//      are cancelled during the verification step (Initial value = empty)

Function Initial_Mapping (M, TA)
{       while all the column have not been explored in M
        {       // Select the most constrained column in M which is not in RC
                Ci= Select_Targeted_Column (M, RC);

                // Generating a valid memory mapping for Ci
                Lci = Mapping_Solution (Ci, TA, TS);

                // Map the solution  Lci in column Ci
                Affect (Ci, Lci);

                // Report the mapping to the related column
                Report (Ci, M, TS);
        }
}
```

Figure 5. Algorithm for the initial mapping

```
Function verification (M, TA, TS)
{       // Verification: columns in M must respect targeted network
        //       and memory constraints
        For all Ci in M
        {       If not (Respect_Constraints(Ci, TA, TS)) then
                {       // dk is a data in column Ci
                        For all dk in Ci
                        {       If IsMapped(dk) then
                                {       Cancel_Mapping(Ci, dk)
                                        RC = Report_Cancellation(M, Ci,  dk)
                                }
                        }
                }
        }
}
```

Figure 6. Verification algorithm

- Conflict solving: This step inputs the resulting mapping matrix from the initial mapping step and then tries to map the unassigned (postponed) data (see Figure 7). The idea is still to map the data in the memory banks as much as possible, according to the memory and network constraints. If some of these data still remain unassigned, then they are removed from the memory banks and additional registers are used to store them.

Finally, the algorithm aims to minimize the number of required additional registers and multiplexers and to reduce the complexity of the associated controller. This binding of the additional registers is performed though a modified Left-Edge algorithm that minimizes the number of registers while taking into account the cost of multiplexers induced by resource sharing.

```
Function  Conflict_Solving
{               // Select a targeted column in RC
                Ci= Select_Target(M, TA, TS, RC)
                For all dk in Ci
                {       // Generation of mapping solution
                        Lci= Final_Mapping_Solution (M, Ci, TA, TS);
                        If  IsEmpty( Lci) then
                                // Affect register to the data
                                Affect_Register (M, Ci)
                        Else
                        {       // Map the solution  Lci in column Ci
                                Affect (Ci, Lci);

                                // Report the mapping to the related column
                                Report (Ci, M, TS);
                        }
                }
}
```

Figure 7. Conflict Solving algorithm

3. PRATICAL IMPLEMENTATION

Let us take as an example the interleaving law presented in figure 2. The targeted architecture is composed of three memory banks (A, B, C) and three processing elements (P1, P2, P3) as illustrated in figure 1. The target interconnection network is a barrel shifter: each column of the final mapped access matrix must be a circular permutation of all others for the network constraint to be satisfied.

Based on the model illustrated in figure 3, our algorithm starts by mapping read accesses of data that belong to the first column i.e. *column 1* (see Figure 8): *data 1, 2 and 3 are respectively* mapped to the memory *banks A, B and C.*

Figure 8. First column mapping

The last write accesses to these three data are then assigned to the same memory banks as it can be seen in Figure 9 where memory mapping matrix has been updated accordingly.

Figure 9. Report the first column mapping

The algorithm then selects the most constrained column i.e. the most constrained instant for read or write accesses. The algorithm tries to map the unmapped data of this column by respecting the memory and the network constraints. In Figure 9, the most constrained column is the last one (write column of *instant 6*). *Data 4* is assigned to the memory *bank C*. Following the memory constraint #1, the cell corresponding to the first read access to *data 4* is updated with the same memory bank label i.e. bank *C* (see Figure 10.a).

At that time, two most constrained columns exist: the read column of *instant 3* or the write column of *instant 5*. In such a case, our algorithm selects the leftmost column i.e. the column which contains the oldest accesses. In our example, the selected column is thus the read column of *instant 3*. In order to follow the network constraint i.e. the circular permutation laid down by the barrel shifter, *data 6* and *data 1* are respectively assigned to the memory *bank B* and *bank A*. According to the memory constraints, the cells corresponding to their previous write accesses are assigned to the memory *bank B* and *A* respectively as presented in Figure 10.b.

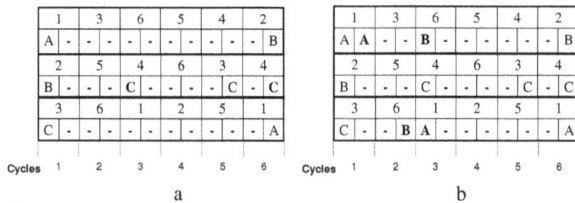

Figure 10. Second column mapping

Our algorithm is performed on the rest of the matrix until it reaches a mapping conflict due the network constraints i.e. *instant 4* in our example (see Figure 11.a) where mapping of *data 5* and *data 2* are conflicting. Indeed, in the write column of *instant 2*, *data 5* has been mapped in the memory *bank A* which implies its next read (at *instant 4*) has to be done in the same memory bank i.e. *bank A*. In addition, in the write column of *instant 1*, *data 2* has also been assigned to memory *bank B* which implies its next read (at *instant 4*) has to be done in the same memory bank i.e. *bank B*. Unfortunately, the memory mapping of read accesses to *data 5* and *data 2* at *instant 4* require a permutation of memory banks that does not comply with the network which can only realize circular permutation of memory banks *ABC, CAB and BCA*. As a consequence, *data 5* and *data 2* not mapped at the corresponding time instances and are postponed. This is represented by grey colored cells in Figure 11.b.

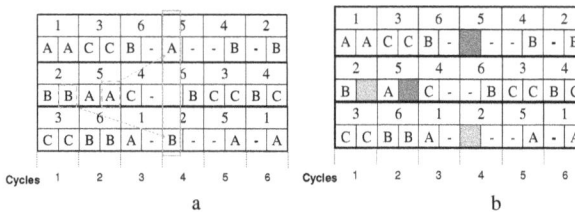

Figure 11. Architectural constraint failure in the read column
column for *instant 4*

The algorithm executes on the rest of the matrix as explained in the previous section. Figure 12 shows the resulting partially mapped matrix obtained by the initial step of our algorithm.

The second step consists in mapping the postponed accesses if empty cells exist in the mapping matrix. For each unassigned access we generate all the possible solutions complying with the memory and the network constraints. If there is no valid memory mapping the data is finally stored in a register. In Figure 12, *data 2* at *instant 1* has one mapping solution with respect to the constraints (i.e. *bank B*).

However, there is no mapping solution for *data 5* at *instant 2*: the required bank at *instant 2* (i.e. *bank A*) cannot be used for the corresponding read access at *instant 4* because the architectural constraint #1 would not be respected. As a consequence, *data 5* is mapped in a register between *instant 2* and *instant 4*. The algorithm executes on the rest of the matrix as explained. Figure 13 shows the first result obtained by our algorithm.

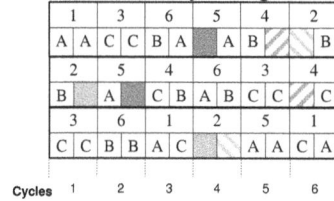

Figure 12. Partially mapped matrix resulting from initial step

Then, once the set of data to be stored in additional registers has been defined, the register binding and architectural optimization algorithms are performed in order to generate the final RTL architecture. Due to non overlapping lifetime, only one register is required in our example (see Figure 14).

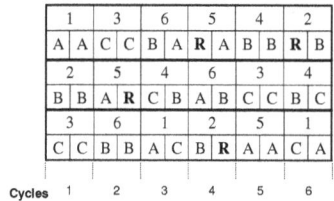

Figure 13. Valid memory mapping solution

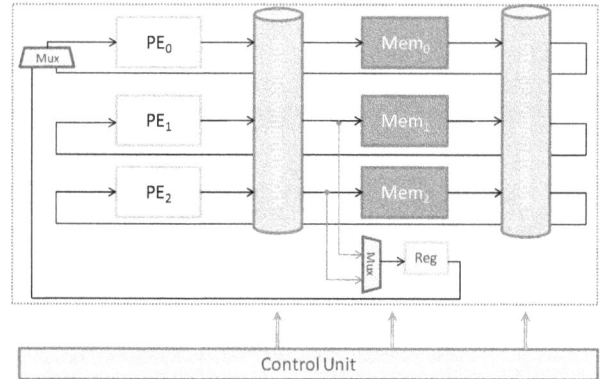

Figure 14. Resulting architecture

4. EXPERIMENTAL RESULTS

Our approach has been applied to an Ultra-WideBand interleaver [21] (UWB) and to a non-binary LDPC applications [22]. Ultra-Wide Band (UWB) is a technology for transmitting information spread over a large bandwidth, which offers high resolution with improved ranging accuracy (multimedia communications, radar applications...) [21]. Non-binary LDPC codes are now recognized as a potential competitor to binary coded solutions, especially when the codeword length is small or moderate [22].
Our results are compared to state of the art approaches [13], [6], [18] and to a *full crossbar* network based architecture. The first approach [13] is based on a simulated annealing mapping algorithm. It is able to find a conflict free memory mapping for any interleaving law, but the interconnection network and the controller complexities are not taken into account. The second approach [6] is able to target user defined network, *iff* the interleaving   law   enables   such   optimization.

Table 1. Synthesis results (area in NAND gate equivalent)

Area	UWB					Non-binary LDPC				
	[5]	[17]	[12]	Full crossbar	Our algorithm	[5]	[17]	[12]	Full crossbar	Our algorithm
Memory	262800	262800	262800	262800	262800	n.a	42048	n.a	42048	35697
Interleaving architecture	1680	2240	2800	4480	1680	n.a	3920	n.a	10080	6020
Control FSM	6042	11970	6954	19266	6042	n.a	1709	n.a	2098	3435
Total Area	270522	277010	272554	286546	270522	n.a	47677	n.a	54226	45152

These two approaches are dedicated to Turbo-Codes like applications only and can not support multiple read and write accesses to the same data like it appears in LDPC decoding algorithms. Third approach [18] proposes a method that is able to deal with Turbo-Codes and LDPC codes.

The synthesis results area have been obtained using a 90 nm technology and are given in NAND-gate equivalent to respect non-disclosure agreement. The experimental results are shown in Table 1. These experiments have been performed on UWB for 1200 data, and the non-binary LDPC for 192 data accessed twice and a parallelism of 6 i.e. 6 processing elements accessing 6 memory banks concurrently. The target interconnection network is a barrel-shifter.

Thanks to the approach proposed in this paper, we have been able to design the UWB interleaver which architecture complexity is the same than the approach proposed in [6]. This is due to the interleaving law which intrinsically supports a barrel-shifter as an interconnection network. This shows that our approach does not degrade the solution compared to [5], when the interleaving rule intrinsically respects the user-defined network constraint. On the contrary, the approach described in [18] and [13] generates an under optimized architecture (i.e. *2,3%* and *0,7%* bigger). This comes from the fact that they do not consider the network cost during the memory mapping process which results in higher complexity architectures: the network is thus composed of dozen of multiplexers and a complex controller. Finally, compared to a *full crossbar* architecture, our solution is *5,6%* smaller. Of course, with higher parallelism this difference would be increased, since the complexity of a barrel shifter evolves linearly while the crossbar evolution is quadratic.

For the non-binary LDPC which requires several accesses to the same data, the approaches presented in [6] and [12] cannot be used anymore because they are not able to deal with multiple data accesses as required in LDPC.

The architecture generated by using our approach is smaller (i.e. *5,6%*) than the architecture generated with the approach proposed in [13]. As already mentioned, [13] is not designed to take into account user-defined network constraint contrary to the method we propose in this paper. This explains why our approach is able to reduce the network complexity. This increases slightly the controller size, but with a marginal impact on the final area. Finally, compared to a *full crossbar* network architecture, our solution is *16,7%* smaller.

5. RELATED WORKS

Interleaving is a solution to arrange data in a non-contiguous way to increase the performances of the coding/decoding approaches like LDPC and Turbo Codes. The interleaver permutes extrinsic values produced by parallel processing elements and writes them into random or quasi random positions in different memory banks. Each decoder has to be fed with extrinsic information which does not originate from itself in order to improve the correctness of its

decisions [2], when two or more of these extrinsic values should be written into the same storage elements at the same clock cycle there is a collision. Moreover within a given standard, different interleaving laws can be used for different modes through varying frame lengths and/or data rates. Indeed, in state-of-the-art parallel turbo-decoding, interleaving is considered as a limiting factor concerning the overall system performance and the architectural cost. To tackle these problems different solutions have been proposed.

A first solution to get rid of conflicts with non-prunable interleavers consists in designing a specific interleaving law. In [2] the authors propose a deterministic methodology to design conflict-free interleavers. In [5] and [8] the authors consider spatial and temporal combinations as a permutation. In [9] the authors simply integrate the conflict-free constraint in the design of their interleaver. However, the multi-modes architectures (depending on the frame length, the data-rate...) cannot be handled by such approaches. Of course, all these solutions are reliable if and only if the designer is free to define his own permutation law.

A second approach consists in adding extra memory elements in the communication network. The aim is to buffer and to postpone the conflicts. In [11] the authors propose, in case of conflict during interleaved order access, to use additional buffers within the network in order to store conflicting data, until the targeted processor can process it. The additional network buffering resources, and the time needed to interleave information, increase with the number of conflicts. This is a suboptimal strategy, in terms of latency and thus throughput that avoids collisions at the expense of area and memory. Moreover, the communication is based on a regular network (i.e. Benes [10]) which might be more complex than a dedicated and optimized architecture.

In [12] the authors propose a solution based on software and/or reconfigurable architecture to obtain the required flexibility, but achieving lower throughput. In [14], an advanced heterogeneous communication network implementation was proposed. Two multistage interconnection network architectures are presented in order to handle on-chip communications in multiprocessor parallel turbo decoders. Both are based on a dedicated network and associated routers. The main feature of these network architectures (Benes and Butterfly based topologies) is their supposed scalability enabling seamless trade-off between hardware complexity and available bandwidth for turbo decoding. The Butterfly network, which lacks of diversity, is a multistage interconnection network with 2-input/2-output routers. There is a unique path between each source and destination. As a consequence, the risk of conflict is increased and the authors have to add queues to store conflicting data. The second proposed network architecture is based on a Benes network. In this case, the latency is constant for all the couples (source, destination), but this network avoids the conflicts if and only if all the paths have a different destination. Unfortunately, it has been shown that it was not true for turbo-decoding applications because interleaving (resp. de-interleaving) ends in potential conflicts. In [15] the authors propose another on-chip interconnection network based on the De Bruijn network. This network allows supporting the communication intensive nature of the application. The conflict accesses are avoided thanks to a dedicated routing algorithm.

In a third approach [13], it has been proved that there is no need for an ad-hoc code design to meet the parallelism requirements because for any scheduling of the reading/writing operations there is a suitable mapping of the variables in the memory that grants a conflict free access. This algorithm is based on two steps: First, a preliminary simple mapping function used to assign data to memory, if no conflict access is generated; if the data is not assigned to any memory bank. The second step consists in filling the blanks corresponding to the unmapped data thanks to a simulated annealing approach. As a consequence, the user cannot predict when the algorithm will end. In addition, the proposed approach neither targets the optimization of the storage elements, nor the optimization of the network.

In [6], the author proposes a methodology which generates a conflict free mapping of the variables in the memory banks and optimizes the resulting interleaving architectures. The main interest of the proposed algorithm is its lower complexity and the possibility to target user defined interconnection network. The resulting designs are compared with architectures generated with the approach proposed in [13], and the authors demonstrate the suitability of their approach. However, the targeted architecture can be achieved only if the interleaved scheme allows it; i.e. if the interleaving law is not intrinsically circular, then the resulting architecture cannot be based on barrel-shifters. Furthermore, this approach is only dedicated to turbo decoder [16]; it is not useable if data are processed more than once like in LDPC codes [17][22].

Some solutions based on a set of elementary memory elements (Registers, FIFO, LIFO), such as [19], have been proposed. But if these solutions are able to generate strongly optimized architectures, they cannot, to this day, target memory block based architecture. Moreover, this approach cannot target a user defined interconnection network. Finally, in [18] a more generic approach has been proposed which is able to deal with multiple data access designs. However this methodology is not able to target dedicated interconnection network architecture.

6. CONCLUSION

In this paper, we proposed an approach to design conflict-free parallel memory system. This approach relies on a formal modeling describing data communication constraints. It systematically allows the generation of a conflict-free memory mapping respecting a given network architecture. Our approach has been compared through industrial test-cases to the state-of-the-art techniques and its interest has been shown.

7. REFERENCES

[1] A. Seznec, J. Lenfant, "Interleaved parallel schemes", *IEEE Trans. Paral. Distrib. Syst., vol.5, no.12, p.1329-1334*, 1994.

[2] A. Giulietti, L. van der Perre, M. Strum, "Parallel turbo coding interleavers :avoiding collisions in accesses to storage elements", *Electro. Leters, vol.38, no.5, pp.232–234*, 2002.

[3] P. Urard, Y. Joonhwan, K. Hyukmin, A. Gouraud, "User needs", in High-Level Synthesis: from Algorithm to Digital Circuit, *Springer*, 2008

[4] A.Norton, E.Melton, "A class of boolean linear transformations for conflict-free power-of-two stride access", *Int. Conf. on Parallel Processing, pp. 247-254*, 1987

[5] D. Gnaedig, E. Boutillon, M. Jezequel, V.C. Gaudet, P.G. Gulak, "On multiple slice turbo codes", in proc.3rd *Int. Symp. TurboCodes, pp. 343-346*, Brest, 2003.

[6] C. Chavet, P. Coussy, E. Martin, P. Urard, "Static Address Generation Easing: a Design Methodology for Parallel Interleaver Architectures", *proc of the International Conference on Acoustics, Speech, and Signal Processing (ICASSP), pp. 1594-1597*, Dallas, 2010

[7] L. Dinoi, S. Benedetto,"Variable-size interleaver design for parallel turbo decoder architecture", *IEEE Trans. Communication, Vol.53, No11*, 2005.

[8] R. Dobkin, M. Peleg, R. Ginosar, "Parallel VLSI architectures and parallel interleaving design for low-latency MAP turbo turbo decoders", *Tech.RepCCIT-TR436*

[9] J. Kwak, K. Lee, "Design of dividable interleaver for parallel decoding in turbo codes," *Electron. Lett., vol. 38, no. 22, pp. 1362–1364*, 2002.

[10] V. E. Benes, "Mathematical Theory of connecting network and telephone trafic", New York, N.Y.: *Academic*, 1965.

[11] N. When, "SOC-Network for Interleaving in wireless Communications", *MPSOC*, 2004.

[12] A. La Rosa, C. Passerone, F. Gregoretti, L. Lavagno, "Implementation of a UMTS turbodecoder on dynamically reconfigurable platform", *DATE*, 2004.

[13] A. Tarable, S. Benedetto, and G. Montorsi, "Mapping interleaving laws to parallel turbo and LDPC decoder architectures", *IEEE Trans. Inf. Theory, vol.50, no.9, pp.2002-2009*, Paris, 2004.

[14] O. Muller, A. Baghdadi, M. Jezequel, "ASIP-based multiprocessor SoC design for simple and double binary turbo decoding", *DATE*, 2006

[15] H. Moussa, A. Baghdadi, M. Jezequel, "Binary de Bruijn on-chip network for a flexible multiprocessor LDPC decoder", 45th *Design Automation Conference, pp.429-434*, 2008.

[16] C. Berrou, A. Glavieux, P. Thitimajshima, "Near-Shannon limit error-correcting coding and decoding: Turbo codes", *Proc. IEEE Int. Conf. Commun., vol.2, pp.1064–1070*, Geneva, 1993.

[17] J.C. MacKay David, R.M. Neal, "Near Shannon limit performance of low density parity check codes", *Electronics letters*, July 1996.

[18] C. Chavet, P. Coussy, "A memory Mapping Approach for Parallel Interleaver design with multiples read and write accesses", *Proc. of the IEEE International Symp. on Circuits and Systems (ISCAS), pp.3168-3171*, Paris, 2010

[19] C. Chavet, P. Coussy, P. Urard and E. Martin, "A Methodology for Efficient Space-Time Adapter Design Space Exploration: A Case Study of an Ultra Wide Band Interleaver", *proc. of the IEEE International Symp. on Circuits and Systems (ISCAS), pp.2946*, New Orleans, 2007.

[20] C. Andriamisaina, P. Coussy, E. Casseau, C. Chavet, "High-Level Synthesis for Designing Multimode Architectures", *IEEE Trans. on Computer-Aided Design of Integrated Circuits and Systems, Vol.29, Issue:11, p.1736-1749*, 2010.

[21] IEEE 802.15.3a, WPAN High Rate Alternative

[22] I. Gutierrez, A. Mourad, J. Bas, S. Pfletschinger, G.Bacci, A.Bourdoux, H.Gierszal, "DAVINCI Non-Binary LDPC codes: Performance and Complexity Assessment", *proc of Future Network & Mobile Summit, Italy*, 2010.

[23] L. Conde-Canencia, E. Boutillon, A. Al-Ghouwayel, "Complexity comparison of non-binary LDPC decoders", *proc of ICT Mobile Summit, Spain*, 2009.

Distributed Sensor Data Processing for Many-cores

Jia Zhao, Russell Tessier and Wayne Burleson

Department of Electrical and Computer Engineering

University of Massachusetts, Amherst MA, USA

{jiazhao, tessier, burleson}@ecs.umass.edu

ABSTRACT

Future many-core systems will rely heavily on a wide variety of sensors which provide run-time information about on-chip environment and workload. In this paper, a new dedicated infrastructure for distributed sensor processing for many-core systems is described. This infrastructure includes a sparse array of dedicated processors which evaluate on-chip sensor data and a two-level hierarchical network-on-chip (NoC) which allows for efficient sensor data collection. This design is evaluated using benchmark driven simulations for a three-dimensional (3D) stack, necessitating inter-layer sensor data communication. The experimental results for up to 1024 cores indicate that for typical sensor data collection rates, one sensor data processor (SDP) per 64 cores is optimal for sensor data latency. The use of a two-level NoC is shown to provide an average of 65% sensor data latency improvement versus a flat sensor data NoC structure for a 256-core system.

Categories and Subject Descriptors

C.4 [**Performance of Systems**]: Reliability, availability, and serviceability

General Terms

Performance, Design, Reliability.

Keywords

Many-core, on-chip monitoring, distributed sensor processing.

1. INTRODUCTION

In the next few years many-core processors containing up to 1000 processor cores will become a reality [1]. Due to performance, power and reliability concerns, these massively parallel computing substrates will be required to evaluate an increasing amount of run-time information pertaining to error, thermal, process variation, wear-out, and supply voltage integrity issues, among others. The emergence of three-dimensional (3D) die stacking will further amplify the need for sensor information and corresponding remediation actions [2]. Currently, real-time system responses for multi-cores, including dynamic voltage and frequency scaling (DVFS), error recovery, and thermal remediation are performed locally and are often isolated within individual cores.

This work was funded by the Semiconductor Research Corporation under Task 2083.001

As system-on-chips (SoC) scale, both local and global techniques are needed to collect, collate, and use the information obtained from on-chip sensors [3]. These actions require multiple processors for deterministic and low latency sensor data processing.

Our approach to managing sensor data for many-cores involves providing architectural support for distributed sensor data collection and processing and system remediation on a chip-wide basis. In many-core systems, system temperature, voltage droop, processor activity, etc. need to be closely monitored and run-time remediation, such as DVFS, is invoked when necessary [4]. Recent advances in the use of sensor data include the use of processor performance signatures and performance counters to predict voltage droop emergencies and prevent thermal emergencies [3][5]. These advances motivate infrastructure for run-time management based on sensor processing components that share and distribute run-time signature, voltage, thermal, and error information. Unlike previous on-chip monitoring infrastructures [6][7], our architecture includes multiple processing components which are dedicated to sensor data analysis. A hierarchical network-on-chip (NoC), which interconnects the sensor data processors (SDPs), allows for both efficient sensor data collection and inter-SDP communication for shared sensor data. The approach is verified for many-core systems for up to 1024 cores in the core layer. A customized interconnect simulator is used to evaluate the communication infrastructure for sensor data. Additionally, the Graphite many-core simulator [1] is used to evaluate the many-core architecture for a collection of accepted benchmarks. A system-level experiment which examines the global distribution of voltage data and thermal information is performed to show the benefit of using our hierarchical infrastructure.

Experimental results show that our hierarchical sensor data communication infrastructure achieves up to an 80% latency reduction compared to a one-layer infrastructure. The system level benefit of our approach is shown using dynamic frequency scaling (DFS) for thermal management and voltage droop compensation. The results show an average many-core performance improvement of 6% using our hierarchical infrastructure, although higher rates of system temperature and supply voltage change will lead to higher benefits.

The remainder of this paper is organized as follows. Section 2 presents a brief background on many-core systems, 3D stacks and on-chip sensor data systems. Section 3 introduces our many-core sensor data collection and processing infrastructure. Section 4 discusses our experimental approach and experimental results are presented in Section 5. Section 6 concludes the paper.

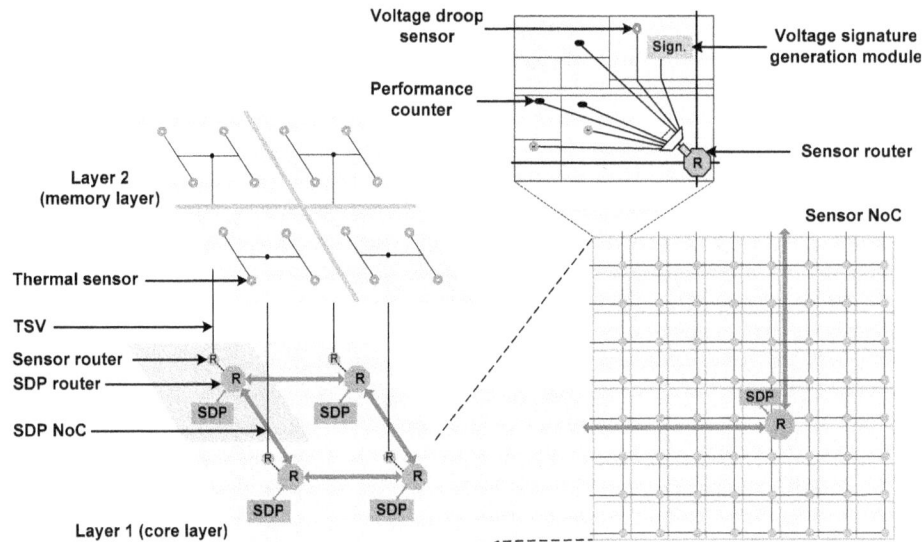

Figure 1. Hierarchical sensor data processing infrastructure for a 256-core system. Two distinct NoCs are shown. The sensor NoC (on the right) connects sensors to SDPs. The SDP NoC (on the left) connects SDPs. Sensors in the memory layer transfer data via TSVs.

2. Background

On-chip sensors are widely used in processors to closely monitor system temperature, performance, supply power fluctuation, and other environmental conditions. For example, the IBM Power7 includes 5 thermal sensors and 31 activity sensors per core [6]. Information from sensors, which is used to perform remediation techniques such as DVFS, voltage droop compensation and error rollback, presents a significant communication and processing workload. In many-cores, the impact of these communication and processing workloads is exacerbated by the highly-distributed nature of hundreds of cores. The dramatic expansion of sensor data necessitates a global and distributed view of sensor data processing.

Numerous remediation approaches based on on-chip sensors have been introduced. DVFS is widely used for processor thermal management in which the system frequency and/or voltage is reduced when a higher-than-threshold temperature is detected [3]. Supply voltage droops pose a threat to multi-core system reliability, thus DFS or voltage boosting needs to be enabled when a significant voltage droop is detected [8]. A recently-developed signature-based voltage droop compensation method detects signatures (a sequence of processor execution events) that are related to significant voltage droops and enables early prediction of incoming droops based on these signatures [8]. System reliability information measured as architectural vulnerability factors (AVF) are monitored at run-time and used to enable redundancy protection (e.g., dual modular redundancy) against soft errors when necessary [9]. Moreover, combinations of on-chip sensor information for multi-core remediation have also been explored [3][9].

Several approaches for on-chip sensor data collection and processing have been introduced. The IBM EnergyScale adaptive energy management approach [6] implements on-chip thermal and critical path sensors and performance counters for each core in an eight-core system. A microcontroller is used for sensor data processing. Intel AMT technology [10] uses a separate communication channel for remote discovery, healing and protection. In Wang, et al. [11], on-chip sensor data is transmitted using the existing NoC for regular inter-processor traffic.

However, none of these techniques is suitable for many-core systems with a large number of distributed on-chip sensors. A previous NoC-based infrastructure [7] for monitoring addressed some of these issues. This system, which is targeted at multi-cores, includes a low-dimensional NoC and up to two microcontrollers for centralized sensor data collection and processing. This earlier interconnect is organized as a flat two-dimensional mesh. No data exchange between the micro-controllers is supported in this infrastructure and the lack of a hierarchical interconnect significantly inhibits the scalability of the interconnect. Although this interconnect is sufficient for multi-cores, the greater throughput demands presented by many-cores motivates a new hierarchical interconnect approach.

The use of 3D stacking technology leads to additional challenges for sensor data collection [2]. Inter-layer communication is facilitated by the use of through silicon vias (TSVs). The total number of TSVs is limited. Most 3D implementations focus on layering memory on top of a processor core layer [2][4]. Although it is expected that multiple stacked core layers will be implemented in the future, this work mainly considers two layer stacks including a memory and a core layer.

3. Distributed Sensor Data Collection and Processing

An overview of the hierarchical infrastructure for distributed sensor data collection and processing is shown in Figure 1 for a two-layer 3D stack many-core implementation. This dedicated interconnect and sensor data processor infrastructure, which only handles sensor data, contains two levels of NoC routers and a series of SDPs. The NoC infrastructure, SDPs, and most of the sensors are implemented in the core layer while thermal sensor data in the memory layer are accessed through TSVs. SDPs can be implemented from available regular cores.

3.1 Hierarchical Sensor Data Interconnect Infrastructure

The sensor data interconnect infrastructure consists of two levels of NoC-style routers. On-chip sensors in each core are connected to a minimalistic packet router through a multiplexer, as shown in the top, right in Figure 1. These routers, called sensor routers, are connected together in a mesh, as shown in the bottom, right in Figure 1. Sensor routers (one per core) send the collected sensor data to an SDP through the sensor NoC. Data from thermal sensors in the memory layer are also collected by the SDP through the sensor NoC in a slightly different fashion. Adjacent thermal sensors (4 in the example shown in Figure 1, not necessarily the number used in real systems) in the memory layer are connected to a multiplexer which sends its output to a serializer. The thermal data are received in the core layer, de-serialized, sent to a sensor router, and subsequently forwarded to the SDP. This approach only uses one TSV for each vertical connection between the serializer in the memory layer and the de-serializer in the core layer. Ten cycles [12] are required to transmit the 8-bit thermal data from the memory layer to the sensor NoC.

The sensor router implemented in this infrastructure has a small data width (sensor data packet width) and a shallow input buffer (e.g., 24-bit width and 4 flits, much smaller than the 256-bit width and 8-16 flits used in standard NoCs). The sensor router supports data packets with two priority levels using two virtual channels. Packets in the priority channel have higher routing priority than those in the regular channel. Emergency sensor data packets (such as an alert for a significant voltage droop) are transmitted in the priority virtual channel to avoid congestion, thus it has lower latency. The widely-used XY routing algorithm is implemented in the sensor router [13]. Each packet generated by the sensor router includes a time stamp which indicates its generation time. An SDP manages sensor NoC-transferred data from a relatively small number of cores (64 in the example shown at the bottom, right of Figure 1) and is physically placed in the center of these cores to reduce sensor data transmission latency.

As processor counts scale to many-cores, there is a need for the SDPs to quickly share data, as will be shown in the next subsection. This need motivates a second interconnect layer between SDPs to reduce the number of hops needed for transmitting packets between SDPs, as shown in the bottom, left in Figure 1. The communication among SDPs is facilitated by the SDP NoC using very low overhead SDP routers interconnected in a mesh, effectively forming a lightweight higher-level network. The SDP router in our infrastructure is interfaced to both the sensor NoC and the SDP NoC. Sensor data packets from sensor routers are processed by an SDP and, when appropriate (determined by the SDP), sent to adjacent SDPs. Additional SDP router details are explained in Section 3.3.

3.2 Packet Transmission in the SDP NoC

In many cases, sensor data does not need to be shared across multiple cores and can be used locally. For example, for thermal management [4], per-core architectural adaptation is employed to reduce individual core temperatures based on thermal sensor information collected in each core.

However, some recent many-core remediation approaches, such as DVFS in response to hotspot detection and voltage boost in response to voltage droop, require the sharing of sensor information on a global scale, which includes aggregations over regions of different scales and broadcasts. For example, global scale hotspot remediation requires the transfer of thermal information and performance counts to a centralized location [3] and voltage droop recovery requires the broadcast of voltage sensor data. Most NoC-based multi-core and many-core systems do not have broadcast support in the NoC, since the NoC is mainly used for accessing shared memory.

In our system for the hotspot case, thermal sensor data packets are sent from all SDPs to a central SDP (aggregation at the chip scale

[3]) which determines frequency change decisions. For the voltage droop case, voltage sensor data packets are broadcast among all SDPs. Voltage droop problems affect every core in a many-core system since they are on the same power grid [8]. Thus, a dangerous voltage droop detected in one core should be known by all other cores so that remediation, such as voltage boost or frequency reduction, can be enabled globally [5].

The SDP NoC facilitates both of these traffic patterns. The XY routing algorithm is implemented in the SDP router. Hotspot traffic is supported by this routing algorithm. The SDP router supports packet broadcast using an accepted approach [14]. A packet is first sent vertically to all nodes (along the Y axis) across the mesh. Then, all the nodes that currently have the packet send it out horizontally (to the left and right along the X axis). In a mesh network of size n×n, the position of a router is represented as (x,y) in which $0<=x,y<n$ and x, y are both integers. There are two scenarios in which new packets need to be generated. At the source SDP router, up to four new packets are generated and sent to router (x,0), (x,n-1), (0,y) and (n-1,y). At routers on the same row as the source router, new broadcast packets are generated and sent to router (0,y) and (n-1,y). These new packets follow the XY routing algorithm to their destinations.

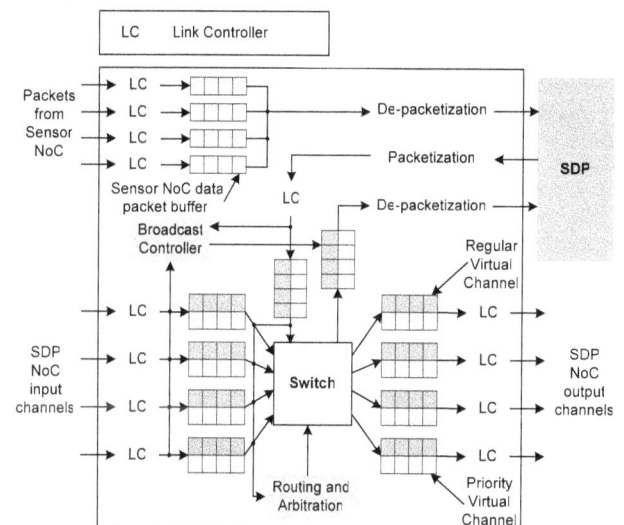

Figure 2. SDP router structure. The structure is simplified from more standard NoCs. Each packet has minimal bit width (24 bit maximum) and storage buffers are shallow (6 flits)

3.3 SDP Router Design

The structure of the SDP router in our infrastructure is shown in Figure 2. The link controller in this figure is responsible for controlling when packets can be sent to/from the buffer based on the usage of the buffer. The SDP router is interfaced to both the sensor NoC and the SDP NoC using shallow buffers (about 6 flits). An SDP write path to the sensor NoC is unneeded so only an input path is provided. Sensor data is extracted from these packets in the de-packetization module and sent to the SDP. Outdated regular sensor data packets, as calculated from the time stamp in the packet, can be discarded (e.g., thermal sensor data packet from the previous sampling period). A crossbar is used to implement the switch.

Similar to the sensor router structure, two virtual channels for regular (e.g., thermal sensor data) and priority packets (e.g. high voltage droop alerts) are implemented in each input and output buffer. The XY routing algorithm is used in the routing and arbitration module. This structure is enhanced with a broadcast controller for performing broadcasts. When a broadcast packet is received at any router buffer, it is sent to the output buffer for the SDP. At the same time, the broadcast control module decides whether new packets need to be generated, based on the method explained in Section 3.2. These new packets have the same sensor data as the original packet but with different destinations.

4. Experimental Approach

A series of simulation and synthesis evaluations were performed to show the benefit of using our infrastructure for 256, 512 and 1024 many-core systems with one memory layer and one core layer in a two layer stack. Two specific sensor data interconnect approaches are considered, the hierarchical approach shown in Figure 1 and a flat sensor data interconnect that consists only of sensor routers (similar to the infrastructure shown in Figure 1 but *without* the SDP NoC). A packet transmitted between two neighboring SDPs in the flat sensor NoC infrastructure needs to go through several sensor routers. The Popnet simulator [15] is heavily modified to model both the new hierarchical infrastructure and the flat sensor NoC infrastructure for many-core systems.

To estimate the overhead of our infrastructure, synthesizable hardware models of the sensor NoC and SDP routers were developed. The hardware models were synthesized by Synopsys Design Compiler using a 45nm standard cell library [16]. The system-level effect of a many-core system with a core layer and a separate DRAM (memory) layer was modeled using the Graphite many-core simulator [1] with a previously-determined memory access latency number for a 3D stack [2]. The performance calculation module in Graphite has been modified to accommodate run-time frequency changes and to report the overall performance of the system with dynamic frequency scaling. The system frequency is set to 1 GHz. The temperature effect of stacking a DRAM layer on top of the core layer is estimated with a highly-accurate many-core temperature estimation method based on the power consumption of all cores. It is assumed that the heat sink is below the core layer. Thus, the temperature in the core layer is proportional to the power consumption in both the core layer and the DRAM layer [2].

The 128-core architecture used in the system level experiment is scaled up from an 8-core UltraSPARC T1 architecture consuming 115mm^2 using 90nm technology [4]. The total area of the 128 core system is estimated to be 460mm^2 using 45nm technology. The power modeling method used in [4] is adopted in our experimentation and scaled to 45nm technology. A maximum value of 5.24 W/core was determined for a 45nm 128-core system. The DRAM layer has the same area as the core layer and hosts 2 GB memory [2]. The DRAM power consumption is set to 1 W/GB [17].

5. Experimental Results

A series of simulations was first performed to find the optimal number of cores per SDP and to show the benefit of using our hierarchical infrastructure versus a flat sensor NoC infrastructure for sensor data distribution. A system-level experiment was then performed for the many-core system using our infrastructure for thermal and voltage droop sensor data transmission.

5.1 On-chip Sensor Setup

In a series of simulations, sensor data from thermal sensors, processor performance counters, and voltage droop sensors were considered based on previously-reported instantiation and sampling rates in multi-core systems. Eight thermal sensors [7], 18 performance counters [3], a voltage droop signature capturing structure and a voltage droop sensor [8] are used in each core. The

total number of thermal sensors in the DRAM layer is 128 (one thermal sensor per 128Mb DRAM [18]). Hardware synthesis indicates that our infrastructure can run at 1 GHz (both the sensor NoC and the SDP NoC).

Thermal data, performance counter data and voltage droop signature data are transmitted in the regular sensor NoC channel. The thermal sensor data injection rate per sensor is based on the maximum temperature rise rate of 10°C/ms [19] and the thermal sensor resolution of 0.1°C [5], which leads to a sample period of 10,000 cycles (1/10,000 cycle injection rate). The performance counter injection rate is 1/3,000,000 cycles [3]. A voltage droop signature injection rate of 1/4,000 cycles is used [8]. Thus, the sensor router regular channel data injection rate is 1/947 cycle based on 8 thermal sensors, 18 performance counters and 1 signature capturing structure. To reduce the total number of TSVs used for transmitting sensor data from the DRAM layer, every 8 thermal sensors in the DRAM are connected to a sensor router. Thus, there are 16 sensor routers that also receive thermal information from sensors in the DRAM layer. These sensor routers are evenly distributed in the core layer. The sensor packet injection rate at these routers is 1/539 cycle based on 16 thermal sensors, 18 performance counters and 1 signature capturing structure. The voltage droop sensor sends out an alert when a dangerous voltage droop happens and this packet is transmitted in the priority channel. A very aggressive injection rate of 1/108 is used for the voltage droop sensor [8].

The SDP NoC traffic includes performance counter and voltage droop broadcast traffic. The performance counter data in the *n* cores managed by a SDP are sent to the central SDP. Every signature used in the core layer is broadcast to all SDPs. Voltage droop alerts are also broadcast in the SDP NoC using the priority channel. The sensor NoC and SDP NoC data widths are set to 24 bits [7]. The regular packet size for thermal sensor and performance counter values is 1 flit [9] while the signature information requires 3 flits [8]. The buffer sizes in the sensor NoC and the SDP NoC are set to 6 flits.

5.2 Core to SDP Ratio Experiment

In the first experiment, we simulated the infrastructure introduced in this paper with varying numbers of cores per SDP. The total number of SDPs decreases as the number of cores per SDP increases. The latency and hardware cost (including wire area) results using varying cores-per-SDP ratios for 256, 512 and 1024 core systems are shown in Table 1. The packet latency numbers shown in this table are for packets transmitted using both the sensor NoC and the SDP NoC. The TSV latency has been included in this experiment.

As shown in Table 1, the average sensor NoC latency (sensor-to-SDP latency) increases as the cores-per-SDP ratio increases since there are more sensors connected to the each SDP through the sensor NoC. The SDP NoC latency (SDP-to-SDP latency) decreases as the ratio increases since the size of the SDP NoC is reduced. The minimum overall latency is located in the middle of the extremes, although the latency differences are relatively small. The SDP NoC hardware cost decreases as the cores-per-SDP ratio increases (fewer SDP routers and shorter interconnect).

A cores-to-SDP ratio of 64 is used in the following experiments since it provides low latency and the highest capacity for sensor packets while requiring moderate hardware cost (less than 6% increase versus sensor NoC-only in the 1024 core system). The hardware cost of both NoCs together is less than 1.5% of the overall many-core hardware area for all the cases, since the data width and buffer size for both NoCs are small and a simplified router structure (versus standard NoC routers) is used, as explained in Section 3.1 and 3.3.

Table 1. Average latency (in cycles) and hardware costs associated with varying core count per SDP

Core num.	SDP num.	Core/ SDP ratio	Sensor NoC lat.	SDP NoC lat.	Total lat.	SDP+sensor NoC to sensor NoC-only HW increase (%)
256	32	8	7.09	18.83	25.92	24.98
	16	16	9.35	13.62	22.97	14.20
	8	32	12.25	11.01	23.26	7.94
	4	64	17.31	7.72	25.03	4.19
	2	128	24.14	5.25	29.39	1.72
512	64	8	7.09	24.39	31.48	25.93
	32	16	9.35	19.34	28.69	15.09
	16	32	12.25	13.37	25.62	8.78
	8	64	17.31	11.42	28.73	4.99
	4	128	24.14	7.88	32.02	2.65
1024	128	8	7.09	39.40	46.49	26.22
	64	16	9.35	26.01	35.26	15.46
	32	32	12.25	20.87	33.12	9.20
	16	64	17.31	15.48	32.79	5.46
	8	128	24.14	11.84	35.98	3.15

Table 2. Average latency (in cycles) comparison

Core and SDP num.	Latency type	Flat sensor NoC	Our method	Latency reduction w.r.t. flat sensor NoC (%)
256 core (4 SDP)	Inter-SDP	45.43	7.72	83.01
	Total	62.82	25.03	60.16
576 core (8 SDP)	Inter-SDP	67.57	11.42	83.10
	Total	84.96	28.73	66.18
1024 core (16 SDP)	Inter-SDP	90.36	15.48	82.87
	Total	107.75	32.79	69.57

5.3 Comparison against the Flat Sensor NoC Infrastructure

In this experiment, the hierarchical approach described in this paper is compared against a flat one-layer (sensor NoC-only) sensor data interconnect. As shown in Table 2, our hierarchical infrastructure achieves 60%, 66% and 70% total latency (SDP NoC latency + sensor NoC latency) reduction versus the flat sensor NoC infrastructure for 256, 512, and 1024 cores, respectively. The sensor NoC latency in both infrastructures barely changes as the total core number increases since each SDP manages the same number of sensors as the total core count increases.

The inter-SDP sensor packet transmission latency makes the difference. The SDP NoC latency in our infrastructure is over 80% lower versus the inter-SDP latency in the flat sensor NoC infrastructure. As the total core number increases from 256 to 1024, the difference becomes larger. As explained in Section 3, SDPs are directly connected in our hierarchical infrastructure while inter-SDP packets in the flat sensor NoC infrastructure needs to go through numerous sensor routers (at least 8 sensor routers for neighboring SDPs in this experiment).

The throughput of the SDP NoC in 256, 512 and 1024 core systems is shown in Figure 3. For comparison, the throughput of the inter-SDP traffic in the flat sensor NoC infrastructure is also shown. The sensor NoC and the vertical communication structure are not included in this simulation since they yield the same throughput in both infrastructures.

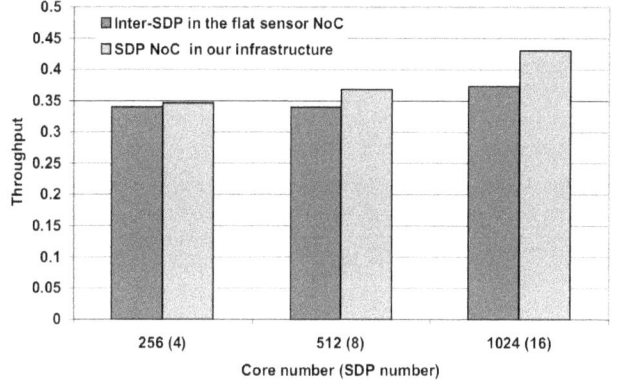

Figure 3. Throughput (packet/SDP router/cycle) comparison

The throughput of the SDP NoC is defined in Equation (1) [20]. P_{total} is the total number of packets received during the simulation, N is the number of routers in the SDP NoC and C is the average number of cycles to route all the packets, which is from the simulation.

$$Throughput = \frac{P_{total}}{N \times C} \quad (1)$$

Figure 3 shows that the SDP NoC has higher throughput for all simulated systems. The difference increases as the number of cores increases since the broadcast requirements of the sensor data negatively impacts the sensor NoC-only case.

5.4 Using DFS for Thermal Management and Voltage Droop Compensation

In a system level experiment, a 2-layer 128-core system with integrated voltage droop sensors and thermal sensors is simulated. A voltage droop alert sent by the voltage droop sensor is broadcast to all SDPs. The frequency of all cores is reduced by half during a voltage droop [5]. Thermal sensor data is processed in the local SDP only. The system frequency is reduced by half when the temperature is over 85°C [4].

The sensor setup includes 1 voltage droop sensor per core [5] and 8 thermal sensors per core. The DRAM layer with 2 GB capacity has 128 thermal sensors, as explained in Section 5.1. No voltage droop sensors are used in the memory layer. The injection rate of the voltage droop sensor and thermal sensor is 1/108 and 1/10,000 cycle respectively, as explained in Section 5.1. In this experiment, each SDP manages sensors in 32 cores, thus 4 SDPs are used in this 128-core system. The hardware cost of adding the SDP NoC is less than 7% of the overall monitoring system hardware cost (sensor NoC, SDP, and SDP NoC). The overall hardware cost of our infrastructure is only 0.94% of the total chip (core layer).

Three cases are considered in this experiment.

1) Case 1: DFS for thermal management only. Without on-chip voltage droop sensors, the system voltage is conservatively set to 1.1V [7].
2) Case 2: DFS for thermal management and voltage droop compensation using the flat sensor NoC.
3) Case 3: DFS for thermal management and voltage droop compensation using the hierarchical infrastructure introduced in this paper.

The latency of sensor data transmission is simulated using the modified Popnet simulator, as described in Section 4. This simulation shows that the latencies for transmitting voltage droop sensor information using the flat sensor NoC infrastructure and our infrastructure are 62 cycles and 21 cycles, respectively. Using the system voltage calculation method described in [7] and scaling it to 45nm technology, the system voltage is set to 1.02V and 1V for case 2 and 3 respectively. Given the same processor activity, the maximum temperature difference from these two voltages is close to 6°C using the temperature estimation method described in Section 4. Case 1 is used as a baseline case for comparison.

Table 3. Results of the system level experiment using DFS for thermal management and voltage droop compensation

Benchmark	Perf. (billion cycles)			Benefit (%)	
	Case 1	Case 2	Case 3	Case 2	Case 3
LU (non-contig)	89.33	81.41	79.46	8.87	11.05
Ocean (non-contig)	3.49	3.26	3.25	6.56	6.74
LU(contig)	24.23	23.70	22.21	2.20	8.32
Ocean(contig)	2.76	2.54	2.51	7.79	9.09
Radix	9.78	9.29	9.08	5.03	7.18
FFT	115.14	115.06	114.73	0.07	0.36
Cholesky	189.66	185.05	182.28	2.43	3.89
Radiosity	121.42	114.71	111.28	5.53	8.35

The results are shown in Table 3. The performance is represented as the total cycles for the 128 core system (sum of execution time in each core) to finish the benchmark. The execution time difference shown in Table 3 for case 2 and 3 are with respect to the execution time in case 1. The system using our infrastructure (case 3) achieves an average 6.8% performance benefit compared to the system without on-chip voltage droop sensors (case 1). The performance benefit is up to 6% higher compared to a system with the flat sensor NoC infrastructure (case 2). Case 3 uses the lowest supply voltage, which leads to the lower system temperature and less DFS enable time using our infrastructure. The benefit is small with the FFT benchmark since the system temperature is almost always below the threshold.

6. Conclusion

A dedicated infrastructure for distributed sensor data collection and processing for many-core systems is introduced. This infrastructure features a hierarchical NoC that supports two types of sensor data traffic. Our infrastructure achieves more than 50% latency reduction versus a flat NoC infrastructure in many-core systems. The system level performance benefit of using our infrastructure is up to 6% versus a nominal system.

7. References

[1] J. Miller, et al., "Graphite: A distributed parallel simulator for multicores," in *Proc. Int'l Symp. on High Performance Computer Architecture*, pp. 1-12, Jan. 2010.

[2] G. Loh, "3D-stacked memory architectures for multi-core processors," in *Proc. Int'l Symp. on Computer Architecture*, pp. 453-464, Jun. 2008.

[3] R. Jayaseelan, et. al., "A hybrid local-global approach for multi-core thermal management," in *Proc. Int'l Conf. on Computer-Aided Design*, pp. 314-320, Nov. 2009.

[4] A. Coskun, et al., "Dynamic thermal management in 3D multicore architectures," in *Proc. Design, Automation & Test in Europe Conf.*, pp. 1410-1415, Apr. 2009.

[5] J. Zhao, et al., "Thermal-aware voltage droop compensation for multi-core architectures," in *Proc. Great Lakes Symp. on VLSI*, pp. 335-340, May 2010.

[6] M. Floyd, et al., "Adaptive Energy Management Features of the POWER7 Processor," [online]. Available: http://www.research.ibm.com/people/l/lefurgy/Publications/hotchips22_power7.pdf.

[7] S. Madduri, et al., "A monitor interconnect and support subsystem for multicore processors," in *Proc. IEEE Design Automation & Test in Europe Conf.*, pp. 761-766, Apr. 2009.

[8] V. Reddi, et. al. "Voltage emergency prediction: A signature-based approach to reducing voltage emergencies," in *Proc. Int'l Symp. on High-Performance Computer Architecture*, pp. 18-27, Feb. 2009.

[9] R. Vadlamani, et al., "Multicore soft error rate stabilization using adaptive dual modular redundancy", in *Proc. Design Automation and Test in Europe Conf.*, pp. 27-32, Mar. 2010.

[10] "Intel Active Management Technology," [online]. Available: http://www.intel.com/technology/platform-technology/intel-amt/

[11] Y. Wang, et al., "Performance evaluation of on-chip sensor network (SENoC) in MPSoC," in *Proc. Int'l Conf. on Green Circuits and Systems*, pp. 323-327, Jun. 2010.

[12] S. Pasricha, "Exploring serial vertical interconnects for 3D ICs," in *Proc. Design Automation Conf.*, pp. 581-586, Jul. 2009.

[13] M. Yang, et. al., "Incremental design of scalable interconnection networks using basic building blocks," in *Proc. IEEE Symp. on Parallel and Distributed Processing*, pp. 252-259, 25-28, Oct. 1995.

[14] E. Modiano, et. al., "Efficient algorithms for performing packet broadcasts in a mesh network," in *IEEE Trans. on Networking*, vol. 4, no. 4, pp. 639-648, Aug. 1996.

[15] L. Shang, et. al., "Dynamic voltage scaling with links for power optimization of interconnection networks," in *Proc. Int'l Symp. on High-Performance Computer Architecture*, pp. 91-102, Feb. 2003.

[16] The Nangate 45nm Open Cell Library [online]. Available: http://www.nangate.com

[17] Samsung Green DDR3, [online]. Available: http://www.samsung.com/global/business/semiconductor/Greenmemory

[18] C. Kim, et al., "CMOS temperature sensor with ring oscillator for mobile DRAM self-refresh control," in *Proc. Int'l Symp. on Circuits and Systems*, pp. 3094-3097, May 2008.

[19] G. Zhao, et al., "Processor frequency assignment in 3D MPSoCs under thermal constraints by polynomial programming," in *Proc. Asia Pacific Conf. on Circuits and Systems*, pp. 1668-1671, Nov. 2008.

[20] A. Weldezion, et al., "Scalability of network-on-chip communication architecture for 3-D meshes," in *Proc. Int'l Symp. on Networks-on-Chip*, pp. 114-123, May 2009.

CMOS Compatible Many-Core NoC Architectures with Multi-Channel Millimeter-Wave Wireless Links

Sujay Deb, Kevin Chang, Miralem Cosic,
Partha Pande, Deukhyoun Heo, Benjamin Belzer
School of Electrical Engineering & Computer Science,
Washington State University, Pullman, WA, USA
{sdeb, jchang, mcosic, pande, dheo, belzer}
@eecs.wsu.edu

Amlan Ganguly
Department of Computer Engineering
Rochester Institute of Technology
Rochester, New York, USA
amlan.ganguly@rit.edu

ABSTRACT

Traditional many-core designs based on the Network-on-Chip (NoC) paradigm suffer from high latency and power dissipation as the system size scales up due to their inherent multi-hop communication. NoC performance can be significantly enhanced by introducing long-range, low power, and high-bandwidth single-hop wireless links between far apart cores. This paper presents a design methodology and performance evaluation for a hierarchical small-world NoC with CMOS compatible on-chip millimeter (mm)-wave wireless long-range communication links. The proposed wireless NoC offers significantly higher bandwidth and lower energy dissipation compared to its conventional non-hierarchical wired counterpart in presence of both uniform and non-uniform traffic patterns. The performance improvement is achieved through efficient data routing and optimum placement of wireless hubs. Multiple wireless shortcuts operating simultaneously provide an energy efficient solution for design of many-core communication infrastructures.

Categories and Subject Descriptors

C.5.4 [**Computer System Implementation**]: VLSI Systems

General Terms

Performance, Design.

Keywords

Many-Core, NoC, Wireless, Small-World.

1. INTRODUCTION

The predicted evolution of many-core *Systems-on-chip* (SoCs) indicates a manifold increase in the number of cores on a single die over the next few years. High performance and low power are crucial for the widespread adoption of such many-core platforms. Achieving these goals cannot be attained by traditional paradigms and we are forced to re-think the basis of designing such systems, in particular the overall interconnect architecture. *Networks-on-Chip* (NoCs) have emerged as communication backbones to enable a high degree of integration in many-core SoCs. Despite their advantages, an important performance limitation in traditional NoCs arises from planar metal interconnect-based multi-hop links, where the data transfer between two distant blocks causes high latency and power consumption. A scalable solution can be achieved by drawing inspiration from the small-

world property possessed by many natural complex networks. Such small-world networks are known for having low average inter-nodal distances with limited resources. This is achieved by adding a few long-range links in a regular lattice, resulting in a significantly lower average hop count. An attempt toward constructing small-world NoCs has been made with metal wires in the past [1]. However, that approach doesn't scale up because multi-hop wired links are required for longer distances. This paper evaluates the performance of a small-world NoC with multiple non-overlapping millimeter (mm)-wave wireless channels as long-range links. These on-chip wireless links are CMOS-compatible and do not need any new technology. But they have antenna and transceiver area and power overheads. Thus, to achieve the best performance, the wireless resources need to be optimally placed and used. To accomplish that goal, a hybrid and hierarchical network where nearby cores communicate through traditional metal wires, but long distance communications are predominantly achieved through high performance single-hop wireless links is implemented.

This paper shows that network performance can be significantly improved by using a hybrid approach and by placing and using multiple simultaneously operating and non-overlapping wireless shortcuts optimally. The wireless link insertion also takes into account the target application or the traffic pattern the chip is designed for. Furthermore, we employ a distributed flow-control-based routing algorithm that ensures an optimum utilization of the wireless links. We demonstrate that the proposed mm-wave wireless NoC (mWNoC) outperforms its more traditional non-hierarchical wire line counterparts in terms of sustainable data rate and energy dissipation. It also performs better than other emerging NoC architectures like, NoC with RF-I and 3D NoC.

2. RELATED WORK

Conventional NoCs use multi-hop packet switched communication. It is shown in [2] that by using virtual express lanes to connect distant cores in the network, it is possible to avoid the router overhead at intermediate nodes, and thereby greatly improve NoC performance. Performance improvements have also resulted by inserting long range wired links following principles of small-world graphs [1].

The design principles of photonic NoC are elaborated in various recent publications [3][4]; photonic NoCs are estimated to dissipate significantly less power than electronic NoCs.

The amalgamation of two emerging paradigms, namely NoCs in a 3D IC environment, allows for the creation of new structures that enable significant performance enhancements over traditional solutions [5]. Despite these benefits, 3D architectures pose new technology challenges and the heat dissipation is a serious concern due to increased power density [6] on a smaller footprint.

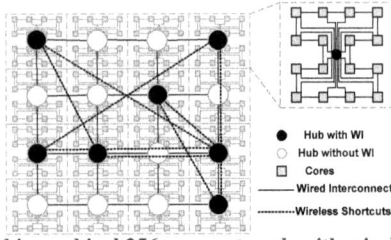

Figure 1. A hierarchical 256-core network with wireless shortcuts.

Another alternative is NoCs with multi-band RF interconnects [7], wherein electromagnetic (EM) waves are guided along on-chip transmission lines created by multiple layers of metal and dielectric stack. As the EM waves travel at the effective speed of light, low latency and high bandwidth can be achieved.

Recently, design of a wireless NoC based on *CMOS Ultra Wideband* (UWB) technology was proposed [8]. In [9], the feasibility of on-chip wireless communication networks with miniature antennas and simple transceivers that operate at the sub-THz range of 100-500 GHz has been demonstrated. With further increase in transmission frequencies the size of antenna can be reduced, occupying much less chip real estate. One possibility is to use nanoscale antennas based on *carbon nanotubes* (CNTs) operating in the THz/optical frequency range [10]. The design of a small-world wireless NoC operating in the THz frequency range using CNT antennas is elaborated in [11]. Though CNT based NoCs offer orders-of-magnitude performance improvements over traditional wire line NoCs, the integration and reliability of CNT devices need more investigation. A preliminary NoC architecture with CMOS-compatible mm-wave wireless links was proposed in [12], where all the communicating nodes share a single wireless channel and hence the performance gain is limited.

This work introduces a comprehensive design methodology for a hierarchical and small-world mWNoC with multiple simultaneously operating wireless shortcuts. Through efficient distributed flow control based routing and optimal placement of wireless hubs, the mWNoC significantly outperforms traditional multi-hop NoCs.

3. PROPOSED NOC ARCHITECTURE

We propose design of a hierarchical NoC architecture with a limited number of wireless shortcuts strategically placed for optimum performance. Our goal is to use the small-world approach to build a highly efficient NoC with both wired and wireless links. The small-world topology can be incorporated in NoCs by introducing long-range, high bandwidth and low power wireless links between distant cores. The system is divided into multiple small clusters of neighboring cores called subnets. These subnets have NoC switches and links as in a standard NoC. The cores are connected to a centrally located hub through direct links and the hubs from all the subnets are connected in a 2nd level network forming a hierarchical structure. This is achieved by interconnecting adjacent hubs with wireline links and introducing a few long range mm-wave wireless links between distant hubs according to the placement scheme. The hubs connected through wireless links require wireless interfaces (WIs). As described in section 3.1, a simulated annealing (SA) algorithm is used to optimally place the WIs so as to establish optimal overall network topology under given resource constraints, i.e., a limited number of WIs. Figure 1 shows a representative hierarchical 256-core network where the subnets have a Ring-Star (a ring with a central hub connecting to every core) topology and it has 16 hubs and 7 WIs. The hubs are connected in mesh architecture with overlaid long-range wireless shortcuts on the 2nd level of the

hierarchy. In this paper Mesh-RingStar architecture is used as an example since it is shown to provide the best performance-overhead tradeoff among several possible mWNoC architectures [13].

3.1 Optimum Placement of WIs

WI placement is crucial for optimum performance as it establishes high-speed, low-energy interconnects on the network. If there are N hubs in the network and n WIs to distribute, the size of the search space S is given by

$$|S| = \binom{N}{n}. \tag{1}$$

Thus, with increasing N, it becomes increasingly difficult to find the best solution by exhaustive search. It is shown in [11] that for placement of wireless links in a NoC, an SA based methodology converges to the optimal configuration much faster than exhaustive search. Hence, in the interest of scalability we adopt SA [14] based optimization for WI placement to get maximum benefits of using the wireless shortcuts. Initially, the WIs of all frequency ranges are placed randomly with each hub having equal probability of getting a WI.

Once the network is initialized randomly, an SA based optimization step is performed. Since only hubs contain WIs, the optimization is performed solely on the 2nd level network of hubs. The number of WIs sharing the same frequency channel is kept equal for different frequency bands along with one gateway hub (which can operate in all frequency channels). Multiple non-overlapping wireless channels are distributed among n WIs and WIs sharing the same channel form a cluster. To perform SA, an optimization metric μ is established, which is closely related to the connectivity of the network. The metric μ is the average distance, measured in number of hops, between all source and destination hubs. A single hop in this work is defined as the path length between a source and destination pair that can be traversed in one clock cycle. To compute μ the shortest distances between all pairs of hubs are computed. The distances are then weighted with the normalized frequencies of communication between hub pairs. The optimization metric, μ can be computed as

$$\mu = \sum_{i,j} h_{ij} f_{ij}, \tag{2}$$

$$h_{ij} = p * d_{i,j\_with\_shortcut} + (1-p) * d_{i,j\_without\_shortcut} \tag{3}$$

where h_{ij} is the distance (in hops) between the ith source and j^{th} destination. The frequency f_{ij} of communication between the ith source and j^{th} destination is the normalized apriori probability of traffic interactions between subnets determined by particular traffic patterns depending upon the application mapped onto the NoC. The probability of getting access to the wireless channel for communication between any source-destination pair is designated by p which is inversely proportional to the number of WIs in a cluster (n_c) sharing the same frequency channel. With the assumption that all the WIs are equally likely to have access to wireless channel in a cluster, p can be computed as

$$p = 1/n_c. \tag{4}$$

The distance ($d_{i,j}$) between source and destination varies depending on whether or not wireless shortcuts are used while routing. In this case, equal importance is attached to inter-hub distance and frequency of communication.

4. COMMUNICATION SCHEME

This section describes the WI components and the adopted routing strategy.

4.1 Wireless Interface

The two principal WI components are the antenna and the transceiver, whose characteristics are outlined below.

4.1.1 On-Chip Antennas

The on-chip antenna for the proposed mWNoC has to provide the best power gain for the smallest area overhead. A metal zigzag antenna [15] has been demonstrated to possess these characteristics. This antenna also has negligible effect of rotation (relative angle between transmitting and receiving antennas) on received signal strength, making it most suitable for mWNoC applications. Zigzag antenna characteristics depend on physical parameters like axial length, trace width, arm length, bend angle, etc. By varying these parameters antennas are designed to operate on different non-overlapping frequency channels in this work.

4.1.2 Wireless Transceiver Circuits

To ensure high throughput and energy efficiency, the mWNoC transceiver circuitry has to provide a very wide bandwidth as well as low power consumption. In designing the wireless transceiver, low power design considerations are taken into account at the architecture level with a design adopted from [16]. The detail description of the transceiver circuit is out of the scope of this paper. Non-coherent on-off keying (OOK) modulation is chosen, as it allows relatively simple and low-power circuit implementation. The transmitter (TX) consists of an up-conversion mixer and a power amplifier (PA). In the receiver (RX), a direct-conversion topology is used, consisting of a low noise amplifier (LNA), a down-conversion mixer and a baseband amplifier. An injection-lock voltage-controlled oscillator (VCO) is reused for TX and RX. With both direct-conversion and injection-lock technology, a power-hungry phase-lock loop (PLL) is eliminated.

4.2 Adopted Routing Strategy

In this proposed hierarchical NoC, intra-subnet data routing depends on the Ring-Star subnet topology. In the subnet, if the destination core is within two hops on the ring from the source, then the flit is routed along the ring; otherwise the flit goes through the central hub to its destination. To avoid deadlock within the subnet, we adopt the virtual channel management scheme from Red Rover algorithm [17]. The ring is divided into two equal sets of contiguous nodes. Messages originating from each group of nodes use dedicated virtual channels. This scheme breaks cyclic dependencies and prevents deadlock.

Inter-subnet data routing requires flits to use the upper level network. By using the wireless shortcuts between the hubs with WIs of the same frequency channel, flits can be transferred in a single hop between them. If the source and destination WIs are tuned to different frequencies, flits are first routed to a gateway hub via wireless links and are then transmitted using the destination WI's frequency channel. If the source hub has no WI, the flits are routed to the nearest hub with a WI via the wired links and are then transmitted through the wireless channel. Likewise, if the destination hub has no WI, then the nearest WI hub receives the

data and routes it to the destination through wired links. Between a source and destination hub pair without WIs, the routing path with a wireless link is chosen if it reduces the total path length compared to the wired path. This can potentially give rise to a hotspot situation in the WIs because many messages try to access wireless shortcuts simultaneously, thus overloading the WIs and resulting in higher latency. Token flow control [18] and distributed routing are used to alleviate this problem. The routing adopted here is a combination of dimension order routing for the hubs without WIs and South-East routing algorithm for the hubs with wireless shortcuts. This routing algorithm is proved to be deadlock free in [1]. Consequently, the distributed routing and token flow control prevents deadlocks and effectively improves performance by distributing traffic though alternative paths.

The wireless hubs are grouped into clusters, each tuned to a particular frequency. As the wireless hubs in a particular cluster use the same frequency and can send or receive data from any other wireless hub in that cluster, an arbitration mechanism must be designed to grant access to the wireless medium to a particular hub at a given instant to avoid interference and contention. To avoid centralized control and synchronization, the arbitration policy adopted is a wireless token passing protocol. (Note that the use of the word token in this case differs from the usage in the above mentioned token flow control.) In this scheme a dedicated token circulates in each cluster. The particular WIs possessing the wireless tokens can broadcast flits into the wireless medium in their respective clusters. The wireless token is forwarded to the next wireless hub in the same cluster after all flits belonging to a packet at the current hub are transmitted.

5. EXPERIMENTAL RESULTS

This section characterizes mWNoC performance through simulation and analysis in presence of various traffic patterns. Characteristics of the on-chip wireless communication channel and selection of the optimum number of WIs for different system sizes are presented, followed by detailed network simulations with various system sizes. We also present performance benchmarking with respect to two other emerging NoC architectures, viz., NoC with RF interconnects (RF-I) and 3D NoC.

5.1 Wireless Channel Characteristics

The metal zigzag antennas described earlier are used to establish the on-chip wireless links. Antenna characteristics are simulated using the ADS momentum tool. High resistivity silicon substrate (ρ=5kΩ-cm) is used for the simulation. To represent the worst case inter-subnet communication range between WIs, the transmitter and receiver were separated by 20 mm. The antenna's forward transmission gains (S21) obtained via simulations are shown in Figure 2. We are able to obtain three different channels with 3 dB bandwidths of 16 GHz and center frequencies of 31, 57.5 and 120 GHz respectively. For optimum power efficiency, the quarter wave antennas use axial lengths of 0.73, 0.38 and 0.18 mm respectively in the silicon substrate. The antenna design ensures that signals outside the communication bandwidth for each channel are

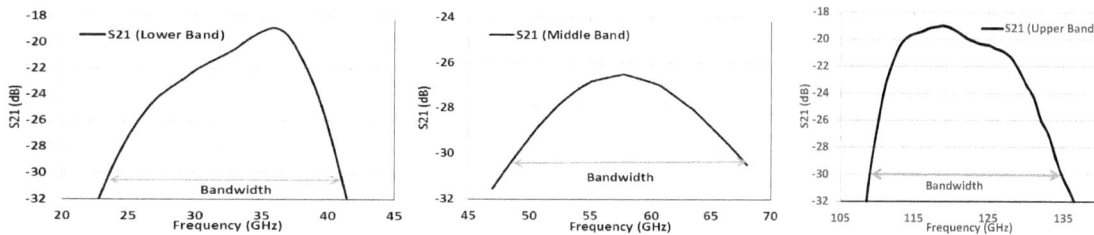

Figure 2. Antenna transmission gain (S21) for three non-overlapping channels.

sufficiently attenuated to avoid inter-channel interference. The wireless transceiver circuitry is designed and simulated using TSMC 65-nm CMOS process. The OOK transceiver can sustain a data rate of 16 Gbps with a power consumption of 43.6 mW.

5.2 Optimal Number of WIs

To reduce hardware overhead, we aim to limit the number of WIs on the chip without significantly compromising the overall performance. We assume round-robin token circulation among WIs. The token is considered to be a single flit transmitted from the WI currently holding it to the next one. The smaller the token return time to a particular WI is, the better the network performance is since wireless medium acquisition delay is minimized. On the other hand, hop-count decreases with more WIs due to higher connectivity as a result of introduction of additional WIs in the network's upper level. Since these are two opposing trends, a tradeoff needs to be established. Hence, we study achievable network bandwidth and packet energy as a function of the number of WIs. The upper level of the network is considered a mesh with three simultaneously operating wireless shortcuts and the subnet architecture is Ring-Star as shown in Figure 1. The WI clusters are equal in size and a single WI with transceivers of all frequencies acts as gateway between different clusters. Figure 3 shows that for a 512-core Mesh-RingStar system (32 subnets with 16 cores per subnet) bandwidth increases with number of WIs until reaching a maximum at 13 WIs (3 clusters of 4 WIs each and a gateway) and then it decreases. Moreover, as the number of WIs increases, the overall energy dissipation from the WIs becomes higher, and it causes the packet energy to increase as well. Considering all these factors, we determine the optimum number of WIs for 512-core mWNoC as 13. Similarly, for 8 and 16 subnet systems optimum performance is achieved with 5 and 7 WIs respectively.

5.3 Performance Evaluation

In this section we analyze mWNoC characteristics and study performance trends as the system size scales up. We consider three different system sizes, namely 128, 256, and 512 cores divided into 8, 16 and 32 subnets respectively. The die area is kept fixed at 20 mm x 20 mm for all system sizes. The NoC switch architecture is adopted from [19]. The hubs and NoC switches in the subnets have 4 virtual channels per port and have a buffer depth of 2 flits. Each packet consists of 64 flits. The WI ports have an increased buffer depth of 8 flits, which ensures that all messages trying to access wireless links are efficiently handled without compromising performance. Increasing the buffer depth beyond 8 gives no further performance improvement for this packet size, but gives rise to additional area overhead [13]. The WI wireless ports are assumed to have antennas and wireless transceivers. A self-similar traffic injection process is assumed.

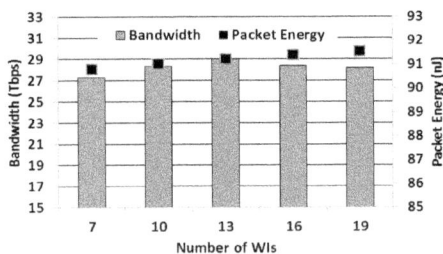

Figure 4. Achievable bandwidth for different system sizes.

The Mesh-RingStar architecture introduced earlier is simulated using a cycle accurate simulator. The subnet switches and the hub digital components are synthesized using 65 nm standard cell library from TSMC at a clock frequency of 2.5 GHz. The delays in flit traversals along the wired interconnects of the hybrid NoC architecture are considered when quantifying the performance. These include the intra-subnet core-to-hub wired links and the inter-hub links in the network's upper level. The delays through the switches and inter-switch wires of the subnets and the hubs are taken into account as well.

Figure 4 shows achievable bandwidth of the proposed mWNoC for three different system sizes considered under a uniform random spatial traffic distribution. We considered mWNoC with one, two and three simultaneously operating wireless channels. For comparison, we also present the bandwidth of three alternative architectures of the same size: (i) a flat mesh; (ii) the same hierarchical architecture as the mWNoC, but without any long-range links; and (iii) hierarchical architecture as the mWNoC, but with shortcuts implemented using buffered metal wires instead of wireless links (BWNoC). The number of wired shortcuts is kept equal to the number of WIs for different system sizes and they are optimally placed using the same SA-based optimization used for the placement of WIs. Each wired shortcut is considered to be 32 bits, which is equal to the width of a flit considered here. The wires are designed with an optimum number of uniformly placed and sized repeaters. The mWNoC with three simultaneously operating channels outperforms all the other alternatives for the three system sizes, except for the system with buffered wired shortcuts. The flat mesh architecture performs the worst due to its high average hop count. The hierarchical architecture improves the performance by reducing hop count, but the best performance is obtained from the hierarchical architecture with multiple shortcuts due to the small-world nature of the network. The hierarchical NoC with buffered wires as shortcuts results in a higher bandwidth as multiple parallel wires can operate together. But it suffers from significant energy dissipation, which is quantified in section 5.4. It can be observed that the bandwidth of the mWNoC with three non-overlapping channels improves compared to the initially proposed mWNoC with single wireless channel [12] for all the system sizes considered. Specifically, for higher system size the performance gain is more.

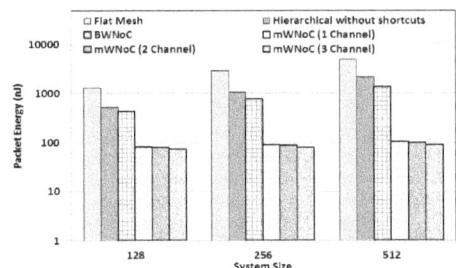

Figure 3. Performance variation with different number of WIs for a 512-core system with 32 subnets.

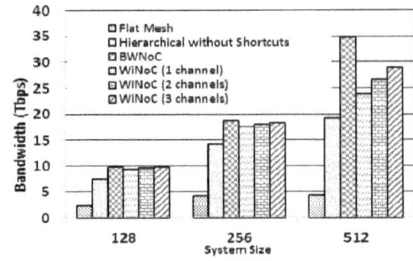

Figure 5. Packet energy for different NoC architectures.

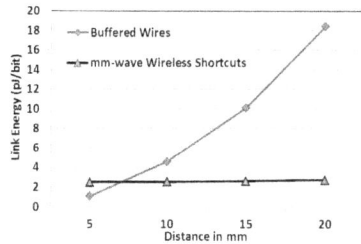

Figure 6. The variation of per bit energy dissipation with distance for a wired and a wireless link.

5.4 Energy Dissipation

We determine the mWNoC's packet energy dissipation. The packet energy is the energy dissipated on average by a packet from its injection at the source to delivery at the destination. The energy dissipations of the switches and hubs are obtained through synthesis using Synopsys tools with 65 nm standard cell libraries from TSMC. The energy dissipated by the wireless transceiver is determined through Cadence simulations. The energy dissipation of the wired links is obtained from the Cadence layout, assuming a 20 mm x 20 mm die area.

Figure 5 shows the packet energy dissipation of the considered architectures for uniform random traffic. The energy dissipation of the hierarchical wired NoCs with or without wireline shortcuts is significantly less than that of the flat mesh architecture. This is because a hierarchical network reduces the average hop count, and hence the latency between the cores. Packets get routed faster and hence occupy resources for less time and dissipate less energy in the process. The mWNoC further improves performance by employing multiple energy efficient long range shortcuts in the hierarchical network. In Figure 6 we show the variation of per bit energy dissipation with distance for a wired and a mm-wave wireless link. From this plot it can be observed that wireless shortcuts are always energy efficient whenever the link length is 7 mm or more. In our implementation, the minimum and maximum distances between the WIs communicating using the wireless channel are 7.07 mm and 18 mm respectively. Therefore, in this design, using the wireless channel is always more energy efficient. The mWNoC with multiple non-overlapping channels has reduced packet energy for all system sizes compared to the single channel mWNoC of [12]. Also, the mWNoC significantly reduces energy dissipation compared to the other two possible wired hierarchical architectures and can reduce the packet energy dissipation by at least an order of magnitude compared to the flat mesh.

5.5 Comparative Evaluation of mWNoC

In this section we perform a comparative analysis between the mWNoC and two other emerging NoCs. The on-chip RF transmission line (RF-I) proposed in [7] is a new interconnect technology that can improve NoC performance. Hence, we designed a small-world NoC (RFNoC) by replacing the wireless communication links of the mWNoC by RF-Is, maintaining the same hierarchical topology. Like the wireless links, these RF links can be used as long-range shortcuts. These shortcuts are optimally placed using the same SA-based optimization used for placing WIs in the mWNoC. As mentioned in [7], in 65 nm technology it is possible to have 8 different frequency channels, each operating with a data rate of 6 Gbps and used for long-range inter-subnet communications. We also considered a 3D mesh-based NoC with four layers as in [5]. Due to the high energy dissipation hierarchical NoC without shortcuts and BWNoC are not considered for this analysis.

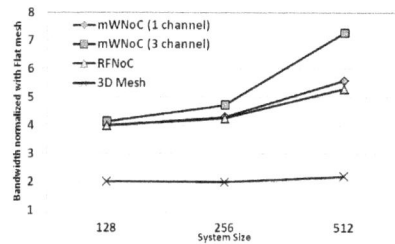

Figure 7. Comparative performance evaluation for different emerging NoCs.

For this comparative evaluation, we first present the normalized bandwidth with respect to flat mesh for different system sizes in Figure 7 for uniform random spatial traffic distribution. From this result it is evident that performance benefits are more prominent for bigger systems and is highest for 512-core. Consequently, a 512-core system is considered for subsequent analysis.

We consider both uniform and non-uniform traffic patterns in this evaluation. For non-uniform traffic patterns we use synthetic and application-based traffic distributions. We considered two types of synthetic traffic. First, a transpose traffic pattern [1] is considered where cores in a certain number of subnet pairs are considered to communicate more frequently with each other. We consider three such pairs and 50% of packets originating from one of these subnets are targeted towards the other in the pair. The other synthetic traffic pattern considered is hotspot traffic [1], where each core communicates with a certain number of subnets more frequently than with the others. We consider three such hotspot subnets to which all other cores send 50% of the packets that originate from them. Transpose and hotspot traffics are mapped in 3D NoC by selecting sets of adjacent cores to form groups (equivalent to subnets of mWNoC). In the transpose traffic three of these groups communicate with each other and also we consider three hotspot groups. We consider two application-based traffics. A 1024-point FFT is considered where each core performs a 2-point radix 2 FFT computation. Multiplication of two 512x512 matrices is used to generate another application-based traffic pattern.

Figure 8 shows the achievable overall network bandwidth for the different NoC architectures in uniform and non-uniform traffic scenarios for a 512-core system. It can be observed that in case of non-uniform traffic, due to the skewed communication pattern, certain interconnects on the path are overloaded and become bottle-neck affecting the overall performance of the NoC. This is most prominent in case of hotspot traffic. The 3D mesh based NoC suffers from the fact that number of layers is limited to four, which results in poor performance for 512-core system size. Though mWNoC and RFNoC have the same hierarchical architecture, mWNoC performs better. In RFNoC once the 8 shortcuts are placed they are fixed for that traffic pattern. But, in mWNoC, 13 WIs are placed depending on the traffic and wireless communication channel can be established between any of those pairs. Moreover, the total long-range link area overhead and the layout challenges of the RFNoC are more significant compared to mWNoC. For example, for a 20 mm x 20

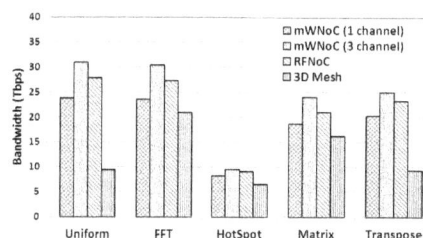

Figure 8. Achievable bandwidth for different traffic

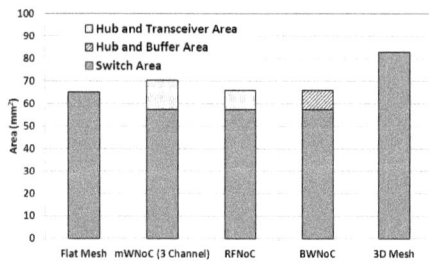

Figure 9. Silicon area overhead for different NoCs of size 512.

mm die, an RF interconnect of approximately 100 mm length has to be allocated for RFNoC following the layout of [7]. This is significantly higher than the combined length of all the antennas used in the mWNoC, which is 6.45 mm for the 512-core system.

6. AREA OVERHEAD

This section, quantifies the area overhead due to the wireless deployment in mWNoCs. The antenna lengths for the three different channels range from 0.18 mm to 0.73 mm with a trace width of 10 μm. The area of each WI transceiver is 0.31 mm^2, 0.32 mm^2 and 0.34 mm^2 for the three selected frequency ranges. The digital part for each WI, which is very similar to a traditional wireline NoC switch, has area overhead of 0.40 mm^2. Therefore, the total area overhead per hub with a WI (inclusive of transceiver and antenna) is determined to be in the range of 0.70 mm^2 to 0.74 mm^2. Gateway hubs have transceivers at all three frequencies so that their total circuit area requirement is 1.3 mm^2. Since the number of WIs is kept limited, the overall silicon area overhead is dominated by the wireline NoC switches. For example, in case of a 512-core mWNoC integrated in a 20 mm x 20 mm die, wireless transceivers consume only 5.9 % of total silicon area. The transceiver area overhead for RFNoC is obtained from [7]. Total silicon area overheads for flat mesh, mWNoC (3 Channel), RFNoC, BWNoC and 3D mesh for a 512-core system are shown in Figure 9. The extra ports in z-dimension for 3D mesh cause the total switch area requirements to increase. The area overheads of the hubs along with the required transceivers (mWNoC, RFNoC) and buffers (BWNoC) are shown separately. The transceiver area overhead for mWNoC is marginally higher than RFNoC and BWNoC. Though the overall silicon area requirements for mWNoC, RFNoC, BWNoC and 3D mesh are higher than flat mesh, the performance benefit of these emerging NoCs clearly outweighs the associated overhead.

7. CONCLUSIONS

This paper demonstrates that by an optimal utilization of long range, high bandwidth, low power, mm-wave wireless channels, significant performance improvements can be achieved in a NoC. By incorporating a hierarchical small-world topology and multiple non-overlapping wireless channels, the proposed mm-wave NoC architecture gives considerable performance gains in presence of various traffic patterns without significant area overhead.

8. ACKNOWLEDGMENTS

This work was supported in part by the US National Science Foundation (NSF) CAREER grant (CCF-0845504) and CRI grant (CNS-1059289) and NSF CAREER Grant (ECCS-0845849).

9. REFERENCES

[1] Ogras, U. Y. and Marculescu R., 2006. It's a Small World After All: NoC Performance Optimization via Long-Range Link Insertion. *IEEE Transactions on Very Large Scale Integration (VLSI) Systems*, vol. 14, no. 7, pp. 693-706.

[2] Kumar A., et al. 2008. Toward Ideal On-Chip Communication Using Express Virtual Channels. *IEEE Micro*, vol. 28, no. 1, pp. 80-90.

[3] Shacham A., et al. 2008. Photonic Network-on-Chip for Future Generations of Chip Multi-Processors. *IEEE Transactions on Computers*, vol. 57, no. 9, pp. 1246-1260.

[4] Joshi A., et al. 2009. Silicon-Photonic Clos Network for Global On-Chip Communication. *Proceedings of the 3rd International Symposium on Networks-on-Chip (NOCS-3)*, pp. 124-133.

[5] Feero B. and Pande P. P., 2009. Networks-on-Chip in a Three-Dimensional Environment: A Performance Evaluation, *IEEE Transactions on Computers*, Vol. 58, No. 1, pp. 32-45.

[6] Davis W. R. et al., 2005. Demystifying 3D ICs: The pros and cons of going vertical. *IEEE Design and Test of Computers*, Vol. 22, Issue 6, pp. 498-510.

[7] Chang M. F., et al. 2008. CMP Network-on-Chip Overlaid With Multi-Band RF-Interconnect. *Proceedings of IEEE International Symposium on High-Performance Computer Architecture (HPCA)*, pp. 191-202.

[8] Zhao D. and Wang Y., 2008. SD-MAC: Design and Synthesis of A Hardware-Efficient Collision-Free QoS-Aware MAC Protocol for Wireless Network-on-Chip. *IEEE Transactions on Computers*, vol. 57, no. 9, pp. 1230-1245.

[9] Lee S. B., et al. 2009. .A Scalable Micro Wireless Interconnect Structure for CMPs. *Proceedings of ACM Annual International Conference on Mobile Computing and Networking*, pp. 20-25.

[10] Kempa, K. et al. 2007. Carbon Nanotubes as Optical Antennae. *Advanced Materials*, vol. 19, pp. 421-426.

[11] Ganguly A. et al., 2011. Scalable Hybrid Wireless Network-on-Chip Architectures for Multi-Core Systems. *IEEE Transactions on Computers (TC)*, vol. 60, issue 10, pp. 1485-1502.

[12] Deb S., et al. 2010. Enhancing Performance of Network-on-Chip Architectures with Millimeter-Wave Wireless Interconnects. *Proceedings of IEEE International Conference on ASAP*, pp. 73-80.

[13] Chang K. et al., 2011. Performance Evaluation and Design Trade-Offs for Wireless Network-on-Chip Architectures. In *ACM Journal on Emerging Technologies in Computing Systems (JETC)*.

[14] Kirkpatrick S. et al., 1983. Optimization by Simulated Annealing. *Science*. New Series 220 (45978): 671-680.

[15] Floyd B. A., et al. 2002. Intra-Chip Wireless Interconnect for Clock Distribution Implemented With Integrated Antennas, Receivers, and Transmitters. *IEEE Journal of Solid-State Circuits*, vol. 37, no. 5, pp. 543-552.

[16] Yu X., et al., 2010. Performance Evaluation and Receiver Front-end Design for On-chip millimeter-wave Wireless Interconnect. *Proceeding of the International Green Computing Conference*, pp. 555-560.

[17] Draper J. and Petrini F., 1997. Routing in Bidirectional k-ary n-cube switch the Red Rover Algorithm. In *Proceedings of the International conference on Parallel and Distributed Processing Techniques and Applications*, 1184-93.

[18] Kumar A., Peh L. S. and Jha N. K., 2008. Token flow control. In *Proceedings of the 41st IEEE/ACM International Symposium on Microarchitecture (MICRO '08)*, 342-353.

[19] Pande P. P., et al., 2005. Performance Evaluation and Design Trade-offs for Network-on-chip Interconnect Architectures. *IEEE Transactions on Computers*, Vol. 54, No. 8, pp. 1025-1040.

Voltage Island-Driven Power Optimization For Application Specific Network-on-Chip Design *

Kan Wang , Sheqin Dong
Department of Computer Science & Technology
Tsinghua University, Beijing, China 100084
wangkan09@mails.tsinghua.edu.cn
dongsq@mail.tsinghua.edu.cn

Satoshi GOTO
Graduate School of IPS
Waseda University
Kitakyushu, Japan 808-0135
goto@waseda.jp

ABSTRACT

In this paper, a voltage island aware framework is proposed for low power design of application specific NoC (LPAS-NoC). Through a three-phase processing including voltage island generation, VI-driven floorplanning and post-floorplan processing, the total power consumption, design cost and total wire length can be optimized. Experimental results show that compared to traditional ASNoC, the proposed method can reduce total core power by about 34.5% and chip area by about 26.8% without increasing communication power.

Categories and Subject Descriptors

B.7.2 [**Integrated Circuits**]: Design Aids

General Terms

Algorithm, Design, Performance

Keywords

Multiple supply voltages, Application Specific NoC, Voltage Island, Network Components, Power Consumption

1. INTRODUCTION

Power consumption has become one of the major challenges in current network-on-chip design. One effective low power technology is *Multiple-Supply Voltage* (MSV), which partitions the whole circuit into domains of different voltage levels. Each voltage domain is called a *Voltage Island* (VI). By assigning lower voltage to some VIs, the total power can be reduced [2].

Some works applied MSV to NoC design to optimize the total power [1, 14, 15]. However, most of the works paid attention to regular NoC design and few of them considered VI technology in the design of *application specific NoC* (ASNoC).

---

*This paper is supported by MOST of china 2011DFA60290.

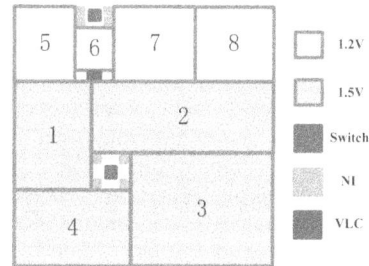

Figure 1: Voltage island aware ASNoC

Different from regular NoC, ASNoC design is more difficult in terms of irregular cores sizes, various core locations and different communication flow requirements [9]. ASNoC integrates IPs of different sizes in a single chip and realizes communication through a kind of mediums: *network components* (NC) [6]. To design a VI-driven ASNoC, it is much more complex since we need to consider not only high-level voltage selection problem, but also physical floorplanning and network component assignment.

Fig. 1 shows an ideational ASNoC containing two voltage islands. Three kinds of NCs are contained in this figure: *switch, network interface* (NI) and *voltage level converter* (VLC). Two switches are used to realize the communication between two islands. In each island, one single NI is allocated for each core to realize data communication and protocol conversion between cores and switches. Besides, switches and NIs share the same voltage level with the communicated cores and voltage level converters [13] can be integrated to realize the communication between NCs in different voltage level islands. In VI-driven ASNoC design, each NC is required to be assigned into the dead-space in bounding box of each voltage island to reduce the communication distance between cores and NCs. To design a low power ASNoC with optimized total power consumption, design cost and total wire length, an effective framework is required.

First, voltage level assignment is the key part in low power ASNoC design, as the voltage level can greatly influence the power consumption and performance of chip. Previous works proposed many methods to assign the best supply voltage from multiple alternative levels of each core according to performance and resource constraint in SoC [2-5]. However, these methods cannot be directly applied to ASNoC design. On one hand, they didn't think of the criticality of cores. [7] implied that performance critical cores such as processors usually require the highest supply voltage level while other functions, such as memories, can operate on lower levels. The criticality of different cores will make the selection of voltage levels more complex. On the other hand, previous voltage assignment didn't think of optimization of communication power, which is also an objective of low power design.

Second, network component assignment is quite important, because NCs are the communication medium and can seriously impact total communication power due to the distance between components determined by physical layout. Therefore, a good floorplan is the key to optimize the communication power. Previous works used ILP formulation and network flow based algorithm to assign NCs [6, 8, 9], but all of them are based on given floorplan or in the post-floorplan stage. On this occasion, the assignment result will be restricted by the floorplan and a bad floorplan will lead to bad assignment and hence large communication power. As a result, a lot of iterations and attempts are needed to get an appropriate floorplan, which will lead to large dead-space or long run-time. Therefore, a floorplanning with NC assignment considered is required for ASNoC design.

What's more, total chip area and power network resource design cost is always used to judge the quality of ASNoC, while the floorplan is one of the most important factors to determine these. [10] implied that the power network resource depends on the area of bounding box of voltage island. Therefore, to minimize the design cost, a VI resource aware floorplanning is necessary.

In this paper, a framework for low power design of ASNoC (LPASNoC) is proposed, which considers all of above factors. Through a three-phase processing, the total power, design cost and total wire length can be optimized.

The contributions in this paper are summarized as follows:

1. A framework is proposed for low power application specific NoC design.

2. Two voltage level assignment algorithms considering communication power are proposed for island generation.

3. VI-driven floorplanning with network component assignment considered is proposed for ASNoC.

The rest of the paper is organized as follows. In Section 2, relative works are introduced. Section 3 describes problems and motivation of this paper. In Section 4, the voltage island-driven framework is introduced, with detailed description of each phase. Experiment results are shown in Section 5 and conclusions are provided in Section 6.

2. RELATED WORK

Two kinds of related works have been proposed: voltage island generation and low power NoC design.

For voltage island generation, [10,11] presented an effective voltage assignment technique based on dynamic programming (DP) while in [12], an ILP formulation is employed to tackle voltage-island generation considering power-network routing resource. However, none of these thought of the relationship between floorplanning and voltage level assignment. [4] integrated the voltage assignment into floorplanning and proposed a core-based voltage island driven floorplanning. [3] presented two ILP formulations to solve voltage level assignment problem for SoC design from chip-level. However, neither of these can be directly used in ASNoC design. With network components assignment considered, the problem will be more difficult with more constraints.

Table 1: Notation of VI driven ASNoC design

v_{ip}	The voltage of core i when it operates at level p
P_{ip}	Power consumption of core i at voltage level p
P_{core}	Total power consumption of cores
P_{vcom}	Total communication power between different VIs
P_{com}	Total power consumption of communication
P_{cons}	Total power consumption of network components
P_{LC}	One bit energy of conversion power on VLC
Com_{ij}	Communication amount between core i and j
CCL	All communication edges in CCG
N_s	Number of network components required in NoC

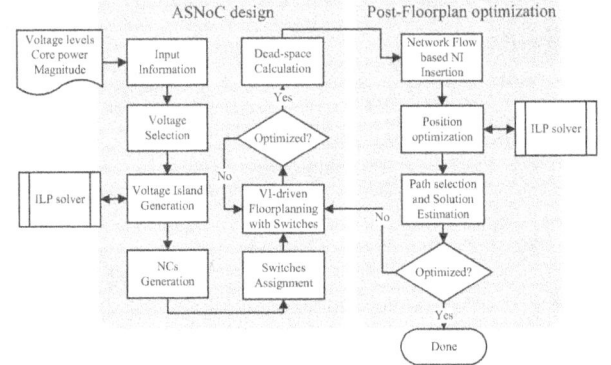

Figure 2: Low power ASNoC design flow

Meanwhile, many other researches paid attention to low power NoC design. [1] integrated MSV into NoC design and proposed a voltage island partitioning method for GALS based NoC. [13] and [14] proposed complete frameworks including portioning, mapping and routing for NoC design. However, all these works are only for regular NoC design. [15] presented a synthesis approach to obtain customized application-specific networks on chips that can support the shutdown of voltage islands. However, this work paid little attention to floorplanning and network components assignment problem, which are both important in ASNoC design.

3. PROBLEM DEFINITION

Given (1) N cores including information of width and height, (2) core communication graph (CCG), (3) legal voltage levels of each core and the corresponding power consumptions at different voltage levels and (4) number of voltage island m, the problems of voltage island generation, NC assignment and m VIs driven Floorplanning are solved respectively under certain constraints such as chip width/height ratio. As the work is applied at the system level, we assume that the inputs and outputs of the cores are registered, therefore no critical path will span across two cores. The objective is to generate a legal layout with optimized total power consumption, total design cost including both chip area and voltage island resource area, and also total wire length. The notations of VI driven ASNoC design are shown in Table 1.

PROBLEM 1. (Voltage Island Generation) Given multiple supply-voltage choices, CCG and the number of voltage levels n, select m levels from n and then assign each core with a supply voltage level to generate m islands so that critical-

ity constraint is satisfied and the total power consumption between VIs are minimized.

PROBLEM 2. *(Network Component Assignment) Given an initial floorplan of NoC, network components, which includes switches, network interfaces and voltage level converters, are required to be assigned into NoC to realize the communication between each core. The objective is to reduce the total communication power of chip.*

PROBLEM 3. *(m VIs Driven Floorplan) Given m voltage islands and the core information in each island, the VI-driven floorplan is to generate a floorplan so that total communication power between VIs, the total power network resource, total chip area and total wire length are optimized.*

4. VOLTAGE ISLAND DRIVEN ASNOC DESIGN

In the proposed framework of LPASNoC, three phases are contained: (1) voltage island generation including voltage selection and assignment; (2) network components assignment and voltage island driven floorplanning with NCs; (3) post-floorplan processing including LP based NC position optimization and path selection.

The design flow is shown as Fig. 2. Given size information, communication requirements and alternative voltage levels of each core, voltage islands are generated by using a DP based voltage level selection and ILP or heuristic based voltage assignment algorithm. Then NCs are generated by partitioning the cores in the same VI into clusters. After that, switches are integrated into voltage island-driven floorplanning to get a good floorplan solution. After min-cost based NI assignment, post-floorplan LP based optimization is employed to improve communication power. Finally, path selection is performed and the solution is evaluated. Since voltage level converters are required for each pair of switches in different VIs, voltage level converters can be integrated into each switch by enlarging the size of switch. During the design flow, voltage level converter assignment is integrated into switch assignment and switches and NIs are assigned separately, which will be described in detail in Section 4.2.

4.1 Voltage Island Generation

For any core, there is a range of possible VDD values, where VDDL is the lowest supply value that meets performance requirement of the core and VDDH is the maximum overall supply voltage.

As a result of different criticality of each core, although it is possible to have a large number of supply voltages on the same chip, only a few of them are practical. The work in [7] used dynamic programming algorithm to choose m islands from given n possible voltage levels that minimizes computation power. In this paper, a similar method is applied to select 3 voltages from 4 given levels. Then two methods are proposed to solve the voltage assignment. The first one is a communication aware ILP formulation while the other is a faster heuristic algorithm.

4.1.1 ILP formulation

Binary decision variables v_{ip} ($1 \le i \le N$, $1 \le p \le m$) are defined such that v_{ip} is 1 if core i is assigned with voltage level p, and v_{ip} is 0 otherwise. Then for each core, one single

voltage is assigned, so:

$$\sum_{p=1}^{m} v_{ip} = 1, \forall 1 \le i \le N \qquad (1)$$

The total core power can be calculated according to:

$$P_{core} = \sum_{i=1}^{N} \sum_{p=1}^{m} P_{ip} \cdot v_{ip}, P_{ip} = R_i C_i V_p^2 + T_i k_i V_i e^{-\frac{V_t}{S_t}} \quad (2)$$

Where R_i is the number of active cycles, C_i stands for the total switched capacitance per cycle, T_i is the number of idle cycles, k_i is a design parameter, and S_t is a technology parameter [1].

The communications between VIs require voltage level converters to realize voltage conversion, which will hence create conversion power. The conversion power can be evaluated by communication amounts between switches, which are also communications between cores in different VIs:

$$P_{vcom} = P_{LC} \cdot \sum_{i=1}^{N} \sum_{j=i+1}^{N} \sum_{p=1}^{m} \sum_{q=p+1}^{m} Com_{ij} \cdot v_{ip} \cdot v_{jq} \quad (3)$$

The formulation can be transformed into linear formulation by replacing v_{ip} and v_{jq} with a binary decision variable $v_{ip,jq}$ according to [3]:

$$v_{ip} + V_{jq} - 1 \le v_{ip,jq} \qquad (4)$$

However, this will bring in lots of variables, which can lead to a long time to solve. However, it's easy to find most of the variables are redundant as there is no communication between them. Therefore, the redundant variables can be reduced and only the ones for the cores which have communication requirement are used:

$$P_{vcom} = P_{LC} \cdot \sum_{(i,j) \in CCL} \sum_{p=1}^{m} \sum_{q=p+1}^{m} Com_{ij} \cdot v_{ip} \cdot v_{jq} \quad (5)$$

With these variables, the objective can be written by:

$$Minimize \ P = P_{core} + P_{vcom} \qquad (6)$$

On the other hand, according to performance criticality, the supply voltages of some critical cores such as processor cores are required to be guaranteed. For this, another constraint is brought in. Let l_i be the lowest available level for core i, and then for each core, the following inequality should be satisfied:

$$\sum_{p=l_i}^{m} v_{ip} \ge 1, 1 \le i \le N \qquad (7)$$

4.1.2 Heuristic greedy algorithm for VLA

Although ILP formulation is exact, it will still take a lot of time if N or m is large enough. For this, a heuristic greedy algorithm is proposed for voltage level assignment in VI-driven ASNoC design.

The objective of voltage assignment is to use as low voltage as possible for cores without increasing communication cost under criticality constraint, and to achieve trade-off among core power and communication cost. With this motivation, the following greedy algorithm for voltage assignment is proposed, as shown in Fig. 3. Cores are first ranked by VDDL incrementally. In each voltage level, the cores are ordered by communication amount decreasingly. Then

```
Function voltageAssignment:
Input: N, m, Li, Comij, Vli   // Number of cores, number of VIs, Lowest level for core i,
communication amount between core i and j, Vector for each VI
Output: Voltage level for each core
Method:
orderCoresByVoltage();   //Order The cores according to the available lowest level of voltage;
orderCoresByComs();   //In the same voltage level, order by total communication of amount
For (i between 1 and m){   //generate voltage islands
    For (j between 1 and N) {
        if(i is ok for core j){
            checkCore(j);   //check the cores communicated with j that already allocated
            For(each neighbor t of j)
            if(Comj >= com_threthold){   //If communication exceed a threat hold
                pulloutofVI(t);   //pull it out of previous island
                insertCoretoVI(t,i);   // insert it into current island
            }
            insertCoretoVI(j,i);   //Insert Core to VI
}   }}
Output wireLengthij;
```

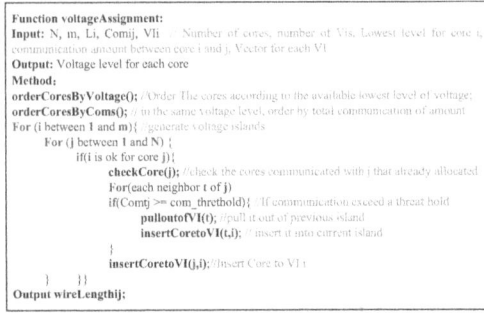

Figure 3: Heuristic voltage assignment algorithm

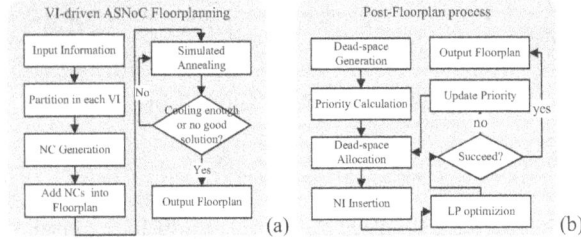

Figure 4: (a)Voltage Island driven floorplan flow (b)Post-floorplan LP optimization flow.

each core will be inserted into the m islands according to the rank. Cores with lower VDDL and lower communication amount will be first inserted to the VI of lower voltage. For each inserted core, all its communicated cores which are already allocated into previous islands will be checked. If the communications are large ones, then the communicated cores will be pulled out of previous island and inserted into the current island.

It's easy to find that the complexity of this algorithm is no more than $O(mMN)$, where M is largest number of neighbor nodes in CCG. It is much faster than ILP formulation when m and N are large.

4.2 Voltage Island Driven Floorplan

In ASNoC design, floorplan has a great effect on NC assignment. However, in previous works, network components were assigned based on given floorplan or in the post-floorplan stage. As a result, the assignment result is restricted by the floorplan and a bad floorplan will lead to large communication power. One feasible method to solve it is to integrate NCs into floorplanning to occupy the objective dead-space. The flow of VI-driven floorplanning is shown in Fig. 4(a). Given the input information, min-cut based partition is performed to generate clusters and NCs. Then NCs are integrated into floorplan and do floorplanning with other cores together. NCs can be pushed into the bounding box of VIs through adding an attractive force to them. After several iterations, a good floorplan can be obtained. During the process, network components are treated as one kind of cores. We call this kind of cores as *dummy cores*, while the actual cores are referred to *real cores*.

4.2.1 Network Component Generation

To generate switches, a similar method as [15] is used to search for every possible number i of partitions in each volt-age island and partition assign cores into i min-cut clusters. Then the best partition solution will be adopted. Those cores with larger communication requirements are assigned to the same cluster and hence use the same switch for communication power optimization. Then NIs can be easily generated for each core according to [8]. Voltage level converters can be clustered into switches by enlarging the size of switches according to communication requirement.

4.2.2 VI-driven Floorplan for NC Assignment

The generated NCs are added into the input of floorplaning and be treated as small dummy cores. However, directly bring in so many cores into floorplan will add to complexity. In this paper, only switches are considered in floorplaning process. NIs can be assigned in the post-floorplan stage. To generate enough dead-space, the sizes of dummy cores can be enlarged to some extent. Then a mechanism is proposed to guide the disturbance during the floorplanning, to eliminate the redundancy in solution space with dummy cores.

The disturbance is partitioned into two stages. The first stage will run when all of the three conditions are satisfied: 1) annealing temperature is larger than $T\_threthold$; 2) solution reject ratio is lower than $reject\_ratio$; 3) dead-space ratio is more than ds_ratio. Otherwise, the disturbance comes to the next stage instead. In the first stage, both real cores and dummy cores are considered in the disturbance, while in the second stage only dummy cores are thought of. The information of small dummy cores is also stored in the whole process in order to distinguish the dummy cores from actual ones.

The cost function in floorplanning is:

$$Cost = \alpha \cdot Area + \beta \cdot VIArea + \gamma \cdot wirelength + \lambda \cdot comAmount \tag{8}$$

where $Area$ represents the floorplan area; $VIArea$ represents the sum of all voltage island bounding resources; and $comAmount$ stands for the total communication amount between clusters. The parameters α, β, γ and λ can be used to adjust the relative weighting of contributing factors.

4.3 Post- Floorplan Processing

The mix-up floorplanning with both real cores and dummy cores in the previous stage make sure that enough dead-space is generated for NCs insertion. However, as a result of the randomness in disturbance, the position of switches may not be the best after floorplanning. In the post-floorplan stage, dead-space re-allocation and NI assignment are performed to fulfill the network components. And then a LP based optimization algorithm is proposed to improve the communication power. The process is shown in Fig. 4(b).

4.3.1 Dead-space Re-allocation

All the dead-space are first searched from a given a floorplan from Section 4.2. This is easy to achieve by using linear scanning method. Then the adjacent dead-spaces are combined to form a bigger one. The final dead-spaces are reallocated for each switch according to a priority which is determined by the communication amount and total wire length of each switch in the dead-spaces. Then switches are inserted into the center of obtained dead-space.

4.3.2 NI assignment

After switches assignment, NIs can be also inserted into

the remaining dead-spaces by using ILP formulation or min-cost network flow algorithm [8].

In this paper, network flow algorithm is applied for NI insertion. Different from [8], different sizes of NIs are considered. Dead-space is divided into small grids and the size of one grid is enough for a NI. The redundant dead-space which is smaller than a grid is merged into nearest grid to form larger grid. On the other hand, larger grids can also be generated by combining two grids together. The network graph consists of an input core node, out-degree core nodes and all available dead-space nodes. Then N min-cost flow is run on the graph to get the best results.

4.3.3 LP optimizaiton for NCs

Given the allocation result from above algorithm, each NC is allocated into the best dead-space. Then a LP optimization is performed to obtain the final best position of each NC. Here, we still refer to NCs as *dummy core*.

First, NCs should satisfy the boundary constraint in the dead-space and cannot exceed the boundary dead-space. Assume (x_{ic}, y_{ic}) and (rx_{ic}, ry_{ic}) are the positions of lower left corner and higher right corner of the dead-space which core i belongs to, respectively. To keep the initial dead-space area, additional inequalities for each core i are needed as follows:

$$x_i \geq x_{ic}, y_i \geq y_{ic}, x_i + w_i \leq rx_{ic}, y_i + h_i \leq ry_{ic} \quad (9)$$

Meanwhile, the cores in the same dead-space are required to satisfy the overlap constraints. The topological relations between core i and core j contains: left, above, right and below.

Let (x_i, y_i), (x_j, y_j) denote the variable positions of the lower left corners of core i and j. From the existing floorplan using representation such as TCG [16], the corresponding relative positions of cores can be captured. If the obtained relation cannot satisfy the dead-space boundary constraint, then other three topological relations will be tried respectively, until one of them satisfies. Given the topological relation, to prevent overlapping between dummy cores i and j, one of the following linear inequalities actually holds:

$$x_i + w_i \leq x_j, \text{ if } i \text{ is left of } j \quad (10)$$

$$y_i + h_i \leq y_j, \text{ if } i \text{ is above of } j \quad (11)$$

The objective of LP is to minimize the total communication cost, which is evaluated by the sum of all the wire lengths weighted by communication amount. The wire length is evaluated by using the half perimeter model. Then following objective can be achieved:

$$Min \sum_{i=1}^{N_s} \sum_{j=1}^{N_s} com_{i,j} \cdot (|x_i - x_j + \frac{w_i - w_j}{2}| + |y_i - y_j + \frac{h_i - h_j}{2}|) \quad (12)$$

The following constraint method can be used to remove the absolute value sign.

$$x_i - x_j + \frac{w_i - w_j}{2} \leq x_{i,j}, -x_i + x_j - \frac{w_i - w_j}{2} \leq x_{i,j} \quad (13)$$

$$y_i - y_j + \frac{h_i - h_j}{2} \leq y_{i,j}, -y_i + y_j - \frac{h_i - h_j}{2} \leq y_{i,j} \quad (14)$$

Then (12) can be re-written by the following formula:

$$Min \sum_{i=1}^{N_s} \sum_{j=1}^{N_s} com_{i,j} \cdot x_{i,j} + com_{i,j} \cdot y_{i,j} \quad (15)$$

Table 2: Results of two algorithms for VLA

Benchmark	V#	E#	ILP		Heuristic	
			Power	Run Time	Power	Run Time
T1	38	46	7.17	0.020	7.42	0.006
T2	76	92	13.87	0.204	15.59	0.015
T3	114	138	21.16	3.185	23.72	0.023
T4	152	184	29.15	5.629	30.93	0.033
T5	190	230	36.53	7.282	38.31	0.044
Average	-	-	21.58	3.264	23.19	0.024
Ratio	-	-	1	1	1.075	0.0074

4.3.4 Path selection

After position optimization, path selection can be applied to get the best communication paths for communication power optimization. In this paper, port number constraint is also taken into account that the port number for each switch must be no more than 5. An improved shortest-path algorithm by dynamically updating capacity of each node can be easily complemented for this problem. We don't discuss in detail here.

5. EXPERIMENTAL RESULTS

Experimental results show that the proposed LPASNoC is effective on power reduction and design cost optimization. All experiments are performed on a workstation with 3.0 GHz CPU and 4GB physical memory. Nine benchmarks are used from [8] in the experiments. A graph partitioning tool [17] is used for switches generation and *lp_solve* 5.5.2.0 [18] is used for ILP of VLA and LP of post-floorplan optimization. To make the comparisons more fairly, the same power model of switch and interconnection is used as [8]. Core power is evaluated according to (2) in Section 4, while the power of voltage level converters is evaluated according to (5).

5.1 Effect of heuristic algorithm for VLA

Five experiments are performed to evaluate our heuristic greedy based algorithm, as shown in Table 2. The five test cases from T_1 to T_5 are generated by duplicating the module and network information from *D_38_tvopd*. voltage requirements are generated randomly. Table IV shows the effect of our our heuristic algorithm. Compared to ILP formulation, our algorithm can save time by about 99.3%, with only 7.5% less power optimization. As the scale of problem increases, the proposed algorithm can save much more time.

5.2 Effect of power optimizaiton of LPASNoC

In order to show the effect of LPASNoC, the work of [8] is complemented and compared to. The results are divided into two patterns: *ASNoC* [8]: application specific NoC design for communication power optimization without voltage island considered; *LPASNoC*: the proposed VI-driven low power design framework. For each benchmark, ten results are selected and the average value is also calculated. Table 3 shows the results on both power consumption and design cost. The core power and communication power of all benchmarks are all normalized values. Fig. 5 shows the generated floorplan of *MPEG4* and *D_38_tvopd* from *LPASNoC*, which are composed of three voltage islands.

From Table 3, with voltage island considered, the core power consumption can be reduced by 34.5% on average.

Table 3: Effects of LPASNoC

Benchmark	V#	E#	ASNoC					LPASNoC					
			Core Power	Com Power	Wire Length	Hops	D.S Ratio	Core Power	Com Power	Wire Length	Hops	D.S Ratio	Run Time(s)
DSP	6	5	12.92	12.5	201	1	3.59	8.11	11.04	218	1	5.52	5.96
PIP	8	8	12.94	16.1	183	1.33	7.37	7.27	15.2	240	1.33	7.02	10.24
MWD	12	12	19.01	40.3	413	1.5	11.82	12.56	34.7	337	1	7.51	40.53
MPED4	12	13	17.2	40.1	402	1	12.25	13.84	30.11	435	1	6.79	11.51
VOPD	12	14	32.4	18.1	578	1	11.70	22.16	20.88	625	1.33	9.18	43.81
263decmp3dec	14	15	54.29	31.1	698	1.5	14.51	32.41	39.6	720	1	9.32	36.24
263encmp3dec	12	12	42.53	70.1	508	1	9.58	29.97	79.9	446	1.25	8.54	30.25
mp3encmp3dec	13	13	34.89	58.02	582	1	12.75	19.34	55.64	480	1.33	9.86	40.62
D_38_tvopd	38	47	65.21	62.7	3078	1.4	15.98	45.23	66.8	2157	1.57	10.93	119.01
Ratio	-	-	1	1	1	1	1	0.655	1.014	0.853	1.007	0.732	37.57

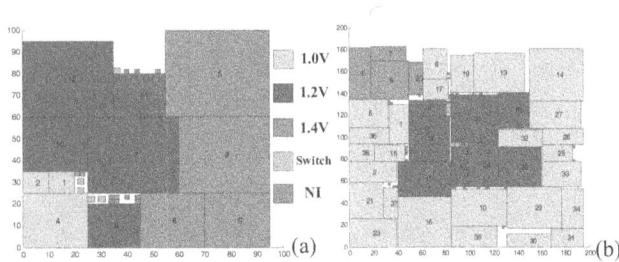

Figure 5: (a)Floorplan for MPEG4 (b)Floorplan for D_38_tvopd.

Compared to *ASNoC*, *LPASNoC* can reduce dead-space by 26.8%. This is because of the strategy to control the floorplan for NCs assignment in the proposed framework instead of blind search for floorplan in *ASNoC*. Furthermore, total wire-length can be improved by 14.7%, due to the optimized position of NCs. Compared to the significant improvement, the little deteriorating on communication power (1.4%) and hop counts (0.7%) is acceptable.

6. CONCLUSION

Voltage level assignment, network components assignment and VI-driven floorplanning are three important phases in low power ASNoC design. In this paper, a framework for low power design of ASNoC is proposed. A heuristic voltage assignment is specially proposed for voltage island generation, which is much faster than ILP algorithm. Then a voltage island-driven floorplaning, which integrates network components planning, is used to generate the ASNoC. In the last phase, a post-floorplan LP optimization is performed for better solution.

7. REFERENCES

[1] Ogras, U.Y., Marculescu, R., Choudhary, P., and Marculescu, D. Voltage-Frequency Island Partitioning for GALS-based Networks-on-Chip. *DAC*, 2007.

[2] Yu, B., Dong, S., Goto, S., and Chen, S. Voltage-island driven floorplanning considering level-shifter positions. *GLSVLSI*, 2009, pages 51-56.

[3] Wai-Kei, M., and Jr-Wei, C. Voltage Island Generation under Performance Requirement for SoC Designs. *ASPDAC*, 2007, pages 798-803.

[4] Ma, Q., and Young, E.F.Y. Voltage island-driven floorplanning. *ICCAD*, 2007, pages 644-649.

[5] Hung, W.L., Link, G.M., Xie, Y., Vijaykrishnan, N., Dhanwada, N., and Conner, J. Temperature-aware voltage islands architecting in system-on-chip design. *ICCD*, 2005.

[6] Srinivasan, K., Chatha, K.S., and Konjevod, G. Linear-programming-based techniques for synthesis of network-on-chip architectures. *IEEE TVLSI*, 2006, 14, (4), pages 407-420.

[7] Sengupta, D., and Saleh, R. Application-driven floorplan-aware voltage island design. *DAC*, 2008.

[8] Yu, B., Dong, S., Chen, S., and Goto, S. Floorplanning and topology generation for application-specific network-on-chip. *ASP-DAC*, 2010.

[9] Murali, S., Meloni, P., Angiolini, F., Atienza, D., Carta, S., Benini, L., De Micheli, G., and Raffo, L. Designing application-specific networks on chips with floorplan information. *ICCAD*, 2006.

[10] Lee, W.P., Liu, H.Y., and Chang, Y.W. Voltage Island aware floorplanning for power and timing optimization. *ICCAD*, 2006.

[11] Ching, R.L.S., Young, E.F.Y., Leung, K.C.K., and Chu, C. Post-Placement Voltage Island Generation. *ICCAD*, 2006.

[12] Wan-Ping, L., Hung-Yi, L., and Yao-Wen, C. An ILP algorithm for post-floorplanning voltage-island generation considering power-network planning. *ICCAD*, 2007.

[13] Wooyoung, J., Duo, D., and Pan, D.Z. A Voltage-Frequency Island aware energy optimization framework for networks-on-chip. *ICCAD*, 2008.

[14] Nishit Kapadia, S.P. VISION: A Framework for Voltage Island Aware Synthesis of Interconnection Networks-on-Chip. *GLSVLSI*, 2011, pages 264-269.

[15] Seiculescu, C., Murali, S., Benini, L., and De Micheli, G. NoC topology synthesis for supporting shutdown of voltage islands in SoCs. *DAC*, 2009.

[16] Jai-Ming, L., and Yao-Wen, C. TCG: A transitive closure graph-based representation for general floorplans. *IEEE TVLSI*, 2005, 13, (2), pages 288-29.

[17] Karypis, G., Aggarwal, R., Kumar, V., and Shekhar, S. Multilevel hypergraph partitioning: applications in VLSI domain *IEEE TVLSI*, 1999, 7, (1), pages 69-79

[18] "Lp_solve". http://www.lpsolve.sourceforge.net/5.5/

Design-Time Performance Evaluation of Thermal Management Policies for SRAM and RRAM based 3D MPSoCs

David Brenner, Cory Merkel, Dhireesha Kudithipudi
Rochester Institute of Technology
Rochester, New York
{dab2704, cem1103, dxkeec}@rit.edu

ABSTRACT

3D-ICs hold significant promise for future generation multi processor systems-on-chip due to their potential for increased performance, decreased power, heterogeneous integration, and reduced cost over planar ICs. However, the vertical integration of these structures exacerbates the heat dissipation and run-time thermal management issues. There have been a number of design- and run-time thermal management policies proposed, but few focus on examining overall system performance. Additionally, the heterogeneity of 3D-ICs allows for the integration of novel technologies, such as resistive random access memories (RRAMs), which offer higher density and lower power than traditional CMOS memory technologies. Our work presents a flexible design-time simulation framework to evaluate system performance and thermal profiles of 3D MPSoCs. We utilize this framework to study the effect of three dynamic thermal management policies (air-cooled load balancing, liquid-cooled load balancing, and air-cooled DVFS) on system performance and die temperature for multi-tiered 3D MPSoCs utilizing SRAM and RRAM-based L2 caches. We find that RRAM-based caches lower overall average maximum temperatures by 120 K and 24 K for air and liquid cooling systems, respectively (when compared to SRAM-based caches), at a worst-case performance delay of 47% and best-case delay of 13% for the parallel shared-memory benchmarks studied.

Categories and Subject Descriptors

B.7 [**Integrated Circuits**]: Advanced Technologies; B.8.2 [**Performance and Reliability**]: Performance Analysis and Design Aids

Keywords

3D-IC, 3D MPSoC, microchannel cooling, thermal management, RRAM cache

1. INTRODUCTION

Three-dimensional integrated circuits (3D-ICs) are rapidly becoming a viable option for overcoming the performance and power bottlenecks of global interconnects in multiprocessor systems-on-chip (MPSoCs). In particular, 3D-ICs offer reduced interconnect lengths, which reduce power, decrease latency, increase bandwidth, and thus increase performance. 3D-ICs also allow for reduced form factor, improved fabrication yield, heterogeneous integration, and, eventually, reduced cost [1]. However, there are still significant challenges to overcome related to the manufacturing and process flow, electrical interconnects, testing, thermal management, etc. [2]. Specifically, the increased power densities, high temperatures, and thermal gradients that result from the increased thermal resistance of the stacked die make thermal management among the key design challenges in the realization of 3D MPSoCs [2]. The heat generated further aggravates thermal gradients as the number of stacks increases. This is due to the device density, complexity in heat dissipation, and limitations in the placement of through silicon vias (TSVs) (also thermal vias).

In 3D-ICs, the thermal gradients span across heterogeneous cores and require hierarchical solutions. Thermal management in 3D-ICs is achieved by using run-time techniques, such as task scheduling, task migration, and dynamic voltage and frequency scaling (DVFS) [3, 4] coupled with interlayer cooling mechanisms using microchannel and/or pin-fin heat sinks [5]. At run-time several of these design knobs (frequency, voltage, liquid flow rate) have to be tuned to minimize thermal gradients in 3D-IC. Ideally, this improves mean time to failure (MTTF) while maintaining system performance, energy consumption, and reliability. Of the cooling systems available – liquid-cooled and air-cooled – the former is practical due to the high heat extraction capabilities (nearly 4 KW/cm^3)[5]. The fluid in the liquid cooled systems is circulated across the stacks using a pump, and a heat exchanger is used to cool the fluid. A significant portion of the cooling system energy is consumed by the pump [6]. An obvious approach to reduce the energy consumption of the MPSoCs and the cooling system is to optimally control the liquid flow rate. To achieve uniform temperature gradients across the layers, it is desirable to combine this with run-time thermal management policies. However, a combination of these techniques can result in overall system performance degradation and/or increased system energy consumption. Existing research work focuses on understanding and minimizing the energy for the different thermal management policies [7, 6].

Our focus in this work is to evaluate the impact on system performance when using the aforementioned thermal management policies, through the design of a simulation framework that allows design-time performance feedback utilizing state of the art modeling tools. This framework allows the designer to customize all aspects of the MPSoC to meet the system requirements. It also allows the user to test real workloads in a full system environment in order to understand how these dynamic thermal management (DTM) policies affect the performance and thermal profile of the system.

As mentioned earlier, one of the advantages with 3D-ICs is the heterogeneous stacks. The advantage is the ability to integrate novel technologies such as resistive random access memories (RRAM's), which offer higher density and lower power over SRAM-based caches [1]. Such technologies have a significant impact on the heat generated and extracted in the different stacks. Therefore, in this work we will also design and study the thermal behavior of 3D MP-SoCs which use RRAM as an L2 cache. Specific reasons to choose RRAM are: experimentally shown to be integrated in a 3D architecture, low static power dissipation, and sensitivity to temperature [8].

The specific contributions of this work are:

- A design-time simulation framework to study the impact on overall system performance with three different dynamic thermal management policies: load balancing, dynamic voltage frequency scaling (DVFS) with interlayer microchannel heat sinks. Previous works have used a similar framework, however the detailed CPU modeling allows for more thorough analysis of the performance impact of the thermal management policies and RAM technologies.

- A comprehensive analysis of the system performance and thermal profile of three DTM techniques on 2-, 4-, and 6-tier (8, 16, and 24-cores respectively) 3D MP-SoCs using traditional CMOS cache implementations with shared memory workloads.

- A comprehensive analysis of the thermal and system performance improvements gained through the use of resistive random access memory for the level 2 cache in the 3D MPSoCs.

The rest of this paper starts with an overview of related work in Section 2. Section 3 describes the integration of RRAM caches into MPSoC architectures. Section 4 explains the DTM policies used and Section 5 presents the simulation framework and methodology. Section 6 explains the results, and Section 7 concludes the paper.

2. RELATED WORK

The application of 3D integration to MPSoCs has been explored in prior work [9]. Black et al. [10] explore 3D processor integration and note decreases in power of 15% and increases in performance of 15% at the cost of a 14K increase in peak temperature compared to a 2D implementation. Puttaswamy et al. [9] show increasing the number of stacked die both increases the performance and the maximum temperature (by 17K and 33K for 2- and 4-tier stacks, respectively) compared to 2D implementations. These prior

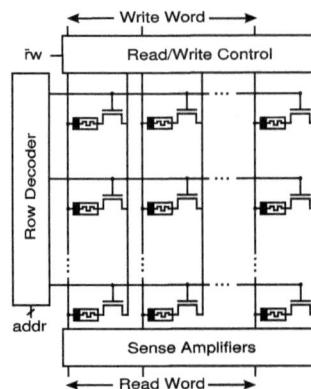

Figure 1: Voltage sensed 1T1R bipolar RRAM block.

works clearly establish the motivation for 3D integration and the need for thermal management techniques of 3D MPSoCs.

Thermal management techniques can be divided into two categories: design-time optimizations and run-time techniques. Prior work on thermal management of 3D-ICs focused on design-time methods including thermal-aware floor-planning [11] and thermal via placement [12]. While floor-planning and thermal via placement alone may not be enough to control the thermal profile [6], it is a necessary step in the design process and is complimentary to this work.

More recent work has emphasized run-time dynamic thermal management techniques including task migration, task scheduling, and DVFS [3, 4]. Sun et al. [3] developed a task scheduling and DVFS thermal management algorithm for air-cooled 3D MPSoCs. In [4], Coskun et al. propose a similar task scheduling algorithm for 2 and 4-tier air-cooled 3D MPSoCs that shows high temperatures in excess of 85°C, motivating the need for alternative cooling mechanisms.

The application of interlayer cooling to 3D MPSoCs has been actively explored [5]. Even more recent work in this area focuses on energy and performance optimization of these techniques [6, 7]. In [6], Sabry et al. utilize a fuzzy logic controller to adjust the liquid flow rate and DVFS. In [7], Zanini et al. frame the solution as an optimization problem, applying a global controller to adjust the liquid flow rate and a local controller policy to adjust the voltage and frequency of individual system components.

Our work offers a detailed analysis on the impact of DTM policies on performance. Specifically, we examine the detailed performance impact of load balancing, and DVFS utilizing the Princeton Application Repository for Shared-Memory Computers (PARSEC) [13] benchmarks. Additionally, this work also expands the scope of 3D MPSoCs to use non-CMOS technologies, such as RRAM technologies to study its effects on thermal management. The potential application of RRAM to 3D MPSoCs is discussed in the next section.

3. INTEGRATING RRAM INTO MPSOCS ARCHITECTURES

An important feature of 3D MPSoCs is that their constituent die may be processed using heterogeneous fabrication methods [1]. This facilitates the integration of conventional CMOS ICs with emerging post-CMOS technologies for improved performance, power, and functionality. In particular, resistive random access memory (RRAM) has been identified as a potential on-chip memory technology for 3D-ICs [14, 15]. RRAM is immune to several of the scaling-related reliability issues associated with SRAM, some of which may limit SRAM's practicality past the 16 nm technology node [2]. Furthermore, RRAM's relatively small static power consumption yields significantly less heat generation, reducing thermal management-related performance degradation (further discussed in Section 6).

A one-transistor, one-resistor (1T1R) RRAM block architecture is shown in Figure 1. Each cell is composed of a single NMOS access transistor and a single resistance switch. The cell size for the 1T1R structure is $\approx 15F^2$. This is a $7\times$ improvement over SRAM's $146F^2$ cell size [16]. During the read operation, a read voltage pulse is applied to each bitline, with the addressed wordline held high. Each resistance switch can be in a low resistance state (LRS) or high resistance state (HRS). During the read operation the sense amplifier detects the switch's state using a voltage or current mode sense amplifier. During the write operation, the addressed wordline is held high, and a positive or negative write voltage is applied to each bitline, depending on the write bit value. Section 6, shows the effects of an RRAM L2 cache architecture on 3D MPSoC performance and heat generation. The effect of the DTM policies on 3D MPSoCs with RRAM L2 cache architectures are discussed in the next section.

4. DTM POLICIES

Three previously studied DTM policies [6] are evaluated in this research. These policies take advantage of air cooling, liquid cooling, load balancing, and DVFS in order to maintain temperature. The specific policies we explored are as follows:

- **Air cooled dynamic load balancing**: This policy incorporates the load balancing feature that aims to evenly distribute tasks to all cores and combines it with an air cooled heat sink. The policy establishes a baseline for system performance, as well as a worst-case thermal profile.

- **Liquid cooled dynamic load balancing**: This policy combines maximum flow rate liquid cooling with load balancing. The interlayer liquid cooled configuration establishes a baseline worst-case for pump energy.

- **Temperature triggered dynamic voltage frequency scaling**: This policy triggers DVFS when the temperature of the core crosses the high thermal trigger threshold. At each evaluation interval, if the core crosses this threshold and there is a lower V, F step, the V, F level will be reduced. Likewise, the V, F level will be increased when the temperature is below the low thermal trigger threshold.

Each of these policies is evaluated for overall system and thermal performance using the simulation framework described in the next section.

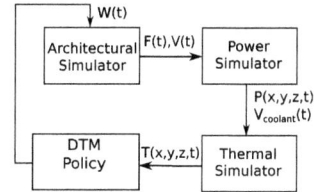

Figure 2: 3D-IC performance evaluation simulation flow.

Resource	Value
Number of cores	8, 16, 24
L2 cache (per 8 cores)	4MB, 32MB (SRAM, RRAM), 8-way @ 10ns, 70ns (SRAM, RRAM) (8 banks)
L2 cache line size	64 B
Per-core:	
CPU Clock	2.6GHz, 1.9GHz, 1.4GHz, 800MHz
Issue	out-of-order
L1 I-cache	16KB, 8-way @ 1ns
L1 D-cache	8KB, 16-way @ 1ns
L1 I & D -cache line size	64B
Issue Width	2
Functional Units	1 Load/Store, 2 INT Exec, 1 FPALU, 1 FPMultDiv
Vdd	1.25, 1.15, 1.05, 0.9 V

Table 1: Architectural model parameters

5. SIMULATION FRAMEWORK AND METHODOLOGY

5.1 Framework

The simulation framework flow used in this research is shown in Figure 2. An architectural simulator is used to obtain performance statistic profiles for the given workload, $W(t)$. For this work, gem5, a modular simulation infrastructure [17], is used to model the CPU, memory, and operating system. We chose to use gem5 due to its ability to quickly and accurately model computer systems, allowing us to capture the performance statistics for the workload at the system level. The architectural description of the MPSoC used in this work, shown in Table 1, is specified in the simulation and the specific workload benchmark, $W(t)$, is executed in gem5. The resulting performance statistics are then used in a power, timing, and area modeling framework (McPAT [18]) to obtain power data, $P(x, y, z, t)$, for each component of the system. McPAT models the gate leakage, subthreshold leakage, and dynamic power. The total power for each component is then combined with the overall system stack configuration and floorplan, shown in Figure 3, and used in a thermal simulator (3D-ICE [19]) in order to obtain the thermal profile, $T(x, y, z, t)$. Depending on the thermal profile, the DTM policy being evaluated triggers to adjust the system parameters: frequency $F(t)$, voltage, $V(t)$, volumetric flow rate $V_{coolant}(t)$, and load balancing according to the policy descriptions given in Section 4. For example, the temperature-triggered DVFS policy adjusts the clock frequency, $F(t)$, and the voltage, $V(t)$, which are inputs to the architectural and power simulators.

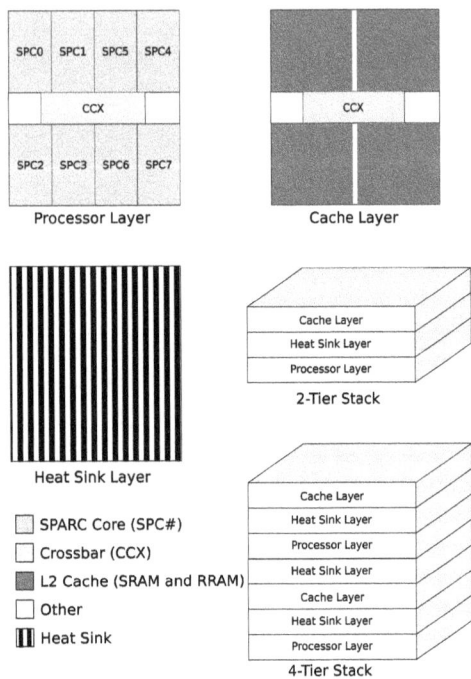

Figure 3: Floorplans and layout of the UltraSPARC T2-based 3D MPSoC. The 6-tier stack is assembled in the same manner as the 4-tier stack.

5.2 Simulation Methodology

The goal of our simulation methodology is to evaluate the performance of our target 3D MPSoC when utilizing each of the DTM policies. We examine the performance impact and the temperature profile of the DTM policies. Additionally, we explore the performance and thermal impacts of using an RRAM cache.

The 3D MPSoCs evaluated by this framework are loosely based on the 65 nm UltraSPARC T2 processor [20]. The architectural model parameters used are shown in Table 1, and the floorplan and stack configuration are shown in Figure 3. The 3D MPSoC is configured as two or more tiers. Microchannel layers between each tier allow for liquid coolant flow. Microchannels have equivalent flow rates and are uniformly distributed in each layer.

In order to emphasize the performance of multithreaded programs, individual benchmarks from the PARSEC benchmark suite [13] are used. Each benchmark is run with each of the DTM policies described in Section 4 for the 2, 4, and 6-tier configurations (8, 16, and 24 core, respectively). While the 6-tier configuration is not shown explicitly in Figure 3, it is assembled in the same manner as the 4-tier configuration. The above configurations are evaluated with both SRAM and RRAM L2 caches (the parameters of which are shown in Table 1). For the RRAM configuration, the entire cache layer is replaced with RRAM cells. The RRAM power consumption was taken as a fraction of the SRAM power using a non-volatile memory simulation tool [21]. RRAM-based L2 cache was also simulated with a 15× larger capacity to account for its higher density (due to the 15× difference in the cell size between SRAM and RRAM). This results in the 32MB cache size shown in the table.

For the air- and liquid- cooled load balancing policies, steady-state thermal analysis of the system at the highest

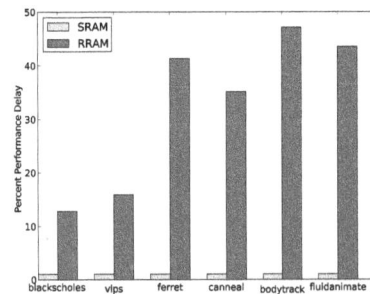

(a)

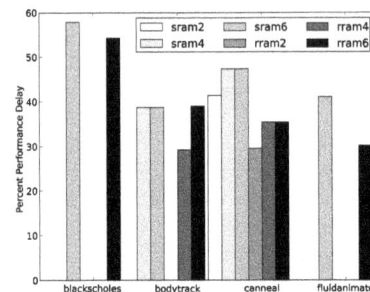

(b)

Figure 5: Performance results. (a) shows the relative RRAM (to SRAM) cache performance impact. (b) shows relative RRAM and SRAM performance impact compared to a system at peak V, F operating points.

V, F setting (peak performance) is used to establish the base case thermal performance. The overall system performance of the RRAM cache is compared to the SRAM, both at the highest V, F setting, in terms of speedup (execution time). The DVFS DTM policy was studied using transient thermal analysis by evaluating the policy at the evaluation interval (0.1 seconds) and triggering the policy. When the policy is triggered and a scaling step is necessary, a modeled DVFS delay of 200us is incurred. For DVFS, the high thermal trigger threshold is set at 80°C and the low threshold is set at 75°C. When the temperature is above the high thermal trigger, the V, F operating point is decreased (if possible), and when the temperature is below the low thermal trigger, the V, F operating point is increased (if possible). The total number of triggered events is recorded in order to evaluate performance impact.

6. RESULTS AND ANALYSIS

Each of the DTM policies in Section 4 was evaluated for the configurations (2, 4,and 6 tier) using SRAM and RRAM L2 caches. The results are summarized in Figure 4. As expected, air cooling with load balancing is not a sufficient policy for controlling the increased power densities of 3D MPSoCs. This is particularly true for high-utilization, shared-memory parallel workloads, such as the PARSEC benchmarks used in this work. Notably, the vips benchmark appears unbounded in steady state thermal analysis, reaching impossibly high temperatures that would certainly trigger thermal shutdown (Figure 4(a)). This is due to the high FP

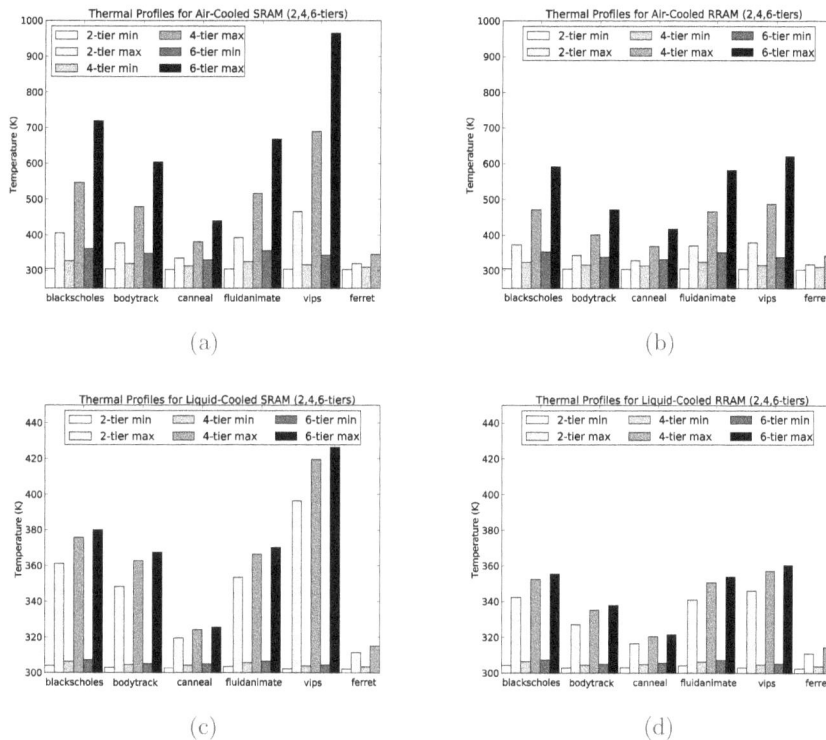

Figure 4: Thermal results for air-cooled ((a) and (b)) and liquid-cooled ((c) and (d)) SRAM and RRAM cache configurations.

utilization in this benchmark. In particular, the FP rename unit is extremely active throughout this benchmark and accounts for the majority of the power consumed per core. The RRAM cache configuration, shown in Figure 4(b), exhibits lower steady state temperatures (average decrease in maximum temperature by 30 K, 70 K, and 120 K for the 2, 4 and 6-tier systems, respectively), but the temperatures are still unacceptably high for the 4- and 6-tier configurations across nearly every benchmark shown. Again, these are expected results that firmly establish the need for additional cooling mechanisms.

The maximum flow rate liquid cooling DTM policy performs better, but still has limitations as the number of tiers grow. The flow rate used could be increased to counteract this, as well as better floorplanning. Again, we see the SRAM temperatures are likely to cause thermal shutdown and serious performance and reliability issues unless combined with additional techniques for the 4 and 6-tier MP-SoCs (Figure 4(c)). The RRAM cache configuration (Figure 4(d)), however, yields a very acceptable range of temperature values (below the DVFS trigger).

The general trend of lower temperatures in RRAM cache structures is not surprising. It is due to the performance slowdown caused by the long latency of inherent in RRAM access times (70 ns). The decrease in static power allows for much higher density circuitry, and thus larger cache sizes, but the latency penalties cancel this performance improvement. Figure 5(a) shows the normalized speedup comparing the SRAM and RRAM caches at peak performance. As shown in the figure, the increase in latency results in a worst case of nearly double the execution time for these benchmarks. This increased idle time is primarily what is responsible for the decreased thermal maxima. While the RRAM

cache static and dynamic power is significantly smaller than that of an equivalent capacity SRAM, the difference in power with a SRAM cache of the same physical area, as the one used in this work, is negligible.

Figure 5(b) shows the performance delay as a result of the DVFS policy. Configurations not shown on the plot were able to maintain temperatures below DVFS thermal triggers. The interesting result here is that due to its lower overall lower power and thermal profile, the RRAM cache configurations are able to outperform the SRAM configurations. It should be noted that if this policy were combined with other methods, such as liquid cooling, the SRAM configurations would not be throttled by the DVFS thermal triggers. However, this does show there may be some potential for use of RRAM cache in low-cost 3D MPSoCs, as the additional complexity of liquid and interlayer cooling is not necessary. That being said, there is still an average 25% performance delay in addition to that of the RRAM cache when compared SRAM cache. Air cooled DVFS proved to be an effective policy in controlling thermal profiles, however there was a great deal of switching between V, F operating points for both the SRAM and some RRAM benchmarks. This switching is reflected in the performance delay.

7. CONCLUSIONS AND FUTURE WORK

Our work has established a functional simulation framework for evaluating system performance and thermal profiles of 3D MPSoCs. We successfully used this framework to evaluate 3 different stacked die configurations, 3 DTM policies, and an RRAM cache configuration. This framework will allow for easier exploration of more advanced DTM techniques and controller optimization.

We find that RRAM based caches lower overall average

maximum temperatures by 120 K and 24 K for air and liquid cooling systems, respectively (when compared to SRAM based caches), at a worst-case performance delay of 47% and best-case delay of 13% for the parallel shared-memory benchmarks studied. Furthermore, we show that through our DVFS policy, the RRAM cache configuration is able to gain a performance advantage over the SRAM configuration.

As RRAM-based devices continue to improve in endurance and performance, we expect these model-based performance numbers to improve as well. Additionally, when combined with more efficient architectures, RRAM caches may prove an excellent fit for 3D MPSoCs.

8. REFERENCES

[1] K. Banerjee, S. Souri, P. Kapur, and K. Saraswat, "3-D ICs: a novel chip design for improving deep-submicrometer interconnect performance and systems-on-chip integration," *Proceedings of the IEEE*, vol. 89, pp. 602 –633, May 2001.

[2] "International Technology Roadmap for Semiconductors," 2010. http://www.itrs.net/reports.

[3] C. Sun, L. Shang, and R. Dick, "Three-dimensional multiprocessor system-on-chip thermal optimization," in *2007 5th IEEE/ACM/IFIP International Conference on Hardware/Software Codesign and System Synthesis (CODES+ISSS)*, pp. 117 –122, Oct. 2007.

[4] A. Coskun, J. Ayala, D. Atienza, T. Rosing, and Y. Leblebici, "Dynamic thermal management in 3D multicore architectures," in *Proceedings of the Conference on Design, Automation and Test in Europe*, pp. 1410 –1415, April 2009.

[5] T. Brunschwiler, S. Paredes, U. Drechsler, B. Michel, W. Cesar, G. Toral, Y. Temiz, and Y. Leblebici, "Validation of the porous-medium approach to model interlayer-cooled 3D-chip stacks," in *IEEE International Conference on 3D System Integration*, pp. 1 –10, Sept. 2009.

[6] M. M. Sabry, A. K. Coskun, D. Atienza, T. S. Rosing, and T. Brunschwiler, "Energy-efficient multiobjective thermal control for liquid-cooled 3-D stacked architectures," *IEEE Transactions on Computer-Aided Design of Integrated Circuits and Systems*, vol. 30, pp. 1883 –1896, Dec. 2011.

[7] F. Zanini, M. Sabry, D. Atienza, and G. De Micheli, "Hierarchical thermal management policy for high-performance 3D systems with liquid cooling," *IEEE Journal on Emerging and Selected Topics in Circuits and Systems*, vol. 1, pp. 88 –101, June 2011.

[8] C. Merkel and D. Kudithipudi, "Towards thermal profiling in CMOS/memristor hybrid RRAM architectures," in *International Conference on VLSI Design*, Jan. 2012.

[9] K. Puttaswamy, "Thermal analysis of a 3D die-stacked high-performance microprocessor," in *In Proceedings of GLSVLSI*, 2006.

[10] B. Black, M. Annavaram, N. Brekelbaum, J. Devale, L. Jiang, G. H. Loh, D. Mccauley, P. Morrow, D. W. Nelson, D. Pantuso, P. Reed, J. Rupley, S. Shankar, J. Shen, and C. Webb, "Die stacking (3D) microarchitecture," in *In Proceedings of MICRO-39*, 2006.

[11] W.-L. Hung, G. Link, Y. Xie, N. Vijaykrishnan, and M. Irwin, "Interconnect and thermal-aware floorplanning for 3D microprocessors," in *International Symposium on Quality Electronic Design*, pp. 6–104, March 2006.

[12] Z. Li, X. Hong, Q. Zhou, S. Zeng, J. Bian, H. Yang, V. Pitchumani, and C.-K. Cheng, "Integrating dynamic thermal via planning with 3D floorplanning algorithm," in *Proceedings of the 2006 International Symposium on Physical design*, ISPD '06, (New York, NY, USA), pp. 178–185, ACM, 2006.

[13] M. Gebhart, J. Hestness, E. Fatehi, P. Gratz, and S. W. Keckler, "Running PARSEC 2.1 on M5," tech. rep., The University of Texas at Austin, Department of Computer Science, October 2009.

[14] D. Lewis and H.-H. Lee, "Architectural evaluation of 3D stacked RRAM caches," in *IEEE International Conference on 3D System Integration*, pp. 1–4, Sep. 2009.

[15] M.-F. Chang, P.-F. Chiu, and S.-S. Sheu, "Circuit design challenges in embedded memory and resistive RAM (RRAM) for mobile SoC and 3D-IC," in *16th Asia and South Pacific Design Automation Conference*, ASP-DAC 2011, pp. 197–203, Jan. 2011.

[16] C. Xu, X. Dong, N. Jouppi, and Y. Xie, "Design implications of memristor-based RRAM cross-point structures," in *Proceedings of the Conference on Design, Automation and Test in Europe*, pp. 1 –6, March 2011.

[17] N. Binkert, B. Beckmann, G. Black, S. K. Reinhardt, A. Saidi, A. Basu, J. Hestness, D. R. Hower, T. Krishna, S. Sardashti, R. Sen, K. Sewell, M. Shoaib, N. Vaish, M. D. Hill, and D. A. Wood, "The gem5 simulator," *SIGARCH Comput. Archit. News*, vol. 39, pp. 1–7, Aug. 2011.

[18] S. Li, J. H. Ahn, R. D. Strong, J. B. Brockman, D. M. Tullsen, and N. P. Jouppi, "McPAT: an integrated power, area, and timing modeling framework for multicore and manycore architectures," in *Proceedings of the 42nd Annual IEEE/ACM International Symposium on Microarchitecture*, MICRO 42, (New York, NY, USA), pp. 469–480, ACM, 2009.

[19] A. Sridhar, A. Vincenzi, M. Ruggiero, T. Brunschwiler, and D. Atienza, "3D-ICE: Fast compact transient thermal modeling for 3D ICs with inter-tier liquid cooling," in *2010 IEEE/ACM International Conference on Computer-Aided Design (ICCAD)*, pp. 463 –470, Nov. 2010.

[20] M. Shah, J. Barren, J. Brooks, R. Golla, G. Grohoski, N. Gura, R. Hetherington, P. Jordan, M. Luttrell, C. Olson, B. Sana, D. Sheahan, L. Spracklen, and A. Wynn, "UltraSPARC T2: A highly-threaded, power-efficient, SPARC SOC," in *Solid-State Circuits Conference, 2007. ASSCC '07. IEEE Asian*, pp. 22 –25, Nov. 2007.

[21] X. Dong, N. Jouppi, and Y. Xie, "Pcramsim: System-level performance, energy, and area modeling for Phase-Change RAM," in *Computer-Aided Design - Digest of Technical Papers, 2009. ICCAD 2009. IEEE/ACM International Conference on*, pp. 269–275, Nov. 2009.

TSUNAMI: A Light-Weight On-Chip Structure for Measuring Timing Uncertainty Induced by Noise During Functional and Test Operations

Shuo Wang and Mohammad Tehranipoor
Dept. of Electrical & Computer Engineering, University of Connecticut
shuo.wang@engr.uconn.edu, tehrani@engr.uconn.edu

ABSTRACT

Noise such as voltage drop and temperature in integrated circuits can cause significant performance variation and even functional failure in lower technology nodes. In this paper, we propose a light-weight on-chip sensor that measures timing uncertainty induced by noise during functional and test operations. The proposed on-chip structure facilitates speed characterization under various workloads and test conditions. Simulation results show that it offers very high sensitivity to noise even under variations. The structure requires negligible area in the chip.

Categories and Subject Descriptors

B.8.1 [**Performance and Reliability**]: Reliability, Testing, and Fault-Tolerance

Keywords

Power supply noise, Temperature, On-chip measurement, Speed characterization, Post-silicon validation

1. INTRODUCTION

Over the past four decades, technology scaling has greatly improved performance and circuit integration density. However, integrated circuits (IC) performance has become less predictable by simulation at design stage due to process and environmental (temperature, crosstalk, and supply voltage noise) variations. As a result, performance limiters, such as noise in the circuit, need to be identified as early as possible during first silicon test and debug and when performing speed characterization during manufacturing test [1][4][5].

Power supply noise (PSN) and temperature's impact on functional operation of the chip as well as test has been extensively investigated in the past several years [1]−[8]. Power supply noise and temperature have shown to have both local and global effects on circuit timing [2][5]. An excessive drop in the power supply voltage can cause a temporary malfunction in the circuit. Additionally, excessive noise could result in miss-binning of the chip under test.

It is beneficial, from design timing stand-point, to measure the noise in post-silicon in functional mode to perform appropriate margining and supply voltage and frequency calibration using adaptive techniques [17]. Understanding the impact of supply noise on circuit timing can also help better correlate structural tests with functional tests for timing characterization and speed binning [9][10].

On-chip measurement architectures have gained significant attention in recent years [11]−[16] to be embedded in the chip for rapid first silicon test, debug, speed characterization, timing margining, IR-drop and temperature measurement, and wear-out mechanism analysis, for which all cause timing uncertainty in the device under test. Such architectures can help record the operation condition in the test mode as well as in the field and help perform post-silicon calibration [3][12][13]. For instance, the authors in [12] proposed SKITTER, an on-chip measurement circuit, to measure timing uncertainty from combined sources in the circuit. Although very effective in capturing noise effects, it requires large area overhead. The on-chip droop detector (ODDD) system designed in [13] enables voltage transient detection as well as a capability to induce voltage transients in a controlled manner to test and debug. The architecture is capable of measuring low frequency noise very accurately in the circuit however it will not be able to measure high frequency voltage noise as effectively, e.g., during launch-to-capture cycle in delay test schemes.

In this paper, we propose a low-cost and light-weight on-chip structure called SUpply Noise- and Temperature-Aware Timing Measurement Instrument (TSUNAMI), taking into account combined effect of supply noise and temperature on clock and on a reconfigurable delay line, to accurately measure the induced timing uncertainty even under process variations. TSUNAMI requires very low area overhead but provides high resolution and sensitivity to voltage noise, especially as technology scales. TSUNAMI consists of two major parts namely: (1) PSN sensor that is based on a reconfigurable delay line, and (2) A control vector unit that configures the PSN sensor and controls the measurement process. Both require negligible area on the chip.

TSUNAMI can operate in both test mode and functional mode. In functional mode, it can measure timing uncertainty (i) during every clock cycle of interest and (ii) within a clock cycle when applying the functional workload. The change in the timing information of the PSN sensor can be converted to the actual noise (e.g., power supply noise) information. In test mode, TSUNAMI can help measure voltage noise during scan as well as launch-to-capture cycle. In addition, it can help measure the noise level depending on the applied launch cycle to analyze the impact of Ldi/dt during

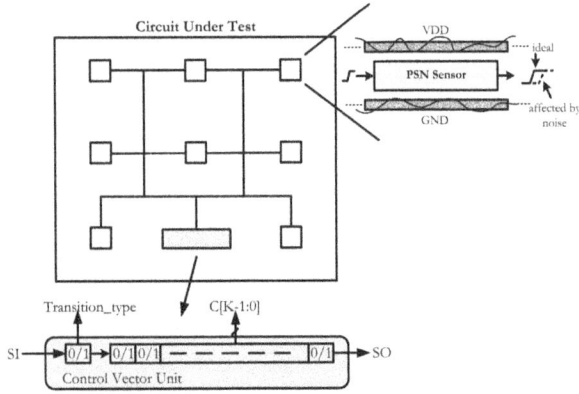

Figure 1: TSUNAMI architecture.

launch-to-capture cycle. Note that, although we call our sensor a PSN sensor, it is in fact able to capture combined timing uncertainty induced by both power supply noise (voltage drop and ground bounce) and temperature in every clock cycle.

The rest of the paper is organized as follows. In Section 2, we introduce the proposed TSUNAMI architecture. Calibration of TSUNAMI is discussed in Section 3. Section 4 presents problem modeling and design flow of TSUNAMI, and Section 5 presents simulation results and analysis. Finally, Section 6 concludes the paper.

2. TSUNAMI ARCHITECTURE

Today's modern designs include very large power distribution network. The voltage noise distribution in the design is not uniform as different blocks in the chip switch differently. Therefore, to take a snapshot of voltage noise distribution in the circuit, we need to insert sensors and measure the noise at various locations of interest under different workloads and test conditions. For instance, one area of interest for sensor insertion is near critical paths. The information can then be analyzed for characterization during post-silicon test, debug, and calibration.

The architecture of TSUNAMI, shown in Fig. 1, consists of PSN sensors, which are distributed across the layout to capture noise at different locations, and a control vector unit, which controls all the sensors.

Transitions are generated at the input of the PSN sensor and propagated through its components. The arrival time and slew of the transitions are affected by the noise on the power/ground lines generated by circuit switching. The PSN sensor is designed to be sensitive to noise. For the sensor to capture noise, it is preferred to be designed as a single macro and placed between power and ground lines. The noise on the power/ground lines will impact the transition propagation time (since it impacts gates' delay in the sensor). The impact of noise can therefore be observed as an additional delay.

Figure 2 shows the detailed implementation of the PSN sensor which consists of the following components: (1) a reconfigurable delay line (RDL), (2) a transition generation (TG) cell, and (3) a transition capture (TC) cell.

1. Reconfigurable Delay Line (RDL): RDL is composed of K reconfigurable stages in addition to an extra fixed stage; each reconfigurable stage contains a multiplexer choosing from an input branch with/without buffer, thereby providing different delay values, while the fixed stage is a series of buffers providing an additional delay that increases beyond what the K reconfigurable stages provide without

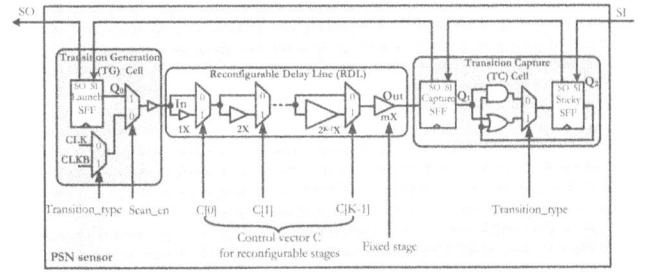

Figure 2: Our proposed PSN sensor.

introducing an additional reconfigurable stage. Thus, by controlling the select signals $C[0]$, $C[1]$, ..., $C[K-1]$, the delay of the entire line becomes reconfigurable. Furthermore, delay of the buffer at each stage is twice the buffer delay in the previous stage. Thus, if each MUX has a delay of t_x and the minimum-sized buffer has a delay of t_b, the minimum and maximum delay that RDL can provide are $t_{min\_RDL}=K*t_x+m*t_b$ and $t_{max\_RDL}=K*t_x+(2^K-1+m)*t_b$, respectively. Since the PSN sensor is designed as a single module, the interconnect delay between the sensor's components is considered negligible.

Different C values create different paths between the input (In) and output (Out) of the RDL. When C="00....00"=0, the path only includes the MUXes, making it the shortest path. However, when C="00....01"=1, the path goes through the first buffer and the remaining MUXes. Finally, when C="11....11"=2^K-1, the path goes through all buffers and MUXes, making it the longest path. The key is to find the C value that allows propagation of a transition from input to output of RDL in just one clock cycle in presence of noise. This C value is further analyzed for understanding the amount of noise incurred by the applied pattern.

It is noteworthy that even for the same control vector C, the total delay varies at different supply voltages. When the delay line is always reconfigured to be constant (e.g., one clock cycle), the variance of the control vector C represents the fluctuation on the power supply, assuming the temperature stays about the same. In this way, the magnitude of power supply noise can be measured and converted to a digital value. The more the noise on the RDL is, the lower the speed of the buffers and MUXes in the sensor is. Thus, the transition takes more time to go through the RDL resulting in an error, signaling that a smaller C value is needed.

2. Transition Generation (TG) Cell: TG cell is inserted at the input of the reconfigurable delay line as shown in Fig. 2, which consists of two MUXes and a launch scan flip-flop (SFF). Transitions at the input of the RDL is generated depending on the circuit operation mode:

- *Functional and Scan Modes*: In these modes, when a workload or a test pattern is applied, the sensor must be able to measure the delay of the RDL in every clock cycle based on the applied C input. Thus, clock signal "CLK", or the inverted clock "CLKB", can be fed into the reconfigurable delay line when a series of rise transitions ("Transition_type=0") or fall transitions ("Transition_type=1") are needed.

- *Launch-to-Capture Mode*: In this mode, initial values can be shifted in during scan mode ("Scan_en=1") to the launch SFF generating the desired transition at launch cycle. In this mode, only one transition is needed when "Scan_en=0" in either launch-off-capture (LOC) or launch-off-shift (LOS) scheme. Figure 2 shows the scan path (SI to SO) going through the FFs in the PSN sensor. Shifting certain values into the three scan FFs will ensure (1) generating a rise/fall at TG cell during launch cycle in delay test and (2) 1/0 at the capture SFF and the sticky SFF as the initial value.

3. Transition Capture (TC) Cell: TC cell is implemented at the end of the reconfigurable delay line to capture transitions. It consists of two SFFs, namely capture SFF and sticky SFF, and a combinational logic to decide whether any of the transitions is not captured properly.

Let us assume the type of transition applied to RDL is rise ("Transition_type=0"). The initial values Q_1 and Q_2 at capture SFF and sticky SFF, respectively, are both set to 1 during scan mode. If a rise transition is not captured, Q_1 becomes 0, and so does Q_2 after one cycle. Then, Q_2 at the sticky SFF stays low even if later Q_1 becomes 1 again after a successful capture of rise transition. Later, when Q_2 is shifted out for analysis, we know that there has been at least one transition failed to be captured during measurement due to a large noise. C value then should be reduced further to make the path shorter; the pattern application is repeated until the path passes the test. The C value that makes the path passes the test thus represents the amount of supply voltage fluctuation on the power/ground lines. Generally, the smaller the C value is, the more noise has appeared on the power/ground lines.

Control vector (CV) unit is the second main component in the TSUNAMI architecture, as shown in Fig. 1. CV unit applies C values to the PSN sensors. Note that hereafter we use C value and control vector intermittently as both represent the select signals for the PSN sensors.

In the CV unit, control signals are applied to all PSN sensors from the scan chain, which includes K-bit control vector C and 1-bit "Transition_type". The scan chain is controlled by the external tester. The C value is shifted into the scan chain which goes to all sensors. Then the workload is applied to the chip and the PSN sensors start capturing transitions generated by the TG cell. After the workload application is over, results in terms of whether the transitions are captured at each sensor are shifted out for analysis. If a new round of measurement is needed, a new control vector (C) is generated and the same procedure is repeated. The collected data is then analyzed for each sensor to obtain the amount of noise each sensor has experienced.

3. CALIBRATION

Calibration is required before noise measurement to establish a mapping relationship between C value and supply voltage. This process begins under no or little background noise condition to identify the C value that makes the path delay one clock cycle. To achieve this, a stable V_{dd} must be applied to the sensor. To analyze the impact of noise on the sensor, new supply voltage V_{dd} is generated and adjusted at a fine granularity and applied to the entire circuit. PSN measurements are performed for each selected $V_{dd} = v_i$. The measurement results (i.e., control vectors) serve as calibration values for given V_{dd}. In other words, a mapping relationship between control vector C and supply voltage V_{dd} is established. Note that, as temperature also plays a role in circuit delay, the calibration process should be performed at similar temperature to that during noise measurement, so that the impact of temperature on delay can be canceled out. This can be achieved by warming up the circuit under test using the workload for measurement before we actually perform calibration. We also assume there are certain on-chip methods [7][8] in place to measure temperature in order to verify that temperature during calibration is indeed close to that during measurement.

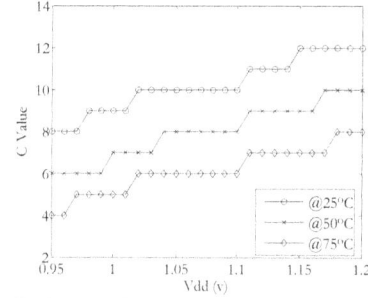

Figure 3: Relationship between V_{dd} and control vector C value (obtained using rise transitions).

An example of HSpice simulation is shown in Fig. 3. A 4-stage (K=4) RDL is implemented in a PSN sensor designed in 90nm technology. Supply voltage is adjusted at granularity of 10mV during calibration. The nominal voltage V_{dd}=1.2V. Decreasing V_{dd} represents voltage noise being generated in the circuit. As V_{dd} decreases, C value also decreases accordingly; a smaller C makes the path delay shorter. For the path to fail for the applied transition, more noise must be applied. This same calibration procedure is tested at different temperatures, i.e., $25^\circ C$, $50^\circ C$, and $75^\circ C$, in this example. However, note that in practice the calibration procedure only needs to be performed once at the equivalent temperature during the measurement mode. Figure 3 clearly shows that measurement results when there is power supply noise in the background can be translated to the magnitude of PSN. Note that rise transitions are used to obtain the relationship shown in Fig. 3. The mapping relationship obtained from fall transitions is slightly different.

It is also noteworthy that measurement results for the same V_{dd} can differ across individual PSN sensors due to process variations. No process variations have been applied to the PSN sensor in this example. However, we will apply variations to the sensor and the results are shown in Section 5 for various process corners.

4. PROBLEM MODELING AND DESIGN FLOW

The design goal of TSUNAMI architecture is to provide the maximum measurement resolution on power supply noise within the budget of area overhead. In this section, we first model the problem of designing the PSN sensor and present the design flow. Then, we analyze the measurement resolution and calibration/measurement time.

1. Problem Modeling: Suppose we need to design a K-stage PSN sensor for a circuit that operates at a clock cycle of T_{clk}. We also assume that the nominal delays of the minimum-sized buffer and the MUXes are t_b and t_x, respectively. In order to ensure that the delay of the RDL meets the one-cycle requirement, the control vector value C must satisfy the equation below:

$$K * t_x + (C + m) * t_b = T - \Delta - \varepsilon * t_b, \quad (1)$$

where Δ is a combinational effect of: (i) the delay of TG cell, (ii) the setup time of the flip-flop in the TC cell, as well as (iii) clock variations. $0 \le \varepsilon < 1$ so that the C value just meets the requirement. m is the number of buffers in the fixed stage within the delay line.

Due to process variations, voltage noise, and temperature, the actual delay can vary in the range of $[t_{b,min}, t_{b,max}]$ and $[t_{x,min}, t_{x,max}]$, respectively. Δ can also vary in a similar way. Consequently, C will also vary from C_{min} to C_{max} as described below.

$$K * t_{x,max} + (C_{min} + m) * t_{b,max} = T - \Delta_{max} - \varepsilon_1 * t_{b,max}, \quad (2)$$
$$K * t_{x,min} + (C_{max} + m) * t_{b,min} = T - \Delta_{min} - \varepsilon_2 * t_{b,min}, \quad (3)$$
$$0 \leq C_{min} < C_{max} \leq 2^K - 1. \quad (4)$$

where $0 \leq (\varepsilon_1, \varepsilon_2) < 1$.

Clearly, the measurement resolution is confined by the number of different observable C values $Diff = C_{max} - C_{min} + 1$, whereas the area overhead is determined by the number of stages K. Therefore, the design problem of finding the optimal K for the PSN sensor design can be expressed below:

$$\text{Maximize: } Diff = C_{max} - C_{min} + 1, \quad (5)$$

$$\text{Subject to: } 0 \leq C_{min} < C_{max} \leq 2^K - 1, \quad (6)$$

$$K \leq K_{budget}. \quad (7)$$

where (5) reflects the efforts to maximize measurement resolution; (6) shows that PSN sensor must be capable of covering the full range of timing variations due to process, supply noise, and temperature, shown also in (2)-(4); (7) is a constraint that overhead budget on stage count K should be met.

Next, we study the measurement resolution and calibration/measurement time:

• *Measurement Resolution:* Delay is determined by supply voltage V_{dd} and threshold voltage V_{th} for a circuit under given temperature. Approximately, *delay* $\propto \frac{V_{dd}}{V_{dd} - V_{th}}$. Without loss of generality, we assume all the devices in the circuit have a universal delay that increases from *delay* to *delay'* = *delay* $* \gamma$, when the circuit experiences a voltage drop from V_{dd} to V'_{dd}. That is,

$$\gamma = \frac{t'_x}{t_x} = \frac{t'_b}{t_b} = \frac{\Delta'}{\Delta} = n * \left(\frac{1-h}{n-h} \right), \quad (8)$$

where $n = \frac{V'_{dd}}{V_{dd}}$ and $h = \frac{V_{th}}{V_{dd}}$.

We further assume that when supply voltage drops from V_{dd} to V'_{dd}, the right C value reduces accordingly from C to C'. Based on (8) and (1), we can obtain the following

$$C - C' = \frac{\gamma - 1}{\gamma} * \frac{T_{clk}}{t_b} - (\varepsilon - \varepsilon'), \quad (9)$$

where $(\varepsilon - \varepsilon') \in (-1, 1)$.

To have two distinguishable voltage levels V_{dd} and V'_{dd}, it requires that C and C' are measured at different values (i.e., $C - C' \geq 1$). Thus, we can derive the lower bound for the measurement resolution that can distinguish V'_{dd} from V_{dd} as

$$min(n = \frac{V'_{dd}}{V_{dd}}) > \frac{1}{1 + 2 * \frac{t_b}{T_{clk}} * (\frac{1}{h} - 1)}. \quad (10)$$

The derived lower bound is in line with our intuition that for a given clock frequency faster buffers can provide better measurement resolution on the power supply noise.

• *Calibration and Measurement Time:* The calibration time T_{calib} and measurement time T_{meas} under the worst case can be obtained as:

$$T_{calib} \leq t_{calib} * (\lceil log_2(C_{max} - C_{min} + 1) \rceil + N - 1), \quad (11)$$
$$T_{meas} \leq W * t_{meas} * (\lceil log_2(C_N - C_1 + 1) \rceil). \quad (12)$$

where t_{calib} (t_{meas}) is the time it takes to shift in the control vector and to check whether it passes the calibration (measurement). N is the number of different voltage levels

considered in the calibration, ranging from $V_{dd,1}$ to $V_{dd,N}$ ($V_{dd,1} < V_{dd,N}$). A binary search can be performed in the range from C_1 to C_N, which are respectively the C values found during calibration for $V_{dd,1}$ and $V_{dd,N}$. W is the number of workload under study. For example, if $C_{min} = 4$, $C_{max} = 13$, $C_1 = 6$, $C_N = 10$, and $N = 26$ different voltage levels, the total time $T_{total} = T_{calib} + T_{meas} \leq 29t_{calib} + W * 3t_{meas}$.

2. Design Flow: According to (5)−(7), we can design the PSN sensor using the flow shown in Algorithm 1. The largest stage count K that is within budget and can provide maximum difference between C_{max} and C_{min} gives the optimal design for the best measurement resolution achievable.

Input:
$T_{clk}, \Delta_{max}, \Delta_{min}, t_{b,max}, t_{b,min}, t_{x,max}, t_{x,min}, K_{budget}$
Output:
K_{opt}, m_{opt}
1 **begin**
2   initialize K with a relatively small value
3       $MAX\_Diff = 0$
4   **while** $K \leq K_{budget}$ **do**
5   calculate m, C_{min}, C_{max} according to (2)−(4)
6   **if** $C_{max} - C_{min} \geq MAX\_Diff$ **then**
7               $K = K + 1$
8   **else**
9   break out of the while-loop
10   **end**
11   **end**
12   return $(K_{opt}, m_{opt}) = (K, m)$
13 **end**

Algorithm 1: PSN sensor design flow.

5. RESULTS AND ANALYSIS

In this section, we evaluate TSUNAMI using HSpice simulations. TSUNAMI is implemented in 90nm technology and the clock frequency for the circuit under test is 1 GHz. We intentionally selected a high frequency for the circuit to demonstrate the efficiency of TSUNAMI for measuring noise for modern designs. To evaluate the sensitivity of TSUNAMI to power supply noise and process variations, we apply various power supply noises to PSN sensors at different process corners and temperatures. Based on the simulation results, we will then estimate the measurement resolution at lower technology nodes. Note that, due to lack of space, in this paper, we only provide results for rise transition; calibration and measurement results from fall transitions are almost the same.

The variations considered in the experiments are: 10% (3 sigma) variation on the effective channel length L_{eff}, and 15% (3 sigma) variation on the threshold voltage V_{th}. As Monte Carlo simulations are extremely time consuming for HSpice, we obtained three corners, i.e., *nominal*, *slow*, and *fast*, and perform simulations on these corners. We perform calibration and measurements for sensors at the three process corners to evaluate the impact of process variations. Meanwhile, three different temperatures $25^\circ C$, $50^\circ C$, and $75^\circ C$ are also used in the simulations to show the impact of temperature as well.

As mentioned earlier, TSUNAMI has two operation modes: calibration and measurement. First, we apply different supply voltages V_{dd}'s to the PSN sensors (K=4) at a granularity of 10mv in the range of [0.95v, 1.20v], where 1.2v is the nominal supply voltage and 0.95v represents more than 20% drop from the nominal voltage level. Note that during calibration background workload is quiet. Mapping relationship between V_{dd} and control vector is then established. The results are shown in Table 1. Each C value corresponds to a specific range of V_{dd} for the given process corner and tem-

Table 1: Mapping relationship between C value and V_{dd} (v) obtained during calibration.

Corner&Temp.		C Value											
		0-3	4	5	6	7	8	9	10	11	12	13	14-15
Slow	25°C	<0.95					0.95-0.97	0.98-1.02	1.03-1.06	1.07-1.14	1.15-1.19	1.20	>1.20
	50°C	<0.95			0.95	0.96-1.03	1.04-1.08	1.09-1.15	1.16-1.20	>1.20			
	75°C	<0.95	0.95-1.00	1.01-1.05	1.06-1.15	1.16-1.20	>1.20						
Nominal	25°C	<0.95					0.95-0.97	0.98-1.01	1.02-1.10	1.11-1.14	1.15-1.20	>1.20	
	50°C	<0.95			0.95-0.99	1.00-1.03	1.04-1.10	1.11-1.16	1.17-1.20	>1.20			
	75°C	<0.95	0.95-0.96	0.97-1.01	1.02-1.10	1.11-1.17	1.18-1.20	>1.20					
Fast	25°C	<0.95						0.95-0.97	0.98-1.05	1.06-1.09	1.10-1.16	1.17-1.20	>1.20
	50°C	<0.95				0.95-0.98	0.99-1.06	1.07-1.11	1.12-1.20	>1.20			
	75°C	<0.95		0.95-0.96	0.97-1.05	1.06-1.12	1.13-1.20	>1.20					

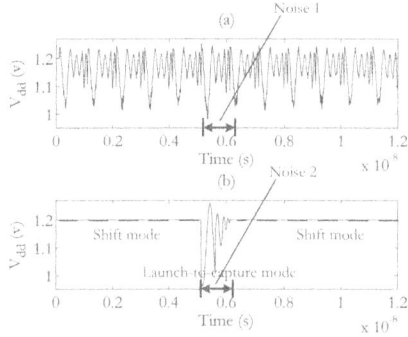

Figure 4: Power supply noise applied during measurements: (a) functional mode and (b) test mode.

Table 2: Measurement results for noise.

Process@Temp.		Noise 1 (actual: ~1.16 v)		Noise 2 (actual: ~1.15 v)	
		C	Meas. (v)	C	Meas. (v)
Slow	25°C	11	1.15-1.19	10	1.07-1.14
	50°C	9	1.16-1.20	8	1.09-1.15
	75°C	7	1.16-1.20	6	1.06-1.15
Nominal	25°C	12	1.15-1.20	11	1.11-1.14
	50°C	10	1.17-1.20	9	1.11-1.16
	75°C	7	1.11-1.17	7	1.11-1.17
Fast	25°C	12	1.10-1.16	12	1.10-1.16
	50°C	10	1.12-1.20	10	1.12-1.20
	75°C	8	1.13-1.20	8	1.13-1.20

perature. This clearly shows that PSN sensors are sensitive to process variations and temperature. As long as the temperature during calibration and that during measurements are similar, they should share the same mapping relationship between V_{dd} and C value.

From Table 1, we can obtain the measurement resolution on average to be about $53mv$ for the 90nm technology. It is calculated from dividing the voltage range of $[0.95v, 1.2v]$, which is $250mv$, by the average count of different C values in each case, which is 4.7. Note that for different temperature the measurement resolution can be different. For example, average measurement resolution when temperature is $75°C$ is calculated to be $58mv$, while resolution when temperature is $25°C$ is $50mv$. We can also verify the measurement resolution bound calculated from (10). For example, when the sensor is at the fast corner, the bound of measurement resolution is calculated to be $\frac{V'_{dd}}{V_{dd}} > \frac{1}{1+\frac{2*41}{1000}*(\frac{1}{0.4}-1)} = 89\%$. This means that for any two voltage levels, as long as they have more than 11% difference in between, their C values should be different. According to the results from Table 1, in the worst case, the indistinguishable voltage levels are $0.97v$ and $1.05v$ when temperature is $75°C$. The difference is $0.97/1.05 = 92\% > 89\%$. This verifies that the bound $\frac{V'_{dd}}{V_{dd}} > 89\%$ indeed holds.

In the next step, we apply power supply noise of different magnitudes to the circuit. Two cases of noise (Noises 1 and 2) are generated to represent functional mode and test mode noises, respectively. Segments of the noises are shown in Fig. 4. Specifically, Noise 1 has a mean value of about $1.16v$ during the period specified in Fig. 4(a), which mimics the power supply noise we have seen from silicon in functional mode; Noise 2 mimics a typical noise during test mode (launch-to-capture cycle), where the average voltage level in a clock cycle drops to about $1.15v$ (and the lowest voltage level reaches at $0.96v$). They are applied as V_{dd} to the PSN sensors. Note that without loss of generality, noises are only applied at V_{dd} while ground line does not experience noise throughout the simulation. However, noise on ground line will affect the results in a similar manner as in power line.

When power supply noise is applied, PSN sensors at dif-ferent process corners and temperatures measure V_{dd} with noise and the results are shown in Table 2. C values are control vector results from the measurements, $Meas.$ values are V_{dd} measured by comparing the C values in measurement and those in calibration (Table 1).

For example, when Noise 1 is applied and the sensor at the slow corner when temperature is $25°C$ reports C value at 11, we can look up this value from Table 1 and find that it represents $1.15 - 1.19v$ during calibration. Therefore, Noise 1 is measured to be somewhere in that range, which is consistent with the average level of $1.16v$. There are a few measurements that are slightly off, such as Noise 1's measurement at nominal corner when temperature is $50°C$ ($10mv$ deviation) and Noise 2's measurement at slow corner when temperature is $25°C$ ($10mv$ deviation). An important reason is that TSUNAMI captures not only a combined effect of noise and temperature, but also clock jitter and setup time variation of the flip-flop in the TC cell. These effects can result in a different mean voltage level within a measurement clock cycle even for the same supply noise. Thus, the noises seem as if they are in a slightly different range compared with other measurements. Nonetheless, they show the same combined effect of timing uncertainty that the functional circuit is experiencing.

In the above simulations for 90nm technology, the average measurement resolution is about $53mv$. For lower technology nodes, we expect the resolution to be higher based on the analysis in Section 4. Although we do not have simulation available for these technology nodes, we would like to provide an estimation on measurement resolution for 45nm and 32nm technologies.

Suppose buffer delay at 45nm and 32nm technologies is 26ps and 18ps for nominal corner at room temperature, respectively. We assume that the ratios of $\frac{t_{b,max}}{t_b}$ and $\frac{t_{b,min}}{t_b}$ at these technology nodes are equal to that at 90nm technology. Thus, we can estimate $t_{b,max}$ and $t_{b,min}$ to be the values listed in Table 3. Using these delay values in the PSN sensor design flow (Algorithm 1) we obtain K, m, C_{max}, and C_{min} for each technology node. Thus, $C_{max} - C_{min} + 1$ indicates totally how many different C values are available for distinguishing power supply noise. In addition, we assume the voltage measurement range is still $250mv$, the same as that in our simulation for 90nm technology. Hence, we can project the measurement resolution for 45nm and 32nm based on the additional different C values rendered by the faster devices, assuming the improvement on measurement

Table 3: Measurement resolution at different technology nodes when $T_{clk} = 1ns$.

Technology	MUX's Delay (ps)			Buffer's Delay (ps)			Sensor Design					resolution (mv)
	t_x	$t_{x,max}$	$t_{x,min}$	t_b	$t_{b,max}$	$t_{b,min}$	K	m	C_{max}	C_{min}	$C_{max} - C_{min} + 1$	
90nm	108	175	104	42	67	41	4	0	13	4	10	53
45nm	67	108	64	26	42	25	5	0	23	8	16	33
32nm	47	76	45	18	29	17	5	8	31	9	23	23

Table 4: Comparison with SKITTER [12]

Technology	SKITTER		TSUNAMI			
	Stage Cnt.	Trans. Cnt.	Stage Cnt.	Trans. Cnt.	$\Delta Area$	ΔT_{meas}
90nm	143	13728	4	262	51X ↓	∼ 2X ↑
45nm	231	22176	5	338	65X ↓	∼ 3X ↑
32nm	334	32064	5	370	86X ↓	∼ 4X ↑

resolution is proportional to the increase of different C values observable. It is noteworthy that the measurement resolution estimated in this way is the same as that estimated by the resolution bound from (10), which only takes into account the nominal buffer delay (t_b). This suggests that one can estimate the achievable measurement resolution quickly according to (10).

Note that in Table 3 the number of buffers in the fixed stage (m) for 90nm and 45nm designs are both 0. In other words, there is essentially no fixed stage in these two designs. However, $m = 8$ for 32nm PSN sensor design, which indicates that there should be a fixed stage consisting of 8 buffers in the RDL. The benefit of this is that without the fixed stage C_{max} will have to be increased to 39, which is beyond what a 5-stage design has to offer - the fixed stage inserted can therefore avoid an otherwise 6-stage design that incurs much larger overhead while the achievable resolution is still equivalent.

Next, we examine the area overhead of the PSN sensor and compare it with SKITTER sensor [12]. The overhead of a K-stage PSN sensor comes from: (i) $K + 3$ MUXes in the RDL and in the TG and TC cells, (ii) $2^K + m$ buffers, (iii) 3 scan flip-flops, and (iv) 2 extra logical gates (AND and OR) in TC cell. In contrast, the original SKITTER sensor requires at least $\frac{3*T_{clk}}{t_{inv}}$ stages to ensure that transition edges of 3 clock cycles can be covered, where t_{inv} is the inverter delay. Thus, it has equivalently $\frac{6*T_{clk}}{t_b}$ stages. Each stage has 1 inverter, 2 FFs, 1 AND, 1 OR, and 1 XOR. When technology scales down into 45nm and 32nm technology, each SKITTER sensor tends to have increasing number of stages, thereby larger area overhead. Thus, stage count and transistor count for TSUNAMI PSN sensor and SKITTER sensor are estimated at different technology nodes as shown in Columns 2 through 5 of Table 4. Note that for 90nm technology we calculate the stage count of SKITTER sensor to be 143, which is different from that in the original SKITTER (i.e., 129 [12]), due to probably different inverter delay and clock cycle. Nevertheless, the trend can be clearly seen that TSUNAMI sensor provides a light-weight solution for power supply noise measurement, at a 51 to 86 times smaller area overhead compared with SKITTER as shown in Column 6 ($\Delta Area$) of Table 4.

Based on (11) and (12), we can also see that TSUNAMI takes about 2 to 4 times longer measurement time than SKITTER as shown in Column 7 (ΔT_{meas}) of Table 4. This is because that TSUNAMI requires multiple runs to find out the right C value. However, it is considered worthwhile to trade some extra time spent during silicon validation for the significantly reduced area overhead on every chip. Take a 32nm one-million-transistor design for example; if we implement 50 sensors on the chip, the area overhead would be 3.2% for SKITTER whereas only a negligible 0.04% for TSUNAMI.

6. CONCLUSIONS

The proposed TSUNAMI architecture provides a low-cost and light-weight solution to measure timing uncertainty induced by voltage drop and temperature in integrated circuits. TSUNAMI can work at different operation modes, hence is helpful for speed characterization under various workloads and test conditions. Simulation results show that TSUNAMI offers high resolution at low technology nodes at significantly reduced area overhead compared to existing work.

7. ACKNOWLEDGEMENTS

This work is supported in part by Semiconductor Research Corporation (SRC) under grants 2053 and 2094, and a gift from Cisco.

8. REFERENCES

[1] J. Saxena, et al., "Case Study of IR-Drop in Structured At-Speed Testing," Intl. Test Conf., pp. 1098–1104, 2003.

[2] M. Tehranipoor and K. Butler, "Power Supply Noise: A Survey on Effects and Research," IEEE Design & Test, vol. 27, no. 2, pp. 51–67, 2010.

[3] Z. Abuhamdeh, et. al., "A Production IR-Drop Screen On a Chip," IEEE Design & Test, vol. 24, no. 3, pp. 216–224, 2007.

[4] J. Wang et al., "Modeling Power Supply Noise in Delay Testing," IEEE Design & Test, vol. 24, no. 3, pp. 226–234, 2007.

[5] P. Pant et al., "Understanding Power Supply Droop During At-Speed Scan Testing," IEEE VLSI Test Symp., pp. 227–232, 2009.

[6] K. Arabi et al., "Power Supply Noise in SOCs: Metrics, Management, and Measurement," IEEE Design & Test, vol. 24, no. 3, pp. 236–244, 2007.

[7] T. Yasuda, "On-chip temperature sensor with high tolerance for process and temperature variation," IEEE Intl. Symp. Circuits and Systems, pp. 1024–1027, 2005.

[8] S. Sharifi and T. S. Rosing, "Accurate Direct and Indirect On-Chip Temperature Sensing for Efficient Dynamic Thermal Management," IEEE Trans. Computer-Aided Design of Integrated Circuits and Systems, vol. 29, no. 10, pp. 1586–1599, 2010.

[9] J. Zeng et al., "On Correlating Structural Tests with Functional Tests for Speed Binning of High Performance Design," Intl. Test Conf., pp. 31–37, 2004.

[10] S. Sde-Paz et al., "Frequency and Power Correlation Between At-Speed Scan and Functional Tests," Intl. Test Conf., pp. 1–9, 2008.

[11] X. Wang et al., "Path-RO: A Novel On-Chip Critical Path Delay Measurement Under Process Variations," Intl. Conf. on Computer-Aided Design, pp. 640–646, 2008.

[12] R. Franch et. al., "On-Chip Timing Uncertainty Measurements on IBM Microprocessors," Intl. Test Conf., pp. 1–7, 2007.

[13] R. Petersen et. al., "Voltage Transient Detection and Induction for Debug and Test," Intl. Test Conf., pp. 1–10, 2009.

[14] A. Muhtaroglu et al., "On-Die Droop Detector for Analog Sensing of Power Supply Noise," IEEE J. of Solid-State Circuits, vol. 39, no. 4, pp. 651–660, 2004.

[15] E. Alon et al., "Circuits and Techniques for High-Resolution Measurement of On-Chip Power Supply Noise," IEEE J. of Solid-State Circuits, vol. 40, no. 4, pp. 820–828, 2005.

[16] Y. Lee et al., "The Impact of PMOS Bias-Temperature Degradation on Logic Circuit Reliability performance," Microelectronics Reliability, vol. 45, no. 1, pp. 107–114, 2005.

[17] J. Tschanz et al., "Effectiveness of Adaptive Supply Voltage and Body Bias for Reducing the Impact of Parameter Variations in Low Power and High Performance Microprocessors," IEEE J. of Solid-State Circuits, vol. 38, pp. 826–829, 2003.

Lazy Suspect-Set Computation: Fault Diagnosis for Deep Electrical Bugs

Dipanjan Sengupta
Dept. Elec. & Comp. Eng.
University of Toronto
Toronto, Canada
dipanjan@eecg.toronto.edu

Flavio M. de Paula
Dept. of Computer Science
University of British Columbia
Vancouver, Canada
depaulfm@cs.ubc.ca

Alan J. Hu
Dept. of Computer Science
University of British Columbia
Vancouver, Canada
ajh@cs.ubc.ca

Andreas Veneris
Dept. Elec. & Comp. Eng.
University of Toronto
Toronto, Canada
veneris@eecg.toronto.edu

André Ivanov
Dept. Elec. & Comp. Eng.
University of British Columbia
Vancouver, Canada
ivanov@ece.ubc.ca

ABSTRACT

Current silicon test methods are highly effective at sensitizing and propagating most electrical faults. Unfortunately, with ever increasing chip complexity and shorter time-to-market windows, an increasing number of faults escape undetected. To address this problem, we propose a novel technique to help identify hard-to-find electrical faults that are not detected using conventional test methods, but manifest themselves as observable functional errors during functional test, system test, or during actual use in the field. These faults are too sequentially deep to be diagnosed using simulation, ATPG, or formal tools. Our technique relies on repeated full-speed chip runs that witness the functional bug, combined with some additional on-chip functional debug support and off-line analysis, to compute a possible set of suspected faults. The technique quickly prunes the suspect set, and for each suspect, it can provide a short test vector for further analysis. Experiments on the ITC'99 benchmarks demonstrate the effectiveness of our approach.

Categories and Subject Descriptors

B.7.2 [**Design Aids**]: Verification; B.7.3 [**Reliability and Testing**]: Testability

General Terms

Algorithms, Performance, Verification

Keywords

Electrical Fault, Post-Silicon Debug, Satisfiability

1. INTRODUCTION

One of the most challenging problems in post-silicon debug is the diagnosis of a fault that has eluded conventional

testing but manifests only later as an observable functional error when the chip is running full-speed during bring-up, system test, or actual use. We dub these bugs "deep electrical bugs" because they have been first sighted only at extreme sequential depth (e.g., many billions of cycles after only seconds of silicon run time). In this paper, we propose a novel technique to help diagnose these bugs.

Existing methods from pre-silicon verification or manufacturing test do not solve this problem. Pre-silicon verification does not consider electrical faults. One could imagine mutating the RTL with a postulated fault and then applying simulation or formal verification — and indeed, this is the primary debugging technique once the set of possible faults is greatly narrowed, and a very short trace demonstrating the bug has been captured — but the vast set of possible faults, the slow speed of simulation, and the capacity limits of formal verification prevent employing these techniques initially for fault diagnosis. ATPG and other methods from manufacturing test explicitly consider faults, but are also inadequate for initial diagnosis of deep electrical bugs. In particular, ATPG algorithms face two fundamental complexity limits. First is the need to cover a fault model. Because the goal is 100% coverage, every possible fault must be analyzed, even if they are irrelevant to a specific, sighted bug. The second limit is even worse: sequential ATPG algorithms [13] blow-up exponentially in the sequential depth (number of time frames). As noted earlier, if we are trying to debug a crash that occurs after a few seconds of the actual silicon running, then it corresponds to billions of cycles of sequential depth — well beyond the capabilities of sequential ATPG methods.

The post-silicon debug team has the luxury of sequential depth, as they are the ones running applications full-speed on real silicon. In our problem scenario, they are the ones who sight the bug. But the almost complete lack of controllability and observability forces them into an extremely challenging, ad hoc debug flow [6]. Some techniques can enhance observability, but these are slow and very limited. For example, physical probing can measure the voltages on a handful of on-chip signals (as long as accessible probe points are available) [5]. However, decreasing feature-size, flip-chip technologies, and growing complexity of chips make such a method cumbersome [7]. Scan chains [11] can aid observability, but typically provide only one-cycle snapshot of the chip's scan chain. Trace buffers [1] provide a recording of hundreds or more cycles, but only of a few key signals. All of these methods help, but it is exceptionally hard to get that scan or trace buffer dump that captures the exact moment that root-causes a bug.

Recently, researchers have started proposing methodologies to harness the on-chip test and debug hardware to provide greater assistance to the post-silicon debug team. For example, scan dumps can be compared automatically between good and bad runs to help root-cause a fault [2], or propagated forward and backward to help improve visibility and diagnosis [10]. A binary-search-based debug method [12] iteratively divides the search space in half until the method identifies the first cycle in which the error is activated and observed. These methods assume the system behavior is deterministic, which is rarely true in practice (due to test case randomness, incomplete control of the operating environment, clock domain crossings, arbitration, etc.). The BackSpace [3] approach can handle non-determinism, but does not consider electrical faults.

Building on those previous works, this paper presents FD-BackSpace (Fault Diagnosis BackSpace) to provide assistance for diagnosing sequentially deep electrical bugs. In particular, the contributions are as follows:

- We propose a novel algorithm that identifies a small set of possible faults that could be responsible for the observed buggy behavior running the actual system in silicon.

- The key characteristic of our method is its laziness. Rather than trying to analyze whether a deep bug could occur for every possible fault, our method starts from the observed bug sighting (e.g., a system crash) and goes backwards, lazily considering only the faults that are relevant to a possible execution leading to the actual bug.

- Experimental results show that our method can reduce the number of suspect faults by an average of 94%. Furthermore, we can eliminate the majority of possible faults within only few clock cycles (going backwards from the bug observation). Even if we cannot run the algorithm to completion, we still greatly reduce the set of possibilities for the debug team to consider.

- Not only does our technique produce a small suspect set of possible faults, it simultaneously reconstructs a trace showing for each fault how that fault can cause the observed buggy behavior.

- Unlike [3,4], our method handles deep electrical faults, being specifically used for fault diagnosis. In contrast to [2,10,12], our method handles non-determinism in the system execution.[1] Unlike [8], our method is not processor-specific and can be applied to any design. Unlike [9], which assumes the existence of test vectors demonstrating faulty behavior before fault diagnosis can be performed, our method simultaneously constructs plausible test vectors along with diagnosing possible faults.

2. PRELIMINARIES

2.1 Notation and Basic Definitions

Let C be a fault-free circuit. We model C as a finite state machine, $M = (Q, I, O, Q_0, \delta, \omega)$, where:

---

[1]Our experimental results are done with deterministic simulation, but we show how to relax this assumption to handle non-determinism.

- $Q = 2^{\mathcal{L}}$ is the set of states, where $\mathcal{L}$ is the set of latches in C;

- I is the input alphabet;

- O is the output alphabet;

- $Q_0 \subseteq Q$ is the set of initial states;

- $\delta \subseteq Q \times I \times Q$ is the transition relation;

- $\omega \in Q \times I \mapsto O$ is the output function.

Notice that we model M as a non-deterministic finite state machine, so the formalism can handle randomness in the bring-up tests as well as transient errors, race conditions, etc.

An execution path (run) on M is a finite sequence of states $\pi = s_0 s_1 s_2 \ldots s_n$, where $n \in \mathbb{N}$. A crash state is a state of the chip where a bug is observable (e.g., a system hang). A path $s_i s_{i+1} \ldots s_n$ is said to be a valid trace leading to the crash state if s_n is the crash state and for each s_j, $i \leq j \leq n-1$, s_j is a predecessor of s_{j+1} and reachable from the initial state.

A signature of a state $s \in Q$ is a projection of s onto a set of latches $\mathcal{S}ig \subseteq \mathcal{L}$, i.e., in this paper, we consider a signature to be just a subset of the bits of a state.

For any node g, we denote by $fanout(g)$ and $fanin(g)$ the set of fan-out and fan-in nodes of g, respectively.

For any fault-free node g, the variable, $\bar{g}$, denotes its faulty counterpart. We denote by $\bar{C}$ a faulty circuit. To simplify our exposition, we will assume a single-stuck-at fault model: the notation $\bar{g}(0)$ ($\bar{g}(1)$) refers to a stuck-at-0 (stuck-at-1) fault at node g. However, our method works for any fault model that can be modeled using the SAT-based technique described next.

2.2 Fault Modeling

We use the SAT-based fault-modeling technique introduced by [9]: we augment the model of the fault-free circuit C (henceforth, C') by adding a mux at the output of each gate g. Each mux has one fault-select and one signal line. We denote the set of fault-select lines by $E = \{e_1, e_2, \ldots, e_n\}$ and the set of corresponding fault-signal lines by $W = \{w_1, w_2, \ldots, w_n\}$, where $n = |\mathcal{G}|$. By setting $e_i = 1$, the node g_i is disconnected from $fanout(g_i)$, and w_i is connected to every node $g_j \in fanout(g_i)$.

This is a very flexible fault-modeling framework, as any gates can be disconnected, with arbitrary faulty values forced into the circuit instead. For example, for the single-stuck-at fault model, we would constrain that exactly one $e_i = 1$ (which enforces "single"), we would assign the corresponding w_i to be 0 (1) for stack-at-0 (stuck-at-1), and we would not allow e_i or w_i to change in different time-frames (which enforces "stuck-at").

Consider the sequential circuit in Fig. 1(a), for example. A possible faulty version is in Fig. 1(b), where a s-a-1 fault is present in g_2. Fig. 1(c) shows the augmented circuit with a set of muxes used to model the fault. (For space reasons, the latches have been removed in Fig. 1(c).) The fault is modeled by setting fault-select line $e_2 = 1$ with all other $e_i = 0$, and fault-signal line $w_2 = 1$.

With the (potentially faulty) circuit in this form, many useful properties are easily phrased as SAT queries. For example, an important basic computation is pre-image: what states/inputs are possible predecessors of a given state? To answer this question, we simply constrain that next-state signals to the given state and ask the SAT solver for solutions for the present-state and input signals. Upon receiving

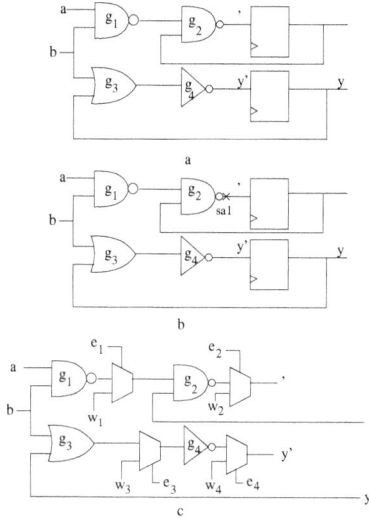

Figure 1: Sequential Circuit (a) fault-free (b) faulty (c) added hardware

a solution, we can force the solver to return additional solutions until we have them all. Returning to Fig. 1, for example, we might ask what states are the predecessors of state 11 under input 01, assuming a fault-free circuit. In that case, we would assert that $x' = y' = b = 1$ and $a = e_i = 0$, and a SAT solver would quickly tell us that this is actually impossible: no solutions for x and y exist. Similarly, we might ask which single-stuck-at faults allow the circuit to transition from state 10 to state 10 under input 01 by asserting $x' = x = b = 1$ and $y' = y = a = 0$ and leaving the fault-select and fault-signal lines partially constrained, and the SAT solver would return the fault in Fig. 1(b) (along with several other solutions). The challenge with using SAT for diagnosing deep bugs, though, is that we don't know the states leading up to the bug, so we don't know the correct constraints for the SAT solver. Our new algorithm solves this problem.

2.3 BackSpace

Since our technique relies on the same formal principles of BackSpace, we briefly review its core assumptions and debug-flow. For a complete presentation, we refer to [3].

The core assumptions of the BackSpace framework are:

- It must be possible to recover the state of the chip when an error has occurred. For example, this could be done via the scan chain with the chip in test mode;

- The silicon implements the RTL (or gate-level or layout or any other model of the design that can be analyzed via formal tools);

- The bring-up tests can be run repeatedly and the bug being targeted will be at least somewhat repeatable (one out of every n tries, for a reasonably small value of n).

The BackSpace framework consists of adding some debug support to the chip: a signature that saves some history information but otherwise has no functional effect on the chip's behavior, and a programmable breakpoint mechanism that allows the chip to "crash" when it reaches a specified state. Given these, the approach repeats the following steps

1. Run the chip until it crashes or exhibits the bug. This could be an actual crash or a programmed breakpoint.

2. Scan out the full crash state, including the signature.

3. Using formal analysis of the corresponding RTL (or other model), compute the set of predecessor-candidates of the crash state. The signature must provide enough information so that the number of predecessor-candidates is reasonably small.

4. For each predecessor-candidate s, let s be the new breakpoint; re-run the chip; if the chip reaches the breakpoint, then s is a valid predecessor.

until it has computed enough of a history trace to debug the design (or Step 3 fails). Each iteration of the loop is like hitting "backspace" on the design – going back one cycle.

Our new algorithm, FD-BackSpace, differs from the original BackSpace in two fundamental ways: 1) we do not assume that the silicon implements the RTL, instead use the techniques from Sec. 2.2 to model faults, and 2) we consider all valid paths to the crash state instead picking just one. The next section presents the FD-BackSpace algorithm.

3. FD-BackSpace ALGORITHM

We start this section by setting forth our assumptions, then we present the algorithm, and afterwards, we prove its correctness.

Similar to the original BackSpace framework, we assume that it is possible to recover the state of the chip when it crashes (or breakpoints) along with a signature of the previous cycle (using scan-chains). In addition we make the following assumptions: 1) we have a fault model that produces a finite set $\mathcal{G}$ of possible faults that might be the cause of the crash and that we can model each fault as in Sec. 2.2; 2) if a fault exists, it will be excited and observed given a reasonably small number of chip-run trials; 3) and the signature is effective at constraining the pre-image computation to yield a small number of predecessor-candidate states. The last two assumptions are in-common with the original BackSpace work, and effective techniques for signature computation are known [3].

Our new FD-BackSpace algorithm is shown in Algorithm 1. It computes a minimal set of the suspect faults, working backwards in time from the crash state towards the initial states. The three main data-structures in this algorithm are two tables, W and H, and the set of suspect faults, $\mathcal{F}$. Each table entry contains a 3-tuple: a state, a signature and a set of faults. We define W as the working-table and H as the history table. The former is a dynamic table where elements are inserted and/or deleted in each iteration. The latter stores all validated states (explained later) and their associated possible faults.

The intuition behind Algorithm 1 is that we are performing a modified graph traversal through the state space under different possible faults. The working table W is the frontier from which we must continue the search, and the history table H accumulates all states we have visited. The key to understanding the algorithm is that we are always maintaining two invariant properties of all triples in W and H: 1) for every state and associated fault, there is an execution path from that state through only states in H to the crash state, assuming that fault; and 2) every state is *validated*, meaning that we actually observed that state occurring on an actual silicon run.

This procedure has 2 main loops: lines $(15 - 29)$ and lines $(30 - 57)$. The first loop (lines $15 - 29$) builds the initial working-table. This loop iterates over all possible faults $\mathcal{G}$. First, it computes a pre-image of the crash state under some fault $g \in \mathcal{G}$ (line 17). If the pre-image is non-empty, Algorithm 1 checks whether each state s in the pre-image is already in W, adding g to the set of possible faults of s, if that is the case. Otherwise, Algorithm 1 validates and inserts s into W by running the actual chip (this step is explained in detail later). At the end of this loop, W contains all valid predecessor-states of the crash state, each of which is associated with its set of possible faults.

The second loop (30 – 57) is essentially just an inductive repetition of the first loop, expanding the graph traversal backward through the state space and building up paths that lead to the crash state under the presence of some fault. First, Algorithm 1 reads the first entry of W (deleting the entry from the table) containing a state s, a signature ξ, and a set of possible faults P_s associated with s. It either copies this entry to the history-table or updates (adding) the set of possible faults for the existing state s. Next, Algorithm 1 iterates over P_s. Different from the previous loop, here we have a few more cases to consider. If a state in the pre-image of s (denoted as s'), under the assumption of a fault p, is in the initial set of states Q_0, then we can safely add it to the final suspect set $\mathcal{F}$. Otherwise, we have three options: a) if s' is already in the history-table, then we can proceed to the next state in the pre-image (thus, we handle paths containing cycles); b) if s' is already in the working-table but p is not in its set, we add p to P_s' (i.e., many faults can explain the same state); c) if s' is in neither table, then it needs to be validated and inserted into W (if valid). Once Algorithm 1 iterates over all faults in P_s, it removes the next entry in W, (i.e., it iterates at line 31). This loop terminates only when the working-table is empty. Thus, $\mathcal{F}$ contains only the faults than can lead the chip from an initial state to the crash state.

Note that Algorithm 1 interleaves off-line software computation with hardware runs exactly as in BackSpace. The hardware interaction is encapsulated by the routine *insertIfValid()*. This routine loads the breakpoint circuitry with the state to be validated, sets a timeout value for each run, runs the chip, and then dumps the signature of the validated state. If the breakpoint is hit, the state is validated; otherwise, the timeout will occur. This routine also handles non-determinism as in the original BackSpace framework: the parameter *ntrials* specifies the number of times the chip must be run before giving up on validating a state. If the chip does not reach the state in any of the *ntrials* run, only then do we conclude that the state cannot be reached by the chip. Setting *ntrials* large enough makes the probability of erroneously not validating a state arbitrarily small.

The termination described in Algorithm 1 is simplified by assuming we can continue until we reach the initial states, but we can easily generalize it. As described, the only termination condition is when W becomes empty, which can only happen if the algorithm reaches the initial set of states (from which point the preImage() returns the empty set). State traversal to any state in Q_0 may take too much time or memory, e.g., if the crash state is too deep from the initial set of states. But it is not difficult to see that if computing a small prefix of the crash state is sufficient to eliminate a large number of faults, then we can simply augment the condition at line 30 to limit to some bounded number of loop iterations. A similar argument could limit to computing a pre-determined number of faults. The last condition not explored in this simplified presentation of the algorithm is the case when the pre-image computation of some state s (lines 17 and 39) grows too big. In this case, we can terminate the computation for s and (conservatively) add its fault to the suspect-set.

To complete the presentation of Algorithm 1, we need to prove its correctness. We start by introducing some definitions. Then, we formally state and prove the correctness properties of the algorithm (under the assumption that

---

Algorithm 1 Suspect-Fault-Set Lazy Computation

```
 1: input   C : circuit-under-debug
 2: input   Q_0 : set of initial states
 3: input   cs : crash state
 4: input   ξ_cs : signature of predecessor state of crash state
 5: input   G : set of all possible faults (as described in
            Sec. 2)
 6: input   timeout, ntrials: parameters to isValid()
 7: output  F: final suspect-fault set
 8: /* Global variables and structures */
 9: s, s′ : states
10: ξ, ξ′ : signatures of the predecessor-state of s and s′
11: P_s : set of possible faults associated with state s
12: W : working-table, where each entry is a 3-tuple
        (s, ξ, P_s)
13: H : history-table, a non-destructive version of table W
14: F := ∅; P, W, H := NULL
15: for each fault g in G do
16:     /* C′ is the faulty version of C with fault g */
17:     if (I := preImage(cs, ξ_cs, C′) ≠ ∅) then
18:         for each state s′ in I do
19:             if isStateMember(W, s′) then
20:                 /* add g to s′ possible-fault set */
21:                 UpdateFaultSetOfState(W, s′, g)
22:             else
23:                 /* validate s′ on-chip assuming g and
24:                 insert new entry into working-table W */
25:                 insertIfValid(W, s′, g, timeout, ntrials)
26:             end if
27:         end for
28:     end if
29: end for
30: while (W ≠ ∅) do
31:     (s, ξ, P_s) = Delete(W[0]) //destructive read
32:     /* Manage history-table H */
33:     if isStateMember(H, s) then
34:         UpdateFaultSetOfState(H, s, P_s)))
35:     else Insert(H, (s, ξ, P_s))
36:     end if
37:     for each fault p in P_s do
38:         /* C′ is the faulty version C with fault p */
39:         for each state s′ in preImage(s, ξ, C′) do
40:             if s′ ∈ Q_0 then
41:                 F := F ∪ {p}
42:             else if isFaultMember(H, s′, p) then
43:                 /*i.e., (s′, p) has already been tested*/
44:                 continue
45:             else if isStateMember(W, s′) then
46:                 if !isFaultMember(W, s′, p)) then
47:                     /* append p to s′ possible-fault set */
48:                     UpdateFaultSetOfState(W, s′, p)
49:                 end if
50:             else
51:                 /* validate s′ on-chip assuming p and
52:                 insert new entry into working-table W */
53:                 insertIfValid(W, s′, p, timeout, ntrials)
54:             end if
55:         end for
56:     end for
57: end while
58: return (F)
```

ntrials is large enough that *insertIfValid()* does not fail to validate any valid state).

DEFINITION 1. *Given a physical chip and a circuit model $\bar{C}$ with the same state bits/flops, a "plausible path" is a finite sequence of states such that every state in the sequence is reachable on the physical chip, each pair of successive states in this sequence is a legal transition in $\bar{C}$, and the sequence ends at the crash state of $\bar{C}$.*

The intuition behind this definition is to capture the best approximation to what we really want, given the information we can gather. What we would really like to compute is an actual execution path of the faulty physical chip that led to the observed crash, but this is impossible, since we do not have full trace-visibility of the chip. A plausible path is an execution path of the *model of* the faulty chip, with the extra information that every state on this path was really reachable on the physical chip. It is possible for a plausible path to connect states of the physical chip together in a way that is possible in the model, but not in the silicon, so plausible paths are an imperfect approximation. But the important point is that every real execution of a faulty chip will also be a plausible path, so considering all plausible paths guarantees not missing possible faults.

DEFINITION 2. *An "initial plausible path" is a plausible path that starts from the initial state of $\mathcal{C}$.*

This definition restricts Defn. 1 to paths that start at an initial state, giving a complete execution from reset to the crash state, assuming some fault. And every state on this execution has been validated as a real state on the silicon.

THEOREM 1. *For every fault $g \in \mathcal{F}$ in Algorithm 1, the computed paths are initial plausible paths.*

Proof: No triple is added to F until the algorithm reaches an initial state, so any path starting from that state is obviously initial. Algorithm 1 maintains invariants that 1) for every state and associated fault, there is an execution path from that state through only states in H to the crash state, assuming that fault; and 2) every state is *validated*. These two invariants are maintained because the only way for a state/fault to be added to W or H requires that they first be in the pre-image of a state in W or H, hence guaranteeing the first invariant; and that they also be validated, hence guaranteeing the second invariant. These two properties combine to show the existence of the plausible path. ∎

THEOREM 2. *For every fault $g \in \mathcal{G}$, if there exists an initial plausible path in $\mathcal{C}$, then $g \in \mathcal{F}$.*

Proof: Since there exists an initial plausible path, as we start from the crash state in the first loop of Algorithm 1, we will find the preceding state of the initial plausible path in the pre-image, under fault g. This will validate (because of our assumption that *ntrials* is sufficiently large), so that triple will be added to W and H. Subsequently, in the second loop, whenever we encounter a state on the initial plausible path, we will find the preceding state of the path in the pre-image, under fault g, and validate it. This will proceed until we reach the initial states, whereupon g will be added to F. ∎

The two theorems state what is promised by our algorithm. Theorem 1 means that every fault returned should be considered seriously by the debug team, because under the assumption of that fault, there is a possible execution from initial state to the crash state, in which every state

Table 1: Suspect Set Computation Results

Circuit Name	No. of Flops	Initial No. of Faults	Final No. of Faults	Runtime	Suspect Set reduction (%)
b01	5	146	10	1m 37s	93
b02	4	70	3	32s	95
b03	30	382	3	4m 42s	99
b04	66	1406	1	54m 40s	99
b05	34	4140	4	34m 41s	99
b06	9	146	39	2m 51s	74
b07	49	886	173	75m 7s	80
b08	21	484	2	66m 17s	99
b09	28	403	3	27m 47s	99
b10	17	426	9	17m 53s	97
b12	121	2802	4	640m 25s	99
AVG					94

on that execution has been validated on the silicon. Theorem 2 means that our algorithm will never miss a fault from G that could have explained the bug (assuming *ntrials* is sufficiently large).

4. EXPERIMENTAL RESULTS

In this section, we present experimental results of our proposed flow for identifying single-stuck-at faults during post-silicon debug. Experiments were conducted on an Intel i5, 3.1 GHz workstation with 16GB of RAM. We report experiments on eleven ITC'99 benchmark circuits. We implemented Algorithm 1 based on the open-source BackSpace v0.3 codebase.[2] To interface with BackSpace, we had to synthesize these benchmarks using BackSpace's cell library. We used Synopsys Design Compiler Version Y-2006.06-SP2. Also, we used Synopsys VCS Version A-2008.09 as our logic simulator.

Recall that BackSpace uses a signature to reduce the size of pre-images. Choosing a good signature depends on knowledge of the circuit [3] or an advanced algorithm for signal selection [7], but this is orthogonal to our work. Thus, for simplicity, we *randomly* chose signatures. Moreover, the signature size was set for each circuit such that the number of possible predecessor states never exceeded 1024.

After synthesizing each benchmark, we randomly inserted either a *s-a-0* or a *s-a-1* in the gate-level netlist. For each experiment, we simulated the gate-level netlist with the accompanying test bench for an arbitrary number of cycles; randomly selected a crash state; and then we started Algorithm 1.

Table 1 shows the results of all our experiments. The first three columns provides details of the circuits. The next two columns show the size of the final suspect set and the runtime of our algorithm. We also report the percentage reduction in the suspect set size (column 6). Note that in the majority of the benchmark circuits the final suspect fault set is extremely small. Thus, this set can now be efficiently handled by ATPG for identifying the actual fault on the chip.

Figure 2 shows the reduction in the set of faults using Algorithm 1 for each circuit. The x-axis represents the clock cycles, starting from the crash state ($x = 0$) to the initial state. (Note that we simulated each circuit for 100 clock cycles before picking a crash state, but due to the presence of loops, some circuits have initial plausible paths of less than 100 states. For example, the initial plausible path for b06 has 39 states. Thus, for some of the circuits, Algorithm 1

---
[2]http://www.cs.ubc.ca/~depaulfm/BackSpace

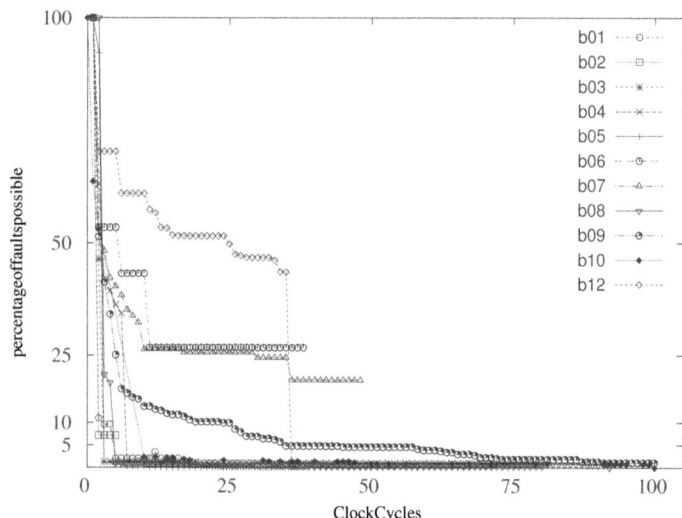

Figure 2: Reduction in possible fault list while tracing from crash state to initial State

Table 2: Suspect-Fault Set Reduction

Circuit Name	$|\mathcal{F}| = 25\%$ of $|G|$	% reduction in 10 clock cycles
b01	2	93
b02	2	95
b03	3	99
b04	7	99
b05	3	99
b06	>100	74
b07	11	74
b08	6	99
b09	5	86
b10	10	97
b12	38	50

computes the shorter path and terminates well below 100 clock cycles.) Notice that we can negate a significant number of faults within the first few iterations. Thus, when tracing to the initial state is impractical, one can stop Algorithm 1 after a few iterations.

Table 2 shows the results where the algorithm terminates prematurely before reaching the initial state. Column 2 shows the minimum number of cycles to be backspaced such that the suspect set size is below 25% of the total faults. In column 3 we show the percentage of fault reduction when the size of the plausible path reaches 10, i.e. maximum number of clock cycles to be iterated is 10. Results show that it is not necessary to trace back to the initial state, in case the crash happens too deep from the initial state. This would lead to considerable reduction in runtime without compromising the effectiveness of our approach.

The plausible path can be considered as the *test vector* in our problem. Unlike ATPG, we do not have any control over these vectors. However, results show that our lazy computation of the suspect set of faults is very efficient in reducing the number of suspect faults. The key insight is that we can use actual "buggy" traces, that lead the chip to a crash state, to prune out suspect faults (instead of using the very expensive, exhaustive methods of automatic test pattern generation).

5. CONCLUSION

We have presented a novel framework to aid debugging deep electrical faults in silicon. Assuming that we can model the faults, our method computes a small set of possible suspects that could explain the silicon's malfunction. In addition, the method also reconstructs, for each fault, plausible paths that lead the silicon to the actual bug. Key to our approach is the laziness of this computation. We avoid the expensive task of computing very long test-vectors for each and every fault. Our experiments show that our method is effective with much less effort. We are able to reduce the suspect set of faults by an average of 94%, and we show that we need only a few cycles from the buggy state to eliminate the majority of the possible faults. The direct line of future work is to target larger designs, experiment with more general fault models, and investigate efficient methods for reducing the runtime of the algorithm by using on-chip circuitry, such as trace buffers.

6. REFERENCES

[1] ARM. *Embedded Trace Macrocell Architecture Specification*, volume 20. July 2007. Ref: IHI0014O.

[2] P. Dahlgren, P. Dickinson, and I. Parulkar. Latch divergency in microprocessor failure analysis. In *ITC'03*, pages 755–763, 2003.

[3] F. M. De Paula, M. Gort, A. J. Hu, S. J. E. Wilton, and J. Yang. Backspace: formal analysis for post-silicon debug. In *FMCAD '08*, pages 5:1–5:10. IEEE Press, 2008.

[4] F. M. de Paula, A. Nahir, Z. Nevo, A. Orni, and A. J. Hu. Tab-backspace: unlimited-length trace buffers with zero additional on-chip overhead. In *DAC '11*, pages 411–416. ACM, 2011.

[5] R. Desplats, F. Beaudoin, P. Perdu, S. Nataraj, T. Lundquist, and K. Shah. Fault localization using time resolved photon emission and stil waveforms. In *ITC'03*. IEEE Computer Society, 2003.

[6] Y.-C. Hsu, F. Tsai, W. Jong, and Y.-T. Chang. Visibility enhancement for silicon debug. In *DAC '06*, pages 13–18. ACM, 2006.

[7] H. F. Ko and N. Nicolici. Automated trace signals identification and state restoration for improving observability in post-silicon validation. In *DATE '08*, pages 1298–1303. ACM, 2008.

[8] S.-B. Park and S. Mitra. IFRA: Instruction footprint recording and analysis for post-silicon bug localization in processors. In *Design Automation Conference*. ACM, 2008.

[9] A. Smith, A. G. Veneris, M. F. Ali, and A. Viglas. Fault diagnosis and logic debugging using boolean satisfiability. *IEEE Trans. on CAD of Integrated Circuits and Systems*, pages 1606–1621, 2005.

[10] V. C. Vimjam, E. Amyeen, R. Guo, S. Venkataraman, M. S. Hsiao, and K. Yang. Using scan-dump values to improve functional-diagnosis methodology. In *VTS'07*, pages 231–238, 2007.

[11] M. J. Y. Williams and J. B. Angell. Enhancing Testability of Large-Scale Integrated Circuits via Test Points and Additional Logic. *IEEE Transactions on Computers*, C-22(1):46–60, January 1973.

[12] C.-C. Yen, T. Lin, H. Lin, K. Yang, T. Liu, and Y.-C. Hsu. Diagnosing silicon failures based on functional test patterns. In *MTV '06*, pages 94–98. IEEE Computer Society, 2006.

[13] L. Zhang, I. Ghosh, and M. Hsiao. Efficient sequential atpg for functional rtl circuits. In *ITC'03*, pages 290–298, 2003.

Influence of Different Layout Styles on the Performance of the Calibration of an On-Chip Programmable Voltage Reference

Dominik Gruber
RIIC - Institute for Integrated Circuits
Altenberger Strasse 69
Linz, Austria
gruber@riic.at

Timm Ostermann
RIIC - Institute for Integrated Circuits
Altenberger Strasse 69
Linz, Austria
oster@riic.at

ABSTRACT

This paper presents an on-chip programmable voltage reference circuit whereat the main block, the switchable resistor array, was realized using four different layout styles. These different layouts, which result in different complexity and chip area consumptions, are analyzed regarding the influence on the calibration performance of the voltage reference circuit.

Although a slight difference can be measured, there is no clear preference for one of the four layout versions. In contrast to the much smaller chip area of the more or less lumped approach (lay3) the distributed approach (lay0) shows only slight advantages in circuit performance.

Categories and Subject Descriptors

B.7.3 [**Integrated Circuits**]: Reliability and Testing; B.8.2 [**Performance and Reliability**]: Performance Analysis and Design Aids

General Terms

Reliability

Keywords

Voltage reference, bandgap, offset voltage, programmable, layout, resistor

1. INTRODUCTION

Voltage references are important building blocks in many of today's electronic circuits, like analog-to-digital converters, radio transceivers or in general any application where a reliable and stable reference voltage is needed.

Due to not only financial but also technical reasons (mismatch, offset errors, additional errors of PCBs etc.) it is useful to calibrate on-chip voltage references.

Usually, the voltage drop V_D over a diode or diode connected bipolar transistor is used to generate voltage reference with low dependency on temperature and supply voltage [5, 3, 1]. Based on this bandgap circuits, more complex references can be realized [4], which can be used to generate voltages that exceed the output voltage of typical bandgap reference circuits (BGR). Those output voltages V_{BG} vary, depending on the used technology and available supply voltage, from $\approx 0.5V$ to $\approx 1.3V$.

If higher reference voltages are required, a circuit like the one shown in Fig. 1 can be used to amplify V_{BG} to nearly any user-defined reference voltage. Depending of the graduation of the resistance ladder ($R_1 \ldots R_3$), different output voltages can be realized. Usually voltage followers are used to decouple the load from the reference circuit itself.

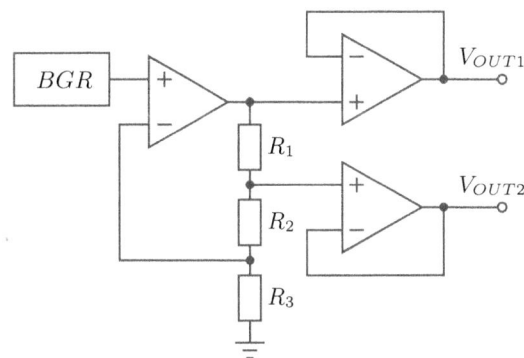

Figure 1: Voltage reference

Temperature errors introduced by the amplifier can be minimized by using an adequately designed resistor ladder with appropriate chosen resistors according to their temperature coefficient.

Due to local (mismatch) and global process variations the initial reference voltages of BGR circuits show deviations of $\pm 10\%$ or more. In addition, offset errors of operational amplifiers and mismatch between resistors make the variations of the reference output voltage even worse.

Even though these extreme deviations are unusual, they have to be considered during the design of the reference circuit in order to keep the production yield as high as possible. A possible method to minimize the effects of process and mismatch variations is to calibrate the output of a volt-

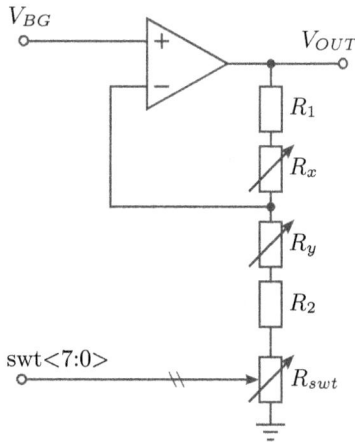

Figure 2: Programmable voltage reference

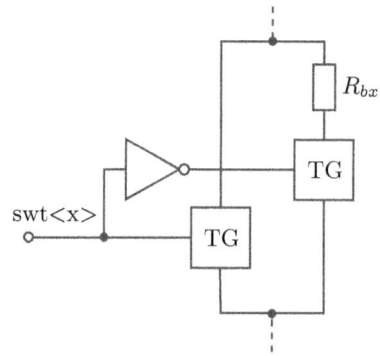

Figure 3: Single resistance element of R_{swt}

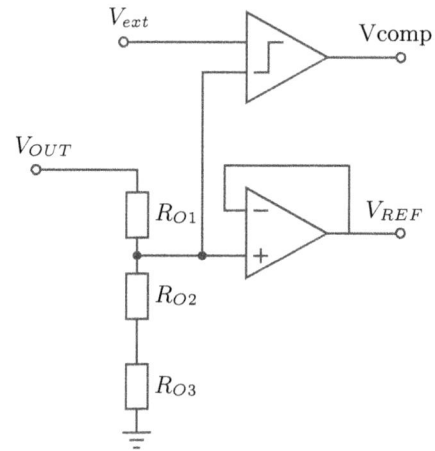

Figure 4: Output structure of the programmable reference

age reference. For this reason we propose a programmable on-chip voltage reference which is described in section 2.

However, even programmable voltage references [4], whose output voltage can be adjusted to balance the imperfections due to process and mismatch, are affected by those problems: The precision of the calibration is reduced if individual elements of the core part (e.g. resistor ladder) of the calibration vary too much.

Therefore different layout styles of the resistor ladder of the calibration circuit are analyzed. The layout styles include complex and simple layout structures of the used calibration resistors. This paper discusses the differences and whether a specific layout style is most suitable.

2. CIRCUIT DESCRIPTION

The programmable voltage reference circuit, which is used to discuss the differences of the layout styles, is shown in Fig. 2. It increases the voltage of the BGR $V_{BG} \approx 1.22V$ to a maximum of $V_{OUT} = 2.5V$ (in given design) and consists of a programmable resistance matrix R_{swt} combined with two additional adjustable resistors R_x and R_y. These two resistors R_x and R_y are used for a rough adjustment of the overall offset of the reference voltage range. They are not required for the analysis of this paper and therefore constantly bypassed by low ohmic transmission gates.

The main and therefore most important part of the on-chip programmable reference circuit in Fig. 2, is the resistance matrix R_{swt} which is used for fine tuning of the reference. The programmable resistance matrix R_{swt} consists of a series connected chain of 8 resistor elements These series connected elements (Fig. 3) are binary weighted due to binary weighted values of the resistors R_{bx}. In fact the resistors R_{bi+1} are not exactly twice as big as R_{bi}, they are slightly smaller to avoid missing calibration codes due to mismatch of the resistors. The overall number of transmission gates (TG), which are used to switch the appropriate single elements on and off and therefore to change the value of the complete resistance matrix R_{swt} and the complete reference ladder, respectively, is always the same. This approach stabilizes the temperature performance of the ladder itself because there are always eight TGs in the current branch of R_{swt}, regardless which resistors R_{bx} are switched on or off.

As an additional measure to improve the performance of such a switchable resistance matrix R_{swt}, the different binary weighted resistors R_{bx} are typically realized through several unit resistors connected in series and parallel to realize the binary weighted values. Even the resistor R_{b0}, acting as LSB resistor typically consists of several (at least) parallel connected unit resistors. Due to series and parallel connected unit resistors the average influences of mismatch effects (see section 4) as well as other gradients, like temperature gradients, are reduced. Dummy resistors (or etching guards) [2] are added to all resistor arrays. Nevertheless an accumulation of these effects to the worst case value is possible, although this is statistically very rare. One important disadvantage of such a parallel/series realization is the increasing chip area consumption.

In addition to the programmable voltage reference itself (Fig. 2), there is also an appropriate output structure which is used to provide any voltage between $2.5V$ and $0V$ (Fig. 4). In the proposed design it consists of a resistance ladder which divides $V_{OUT} = 2.5V$ to $V_{REF} = 2V$. A voltage follower is used to decouple the load of the reference from the reference itself. A comparator is used to automatically calibrate the given voltage reference [4].

With such a programmable programmable voltage reference variations of the output voltage V_{REF} due to e.g. process variation and mismatch can be reduced to values below $\pm 2mV$.

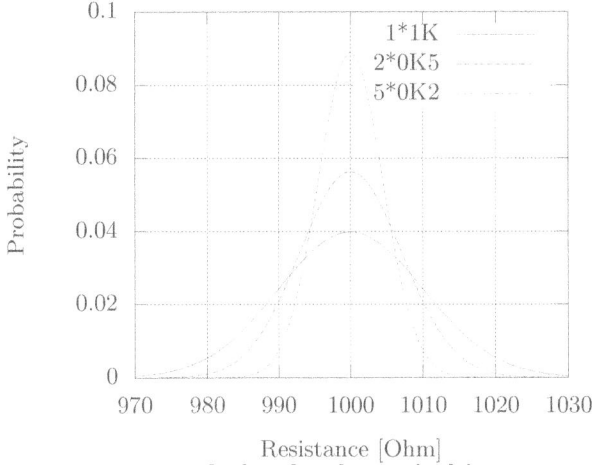

Figure 5: Example for the theoretical improvement of the probability to manufacture the target value of $R = 1k\Omega$ due to segmenting the original resistor

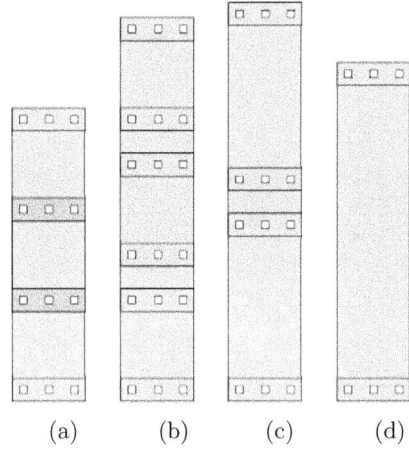

Figure 6: Schematic representation of the different layout styles. (a) lay0, (b) lay1, (c) lay2, (d) lay3

3. DIFFERENT LAYOUT STYLES

Local (mismatch) and global process variation of all involved elements (BGR, OPAMP, Resistors, etc.) directly influence the output voltage of voltage references. The proposed digitally programmable voltage reference is designed to compensate for these effects.

However, the precision of the calibration is reduced if there is too much deviation between the individual resistors R_{bx} of the programmable resistor R_{swt}.

Serial and parallel connection of resistors should increase the probability of the accuracy of the resistance value, if the random error of each single resistor is Gaussian distributed. An example of the theoretical improvement of the probability to manufacture the target value of $1k\Omega$ due to segmenting the original resistor is shown in Fig. 5.

In reality this assumption is not always true. Short resistor segments tend to vary more because of dominating contact resistance effects [2]. To find out which layout style shows the best result, simulation is not always suitable. In order to analyze the effects of different layout measures on the precision of the calibration, four different versions of the programmable voltage reference were produced. R_{b0} is the least significant resistor, $R_{b7} \approx 128 R_{b0}$ the most significant. The differences in the layout of Rbx are:

- Layout version 0 (lay0, see (a) in Fig. 6): Lay0 is realized in the classical way of a distributed layout. Each resistor R_{bx} consists of ten unit resistors in parallel. The binary weighted values are realized by connecting the appropriate number of unit resistors in series. Therefore in total R_{b0} consists of ten and R_{b7} of 1280 unit resistors, respectively. The series connected (polysilicon) resistors are placed without any space between the single resistors; the contacts are placed on top of each other.

- Layout version 1 (lay1, (b) in Fig. 6): Lay1 is similar to lay0 with the exception that the poly-silicon material of each resistor is separated by approximately $0.5um$

- Layout version 2 (lay2, (c) in Fig. 6): For lay2 the number of single resistors to realize R_{bx} is reduced. R_{b7} for example is simplified to 23 resistors in series and 5 in parallel.

- Layout version 3 (lay3, (d) in Fig. 6): In this version all binary weighted resistors R_{bx} are simplified to single resistors (but each with 5 in parallel)

The difference in chip area consumption of all four layout versions regarding the most significant resistor R_{b7} is:

- Lay0: $\approx 60um$ x $\approx 180um (100\%)$

- Lay1: $\approx 60um$ x $\approx 250um (\approx 138\%)$

- Lay2: $\approx 32um$ x $\approx 121um (\approx 35\%)$

- Lay3: $\approx 14um$ x $\approx 206um (\approx 27\%)$

Regarding the complete area of the reference circuit, the chip area consumption is:

- Lay0: $\approx 530um$ x $\approx 370um (100\%)$

- Lay1: $\approx 530um$ x $\approx 430um (\approx 115\%)$

- Lay2: $\approx 530um$ x $\approx 335um (\approx 90\%)$

- Lay3: $\approx 530um$ x $\approx 315um (\approx 84\%)$

4. MEASUREMENT RESULTS

The complete circuit was designed and manufactured using a $250nm$, $3.3V$ CMOS process. In order to verify which layout version of the resistors R_{bx} provides the most accurate calibration behavior, all four versions of the programmable voltage reference as well as a test resistor (the most significant resistor of R_{swt}, R_{b7}) were manufactured.

A corner lot of nine wafers was produced. Each chip of the corner lot includes all four voltage references and four test resistors, which all differ in the layout of the resistors R_{bx}.

One wafer is manufactured at nominal process conditions, the remaining eight wafers are from a corner lot where every wafer was intentionally produced at selected process corners:

- NOM: nominal case

197

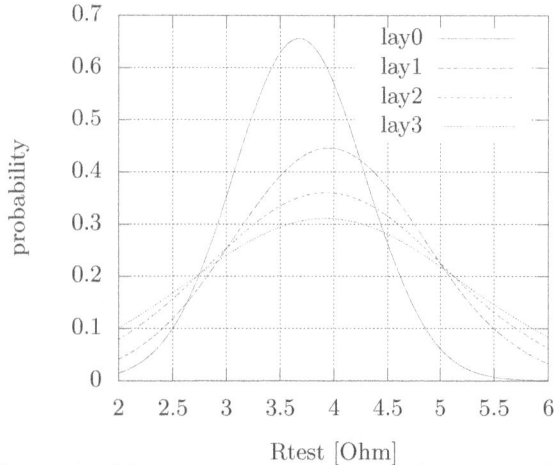

Figure 7: Measurement results of the test resistor R_{b7}

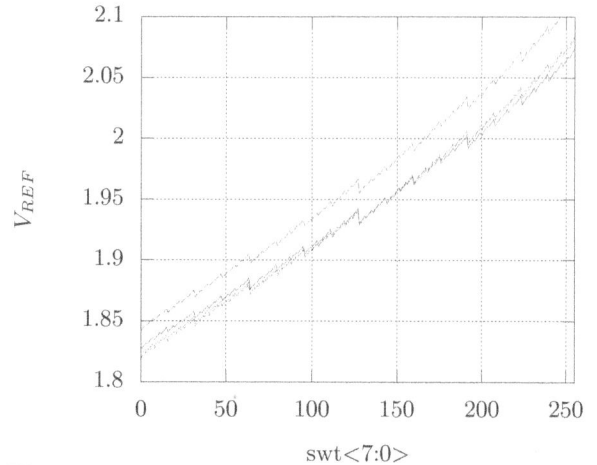

Figure 8: Comparison of all four layout versions: Measurement of the output voltage V_{REF} **using the complete calibration range of** swt<7:0> = 0 ... 255

- FAST: all MOSFET have minimal threshold voltages

- SF: all n-type MOSFET have maximal and all p-type MOSFET have minimal threshold voltages

- FS: all n-type MOSFET have minimal and all p-type MOSFET have maximal threshold voltages

- SLOW: all MOSFET have maximal threshold voltages

- HRES: all used resistors have maximum values

- LRES: all used resistors have minimum values

- LGL: gate length larger than nominal value

- SGL: gate length smaller than nominal value

All measurements are done using nominal supply voltage $V_{DD} = 3.3V$ at room temperature ($\approx 25°C$).

4.1 Test Resistor

Each chip includes four test resistor which was directly connected to test pins.

All wafers but two have no direct effect on the resistance of the manufactured dies. Only CMOS parameters are changed, so resistor variations are only local (die to die) mismatch variations. Only two wafers were produced to change the resistance of poly-silicon-resistors to the maximal and minimal worst case.

The resistance of R_{b7} is measured at room temperature with a measurement current of $30uA$ which corresponds to the current in the application. The average value and the standard deviation of the four different layout styles are determined whether there are statistical differences detectable or not. Fig. 7 shows the results. The plot shows the distribution function of the different layout styles.

The standard deviation of the resistance value changes depending on the complexity of the resistor array. The more elements are connected in series/parallel, the smaller the average deviation. The arithmetical mean value of layout version 0 (lay0) slightly shifts towards smaller resistance value, but this behavior has no disruptive influence on the accuracy of the calibration.

4.2 Voltage Reference

Fig. 8 shows an example of all possible output voltages of four different programmable voltage references (with different layout styles lay0 ... lay3). All of them are located on the same die. As mentioned in section 2, R_{bi+1} is slightly smaller than $2R_{bi}$. This can be recognized by the degreasing output voltage at e.g. the change from swt<7:0> = 127 to swt<7:0> = 128. Because of this, no missing calibration codes occur.

In addition the figure shows how the programmable reference circuit is able to compensate for varying offset voltages. All four voltage references can be calibrated to the target value of $V_{REF} = 2V$ Figure 9, 10, 11 and 12 show histograms of the average step size of all different layout versions (lay0: Fig. 9, lay1: Fig. 10, lay2: Fig. 11 and lay3: Fig 12) during calibration. The average step size of layout version 0 is smaller than the average step size of the other layout versions. This finding correlates with the measurements from section 5.1. Smaller resistors result in smaller steps. But the absolute value of the step size is less important than the deviation from the average value.

Figure 13, 14, 15 and 16 show histograms of the standard deviation of the step size of all different layout versions (lay0: Fig. 13, lay1: Fig. 14, lay2: Fig. 15 and lay3: Fig. 16) during calibration.

Measurements show that the standard deviation of layout version 0 is slightly smaller than the other layout version. But the advantage over layout versions 1-3 is marginal.

Fig. 17 shows a comparison of the temperature behavior between all four layout versions (all on the same die). They behave very similar, there is no difference detectable. Several other temperature measurements approve the example of Fig. 17.

5. CONCLUSIONS

An on-chip programmable voltage reference circuit was demonstrated. The main building block regarding the calibration, the switchable resistor array was realized with four different layout versions. This paper compares those four versions:

The different layout versions differ in complexity and chip

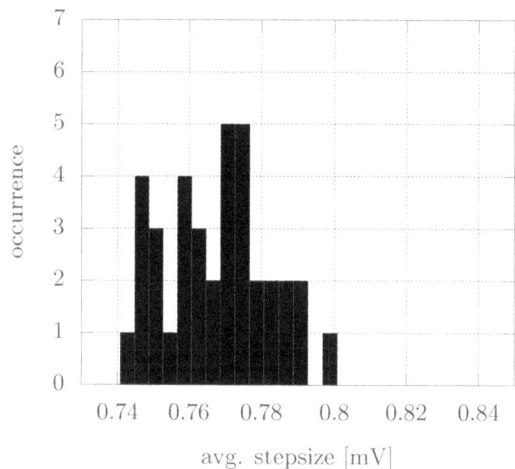

Figure 9: Average step size of layout version 0 (lay0)

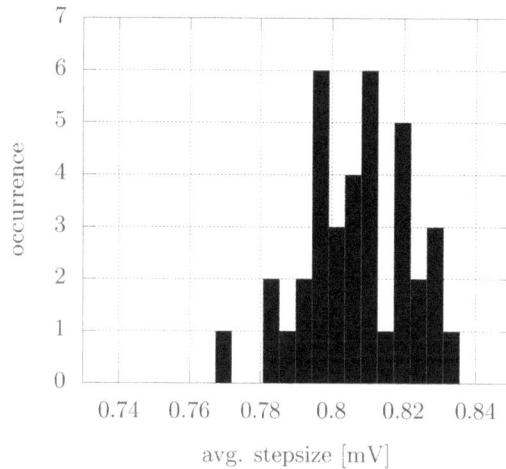

Figure 12: Average step size of layout version 3 (lay3)

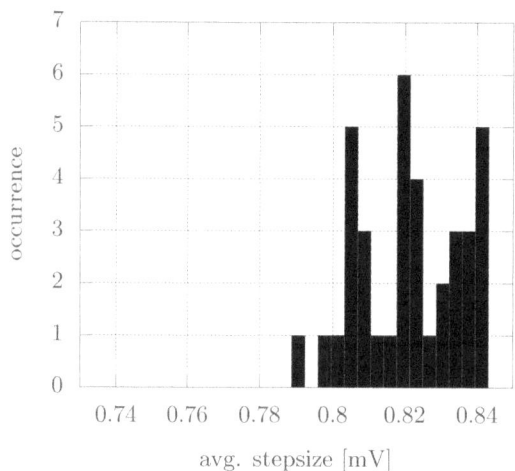

Figure 10: Average step size of layout version 1 (lay1)

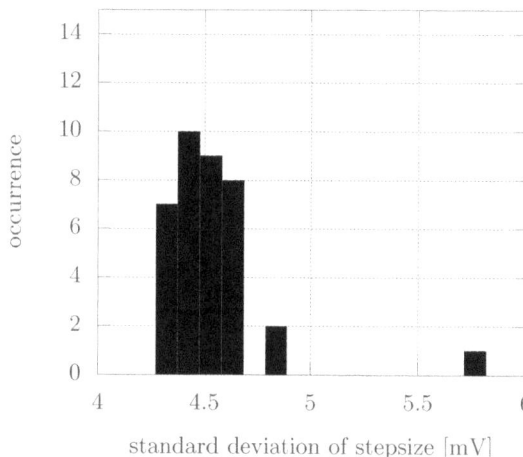

Figure 13: Standard deviation of the step size of layout version 0 (lay0)

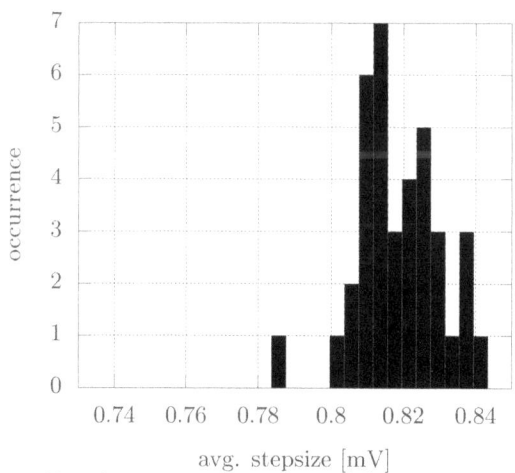

Figure 11: Average step size of layout version 2 (lay2)

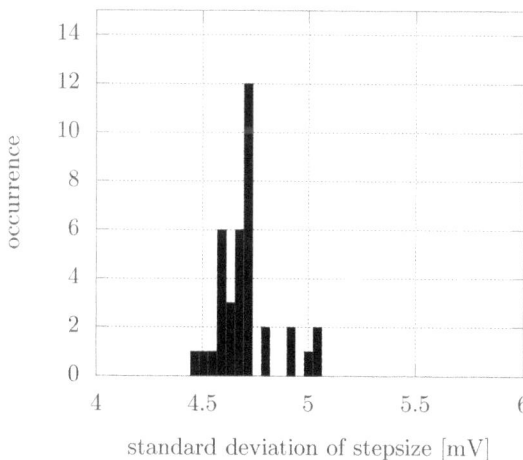

Figure 14: Standard deviation of the step size of layout version 1 (lay1)

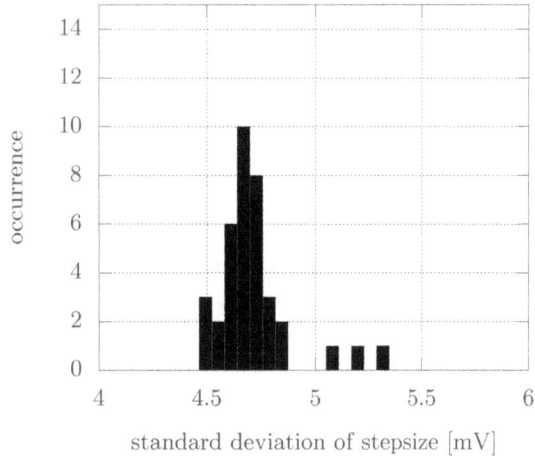

Figure 15: Standard deviation of the step size of layout version 2 (lay2)

area consumption. Although a slight difference can be measured, there is no clear preference which of the four layout versions should be used. Layout version 0 shows a slightly higher accuracy, but version 2 and 3 are also acceptable because of the reduced chip area and complexity.

6. REFERENCES

[1] P. Brokaw. A simple three-terminal IC bandgap reference. *IEEE Journal of Solid-State Circuits*, 9(6):388–393, 1974.

[2] R. A. Hastings. *The Art of Analog Layout*. Prentice Hall, 2000.

[3] K. Kuijk. A Precision Reference Voltage Source. *IEEE Journal of Solid-State Circuits*, 8(3):222–226, 1973.

[4] R. Spilka, M. Hirth, G. Hilber, and T. Ostermann. On-Chip Digitally Trimmable Voltage Reference. *NORCHIP*, Aug. 2007.

[5] Widlar. New Developments in IC Voltage Regulators. *IEEE Journal of Solid-State Circuits*, pages 1–6, 1971.

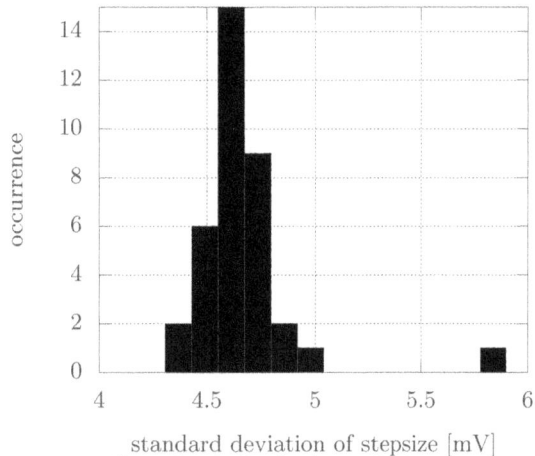

Figure 16: Standard deviation of the step size of layout version 3 (lay3)

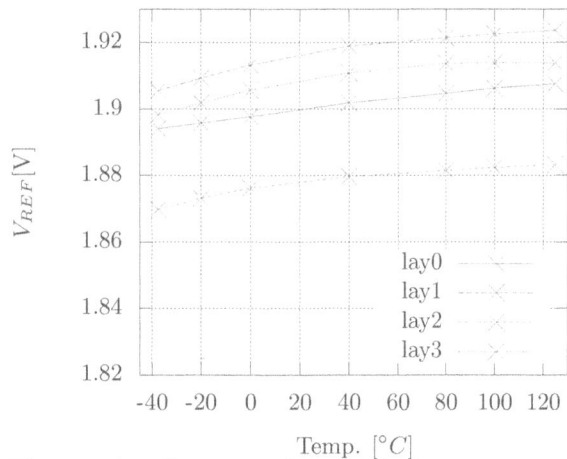

Figure 17: Output voltage of the programmable voltage reference versus temperature

Input and Transistor Reordering for NBTI and HCI Reduction in Complex CMOS Gates

Saman Kiamehr
Karlsruhe Institute of Technology
Karlsruhe, Germany
kiamehr@kit.edu

Farshad Firouzi
Karlsruhe Institute of Technology
Karlsruhe, Germany
firouzi@ira.uka.de

Mehdi B. Tahoori
Karlsruhe Institute of Technology
Karlsruhe, Germany
mehdi.tahoori@kit.edu

ABSTRACT

As CMOS feature size scales to the nanometer regime, transistor aging mostly due to Negative Bias Temperature Instability (NBTI) and Hot Carrier Injection (HCI), has emerged as a major reliability concern. Threshold voltage shift causes the circuit to fail, once the post-aging delay exceeds the timing constraint. In this paper, we investigate the stacking effect of transistors on aging and propose a novel input/transistor reordering approach to alleviate the effect of NBTI and HCI during the active mode operation of the circuit. According to the results, the circuit failing due to aging effect is postponed by increasing the operational lifetime for ISCAS benchmarks by 23.6%, in average, while it has a negligible effect on delay, area, and power compared to the original cell input ordering.

Categories and Subject Descriptors

B.8.1 [**Performance and Reliability**]: Metrics—*Reliability, Testing, and Fault-Tolerance*

General Terms

Reliability

Keywords

Transistor aging, NBTI, HCI

1. INTRODUCTION

As CMOS technology approaches nanometer scales, transistor aging is becoming a major reliability issue for the lifetime operation of VLSI chips. Transistor aging in scaled CMOS technologies mostly arises from *Negative Biased Temperature Instability* (NBTI) and *Hot Carrier Injection* (HCI) [1]. NBTI and HCI increase the absolute value of the PMOS and NMOS threshold voltages respectively and as a result, the performance of the circuit decreases over time. Eventually, when the delay of the circuit exceeds the timing constraints, the circuit fails. Consequently, the operational lifetime of CMOS VLSI chips is reduced.

There is a considerable amount of work for alleviating the effect of transistor aging at various design levels. To name just a few, gate sizing [2], Dynamic Voltage and Frequency Scaling (DVFS) [3], body biasing [4], power gating [5], Input Vector Control (IVC) [6] and Internal node control [7] are some of the existing aging mitigation methods mostly due to NBTI.

Connection of multiple transistors in series is referred to as transistor *stacking*. The way the inputs are connected to different transistors in the stack has a considerable impact on the NBTI and HCI effects. The NMOS (PMOS) transistor stacking increases (decreases) the source voltage of the upper (lower) transistors in the stack as well as lowers the absolute value of the gate-source voltage of these transistors. Since the amount of aging due to NBTI and HCI depends on the initial value of transistor threshold voltage as well as gate-source voltage, stacking has a strong impact on these effects.

A stacking aware gate-level NBTI delay degradation model is presented in [8]. This work was limited to NBTI and does not consider HCI. However, as the channel length is scaling aggressively beyond 65nm, the effect of HCI should be considered as well [9]. In [10, 11], a stacking-aware pin reordering approach is proposed to reduce the effect of NBTI. However, it was limited only to simple stacking cases (i.e. not complex CMOS gates) and also it did not take HCI effect into account.

In this paper, we propose a new simplified stacking-aware NBTI and HCI aging model. Based on this model, we present an algorithm to precisely calculate NBTI and HCI considering stacking effect. Next, we present an *Input and Transistor Reordering* technique to reduce the transistor aging effect during circuit operation (active mode). By exploiting this approach, the number of transistors experiencing NBTI and HCI can be reduced while providing the same output. This method can be applied to circuits containing both simple gates (e.g. NAND, NOR) and complex gates (i.e. the gates which contain various combination of transistors in series and parallel).

The rest of this paper is organized as follows. Section 2 presents the background on NBTI and HCI. The stacking effect for NBTI and HCI, and stacking-aware aging model are presented in Section 3. The proposed input reordering technique is described in Section 4. The experimental results are discussed in Section 5. Finally, Section 6 concludes the paper.

2. TRANSISTOR AGING

2.1 NBTI

A PMOS transistor is under stress when $V_{gs} = -V_{DD}$. In this phase, some interface traps are generated at the interface of $Si - SiO2$. These interface traps cause the absolute value of threshold voltage to increase [12]. During the recovery mode, $V_{gs} = 0$, the threshold voltage is partially recovered and decreased toward its initial value by removing some of the generated traps. However, the shift cannot be compensated totally and hence does not reach to its original value before stress [12]. The state of PMOS transistors frequently changes between stress and recovery modes due to different input vectors, during a normal circuit operation (dynamic NBTI). Since recovery cannot completely relieve the effect of stress on the threshold voltage, there is an overall positive shift in the magnitude of PMOS transistor threshold voltage over time, as illustrated in Figure 1(a).

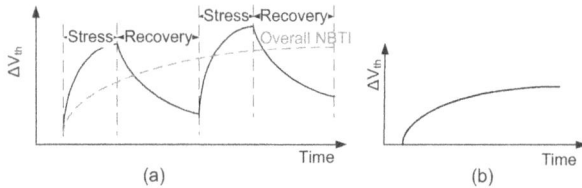

Figure 1: V_{th} change due to a) NBTI b) HCI

The overall NBTI effect can be modeled as follows [13]:

$$\Delta V_{th} = A \times D_c^n \times t^n \quad (1)$$

where A is technology dependent factor, n is a constant which depends on the fabrication process, D_c is the duty cycle (the ratio of stress period to the total operational time), and t is the total operational time.

2.2 HCI

HCI affects mainly NMOS transistors when the gate of NMOS is making a transition. Carriers in the channel are subjected to different electric fields when traveling between the source to the drain. If these hot carriers collide with the gate oxide interface, some electron-hole pairs are generated. Some of these generated electrons are energetic enough to accelerate and get trapped in the gate oxide. These interface traps are generated at the $Si - SiO_2$ interface near the drain causing the threshold voltage to increase [14] (Figure 1(b)). Since hot electrons are generated when the gate of the NMOS switches, the threshold voltage change due to HCI has a direct dependency with the operational frequency. The threshold voltage change can be estimated by Equation (2) [15].

$$\Delta V_{th} = A_{HCI} \times \alpha \times f \times e^{\frac{V_{DD} - V_{th}}{t_{ox} E_1}} \times t^{0.5} \quad (2)$$

where A_{HCI} is a technology dependent constant, α is the activity factor, and f is the clock frequency. V_{th} and V_{DD} are the threshold voltage and supply voltage, respectively. t_{ox} is the oxide thickness, E_1 is a constant equal to $0.8V/nm$ [16] and t is the total time.

3. TRANSISTOR STACKING AND AGING

When multiple transistors are connected in series, the drain-source voltage (V_{ds}) of each transistor is smaller than V_{dd}. As a result, the magnitude of the threshold voltage increases due to *Drain Induced Barrier Lowering* (DIBL). Furthermore, the absolute gate-source voltage (V_{gs}) of upper

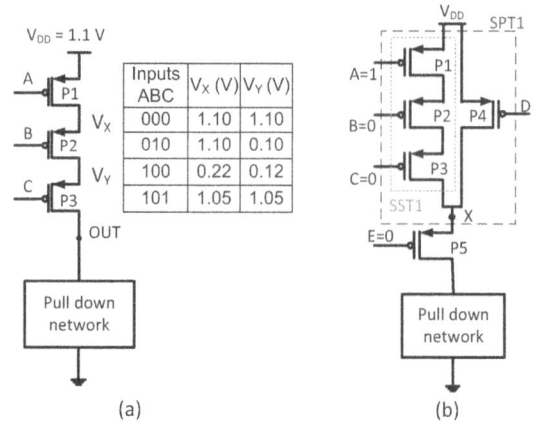

Figure 2: Stacking effect in a) 3 input NOR gate b) complex gate

(lower) transistors is smaller in a series structure of a pull-down (pull-up) network. This phenomenon is called stacking effect. Since there is an exponential relation between aging effects and both gate-source and threshold voltages of transistors, the stacking effect has a noticeable impact on both NBTI and HCI.

Complex gates consist of a combination of parallel and serial transistors in pull-up/pull-down network. Based on this definition, simple gates (NAND, NOR) are a sub-set of complex gates. We define a *Super Transistor* (ST) as a virtual transistor that might be either a *Super Series Transistor* (SST), *Super Parallel Transistor* (SPT), NMOS transistor (NT) or PMOS transistor (PT). A SST is a virtual transistor which is an ordered list of several ST in series. A SST is ON when all its internal STs are ON. We also define a *Super Parallel Transistor* (SPT) as a virtual transistor which consists of several STs in parallel. A SPT is ON when at least one of its internal parallel STs is ON.

$$ST = SST \mid SPT \mid NT \mid PT$$
$$SST_i = (ST_i^1, ..., ST_i^n)$$
$$SPT_j = \{ST_j^1, ..., ST_j^m\} \quad (3)$$

Using aforementioned definitions, all complex gates can be reduced to a simple structure which has only serial or parallel STs. We also refer to simple-SST as a SST which contains only transistors in series. On the other hand, complex-SST is that containing transistors and super transistors. Based on these definitions, since the transistors in a simple-SST are symmetrical, reordering the inputs of transistors, does not affect the functionality of simple-SST. In contrast, input reordering (connecting different inputs to different transistors) changes the functionality of complex-SSTs, because they are not necessarily symmetric.

3.1 Stacking effect on NBTI

Since NBTI only affects PMOS transistors, stacking effect on NBTI should be considered only for the pull-up network. For better understanding of the stacking effect in serial structures, we start with a 3-input NOR gate as shown in Figure 2(a). In a 3-input NOR gate, 8 different cases may occur. Here we investigate the four most important cases.

1. $ABC = 000$: $V_X = V_{DD}$ and $V_Y = V_{DD}$. Therefore, $V_{GS}^{P1} = V_{GS}^{P2} = V_{GS}^{P3} = -V_{DD}$. In this case, the three PMOS transistors are in stress mode.

2. $ABC = 010$: $A = 0$, hence $V_X = V_{DD}$. Since $V_g^{P2} =$

V_{DD}, $V_{gs}^{P2} = 0$, and $P1$ is in stress mode and $P2$ is in recovery mode. Moreover, since $V_Y \approx 0$, $V_{gs}^{P3} \approx 0$. Therefore, although its input is zero, it is in recovery mode.

3. $ABC = 100$: The input of transistor $P1$ is one and the other inputs are zero. Because of the stacking effect, the resistance of $P1$ is much larger than the resistance of $P2$ and $P3$. As a result, $V_X \approx 0$ and hence $V_{gs}^{P2} = V_{gs}^{P3} = 0$. Therefore, both lower transistors are in the recovery mode although $V_{gs}^{P2} = V_{gs}^{P3} = 0$.

4. $ABC = 101$: In this case, $P1$ and $P3$ are OFF and $P2$ is ON. The resistance of $P3$ is much bigger than the resistance of $P1$ and $V_X = V_Y \approx V_{DD}$. As a result, $V_{gs}^{P2} = -V_{DD}$ and hence $P2$ is under stress. The other transistors are in recovery mode.

It can be concluded that the state of a transistor (stress or recovery) depends not only on its input, but also on the state of its upper and lower transistors. For example transistor $P2$ (Figure 2(a)) is in recovery and stress mode for cases $(ABC = 100)$ and $(ABC = 101)$ respectively, although the input of $P2$ is the same.

To investigate the stacking effect on a complex gate, consider an example illustrated in Figure 2(b). In this example $SST1$ consists of transistors $P1$, $P2$ and $P3$ and $SPT1$ consists of $SST1$ and $P4$. First, we investigate the stacking effect on the transistors in $SST1$. The situation in $SST1$ is almost the same as the pull-up network of a 3-input NOR gate. However, $SST1$ has an interaction with other transistors in the complex gate example. To investigate the interaction between $SST1$ and $P4$ consider Figure 2(b). If $D = 1$, $P1$ and $P4$ are OFF and the other transistors are ON. In this situation the parallel transistor of SST1 ($P4$) has no effect on $SST1$ and similar to the 3-input NOR (Figure 2(a) with $ABC = 100$), all transistors inside $SST1$ are in recovery mode. On the other hand, if transistor $P4$ is ON, the voltage of node X is equal to V_{DD}. As a result, the voltages of all internal nodes in $SST1$ are equal to V_{DD}. Therefore, in contrast to previous case, $P2$ and $P3$ are in stress mode. Finally we analyze the stacking effect on transistor $P5$. The pull-up network of this complex gate can be considered as a two-series transistors ($P5$ and $SPT1$). As a result, $P5$ is in stress mode if only the corresponding input is zero and $SPT1$ is ON. The Stress-Recovery (SR) flowchart illustrated in Figure 3 summarizes the above cases for an N-input complex gate.

3.2 Stacking effect on HCI

Since HCI affects NMOS transistors, the stacking effect on HCI should be considered only in a pull-down network, such as NAND gate. Consider a simple two input NAND gate shown in Figure 4. Here, we assume that a falling transition occurs at the input of transistor. The situation for the rising transition is similar.

For the lower NMOS transistor (N2), because its source voltage is zero, each falling transition of its gate input results in a falling transition in V_{gs}. Consequently, the effective activity factor of N2 used in Equation (2) is equal to the activity factor of the input connected to it (Figure 4(a)). However, for the upper transistor (N1) two different situations can occur depending on the status of N2 (Figures 4 (b) and (c)). When the input of N2 is connected to V_{DD}, this transistor is ON and hence $V_{S1} = 0$. Therefore each falling transitions of IN2 leads to a falling transition in V_{gs}^{N2}.

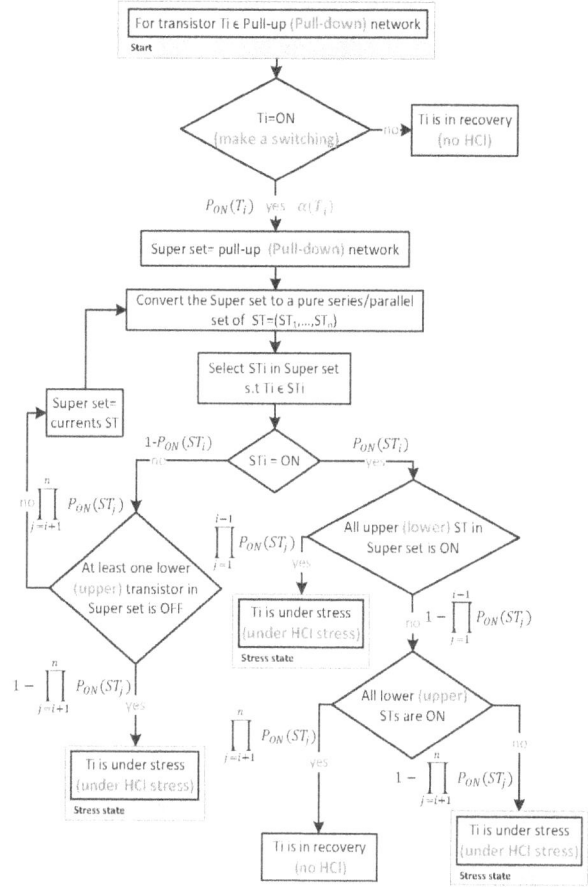

Figure 3: Stress-Recovery (SR) flowchart for NBTI (HCI) in an N input complex gate. For $ST = (ST_1, ..., ST_n)$ if $i < j$, ST_i is closer to power-line

On the other hand when $IN2 = 0$, V_{gs}^{N1} remains constant even when there is a falling transition on IN1. During a falling transition of IN1, the initial value of IN1 is one and IN2 is zero. In this situation, N1 is ON and has a lower resistance in comparison to N2 which is OFF. By this observation, $V_{S1} = V_{DD}$ and consequently the initial value of V_{gs}^{N1} is zero. When IN1 switches to zero, both transistors become OFF. Since the resistance of N1 is much larger than the resistance of N2 (due to the body effect), V_{S1} becomes zero. This means V_{gs1} remains constant and does not switch. It can be concluded that, when IN2 is zero, the switching of IN1 does not result in a switching of V_{gs}^{N1}.

Consequently, the effective switching activity of the transistor N1 is less than the switching activity of its input. In other words, the lower transistor (N2) masks some switchings of the upper transistor which causes N1 to suffer less from HCI. Considering the above cases and by using the definitions of the super-transistors (SST and SPT), the flowchart of Figure 3 (in bold red) shows how the stacking affects HCI in an N input complex gate.

4. REORDERING METHODOLOGY

Since transistor aging is considerably affected by the order in which the transistors are placed in a complex gate and their relative input values, our key idea is to reorder the inputs of each gate (which input is connected to which transistor in the gate connection) without changing the functionality of the gate to decrease transistor aging during the active mode. Another technique which can improve the efficiency of the input reordering approach is Super-Transistor (ST)

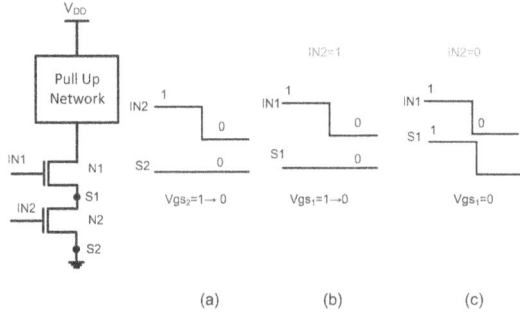

Figure 4: Stacking effect in a NAND gate

reordering. In this work, this is done for pull-up network to reduce NBTI, as well as pull-down network to reduce HCI. For input and transistor reordering we need to calculate the ON/OFF probability of each ST based on the Signal Probabilities (SP) of the inputs (i.e the probability of being one). According to the definitions presented in Equation (3) and the definitions of ST, SST, and SPT the following equations are extracted:

$$P_{ON}(NT) = SP(NT), \quad P_{ON}(PT) = 1 - SP(PT)$$

$$P_{ON}(SST_i) = \prod_{k=1}^{n}(P_{ON}(ST_i^k))$$

$$P_{ON}(SPT_j) = 1 - \prod_{l=1}^{m}(1 - P_{ON}(ST_j^l)) \quad (4)$$

4.1 NBTI reduction

To describe our NBTI reduction method, a simple example is shown in Figure 5. We assume that the input vector is $ABC = 100$. If the inputs are connected in a way depicted in Figure 5(a), then based on the rules described in the previous section, all three transistors are in recovery mode (see Figure 2(b) $ABC = 100$). On the other hand, if the inputs are connected as in Figure 5(b), only the two lower transistors are in recovery mode (see Figure 2(b) $ABC = 010$). It can be concluded that for the input $ABC = 100$, the first interconnection between the inputs and transistors is more favorable, in terms of NBTI, than the second one.

Based on this observation, in our aging-aware input reordering approach, all possible combinations of the input connections of series transistors in each simple-SST are considered and the order resulting in the minimum aging due to NBTI is chosen (reordering inputs of transistors $P1$, $P2$, and $P3$ in Figure 2). Since, the series transistors in a simple-SST are symmetrical, the order of the inputs does not affect the functionality of the gate. This can be further enhanced with ST reordering. In this technique, all possible permutation of ST in each complex-SST are considered and the ordering leading to minimum NBTI effect is selected. Since the input reordering changes the functionality of complex-SST, in ST reordering method, the ST (not their inputs) are reordered. For example in Figure 2 reordering $SPT1$ and $P5$ may change the overall NBTI effect on the complex gate. The following steps show the general methodology of the NBTI-aware input/ST reordering for an N-input complex gate:

1. The pull-up network is converted to a set of SSTs and SPTs.

2. For each complex-SST, all permutations of ST ordering are considered

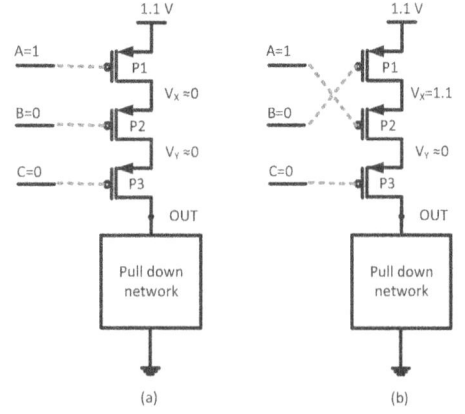

Figure 5: Input reordering in a 3 input NOR gate

3. For each permutation, all possible orders of inputs are considered

4. For each case, the effective duty cycle (the time ratio between stress time to the total time) of each transistor is calculated

5. The case which has the minimum $\sum_i (D_{c-eff}(P_i))^n$ is selected

The effective duty cycle (D_{c-eff}) represents the NBTI effect on a transistor using the Equation (1). The NBTI status of a transistor (stress/recovery) can be determined according to the ON/OFF status of this transistor and the other transistors in the gate. The following steps show the process of the effective duty cycle calculation of each transistor in a complex gate:

1. Construct a Stress-Probability tree from the stress-recovery (SR) flowchart. The nodes and the edges of this tree are defined as follow:

   - Node: traversed ST in the SR flowchart
   - Edge: branch probability in the SR flowchart

2. Remove leaves (end edges) leading to recovery

3. $Prob\_leaf$ is equal to product of probability of edges on the path from the root to that leaf

4. $D_{c-eff} = \Sigma Prob\_leaf$

As an example, Figure 6 shows the steps of the effective duty-cycle calculation for transistor $P2$ in Figure 2. It should be noted that, only for input reordering, the minimum overall effective duty cycle (minimum NBTI) is obtained by connecting the input with a higher SP to the transistor with a higher position in series structure.

4.2 HCI reduction

As described in Section 3.2, in a 2-input NAND gate when the input of the lower NMOS transistor is zero, the falling transition of the gate-source voltage of the upper transistor is masked. This phenomenon leads to an HCI reduction of the upper transistor. The effective switching activity, α_{eff}, of a 2-input NAND gate (see Figure 4) can be calculated based on Table 1. The key idea is to reorder the inputs/ST to minimize the total gate delay degradation due to HCI. According to the Equation (2), V_{th} shift due to HCI has a direct relation with the switching activity. Consequently, the HCI minimization problem of a gate is equivalent to a minimization of the total effective switching activity.

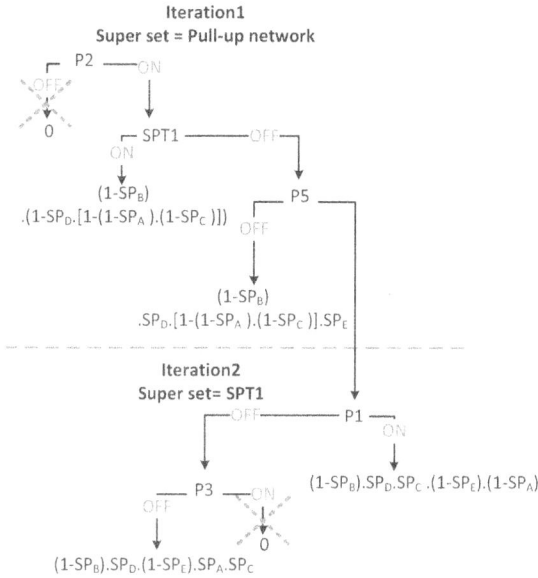

Iteration1
Super set = Pull-up network

P2 — ON
OFF
0

⌐ SPT1 —— OFF
ON

$(1-SP_B)$
$.(1-SP_D.[1-(1-SP_A).(1-SP_C)])$

P5
OFF

$(1-SP_B)$
$.SP_D.[1-(1-SP_A).(1-SP_C)].SP_E$

Iteration2
Super set= SPT1

OFF —— P1 — ON

P3 — ON
OFF
0

$(1-SP_B).SP_D.SP_C.(1-SP_E).(1-SP_A)$

$(1-SP_B).SP_D.(1-SP_E).SP_A.SP_C$

Figure 6: Stress-probability tree: effective duty cycle calculation of $P2$

Based on this fact, the input reordering of a two-input NAND gate for HCI minimization can be formulated as shown below.

$$Minimize(\alpha_{eff}(N1) + \alpha_{eff}(N2)) =$$
$$Minimize(\alpha_A \times SP_B + \alpha_B)$$

where α is the switching activity factor of an input and SP is the signal probability. Therefore, input reordering is done in a way that $\alpha_{eff}(N1) + \alpha_{eff}(N2)$ is minimized.

The general flow of the HCI-aware input/ST reordering for an N-input complex gate is similar to the one for NBTI reduction (Section 4.1), except that here, the objective is to find the reordering which has the minimum $\sum_i \alpha_{eff}(N_i)$. The α_{eff} of each transistor is a representative of its HCI effect based on the Equation (2). When a transistor makes a switching, depending on the ON/OFF status of the other transistors in the gate, this transistor might experience HCI effect. The α_{eff} of each transistor is calculated by a similar flow as the effective duty cycle (D_{c-eff}) calculation explained in Section 4.1.

5. EXPERIMENTAL RESULTS

We evaluate the efficiency of the proposed methods with ISCAS'85 and ISCAS'89 benchmark circuits. The overall proposed methodology for NBTI and HCI reduction by means of stacking-based input reordering is summarized below.

First, a logic synthesis tool is used to map a benchmark circuit to a gate-level netlist. In our experiments, we used Synopsys Design Compiler as the synthesis tool to map the circuits to SAED 90nm standard cell library. The gate delays are extracted from the standard cell library as well. Afterwards, the extracted gate-level description of the circuit is given to a Logic Simulator and signal probabilities and

Table 1: Effective Activity Factor of a 2-input NAND gate

Transistor	Effective Activity Factor
$N1$	$\alpha_{IN1}.SP_{IN2}$
$N2$	α_{IN2}

activity factors of all internal nodes are calculated based on a Monte Carlo (MC) approach. It should be noted that, the signal probability changes from one workload to another. Therefor, for accurate signal probability estimation a real trace of the application can be used to consider workload dependencies. To alleviate the effect of NBTI and HCI, the input/transistor of pull-up and pull-down network are reordered based on the methodology proposed in the Section 4 to minimize the total effective duty cycle and activity factor, respectively. Finally, the obtained aging-aware netlist and original netlist are further processed to analyze the effectiveness of the proposed technique on lifetime of the circuit and its overheads as well. The overall runtime of this methodology depends on the size of the circuit and types of its primitive logic cells. In our experiments this process takes several minutes in average.

Table 2 illustrates the effect of considering stacking on NBTI and HCI delay degradation models. Based on these results, traditional methods which do not consider the stacking effect overestimate the delay degradation more than 16.6-% on average. This overestimation leads to pessimistic analysis in aging mitigation techniques resulting in over-design in terms of unnecessary power consumption, performance and area overhead.

Table 2: Delay degradation of benchmark circuits

Circuit	No Stacking $\Delta delay(ps)$	Stacking $\Delta delay(ps)$	Error
C432	146.03	135.42	14%
C880	155.57	131.31	23%
C1355	81.11	71.83	19%
C2670	100.89	91.17	18%
C3540	201.94	185.28	14%
C5315	40.49	36.31	16%
C6288	623.99	589.06	12%
Average			16.6%

The experimental results of our proposed input/transistor reordering technique are shown in Table 3. The new lifetime after reordering is calculated and the percentage increase is reported in this table. According to this table, the choice of suitable order of inputs can significantly impact the delay degradation due to NBTI and HCI. Based on these results, in average, the lifetime can be extended by 23.6%. According to the results, our proposed model has better improvement in sequential logics. This better improvement is due to a wider range of signal probabilities in sequential circuits. Furthermore, based on the results, reordering the inputs/transistors for both NBTI and HCI improves the lifetime of the circuit by 22.3% in average comparing to reordering for only NBTI effect.

It should be noted that, this method may change the pre-aging delay of the circuit. It is due to the fact that, the delay optimization tools use reordering based on the arrival time of the signal to minimize the pre-aging delay. However, since we reorder inputs/transistors to minimize aging, it may have a side-effect on the pre-aging delay. The delay overhead of the proposed technique is reported in the table as well. According to the results, the proposed method does not significantly affect the pre-aging delay of the circuits and the introduced overhead is negligible. In addition, area and power overhead of the proposed methodology of reordering are reported which indicates that our method does not incur significant area and power overhead. The area overhead is estimated by Design-compiler that does not consider the

205

Circuit	Lifetime Improvement NMOS Reordering	Lifetime Improvement PMOS Reordering	Total lifetime Improvement	Delay Overhead	Power Overhead	Area Overhead
C432	5.3%	5.5%	10.8%	0.85%	0.12%	-0.10%
C880	7.8%	4.7%	11.3%	0.00%	-0.05%	0.00%
C1355	10.2%	4.8%	13.2%	-0.38%	-1.70%	0.00%
C2670	5.9%	11.9%	17.3%	0.12%	-0.01%	0.00%
C3540	3.6%	6.3%	10.2%	-0.06%	-0.19%	0.00%
C5315	2.8%	2.3%	4.6%	-0.43%	-0.01%	0.00%
C6288	2.9%	6.1%	7.4%	0.00	0.00%	0.00%
S05378	0.4%	1.9%	2.1%	0.45%	0.12%	0.00%
S09234	11.1%	3.1%	11.4%	0.37%	-0.45%	0.00%
S35932	7.5%	27.1%	31.4%	-0.67%	0.23%	0.00%
S15850	10.7%	26.6%	35.1%	-.08%	0.13%	0.00%
S13207	1.5%	127.5%	128.2%	-.01%	-0.14%	0.00%
S38417	0.1%	23.7%	23.7%	-.09%	0.03%	0.00%
Average	5.7%	19.3%	23.6%	0.00%	-0.15%	0.00%

layout information. However, due to input reordering, the routing might be affected and hence the actual area overhead depends on the underlying layout and technology information.

6. CONCLUSIONS

By scaling the CMOS technology, transistor aging mostly due to Negative Bias Temperature Instability (NBTI) and Hot Carrier Injection (HCI) is becoming one of the major reliability issues. NBTI and HCI result in a threshold voltage shift for PMOS and NMOS transistors, respectively. These two phenomena, lead to runtime failures and reduced operational lifetime eventually. In this paper, we exploited the stacking effect on transistor aging and proposed an input and transistor reordering approach to reduce the effect of transistor aging during the active mode. This was done for both PMOS and NMOS stacks as well as for complex CMOS gates, to reduce the effect of NBTI and HCI, respectively. This approach comes at negligible area, delay, or power overhead. Our experiments for ISCAS benchmark circuits show that this method can increase the lifetime by 23.6%, in average.

7. REFERENCES

[1] K. Bernstein, D. J. Frank, A. E. Gattiker, W. Haensch, B. L. Ji, S. R. Nassif, E. J. Nowak, D. J. Pearson, and N. J. Rohrer. High-performance CMOS variability in the 65-nm regime and beyond. *IBM Journal of Research and Development - Advanced silicon technology*, 50:433–449, July 2006.

[2] X. Yang and K. Saluja. Combating nbti degradation via gate sizing. In *Proc. Int'l SympQuality Electronic Design*, pages 47–52.

[3] M. Basoglu, M. Orshansky, and M. Erez. NBTI-aware DVFS: a new approach to saving energy and increasing processor lifetime. In *ACM/IEEE int'l symposium on Low power electronics and design*, pages 253–258, 2010.

[4] Z. Qi and M.R. Stan. NBTI resilient circuits using adaptive body biasing. In *Proc. ACM Great Lakes Symp. on VLSI*, pages 285–290, 2008.

[5] A. Calimera, E. Macii, and M. Poncino. NBTI-Aware power gating for concurrent leakage and aging optimization. In *Proc. Low power electronics and design int'l Symp.*, pages 127–132, 2009.

[6] F. Firouzi, S. Kiamehr, and M.B. Tahoori. A linear programming approach for minimum nbti vector selection. In *Proceedings of the 21st edition of the great lakes symposium on Great lakes symposium on VLSI*, pages 253–258. ACM, 2011.

[7] D.R. Bild, G.E. Bok, and R.P. Dick. Minimization of NBTI performance degradation using internal node control. In *Proc. Design, Automation and Test in Europe Conf.*, pages 148–153, 2009.

[8] H. Luo, Y. Wang, K. He, R. Luo, H. Yang, and Y. Xie. A novel gate-level NBTI delay degradation model with stacking effect. *Integrated Circuit and System Design. Power and Timing Modeling, Optimization and Simulation*, pages 160–170, 2007.

[9] K.K. Kim, H. Nan, and K. Choi. Adaptive HCI-aware power gating structure. In *ISQED Int'l Symposium*, pages 219–224. IEEE, 2010.

[10] K.C. Wu and D. Marculescu. Joint logic restructuring and pin reordering against nbti-induced performance degradation. In *Proceedings of the Conference on Design, Automation and Test in Europe*, pages 75–80, 2009.

[11] S.V. Kumar, C.H. Kim, and S.S. Sapatnekar. Nbti-aware synthesis of digital circuits. In *Proceedings of the 44th annual Design Automation Conference*, pages 370–375. ACM, 2007.

[12] S. Bhardwaj, W. Wang, R. Vattikonda, Y. Cao, and S. Vrudhula. Predictive modeling of the NBTI effect for reliable design. In *Proc. Custom Integrated Circuits Conf.*, pages 189–192, 2006.

[13] W. Wang, S. Yang, S. Bhardwaj, S. Vrudhula, F. Liu, and Y. Cao. The impact of NBTI effect on combinational circuit: modeling, simulation, and analysis. *Very Large Scale Integration (VLSI) Systems, IEEE Trans.*, 18(2):173–183, 2010.

[14] W. Wang, V. Reddy, A.T. Krishnan, R. Vattikonda, S. Krishnan, and Y. Cao. Compact modeling and simulation of circuit reliability for 65-nm CMOS technology. *Device and Materials Reliability, IEEE Transactions on*, 7(4):509–517, 2007.

[15] A. Tiwari and J. Torrellas. Facelift: Hiding and slowing down aging in multicores. In *Microarchitecture, IEEE/ACM Int'l Symposium*, pages 129–140, 2008.

[16] http://ptm.asu.edu/reliability/.

Memristor: The Illusive Device

Khaled Nabil Salama
King Abdullah University of Science and
Engineering
4700 King Abdullah University of Science
and Technology
Thuwal 23955-6900
Kingdom of Saudi Arabia
khaled.salama@kaust.edu.sa

ABSTRACT

The memristor (M) is considered to be the fourth two-terminal passive element in electronics, alongside the resistor (R), the capacitor (C), and the inductor (L). Its existence was postulated in 1971 but its first implementation was reported in 2008. Where was it hiding all that time and what can we do with it? Come and learn how the memristor completes the roster of electronic devices much like a missing particle that physicists seek to complete their tableaus.

Categories and Subject Descriptors

B.3 [Hardware]: Memory Structures

General Terms

Design, Experimentation, Measurement, Performance

Keywords

Passive Element, Electronic Devices

Bio

Dr. Salama received his bachelor's degree with honors from the Electronics and Communications Department at Cairo University in Egypt in 1997, and his master's and doctorate degrees from the Electrical Engineering Department at Stanford University in the United States, in 2000 and 2005 respectively. He was an assistant professor at RPI between 2005 and 2009. He joined KAUST in January 2009 and was the founding program chair until August 2011. His work on CMOS sensors for molecular detection has been funded by the National Institutes of Health (NIH) and the Defense Advanced Research Projects Agency (DARPA), awarded the Stanford-Berkeley Innovators Challenge Award in biological sciences and was acquired by Lumina Inc. He is the author of 90 papers and 8 patents on low-power mixed-signal circuits for intelligent fully integrated sensors and non linear electronics specially memristor devices. He is a senior member of IEEE.

GLSVLSI'12, May 3–4, 2012, Salt Lake City, Utah, USA.
ACM 978-1-4503-1244-8/12/05.

InMnAs Magnetoresistive Spin-Diode Logic

Joseph S. Friedman[1], Nikhil Rangaraju[2], Yehea I. Ismail[1], and Bruce W. Wessels[1,2]
[1]Department of Electrical Engineering & Computer Science
[2]Department of Materials Science & Engineering
Northwestern University
Evanston, IL, USA
jf@u.northwestern.edu, nikhilrangaraju2011@u.northwestern.edu, ismail@eecs.northwestern.edu,
b-wessels@northwestern.edu

ABSTRACT

Electronic computing relies on systematically controlling the flow of electrons to perform logical functions. Various technologies and logic families are used in modern computing, each with its own tradeoffs. In particular, diode logic allows for the execution of logic with many fewer devices than complementary metal-oxide-semiconductor (CMOS) architectures, which implies the potential to be faster, cheaper, and dissipate less power. It has heretofore been impossible to fully utilize diode logic, however, as standard diodes lack the capability of performing signal inversion. Here we create a binary logic family based on high and low current states in which the InMnAs magnetoresistive semiconductor heterojunction diodes implement the first complete logic family based solely on diodes. The diodes are used as switches by manipulating the magnetoresistance with control currents that generate magnetic fields through the junction. With this device structure, we present basis logic elements and complex circuits consisting of as few as 10% of the devices required in their conventional CMOS counterparts. These circuits are evaluated based on InMnAs experimental data, and design techniques are discussed. As Si scaling reaches its inherent limits, this spin-diode logic family is an intriguing potential replacement for CMOS technology due to its material characteristics and compact circuits.

Categories and Subject Descriptors

B.7.1 [Integrated Circuits]: Types and Design Styles—*Advanced technologies*

Keywords

Spin-diode, spintronics, diode logic, magnetoresistance.

1. INTRODUCTION

As transistor size continues to decrease, predictions for the end of scaling Si transistors continue to be postponed. It is of critical importance, however, to develop new device technologies that will provide further computing improvements over the long term. Furthermore, these new technologies complement CMOS in niche applications such as high performance circuits. These technologies include devices derived from single-electron transistors [1], carbon nanotubes [2] and related graphene

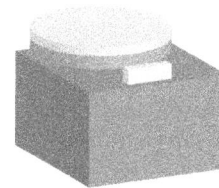

p-III-Mn-V   n-III-V   Contact

Figure 1. Magnetoresistive spin-diode.

structures [3], [4], nanowires [4], [5], and molecular switches [6]. Additionally, there has been significant interest in devices and logic design techniques that utilize electron spin [7]–[17].

One spintronic device of particular interest is the magnetoresistive spin-diode, which is produced by doping a semiconductor p-n junction with an element that interacts strongly with a magnetic field. A common dopant element is Mn. In this semiconductor heterojunction, shown in Fig. 1, Mn is added to form the paramagnetic p-type layer. The spin-diode acts as a conventional diode in the presence of zero or low magnetic fields, with a high ratio of forward current to reverse current. However, when a magnetic field is applied across the junction, there is an increase in resistivity [12]. Thus, under forward bias, it is possible to define two distinct states: a resistive state in the presence of a magnetic field, and a conductive state in its absence.

Logic styles and circuit architectures should be reconsidered in order to fully utilize new materials and devices. While CMOS transistors and logic have dominated Si-based circuits [18], other devices and logic families exhibit significant advantages. Diode logic is elegant in several respects, such as simple OR gates and single junction devices that allow for compact circuit structures. Circuits based on diode logic use fewer devices than their CMOS counterparts, and therefore potentially consume less power and area while operating at higher speeds. Diode logic, however, has historically been impractical due to the inability of a diode to act as an inverter [19]. As inversion is a necessary function of a complete logic family, standard diodes can only perform complex logic functions in concert with transistors.

The recent invention of the magnetoresistive spin-diode solves this problem, as it allows for the creation of a complete logic family composed solely of spin-diodes, including an inverter [20]. This diode logic family performs logical functions with significantly fewer devices than CMOS, and is therefore a potential replacement for Si CMOS computing. The rest of this paper is organized as follows: the logic family is explained in section 2. In section 3, compact spin-diode logic circuits are discussed, and simulation results are presented in section 4. The computing implications of this diode logic family are discussed in section 5. The paper is concluded in section 6.

2. SPIN-DIODE LOGIC FAMILY

The spin-diode logic family presented in [20] executes logic functions by passing an electric current through spin-diodes to control a magnetic field through other spin-diodes. This logic family is composed solely of spin-diodes; no transistors are required. It is therefore the first example of a complete diode logic family. As such, it provides the simplicity advantages of a two-terminal device structure while exploiting the efficiencies of diode logic.

2.1 Digital States

In this spin-diode logic family, a digital '1' is defined as the "high" current produced by the spin-diode conductive state and a digital '0' is defined as the "low" current produced by the resistive state. Each spin-diode is forward biased, with the positive and negative terminals connected, respectively, to the power supply, V_{DD} and ground. The wires that form these connections are routed to control the magnetoresistance of other spin-diodes. A spin-diode propagating a '1' thus creates a magnetic field in another spin-diode, while a spin-diode propagating a '0' does not.

A metal wire or wires can be placed parallel to the plane of the junction, isolated by an insulator. These wires control the magnetoresistive state of the diode, as currents through the wires create magnetic fields perpendicular to the plane of the junction. Under zero or small field, the diode is in its conductive state; a large field asserts the diode's resistive state. Depending on the relative direction of the currents, these currents interfere either constructively or destructively. If the currents in the two wires travel in opposite directions, the fields will add, doubly suppressing the diode current; if the currents are in the same direction, the fields will cancel, allowing current to flow through the diode. These features form the building blocks of spin-diode logic.

2.2 Basis Logic Elements

In Fig. 2, each of this family's basis logic elements is illustrated using at most a single diode: an inverter, NOR gate, XNOR gate, and OR gate. In each of these configurations, the positive terminal of the diode is connected to V_{DD}, while the negative terminal is routed through the circuit before eventually being grounded. Therefore, in the absence of current through the various control wires, each of these diodes propagates a '1', and the configuration of the control wires dictates the logical function of each gate.

An inverter, the simplest gate, is shown in Fig. 2a. The input current I_A is routed alongside the diode, and induces a field proportional to its current. If I_A is a '1', the current creates a large magnetic field, thereby reducing the current through the diode, and causing the output current I_O to propagate a '0'. If I_A is a '0', it does not create a sufficient magnetic field through the junction, and a '1' is propagated.

The addition of a second input current I_B results in a NOR or XNOR gate, depending on its relative placement in the circuit, as shown in Figs. 2c and 2e. If the two currents flow in opposite directions, the magnetic fields produced by the currents will be oriented in the same direction. Therefore, if at least one of the two NOR inputs is a '1', the output propagates a '0'; otherwise, the output is a '1'. In the case of the XNOR, each current creates a magnetic field through the diode in the opposite direction. Therefore, if both inputs are '1', there is no net magnetic field through the diode, and a large current flows, propagating a '1'. This single device XNOR gate is significantly more compact than the standard ten device CMOS implementation.

As the magnitude of the currents defines the digital states, an OR gate is constructed simply by combining two wire currents, as shown in Fig. 2g. Therefore, if at least one of the inputs is a '1', the output is a '1'. By placing an OR gate as an input to another

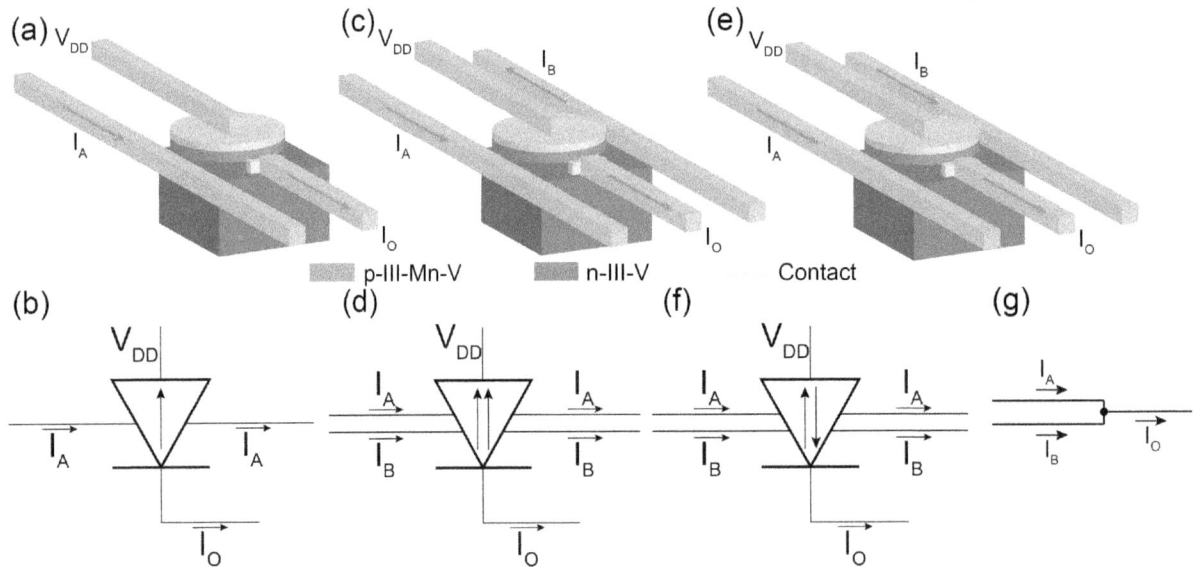

Figure 2. (a) Inverter consisting of input wire with current I_A routed alongside the diode to control the output current I_O. The positive terminal is connected to V_{DD} and the negative terminal is connected to ground. A high input current results in a magnetic field that suppresses the output current. (b) Symbol for spin-diode inverter. (c) Additional input current I_B results in a NOR gate. A high current on either input suppresses the output current. In the case of high currents on both inputs, the output current is doubly suppressed. (d) Symbol for spin-diode NOR gate. (e) Inverting the direction of I_B produces an XNOR gate. In this configuration, a high current on one of the input wires suppresses the output current. Unlike the NOR gate, high currents on both inputs cancel the magnetic fields, allowing a high current to flow through the output. (f) Symbol for spin-diode XNOR gate. (g) OR gate formed by the summation of currents, satisfying the principle of conservation of charge.

gate, a variety of logic functions with more than two inputs can be produced.

3. SPIN-DIODE CIRCUIT DESIGN

These four basis logic elements can be arranged to implement any logical circuit. While the functions are logically equivalent to traditional CMOS circuits, this spin-diode logic family has a starkly different structure.

3.1 Adder

A common circuit is a one-bit full adder, shown in Fig. 3a, which calculates a one-bit sum and carry-out based on the addition of two bits and a carry-in [21]. There are three inputs (A, B, and C_{IN}) and two outputs (C_{OUT} and Sum), optimized for this logic style as

$$C_{OUT} = (A \wedge B) \vee (A \wedge C_{IN}) \vee (B \wedge C_{IN}) \qquad (1)$$
$$= \overline{\overline{A \vee B} \vee \overline{A \vee C_{IN}} \vee \overline{B \vee C_{IN}}}$$
$$Sum = A \oplus B \oplus C_{IN} = \overline{\overline{A \oplus B \oplus C_{IN}}} \qquad (2)$$

As illustrated in the figure, Sum is generated by cascading two XNOR gates. In XNOR1, A and B are inputs, and in XNOR2, one input is C_{IN}, and the other input is the signal propagated by XNOR1. In both gates, if the two inputs are the same, a '1' is propagated. If the two inputs are different, a '0' is produced. Similarly, C_{OUT} is achieved with three NOR gates, an OR gate, and an inverter. NOR1, NOR2, and NOR3 each propagate a '0' if either input is a '1'; when these propagated currents are summed

in OR1, a '0' is propagated unless at least two of the initial inputs are '0'. This value is inverted to generate C_{OUT}.

3.2 Multiplexer

A two-to-one multiplexer is another useful circuit that can be built with this logic family. As shown in Fig. 3b, INV1 produces an inverted select signal. The select and inverted select signals are routed, respectively, through NOR gates alongside A and B. Each NOR gate propagates a '0' unless a signal with a '0' has been selected. These signals are combined and inverted such that a '1' is propagated unless a signal with a '0' has been selected.

3.3 Latch

This logic family also enables an intriguing and simple latch, shown in Fig. 3c. As seen in the diagram, each NOR latch output is routed to the other NOR input, forcing opposite values to be propagated. When one of the diodes propagates a high current, the current is suppressed in the second diode, which allows the first diode to propagate a high current, thereby maintaining a self-consistent state. To set the value stored in the latch, an external current is passed through one of the diodes. To set a '1', a current is passed through NOR1. To set a '0', a current is passed through NOR2. This circuit is a bistable inverter chain, and can be used to create stable memory storage.

4. CIRCUIT CHARACTERIZATION

This logic family integrates the capability of a current to create a magnetic field with the effect of this field on spin-diode

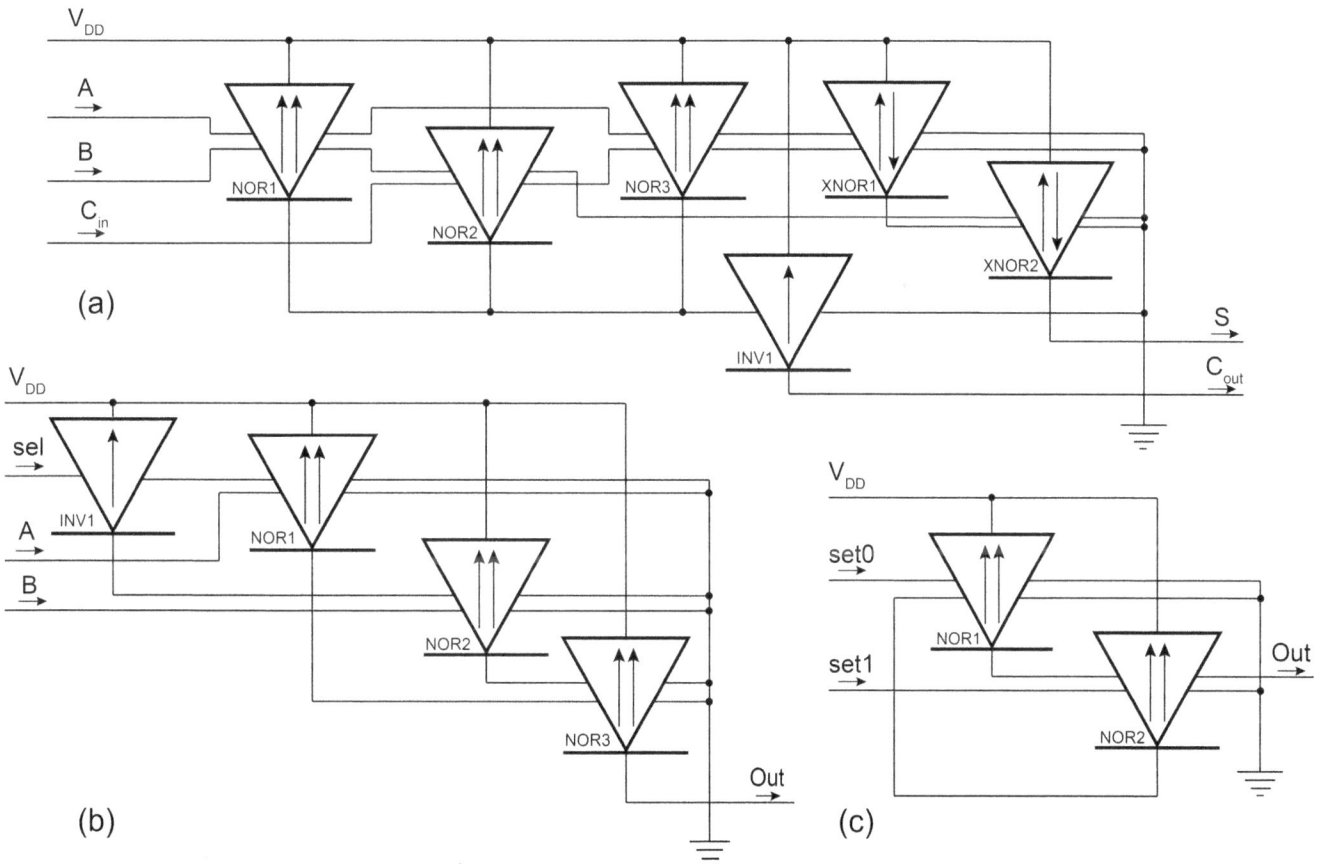

Figure 3. Spin-diode functions: (a) Adder (b) Multiplexer (c) Latch.

current. These phenomena have been evaluated in the context of the circuits discussed in section 3.

4.1 Magnetic Field Strength

Ampere's law can be used to calculate the field created by a current. For a circular line integral, the ratio between the magnetic field B and the current I, in units of T/A, is

$$\frac{B}{I} = \frac{\mu_r \mu_0}{2\pi R} = \frac{2 \times 10^{-7} \mu_r}{R} \qquad (3)$$

where R is the distance in meters between the wire and the diode. This ratio is important for signal integrity, as a higher ratio allows for greater control of the magnetoresistive effects of the diode. Prior art indicates that B/I ratios on the order of 25 T/A has previously been achieved, suggesting a radius on the order of 10 nm. This ratio is sufficient for the current from one of these diodes to control the magnetoresistance of another diode.

4.2 InMnAs Diode Characteristics

Magnetoresistive III-Mn-V heterojunction spin-diodes have been fabricated by depositing III-Mn-V magnetic thin films on III-V semiconductor substrates followed by standard photolithographic processes [13], [15]. The response of this diode to the magnetic field exhibits the characteristics necessary to create magnetoresistive diode logic. The diode has the additional outstanding feature of a positive magnetoresistance that persists even at room temperature. As the measurements in Fig. 4 make clear, the current I decreases significantly as the magnetic field is increased. When the voltage is increased, the I_{on}/I_{off} ratio increases, where I_{on} is the zero-field current and I_{off} is the current at a higher field (e.g., 5 T). A high ratio is useful to differentiation between the two digital states. With developments in nanowire technology and other charge-carrying structures, it should be possible to place these currents sufficiently close to the diodes to produce the magnetic fields necessary to cascade diode logic. It is expected that increasing the Mn content of the heterojunction will lead to increased sensitivity to magnetic fields (i.e., increased g-factor), thereby reducing the current, field, power, and area requirements [22]. Operation below room temperature will also reduce the structural requirements of the system [17].

4.3 Circuit Simulation

To implement this logic family, it is necessary to define the magnitude of the '1' and '0' currents as well as V_{DD}. As the InMnAs device functions non-linearly, it is difficult to determine

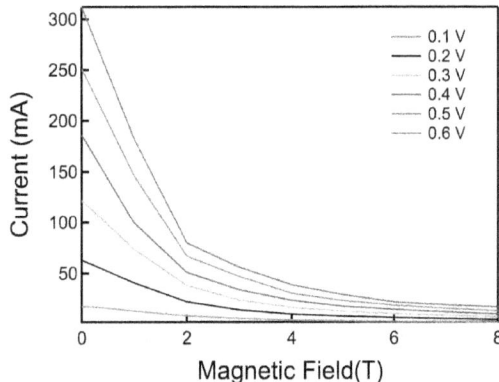

Figure 4. Spin-diode current as a function of magnetic field at various voltage biases [16].

a precise current value to represent '0' and '1', but an approximate value of 32 mA and 160 mA, respectively, is chosen for a V_{DD} of 0.4 V. SPICE models for the basis logic elements have been developed based on the magnetic field calculations and experimental data characterizing the diode. Simulations for the adder, multiplexer, and latch using Synopsis HSPICE are discussed below.

4.3.1 Adder

The signal integrity of this logic family is illustrated in Table 1 for the adder shown in Fig. 3a. As seen in the table, there is a clear differentiation between high and low output currents. Due to the characteristics of the circuit, there are slight variations among the '0' and '1' signals. The values of C_{OUT} make the relationship between the circuit structure and output values particularly clear; the magnitude of C_{OUT} is directly related to the number of '1' inputs. This result agrees with the expected function, as the parallel inputs to INV1 are determined by the sum of the inversion of the input signals. Thus, an increased number of '1' inputs results in a decreased input to INV1 and therefore an increased C_{OUT} current.

Table 1. Adder signal integrity (all currents in mA)

A	B	C	Sum	C_{OUT}
32	32	32	24	15
32	32	160	140	36
32	160	32	186	36
32	160	160	31	97
160	32	32	186	36
160	32	160	31	97
160	160	32	24	97
160	160	160	140	128

4.3.2 Multiplexer

The simulation results for the multiplexer are shown in Table 2. Similar to the adder, the differentiation between the '0' and '1' outputs is sufficient to drive spin-diode logic circuits. In this circuit, the variation in output currents is instructive to the series nature of signal propagation. The input currents to NOR3 are the output currents from NOR1 and NOR2, and NOR2 in turn receives one of its inputs from INV1. The nonlinear behavior of the spin-diode is therefore accentuated by the series connection of these diodes, resulting in an irregular set of output currents.

Table 2. Multiplexer signal integrity (all currents in mA)

Sel	A	B	Out
32	32	32	44
32	32	160	55
32	160	32	106
32	160	160	129
160	32	32	41
160	32	160	112
160	160	32	47
160	160	160	130

4.3.3 Latch

In the simulation shown in Fig. 5, the latch starts in the metastable state, where the outputs of the two NOR diodes are equivalent. As these two diodes drive each other, this situation is stable. When set0 is asserted, NOR1 turns off, allowing NOR2 to turn on. When set0 reverts to '0', the latch continues in a bistable

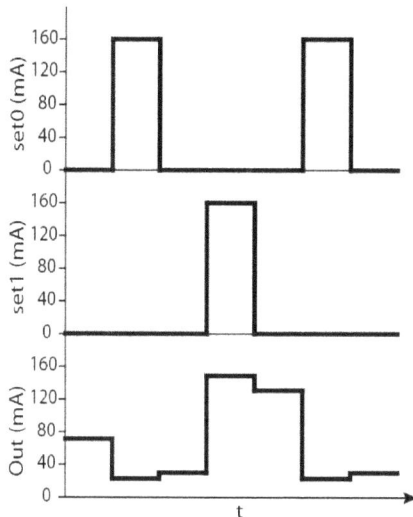

Figure 5. Setting and resetting the spin-diode latch.

state, with NOR1 off and NOR2 on. If set1 is asserted, NOR2 turns off and NOR1 turns on. Once the latch has been switched from the metastable state to a bistable state, the latch will maintain its state until a set signal is asserted.

5. ANALYSIS OF LOGIC FAMILY

5.1 Circuit Design

As the spin-diode has unique properties, such as its power dissipation and fan-out characteristics, the circuit design process must fundamentally change. An analysis of several example circuits demonstrates these design techniques.

5.1.1 Fan-in

Having demonstrated the feasibility of OR and NOR gates with multiple inputs in section 4, it is tempting to consider a gate with an arbitrarily large number of inputs. This scheme, however, is not currently practical, as the '0' current input to the diode is proportional to the number of inputs, and with a sufficiently large number of inputs acts as a '1' current. A device of this sort should therefore be implemented with multiple levels of logic, although this circuit remains far more efficient in terms of area, speed, and power than a comparable CMOS device.

In the adder mentioned above, the output '1' currents produced by *Sum* are significantly greater than those produced by C_{OUT}. This is caused by the fact that whereas *Sum* is driven by an XNOR gate, C_{OUT} is driven by the inversion of the sum of three currents. Therefore, since the current in a '0' signal is not sufficiently small to be negligible, the minimum field through INV1 is of a moderate value, whereas the minimum field through XNOR2 is close to zero. The maximum output current from XNOR2 is thus significantly higher than INV1. While this does not cause problems in most spin-diode circuits, signal integrity may be compromised when implementing XOR gates.

5.1.2 Leakage Current

The aforementioned issue is caused by the leakage characteristics of the spin-diodes that have been produced to date [12], [14], [17]. Even in the presence of high magnetic fields, a small leakage current is caused by the imperfect scattering of holes at the p-n junction. Efforts are currently underway to

decrease the leakage current, but it currently poses a complication for spin-diode circuits. Two solutions include:

1) Multiple power supplies: If multiple values of V_{DD} are used and applied separately to different diodes, leakage effects can be mitigated. As the current through a diode is directly related to the voltage across the diode, the current can be enlarged by increasing the V_{DD} connected to a diode. A large V_{DD} should therefore be applied to diodes with a large fan-in, and a small V_{DD} to diodes with a small fan-in. This technique comes at the cost of additional power grid circuitry, but guarantees that the output current of the various diodes will be similar. In the case of the adder circuit, changing V_{DD} for INV1 from 0.4 V to 0.6 V results in output currents more similar to those produced by NOR2.

2) Limited design space: Even after accounting for leakage current, it is possible to construct any logical function with this spin-diode logic family. By creating design rules that prevent certain circuit structures, signal integrity can be ensured. For example, by prohibiting a multiple-input NOR gate from driving an XOR gate, many leakage current issues can be avoided. Another design rule that suppresses the impact of leakage current is to limit the effective number of gate inputs to two.

5.1.3 Fan-out

Another unique aspect of this logic family is the concept of "fan-out." As an "input" to a device is merely a wire close to a diode, this wire can be used as an input to multiple devices without degrading the signal. This structure is possible because the positive and negative terminals of each diode are directly connected, respectively, through wires to V_{DD} and ground, with no constraint on the wire path or length.

5.1.4 Power Dissipation

The mechanism for consuming power is profoundly different in this logic family. In CMOS circuits, there are two sources of power dissipation: the dynamic power used to switch the state of a gate, and the static power used when a gate is at steady-state. In this spin-diode logic family, there is only static power dissipation, and no dynamic power dissipation. This behavior is because the anode and cathode of each spin-diode are always connected, respectively, to V_{DD} and ground, and no charge needs to accumulate to switch a voltage state.

5.2 Comparison to CMOS

While spin-diode logic is in its early stages of development, it shows promise as a potential replacement for CMOS. In particular, spin-diodes have a more compact logic structure and advantageous material properties.

5.2.1 Compact Logic

As implied in section 2, the spin-diode is most effective when implementing inverters, NOR gates, XNOR gates, and OR gates. While other logic functions, such as AND and NAND, can be implemented with a combination of these basis gates, these logic functions require additional gates, resulting in greater area, delay, and power consumption. It is thus worthwhile to consider the implementation of larger logic blocks, as it may frequently be difficult to optimize smaller logical functions. Most logical functions can be implemented with many fewer devices than its Si CMOS counterpart, as each diode has the equivalent functionality of at least two transistors. For example, an adder is here created using only six spin-diodes, which is far more compact than the typical CMOS 28-transistor implementation. Furthermore, the use of a single device type simplifies the fabrication process. The multiplexer, however, is not easily optimized in this logic family. When a multiplexer is an element of a combinational logic circuit, it should be considered as part of a larger circuit.

5.2.2 Power-Delay Product

An important performance metric of a device is the power-delay product (PDP). The *PDP* of the current heterojunction diode is

$$PDP = \frac{I_{ON} + I_{OFF}}{2} V_{DD} t_D, \qquad (4)$$

where I_{ON} and I_{OFF} are the '1' and '0' currents, respectively. At 0.3 V, preliminary analysis of the switching time of the device exhibits a worst case propagation delay t_D on the order of 10 ns. The *PDP* is thus approximately 2×10^{-10} J. With scaling, it is expected that this metric will be greatly improved. In particular, if the junction diameter is decreased from 300 um to 30 nm, the junction area and current will decrease by a factor of 10^8, and if the length is decreased from 400 um to 40 nm, the delay will decrease by a factor of 10^4. A *PDP* on the order of 10^{-22} J is therefore possible, although other effects are likely to increase the power and delay. As the mobilities in InAs are significantly higher than in Si, these estimates compare favorably with standard Si CMOS technologies, which exhibit a *PDP* on the order of 10^{-15} J.

6. CONCLUSIONS

This logic family is an effective method for exploiting the InMnAs magnetic semiconductor heterojunction. In its current state of development, the diode characteristics are suitable for this logic family, and the performance characteristics are expected to improve with future work. In particular, by increasing the Mn content of the heterojunction, the diode is predicted to have increased magnetoresistance. Experimentation with dielectrics and other materials will likely also result in a higher *B/I* ratio. Additionally, as the device is scaled to nanometer dimensions, the delay and power consumption will decrease. In concert with an expanded design space including the use of recently demonstrated InMnAs transistors and the application to memory and reconfigurable logic, this device and accompanying logic family has the potential to replace CMOS and thereby have a significant impact on the future of computing.

7. REFERENCES

[1] R. H. Chen, A. N. Korotkov, and K. K. Likharev, "Single-electron transistor logic," *Appl. Phys. Lett.*, vol. 68, pp. 1954-1956, Apr. 1996.

[2] A. Bachtold, P. Hadley, T. Nakanishi, and C. Dekker, "Logic circuits with carbon nanotube transistors," *Science*, vol. 294, pp. 1317-1320, Nov. 2001.

[3] A. Geim and K. Novoselov, "The rise of graphene," *Nature Materials*, vol. 6, pp. 183-191, Mar. 2007.

[4] L. Liao *et al.*, "High-speed graphene transistors with a self-aligned nanowire gate," *Nature*, vol. 467, pp. 467, 305-308, Sep. 2010.

[5] Y. Huang *et al.*, "Logic gates and computation from assembled nanowire building blocks," *Science*, vol. 294, pp. 1313-1317, Nov. 2001.

[6] C. P. Collier *et al.*, "Electronically configurable molecular-based logic gates," *Science*, vol. 285, pp. 391-394, July 1999.

[7] S. Datta and B. Das, "Electronic analog of the electro-optic modulator," *Appl. Phys. Lett.*, vol. 56, pp. 665-667, Feb. 1990.

[8] H. Akinaga and H. Ohno, "Semiconductor spintronics," *IEEE Trans. Nanotechnology*, vol. 1, pp. 19-31, Mar. 2002.

[9] A. Ney, C. Pampuch, R. Koch, and K. H. Ploog, "Programmable computing with a single magnetoresistive element," *Nature*, vol. 425, pp. 485-487, Oct. 2003.

[10] D. A. Allwood *et al.*, "Magnetic domain-wall logic," *Science*, vol. 309, pp. 1688-1692, Sep. 2005.

[11] A. Imre, G. Csaba, L. Ji, A. Orlov, G. H. Bernstein, and W. Porod, "Majority logic gate for magnetic quantum-dot cellular automata," *Science*, vol. 311, pp. 205-208, Jan. 2006.

[12] S. J. May and B. W. Wessels, "High-field magnetoresistance in p-(In,Mn)As/n-InAs heterojunctions," *Appl. Phys. Lett.*, vol. 88, pp. 072105-1-3, Feb. 2006.

[13] A. Khitun, M. Bao, and K. L. Wang, "Spin wave magnetic nanofabric: a new approach to spin-based logic circuitry," *IEEE Trans. Magnetics*, vol. 44, pp. 2141-2152, Sep. 2008.

[14] N. Rangaraju, P. C. Li, and B. W. Wessels, "Giant magnetoresistance of magnetic semiconductor heterojunctions," *Phys. Rev. B*, vol. 79, pp. 205209-1-5, May 2009.

[15] S. Sugahara and J. Nitta, "Spin-transistor electronics: an overview and outlook," *Proc. IEEE*, vol. 98, pp. 2124-2154. Dec. 2010.

[16] S. A. Wolf, J. Lu, M. R. Stan, E. Chen, and D. M. Treger, "The promise of nanomagnetics and spintronics for future logic and universal memory," *Proc. IEEE*, vol. 98, pp. 2155-2168, Dec. 2010.

[17] J. A. Peters, N. Rangaraju, C. Feeser, and B. W. Wessels, "Spin-dependent magnetotransport in a p-InMnSb/n-InSb magnetic semiconductor heterojunction," *Appl. Phys. Lett.*, vol. 98, pp. 193506-1-3, May 2011.

[18] F. M. Wanlass, "Low stand-by power complementary field effect circuitry," US Patent # 3,356,858, Dec. 1967.

[19] R. H. Katz, *Contemporary Logic Design*. Redwood City, CA: Benjamin/Cummings, 1994, pp. 669-670.

[20] J. S. Friedman, N. Rangaraju, Y. I. Ismail, and B. W. Wessels, "A Spin-Diode Logic Family," *IEEE Trans. Nanotechnology* (in review).

[21] J. Sklansky, "Conditional-sum addition logic," *IRE Trans. Electronic Computers*, vol. 9, pp. 226-231, Jun. 1960.

[22] B. W. Wessels, "Ferromagnetic semiconductors and the role of disorder," *New J. Phys.*, vol. 10, pp. 055008-1-17, May 2008.

An Efficient Approach for Designing and Minimizing Reversible Programmable Logic Arrays

Sajib Kumar Mitra
Samsung Bangladesh R&D
Center Ltd.
57&57/A Gulshan Avenue (I)
Dhaka-1212, Bangladesh
sajib.mitra@samsung.com

Lafifa Jamal
Dept. of Computer Sci. and
Engineering
Faculty of Engg. and Tech.
University of Dhaka
Dhaka-1000, Bangladesh
lafifa@yahoo.com

Mineo Kaneko
School of Information Science
Japan Advanced Institute of
Sci. and Tech.
1-1 Asahidai, Ishikawa
932-1292 Japan
mkaneko@jaist.ac.jp

Hafiz Md. Hasan Babu[*]
Dept. of Computer Sci. and
Engineering
Faculty of Engg. and Tech.
University of Dhaka
Dhaka-1000, Bangladesh
hafizbabu@hotmail.com

ABSTRACT

Reversible computing dissipates zero energy in terms of information loss at input and also it can detect error of circuit by keeping unique input-output mapping. In this paper, we have proposed a cost effective design of Reversible Programmable Logic Arrays (RPLAs) which is able to realize multi-output ESOP (Exclusive-OR Sum-Of-Product) functions by using a cost effective 3×3 reversible gate, called MG (MUX Gate). Also a new algorithm has been proposed for the calculation of critical path delay of reversible PLAs. The minimization processes consist of algorithms for ordering of output functions followed by the ordering of products. Five lower bounds on the numbers of gates, garbages and quantum costs of reversible PLAs are also proposed. Finally, we have compared the efficiency of proposed design with the existing one by providing benchmark functions analysis. The experimental results show that the proposed design outperforms the existing one in terms of numbers of gates, garbages, quantum costs and delay.

Categories and Subject Descriptors

B.6.1 [**Logic Design**]: Design Styles—*combinational logic, logic arrays*

General Terms

Design, Algorithm, Performance

*Corresponding Author

Keywords

Reversible Logic, Programmable Logic Arrays, MUX Gate

1. INTRODUCTION

Programmable Logic Devices such as PLA, PAL or GAL etc use the array of conventional gates which are not reversible except NOT. Such type of irreversible circuit dissipates $kT*log2$ joules of heat energy to reload information per bit [1, 2], where k is the Boltzmann's constant and T is the absolute area temperature. The performance is not so pleasant rather reversible computing drives multiple operations in a single cycle [3]. Reversible circuits are of particular interest in low power CMOS design [4], optical computing [5], quantum computing [6] and nanotechnology [7].

Array Logic was introduced by Fleisher and Maissel [8] based on AND, OR and NOT synthesis to implement SOP or POS whereas Reversible Logic prefers EX-OR operation as well as Exclusive Sun-Of-Product (ESOP) synthesis. ESOP synthesis gives out better result than SOP realization where many useful methods are proposed for minimizing multi-output Boolean functions into ESOP form [9], [10]. A regular structure of reversible wave cascade of ESOP synthesis is proposed in [11]. Cascade realization of reversible functions and garbage minimization technique is proposed in [12]. The generalized structure of Reversible PLA was first proposed in [13] based on ESOP realization of multi-output functions. Finally, this paper has proposed a new approach of designing Reversible Programmable Logic Arrays as well as compared the proposed design with existing [13] one.

Our work significantly advances the design of cost effective Reversible Programmable Logic Arrays by combining the overview of the design of 3×3 MUX gate [14]; introducing the architecture of RPLAs by using Feynman and MUX gates; novel approach for calculating delay of RPLAs; comparison with existing design and performance analysis by using different benchmark functions.

2. BASIC DEFINITIONS AND PROPERTIES

In this section, we have discussed the basic phenomena of reversible logic, quantum realization and cost calculation of reversible circuits including the architecture of Programmable Logic Arrays (PLAs).

2.1 Reversible Logic

Reversible Computing is the only one method to recover bit loss by using unique mapping between input and output vectors. Frequently used logical operations are composed into gate level called ***Reversible Gate*** where the number of inputs is equal to the number of outputs and also preserves an unique mapping between input and output vectors [15].

Let, the input vector be I_v $\{I_1, I_2, I_3, ..., I_n\}$ and the output vector be O_v $\{O_1, O_2, O_3, ..., O_n\}$ of any Reversible Gate then according to the above definition the relation between two vectors is, $I_v \leftrightarrow O_v$. The input vector, I_v and output vector, O_v for 2×2 ***Feynman Gate (FG)*** [16] are (a, b) and $(a, a \oplus b)$ respectively. Fig. 1 shows the block diagram of Feynman gate and the unique mapping between input-output vectors. The unused outputs of any reversible gate or circuit is called ***Garbage Output*** which will never be used in future rather than to check reversibility [15]. Feynman gate can be used to implement reversible EX-OR operation which generates a dummy output along with its principle output signal to preserve reversibility. The garbage output is denoted by p in Fig. 1. Every reversible circuit has a lower bound of total number of garbage outputs. ***Critical Path Delay*** is the another measurement of the circuit efficiency which is the maximum number of gates from any input to any output [15]. Reversible EX-OR operation requires one gate and the corresponding delay is one.

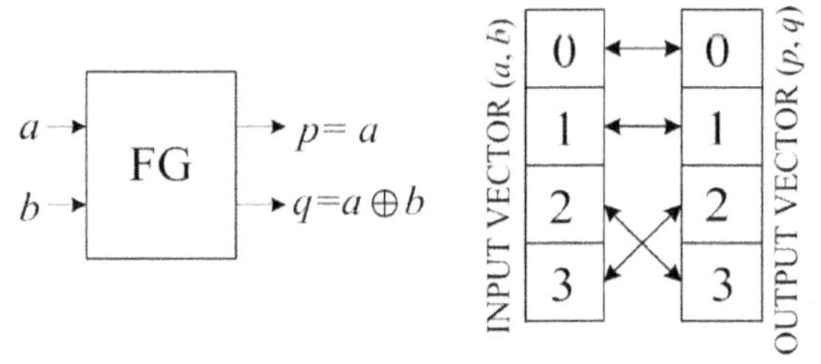

Figure 1: 2×2 Feynman gate and its corresponding input-output mapping

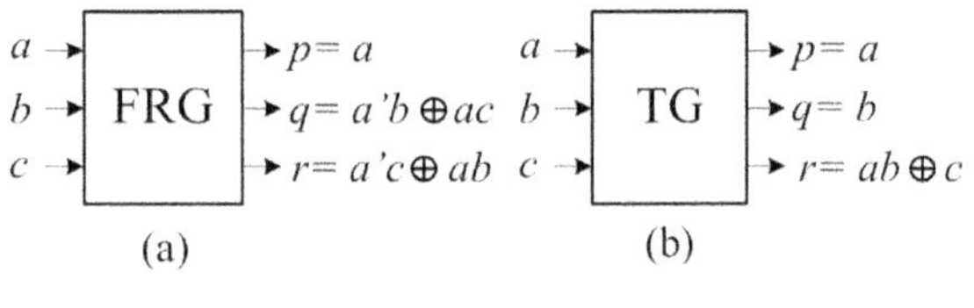

Figure 2: (a) Fredkin gate and (b) Toffoli gate

The input vector, I_v and output vector, O_v of 3×3 ***Fredkin gate (FRG)*** [17] can be defined as: $I_v = (a, b, c)$ and $O_v = (a, a'b \oplus ac, a'c \oplus ab)$, respectively. The pictorial representation of FRG is shown in Fig. 2(a). The input and output vector of 3×3 ***Toffoli gate (TG)*** [18] (shown in Fig. 2(b)) are (a, b, c) and $(a, b, ab \oplus c)$, respectively.

2.2 Quantum Realization of Reversible Circuit

Quantum Computation is gaining popularity as some exponentially hard problem can be solved in polynomial time [19] and reversibility can be used to construct Quantum circuits [15]. Quantum computation uses matrix multiplication rather than conventional Boolean operations and the information measurement is realized using qubits rather than bits. The matrix operations of qubits are performed by using quantum primitives. The value of qubits is the probability factor of 0 and 1 which are represented as $|0\rangle$ or $|1\rangle$ where

$$|0\rangle = \alpha|0\rangle + \beta|1\rangle \ \ and \ \ |1\rangle = \alpha|1\rangle + \beta|0\rangle$$

The **Quantum Cost (QC)** of any reversible circuit is the total number of 2×2 quantum primitives which are used to form equivalent quantum circuit [15]. The Quantum Cost of reversible Feynman gate (shown in Fig. 3(a)) is 1 because the single 2×2 Quantum EX-OR gate can realize the operation of Feynman gate. Fig. 3(b) shows the quantum circuit representation of Toffoli gate where the quantum cost is 5.

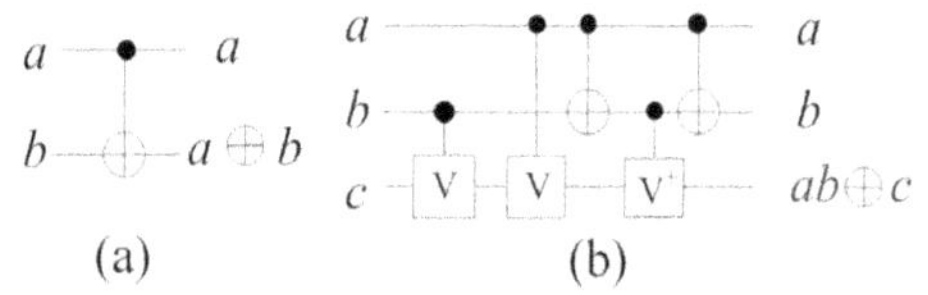

Figure 3: Quantum circuit realization of Reversible gates: (a) Feynman gate (Quantum EX-OR) and (b) Controlled Controlled NOT or Toffoli gate

3. REVERSIBLE PROGRAMMABLE LOGIC ARRAYS

In Section 3.1, we have discussed about the existing design of reversible PLA [13] and its limitations. Rest of the part has described the proposed idea of reversible PLAs based on ordering of output functions and input variables.

3.1 Existing Design of RPLAs

The design of Reversible PLA, is proposed in [13], has used Feynman and Toffoli gates to realize Reversible PLA for multi-output ESOP operation where Toffoli gate is used for AND operation and Feynman gate is used for EX-OR operation. But the existing design has following limitations:

a. Has not treated primary input as garbage when it becomes as an output (But according to [15], unused outputs of any circuits are considered as garbage)

b. Used Conventional Architecture (Complement and non-complement lines for copying input variables)

For example, Equation (I) shows the multi-output ESOP function, where $F = \{f_1, f_2, f_3, f_4, f_5\}$. The design used three templates of Feynman gate for implementing COPY, EX-OR and NOT operations (shown in Fig. 4(a)) and single template of Toffoli gate for doing AND operation (shown in Fig. 4(b)). AND plane has used both Feynman and Toffoli gate where as EX-OR plane used only Feynman gate. The realization of multiple-output function in Equation (I) based on existing algorithm is shown in Fig. 4c. The existing design used Toffoli gate AND operation which generates

$$F = \begin{cases} f_1 = ab' \oplus ab'c \\ f_2 = ac \oplus a'b'c \\ f_3 = ab' \oplus ab'c \oplus bc' \\ f_4 = ac \\ f_5 = ab' \oplus ac \oplus bc' \end{cases} \cdots \cdots \cdots (I)$$

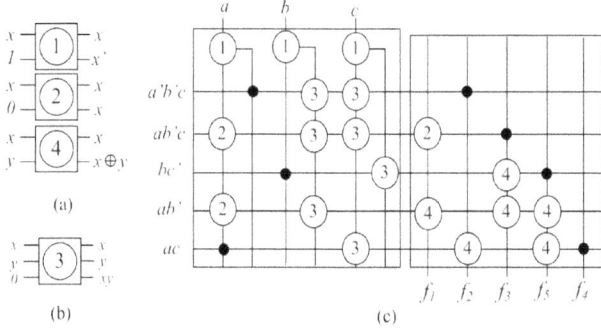

Figure 4: (a) Different templates of Feynman gate for different purposes, (b) Template of Toffoli gate and (c) Design of Reversible PLAs according to [13].

products without changing the form of input literals (complement or non-complement). On the other hand, Toffoli gate produces huge number of unused outputs which are same as primary inputs. But in proposed design of PLAs there is no scope to use those unused outputs.

3.2 Proposed Design of Reversible PLAs

Proposed design is based on the ordering of input variables which depends on the corresponding order of Products. But the order of Products will be generated after the optimization of EX-OR plane. In this subsection, we have proposed two algorithms on the construction of EX-OR plane followed by the realization of AND plane for generalizing the proposed design. The 3×3 reversible MUX [14] or MG gate is used to design proposed Reversible PLAs which has minimum quantum cost i.e 4 (shown in Fig. 5). MG gate can realize the operation of (2 to 1) multiplexer circuit and able to produce half of minterms generated by two variables. Fig. 6 shows the templates (MG-5 and MG-6) of a MUX gate which have been used in proposed design. We have used three modes (FG-1, FG-2 and FG-4) of Feynman gate proposed in [13] (shown in Fig. 4(a)) and other two modes (MG-5) and (MG-6) of MUX gate shown in Fig. 6. In our further discussion, we have used symbol 1, 2, 4 (Fig. 4) and symbol 5, 6 (Fig. 6) rather than full name (FG-1 or MG-5 etc.) to represent the particular modes. The cross point in RPLA, in which no gate is used, is termed as DOT [13].

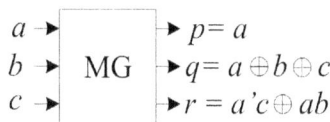

Figure 5: 3×3 Reversible MUX gate

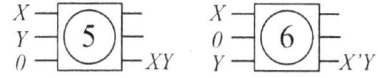

Figure 6: Two different templates of MUX gate

Table 1: Size of each function of Equation (I)

function	f_1	f_2	f_3	f_4	f_5
SizeOf (f_i)	2	2	3	1	3

Definition 1. **Size of Function (SizeOf (f_i)):** Total number of products has been used by f_x function. For example, in Equation (I), SizeOf(f_1)= 2 because the total number of products of f_1 is 2 (shown in Table 1).

Definition 2. **Product Lines** are the horizontal lines corresponding to the products of an AND plane. These product lines are used in the EX-OR plane to generate the output of a particular function consisting of EX-OR operations. The number of product lines is equal to the number of total products. There are 5 product lines for 5 products of function F (see Equation (I)).

In the proposed design, the ordering of output functions is related to SizeOf (f_x). Functions are generated in ascending order based on this criterion. We have optimized EX-OR Plane followed by AND Plane minimization by using MUX and Feynman gates.

Algorithm 1: Construction of EX-OR Plane

---
1. **START** $TDOT := 0$ [$TDOT =$ Total number of DOT]
2. Sort output Functions according to Sizeof (f_i)
3. **REPEAT** Step 4 for each output function
4.   **IF** Sizeof (f_i) of f_i is one **THEN**
5.   **IF** product p_j exists **THEN** use FG-2
6.   **ELSE** assign a line for product (p_j) and use DOT
7. $TDOT := TDOT + 1$
8.   **END IF**
9.   **ELSE**
10.   **IF** all product(s) p_j exist **THEN** use FG-2 for top most line and FG-4 for others
11.   **ELSE** assign the upper lines for products p_j and use DOT for top most and FG-4 for existing
12. $TDOT := TDOT + 1$
13.   **END IF**
14.   **END IF**
15. **END**

By using the proposed algorithm, the realization of EX-OR plane for Equation (I) is shown in Fig. 7.

THEOREM 1. *Let n be the number of EX-OR operations of m output functions and $TDOT$ be the number of crosspoints, then the minimum number of Feynman gates to realize EX-OR plane is $n + m - TDOT$.*

PROOF. When there are $TDOT$ cross-points for m functions, the number of additional Feynman gates in EX-OR plane of RPLA is $m - TDOT$. As there are n Ex-OR operations by n Feynman gates, the total number of Feynman gates in the EX-OR plane of RPLA is $n + m - TDOT$. $\square$

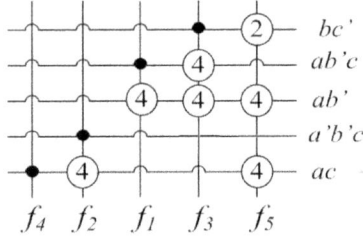

Figure 7: EX-OR plane realization of Equation (I) based on proposed Algorithm 1

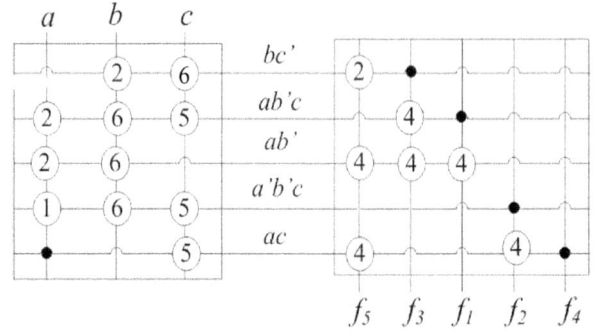

Figure 8: Proposed reversible PLAs design of multi-output function F in Equation (I)

For multi-output function F in Equation (I), the number of outputs (m) is 5 and the number of EX-OR operations is 6. The number of $TDOT$ is 4 in Fig. 7. So the number of Feynman gates $= n + m - TDOT = 6 + 5 - 4 = 7$.

THEOREM 2. *Let p be the number of products and $TDOT$ be the number of cross-points in the EX-OR plane of RPLA. The minimum number of garbages to realize EX-OR plane of RPLA is $p - TDOT$.*

PROOF. As there are $TDOT$ cross-points and p products for m output functions, the total number of garbage outputs in the EX-OR plane of RPLA is $p - TDOT$. □

Consider Fig. 7 for multi-output function F in Equation (I). In Fig. 7, number of products (p) is 5 and number of cross-points ($TDOT$) is 4. So the number of garbage outputs is $p - TDOT = 5 - 1 = 4$. The realization of EX-OR plane generates the order of Products shown in Fig. 7 and AND plane will be constructed according to this order by using MUX and Feynman gates.

Algorithm 2: Construction of AND Plane
---

1. **START** $TDOT := 0$ [$TDOT$= Total number of DOT]
2. **REPEAT** Step 3 for each product (p_j)
3.   **IF** l_j is the first literal of p_j **THEN**
4.   **IF** l_j is in complemented form THEN apply FG-1
5.   **ELSE**
6.   **IF** l_j is further used **THEN** apply FG-2
7.   **ELSE** use DOT and $TDOT := TDOT + 1$
8.   **END IF**
9.   **END IF**
10.   **ELSE IF** l_j in complemented form **THEN** apply MG-6
11.   **ELSE** apply MG-5
12.   **END IF**
13. **END**

In AND plane, the Feynman gates are used to copy or recover fan-out problem and the MUX gates are used for AND operations. The generation of complementary forms of input literals are unnecessary for the proposed AND plane because MUX and FG are used together to generate all the minterms of two variables without having any dedicated lines of complemented forms of input variables. By using Algorithms 1 and 2, the realization of proposed Reversible PLA is shown in Fig. 8.

THEOREM 3. *Let p be the number of products and $TDOT$ be the number of cross-points in the AND plane of RPLA. The minimum number of Feynman gate to realize AND plane of RPLA is $p - TDOT$.*

PROOF. Cross-points reduce the usability of Feynman gates for any particular product line. So, in response to $TDOT$ cross point and p products for m output functions, the total number of Feynman gates in the AND plane of RPLA is $p - TDOT$. □

Consider Fig. 8 for multi-output functional Equation (I). In Fig. 8, number of products (p) is 5 and number of cross points ($TDOT$) is 1. So the number of Feynman gates is $p - TDOT = 5 - 1 = 4$.

THEOREM 4. *Let q be the number of AND operations among products in the AND plane of RPLA. The minimum number of MUX gate to realize AND plane of RPLA is q.*

PROOF. As there are q AND operations for p products of m output functions, the total number of MUX gates to realize the AND plane of RPLA is q. □

THEOREM 5. *Let l be the number of inputs and q be the number of AND operations among products and $TDOT$ be the number of cross-points in the AND plane of RPLA. The minimum number of garbages to realize AND plane of RPLA is $l + q - TDOT$.*

PROOF. As there are q AND operations for p products of m output functions F in Equation (I), l inputs and $TDOT$ cross-point, the total number of garbages in the AND plane of RPLA is $l + q - TDOT$. □

3.3 Delay Calculation

In this paper, we have calculated the delay of reversible PLAs in greedy approach and the proposed algorithm generates better throughput. We divide the calculation into two phases: a. AND Plane Delay (APD (p_i)) and b. EX-OR Plane Delay (XPD (p_i)) in terms of product lines (horizontal lines). Then we have merged both of the delay respect to both planes. We have used Equation (I) to calculate the delay. First we calculate the delay of AND plane followed by EX-OR plane. Fig. 9 and Fig. 10 show the delay calculation of AND plane and EX-OR plane respectively. In further realization of delay calculation, we consider the following things:

a. Gate (Via) is represented as circle (DOT).

b. Delay of any gate is 1 and via (DOT) denotes 0.

c. Decimal values show the delay of corresponding circle

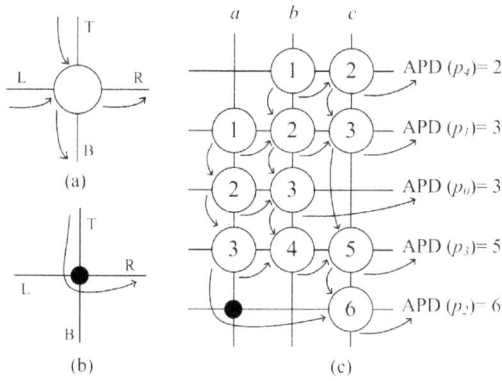

Figure 9: Delay calculation of AND plane: (a-b) Delay propagation path of a gate and a cross-point respectively in AND plane and (c) Overall delay propagation path for AND plane

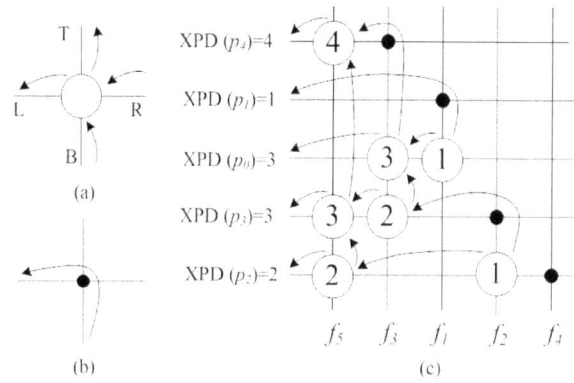

Figure 10: Delay calculation of EX-OR plane: (a-b) Delay propagation path of a gate and a cross-point respectively in EX-OR plane and (c) Overall delay propagation path for EX-OR plane

3.3.1 Delay Calculation of AND Plane

For AND plane, every gate updates its Delay by comparing the delay of neighboring gates at Left (L) and Top (T) and then, it propagates the updated delay to neighboring gates placed in Right (R) and Bottom (B) sides as shown in Fig. 9. Each Circle in Fig. 9 represents the delay of particular point and Arrows show the path of delay propagation. The size of AND plane of proposed design is less then existing design because proposed design does not need to generate complement lines of corresponding input lines.

3.3.2 Delay Calculation of EX-OR Plane

For EX-OR plane, every gate updates its Delay by comparing the delay of neighboring gates at Right (R) and Bottom (B) and then, it propagates the updated delay to neighboring gates placed in Left (L) and Top (T) sides as shown in Fig. 10. The size of EX-OR plane of proposed design is same as existing design and this plane is optimized in terms of cost analysis as Theorem 1 & 2.

3.3.3 Delay of Overall Design

After calculating the delay of both planes, the delay of product lines having maximum value is the final delay of the overall design of reversible PLAs. We proposed the following algorithm for the calculation of delay of reversible PLAs. According to proposed design, the delay propagation of AND (EX-OR) plane is Top-Bottom-Right (Bottom-Top-Left).

Algorithm 3: Delay Calculation of Reversible PLAs

---

1. **START** Calculate APD (p_i) (XPD (p_i)) of each product lines of AND (XOR) plane
2. Delay:= MAX $\{APD\ (p_i) + XPD\ (p_i)\}$ where $i = 1$ to n (n= total number of product)
3. **END**

3.4 Experimental Results

We have realized the calculation of all proposed algorithms by using programming language Java (J2SE 1.6.0_17) on Netbeans IDE (6.8) in Linux Workstation. All the experimental results are tested on Intel(R) Core(TM)2 Duo CPU

E7300 2.66GHz edition with 2 GB RAM. Table 2 shows the experimental results for different benchmark functions and the comparison with the existing method [13].

4. CONCLUSIONS

In this paper, we proposed a regular structure of Reversible Programmable Logic Arrays (RPLAs) based on MUX-Feynman logic and also we presented the minimization techniques for both AND and EX-OR planes of reversible PLAs. We used the garbage outputs as operational outputs that reduced the number of AND operations in RPLAs. The minimization of AND plane based on the ordering of input variables gives an excellent throughput of the overall design. Finally, we figured the performance of the proposed design over the existing one. The experimental results show that the proposed design outperforms the existing one in terms of numbers of gates, garbages and quantum costs. The proposed algorithm also required less time than the existing one. We also presented five lower bounds on the numbers of gates, garbages and quantum cost of RPLAs. RPLAs are useful in embedded circuits and others technologies for power consumption and fault tolerant [8], [4], [18].

5. ACKNOWLEDGMENTS

We would like to thank Mr. A. R. Chowdhuary, Assistant Prof. of Department of Computer Science and Engineering, University of Dhaka, Dhaka-1000, Bangladesh. He is currently pursuing PhD at the Monash University, Australia.

6. REFERENCES

[1] R. Landauer. Irreversibility and heat generation in the computing process. *IBM J. Research and Development*, 5(3):183–191, 1961.

[2] C. H. Bennett. Logical reversibility of computation. *IBM J. Research and Development*, 17:525–532, 1973.

[3] I. Hashmi and H. M. H. Babu. An efficient design of a reversible barrel shifter. In *IEEE 23rd International Conf. on VLSI Design*, pages 93–98, Banglore, India, 2010.

[4] G. Schrom and S. Selberherr. Ultra-low-power cmos technology. In *Semiconductor Conf.*, Romania, 2002.

Table 2: Experimental results using different benchmark functions

	Proposed Design					Existing Design [13]				
	*GA	*GB	*DL	*QC	*CT[ms]	GA	GB	DL	QC	CT[ms]
5xp1	166	112	36	418	1.689	170	118	30	508	3.648
9sym	427	385	59	1405	0.844	439	391	62	1737	1.305
adr3	67	48	19	157	0.153	69	50	19	189	0.268
apex3	3998	1799	279	8654	4.376	4047	1848	234	10255	8.678
b12	159	132	35	453	0.499	170	140	29	562	0.652
bw	305	64	44	446	0.499	350	70	45	499	0.572
cordic	12162	10640	792	41694	19.737	18184	10662	792	51560	75.955
duke2	941	667	99	2735	1.305	931	695	93	3361	1.804
e64	2170	2148	114	8410	2.572	2228	2206	116	10548	3.072
inc4	16	10	9	34	0.192	16	10	6	40	0.153
inc5	23	16	8	53	0.153	24	17	8	64	0.115
misex1	88	51	22	193	0.192	95	58	20	235	0.192
misex2	199	176	36	622	0.384	227	200	34	787	0.460
pdc	3801	3006	275	12096	7.372	3825	3030	273	14885	13.939
rd53	56	42	17	134	0.152	55	44	18	162	0.192
rd73	187	141	44	487	0.345	188	144	37	590	0.499
rd84	328	265	68	928	0.844	321	269	58	1132	1.382
sao2	284	236	41	890	0.691	291	243	41	1099	0.768
sasao	14	13	9	29	0.115	15	14	8	35	0.115
t481	49	53	17	133	0.192	64	68	17	176	0.230
table3	2602	1814	190	7537	2.956	2619	1831	173	9199	4.569
table5	2539	1819	182	7516	2.841	2559	1839	148	9195	4.416
vag2	2018	1836	188	6926	3.955	2038	1856	173	8582	7.680
xor5	8	8	5	8	0.115	9	9	5	9	0.075
z5xp1	167	114	35	425	0.345	171	122	29	519	0.422
z9sym	427	385	59	1405	0.960	433	391	62	1737	1.459

*GA= Total Gates; *GB= Total Garbages; *QC= Quantum Cost; *DL= Delay of the Circuit; *CT= CPU Time in millisecond (ms)that required to calculate the numbers of gates and garbages as well as delay counting and quantum cost analysis.

[5] E. Knill, R. Laamme, and G. J. Milburn. A scheme for e-cient quantum computation with linear optics. *Nature*, 409:46–52, January 2001.

[6] M. A. Nielsen and I. L. Chuang. Quantum computation and quantum information, 2001.

[7] R. C. Merkle. Two types of mechanical reversible logic. *Nanotechnology*, 4(2):114–131, January 1993.

[8] H. Fleisher and L. I. Maissel. An introduction to array logic. *IBM J. of Research and Development*, 19, 1975.

[9] T. Sasao. Exmin2: A simplification algorithm for exclusive-or-sum-of-products expressions for multiple-valued input two-valued output functions. *IEEE Transactions on Computer-Aided Design of Integrated Circuits and Systems*, 12(5):621–632, 1993.

[10] A. Mishchenko and M. Perkowski. Logic synthesis of reversible wave cascades. In *International Workshop on Logic Synthesis*, pages 197–202, June 2002.

[11] M. Perkowski, A. B. P. Kerntof and et al. Regularity and symmetry as a base of efficient realization of reversible logic circuits. In *International Workshop on Logic Synthesis*, pages 245–252, June 2001.

[12] D. Maslov and G. Dueck. Reversible cascades with minimal garbage. *IEEE Transactions on CAD*, 23(11):1497–1509, November 2004.

[13] A. R. Chowdhury, R. Nazmul, and H. M. H. Babu. A new approach to synthesize multiple-output functions using reversible programmable logic arrays. In *IEEE 19rd International Conf. on VLSI Design*, pages 311–316, Hyderabad, India, 2006.

[14] A. S. M. Sayem and S. K. Mitra. Efficient approach to design low power reversible logic blocks for field programmable gate arrays. In *IEEE International Conference on Computer Science and Automation Engineering*, pages 251–255, July 2011.

[15] A. K. Biswas, M. M. Hasan, A. R. Chowdhury, and H. M. H. Babu. Efficient approaches for designing reversible binary coded decimal adders. *Elsevier Microelectron. Jounrnal*, 39(12):1693–1703, December 2008.

[16] R. P. Feynman. Quantum mechanical computers. *Opt. News*, 11(2):11–20, 1985.

[17] E. Fredkin and T. Toffoli. Conservative logic. *International J. Theoretical Physics*, 21:219–253, 1982.

[18] T. Toffoli. Reversible computing. *MIT Lab for Comp. Sci.*, 85:632–644, 1980.

[19] V. V. Shende, A. Prasad, I. Markov, and J. Hayes. Reasoning about naming systems. *IEEE Trans. CAD*, 22(6):723–729, 2003.

Modeling a Single Electron Turnstile in HSPICE

Wei Wei
ECE Dept, Northeastern University
Boston, MA 02115, USA
617-373-7780
wei.w@husky.neu.edu

Jie Han
ECE Dept, University Of Alberta
Edmonton, Canada
780-492-1361
jhan8@ualberta.ca

Fabrizio Lombardi
ECE Dept, Northeastern University
Boston, MA 021152, USA
617-373-4854
lombardi@ece.neu.edu

ABSTRACT

This paper presents a novel HSPICE circuit model for designing and simulating a Single-Electron (SE) turnstile, as applicable at the nanometric feature sizes. The proposed SE model consists of two nearly similar parts whose operation is independent of each other; this disjoint feature permits to accurately model the sequential transfer of electrons through the turnstile in the storage node (modeled on a voltage level basis). It therefore avoids the transient (current-based) nature of a previous model. The model has been simulated and results show that it can correctly operate at 32 nm with excellent stability in its operation. Extensive simulation results are presented to substantiate the advantages of using the proposed model with respect to changes in the circuit model parameter.

Categories and Subject Descriptors

B.7.1 [**Integrated Circuits**]: Types and Design Styles – *Advanced Technologies, Memory Technologies, VLSI (Very Large Scale Integration).*

General Terms

Design

Keywords

Single-Electron (SE) Turnstile, HSPICE, Single Electron Tunneling Transistor

1. INTRODUCTION

The last decade has seen the development of nanometric circuits as MOSFET-based design has moved to feature sizes well below 100 nm. However, concerns over short channel effects, ultrathin gate leakage, doping fluctuations, and large power dissipation have been widely reported for nanoscale integrated. So-called emerging technologies have been advocated to surmount the physical and economic barriers of current technology [1-7]. However emerging technologies have encountered only limited success; among them, single-electron devices have been attracting considerable attention for application to arithmetic and memory circuits as well as sensors. Modeling and fabrication techniques for single-electron (SE) devices have been reported [8-13].

SE transfer requires the utilization of specific devices, such as pumps and turnstiles [19]. SE pumps and turnstiles have been proposed and experimentally demonstrated by using multiple metal islands separated by metal-oxide tunnel junctions.
An accurate transfer with an error rate of 10^{-8} has been achieved by using a seven-tunnel junction pump, so the operation frequency is still limited to the order of MHz due to the resistance of the tunnel junctions. [18] has reported an operating frequency of 166MHz. Although the operating frequency is still limited, modulated tunnel barriers have been proposed to improve performance as well as fabrication A HSPICE model has been presented in [17] using a charging pulse to represent the electron transfer event; however with the development of new fabrication techniques and SE devices, modification and improvements to this simulation model are required. The objective of this paper is to propose a new HSPICE compatible model for a SE turnstile; the proposed model takes advantage of a nearly symmetric circuit in which each of its two parts operates almost independently. The voltage driven outcome of the model circuit avoids the transient nature of the current-driven output of a previous model [17]. The operation of the HSPICE model is evaluated and simulated at 32 nm, because several scaling challenges have been encountered at this feature size [20]. Extensive simulation results are provided as assessment of the proposed model with respect to its operational functionality in the presence of parameter changes as well as the experimental data of [19].

2. REVIEW

The MOSFET based SE turnstile is a promising device that can accurately transfer Single Electrons (SEs) at high speed even at room temperature [14–16]. Its equivalent circuit is shown in Figure 1(a). The SE turnstile consists of the following elements: a source S, a drain D, a gate voltage terminal G, a bias voltage terminal B, and two clock voltage terminals (clk1 and clk2). The source is connected to a supply voltage (V_{SS} or $-V_{SS}$). The drain is connected to an electron storage node (SN). Electrons can be injected into the SN or ejected from the SN by the SE turnstile. The turnstile consists of two MOSFETs (FET1 and FET2). Single electrons are transferred from the source to the drain one by one (i.e. sequentially) by turning FET1 and FET2 ON and OFF alternately [16]. The circuit symbol shown in Figure 1(b) is commonly used to represent the SE turnstile.

Compared with [16], an additional bias terminal B is added to the circuit. Terminals G and B are the upper or side gates and they control the number of electrons transferred per cycle. G and B are connected to the input voltage V_G and the bias voltage V_B, respectively. Figure 1(c) shows the pulse sequences for Vclk1 and Vclk2. Figure 1(d) shows the process by which SEs are

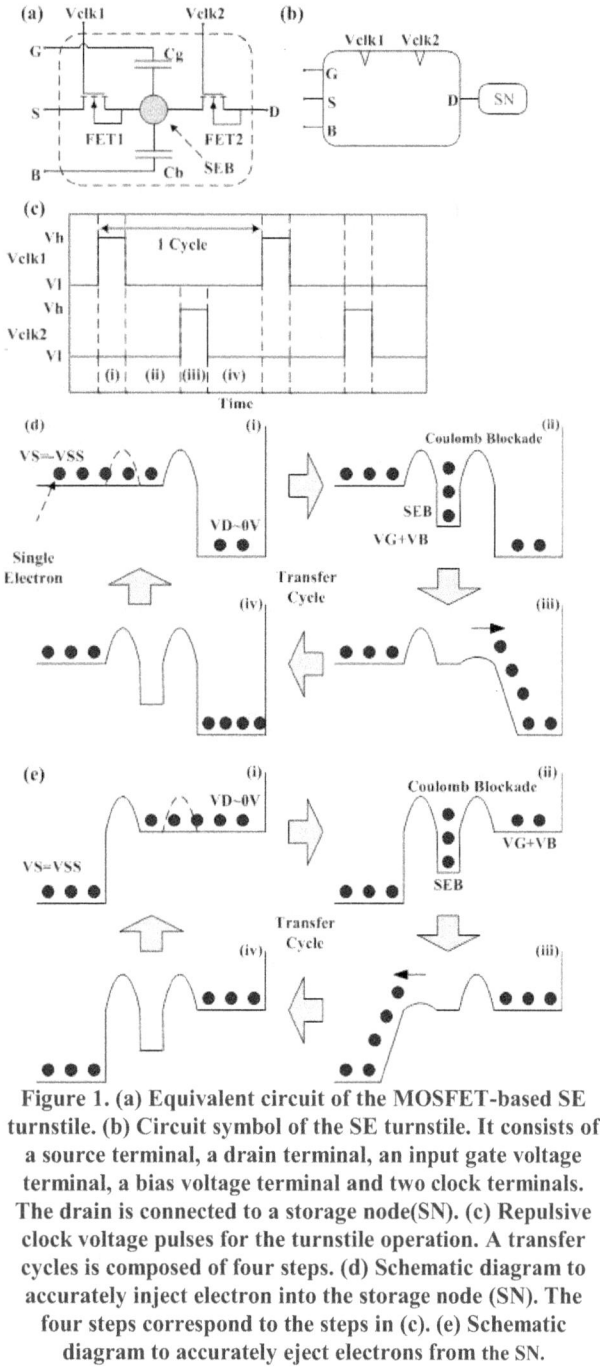

Figure 1. (a) Equivalent circuit of the MOSFET-based SE turnstile. (b) Circuit symbol of the SE turnstile. It consists of a source terminal, a drain terminal, an input gate voltage terminal, a bias voltage terminal and two clock terminals. The drain is connected to a storage node(SN). (c) Repulsive clock voltage pulses for the turnstile operation. A transfer cycles is composed of four steps. (d) Schematic diagram to accurately inject electron into the storage node (SN). The four steps correspond to the steps in (c). (e) Schematic diagram to accurately eject electrons from the SN.

transferred from the source to the SN as per steps (i)–(iv) (shown in Fig. 1(c)). The source of the SE turnstile is connected to $-V_{SS}$.

The operation of the SE turnstile can be described by a *transfer cycle* made of four steps as follows. When both FET1 and FET2 are turned OFF, a single-electron-box (SEB) is electrically formed. The potential of the SEB is controlled by electrically coupling the voltages of V_G and V_B [14]. When FET1 is turned ON, the electrons enter the SEB from the source [step (i)]. After FET1 is turned OFF again, the electrons are retained in the SEB [step (ii)]. The number of captured electrons (N_S) is determined by the potential difference between the SEB and the source.

The electron injection process is given as follows. When FET2 is turned ON, the SEB is connected to the SN [step (iii)]. The capacitance of SN is denoted by C_{SN} and is much larger than the capacitance of the SEB; therefore, when the electrons enter the SN, its potential change cannot affect the behavior of the SE turnstile, and thus, can be neglected. Therefore, it is assumed for simplicity that the potential of the SN is always 0V. When $V_G + V_B < 0$, the potential of the SEB is higher than the SN, such that all SEs flow into the SN [14]. In this case, the number of electrons (given by N = Ns) transferred depends exclusively on V_G. However, when $V_G + V_B > 0$, not all electrons flow out of the SEB; in this case, N only depends on the potential difference between the source and the SN. Finally, FET2 is turned OFF [step (iv)], and the transfer cycle is completed.

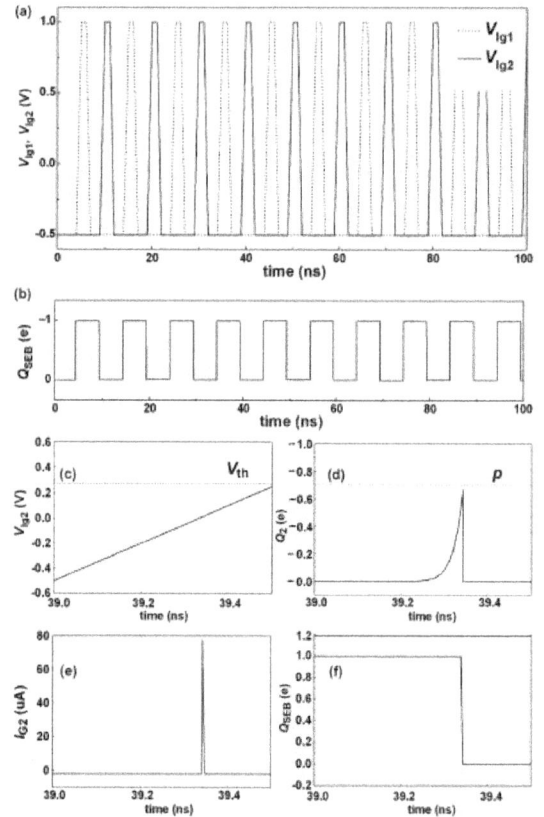

Figure 2. Simulation results of the HSPICE SE turnstile model of [17]. (a) Applied gate voltage pulses for turnstile operation. (b) Amount of charge in SEB as function of time. (c)-(f) have same time scale. (c) FET2 is gradually opened. (d) Amount of charge as function of time. SEB is gradually discharged.. (e) Transient current through G2 to show a SE transfer event. (f) After the transfer event, there is no electron in SEB.

Correspondingly, single electrons can also be ejected from the SN, as shown in Figure 1(e). In this case, the source of the SE turnstile is connected to V_{SS} and V_G is positive. The number of electrons (N_d) captured in the SEB is determined by the potential difference between the SEB and SN. In steps (iii) and (iv), electrons flow out of the SEB to the source. When $V_G + V_B < V_{SS}$, the potential of the SEB is higher than the source, and all electrons flow out of the SEB. So N = N_d also depends exclusively on V_G. However, when $V_G + V_B > V_{SS}$, the potential of the SEB is lower than the source, and not all electrons flow out of the SEB, i.e. N only depends on V_{SS}. This characteristic makes the SE turnstile flexible for many applications, such as multi-valued memory cells and threshold logic circuits [15] [16]. For circuit design, it is likely that the SE turnstile will be used together with CMOS devices (due to compatible fabrication [19]); hence, an electrical model of its operation is highly desirable. A HSPICE model for the SE turnstile has been previously proposed in [17]. Figure 2 shows the simulation result of this HSPICE model for the SE turnstile [17] and in particular shows the transient nature of [17] with respect to its current-driven mode of operation.

3. PROPOSED HSPICE MODEL

Figure 3 shows the proposed HSPICE model of a SE turnstile. The goal of this circuit model is to reduce the transient effects from the input signals (source) to the output node (drain) as occurring in [17]; moreover stability should be achieved also at nanometric feature size.

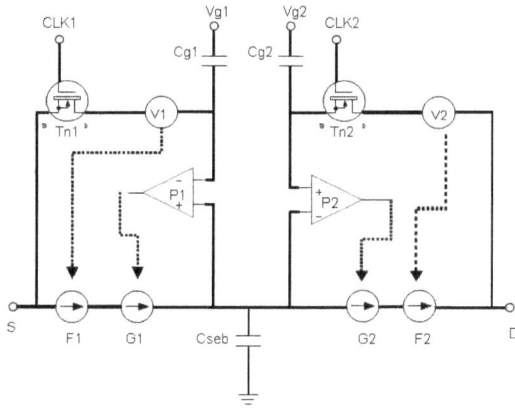

Figure 3. Proposed HSPICE Model of a SE Turnstile

The SEB and the SN are modeled as ideal capacitors with capacitance C_{SEB} and C_{SN}, respectively. An electron transfer event is accomplished as follows: when the two transistors Tn1 and Tn2 are turned ON, two separate reference voltage nodes (Vg1 and Vg2) are utilized. There is no direct connection between the transistors, when both the input signal and CLK1 are high, the output will not experience a voltage change; therefore, the two transistors can separately control the two parts of this model.

Consider initially the left part of the proposed model; when CLK1 and the input signal are both "1", the path of the current mirror F1 is open by controlling (turning ON) the transistor Tn1. Meanwhile, an electron is transferred from the source node into the SEB. Then, C_{SEB} is charged and its voltage V_{SEB} is fed back to the comparator P1. P1 is then used to compare the voltage difference between V_{SEB} and Vg1. When $V_{SEB} > N*e/ C_{SEB}$, the output of the comparator P1 shuts off the voltage-controlled current source G1. Therefore, the electron that has been transferred into SEB, can be stored and not returned back to the input source node. This is a very stable and precise process. The value of the voltage of the SEB will not decrease back to its initial value, so the equivalent charge stored in the SEB is exactly N*e. This structure can eliminate the effects of CLK1 when it changes back to "0", causing a negative current pulse in the controlling path [17]. The number of single-electrons stored in the SEB (N) is controlled by Vg1; using CLK1, N electrons are stored in C_{SEB}.

Table 1. Device Parameters for HSPICE Simulation

Feature Size		32 nm
Temperature		300 K
Power Supply	Vdd	0.9 V
	-Vdd	-0.9 V
SE turnstile	W_{FET1}, W_{FET2}	32 nm
	L_{FET1}, L_{FET2}	32 nm
	Vg1, Vg2	0.9 V
	V_h	0.9 V
	V_l	-0.9 V
	C_{SEB}	0.5 aF
	C_{SN}	10 aF
	C_0	1 aF

As for the right part of Figure 3, a circuit similar to the left part is used in the model. The right part operates as follows. With the rise edge of CLK2, the current mirror F2 generates a current pulse from the drain to the SEB; this indicates that an electron is transferred from the SEB to the drain. Similarly, a voltage-controlled current source G2 is used for the voltage of the drain node, i.e. the voltage of the storage node (SN). All single-electrons are transferred into the SN using a series of pulses on CLK2.

4. SIMULATION RESULTS

The proposed SE turnstile model has been simulated at 32 nm; Table 1 shows the values of all device parameters. Hereafter in the evaluation, the operational frequency of the turnstile has been set to 50 MHz; this value was selected by considering the transfer rate for the correct operation of the SE turnstile [18]. Figure 4 shows the simulation results for the SE turnstile at 32 nm. This figure shows the two clocks (CLK1 and CLK2) as well as the input (at the source) and the drain (SN). In both cases, the electrons transfer into the SN occurs sequentially, as shown by the decrease of the voltage level of the SN. The number of electrons transferred by the SE turnstile (N) depends on the value of Cg, i.e. the capacitance between the gates and the SEB. In this paper it is assumed that $Cg/C_0 = x$, where $C_0 = 1$ aF as a unit capacitance and x is an integer as simulated device parameter.

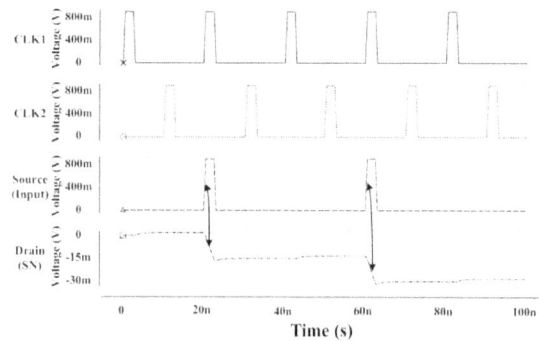

Figure 4. Simulation Timing Diagram of SE Turnstile at 32nm node

Figure 5 shows the number of electrons transferred by the SE turnstile in which the voltage level can be controlled by different values of the capacitance Cg. Besides, the N transferred electrons correspond to a negative voltage level. By increasing the input pulses, the voltage level of the SN decreases and N is linear to x.

The simulation results in Figure 5 are average values (over four trials) for the voltage levels (linear to N) as function of time when x=1, 2, 4, 8 and V_B=0. The voltage-time function has several plateaus with a width given by the transfer cycle. The first plateau is the highest as corresponding to N=0. For example when x=2, the number of electrons transferred by the SE turnstile is two and the plateau has a negative voltage level and in each transfer cycle, two electrons are transferred into the SN through the SEB.

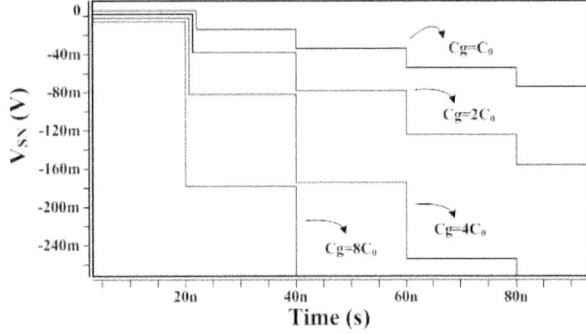

Figure 5. Number of electrons transferred (up to 4) by the SE turnstile with different Cg at 32 nm node; the bias voltage V_h is given by 0V

For a comprehensive and complete evaluation of the proposed model of the SE turnstile, the main operational modes of the circuit are assessed at 32 nm. In Figure 6, the effects of having a disjoint circuit structure for the proposed HSPICE model (made of two parts) are evidenced as V_{SN} is affected only by the electron transfer process.

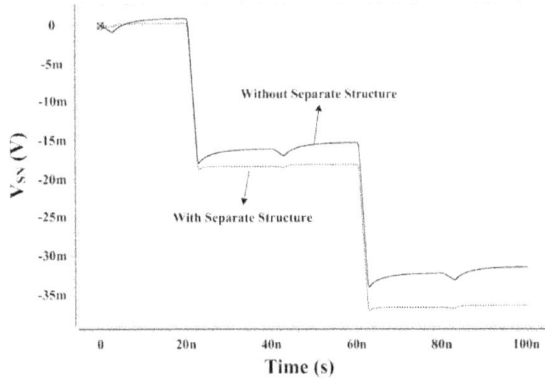

Figure 6. Effect of single/disjoint parts in the proposed model on V_{SN} at 32 nm

Figure 7 is obtained under different values for the capacitance C_{SN}, i.e. from 5 aF to 14 aF. The voltage levels corresponding to the numbers of transferred single-electrons are distinct, thus showing the effectiveness of the proposed circuit to model this process. The (absolute) value of the SN voltage decreases with an increase of C_{SN}; this confirms the original design objective to use voltage levels to represent the stored electron(s) according to the relation

between the capacitance C_{SN} and the SN voltage V_{SN} (i.e. $Q=N*C_{SN}*V_{SN}$).

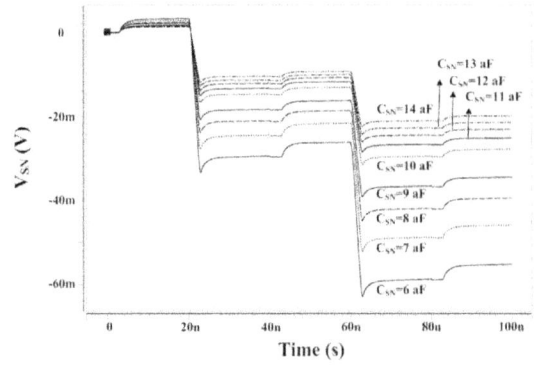

Figure 7. V_{SN} by varying C_{SN} at 32 nm

5. COMPARATIVE ASSESSMENT

The turnstile of [19] uses two one-dimensional field effect transistors (FETs) to sequentially transfer electrons into the Memory Node (MN). The MN is equivalent to the Storage Node (SN) as outlined in previous discussion. As part of the evaluation of the proposed model, the experimental parameters of the fabricated MOSFET-based SE turnstile (as reported in [19]) have been utilized as data. These parameters are used to compare the experimental and simulated turnstiles, which are given in Table 2.

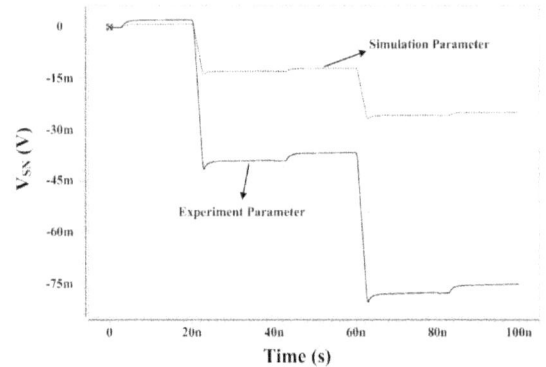

Figure 8. Simulation results using experimental parameters of [19]

Figure 8 shows the experimental and the simulated results using the values of [19]; it shows that the proposed model can capture the SE transfer process quite accurately in terms of timing performance as well as the SN voltage (at all three levels). The difference in SN voltage levels between experimental and simulated results reflects a constant DC voltage that does not affect the correctness of the proposed model. This is mostly related to the difference in SN capacitance, i.e. the absolute value of the SN voltage increases by reducing the SN capacitance (in the proposed model, an ideal capacitor is used for simulating its characteristics). However, the reduced MOSFET feature size (affecting mostly the gate width) and the lower supply voltage are also contributing factors. The supply voltage affects the output voltage swing, while the lower feature size of the MOSFET affects the rate of the charging/discharging process.

5.1 Functional Operations

TABLE 2. Device Parameters for Comparison

Feature		Experimental Data	Simulation Model
Temperature		26 K	26 K
Power Supply	Vdd	1 V	0.9 V
	-Vdd	-1 V	-0.9 V
SE turnstile	W_{FET1}, W_{FET2}	80 nm	32 nm
	L_{FET1}, L_{FET2}	30 nm	32 nm
	Vg1, Vg2	1 V	0.9 V
	V_h	1 V	0.9 V
	V_l	-1 V	-0.9 V
	C_{SEB}	0.5 aF	0.5 aF
	C_{SN}	4 aF	10 aF

The modes of the proposed circuit have been analyzed with respect to the functional operation of the SE turnstile. Figure 9 shows the evaluation over two cycles for those pulses required for its operational modes. Figure 9(a) shows the gate voltage pulses for the turnstile at an operational frequency of 50 MHz; together with these two control pulses, an input pulse is then utilized to start the electron transfer (Figure 9(b)). Figure 9(c) shows the amount of charge in the SEB as function of time; this is consistent with the transfer event cycles. After a transfer event, the voltage of the SEB rises back to 0, as corresponding to the number of electrons.

Therefore as expected, simulation proves that there is no electron stored in the SEB after the electron transfer. Using the same scale for the x-axis (time), Figure 9(d) shows that when the control pulse on CLK1 is high, the voltage-controlled current source G1 is open and a transient current flows in the turnstile, leading to an electron to be transferred into the SEB. In the first cycle for the operation of the turnstile in Figure 9(d), the current does not change without a variation in the value of the input signal, i.e. no electron is transferred through G1. As no electron is transferred, then the SEB is still empty (Figure 9(c)). Differently from the first cycle, when the input signal is high, the control pulse on CLK1 opens the current source G1; hence, an electron is transferred into the SEB during the second cycle. So, changes in I_{G1} and the number of electrons stored in the SEB (Figures 9(c) and 9(d)) characterize the SE transfer process. This shows that the proposed HSPICE model works correctly by utilizing the control pulses on CLK1 and CLK2.

5.2 Comparison with [17]

Based on the above discussion of the operational modes of the proposed model, a comparison can be pursued with respect to [17]. The simulated operation of [17] is shown in Figure 10.

Compared to the previous HSPICE model [17], the proposed model uses a symmetric design (made of two separate and almost identical parts) in the circuit to simulate the operation of the SE turnstile. The difference in circuit model structure has been shown previously in Figure 6; the model without disjoint and separate parts (i.e. a single circuit structure) shows at the plateaus a

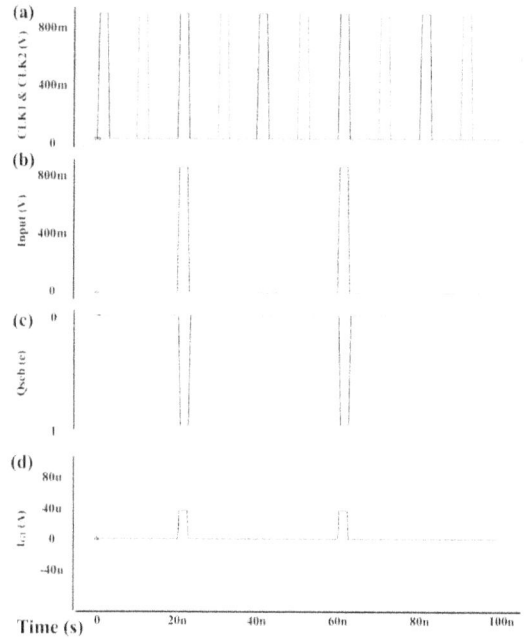

Figure 9. Simulation results for operational modes of the proposed model at 32 nm. (a) Applied gate voltage control pulses. (b) Input pulse to start a SE transfer cycle. (c) Amount of charge in SEB as function of cycle time. (d) Transient current through G1 for a SE transfer.

relatively larger variation in SN voltage than for the one having a circuit model with two disjoint parts. This effectively stabilizes the output voltage, thus modeling closely the operations of the SE turnstile.

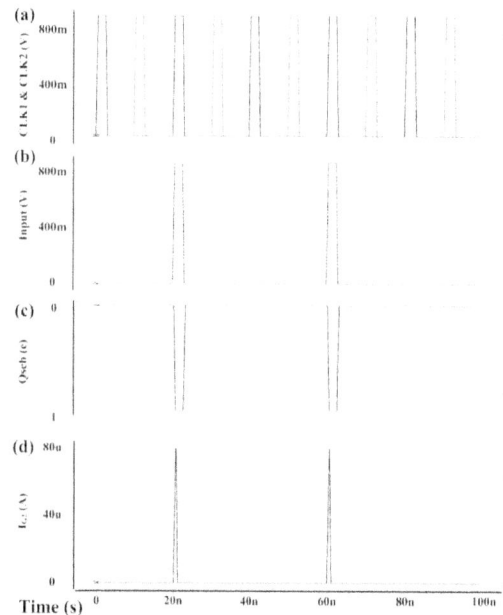

Figure 10. Simulation results for the model of [17] at 32 nm. (a) Applied gate voltage control pulses. (b) Input pulse to start a SE transfer process. (c) Amount of charge in SEB as function of time. (d) Transient current through current source to represent a SE transfer.

225

Moreover, the output signal operates differently in [17] compared to the proposed model. [17] utilizes a current pulse to represent the SE transfer. When the charge on Ce is larger than e (i.e. Ne/Ce> V_{SEB}), a comparator sets the charge on Ce to "0". On a transient basis, the output of the comparator enables a voltage-controlled current source, such that it outputs a sharp current pulse (Figure 10). Differently from the current pulse for representing the electron transfer process, the output of the proposed model is given by the voltage in the SN (whose level corresponds to the number of transferred electrons as in Figure 9). As evidenced by simulation, the proposed model operates correctly in terms of logic and timing. This is in agreement with the expected output (as shown in Figure 9(c) for the discharge event of a SEB).

6. CONCLUSION

This paper has presented a HSPICE circuit model for a single-electron (SE) turnstile. The proposed model captures the sequential transfer of electrons through the turnstile using a nearly symmetric circuit that shows stability at nanometric feature sizes using a voltage level output as mode of operation.. The proposed model is HSPICE compatible and its assessment has shown that it can operate at nanometric scales, while correctly characterizing the process of single-electron transfer. The proposed circuit model has been compared with [17]. It has been shown that the nearly disjoint operation of the proposed circuit model (consisting of two nearly independent parts) results in a stable output; stability has also been accomplished when changing many parameters such as capacitance, feature size and voltages.

7. REFERENCES

[1] Yu, B., Chang, L., Ahmed, S. et al. 2002. FinFET scaling to 10 nm gate length. *Int. Electron. Devices Meet.* 251–254.

[2] Martel, R., Derycke, V., Appenzeller, J. et al. 2002. Carbon nanotube fieldeffect transistors and logic circuits. *ACM SIGDA DAC.*

[3] Redwing, J., Mayer, T., Mohney, S. et al. 2002. Semiconductor nanowires: building blocks for nanoscale electronics. *NSF Nanoscale Science and Engineering Grantees Conference.*

[4] Peatman, W.C.B., Brown, E.R., Rooks, M.J. et al. 1994. Novel resonant tunneling transistor with high transconductance at room temperature. *IEEE Electron. Device Lett.* 15 (7).

[5] Hadley, P. 1998. Single-electron tunneling devices. *AIP Conference Proceedings.* 427, Woodbury, New York, 256–270.

[6] Amlani, I., Orlov, A.O., Toth, G. et al. 1999. Digital logic gate using quantum-dot cellular automata. *Science* 284, 289–291.

[7] Chen, Y., Jung, G. Y., Ohlberg, D.A.A. et al. 2003. Nanoscale molecularswitch crossbar circuits. *Nanotechnology* 14, 462–468.

[8] Kim, D. H. et al. 2002. Fabrication of single-electron tunneling transistors with an electrically formed coulomb island in a silicon-oninsulator nanowire. *J. Vac. Sci. Technol.* 20 (4) 1410–1418.

[9] Klein, D.L., Roth, R. and Lim, A.K.L. 1997. A single-electron transistor made from a cadmium selenide nanocrystal. *Nature.* 389, 16.

[10] Soldatov, E.S., Khanin, V.V. and Trifonov, A.S. 1998. Room temperature molecular single-electron transistor. *Phys. Usp.* 41 (2) 202–204.

[11] Wasshuber, C., Kosina, H. and Selberherr, S. 1997. SIMON—a simulator for single-electron tunnel devices and circuits. *IEEE Trans. Comput. Aided Des.* Vol. 16, No. 9, 937–944.

[12] Mahapatra, S., Ionescu, A.M. and Banerjee, K. 2002. A quasi-analytical set model for few electron circuit simulation. *IEEE Electron. Device Lett.* 23 (6) 366–368.

[13] Yu, Y.S., Hwang, S.W. and Ahn, D. 1999. Macromodeling of single-electron transistor for efficient circuit simulation. *IEEE Trans. Electron. Devices.* 46 (8) 1667–1671.

[14] Oya, T., Asai, T. and Amemiya, Y. Stochastic resonance in an ensemble of single-electron neuromorphic devices and its application to competitive neural networks. *Chaos, Solitons Fractals.* Vol. 32, 855–861.

[15] Smith, K. C. 1981. The prospects of multi-valued logic: A technology and applications view. *IEEE Trans. Comput.* Vol. AC-30, No. 9, 619-634.

[16] Inokawa, H., Fujiwara, A. and Takahashi, Y. 2003. A multiple-valued logic and memory with combined single-electron and metal-oxide-semiconductor devices. *IEEE Trans. Electron Devices.* Vol. 50, No. 2, 462-470.

[17] Zhang, W. C. and Wu, N. J. 2008. Nanoelectronic Circuit Architectures Based on Single-Electron Turntiles. *2nd IEEE International Nanoelectronics Conference.* 978-1-4244-1573-1.

[18] Zimmermana, N. M., Hourdakis, E., Ono, Y., Fujiwara, A. and Takahashi, Y. Error mechanisms and rates in tunable-barrier single-electron turnstiles and charge-coupled devices. *J. Appl. Phys.* Vol. 96, 5254–5266.

[19] Nishiguchi, K., Inokawa, H., Ono, Y., Fujiwara, A. and Takahashi, Y. 2004. Multilevel memory using single-electron turnstile. *Electronics Letters.* Vol. 40, No. 4.

[20] Nara, Y. 2009. Scaling challenges of MOSFET for 32 nm node and beyond. *VLSI Technology, System, and Application.* 72-73.

Limits of Writing Multivalued Resistances in Passive Nanoelectronic Crossbars Used in Neuromorphic Circuits

Arne Heittmann and Tobias G. Noll
Chair of Electrical Engineering and Computer Systems
RWTH Aachen University
Schinkelstr. 2, D-52062 Aachen, Germany
{heittmann,tgn}@eecs.rwth-aachen.de

ABSTRACT

In this paper, limits of writing multivalued resistances in passive nanoelectronic crossbars are examined. The results are based on circuit simulation including device models for resistive switches based on the electrochemical metallization effect. The write operation is performed using a current mirror based on 40nm CMOS technology which operates in subthreshold mode. The results show that only sparsely coded pattern with low mutual overlap can be robustly brought into the matrix which limits the use of passive crossbar to applications that feature particular spatial distributions of resistive weights.

Categories and Subject Descriptors

B.7.1 [Integrated Circuits]: advanced technologies; C.1.3 [Processor Architectures]: neural nets

General Terms

Theory, Performance

Keywords

passive crossbar, multilevel writing, nanoelectronics, electrochemical metallization, neuromorphic circuits, hybrid circuits, resistive switches

1. INTRODUCTION

Systems which are based on artificial neural networks (ANNs) are inherently parallel architectures. Since parallelism is one of the most powerful design principles to realize energy efficient systems while keeping performance high, ANNs are attractive candidates to be used in cognitive applications where energy is a limited resource and real time capabilities are mandatory. However, parallelism comes at the price of space. In neuromorphic architecture synapses provide the substantial amount of computing resources [1].

Since synapses are generally dynamic elements, i.e. they have to be programmed or adapted in order to optimize system performance, a particular challenge is given by providing adaptive

elements with low space requirements. For most networks, the synaptic strength has to be adaptive with respect to a learning law or with a special dynamics modeling short term plasticity. In either case, storage of the concurrent synaptic strength is required. If multilevel strengths are necessary, multiple bits in memory (for a digital implementation) or analog memories such as capacitors (for analog implementations) are traditionally used which constitute an essential portion of the space demands for artificial synapses, see [1,2] for instance.

In this paper, the focus is set on nanoelectronic devices, resistive switches (RS) or memristive devices, which can be used as adaptive and/or non-volatile multi-level storage elements. RS resemble the functionality of memristors [3]. In principle, most types of resistive switches can be implemented with an area occupation of $4F^2$ (F: lithographic feature size) and hence outperform most established storage techniques with regard to area [4]. This is one reason why many researchers have proposed neuromorphic architectures comprising RSs as synapses [5,6,7,8,9,10]. Also adaptive filter circuits were recently proposed as an application [11]. However, there are two additional considerations that have to be made.

First, resistive switches are passive devices. Therefore, they cannot be used for signal amplification. Consequently, circuits of resistive switches have to be connected to an interface of active circuits in order to realize useful functions. If, for instance, scaled CMOS is brought together with memristive devices, circuit designers are faced with the problem that voltages as well as current levels, which are necessary for robust RS operation, have to match with those provided by standard CMOS devices. Although electrical prospects for CMOS are known [5], there is still a considerable gap for the electrical requirements between particular RS device technology and CMOS. Low resistances per device as well as high voltages definitely pose a serious problem of power consumption on circuit level (for drivers) as well as on architecture level, see [3] or [12] for typical memristive device parameters. In addition, the design of hybrid RS-CMOS circuits is often hindered due to the absence of physical models which can be used for circuit simulation. Typically, phenomenological device models are used instead [13]. This poses the risk of missing considerable device dynamics during simulation. So far, for particular material systems models exist that provide physical modeling of switching processes combined with empirical functions and parameters that are fitted against measured data [14,15].

Second, the highest possible storage density of memristive devices is achieved by *passive* crossbar arrays, i.e. the memristive

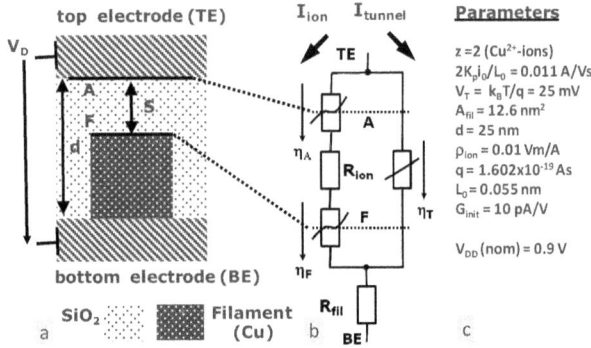

Figure 1: a) Sketch of a device cross section. b) Equivalent circuit diagram with nonlinear elements used in SPICE. c) Representative parameters used for simulation, from [22].

devices are arranged in a matrix-like layout architecture comprising bitlines as well as wordlines, and are connected without additional devices (such as transistors or diodes acting as selector devices). Here, the smallest possible footprint per memristive device is achieved.

But there are costs associated with passive crossbars: the sneak-path problem [16] seriously limits the performance of passive crossbars. For general memory applications where a resolution of 1 bit per cell was assumed, a fundamental analysis for reading the bit information has been done in [16]. Also for bit-based write operations limits were derived in [17].

Along with the discovery of *multilevel* capabilities of memristive devices circuits have been proposed, which were able to tune the state of a memristive device very precisely [18,19,20]. However, only single, isolated memristive devices were analyzed so far. Since passive crossbars are once more attractive as a physical architecture for multilevel information storage with application in neuromorphic systems, the impact of sneak paths on the adjustability of conductances has to be considered.

In this paper, analysis is done for a particular class of memristive devices which are based on the Electrochemical Metallization effect (ECM) [21]. Since a closed physical device model for ECM cells has been developed recently [22] detailed dynamic simulations including CMOS circuits can be performed.

2. The ECM Device Model

So far, nine different electrically induced resistive switching effects have been identified [21]. The electrochemical metallization cells (ECM) provide multilevel resistance programmability where resistances range from 10^{11} V/A down to 10^6 V/A [21]. ECM devices are particularly interesting since the involved device currents can be hold in the nA-domain resulting in ultra-low power consumption during operation.

In Fig.1a a sketch of an ECM device based on Cu-SiO$_2$ is shown, which has been used for the analysis. The device consists of a bottom electrode as well as a top electrode which define the nodes for electronic access. Both electrodes are separated by an electronically insulating material (e.g. SiO$_2$ for Cu-SiO$_2$ cells) which also acts as an ion conducting layer (electrolyte).

If a voltage V_D is applied, electrochemical active ions move into the ion conducting layer and drift towards the counterelectrode which is separated from the top electrode by a gap of size S. The deposition of ions on the counterelectrode results in the continuous growth of a filament.

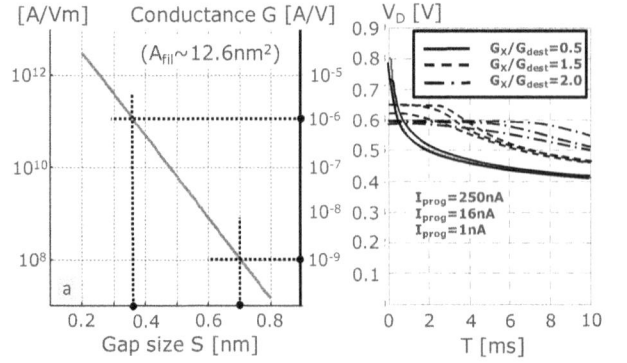

Figure 2: a) S-G relation of the tunnel gap assuming a effective filament area of 12.6nm^2 b) Device voltage V_D vs. time for different G_X/G_{dest} ratios for different I_{prog}.

The current density for the charge transfer across the electrolyte-electrode interface during the cathodic reduction is described by the Butler-Volmer-equation (1).

$$I_{ion} = A_A \cdot 2 \cdot i_0 \sinh\left(\frac{z\eta_A}{2V_T}\right) = A_F \cdot 2 \cdot i_0 \sinh\left(\frac{z\eta_F}{2V_T}\right) \quad (1)$$

Since the ionic charge transport is associated with material transport towards the filament, the filament growth is in direct proportion to the ion current density and is reflected by a shrink in the gap size S (2):

$$\frac{dS}{dt} = -K_p \cdot 2 \cdot i_0 \sinh\left(\frac{z\eta_A}{2V_T}\right) = -K_p \cdot 2 \cdot i_0 \sinh\left(\frac{z\eta_F}{2V_T}\right) \quad (2)$$

In (1) η_A (η_F) describes the voltage across the interface A (F), z is the number of charges per ion and V_T the temperature voltage. A_F as well as A_A are effective active areas of filament and top electrode, K_p models the relation between electronic charge transport and matter deposition. For small gap sizes S the resistance R_{ion} can be neglected and the voltage across the interfaces (A and F) is approximately half of the device voltage V_D, as the effective areas A_F as well as A_A are almost equal in this case. As long as S becomes smaller than approximately 1nm the electronic current becomes dominated by a tunneling current which strongly depends on the gap size S [22,24]. The total device current is the sum of tunnel current and ionic current:

$$I_D = I_{ion} + I_{tunnel} \approx I_{tunnel} \quad (3)$$

For voltages V_D below 1V the overall device conductance (established mainly by I_{tunnel}) can be approximated by (4) which constitute a linear I-V relationship

$$G \approx \frac{q^2}{4 \cdot \pi \cdot h} \cdot e^{-S/L_0} \cdot \frac{A_{fil}}{S \cdot L_0} \quad (4)$$

In (4) the definition (5) was used

$$L_0 = \frac{1}{\sqrt{2m_e \cdot q \cdot \phi_0}} \cdot \frac{h}{4 \cdot \pi} \quad (5)$$

Fig.2a shows the conductance $G = I_D/V_D$ with regard to the gap size S. Between S=0.3nm and S=0.8nm, the conductance shows a variation over more than five orders of magnitude which is the desired operating range for the envisioned neuromorphic application [1,7].

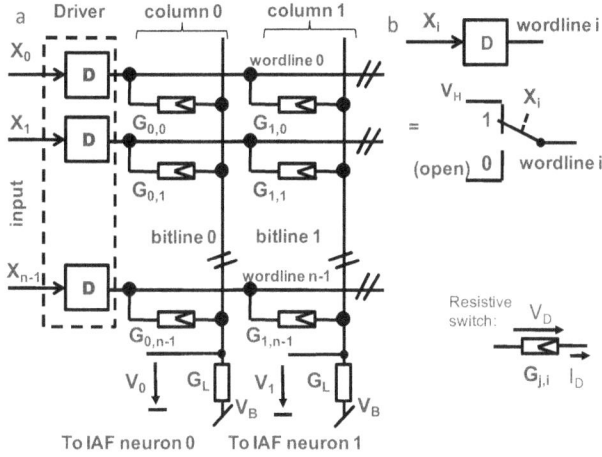

Figure 3: a) Synaptic array. b) Driver circuit.

Combining (2) with (3) the conductance adaptation rate is obtained (6).

$$\frac{dG}{dt} = \frac{\partial G}{\partial S} \cdot \frac{dS}{dt} \approx \frac{2 \cdot K_p \cdot i_0}{L_0} \cdot G \cdot \sinh(\frac{z \cdot V_D}{4 \cdot V_T}) \qquad (6)$$

The right hand side of (6) provides the fundamental relationship between the conductance adaptation rate, the applied device voltage V_D and the conductance G. Positive device voltages let the conductance increases while negative voltages result in a decrease of the conductance G. Note, that a synaptic adaptation function similar to (6) has been proposed in [6].

The model presented in [22] was fully incorporated in our circuit simulation flow using SPICE allowing for the hybrid simulation of CMOS transistor devices and ECM devices with representative parameters (see Fig.1c or [22]). In particular, a circuit using TSMCs 40nm CMOS technology was designed and some expected performance figures were derived from simulation. As a welcomed by-product, the results show that the electrical device properties match with the requirements of scaled CMOS. Hence, it could be used as a prospective device for future neuromorphic systems.

3. Synaptic Array

A fundamental building block which uses resistive switches as programmable synapses is illustrated in Fig.3a. In the envisioned target architecture [1,7] integrate-and-fire neurons (IAF) are used as the neuron model, but the results presented here are not limited to that particular signal processing scheme.

The synaptic array receives n pulsed input signals X_i from a receptive field which are connected to an adjacent driver circuit D. This circuit transforms pulses into voltage levels for driving the resistive switches $G_{j,i}$. The wordlines are connected to the resistive switches and give rise to currents to the bitlines where the currents are summed up. At the load conductances G_L output voltages V_j are generated which are used as the input signals for IAF neurons. In order to keep the voltage swing for V_j large and power consumption low, the driver circuits operate in the so-called open-mode: only conductances associated with active pulses ($X_i=1$) are connected to the activation voltage V_H, see Fig.1b. As the ECM devices provide a linear I-V relationship the output voltage V_j (of column j) is given by the set of active pulses, the activated conductances and the load conductance G_L (7). In (7) the impact of cross current flowing through open wordlines from

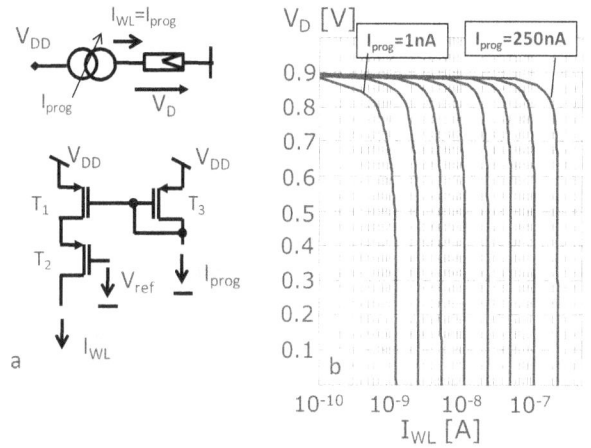

Figure 4: a) Current source and current mirror; b) I-V charac-teristics of the circuit shown in Fig.4a for different currents I_{prog} ranging from 1nA up to 250nA (shown on a logarithmic scale).

$$V_j(t) - V_B = (V_H - V_B) \cdot (\sum_{X_i=1} G_{j,i})/(G_L + \sum_{X_i=1} G_{j,i}) \qquad (7)$$

one bitline to the other was neglected. This can be assured if conductances are distributed in such a way that at least one conductance per wordline is in a high resistive state (HRS). However, due to the feedback of the load conductance G_L the transfer characteristics given by (7) is nonlinear [7]. It becomes linear if the probability of finding synchronous input pulses is low. For instance, this is given for uncorrelated input pulses at low pulse frequencies. Then, an input pulse X_i is exactly weighted by (8) for column j, and the impact on the membrane of the receiving neuron is additive for any pulse at the input and establishes the well-known weighted-sum-of-inputs for the synaptic array.

$$W_{j,i} \sim G_{j,i}/(G_L + G_{j,i}) \qquad (8)$$

Particular weights within a neuromorphic architecture are known in advance (from simulation or from calculation), such as filter coefficients of low-level feature detectors [1,7]. If functional layers (comprising known weights) are implemented by means of a passive crossbar array, the challenge is to program/adapt $G_{j,i}$ in order to maintain the relation (8) up to a certain level of precision. For particular neuromorphic architectures moderate variations of $G_{j,i}$ can be tolerated [1,6]. But even when the right order of magnitude for $G_{j,i}$ has to be ensured, passive crossbars provide strong limitations on the possible distribution of weights. In this paper, some limitations will be shown.

4. Tuning of Conductances

In literature, several programming schemes for memristive devices have been presented [17,18,19]. Generally, the used procedure to program a memristive device is to (i) drive the memristive device into an operating point with high adaptation rate by applying a sufficiently large driving force (i.e. a current or a voltage), (ii) monitor the obtained resistance of the memristive device by measuring the device voltage under defined currents (or vice versa) and (iii) cut-off the driving force if the desired resistance value has been obtained. For single devices the processes (i)-(iii) were successfully realized by either active feedback circuits or by voltage-sources with current compliance [18,19,20].

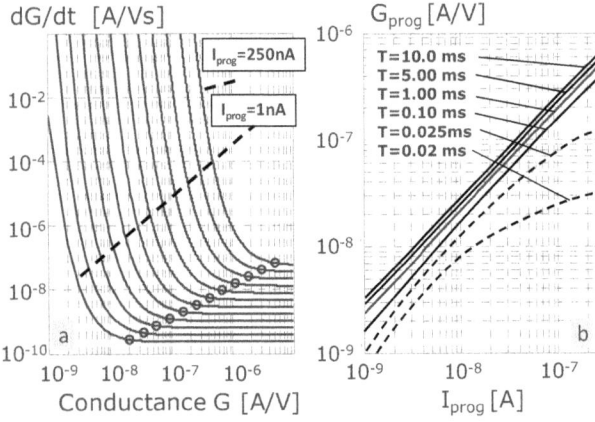

Figure 5: a) Conductance adaptation rate dependent on the conductance G for different currents I_{prog}.. Currents are provided on a logarithmic scale. Circles mark the turing point where the exponential dependency turns into an almost linear relationship; b) Attained conductance G_{prog} dependent on I_{prog} for different write times T. For small write times T the adaptation dynamics has not been settled which makes the I_{prog}-G_{prog} relation non-linear.

However, in passive crossbar circuits (i.e. in crossbars without additional selector devices), process (ii) is crucial as on one hand sneak paths seriously disturb the measuring process and on the other hand even may inhibit the adaptation of resistance itself due to lowering the driving force, then making any adaptation scheme practically non-effective.

4.1 Voltage Source with Current Compliance

In the following, a voltage source with current compliance is used to adapt the conductance of a memristive device. In order to account for a realistic scenario a circuit was simulated which is based on 40nm CMOS devices. Although the circuit is very simple compared to other hybrid CMOS/RS implementations the fundamental problems become visible and are hard to overcome in passive crossbars, even with complex circuits in the crossbar periphery.

The circuit (Fig.4a) is an implementation of a current mirror with an additional transistor T_2 which stabilizes the drain potential of T_1 in order to precisely mirror I_{prog} to I_{WL}. The circuit operates in subthreshold mode [23] and is able to deliver currents in the range of 1nA < I_{prog} < 250nA using minimal sized transistors, see Fig.4b.

4.1.1 Adaptation Dynamics and Switching

The supply voltage was fixed to V_{DD}=0.9V. If I_{WL} gets delivered to an isolated memristive device (i.e. without additional devices, cf. Fig. 4a) and the initial conductance is low, the device voltage V_D saturates to V_{DD} which lets the circuit initially act as an voltage source. If the tunnel current is already dominant, a very fast increase of G is the result (9).

$$G(t) = G(0) \cdot \exp\left(t \cdot \frac{K_p \cdot i_0}{L_0} \cdot \sinh\left(\frac{z \cdot V_{DD}}{4 \cdot V_T} \right) \right) \quad (9)$$

As G rises and I_{prog} is constant the condition for saturation fails very soon and the device voltage V_D is a function of G and I_{prog}: V_D=I_{prog}/G, see also Fig.8a. Then, the adaptation rate is continuously slowed down (10), see Fig.5a

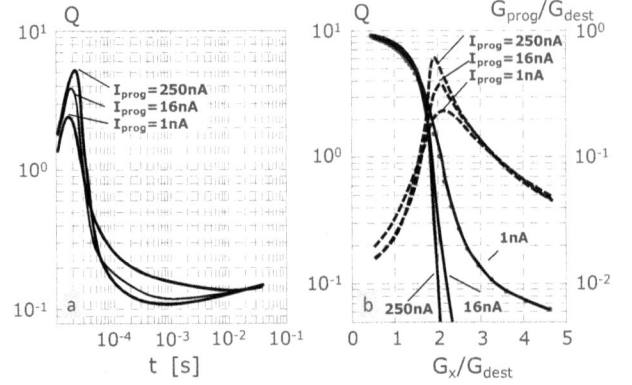

Figure 6: a) Course of Q dependent on time for different I_{prog}. Local maxima of Q indicate the onset of conductance adaptation b) Attained Q for write time T=5ms (dotted lines) for different I_{prog}; Ratio attained G_{prog} : desired G_{dest} (straight lines) for write time T=5ms.

$$\frac{dG}{dt} = \frac{2 \cdot K_p \cdot i_0}{L_0} \cdot G \cdot \sinh\left(\frac{I_{prog}}{G} \frac{z}{4 \cdot V_T} \right) \quad (10)$$

For a specific G dependent on I_{prog} the adaptation rate becomes almost linear. Roughly, the exponential adaptation rate turns into a linear relationship at the condition (11), see also Fig.5a, marked with circles.

$$G = G_{prog} = I_{prog} \frac{z}{4 \cdot V_T} \quad (11)$$

The obtained conductance is a linear function of I_{prog}, which is the desired result: processes (i)-(iii) are provided by the current mirror.

However, the time needed to drive G towards this specific turning point is quite large and is not maintainable under economical considerations. In fact, the device voltage V_D applied to the device at that point is V_D=4V_T/z which is in the range of the voltage used to *read* the device state (and is associated with very slow adaptation rates). An almost linear relationship between I_{prog} and G_{prog} can be obtained yet earlier.

In order to determine a usefull time margin for writing, the switching process has to be separated from the operating points with comparatively low dynamics.

On one hand, the write current should be applied long enough in order to overcome the main adaptation process of the device. On the other hand, the write current is considered to be non-effective if the adaptation rate of the conductance compared to the currently attained conductance is smaller than the relative progress made in time (the latter is an economical argument). All in all, a function which describes this tradeoff is given by (12).

$$Q := \frac{dG}{dt} \frac{t}{G} \quad (12)$$

Fig.6a shows the course of Q dependent on time for different I_{prog} and G_{init}=10^{-11} A/V. The strong maximum of Q characterizes the onset of the adaptation process. Then, as Q decreases the adaptation process has decayed. If Q is low enough (i.e. Q << 1), the adaptation can be considered to last sufficiently long. Here, by taking (6) into account, the device voltage V_D develops only logarithmically with time.

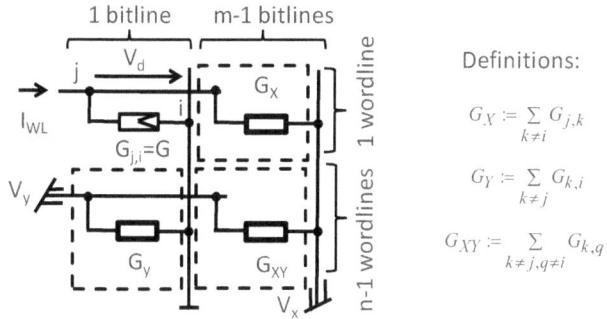

Figure 7: Equivalent circuit of a passive crossbar during writing to memristive device $G_{j,i}$. The crossbar comprises n x m ECM devices The write current is delivered to wordline j. Bitline i is shortened to ground. Remaining wordlines as well as bitlines are tied to voltage levels V_x and V_y respectively. G_X, G_Y and G_{XY} model effective conductances of remaining elements which partially act as a parasitic load to I_{WL}.

Definitions:

$$G_X := \sum_{k \neq i} G_{j,k}$$

$$G_Y := \sum_{k \neq j} G_{k,i}$$

$$G_{XY} := \sum_{k \neq j, q \neq i} G_{k,q}$$

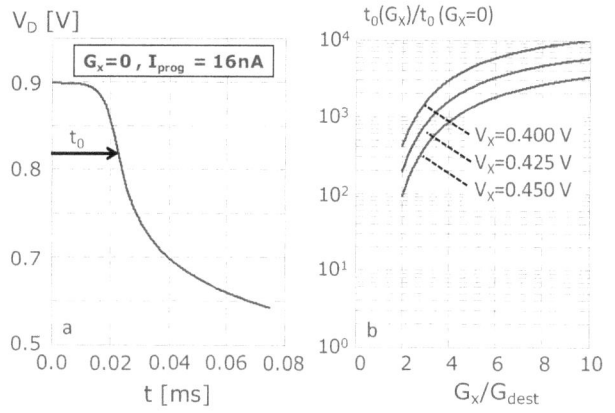

Figure 8: a) V_D-characteristics for an isolated ECM device (I_{prog}=16nA). Until t=t_0 the current mirror acts as a voltage source. b) Increase of t_0 due to additional load conductance G_X. The increase is shown with respect to the ratio G_X/G_{dest} (parasitic load:desired conductance).

From Fig.5, a reasonable write time T can be found. Fig.5b shows the attained conductance G_{prog} for different write currents I_{prog} dependent on the write time T. As the switching process is still on for T < 0.025ms the G_{prog}-I_{prog} relationship is strongly nonlinear (dotted lines). In order to avoid falling into the rapid adaptation regime of the device the time margin has to be made large enough. Therefore, the minima of Q (see Fig.5a) may serve as an orientation for defining the write time which has been set to T=5ms for the subsequent examinations. By combining (12) with (6) it becomes obvious that Q depends on the concurrent device voltage as well as on the time passed by. Note that the maxima of Q are not defined by absolute threshold voltages but by the combined dynamics of load circuit and the switching device.

4.1.2 Equivalent Circuit of a Crossbar

For the following analysis, ECM devices are embedded into a passive crossbar architecture, see Fig.3a. An array of n x m devices is assumed (note: in Fig.3a the crossbar contains only 2 bitlines and n wordlines). First, it is assumed that a single device is programmed at any time. Hence, the impact of programming to other embedded devices with regard to adaptation should be made as low as possible, i.e. the voltages across the remaining devices should be as low as possible. In order to achieve this condition, the remaining wordlines as well as bitlines are tied to predefined voltage levels. As there is no particular preference all bitlines apart from bitline i are connected to the same potential V_X while all wordlines apart from worline j are connected to the potential V_Y. Under this particular connection scheme an equivalent circuit emerges which is shown in Fig.7. The non-addressed resistive switches are merged into one of three substitutional conductances in an additive way. As V_X as well as V_Y are fixed, only G_X may disturb the writing process of $G_{j,i}$.

In order to obtain usefull values for V_X and V_Y the characteristic of an isolated device has to be considered. If the I_{WL} gets delivered to an ECM cell (Fig.4a) the device voltage V_D follows a falling characteristics, see. Fig.8a. As soon as the write time is reached, the voltage level V_D is almost independent from the write current I_{prog} (or I_{WL}), cf. Fig.2b., and the attained G is a function of $I_{prog.}$. Finally, V_X should be approximately in the range of V_D obtained at write time, which is in our case approximately half of V_{DD}. Note, that larger values of V_X let G_X act as an additional current source which eventually results in a larger

conductance for G_{ij} after writing. In such a case, any method which potentially could be used to correct such errors would require to apply negative device voltages to G_{ij} which is out of scope of this paper.

Now, the device to be programmed is connected to bitline i (shared by n-1 devices) and wordline j (shared by m-1 devices). The programming current I_{WL} is delivered to wordline j and bitline i is connected to ground. G_X is now an additional load to the current I_{WL} and (for large G_X compared to the desired G for G_{ij}) may short V_D. For different G_X/G ratios the voltage V_D (measured at write time) gets larger compared to the case considering an isolated device. On the other hand, in the initial phase where the current mirror drives the device in voltage mode, the device voltage is considerably lower than for the isolated case, cf. Fig. 2b. As the initial adaptation gets slower, the device stays longer in the (initially) high-resistive state. The effective load of the current source is given by G_X only. If G_X is as large as the desired G_{dest} the associated current I_{prog} is large enough to rise V_D to $2V_X$ which lets the device still adapt sufficiently fast, cf. Fig.6b. The relative loss in the attained G is less than 10%.

However, larger loads G_X result in a rapid decrease in adaptation speed. In Fig.6b the actual value of Q in dependence of the ratio G_X/G_{dest} is shown. For G_X/G_{dest}=2 it can be concluded that the memristive device has initially started its adaptation at write time only (Q hat a maximum there) which is reflected in a very high loss of more than 99% comparing the attained G with the desired value. Even larger G_X/G_{dest} ratios let the memristive device stay in an almost unreactive state (Q is low, while the loss is almost 100%).

However, as the device voltage V_D is different form 0V, adaptation is (theoretically) given. Consequently, a larger time frame potentially could help in getting the desired results of conductance adaptation. In order to get an impression of the required (additional) time frame, a formula is provided (13) that approximates the required time to overcome the voltage-mode driven state. Taking (9) into account the required time t_0 can be related to the unloaded case of an isolated device, see also Fig.8b. The relation (13) predicts (see also Fig 8b), that the adaptation becomes almost ineffective if G_X becomes larger than twice of G_{dest} since V_D becomes increasingly shorted to V_X.

$$t_0(G_X) \approx t_0(G_X = 0) \cdot \frac{\sinh\left(\dfrac{z \cdot V_{DD}}{4 \cdot V_T}\right)}{\sinh\left(\dfrac{z \cdot V_X}{4 \cdot V_T} \cdot \left(1 + \dfrac{G_{dest}}{G_X}\right)\right)} \qquad (13)$$

The required additional time becomes more than 2 orders of magnitude larger which is not maintainable under an economical point of view.

5. CONCLUSION

For an economic programming scheme the following considerations have to be taken into account. As long as the parasitic load given by G_X is smaller than the desired G_{prog} programming-by-reference in a single step is possible, if a precision of 10% can be tolerated (see Fig.6b). For fixed write time, larger G_X result in considerable errors making further adaptation necessary which is potentially possible as long as the memristive device sufficiently reacts to the applied stimulus. However, the additional margin is not very large. Since the device kinetics depends in an exponential way from the applied stimulus yet a small decrease in the device voltage results in a considerable reduction of the adaptability for the device conductance. If the driving circuit (in our case: the current mirror) has limited output capabilities (e.g. a maximum current the circuit is able to deliver) the sum of conductances per wordline is strictly bounded. Under this consideration only those applications benefit from passive resistive crossbars which show particular weight distributions that are highly nonoverlapping with regard to either the wordline contents or the bitline contents (as the results still apply if wordline and bitline are interchanged). One possible application is given by a (n x 2) crossbar architecture as proposed in [7] which requires that only one conductance per wordline is in the low resistive state, see also Fig.3a.

The analysis of this simple programming scheme has also consequences for non-neuromorphic resistive crossbar applications, such as small-sized memories [16] comprising resistive switches which are either in a high resistive state or a low resistive state. If a current source is used to limit the on-resistance of a particular memory element during programming, the parasitic load G_X has to be known in advance in order to size the write current correctly. If the memory contents is known upfront (this could be given for look-up tables in configurable logic applications, or if all cells connected to a particular wordline are modified at once), the required write current I_{load} can be predicted. On the contrary, for a random write access scheme where individual memory cells change their contents at random, the wordline load G_X has to be measured *before* any write action takes place in order to provide the right write current for the modification of a single bit. Here, the cycle time of a random write access could be affected in a significant way and the write circuit gets more complicated.

6. ACKNOWLEDGMENTS

The authors like to thank Stephan Menzel, Eike Linn and Rainer Waser for providing the fundamental ECM device model including representative parameters for simulation.

7. REFERENCES

[1] U.Ramacher, C.v.d.Malsburg, "On the Construction of Artificial Brains", Springer, 2010.

[2] C.Bartolozzi, G.Indiveri, "Synaptic Dynamics in Analog VLSI", Neural Computation 19, pp 2581-2603, 2007.

[3] D.Strukov, G.Snider, D.Steward, R.Williams, "The Missing Memristor Found", Nature 53, 80, 2008.

[4] ITRS roadmap, http://www.itrs.net

[5] S. H.Jo, T.Chang, I.Ebong, B. B.Bhadviya, P.Mazumder and W.Lu, "Nanoscale memristor device as synapse in neuromorphic systems", Nano Lett. pp. 1297-1301, 2010.

[6] G.Snider, "Self-organized computation with unreliable, memristive nanodevices", Nanotechnology, vol. 18, no. 36, pp. 1-13, 2007.

[7] A.Heittmann, T.G.Noll, "Sensitivity of neuromorphic circuits using nanoelectronic resistive switches to pulse synchronization", in Proc. ACM GLSVLSI, pp.375-378, 2011.

[8] T.Hasegawa et.al, "Learning Abilities Achieved by a Single Solid-State Atomic Switch", Advanced Materials, Vol.22, Iss. 16, pp. 1831-1834, 9.Feb.2010.

[9] M.Versace, B.Chandler, "MoNETA: A Mind Made from Memristors", IEEE Spectrum, Cover page fearured article, December 2010.

[10] S.Yu, Y.Wu, R.Jeyasingh, D.Kuzum, H.-S. P. Wong, "An Electronic Synapse Device Based on Metal Oxide Resistive Switching Memory for Neuromorphic Computation", IEEE Transactions on Electron Devices, Vol. 58, No.8, pp 2729-2737, August 2011

[11] T.Driscaoll, J.Quinn, S.Klein, H.T.Kim, B.J.Kim, Y.V.Pershin, M.DiVentra, D.N.Basov, "Memristive Adaptive Filters", Applied Physics Letters 97, pp. 093502 - 093502-3, 2010

[12] M.Aono, T.Hasegawa, "The Atomic Switch", Proc. of the IEEE, vol.98, no.12, pp.2228-2236, Dec. 2010.

[13] S.Shin, K.Kim, and S.-M.Kang, "Compact Models for Memristors Based on Charge-Flux Constitutive Relationships," IEEE Transactions on Computer-Aided Design of Integrated Circuits and Systems, vol. 29, no. 4, pp. 590–598, 2010.

[14] M.D.Pickett, D.B.Strukov, J.L.Borghetti, J.J.Yang, G.S. Snider, D.R.Stewart, R.S.Williams, "Switching dynamics in titanium dioxide memristive devices", Journal of Applied Physics 106, pp. 074508 - 074508-6, 2009

[15] H.Abdalla, M.D.Pickett, "SPICE Modeling of Memristors", IEEE International Symposium on Circuits and Systems (ISCAS 2011), pp 1832-1835, May 2011

[16] A.Flocke, T.G.Noll, C.Kügeler, C.Nauenheim, R.Waser. "A Fundamental Analysis of Nano-Crossbars with Non-Linear Switching Materials and its Impact on TiO2 as a Resistive Layer". 8th IEEE Conference on Nanotechnology, pp. 319–322, 2008.

[17] A.Flocke, T.G.Noll, "Fundamental Analysis of Write-Operations in Resistive Crossbar Arrays", nanoelectronic days 2008

[18] F.Alibart, L.Gao, B.Hoskins, D.Strukov: "High-Precision Tuning of State for Memristive Devices by Adaptable Variation-Tolerant Algorithm", In: CoRR, abs/1110.1393, 2011

[19] W.Yi, F.Perner, M.S.Qureshi, H.Abdalla, M.D.Pickett, J.J.Yang, M.-X. M.Zhangm G.Medeiros-Ribeiro, S.Williams: "Feedback write scheme for memristive switching devices", Applied Physics A 102, pp. 973-982, 2011

[20] N.Papandreou, H.Pozidis, A.Pantazi, A.Sebastian, M.Breitwisch, C.Lam, E.Eleftheriou, "Programming Algorithms for Multilevel Phase-Change Memory", Proc. IEEE International Symposium on Circuits and Systems, pp. 329-332, May 2011

[21] R.Waser, "Electrochemical and Thermochemical Memories", IEDM, pp. 1-4, 2008.

[22] S.Menzel, U.Böttger, and R.Waser, "Simulation of Multilevel Switching in Electrochemical Metallization Memory Cells", Journal of Applied Physics Vol. 111, pp.014501-014501-5, Jan. 2012

[23] S.C.Liu, J.Kramer, G.Indiveri, T.Delbrück, R.Douglas, "Analog VLSI: Circuits and Principles", Cambridge, MA: MIT Press, 2002

[24] J.G.Simmons, "Generalized Formula for the Electric Tunnel Effect between Similar Electrodes Separated by a Thin Insulating Film", Journal of Applied Physics, Vol. 34, No.6, pp. 1793-1803, June 1963

Stepwise Sleep Depth Control
for Run-Time Leakage Power Saving

Seidai Takeda [†], Shinobu Miwa [‡], Kimiyoshi Usami [§], Hiroshi Nakamura [†,‡]

[†]Graduate School of Engineering, The University of Tokyo
[‡]Graduate School of Information Science and Technology, The University of Tokyo
[§]Department of Information Science and Engineering, Shibaura Institute of Technology

[†‡]{takeda, miwa, nakamura}@hal.ipc.i.u-tokyo.ac.jp, [§] usami@shibaura-it.ac.jp

ABSTRACT

Recently, run-time sleep control scheme using multiple sleep modes have been studied. In those studies, each sleep mode has its own sleep depth. Deeper sleep mode provides higher leakage saving but incurs larger overhead energy. Use of multiple modes is helpful for further leakage saving if an appropriate mode is selected, but the best mode depends on the idle period whose length cannot be told in advance. Although the implementations how to realize different sleep depths have been well studied, few attention has been paid to the method of how to select the best sleep depth dynamically during execution. This paper proposes a simple but novel sleep control scheme, called stepwise sleep depth control, which aims to select the best depth among provided multiple sleep depths. Our scheme automatically applies deeper depth in a step-by-step manner after an idle state starts. It successfully reduces leakage energy while only a small modification is required for circuit implementation. This paper also proposes a methodology for optimizing control parameters of our sleep control scheme according to program behavior and temperature. Experimental result shows that stepwise sleep depth control applied to body biasing circuit improves net leakage saving of up to 43% for FPAlu at 1.0GHz, 75°C compared to conventional reverse body biasing.

Categories and Subject Descriptors:
B7.1 [Integrated Circuits]: Types and Design Style
General Terms: Design

1. INTRODUCTION

Since process technology has been in the deep sub-micron era, leakage power reduction is one of the major concerns in modern VLSI circuit design. Body Biasing (BB) [1] and Power Gating (PG) [2] are promising techniques for leakage saving and are commonly used in products from high-performance server processors to low-end micro-controllers. Nowadays, their leakage saving opportunities are confined

to standby-time because their sleep control granularities are much coarse, e.g. processor core or IP macro levels [3, 4].

To achieve further leakage saving, approaches which introduce finer sleep granularities and save leakage power even in run-time are studied [5, 6, 7]. For efficient run-time leakage saving, the overhead energy associated with sleep control is a critical concern. Unfortunately, due to their large overhead energies, those approaches exacerbates energy dissipations if circuits enter into sleep mode during short idle periods. In our evaluation, the minimum sleep time to pay off the overhead energy is more than 200 cycles. For some programs, there is no chance to achieve net leakage saving because the majority of idle periods appeared in the execution is less than the time. Thus, reducing overhead energy is necessary to achieve successful leakage saving in run-time.

Recently, advanced BB and PG using multiple sleep modes are studied [8, 9, 10, 11, 12]. In those studies, each sleep mode has its own sleep depth. A shallower sleep mode provides smaller overhead energy than a deeper sleep mode; hence it achieves net leakage saving in shorter idle intervals. In contrast, a deeper sleep has more leakage saving efficiency; hence it is effective for longer idle intervals. These circuits are able to achieve efficient leakage saving by applying appropriate sleep mode to various idle intervals.

Circuit techniques to produce multiple sleep depths are well studied. However, how to control their sleep modes have not been established. The best sleep mode depends on the length of the idle period. But it is difficult to select the best mode at the beginning of each idle state because the length of the period is unpredictable.

In this paper, we propose a novel sleep control scheme called *stepwise sleep depth control* for multiple sleep modes. Our scheme automatically applies deeper sleep depth in a step-by-step manner. It applies the next sleep depth when the elapsed time reached to each pre-set time, called threshold, after an idle period starts. This scheme is simple but effective to achieves run-time leakage saving for both short and long idle intervals. Moreover, proposed sleep scheme is easily realized with small additional circuits, including a counter and some registers.

Meanwhile, leakage saving efficiency of proposed sleep control scheme receives significant influence from run-time factors, such as application behavior and temperature. For time-out based sleep control, considering the nature of an application, occurrence of idle lengths, is important to avoid exacerbation of power dissipation [13]. Intervals whose length are longer than a threshold but shorter than a time to pay off an overhead energy incur total energy dissipation. In addition, temperature variation exponentially changes leakage

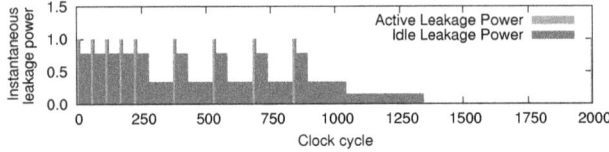

Figure 1: Leakage saving by proposed scheme.

currents; hence, a time to pay off the overhead energy for each sleep depth and the best sleep depth for an idle period greatly vary with temperature variation [12, 14].

We propose a methodology for optimizing control parameters, depths and thresholds, of stepwise sleep depth control to maximize net leakage saving. This optimization utilizes temperature and distribution of idle intervals for an application. Note that idle interval distribution is statically obtained by simple architecture simulation.

The contributions of this paper are as below.

- We propose a novel sleep control scheme for circuits having multiple sleep modes. It efficiently enlarges net saving of leakage power in run-time with small circuit modifications.
- We propose an implementation of our sleep control scheme using Body Biasing circuits.
- We propose optimization method for control parameters of proposed sleep scheme considering application behavior and temperature.

Rest of this paper is organized as follows. In section 2, we explain our stepwise sleep depth control scheme and its implementation using BB. In section 3, control parameter optimization for proposed sleep control scheme are detailed. Experimental results are indicated in section 4. Finally, we summarize this paper in section 5.

2. STEPWISE SLEEP DEPTH CONTROL

Our sleep control scheme applies deeper sleep mode in a step-by-step manner according to elapsed time since an idle state starts. We label a duration where the circuit is applied a same sleep depth as *stage*.

Figure 1 indicates a behavior of proposed sleep control scheme for various idle intervals. *Active Leakage Power* and *Idle Leakage Power* represent instantaneous leakage power of an active state and an idle state, respectively. As previously mentioned, a deeper sleep can not achieve net leakage saving in short idle periods due to its large overhead energy. In our scheme, shallower sleep is applied for leakage saving even in a short idle period until the elapsed time reaches the next threshold. However, shallower sleep has less leakage saving efficiency compared to deeper sleep. When the elapsed time reaches the next threshold, the circuit applies the next deeper sleep mode and to achieves more leakage saving efficiency. Since then, as shown in figure 1, further sleep depths are applied in the same way. Note that each leakage power used in this figure comes from our evaluation.

Rest of this section, we describe an implementation of proposed sleep control scheme using BB and a characterization of its leakage power and overhead energy.

2.1 Implementation using Body Biasing

A. Circuit implementation

Figure 2 indicates the implementation of our stepwise sleep depth control utilizing the body biasing technique.

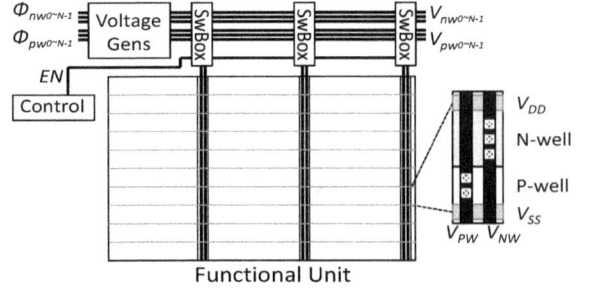

Figure 2: Implementation of stepwise sleep depth control using Body Biasing circuits.

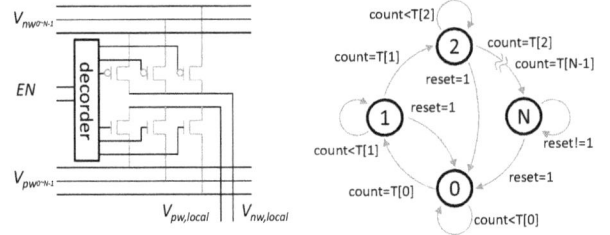

Figure 3: (Left) Structure of switch box. (Right) State transition diagram of control unit.

Proposed circuit is comprised of following components, a basic functional unit, voltage generators, switch boxes and a control unit.

In proposed circuit, stages are switched by switch boxes by selecting N-well and P-well biasing voltages from multiple generated voltages. When the number of stages is N, $2 \times N$ voltage generators for N-well and P-well are necessary. We use charge pump circuits for these voltage generators, the same as conventional BB [1]. These generators are able to change their output voltages by changing control frequencies (ϕ_{nw_i}, ϕ_{pw_i}). However, changing output voltages is much slower than selecting a voltage by switch boxes. Thus, output voltages are not varied when an application is running.

The switch box consists of a decoder and nMOS and pMOS switches to select P-well and N-well biasing voltages, respectively. The left side of figure 3 indicates the structure of the switch box. The decoder generates the control signal from the enable signal (EN). nMOS and pMOS switches have sufficient widths which permit to transfer between awake state and the deepest sleep state within one cycle. Note that charge pumps cannot change their output load current in short time; hence, decoupling capacitances at their outputs are needed for mode switching within one cycle.

The control unit consists of a counter (count) which counts up elapsed time of an idle interval and N registers (T[0]~T[N-1]) which store threshold times for stages. The right side of figure 3 indicates the state transition diagram of the control unit. The control unit compares the value of count and the value of a register corresponding to current stage every cycle. If these values are the same, the control unit switches to further stage and updates EN. When an idle interval is ended, reset signal (reset) is asserted and state becomes 0 which corresponds to the awake mode.

B. Leakage power and overhead energy

In this subsection, we describe leakage power and overhead energy of BB circuits. For BB circuits, sleep control

234

(a) Cross section of Triple-well structure

(b) Equivalent circuit

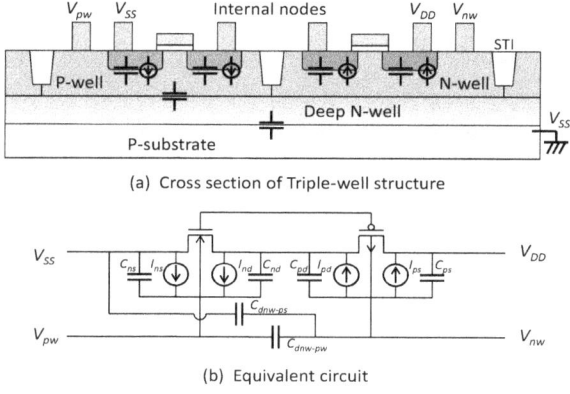

Figure 4: Cross section of Body Biasing circuit and its equivalent circuit.

is realized by changing N-well and P-well biases (V_{nw}, V_{pw}) independently. V_{nw} and V_{pw} are represented as following equations using their standard values (V_{DD}, V_{SS}) and displacements ($\Delta V_{nw}, \Delta V_{pw}$), respectively.

$$V_{nw} = V_{DD} - \Delta V_{nw}, \quad V_{pw} = V_{SS} + \Delta V_{pw}$$

The case $\Delta V_{nw}, \Delta V_{pw} > 0$ and $\Delta V_{nw}, \Delta V_{pw} < 0$ represent forward biasing and reverse biasing situations, respectively. In our work, ΔV_{nw} and ΔV_{pw} has discrete values and their unit value is V_s. In the rest of this paper, we represent a mode of BB circuit as the combination of displacements.

For BB circuits, two unique factors, Band to band tunneling (BTBT) leakage current [15] and charging energies of well capacitances must be taken into account for. Figure 4 indicates the cross section and the equivalent circuit of a BB circuit. $I_{ns}, I_{nd}, I_{ps}, I_{pd}$ represent BTBT leakage currents of source and drain diffusion areas for nMOS and pMOS transistors. $C_{ns}, C_{nd}, C_{ps}, C_{pd}$ represent capacitances between diffusion areas and N/P-wells. C_{dnw-ps} and C_{dnw-pw} represent a capacitance between deep N-well and P-substrate and a capacitance between deep N-well and P-well, respectively. Those values are calculated from doping densities of layers and geometric area information of the circuit.

Leakage power of BB circuit is represented as summation of sub-threshold leakage, gate leakage and BTBT leakage powers. In our evaluation, sub-threshold leakage and gate leakage powers are obtained by SPICE simulation and BTBT leakage power is calculated using $I_{ns}, I_{nd}, I_{ps}, I_{pd}$.

On the other hand, overhead energy is represented as summation of switching energy associated with switch boxes and the well charging energy. Under a condition, the well charging energy is represented as the difference of stored energies within well capacitances between the previous and the next modes. The condition is that either increasing or decreasing of both of N-well and P-well biasing occurs between two modes. If increasing and decreasing simultaneously occurred, larger overhead energy is necessary than the difference of stored energies. Well charging energy occupies the large ratio of overhead energy. Thus, mode switchings which cause such extra energy dissipation should be prohibited.

3. OPTIMIZATION OF STEPWISE SLEEP DEPTH CONTROL

Our sleep depth control achieves successful leakage saving by adjusting control parameters, depth and threshold of each

stage. This section, we propose an optimization algorithm to maximize net leakage saving by adjusting sleep control configuration with run-time factors, application and temperature. Firstly, we introduce the definition of this problem. Secondly, we explain how to select a set of modes. Finally, we describe the simulated annealing based optimization algorithm. The algorithm finds out optimal sleep depth and threshold for each stage.

3.1 Problem definition

The objective of our optimization problem is to maximize net leakage saving in run-time. In active intervals, leakage power cannot be reduced and leakage power does not change before and after the optimization. Thus, this optimization problem is redefined as to minimize a summation of net energy consumptions for each idle interval.

The cost of this problem is calculated by distribution of idle intervals and net leakage energy consumption for each length of idle interval. A function of net leakage energy for a given idle interval is represented as following equation.

$$\varepsilon(D, T, t) = \sum_{k=0}^{j-1} \{p_k(t_{k+1} - t_k) + e_{k+1} - e_k + e_{sw}\} + p_j(t - t_{j-1})$$

$$D = \left(\begin{array}{ccc} m_0 & \cdots & m_{n-1} \end{array} \right) = \left(\begin{array}{ccc} p_0 & \cdots & p_{n-1} \\ e_0 & \cdots & e_{n-1} \end{array} \right)$$

$$T = \left(\begin{array}{ccc} t_0 & \cdots & t_{n-1} \end{array} \right)$$

D and T represent depth and threshold for each stage. n represents the number of stages. e_{sw} represents energy consumption of switch boxes. A duration of stage i is $(t_i, t_{i+1}]$ excepting the last stage; and that of the last stage is $(t_i, \infty]$. At the beginning of stage i, mode switching energy ($e_{sw} + e_i - e_{i-1}$) is consumed. And gradient is p_i in this stage.

The problem definition is represented as follows by using $\varepsilon(D, T, t)$ and a distribution of idle intervals ($\nu(t)$).

$$\text{minimize} \quad E_{total} = \sum_{t=1}^{\infty} \varepsilon(D, T, t)\nu(t) \quad (1)$$

subject to

$$t_0 = 0, \qquad t_i \in (0, \infty], \qquad t_{i-1} < t_i$$

$$m_0 = \widehat{m}_{awake}, \qquad m_i \in \widehat{M}, \qquad m_{i-1} < m_i$$

$\widehat{M}$ represents a set comprised of modes produced by a circuit. It includes not only sleep modes but also an awake mode. Depth of each stage is selected from $\widehat{M}$. For the first stage, the depth is selected as that of awake mode ($\widehat{m}_{awake}$) and threshold is 0. These values are unchanged by optimization. Depth and threshold of each stage must be deeper and larger than those of previous stage and must be shallower and smaller than those of next stage, respectively.

3.2 Depth ordering for each sleep mode

In this subsection we describe how to select depth ordered mode set ($\widehat{M}$) from all modes produced by BB circuits. Depth ordering using such definition causes two problems. One problem is that there are useless modes from the perspective of both leakage saving and overhead energy compared to other modes. Such useless modes are not useful to minimize total energy consumption. Moreover, adding these modes into $\widehat{M}$ diminishes optimization efficiency because search space is extended. In ideal, the set comprised of

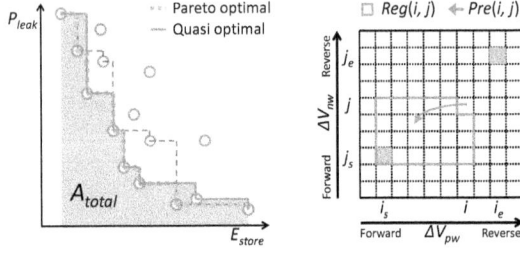

Figure 5: (Left) Cost for depth ordering. (Right) Internal step of Dynamic Programming.

Pareto-optimal solution for all producible modes indicated in the left of figure 5 is desirable for $\widehat{M}$. The other problem is that some mode-to-mode switchings consume not only the difference of two stored energies but also an extra energy described in section 2.1.B. If extra energy consumption is taken into account for, optimization problem will be complicated because a depth of each mode dynamically varies according to the previous mode. To overwhelm complication, we defined that $\widehat{M}$ never include a element which consume extra energy with other elements. However, there are many sets of depths which satisfy two conditions, Pareto-optimality and extra energy free. The left side of figure 5 indicates two sets with a solid line and a dashed line. It is difficult to judge which set achieves the best net leakage saving.

We introduce a heuristic to select the best set of depths. Our heuristic uses an area(A_{total}) constructed by elements of a set of depths as shown in the right side of figure 5. A set having minimum area is adopted as the best set of depths. The area minimization problem is defined as follows.

$$\text{minimize} \quad A_{total} = \sum_{i=0}^{m-1} P_{x_i y_i} \left(E_{x_{i+1} y_{i+1}} - E_{x_i y_i} \right)$$

subject to

$$x_{m-1} \leq x_{i+1} \leq x_i \leq x_0, \quad y_{m-1} \leq y_{i+1} \leq y_i \leq y_0,$$
$$P_{x_i y_i} > P_{x_{i+1} y_{i+1}}, \quad E_{x_i y_i} < E_{x_{i+1} y_{i+1}}$$

(x_0, y_0) and (x_{m-1}, y_{m-1}) represent the awake and the deepest sleep mode, respectively. m represents the number of elements of the set. To ensure Pareto-optimality, two elements have to satisfy $P_{x_i y_i} > P_{x_{i+1} y_{i+1}}$ and $E_{x_i y_i} < E_{x_{i+1} y_{i+1}}$.

We develop dynamic programming algorithm to solve above problem. The algorithm is represented as follows.

$$A(i,j) = \min_{k,l \in Reg(i,j)} \left\{ A(k,l) + \Delta A(k \to i, l \to j) \right\}$$

$$Reg(i,j) = \left\{ k,l \,\middle|\, \begin{array}{l} i_s \leq k \leq i_e, j_s \leq l \leq j_e, (k \neq i \text{ or } l \neq j), \\ P_{kl} > P_{ij}, \ E_{kl} < E_{ij} \end{array} \right\}$$

$$\Delta A(k \to i, l \to j) = P_{kl}(E_{ij} - E_{kl})$$

$$Pre(i,j) = \{(k,l) \,|\, \min\{A(i,j)\}\}$$

In this algorithm, the awake mode and the deepest sleep mode are given as the start-point (i_s, j_s) and the end-point (i_e, j_e). The algorithm solves the minimum area from the start-point to end-point by using other minimum areas which already be solved. Firstly, minimum area of the start-point is set as 0. Subsequently, we solve the minimum areas of each point (i,j) in $i_s \leq i \leq i_e$ and $j_s \leq j \leq j_e$. A partial area, $A(i,j)$, is calculated as below. For each point (k,l) included in $Reg(i,j)$, we calculate a summation of minimum area for (k,l) and incremental area constructed by (k,j) and

---

Algorithm : $SA\ (temp, \alpha, len, \beta, maxL, n, \nu, \widehat{M})$
1: $curS = Initial(n, \widehat{M})$
2: $curC = E_{total}(curS, \nu)$
3: $bestS = curS$; $bestC = curC$
4: **for** $i = 0$ to $maxL$ **do**
5: $\quad ML(temp, len, curS, curC, bestS, bestC, \nu, \widehat{M})$
6: $\quad temp *= \alpha$
7: $\quad len *= \beta$
8: **end for**
9: **return** $bestS$

Algorithm : $ML(temp, len, curS, curC, bestS, bestC, \nu, \widehat{M})$
1: **for** $i = 0$ to len **do**
2: $\quad$ **if** $RAND < 0.5$ **then**
3: $\quad\quad newS = NeighborT(curS, \widehat{M})$
4: $\quad$ **else**
5: $\quad\quad newS = NeighborD(curS, \widehat{M})$
6: $\quad$ **end if**
7: $\quad newC = E_{total}(newS, \nu)$
8: $\quad$ **if** $newC < curC$ **then**
9: $\quad\quad curS = newS$; $curC = newC$
10: $\quad\quad$ **if** $curC < bestC$ **then**
11: $\quad\quad\quad bestS = curS$; $bestC = curC$
12: $\quad\quad$ **end if**
13: $\quad$ **else if** $RAND < exp(-(newC - curC)/temp)$ **then**
14: $\quad\quad curS = newS$; $curC = newC$
15: $\quad$ **end if**
16: **end for**

Figure 6: Pseudo-code for optimization.

(i,j). $Reg(i,j)$ represents a set which includes points shallower than (i,j) and satisfying conditions, Pareto-optimality and extra energy free. The minimum value of those summations is adopted as the minimum area of (i,j). And the point in $Reg(i,j)$ which provides the minimum area is registered in $Pre(i,j)$. Consequently, we trace $Pre(i,j)$ from the end-point to the start-point. The set of traced points becomes quasi optimal depth ordered mode set.

3.3 Algorithm based on simulated annealing

We implement an algorithm based on simulated annealing [16] to solve the problem defined in section 3.1. The cost function is represented as equation 1. We define neighbor solutions as a solution whose depth or threshold of a stage is changed from current one. Changing depth or threshold is chosen in random manner.

Figure 6 indicates the pseudo code of our algorithm. Function SA schedules cooling and call a sub-routine (ML) which searches solutions in the same temperature. Inputs of SA are as follows, $temp$: initial temperature, α : scheduling factor for temperature, len : initial number of attempt in the same temperature, β : scheduling factor for attempt number, $maxL$: the maximum number of scheduling, n : the number of stage, ν : idle interval distribution, $\widehat{M}$: ordered mode set obtained in section 3.2. Output is that, $bestS$: the best solution in the searching. Firstly, SA generates an initial solution. Its number of stages is n. Depths and thresholds for each stage are randomly chosen with satisfying constraints. Secondly, repeats calling ML and cooling $maxL$ times. Consequently, outputs the best solution which has ever found as the optimal solution.

Inputs of ML are as follows, $temp$: temperature, len : number of attempts, $curS$: current solution, $curC$: cost of current solution, $bestS$: best solution in the past search,

$bestC$: cost of the best solution, ν : idle interval distribution, $\widehat{M}$: meaningful mode set. $curS$, $curC$, $bestS$, $bestC$ are also used as outputs. ML repeats following tasks len times. ML generates a neighbor solution from current solution (line.2~6). If the cost of neighbor solution is better than that of current one, do update process (line.8~12). Even the cost is worse, the neighbor solution is accepted according to the acceptance probability function with current temperature (line.12~15).

4. EVALUATION

A. Leakage power, stored energy, switch box energy and idle interval analyses

We designed physical layout of 32-bit parallel prefixed adder with 90nm process technology to obtain leakage powers and stored energies for our evaluation. We defined that ΔV_{PW} and ΔV_{NW} vary from 0.5[V] to -2.0[V] with 0.1[V] as a step. Thus, the total number of unordered modes are $26 \times 26 = 676$. We calculated leakage powers and stored energies of each unordered mode under various temperatures (75, 100, 125°C). Figure 7 indicates these values at 75°C. Each value is normalized to the value of the zero body biased mode (ZBB). In general, leakage power is increased by larger forward biasing and decreased by larger reverse biasing. However, larger reverse biasing for P-well also increases leakage power due to increasing of BTBT leakage power. On the other hand stored energy monotonically increases from larger forward biasing to larger reverse biasing.

For depth ordering, we determined the active mode and the deepest sleep mode. The active mode is selected as the mode which has the smallest stored energy among modes whose leakage power are less than three times of that of ZBB. The deepest sleep mode is selected as the mode which has the smallest leakage power among all unordered modes.

Transistor widths of switch boxes are determined as sufficient width to switch between the active and the deepest sleep modes within one cycle. We calculated the switch box energy from the gate capacitances of these transistors.

In addition, we obtained distributions of idle intervals and proportions of total idle times to the total execution time for three floating point functional units, ALU, multiplier and divider (FPAlu, FPMul, FPDiv) by an architectural simulator, Onikiri 2 [17], with 12 and 17 applications chosen from SPEC 2006 CINT and CFP benchmark suits. The configuration of the simulator is shown in table 1.

Note that circuit sizes of 32bit adder and target functional units are different. However, leakage power and stored energy are proportional to the circuit area ($\fallingdotseq$ size) and transistor widths of switch boxes are proportional to the stored energy with first-order approximation. Therefore, ratio of leakage energy, stored energy and switch box energy is constant regardless of the size at the same frequency.

Table 1: Configuration of architectural simulator.

ISA, Pipeline	Alpha64, Out of order, 4way fetch
Functional unit	IntAlu×2, IntMul×1, IntDiv×1, FPAlu×1, FPMul×1, FPDiv×1, Address×2, Load/Store×2
L1 I/D cache	32KB, 4way, 3cycle latency
L2 cache	1MB, 8way, 15cycle latency
Main memory	200cycle latency

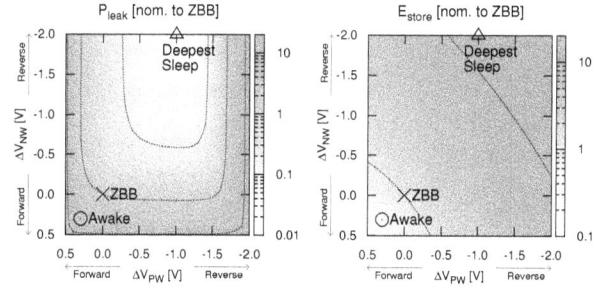

Figure 7: Leakage power and stored energy at 75°C.

B. Extraction of depth ordered mode set

Table 2 indicates the summary of depth ordered mode set produced for each temperature. A_{DP} and A_{Ideal} represent the costs of the mode set produced by our algorithm and the mode set comprised of Pareto-optimal solution, respectively. The errors of A_{DP} compared to A_{Ideal} are very small. Therefore, optimization using our ordered mode set is expected to achieve as much net leakage saving as the optimization using unordered mode set. In addition, our algorithm reduces the number of modes before optimization to get good solution with less searching time. Our algorithm achieves approximately 95% reduction of modes.

C. Total leakage energy consumption

Figure 8 indicates total leakage energy consumption for FPAlu at 1.0GHz, 75°C with various sleep control schemes. $Deep$ and $minBET$ represent circuits having only one sleep mode. Deep has the deepest sleep mode and minBET has the shallowest sleep mode which provides minimum BET which is time to compensate overhead energy [18]. These schemes adopt time-out scheme to enter sleep mode. Their thresholds are adjusted to maximize net leakage saving for each application. Each $N=x$ represents stepwise sleep depth control scheme and x represents the number of stages. Their depths and thresholds have been optimized. Act represents leakage energy of active operation cycles.

SPEC 2006 benchmark suits are divided into CINT and CFP which mainly composed of integer operations and floating point operations, respectively. In the case of integer benchmarks, most of applications have low usage frequency of FPAlu and long idle intervals frequently appear. Thus, conventional deepest sleep are effective but shallowest sleep are not effective due to its less leakage saving efficiency. On the other hand, in the case of floating point benchmarks, usage frequencies of FPAlu are high and short idle intervals frequently appear. Thus, conventional deepest sleep is inefficient for some applications and shallowest sleep achieves more leakage saving than deepest sleep for some applications, such as bwaves, milc and poray.

Proposed stepwise sleep depth control scheme achieves further leakage saving compared to conventional deepest sleep and shallowest sleep for all applications. Additional leakage saving efficiency varies with applications. For integer benchmarks, conventional deepest sleep usually achieves sufficient leakage saving. However, our scheme achieves further saving

Table 2: Summary of depth ordering.

Temperature	[°C]	75	100	125
$A_{DP}/A_{Ideal} - 1.0$	[%]	0.48	1.26	1.25
# of ordered modes		36	36	37
red. of modes	[%]	94.76	94.76	94.52

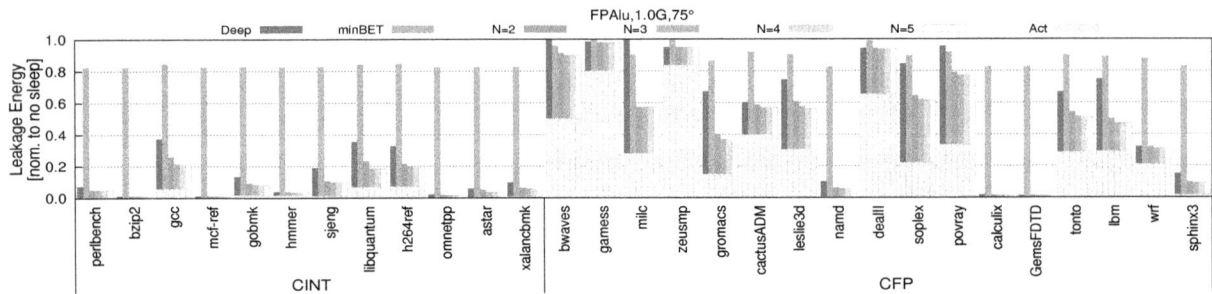

Figure 8: Total leakage energy consumption of FPAlu at 1.0GHz, 75°C.

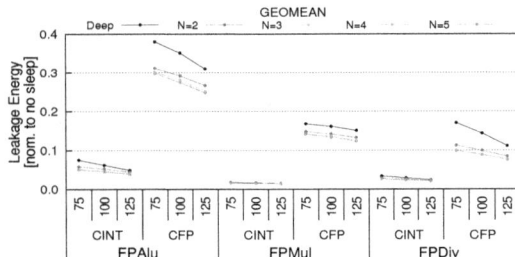

Figure 9: Geometric means of total leakage consumption for each functional unit.

by exploiting short idle intervals in some applications, like gcc, sjeng, libquantum and h264ref. In contrast, for floating point benchmarks, our scheme achieves further saving than deepest sleep and shallowest sleep in most applications. In particular, for milc, deepest sleep scheme is totally incapable of leakage saving, but our scheme achieves 43% net saving.

As with Deep and minBET, N=2 utilizes only one sleep mode at a time. Comparison among Deep, minBET and N=2 indicates that adjusting sleep depth according to application has a major impact on leakage saving efficiency. Meanwhile, increasing the number of stages provides small improvement for leakage saving efficiency and improvement ratio diminishes with each increment.

Figure 9 indicates geometric means of leakage energy consumption for various functional units and temperatures. Integer applications does not use each functional unit so frequently. Hence, achievements of our scheme for each functional unit are rarely different from that of conventional deep sleep scheme. Proposed scheme achieves a geometric mean of 8% further leakage saving than conventional deep sleep scheme for FPAlu at 75°C. Our scheme achieves further leakage saving in lower temperatures because BETs for each mode become longer and leakage saving chances for conventional deep sleep decrease.

5. CONCLUSION

In this paper we proposed a novel sleep control scheme called stepwise sleep control scheme for circuits employing multiple sleep modes. Our scheme applies deeper sleep mode in a step-by-step manner according to elapsed time since idle state started. In run-time, it achieves efficient net leakage saving including overhead energy by applying appropriate sleep depth for various idle intervals.

Leakage saving efficiency of proposed sleep scheme is significantly changed by adjusting sleep depth and starting time of each step. To maximize leakage saving, we proposed a methodology for optimizing depths and starting times of

steps by adjusting depths and starting times according to run-time factors, temperature and idle interval distribution.

We proposed an implementation of body biasing circuit applying stepwise sleep control scheme. Evaluation result shows that our sleep control scheme improves net leakage saving of up to 43% with a geometric mean of 8% within floating point benchmarks for FPAlu at 1.0GHz, 75°C compared to conventional deepest sleep scheme.

6. ACKNOWLEDGMENTS

This work is supported by the Japan Science and Technology Agency (JST) as a CREST Research program, VLSI Design and Education Center (VDEC) and Synopsys, Inc.

7. REFERENCES

[1] T. Kuroda, et al., "A 0.9-V, 150-MHz, 10-mW, 4mm², 2-D Discrete Cosine Transform Core Processor with Variable Threshold-Voltage (VT) Scheme," JSSC, vol.31, no.11, pp.1770–1779, 1996.

[2] S. Mutoh, et al., "1-V Power Supply High-Speed Digital Circuit Technology with Multithreshold-Voltage CMOS," JSSC, vol.30, no.8, pp.847–854, Aug. 1995.

[3] R. Kumar, G. Hinton, "A family of 45nm IA processors," ISSCC, pp.58–59, 2009.

[4] Y. Kanno, et al., "Hierarchical Power Distribution with 20 Power Domains in 90-nm Low-Power Multi-CPU Processor," ISSCC, pp.2200–2209, 2006.

[5] J. Tschanz, et al., "Dynamic Sleep Transistor and Body Bias for Active Leakage Power Control of Microprocessors," JSSC, vol.38, no.11, pp.1838–1845, 2003.

[6] Z. Hu, et al., "Microarchitectural Techniques for Power Gating of Execution Units," ISLPED, pp.32–37, 2004.

[7] K. Usami, N. Ohkubo, "A Design Approach for Fine-grained Run-Time Power Gating using Locally Extracted Sleep Signals," ICCD, pp.155–161, 2006.

[8] K. Agarwal, et al., "Power Gating with Multiple Sleep Modes," ISQED, pp.633–637, 2006.

[9] E. Pakbaznia, M. Pedram, "Design and Application of Multimodal Power Gating Structures," ISQED, pp.120–126, 2009.

[10] H. Xu, et al., "Novel Vth Hopping Techniques for Aggressive Runtime Leakage Control," VLSID, pp.51–56, 2010.

[11] Z. Zhang, et al., "A Robust and Reconfigurable Multi-mode Power Gating Architecture," VLSID, pp.280–285, 2011.

[12] S. Takeda, et al., "Efficient Leakage Power Saving by Sleep Depth Controlling for Multi-mode Power Gating," ISQED, 2012.

[13] A. Lungu, et al., "Dynamic Power Gating with Quality Guarantees," ISLPED, pp.377–382, 2009.

[14] H. Xu, et al., "Stretching the Limit of Microarchitectural Level Leakage Control with Adaptive Light-Weight Vth Hopping," ICCAD, pp.632–636, 2010.

[15] Y. Taur, T. H. Ning, Fundamentals of modern VLSI devices, Cambridge University Press, 1998.

[16] S. M. Sait, H. Youssef, Iterative Computer Algorithms with Applications in Engineering, IEEE Computer Society Press, 1999.

[17] Pipeline simulator onikiri 2, http://www.mtl.t.u-tokyo.ac.jp/~onikiri2/.

[18] H. Xu, et al., "Run-time Active Leakage Reduction by Power Gating and Reverse Body Biasing: An Energy View," ICCD, pp.618–625, 2008.

An Efficient Power Estimation Methodology for Complex RISC Processor-based Platforms

Santhosh Kumar
Rethinagiri
Inria Lille - Nord Europe
59650 Villeneuve d'Ascq -
France
santhosh-
kumar.rethinagiri@inria.fr

Rabie Ben Atitallah
Université de Valenciennes et
du Hainaut Cambrésis,
Valenciennes
59313 Valenciennes - France
rabie.benatitallah@univ-
valenciennes.fr

Jean-Luc Dekeyser
Laboratoire d'Informatique
Fondamentale de Lille (LIFL)
Université de Lille 1
59655 Villeneuve d'Ascq
Cédex - France
dekeyser@lifl.fr

Smail Niar
Université de Valenciennes et
du Hainaut Cambrésis,
Valenciennes
59313 Valenciennes - France
smail.niar@univ-
valenciennes.fr

Eric Senn
Université de Bretagne Sud
Centre de Recherche
56321 LORIENT Cédex -
France
eric.senn@univ-ubs.fr

ABSTRACT

In this contribution, we propose an efficient power estimation methodology for complex RISC processor-based platforms. In this methodology, the Functional Level Power Analysis (FLPA) is used to set up generic power models for the different parts of the system. Then, a simulation framework based on virtual platform is developed to evaluate accurately the activities used in the related power models. The combination of the two parts above leads to a heterogeneous power estimation that gives a better trade-off between accuracy and speed. The usefulness and effectiveness of our proposed methodology is validated through ARM9 and ARM CortexA8 processor designed respectively around the OMAP5912 and OMAP3530 boards. This efficiency and the accuracy of our proposed methodology is evaluated by using a variety of basic programs to complete media benchmarks. Estimated power values are compared to real board measurements for the both ARM940T and ARM CortexA8 architectures. Our obtained power estimation results provide less than 3% of error for ARM940T processor, 3.5% for ARM CortexA8 processor-based system and 1x faster compared to the state-of-the-art power estimation tools.

Categories and Subject Descriptors

I.6 [**Simulation and modeling**]: Model validation, analysis

General Terms

System Level Power Estimation

Keywords

Power Estimation, Virtual Platform, SystemC, TLM, FLPA, ASIC

1. INTRODUCTION

Today's embedded industries focus more on manufacturing RISC processor-based platform as they are cost and power effective. On the other side, modern embedded applications are becoming more and more sophisticated and resource demanding. Examples of the concerned applications are numerous such as software defined radio, GPS, mobile applications, etc. The computation requirements of such systems are very important in order to meet real-time constraints and high quality of services. At the same time, the recent advances in silicon technologies offer a tremendous number of transistors integrated on a single chip. For this reason, embedded hardware designers are directed more and more towards complex RISC architectures, which may contain several pipeline slots, hierarchical memory system (L1 and L2 cache level), and specific execution units such as NEON architecture for ARM CortexA8 processor as a promising solution to deal with the potential parallelism inherent from modern applications. Recently, the ITRS [8] and HiPEAC [1] roadmaps promote *power defines performance* and *power is the wall*. In fact, power consumption is becoming a critical pre-design metric in complex embedded systems. An efficient and fast design space exploration (DSE) of such systems needs a set of tools capable of estimating performance and power at higher abstraction level in the design flow. Today, virtual platform power estimation is considered as an important hypothesis to cope with the critical design constraints. However, the development of virtual platform tools for power estimation and optimization

---

[1] http://www.hipeac.net/system/files/hipeacvision.pdf

is in the face of extremely challenging requirements such as the seamless power-aware design methodology.

At the virtual platform level, the power estimation process is centred around two correlated aspects: *the power model granularity* and *the system abstraction level*. The first aspect concerns the granularity of the relevant activities on which the power model relies. It covers a large spectrum that starts from the fine-grain level such as the logic gate switching and stretches out to the coarse-grain level like the hardware component events. In general, fine-grain power estimation yields to a more correlated model with data and to handle technological parameters, which is tedious for virtual platform designers. On the other hand, coarse-grain power models depend on micro-architectural activities that cannot be determined easily. The second aspect involves the abstraction level on which the system is described. It starts from the usual Register Transfer Level (RTL) and extends up to the algorithmic level. In general, going from low to high design level corresponds to more abstract description and then coarser activity granularity. The power evaluation time increases as we go down through the design flow and the accuracy depends on the extraction of each relevant activity and the characterization methodology to evaluate the related power cost. In order to have an efficient power estimation methodology, we should find a better trade-off between these two aspects.

To answer the above challenges, we propose an efficient power estimation methodology for consumption estimation of complex RISC processor-based systems. The idea here is to develop a power estimation virtual platform, which combines Functional Level Power Analysis (FLPA) for hardware power modeling and a system-level simulation technique for rapid prototyping and fast power estimation. The functional power estimation part is coupled with a OVPSim [2] simulator in order to obtain the needed functional-unit activities for the power models, which allows us to reach a superior bargain between accuracy and speed.

This paper is organized as follows. After Section 2 which, presents the related works, Section 3 exposes the proposed power estimation methodology. In Section 4, the power modeling methodology is applied to 2 complex RISC processors designed around OMAP5912 and OMAP3530 boards. To evaluate our methodology in terms of accuracy and speed, experimental results are presented in Section 5.

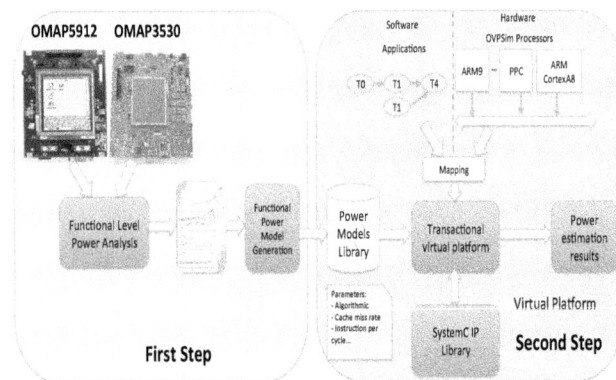

Figure 1: Hybrid power estimation methodology

[2]http://www.ovpworld.org/

2. RELATED WORKS

In order to make a better trade-off between power estimation time and accuracy, several studies have proposed evaluating system power consumption at higher abstraction levels. Almost of these tools use a micro-architectural simulators to evaluate system performance and with the help of analytic power models to estimate consumption for each component of the platform. Wattch [3], SimplePower [17] and Skyeye [4] are example of available tools. In general, these tools rely on Cycle-Accurate (CA) simulation technique. Usually, to move from the RTL to the CA level, hardware implementation details are hidden from the processing part of the system, while preserving system behavior at the clock cycle level. The power consumption of the main internal units is estimated using power macro-models, produced from lower-level characterizations. The contributions of the unit activities are calculated and added together during the execution of the program on the cycle-accurate micro-architectural simulator. Though using CA simulators has allowed accurate power estimation, evaluation and simulation time are very significant for the off-the-shelf processor.

In an attempt to reduce simulation time, recent efforts have been done to build up fast simulators using *Transaction Level Modeling* (TLM) [2]. SystemC [14] and its TLM 2.0 kit have become a de facto standard for the system-level description of Systems-on-Chip (SoC) by the means of offering different coding styles. Nevertheless, power estimation at the TLM level is still under research and is not well established. In [12] and [13], a methodology is presented to generate consumption models for peripheral devices at the TLM level. Relevant activities are identified at different levels and granularities. The characterization phase is however done at the gate level: from where they deduce the activity and power consumption for the higher level. Using this approach for recent processors and systems is not realistic. Dhawada et al. [5] proposed a power estimation methodology for a monoprocessor PowerPC and CoreConnect-based system at the TLM level. Their power modeling methodology is based on a fine-grain activity characterization at the gate level, which needs a huge amount of development time. Due to a high correlation with data, a power estimation inaccuracy of 11% is achieved. Compared to the previous works, our proposed methodology for power estimation also partially uses SystemC/TLM simulation with coarse grain power models. Today, the Open Virtual Platform by Imperas Inc. [1] uses the same level of simulation but also tackles the simulation speed problem by proposing the OVPSim simulator which is very fast since processors are not ISS but use code morphing and just-in-time (JIT) compilation. This technique will be also used in our framework.

For the functional level, Tiwari et al. [16] have introduced the concept of Instruction Level Power Analysis (ILPA). They associate a power consumption model with instructions or instruction pairs. The power consumed by a program running on the processor can be estimated using an Instruction Set Simulator (ISS) to extract instruction traces, and then adding up the total cost of the instructions. This approach suffers from the high number of experiments required to obtain the power model. In addition, significant effort is required to obtain the ISS of the target processor. To overcome this drawback the *Functional Level Power Analysis* (FLPA) was proposed [10] [9], which relies on the identification of a set of functional blocks that influence the power consumption of the target component. The model is represented by a set of analytical functions or a table of consumption values which depend on functional and architectural parameters. Once the model is build, the estimation process consists of extracting the appropriate parameter values from the design, which will be injected into the model to compute the power consumption. Based on this methodology, the tool SoftExplorer [6] has been developed and included in the recent toolbox CAT [15]. It includes a library of power models for simple to complex processors. Only a static analysis of the code, or a rapid profiling is necessary to determine the input parameters for the power models. However, when complex hardware or software components are involved, some parameters may be difficult to determine with precision. This lack of precision may have a non-negligible impact on the final estimation accuracy. In order to refine the value of sensible parameters with a reasonable delay, we propose to couple the OVPSim simulator with the functional level power models which offers us the reasonable trade-off between estimation speed and accuracy.

3. THE HYBRID POWER ESTIMATION METHODOLOGY

This section exposes our proposed power estimation methodology that is divided into two steps as shown in Fig. 1. The **first step** concerns the power model elaboration for the system hardware components. In our framework, the FLPA methodology is used to develop generic power models for different target platforms. The main advantage of this methodology is to obtain power models which rely on the functional parameters of the system with a reduced number of experiments. As explained in the previous section, FLPA comes with few consumption laws, which are associated to the consumption activity values of the main functional blocks of the system. The generated power models have been adapted to the system level, as the required activities can be obtained from the virtual platform. For a given platform, the generation of power models is done at once. To do so, the first part is to divide the architecture of the corresponding processor into different functional blocks and then to cluster the components that are concurrently activated when the code is running.

There are two types of parameters: *algorithmic parameters* that depend on the executed algorithm typically the cache miss or instruction per cycle rates and *architectural parameters* that depend on the component configuration set by the designer typically the clock frequency. For instance, Table 1 presents the common set of parameters of our generic power model. These sets of parameters are defined for a general class of RISC processors. Additional parameters can be identified for specific processors based-architecture such as Superscaler. The next step is the characterization of the embedded system power consumption when the parameters vary. These variations are obtained by using some elementary assembly programs (called scenario) or built-in test vectors elaborated to stimulate each block separately. In our work, characterization is performed by measurements on real boards. Finally, a curve fitting of the graphical representation will allow us to determine the power consumption models by regression. The analytical form or a table of values expresses the obtained power models. This power modeling approach was proven to be fast and precise.

The **second step** of the methodology defines the architecture of our power estimator that includes the *functional power estimator* and *fast Just In Time (JIT) compilation simulator* as shown in Fig.1. The functional power estimator evaluates the consumption of the target system with the help of the elaborated power models from the first step. It takes into account the architectural parameters (e.g. the frequency, the processor cache configuration, etc.) and the application mapping. It also requires the different activity values on which the power models rely. In order to collect accurately the needed activity values, the functional power estimator communicates with a fast JIT OVPsim at the TLM level. The combination of the two components above described at different abstraction levels (functional and TLM) leads to a hybrid power estimation that gives a better trade-off between accuracy and speed.

The vital function of the this power estimation methodology is to offer a detailed power analysis by the means of a complete simulation of the application. This process is initiated by the functional power estimator through *mapping interface* (Fig. 2). In this way, the mapping information is transmitted to the fast OVPSim simulator. Our simulator consists of several hardware components which are instantiated from the Open Virtual Platform (OVP) [7] library in order to build a virtual prototype of the target system. We highlight that processors are described using JIT simulator that works by coding morphing concept and is instruction accurate.

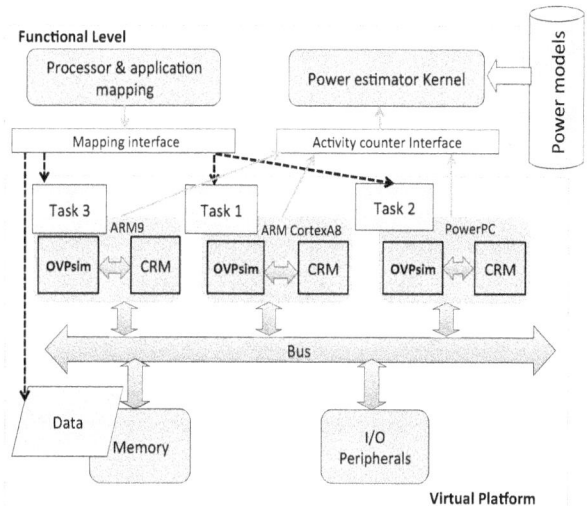

Figure 2: Power estimation methodology functioning

In the power estimation step, the simulator collects the activities that are influenced by the application and the input data. At the end of the simulation, the values of the activities are transmitted to the power consumption models or power estimator kernal using *the activity counter interface* in order to calculate the global power consumption as illustrated in Fig. 2. As we have stated before, the following section will discuss the **first step**; the elaboration of the power model for the OMAP5912 and OMAP 3530 platform by using FLPA methodology.

Table 1: Generic power model parameters

	Name	Description
Algorithmic	τ	External memory access rate
	γ	Cache miss rate
	IPC	Instruction per cycle rate
Architectural	$F_{processor}$	Frequency of the processor
	F_{bus}	Frequency of the bus

Table 2: Generic power models for different processors

Processor	Power models
ARM9	P(mW)=1.03 $F_{Processor}$ + 0.6 (γ) + 5.3
ARM CortexA8	P(mW)=0.79 $F_{Processor}$ + 18.65 IPC +0.26 ($\gamma 1 + \gamma 2$) + 10.13
PowerPC	P(mW)=4.1 (γ) + 6.3 F_{bus} + 1599

4. POWER MODEL ELABORATION

In order to prove the usefulness and the effectiveness of the proposed power estimation methodology, we used an ARM9 architecture implemented into the OMAP5912 and an ARM CortexA8 -based architectures implemented into the OMAP 3530 platform. The OMAP5912 contains an ARM926EJ-S processor (16KB instruction cache and 8KB data cache). The OMAP3530 contains an ARM Cortex A8 processor (16KB, 2-way set associative instruction and data caches and 256KB L2 cache). Each processor has access to the off-chip memory (SDRAM) via the processor bus interconnect. As explained above, we used the FLPA methodology to generate a power model for each target system. As a first step, we divided the architecture into different functional blocks such as the core clock system, the memory system, and the functional unit for ARM CortexA8 processor as shown in the Fig. 3. A parameter is denoted for each functional block such as $\gamma 1$ and $\gamma 2$ respectively for L1 cache miss rate and L2 cache miss rate. The second step is the characterization of the power model by varying the parameters. These variations are obtained by using some elementary assembly programs (called scenario) or built in test vectors elaborated to stimulate each block separately. In our work, characterization is performed by measurements on real boards. Finally, a curve fitting of the graphical representation will allow us to determine the power consumption laws by regression. The analytical form or a table of values expresses the obtained power laws. This power modeling approach was proven to be fast and precise.

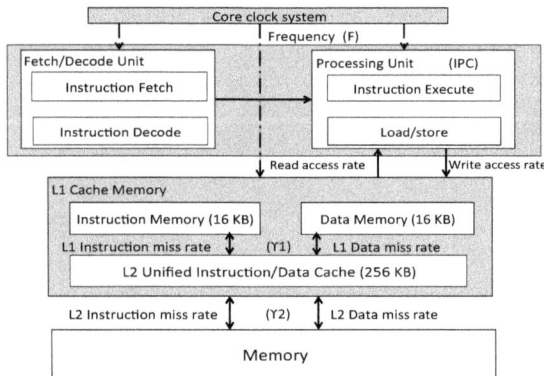

Figure 3: Main functional blocks of ARM CortexA8 processor

Table 2 shows the power consumption models for the ARM9, ARM CortexA8, and PowerPC405 by using FLPA methodology. The input parameters on which the power models rely are the frequency of the processor ($F_{processor}$(MHz)), Instruction Per Cycle (IPC), and the cache miss rate ($0 < \gamma < 100$ (%)). The system designer

chooses the frequency of the processor and the bus while the cache miss rate and the IPC are considered as an activity of the processor, which could be extracted from the simulation environment.

5. SYSTEM LEVEL POWER ESTIMATION RESULTS

5.1 Power estimation accuracy

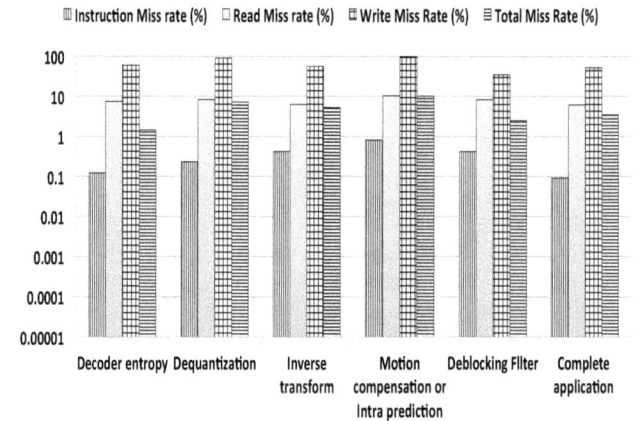

Figure 4: H264 application cache miss rates for ARM9 processor

For the **second step** of our power estimation methodology, a virtual platform prototype of an ARM9 and an ARM CortexA8 based architecture has been developed. This prototype uses different virtual hardware models, a cache ratio monitor (CRM) provided with the virtual platform for cache miss rate, and the JIT for the target processor as shown in the Fig. 2. Furthermore, the cache parameters and the bus latencies are set to emulate the real platform behaviour. From the CRM, we are able to determine the occurrences of the main activities. For the ARM9 processor the following counters are used for different cache miss rates: read data miss, write data miss and read instruction miss. As a main application, we used the H.264/AVC baseline profile decoder that supports intra and inter-coding, and entropy coding with Context-Adaptive Variable-Length Coding (CAVLC) as a benchmark for ARM9 processor. The H264 decoder application consists of 5 main tasks: decoder entropy, dequantization, inverse transform, motion compensation or intra prediction and deblocking filter.

Fig. 4 shows the detailed results of the activities fetched by the fast JIT-SystemC simulator for each task of the H264 application for ARM9 processor. From these results several remarks can be drawn. First, we can notice that instruction cache miss rates and read data miss rates are very low when compared with write data miss rates. This is due to the reduced task kernel and data pattern sizes that are very low compared to the cache size (16 KB), which decreases the access to the external memory and thus having a minimal effect on the dynamic power consumption. Second, the data write miss rates have a high impact on the total power consumption of the system. This is because of the algorithm's structure, which does not favour the reuse of data output arrays and the usage of cache policy. Therefore, the statistics collected in Fig. 4 could help in tuning the application structure for a better optimization of the system power consumption. In a similar fashion, we extracted the activities for the ARM CortexA8 processor.

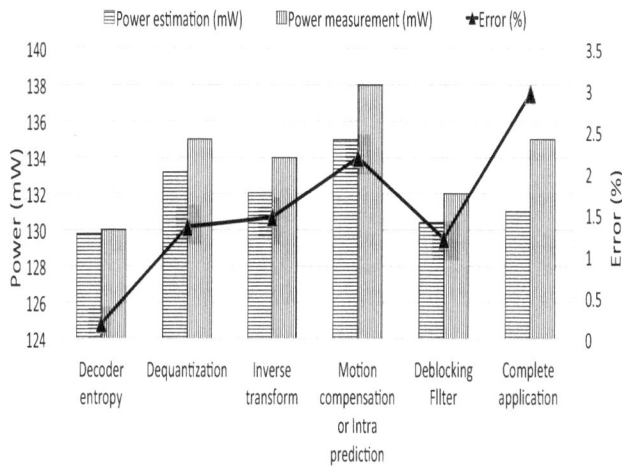

Figure 6: Power estimation accuracy vs real board measurement (ARM9 at 120 MHz)

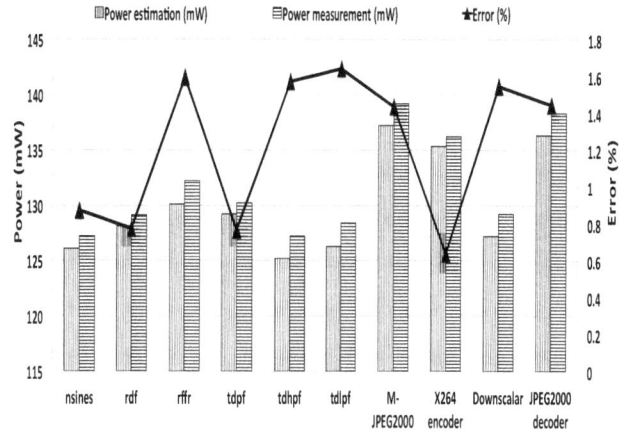

Figure 5: Power estimation accuracy for the H264 application (ARM9 at 120 MHz)

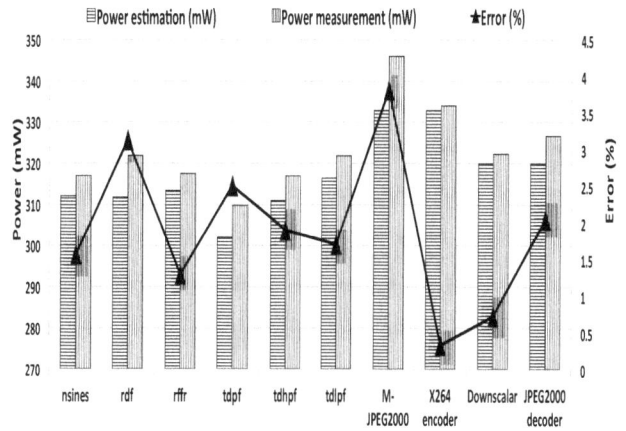

Figure 7: Power estimation accuracy vs real board measurement (ARM CortexA8 at 500 MHz)

In the next step, we estimated the total power consumption of each task using the power models shown in Table 2. Fig. 5 illustrates the results and shows the comparison between the proposed methodology and the real board measurements. First, our power estimator has a negligible average error of 2%. This study offers a detailed power analysis for each task in order to help designers to detect peaks of consumption and thus to propose efficient mapping or optimization techniques. In order to evaluate the accuracy of our methodology, we carried out power estimation on several image & signal processing benchmarks. Fig. 6 and Fig. 7 illustrates the power results by showing the estimation accuracy between the proposed power estimation methodology and the real board measurements. Our proposed methodology has an negligible average error of 1.24% and 2.4% respectively for ARM9 and ARM Cortex A8 processor, which is considered as a high accuracy level when compared to SoftExplorer's [11] average error of 8% for RISC processors.

5.2 Power estimation speed

In this section, we will compare the efficiency of the proposed methodology in term of estimation speed with the SimplePower(Cycle-Accurate), TLM with ISS based simulation and SoftExplorer (functional level simulator) ap-

proaches as introduced in Section 2. This comparison is for the quantification of our proposed methodology to the state-of-art power estimation tools used in current industrial and academic practices. SoftExplorer, TLM with based simulation, and our proposed methodology are executed on a PC (Intel, 1.8 GHz, 2 Go of RAM), whereas SimplePower on a Workstation (Ultra Sparc T2+, 1.6 GHz, 2 Go of RAM). In order to compare the result, computer benchmarking has been done to confirm that the workstation is always faster compared to the PC for all kind of applications. Power estimation has been carried out with a set of image & signal processing applications and also with SPEC 2008 [3] benchmarks.

From Fig. 8, we can notice that SoftExplorer and TLM with ISS based simulation have an average estimation time of 5 seconds, which is faster when compared to the Simple-Power's average estimation time of 20 seconds. Our proposed methodology has a average estimation time of 2.45 seconds, which is faster compared to the other tools. Our methodology works by running the application on the virtual

---

[3]http://http://www.spec.org/benchmarks.htmlpower/

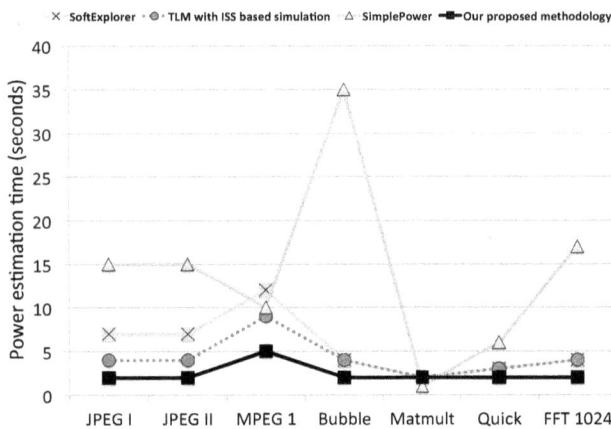

Figure 8: Comparison of estimation time for our proposed methodology, SimplePower, TLM with ISS based simulation and SoftExplorer tools

platform thereby collecting the dynamic activities. Simple-Power uses cycle accurate specifications to collect the necessary power data, whereas SoftExplorer realizes a static profiling of the code, which results in reduced execution time and thus resulting in a low power consumption estimation time. Static profiling of the C code is not sufficient to determine the average execution time and the global energy consumption, for this reason we need to run the application on the virtual platform in order to collect the activities accurately and efficiently. Experimental results prove that our proposed methodology is efficient, fast and accurate.

6. CONCLUSIONS

This paper presents a efficient system level power estimation methodology for RISC processor-based embedded system. Indeed, power/energy constraints are considered as a major challenge when the system runs on batteries. Thus, designers must take these constraints into account as early as possible in the design flow. First, a power modeling methodology has been defined to address the global system consumption that includes core clock system, memory, etc. Secondly, the functional power modeling part is coupled with a fast virtual platform to obtain the needed micro-architectural activities for the power models, which allows us to reach accurate estimates. With such proposed methodology, the designer can explore several implementation choices on different processors. The future works of this project will focus on more complex heterogeneous platforms. Furthermore, in order to obtain more accurate power estimations, some power model refinements must be realized.

7. REFERENCES

[1] B. Bailey. System level virtual prototyping becomes a reality with OVP donation from imperas. Technical report, EDA, June 2008.

[2] G. Beltrame, L. Fossati, and D. Sciuto. ReSP: A Nonintrusive Transaction-Level Reflective MPSoC Simulation Platform for Design Space Exploration. *Computer-Aided Design of Integrated Circuits and Systems, IEEE Transactions on*, 28(12):1857–1869, Dec. 2009.

[3] D. Brooks, V. Tiwari, and M. Martonosi. Wattch: A framework for architectural-level power analysis and optimizations. In *Proc. International Symposium on Computer Architecture ISCA'00*, pages 83–94, 2000.

[4] Y. Chen. *The Analysis and Practice on Open Source Embedded System Software–Based on SkyEye and ARM Developing Platform*. Beihang University Press, 2004.

[5] N. Dhanwada, R. A. Bergamaschi, W. W. Dungan, I. Nair, P. Gramann, W. E. Dougherty, and I.-C. Lin. Transaction-level modeling for architectural and power analysis of powerpc and coreconnect-based systems.

[6] S. Dhouib, J.-P. Diguet, D. Blouin, and J. Laurent. Energy and power consumption estimation for embedded applications and operating systems. *Journal of Low Power Electronics (JOLPE)*, 5(3), 2009.

[7] Imperas Inc. OVP World home page. http://www.ovpworld.org/.

[8] ITRS. Design, 2010 edition. http://public.itrs.net/, 2010.

[9] J. Laurent, N. Julien, and E. Martin. High level energy estimation for DSP systems. In *Proc. Int. Workshop on Power And Timing Modeling, Optimization and Simulation PATMOS'01*, pages 311–316, 2001.

[10] J. Laurent, N. Julien, and E. Martin. Functional level power analysis: An efficient approach for modeling the power consumption of complex processors. In *Proceedings of the Design, Automation and Test in Europe Conference*, Munich, 2004.

[11] J. Laurent, N. Julien, and E. Martin. Softexplorer: estimation, characterization and optimization of the power and energy consumption at the algorithmic level. In *Fourteenth International Workshop on Power and Timing Modeling (PATMOS 2004)*, pages 15–17, Santorini, Greece, September 2004.

[12] I. Lee, H. Kim, P. Yang, S. Yoo, E. Chung, K.Choi, J.Kong, and S.Eo. Powervip: Soc power estimation framework at transaction level. In *Proc. ASP-DAC*, 2006.

[13] N.Dhanwada, I. Lin, and V.Narayanan. A power estimation methodology for systemc transaction level models. In *International conference on Hardware/software codesign and system synthesis*, 2005.

[14] Open SystemC Initiative. Systemc, 2008. World Wide Web document, URL: http://www.systemc.org/.

[15] J. D. S. Douhib. Model driven high-level power estimation of embedded operating systems communication and synchronization services. In *Proceedings of the 6th IEEE International Conference on Embedded Software and Systems*, China, May 25-27 2009.

[16] V. Tiwari, S. Malik, and A. Wolfe. Power analysis of embedded software: A first step towards software power minimization. In *Transactions on VLSI Systems*, 1994.

[17] W. Ye, N. Vijaykrishnam, M. Kandemir, and M. Irwin. The design and use of simplepower: a cycle accurate energy estimation tool. In *Proc. Design Automation Conference DAC'00*, June 2000.

ADAM: An Efficient Data Management Mechanism for Hybrid High and Ultra-Low Voltage Operation Caches

Bojan Maric[1,2], Jaume Abella[1], Mateo Valero[1,2]
[1]Barcelona Supercomputing Center (BSC-CNS)
[2]Universitat Politecnica de Catalunya (UPC)
{bojan.maric, jaume.abella, mateo.valero}@bsc.es

ABSTRACT

Semiconductor technology evolution enables the design of ultra-low-cost chips (e.g., below 1 USD) required for new market segments such as environment, urban life and body monitoring, etc. Recently, hybrid-operation (high Vcc, ultra-low Vcc) single-Vcc-domain cache designs have been proposed to tackle the needs of those chips. However, existing data management policies are far from being optimal during high Vcc operation.

This paper presents ADAM, a new and extremely simple Adaptive Data Management mechanism, which is tailored to detect hit distribution and changing application conditions dynamically at fine grain with negligible hardware overhead. ADAM is proven to save significant energy (29% on average) in L1 caches with negligible performance degradation (1.7% on average), thus improving the energy-delay product (EDP) noticeably across different cache configurations with respect to *all* existing data management approaches.

Categories and Subject Descriptors

B.3.1 [**Memory structures**]: Semiconductor Memories—*Static memory (SRAM)*; B.3.2 [**Memory structures**]: Design styles—*Cache memories*

General Terms

Design, Performance

Keywords

Cache Memories, High Performance, Low Energy

1. INTRODUCTION

Higher semiconductor technology integration due to geometry scaling opens the door to new market segments. In particular, technology evolution enables adding some degree of intelligence to any control or measuring engine such as environment sensor applications to monitor wind, sea level, temperature, tsunamis, the body, etc., by means of battery-powered ultra-low-cost (e.g. below 1 USD) computing devices. The main requirements for this new market segment are: (i) ultra-low energy consumption in order to extend battery lifetime and (ii) very simple system design for increased yield and reduced cost.

Those new types of applications are very low duty cycle applications. All of them require a sensor to be read and to process the data and to react quickly [14] on a relatively infrequent basis (which varies for different applications) [19]. Typically, those computing systems have two operation modes with different needs and different optimal supply voltages (Vcc): (i) high-performance and low-power operation mode under high or moderate voltage (HP mode for short) during relatively short periods of time to react to some infrequent particular events and (ii) ultra-low energy and reliable operation mode under near-/sub-threshold voltage (ULE mode for short) during most of the time until infrequent events arise.

New cache designs with a single-Vcc domain have been recently proposed [16] to satisfy the stringent needs of this market. Such cache architectures guarantee reliable ULE operation at near-/sub-threshold Vcc and HP operation at high Vcc. The cache is designed in such a way that different cache ways may use different SRAM cell types (hybrid cache designs). Some of the cache ways are optimized to satisfy high performance requirements during high voltage operation (HP ways) whereas the rest of the ways provide ultra-low energy and reliability during near-/sub-threshold voltage operation (ULE ways). HP ways are disabled at ULE mode since they would experience many faults. However, ULE ways are used at HP mode despite their energy inefficiency at high Vcc because they reduce the number of off-chip accesses, which are much more expensive in terms of performance and energy. Data management policies are required to access HP and ULE ways at HP mode to minimize the number of accesses to ULE ways. Unfortunately, existing policies are far from being efficient across all applications.

This paper proposes an efficient, but simple data management mechanism for HP operation on single-Vcc domain caches for hybrid operation. Our mechanism is called Adaptive Data Management (ADAM). ADAM is tailored to detect hit distribution dynamically across the cache lines during program execution and adapts to different application behaviors to optimize performance and energy consumption by means of an extremely simple hardware mechanism. ADAM achieves average energy savings between 12% and

29% with respect to all state-of-the-art approaches at HP mode with negligible performance impact (1.7% w.r.t. the best state-of-the-art approach). ADAM largely outperforms all mechanisms in terms of energy-delay product.

The rest of the paper is organized as follows. Section 2 discusses related work. Our approach and state-of-the-art mechanisms are presented in Section 3. The methodology used for evaluation as well as results are presented in Section 4. Finally, Section 5 summarizes the main conclusions of this work.

2. RELATED WORK

Literature on low-power techniques for caches is abundant. Double-ended 6T (6 transistors) SRAM cells have been widely deployed for high voltage operation. Numerous SRAM cell designs such as 8T [11], Schmitt-Trigger 10T (ST 10T) [13], etc. target different voltage and robustness scenarios. However, those SRAM cells introduce significant area and energy overheads w.r.t. 6T cells at high voltage, which is unaffordable in embedded cache design if used extensively.

Some authors present techniques to save energy by reconfiguring cache characteristics such as cache size and associativity or lowering cache Vcc [2, 9] (or even gating it [18]) for some cache sections or the whole cache, which is orthogonal to data management techniques.

Some authors propose splitting the cache into different modules. Kin et al. [12] propose putting a small cache in front of the L1 cache to filter accesses to the L1 cache. This technique is known as a filter cache. A similar approach is explored by Abella and Gonzalez [1] and Fujii and Sato [10]. Maric et al. [16] propose hybrid, single-Vcc domain cache designs, which suit our target market and therefore, are used as our baseline cache design.

Simple data management policies fitting our target cache architecture have been proposed recently [8, 7]. However, as shown later, they are largely inefficient in our target market because they require many accesses to the ULE ways either because they swap data unnecessarily [8] or because they do not swap it when needed [7]. As shown later, ADAM outperforms those techniques consistently across all applications.

3. ADAPTIVE DATA MANAGEMENT

Hybrid cache designs offer significant room to trade off between energy and performance at HP mode to fit application requirements dynamically by using different data management mechanisms to access HP and ULE ways. This section describes existing state-of-the-art data management mechanisms, analyzes their drawbacks and proposes ADAM, an Adaptive Data Management mechanism, which achieves significant energy savings with negligible performance degradation at HP mode for hybrid caches with single Vcc domain.

3.1 Baseline Cache Architecture

Cache memories are the main energy contributor in our target market, thus they must be carefully designed and managed. For instance, L1 caches occupy 50% chip area and are one of the main energy contributors in the ARM Cortex A5 processor [3] in order to improve energy efficiency at HP mode. In particular, our approach is built on top of a 6T+10T hybrid cache although any other hybrid configurations proposed in [16] may be used.

Figure 1 depicts the 6T+10T cache design used [16]. We have considered different hybrid configurations with similar total area to that of a cache where all cache ways are implemented with 6T cells, although such a design would not work at ultra-low voltage. However, our data management technique is not restricted to any particular configuration. The first considered hybrid configuration consists of an 8KB 8-way cache where 7 ways are implemented with 6T cells and 1 way with ST 10T cells (7+1 for short). Due to the additional 45% area of the ST 10T cells with respect to 6T cells, this configuration has an area overhead around 5% with respect to the case where all cache ways are implemented with 6T cells (8+0). We also consider a 7KB 7-way cache implemented with a 5+2 hybrid configuration. Such configuration has similar area to that of an 8+0 one.

3.2 State-of-the-art Data Management Mechanisms

Existing state-of-the-art data management mechanisms are not particularly devised for hybrid caches implemented with a single Vcc domain, but we use them for comparison purposes. They are extensively evaluated against our proposed mechanism in Section 4. In general, they are built on top of the LRU replacement policy, so we assume LRU as the replacement policy in the rest of the paper.

3.2.1 Parallel Mechanism

The first mechanism is the simplest one and accesses all cache ways in parallel. This technique is the fastest one, but very inefficient in terms of energy since 10T ways are accessed always. We refer to this mechanism as *Parallel* in the rest of the paper.

3.2.2 Sequential Mechanism

The second mechanism [7] (referred to as *Sequential*) improves energy efficiency upon the *Parallel* one by accessing 6T ways first since they are the most energy efficient ways at HP mode. In case of a miss in the 6T ways, the 10T ways are accessed next. Energy efficiency is achieved only if hits concentrate mostly in the 6T ways. Otherwise, this mechanism will be slower than the *Parallel* one due to the extra delay to access the 10T ways.

3.2.3 Swap Mechanism

The *Swap* [8] mechanism accesses cache ways sequentially as the *Sequential* one does (first 6T ways and, in case of a miss, 10T ways), but on a 10T hit the hit line is swapped with the 6T line closest to the Least Recently Used (LRU) position and becomes the Most Recently Used (MRU). On a miss, the line in the LRU of the 10T ways is evicted, the LRU line of the 6T ways is moved to the 10T ways and the fetched line is then placed in the 6T ways. Therefore, the most recently used lines reside in the 6T ways and the LRU ones in the 10T ways. Figure 2 depicts the hardware support required for swapping. Swapping is done in four phases: (i) LRU 10T data (rd10T) read out and written into the fetch line buffer (this buffer is used also to transfer data to/from memory), (ii) LRU 6T data (rd6T) read out and written into the swap buffer, (iii) write swap buffer contents into the 10T LRU position (wr10T) and (iv) write fetch line buffer contents into the 6T LRU position (wr6T). In order to perform swapping successfully, we adopt a fairly simple solution: *stall any access to cache during swapping, which*

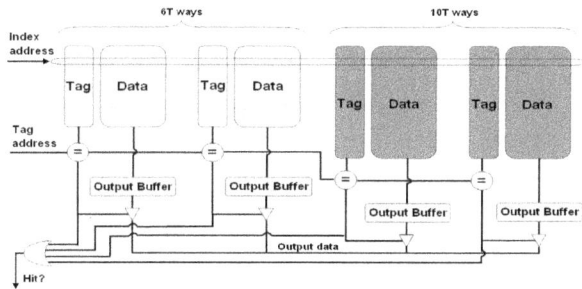

Figure 1: Hybrid 6T+10T cache design.

lasts for the latency of 4 cache accesses. Typically, preventing further accesses during some cycles is as easy as keeping the cache ports busy to prevent the port arbiter from issuing new accesses. Besides the latency of 4 accesses, swapping overhead includes extra read/write operations energy.

The *Swap* mechanism is particularly efficient for irregular applications where hits concentrate in few cache lines during long intervals. Thus, if any of those cache lines resides in a 10T way, it is quickly swapped to the 6T ways so that the energy spent swapping lines is quickly offset by the hits in such cache line. In this case, most of the hits concentrate in the 6T ways.

However, for regular applications hit distribution is quite homogenous across the cache lines. Moreover, if the working set matches quite well cache size (the ideal case to exploit cache capacity), hits concentrate mainly in the LRU positions of each cache set, which would reside in the 10T ways if swapping is performed. In this case, swapping can introduce additional energy consumption due to many unnecessary swaps. The energy overhead does not pay off because once swapped those lines are barely hit in the 6T ways. A good example is the continuous traversal of a vector or array that occupies most of the cache. In this example many hits concentrate in the LRU ways (thus, 10T ways when *Swap* is used). If no swapping was performed, most of the hits would concentrate in the 6T ways, performance would be higher and swapping overheads would not be incurred.

Based on the inefficiencies observed for both *Sequential* and *Swap* mechanisms, we propose ADAM, a very simple, adaptive mechanism to swap only when it pays off.

3.3 ADAM: Adaptive Data Management Mechanism

ADAM tracks the hit distribution across the different cache regions (6T and 10T regions) during program execution and dynamically activates/deactivates swapping to optimize performance and energy consumption. Hence, our mechanism has two distinct modes: swap and no-swap. Swap mode allows swapping between the LRU cache location of 10T ways and the LRU cache location of 6T ways similarly to [8]. Information about LRU position in the 6T and 10T ways is obtained from the global LRU stack (see Figure 3) since our technique forces the LRU locations to remain into the 10T ways during swap mode. This greatly simplifies the implementation and avoids using extra control logic to create three separated LRU stacks (one for 6T ways, one for 10T ways and a global one).

Swapping is then performed by using existing fetch line buffer and additional swap buffer to switch the cache line contents as for the *Swap* mechanism. No-swap mode stops this functionality.

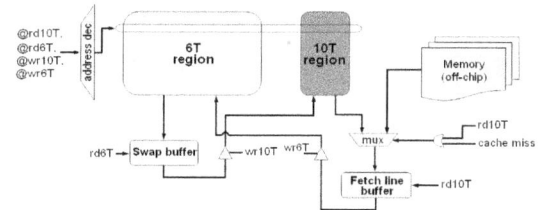

Figure 2: Hardware support for Swap operation.

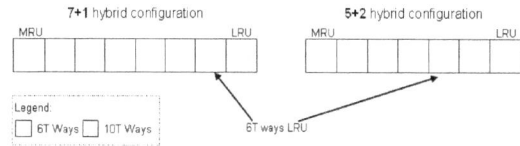

Figure 3: Global LRU stacks for 7+1 and 5+2 hybrid cache configurations.

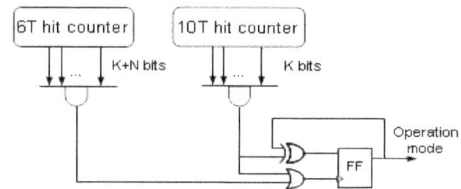

Figure 4: ADAM control logic.

ADAM tracks the number of hits in each region (6T and 10T). Whenever the number of hits in the 10T region is high, the operation mode is changed. Thus, if swap was activated it is deactivated and vice versa. This way, if swap is deactivated and many hits occur in the 10T ways, ADAM activates swap so that those lines are swapped into the 6T ways for higher efficiency. Conversely, if swap is activated, but many hits occur in the 10T ways, ADAM deactivates swap to stop useless swapping. By doing so, ADAM combines the advantages of the *Swap* and *Sequential* approaches, and moreover, reacts in front of the different phases of a program. Next we describe how to implement such a mechanism and when the operation mode must be changed.

3.3.1 Implementation Details

Figure 4 depicts the control logic to change the operation mode. Counters are employed to track the numbers of hits to the 6T and 10T ways respectively. The size in bits of each counter is set based on the expected hit probabilities in 6T and 10T cache ways.

Hit distribution largely depends on the program. We assume the worst case where the hit distribution is uniform across the cache lines. In this scenario each cache line has similar hit probability in each cache set, so it does not matter which particular cache lines reside in each cache way. Therefore, the size in bits of the hit counters depends on the fraction of cache space devoted to 6T and 10T ways. The size of the 10T counter is smaller due to a smaller number of 10T ways with respect to the 6T ways. If the size of 10T counter is K bits, then the size of the 6T counter must be $K+N$ bits. This means that by setting up the 6T hit count threshold 2^N times larger than that for 10T, we can detect with high accuracy whether 10T hits are above or below their expected value. In other words, the 10T hit counter saturates first if and only if more than $\frac{1}{2^N+1}$ of total hits occur in the 10T ways. Therefore, N determines the acceptable fraction of hits in the 10T ways. The value of K determines at which granularity decisions are taken.

For instance, for the 7+1 hybrid configuration, if the hit distribution is uniform across the cache lines, then around 7/8 of the total hits concentrate in the 6T ways and the remaining 1/8 of hits in the 10T ways. This means that by setting up the 6T hit count threshold seven times greater than that for 10T, we can detect whether 10T hits are above or below their expected value. Using a counter for 6T hits with $N = 3$ bits more than for 10T hits, suffices to detect this scenario. However, we decided to be more aggressive to reduce the number of accesses to the 10T way (10T ways are more power hungry than 6T ones), and thus, the 6T hit counter has $N = 4$ bits more than the 10T hit counter. Thus, the 10T hit counter saturates first if and only if more than $\frac{1}{2^4+1} = \frac{1}{17}$ of the hits occur in the 10T ways. Since changing the operation mode (swap or no-swap) has negligible cost, we take decisions at very fine grain. Thus, we use $K = 7$ bits for the 10T hit counter (up to 127 hits) and $K+N = 11$ bits for the 6T hit counter (up to 2047 hits). Based on the same methodology, we use 7 bits for the 10T hits counter (up to 127 hits) and 10 bits for the 6T hits counter (up to 1023 hits) for the 5+2 hybrid configuration. However, our mechanism is not restricted to those particular values. Threshold sensitivity is studied in the evaluation section.

Figure 5 shows the algorithm for the ADAM mechanism. Initially, the operation mode can be either swap or no-swap. If there is a 6T hit, the 6T hit counter depicted in Figure 4 is incremented. If this counter is saturated, the operation mode is not changed (it is kept as it was before a hit) and 6T and 10T hit counters are reset. If there is a 6T miss, the 10T cache ways are accessed next. In case of a hit, the 10T hit counter is incremented. If this counter saturates, the operation mode is changed and 6T and 10T hit counters are also reset. Regardless of whether the 10T hit counter saturates, if swap is activated, the 10T line hit and the 6T LRU line are swapped.

In case of a miss in both 6T and 10T ways, we must evict the 10T LRU line, move the 6T LRU line to the available 10T entry and fetch the new data into the available 6T entry. Such process increases latency by the latency of an extra access to write data read from 6T ways into 10T ways. Note that we could simply evict 6T lines if the swap mode is not activated. However, we empirically observed that cache lines not in cache experience a burst of hits after being fetched and it is better fetching them to the 6T ways. Nevertheless, fetching those lines to the absolute LRU location if swap is deactivated produces negligible changes in the results.

As shown in Figures 2 and 4, hardware complexity is low.

4. EVALUATION

This section presents the evaluation methodology, performance and energy results for proposed mechanism as well as for those used for comparison purposes, and threshold sensitivity study for the hit counters.

4.1 Methodology

We have chosen a very simple processor architecture with one core and in-order execution because energy and power efficiency are the main goals for the ultra-low-cost market segment. Data management mechanisms have been implemented for both on-chip L1 data and instruction caches.

```
If access to the 6T ways is a hit:
1.     Increment 6T hit counter;
2.     If(6T hit counter is saturated)
2a.         Operation mode = Keep the same operation mode as
                 it was before hit;
2b.        Reset 6T and 10T hit counters;
3.     Return data.
Else if access to the 10T ways is a hit:
4.     Increment 10T hit counter;
5.     If(10T hit counter is saturated)
5a.         Operation mode = Change the operation mode that
                 was before hit;
5b.        Reset 6T and 10T hit counters;
6.     If(Operation mode = swap)
6a.        Swap hit 10T line with 6T LRU line;
7.     Return data.
Else (we have a miss here):
8.     Evict the LRU data from 10T ways;
9.     Move the LRU data from 6T ways into the evicted 10T
       location;
10.    Fetch new data into the free 6T location.
```

Figure 5: ADAM algorithm implementation.

The relative memory latency is low (in the order of 20 cycles) given the simplicity required in those systems, its small size (typically few MBs) and its high integration with the processor itself. Nevertheless, other memory latencies do not change the trends reported later since all data management mechanisms keep exactly the same contents in cache at any time and, therefore, they neither increase the number of misses nor change the order of memory requests. *Therefore, off-chip behavior remains unchanged.*

4.1.1 Benchmarks

So far, a set of benchmarks specific for the domain that we target basically does not exist. We have chosen Media-Bench [15] because they fit very well the expected needs of the ultra-low-cost segment: abundant data processing during HP mode and small workloads. For instance, we expect that sensor applications which monitor wind, sea level, temperature, tsunamis, etc., to be data intensive.

4.1.2 Operating Modes

Our system has two distinct operating modes: HP and ULE. We have set Vcc to 1V and 350mV for HP and ULE mode respectively. Operating frequencies are set to 1GHz for HP mode, and 5MHz for ULE mode, which is in line with state-of-the-art results [5, 6].

4.1.3 System Modeling

The technology node considered is 32nm. In order to support two different operating modes, we have created accurate energy models for ST 10T SRAM cells when operating at high and near-threshold Vcc. We use the analytical methodology proposed in [5] coupled with iso-robustness scaling analysis proposed by Chen et al. [6] to determine appropriate transistors sizes to meet reliability constrains, basically introduced by process variations at near-/sub-threshold voltage. We have sized ST 10T cells when operate at ultra-low voltage (350mV) to match the same failure rate as for 6T cells when operate at high voltage (1V). These models are incorporated into CACTI 6.5 [17] to determine dynamic and leakage energy of hybrid 6T+10T cache designs. CACTI is a flexible and accurate cache delay, power and area simulator. Similarly, we have enabled further degrees of freedom in CACTI to implement hybrid cache memories where different cache ways may use heterogeneous cell types (6T and 10T). We have also incorporated our custom-modified CACTI tool

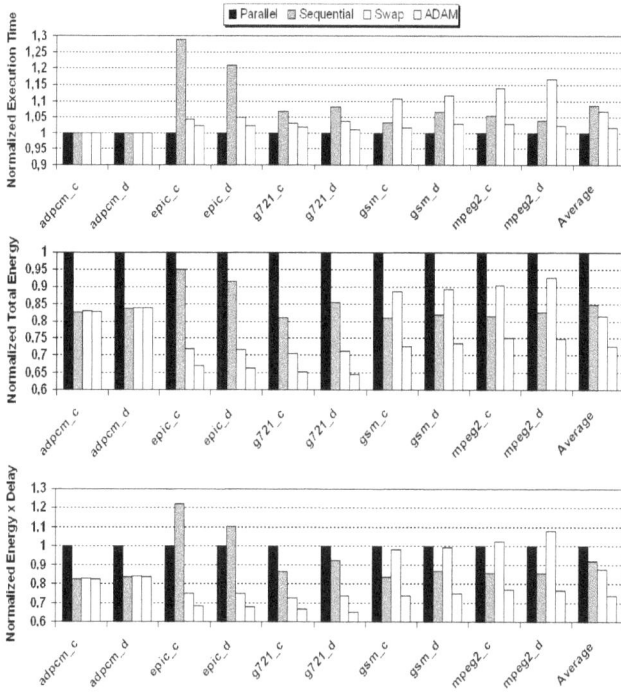

Figure 6: Normalized execution time, total energy and EDP for 7+1 hybrid configuration.

into a full-chip simulator. As full-chip simulator, we have used an enhanced version of SMTSim [20] extended with power models analogous to those of Wattch [4], but using our enhanced CACTI version to model all SRAM array-like structures (Caches, TLB, register files, etc.). We account for the extra energy introduced by swapping as well as additional latencies to perform swaps correctly. All SRAM arrays except L1 caches have been implemented using ST 10T cells so that they operate properly at any voltage level considered.

Off-chip main memory is not modeled in detail. We have presented execution time and energy for the processor as the measure of the overall system performance. All mechanisms produce the same off-chip memory pattern (exactly the same addresses and in the same order with similar inter-access time gaps). Thus, off-chip behavior should not change. Moreover, different memory technologies (e.g., ROM, DRAM, SRAM, PCMs) offer strongly different behavior, so we left it out of the scope of this paper.

4.2 Results

In this subsection we present execution time, total energy and energy-delay product (EDP) results for the different data management mechanisms at HP mode.

4.2.1 Performance and Energy

Execution time and energy vary across benchmarks. Thus, for the sake of clarity, results have been normalized with respect to the *Parallel* mechanism, which is the most optimal mechanism in terms of performance.

Figure 6 shows the execution time, total energy and EDP when running the 10 benchmarks on our 7+1 hybrid configuration. As stated before in Section 3, there are some pathological (and quite frequent) situations where *Sequential* and *Swap* lose significant performance. For instance, *Sequential*

achieves the worst performance for epic_c, epic_d, g721_c and g721_d applications, since those programs use few cache lines intensively. By chance, some of those lines reside in the 10T ways, which degrades performance and does not allow saving dynamic energy. In fact, energy consumption increases due to the extra leakage (higher execution time implies higher leakage). Epic_c experiences the highest performance degradation for this mechanism (29%). Conversely, *Swap* shows very low performance degradation for those "swap-friendly" applications (around 4%). *Sequential* achieves high performance (only 5% degradation) for regular applications such as gsm_c, gsm_d, mpeg2_c and mpeg2_d since those programs use cache lines quite homogeneously. However, *Swap* shows a high performance loss (up to 17% for mpeg2_d) because it performs many unnecessary swaps that end up moving to the 10T way those cache lines to be reused soon.

It can be seen that ADAM shows the best results in terms of execution time for all regular and irregular applications (up to 2.9% performance degradation for mpeg2_c and 1.7% on average). ADAM improves results even for those applications where *Sequential* and *Swap* perform well because of its adaptive nature that allows ADAM to adapt dynamically to a different program phases.

All mechanisms behave nearly the same for adpcm_c and adpcm_d because those applications have very small workloads. Since we start fetching data into the 6T ways, even *Sequential* works well for those benchmarks.

Relative trends for total energy savings match quite well performance for *Sequential*, *Swap* and ADAM mechanisms. The *Parallel* mechanism provides the highest performance, but also the highest energy consumption because it accesses all cache ways on every request. For the remaining mechanisms, the higher execution time, the higher energy consumption is. Whenever any of those mechanisms hits the 10T way often, it saves little energy (both 6T and 10T ways are accessed) and performance is degraded due to the sequential access to 6T and 10T ways. Since ADAM is the most effective data management mechanism concentrating hits in 6T ways, it is the most effective mechanism in terms of both performance and total energy. In particular, ADAM decreases energy consumption by 29% on average.

As expected, ADAM is the best performing mechanism in terms of EDP. ADAM reduces EDP by 27%, 21% and 16% with respect to *Parallel*, *Sequential* and *Swap* mechanisms respectively.

Similar trends are observed for the 5+2 configuration. Details are not reported due to lack of the space.

4.2.2 Disabling 10T Ways at HP Mode

We have also compared ADAM mechanism with the power gating technique [18] at HP mode. We have analyzed the case when 10T ways are turned off at HP mode, keeping only 6T ways enabled (7KB/5KB, 7-way/5-way, for 7+1/5+2 configuration). Results are compared with respect to the same baseline (7+1/5+2 configuration with *Parallel* access). In this scenario, execution time is increased up to 8% (for gsm_c, gsm_d, mpeg2_c and mpeg2_d applications) due to reduced cache size and additional cache misses, while total energy is higher (17% on average for all applications) than when ADAM mechanism is in the place for 7+1 configuration. Moreover, total energy consumption will be even higher if extra off-chip energy is accounted due to additional off-chip accesses occured in this scenario.

Table 1: Normalized execution time, total energy and EDP for different threshold values of hit counters for ADAM for 7+1 configuration.

	Configurations (K,N)				
	(5,4)	(9,4)	**(7,4)**	(7,3)	(7,5)
Execution time	0.988	1.018	1	1.004	1.007
Total energy	0.983	1.024	1	1.007	1.015
EDP	0.984	1.038	1	1.012	1.023

Those facts confirm and justify our decision to keep ULE ways operating at HP mode and further prove the efficiency of ADAM mechanism. Similar trends are observed for the 5+2 configuration.

Note that only 10T ways are active during ULE operation, so no data management policy is required. Details about ULE operation can be found in [16].

4.3 Hit Counters Size Sensitivity Study

Default saturation values of hit counters (see Figure 4) are determined by setting parameters K (i.e. 7 bits) and N (i.e. 4 bits). A lower threshold makes more aggressive choices whereas choosing a larger threshold may be so conservative and may miss more opportunities.

To better understand the effects of thresholds, we vary K (5, 7 and 9 bits), while keeping $N = 4$ bits. Similarly, we also vary N (3, 4 and 5 bits), while keeping $K = 7$ bits. Table 1 presents performance and energy results averaged across all 10 applications for the 7+1 configuration for those 5 different (K,N) configurations. All results are normalized with respect to the default (7,4) configuration. From the performance perspective, results show that all applications are highly insensitive to the thresholds (up to 1.8% performance variation across different thresholds). Whenever N is high ($N=5$) the operation mode is changed too aggressively. Conversely, whenever N is low ($N=3$) the 6T hit counter saturates quickly and the operation mode is changed in an excessively conservative manner. Since changing operation mode has negligible cost, all applications prefer to take decisions at very fine grain ($K=5$), but difference in performance between this case and the default one ($K=7$) is very small (1.2%). On the other hand, if decisions are taken at coarse grain ($K=9$), performance degradation is negligible (1.8%).

From the energy perspective, the trends are similar as they are for performance. The reason is that those configurations inefficient for performance are also inefficient for energy due to either unnecessary swaps or unnecessary 10T accesses.

Similar trends are observed for 5+2 configuration, but results are not reported due to lack of the space.

5. CONCLUSIONS

In this paper, we have proposed a new, efficient and very simple Adaptive Data Management mechanism (ADAM) for hybrid voltage operation caches designed particularly for ultra-low-cost (e.g. below 1 USD) battery-powered systems, which are a new market. ADAM is a counter-based data management mechanism which dynamically detects hit distribution across the cache lines during program execution and activates/deactivates swapping to optimize performance and energy consumption at fine grain with negligible hardware overhead. ADAM saves significant energy (29% on average) in L1 caches with negligible performance degradation (1.7% on average), thus outperforming *all* existing data

management approaches noticeably in terms of energy-delay product (EDP) across different cache configurations.

Moreover, ADAM meets the golden requirement of this new market segment, which is *simplicity of design*, as required to achieve reduced fabrication cost, increased integration, energy efficiency and high performance.

Acknowledgments

This work has been partially supported by the Spanish Ministry of Education and Science under grant TIN2007-60625, HiPEAC, the UPC under grant FPI-UPC and the Generalitat of Catalunya under grant Beatriu Pinós 2009 BP-B 00260.

6. REFERENCES

[1] J. Abella and A. Gonzalez. Power efficient data cache design. In *ICCD*, 2003.

[2] D. H. Albonesi. Selective cache ways: On-demand cache resource allocation. In *MICRO*, 1999.

[3] ARM. Cortex-A5™ Technical Reference Manual.

[4] D. M. Brooks, V. Tiwari, and M. Martonosi. Wattch: A framework for architectural-level power analysis and optimizations. In *ISCA*, 2000.

[5] J. Chen, L. Clark, and T.-H. Chen. An ultra-low-power memory with a subthreshold power supply voltage. *IEEE JSSC*, 41, 2006.

[6] G. Chen et al. Yield-driven near-threshold SRAM design. In *ICCAD*, 2007.

[7] R. Dreslinski et al. Reconfigurable energy efficient near threshold cache architectures. In *MICRO*, 2008.

[8] S. Dropsho et al. Integrating adaptive on-chip storage structures for reduced dynamic power. In *PACT*, 2002.

[9] K. Flautner et al. Drowsy caches: Simple techniques for reducing leakage power. In *ISCA*, 2002.

[10] S. Fujii and T. Sato. Non-uniform set-associative caches for power-aware embedded processors. In *LNCS'04: Embedded and Ubiquitous Computing*, 2004.

[11] S. Jain and P. Agarwal. A low leakage and SNM free SRAM cell design in deep sub micron CMOS technology. In *VLSID*, 2006.

[12] J. Kin, M. Gupta, and W. Mangione-Smith. The filter cache: An energy efficient memory structure. In *MICRO*, 1997.

[13] J. Kulkarni, K. Kim, and K. Roy. A 160 mV, fully differential, robust schmitt trigger based sub-threshold SRAM. In *ISLPED*, 2007.

[14] F. Kusumoto and N. Goldschlager. *Cardiac Pacing for the Clinician*. Springer Science+Business Media LLC, 2008.

[15] C. Lee et al. Mediabench: A tool for evaluating and synthesizing multimedia and communication systems. In *MICRO*, 1997.

[16] B. Maric, J. Abella, F. J. Cazorla, and M. Valero. Hybrid high-performance low-power and ultra-low energy reliable caches. In *ACM CF*, 2011.

[17] N. Muralimanohar, R. Balasubramonian, and N. Jouppi. CACTI 6.0: A tool to understand large caches. *HP Tech Report HPL-2009-85*, 2009.

[18] M. Powell et al. Gated-vdd: A circuit technique to reduce leakage in deep-submicron cache memories. In *ISLPED*, 2000.

[19] R. Szewczyk et al. Lessons from a sensor network expedition. In *Proceedings of the 2004 European Workshop on Sensor Networks*, 2004.

[20] D. Tullsen, S. Eggers, and H. Levy. Simultaneous multithreading: Maximizing on-chip parallelism. In *ISCA*, 1995.

A Low Stand-by Power Start-up Circuit for SMPS PWM Controller

In-Seok Jung
Department of Electrical and Computer Engineering
Northeastern University
Boston, MA, USA
1-617-373-7780
ijung@ece.neu.edu

Yong-Bin Kim
Department of Electrical and Computer Engineering
Northeastern University
Boston, MA, USA
1-617-373-2919
ybk@ece.neu.edu

ABSTRACT

In this paper, a novel start-up circuit with a simple topology and low stand-by power during under voltage lockout (UVLO) mode is proposed for SMPS (switching mode power supplies) application. The proposed start-up circuit is designed using only a few MOSFETs, LDMOSs, and one JFET based on the analysis of the existing start-up circuits to address the power consumption and input voltage range issues of the conventional start-up. Simulated results using 0.35um BCDMOS process demonstrate that the leakage current of the proposed circuit is less than 1uA after UVLO signal turns on. Setting time is less than 1ms when the load current changes from 10mA to 20mA and vice versa.

Categories and Subject Descriptors

B.7.0 [**Integrated Circuits**]: General

General Terms

Performance, Design, Economics,

Keywords

Start-up circuit, SMPS, PWM, AC/DC

1. INTRODUCTION

The flyback topology is one of the popular choices as SMPS (switching mode power supplies) for low power applications such as adapter for mobile devices and auxiliary power supplies of the personal computer [1] and electrical home appliances [2]. Due to its insulating properties and simplicity, the flyback can be implemented using one switching component and one transformer. The switching component can be implemented using discrete or integrated power MOSFET, and control circuits are required to turn it on or off.

Most of the control integrated circuits for SMPS using the flyback topology require their starting current to charge a capacitance of the self-supply circuit in the integrated circuit to a certain level [3]. Such current comes from the starting circuit known as the bootstrap circuit (start-up), which is constituted in the simplest case by a resistance connected to the supply line [4].

When the voltage on the capacitance achieves a preset voltage

value called start-up voltage, the self-supply circuit supplies the control circuit. The self-supply circuit is generally constituted by an additional winding performed on the main transformer of the switching power supply to which a suitable rectification and filtration circuit is connected.

Since there are ever increasing demand for the reduction of the supply consumption lately, the simple aforementioned resistance is replaced by the circuits that are active during the starting phase and inactive during the normal operation.

Lately there are also demands for monolithic devices for the SMPS that are comprised of both control circuit and the active starting circuit. However, the technologies used for these integrated circuits have limitation regarding the maximum sustainable voltage because the start-up circuits are hard to be integrated due to its high and wide input voltage range such as 80~275 VAC for universal operation. Therefore, there is a need for a bootstrap start-up circuit that has lower power consumption and accommodates a wide bias voltage range. To address these problems, 700V DMOS or BiCMOS process technology is often used for products.

This paper proposes a novel start-up circuit using only a few MOSFETs, LDMOSs, and one JFET based on the analysis of the existing start-up circuits to address the power consumption and input voltage range problems of the conventional start-up circuits.

The remainder of this paper is organized as follows. Section II introduces the basic concept of the start-up circuit by showing the conventional topologies as well as the state-of-the-art topologies. Section III describes the proposed start-up circuit, and the simulated results are discussed in Section IV to prove the proposed design, followed by the conclusion in Section V.

2. CONVENTIONAL SART-UP CIRCUITS

2.1 Prior Arts

Figure 1(a) shows a prior art SMPS using flyback topology that includes a full-wave bridge rectifier, a transformer, and control circuit using pulse width modulation (PWM) mode.

M_P (Power MOSFET) controls the power switching of the primary winding of the transformer, and high input voltage is applied to the transistor during initial power-up.

The power to operate the control circuit is produced by the secondary winding side and is referred to as Vbias in Figure 1(a). However, at power start-up, controller will be without power because primary winding will be open and no voltage will be induced into the secondary winding because the controller is not switching during this time. To initiate such switching, node A with high DC voltage is tapped on bridge rectifier by the resistor, and the current is filtered by the capacitor. This tap supplies just enough current through the resistor to start chip because significant amounts of power can be dissipated in the process of dropping the high voltage of the line to the low voltage required by chip. This waste of power is continuous, even after Vbias is available. This problem is severe particularly after Vbias is available, which makes the resistor dissipate all the more power at the highest expected voltage. Therefore, high wattage resistor type is inevitable and it requires extra space and air circulation.

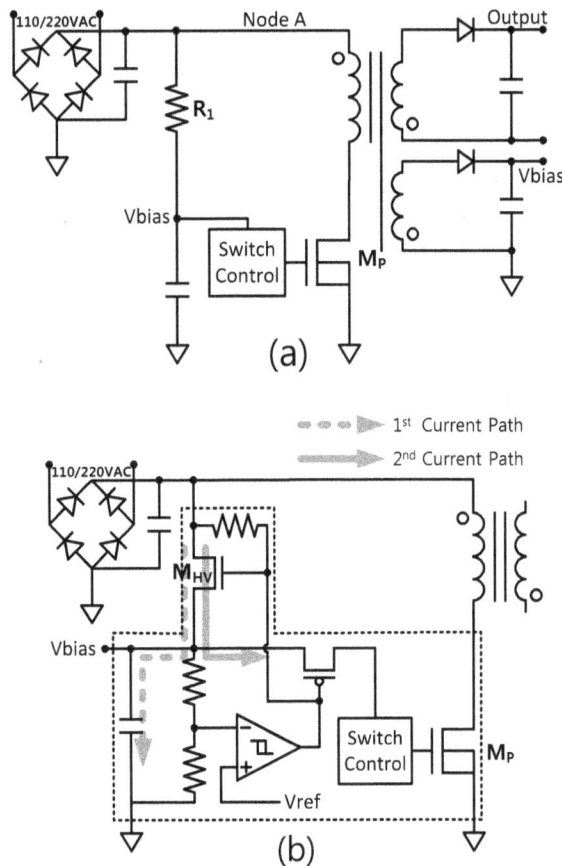

Figure 1 Conventional start-up circuits

Figure 1(b) shows an evolved SMPS that is similar to the circuit shown in Figure 1(a), where the resistor R_1 has been eliminated and several components have been included.

A voltage regulator inside the chip (dashed line in Figure 1(b)) allows the elimination of the high watts resistor and includes a high voltage pre-regulator transistor and supply power to PWM. A voltage is developed across the primary winding and induces a voltage in the secondary winding that supplies Vbias. With Vbias being supplied, the comparator operates to maintain the transistor off and no further high voltage power is required. Turning off high voltage pre-regulator saves power after start up, but such a function is more expensive to implement as it requires a high voltage transistor and extra pin with high voltage safety spacing. The high voltage transistor used in the pre-regulator of SMPS chips is usually relatively small device. The transistor's immunity to line transients is therefore limited. Thus, the pin associated with the transistor becomes a limiting factor for electrical static discharge (ESD) and safe operating area (SOA) rating of the switching regulator chip.

2.2 The-State-of-the-arts

Figure 2 shows SMPS that includes a start-up circuit, which is very similar to Figure 1(b) but uses one JFET. The voltage of the drain node of the transistor M_P is lower than the voltage at the output of the primary winding (drain node of JFET) because the voltage at the output of the primary winding is dropped through the JFET, and the JFET supplies a regulator with power either temporarily or continuously to operate a PWM in the chip.

Advantages of the circuit are that the circuits not only save power but also reduce the number of pins of the chip while ESD and SOA concerns are reduced by using JFET [5].

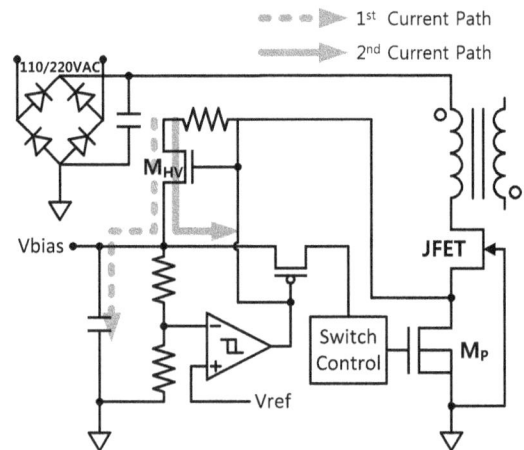

Figure 2 Schematic of the start-up circuits that uses JFET

When a line voltage is applied to the drain of the JFET, the JFET's body is fully depleted with a typical source to substrate pinch-off voltage of 50V wherein the body of the JFET is often referred to as the drift region of the M_P. The JFET is operated in the saturation region. Since the source voltage of the JFET is limited to approximately 50V, most of the line voltage applied at a terminal of the transformer is dropped across the body of the JFET thereby providing a buffer from the line voltage. This protects M_V and the resister connected to this node.

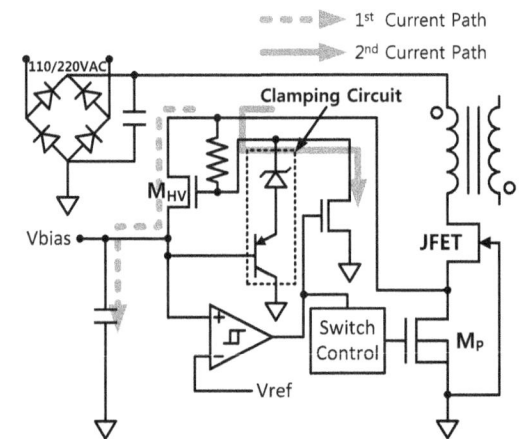

Figure 3 Schematic of the start-up circuit using clamping diode

Figure 3 is the schematic of another conventional start-up circuit. Clamping circuit limits the voltage appearing at the control electrode of M_V thereby limiting the current flowing between a terminal of the transformer and an input terminal of the comparator. Clamping circuit includes an avalanche diode having a typical breakdown voltage of 10V, and a PNP transistor. The Clamping voltage is typically below 20V to prevent a long term degradation of the gate oxide of the M_{HV} (MOSFET for High Voltage). The 10V avalanche diode was selected for its low leakage, providing a sharp knee at low bias currents.

The advantages of the circuit are that it provides a high ESD breakdown voltage and highly SOA. Furthermore, the start-up charging current can be adjusted from 5mA ~ 25mA by controlling the gate voltage of M_{HV}.

3. The PROPOSED START-UP CIRCUIT DESIGN

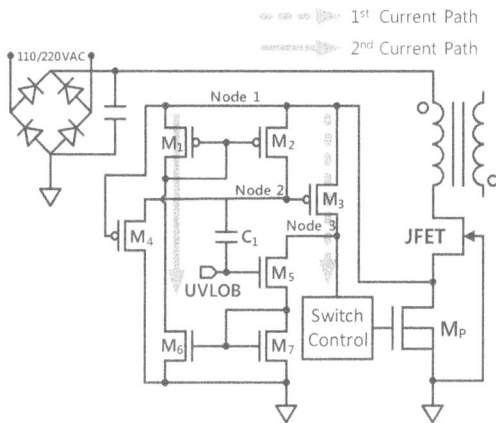

Figure 4 Schematic of the proposed start-up circuit.

Figure 4 shows the schematic of the proposed start-up circuit. The circuit consists of several MOSFETs, one capacitor, and one JFET. Because the role of capacitor C1 is to change the voltage of Node 2, C1 can be smaller size than 100fF. In the initial mode, the current through JFET goes to high voltage P-channel LDMOS transistors (M1 to M4). Initial voltage of Node 1 is zero and it makes M4 turn on initially, making M3 turn on in turn. As the voltage at Node 1 rises, M4 is turned off. Thus, the current flows through M3, M5, M6, and M7. This is the first phase of the start-up circuit operation, and this current path in indicated as 1st current path in the Figure 4. When UVLOB (complement of UVLO) is applied to the gate of M5 (when the voltage of UVLO goes to under 5V), the current flows through the 2nd current path shown in Figure 4. At this time, the voltage of Node 2 becomes high, making the current of M3 smaller. The total leakage current is expected to be very low because M4 and M3 are in linear region and there is no short circuit path. By switching the voltage of the gate of M5 and Node 2, start-up circuit is able to supply current to the controller. The Maximum voltage range of these nodes is equal to UVLO voltage so that the stress of M3 can be reduced.

As shown in Figure 5, the proposed circuit needs an internal signal called UVLO [6] to turn off the start-up circuit in order to save power and to protect the switch controller circuit when the input voltage or the voltage of the internal circuit of the controller is under specific voltage. In the proposed circuit, the signal of UVLO comes from the switch control block as shown in Figure 5, which shows the block diagram of the entire integrated circuit of controller including start-up, switch controller, and IO pin. The voltage of UVLO is controlled by the start-up circuit in the loop

for automatic operation of the circuit. UVLO voltage swings between 0V to 5V.

Not only UVLO voltage but also the reference voltage Vref and bias current I_{BIAS} are generated by the voltage supplied from the proposed start-up circuit. Therefore, these signals should be independent of the supply voltage.

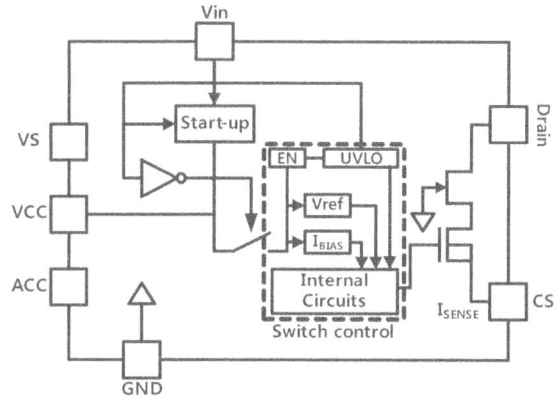

Figure 5 Block diagram of the integrated circuit including the proposed start-up circuit and controller.

The proposed start-up circuit is designed using only a few MOSFET, LDMOS, and one small capacitor for cost effective integration. However, in the case of PMOS such as M1, M2, M3 and M4, the drain voltages are very high with large voltage swing. Especially, M4 gate voltage can be up to 30V and it controls the gate voltage of M3 (Node2). Transistor M4 increases the circuit immunity against the change of Node 1 voltage to guarantee the start-up operation. Therefore, M4 transistor should have high breakdown voltage and its size should be carefully selected

4. SIMULATION RESULTS

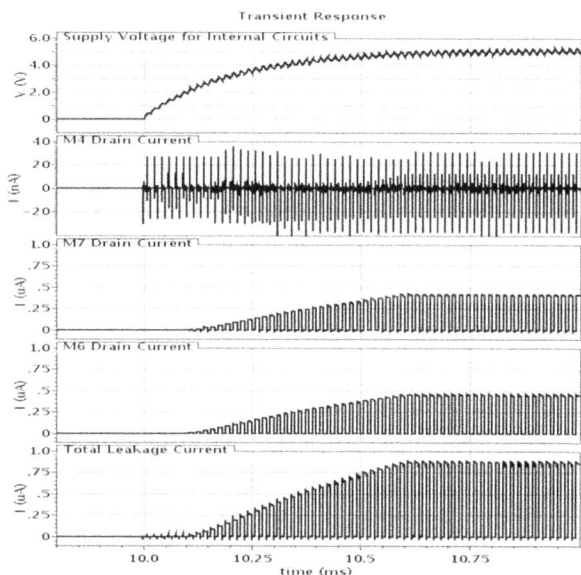

Figure 6 Simulation waveforms of the proposed circuit.

Figure 6 shows the waveforms of each nodes of the proposed start-up circuit. The fabrication process technology used for the proposed design is 0.35um 60V BCDMOS process and the breakdown voltage of the discrete JFET transistor is 700V. The parameters of the BCDMOS process are shown in Table I. A variety of devices are available in the BCD process including normal MOSFET and LDMOS operating between 5V and 60V as well as high gain BJT. M5, M6 and M7 MOFETs in Figure 4 are designed using the transistor model with 8V supply voltage considering its operating voltage regions, and other P-channel MOS are designed using 36V Low Vgs LDMOS transistor model. The output voltage of the start-up circuit goes to around 5V. Therefore, it is possible to integrate the JFET with controller using 700V BCDMOS process. The simulated average leakage current is 0.5uA and the possible operation current is up to 29mA, which satisfies the target specification. Figure 7 shows the transient responses when the load current changes from 10mA to 20mA (first dip) and 20mA to 10mA (second dip). Setting time is less than 1ms as shown in the Figure.

Table I. Parameters of BCDMOS process

Parameter	8V nMOS	36V LDpMOS
Electrical Tox [A]	300	130
Vt_lin @ L=Min. [V]	0.6	0.8
Gate-scource breakdown voltage [V]	12	5.0
Drian-source breakdown voltage [V]	8	36
Min. L [um]	2	1.8

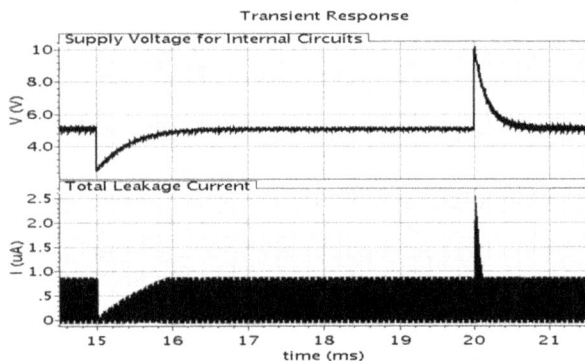

Figure 7 Transient response as changing required load current

Table II. Comparison of the start-up circuits' characteristics

	[7]	[8]	[9]	[10]	This work
JFET	O	O	O	X	O
Start-up On/off Logic	UVLO	UVLO	Time Delay Circuit	UVLO	UVLO
Leakage Current	Under 10uA	Under 10uA	Under 10uA	Under 15uA	Under 1uA
Cost	Mid	High	High	High	Low
2nd or 3rd winding feedback	Need	Need	Need	Need	None Need
Other Control Signal	None Need	None Need	None Need	Need	Need

Table II shows the comparison of the proposed start-up circuits with the ones that are commercially available. The proposed start-up circuit uses smaller area, which means low cost and low leakage current compared to other circuits because junction leakage is proportional to area as can be seen from the Table II.

5. CONCLUSIONS

This paper presents a novel start-up circuit for SMPS using flyback topology. The circuit is composed of only a few MOSFETs, LDMOSs, one capacitor, and one JEFT to make it possible to integrate with lower cost and smaller area. In addition, ESD and SOA concern is reduced by using JFET and it can be integrated into a single integrated circuit using commercially available 700V BCDMOS foundry process. Therefore it is possible to reduce the number of pins of the controller chip by integrating start-up circuit, PWM, JFET, and Power MOSFET except transformer, which means the significant reduction of the cost of the controller. Simulation results show that the proposed circuit consumes a small current less than 1uA after UVLO signal turns on. The design demonstrates a novel and viable solution of the modern start-up circuit design.

6. REFERENCES

[1] Amarasinghe, A. and Mason, S.C. 1992. Soft switching in traction auxiliary power supplies. *Auxiliary Power Supplies for Rolling Stock, IEE Colloquium on.* 7/1-7/4 (Feb. 1992).

[2] T. Qian and B. Lehman. 2008. Coupled input-series and output-parallel dual interleaved flyback converter for high input voltage application. *IEEE Transactions on Power Electronics*, vol. 23, no. 1 (Jan. 2008), 88–95.

[3] Tamyurek, B. and Torrey, D.A. 2011. A Three-Phase Unity Power Factor Single-Stage AC–DC Converter Based on an Interleaved Flyback Topology. *IEEE Transactions on Power Electronics*, vol.26, no.1 (Jan. 2011), 308-318.

[4] Yuan Bing, Lai Xinquan, Wang Hongyi and Wang Yi. 2005. The design of a start-up circuit for boost DC-DC converter with low supply voltage. *ASICON 2005*, vol.1, (Oct. 2005), 483-487.

[5] Mousa, R., Planson, D., Morel, H., Allard, B. and Raynaud, C. 2008. Modeling and high temperature characterization of SiC-JFET. *Power Electronics Specialists Conference, 2008. IEEE*, 15-19 (June 2008), 3111-3117.

[6] Li Fuhua, Wang Wei, Hang Qiuping, Xie Weiguo and Lu Zhenghao. 2009. Design of a under voltage lock out circuit with bandgap structure. *Integrated Circuits, ISIC '09*, 14-16 (Dec. 2009), 224-227.

[7] Balu and Balakrishman. 1994. Switched mode power supply integrated circuit with start-up self-biasing. U.S. Patent 5,285,369, Feb. 8, 1994

[8] Eric W. Tisinger and David M. Okada. Off-line bootstrap startup circuit. U.S Patent 5,477,175, Dec. 19, 1995

[9] Hang-Seok Choi. Switching mode power supply and method for generation a bias voltage. U.S. Patent 7,525,819 B2, Apr. 28, 2009

[10] Mauro Fagnani, Albino Pidutti and Claudio Adragna. Starting circuit for switching power supplies. U.S. Patent 7,319,601 B2, Jan. 15, 2008

Particle Swarm Optimization over Non-Polynomial Metamodels for Fast Process Variation Resilient Design of Nano-CMOS PLL

Oleg Garitselov[1], Saraju P. Mohanty[2], Elias Kougianos[3], Geng Zheng[4]
NanoSystem Design Laboratory (NSDL)[1,2,3,4]
Department of Computer Science and Engineering[1,2,4]
Department of Electrical Engineering Technology[3]
University of North Texas, Denton, TX 76203.[1,2,3,4]
omg0006@unt.edu[1], saraju.mohanty@unt.edu[2], eliask@unt.edu[3]

ABSTRACT

An automated top-down design flow to achieve physical design of Analog/Mixed-Signal Systems-on-Chip (AMS-SoCs) is difficult, especially for nano-CMOS. Process variation effects have profound impact on the performance of silicon versus layout design. In this paper metamodels, (surrogate models) and Particle Swarm Optimization (PSO) have been combined in an automated physical design flow for fast design exploration of AMS-SoCs. Neural network based non-polynomial metamodels that handle large numbers of design parameters, are used to predict the statistical process variation effects instead of exhaustive Monte Carlo simulations. The PSO algorithm is used for optimization of the AMS-SoC components using their metamodels instead of the actual circuit. The PSO algorithm followed a two step approach: local and global. The physical design of a Phase Locked Loop (PLL) is considered as a case study circuit. The proposed design flow is approximately 5 times faster while the error is under 2% compared to the Monte Carlo analysis.

Categories and Subject Descriptors

B.7.1 [**Integrated Circuits**]: Types and Design Styles—VLSI (very large scale integration)

Keywords

Neural Network Metamodel, Mixed-Signal Design, PLL

1. INTRODUCTION AND MOTIVATION

Modern consumer electronic devices are Analog/Mixed-Signal Systems-On-Chips (AMS-SoCs). Design of these AMS-SoCs presents specific challenges as the design of analog and digital circuits involves two distinct approaches. For example, the digital design is performance driven, whereas the analog design is specification driven. When AMS-SoCs are fabricated using nano-CMOS, their circuits are strongly impacted by the imperfections of manufactur-

ing processes [1]. However, design decisions are often based on nominal (rather than statistical) values of circuit attributes under the assumption that all transistors are alike. Thus, design decisions based on nominal data are not correct because the used data are either overestimations or underestimations of actual (silicon) data (i.e. design-to-silicon gap).

Process variations have an increasing effect on circuits as the technology is scaling past 100 nm [3],[2] with more dramatic effect on analog circuits [5]. It is essential to produce a process variation resilient design to increase the production throughput and reduce the chip cost.

The simulation analysis for process variation is usually done using a large number of Monte Carlo simulations. To reduce the amount of time it takes to produce the analysis, different techniques have been introduced such as symbolic analysis, hierarchical statistical analysis, and regression based approaches. For a simple circuit, it is possible to produce a process robust design just from the understanding of the behavior of that circuit. For large circuits it is time consuming to run large amount of simulations for process variation analysis due to high complexity of the circuit. The primary goal of this paper is to reduce the time-to-market constraints that are enforced on the design time to create a process resilient design of the circuit. This paper proposes the use of neural network based metamodels that can capture the circuit output in small and large ranges of the design parameters which can then be used for process variation analysis and also for circuit optimization. Then, a Particle Swarm Optimization (PSO) algorithm performs the design exploration over the non-polynomial metamodels to quickly converge to a target design.

2. CONTRIBUTIONS & PRIOR RESEARCH

The **novel contributions of this paper** are:

- A novel design flow that combines non-polynomial metamodels and particle swarm optimization for fast nanoscale process variation resilient mixed-signal design exploration.

- Accurate neural network based metamodels are proposed for frequency, power, jitter and locking time of the PLL system.

- Statistical process variation analysis over the metamodels instead of the actual circuit are performed and are shown to be much faster without significant loss of accuracy.

- A particle swarm optimization algorithm is presented for global optimization and process variation analysis.

Design optimization approaches to produce process variation tolerant design have being proposed in the existing literature. In [4], variability is estimated for frequency acquisition in digitally controlled oscillators. In [10], PVT-tolerant PLLs are proposed. A PVT-tolerant amplitude controller is proposed in [7] to minimize the phase noise of an LC-VCO. A PVT-tolerant low-jitter digital PLL is presented in [6]. The above circuit-specific approaches are not top-down AMS-SoC design flows and cannot handle large circuits with full-blown parasitics as the exclusive use of SPICE does not make them easily scalable. The current paper will thus significantly advance the state-of-the-art in AMS-SoC design exploration. Particle swarm optimization (PSO) has been successfully used on op-amp optimization in [9]. The design of an RF CMOS distributed amplifier is done in [8] using the PSO algorithm.

3. PROPOSED DESIGN FLOW USING NON-POLYNOMIAL METAMODELS AND PARTICLE SWARM OPTIMIZATION (PSO)

3.1 The Proposed Fast/Accurate Design Flow

This section briefly discuses the proposed fast design flow for fast and yet accurate process variation resilient optimization of the mixed-signal systems. After the creation of the physical design, the design characteristics of the PLL has changed from the schematic, as expected. Usually comprehensive manual design iterations follow to adjust the physical design to bring the circuit back to the desired specifications. Then Monte Carlo or corner analysis follows to conduct process variation study. If the circuit does not pass the process variation specifications then more manual labor is needed to adjust the design and more simulations are required for this iterative flow. The proposed design flow uses neural network based metamodels that predict the characteristics of the PLL. The physical design parametric netlist is parameterized and Latin Hypercube Sampling (LHS) is used to create two data sets for training and verification of the neural network model. The neural network based non-polynomial metamodels are created based on the training data set and then verified. The accuracy of the metamodels is required to be very high to be able to predict process variation effects. For each particle in the particle swarm optimization algorithm, a Monte Carlo analysis is run on the model if the output of the model is within the optimization constraints of frequency. This optimization effectively can optimize the circuit on a global scale and also account for process variation of the circuit. The final parameters are then used to adjust the physical design. This design flow reduces the amount of manual labor down to two physical design iterations considerably reducing the design process time.

3.2 The Case Study Circuit PLL in Brief

The PLL, which is shown in Fig. 1, provides a good example of a mixed signal system and is a good candidate for the application of our methodology.

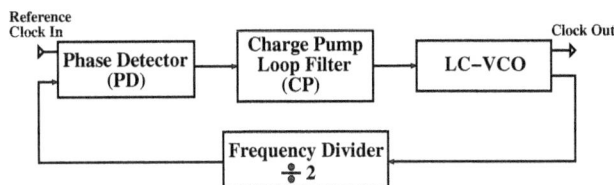

Figure 1: Block diagram of a phase locked loop.

The PLL has distinct components performing different functions and hence their parameters need to be tuned for process variation analysis and optimization. The following design and process parameters were considered: For the LC-VCO, W_{nLCVCO} and W_{pLCVCO}, the width of NMOS and PMOS, L the common length of both transistors, and T_{oxn} and T_{oxp} for the oxide thickness of the NMOS and PMOS. For the divider, Wn_{DIV} and Wp_{DIV}, the width of NMOS and PMOS, L the common length of both transistors, and T_{oxn} and T_{oxp} for the oxide thickness of the NMOS and PMOS. For the charge pump, Wn_{CP} and Wp_{CP} the width of NMOS and PMOS, and L the common length of both transistors. For the phase detector, Wn_{PD} and Wp_{PD} the width of NMOS and PMOS, and L the common length of both transistors.

4. PROCESS VARIATION ANALYSIS USING NEURAL NETWORK METAMODELS

4.1 Non-polynomial metamodeling using feed forward neural networks

Neural network models are composed of a mass of fairly simple computational elements and rich interconnections between them. They operate in a parallel and distributed fashion which resembles biological neural networks. Most neural networks have some form of initial "training" rule in which the weights of connections are adjusted on the basis of presented patterns.

A multiple layer neural network consists of an input, a nonlinear activation function in a hidden layer, and a linear activation function in the output layer. This makes multilayer networks very flexible and powerful due to their ability to represent nonlinear as well as linear functions. The multilayer network needs to have at least one non-linear function layer, otherwise a composition of linear functions becomes just another linear function. The linear layer function output is:

$$v_i = \sum_{i=1}^{s} w_{ji} x_i + w_{j0}, \qquad (1)$$

where w_{ji} is the weight connection between the jth component in the hidden layer and the ith component of the input x_i. The nonlinear tanh activation function used for the hidden layer has the following format: $b_j(v_j) = \tanh(\lambda v_j)$. The network training is performed to minimize the least squares criterion between the predicted ($\hat{y}_k$) and actual (y_k) responses: $E = \sum_{k=1}^{n} (y_k - \hat{y}_k)^2$.

The input data set is generated from SPICE simulations, is the same for every metamodel and is generated using LHS. LHS supports any amount of planes and is proven to work better than Monte Carlo due to the more even distribution of points with still the random factor that helps to detect nonlinearity. LHS divides each plane (parameter) into Latin squares and randomly picks a point from each square. Output is generated for each run from a SPICE simulation saving each needed value to its own data set. Hence, each metamodel will have its own target data set. Since the input data set has a large dynamic range, either normalization or standardization of the input data can improve numerical stability. If not, the training of higher values can outweigh the lower and neural network will not train properly. In this paper, the data sets are normalized to mean 0 and standard deviation 1, as experimental studies showed that the normalization is easier to handle than standardization of data even though neural networks performed much better with either than without one.

The validation and test data must be also normalized using the statistics (μ and σ) that were computed from the training data. Since a neural network is created for each desired output there is no need to standardize the output. The output standardization is usually used if there are more than one output and they are in different order, hence affecting the way weights converge during the learning process. The statistical data is then collected to calculate root mean square error (RMSE) and coefficient of regression (R^2) values for both sets. Since there may be numerous metamodels created from the same sample set. RMSE and R^2 are the metrics used for goodness of fit. The created model may fit perfectly the training data set even though it may not qualify as a good model to represent the output for the given process at other points. Hence the verification data set is created so that the points are at the different locations than the sample data. If the verification dataset RMSE and R^2 values do not differ much from the training values, then the model has trained correctly, otherwise it must have been overfitted or trained improperly. If the neural model did not train correctly, the training parameters of the model can be adjusted or additional sample points can be collected from the circuit simulation. Otherwise the neural network can act as an accurate metamodel for the PLL circuit. A total of 200 simulations were needed to create the metamodels.

4.2 Proposed Method for Statistical Process Variation Analysis

In a typical approach, Monte Carlo (or its derivatives) are used *on the actual circuits (i.e. netlists)*. This is prohibitively slow. In our approach, the non-polynomial metamodels are used instead, thus speeding up significantly the statistical process variation analysis.

A Monte Carlo (MC) analysis of the PLL circuit containing full-blown parasitics is performed first for comparison purposes. The mean (μ) and standard deviations (σ) of the 4 Figures-of-Merit (FoMs) are presented in Table 1. It may be noted that **the processing time for 1000 MC runs on the actual circuit (i.e. netlist) was approximately 5 days**.

In the current non-polynomial metamodel-based design flow, the process variation analysis is conducted over the metamodels. The MC analysis is performed on the metamodels that show the best fit. The results are compared with the 1000 circuit Monte Carlo simulations. The PDFs extracted from the Monte Carlo analysis on the neural networks is shown in Fig. 2. The statistical parameters of these PDFs are also presented in Table 1.

5. PROPOSED PARTICLE SWARM OPTIMIZATION (PSO) ALGORITHM

The PSO algorithm, uses multiple particles to find a solution based on the cost function. The particle movement is calculated based on the local intelligence of each particle which is offset using global knowledge. The steps of the approach are shown in Algorithm (1). Each particle location information holds a 35 dimensional location, where each dimension corresponds to a parameter. The algorithm starts at a random location of each parameter for each particle, with random velocity. When the information is acquired from the cost function $f(p_i)$ the minimum global position is retained in $f(g)$, while the local position is saved for each particle in p_i. With each iteration of the loop, while the amount of iteration is not reached, the particles keep searching for a minimum solution in the design space by updating the particle velocity vector v_i.

For the calculation of cost function, the Monte Carlo analysis is done around the x_i parameter points. To minimize the amount of calculations the analysis is only done on the frequency metamodel first. If the mean and standard deviation is within the specifications, then the rest of the metamodels are used to calculate the mean and standard deviation for other FoMs. Before the calculation of the

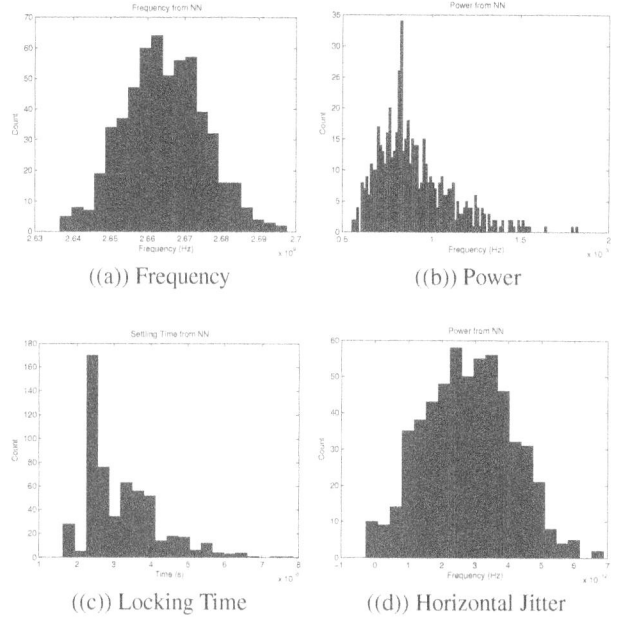

((a)) Frequency   ((b)) Power

((c)) Locking Time   ((d)) Horizontal Jitter

Figure 2: Statistical analysis of the FoMs of the PLL using the *neural network metamodels*.

cost value is done, all the values are brought to the same power to even their weight composition on the circuit.

6. RESULTS AND CONCLUSIONS

PSO is used for optimization using neural network metamodels. Two optimization cases are considered in this paper. The first case optimizes the average case, while the worst case scenario is considered in the second case. A total of 35 parameters were used for both optimization cases. For Monte Carlo analysis each parameter is varied by 5% around the mean for process variation. The optimization results are shown in Table 2.

In this paper it is shown that neural network based metamodels can closely follow the behavior of a circuit for process variation and global optimization. On an example PLL system, the models were generated from the physical layout netlist for 4 different components. An error under 2% has been observed in the models for process variation analysis for mean and standard deviation. The PSO algorithm has been successfully used to bring the circuit back to specifications, even though the physical layout did not meet them. The processing time for running 1000 Monte Carlo analysis on the PLL system was approximately 5 days, in comparison to 200 simulations needed to create the metamodels and running the algorithm. The proposed design flow has reached a speedup of roughly 5 $\times$ over the Monte Carlo analysis without any iterative modifications to the physical layout.

Acknowledgements

This research is supported in part by SRC award P10883 and NSF awards CNS-0854182 and DUE-0942629.

7. REFERENCES

[1] K. Bernstein, D. J. Frank, A. E. Gattiker, W. Haensch, B. L. Ji, S. R. Nassif, E. J. Nowak, D. J. Pearson, and N. J. Rohrer. High-Performance CMOS Variability in the 65nm Regime

Table 1: Before Optimization: Statistical Figures of Merit of the PLL.

	Circuit Monte Carlo		Neural Network Monte Carlo		Error	
	Mean (μ)	Standard Deviation (σ)	Mean (μ)	Standard Deviation (σ)	Mean (μ)	Standard Deviation (σ)
Frequency	2.66 GHz	10.95 MHz	2.66 GHz	10.96 MHz	0.0%	0.11%
Power	0.90 mW	0.21 mW	0.90 mW	0.21 mW	0.14%	1.3%
Locking Time	3.24 μs	1.07 μs	3.22 μs	0.99 μs	0.7%	6.93%
Horizontal Jitter	2.79 ps	1.32 ps	2.80 ps	1.32 ps	0.12%	0.5%

Table 2: After Optimization: Statistical Figures of Merit of the PLL.

	$\mu + \sigma$ Optimization		$\mu + 3\sigma$ Optimization	
	Mean (μ)	Standard Deviation (σ)	Mean (μ)	Standard Deviation (σ)
Frequency	2.75 GHz	28.64 MHz	2.74 GHz	29.14 MHz
Power	0.99 mW	0.28 mW	0.98 mW	0.27 mW
Locking Time	4.69 μs	1.15 μs	4.61 μs	1.13 μs
Horizontal Jitter	5.82 ps	3.42 ps	5.97 ps	3.34 ps

Algorithm 1 The Proposed Particle Swarm Optimization (PSO) for the PLL Components.

1: **Set** N - number of particles
2: **Start** at a random location with uniform distribution
3: **Get** current position x_i and use it initially for best particle position $f(p_i)$ and $f(g) = min(p_i)$
4: v_i $U(min_{p\_i}, max_{p\_i})$
5: **Initialize** iter=0
6: **Initialize** weight for swarm effect ϱ_p
7: **Initialize** weight for swarm effect ϱ_p
8: **Initialize** weight for velocity effect (acceleration/inertia) w
9: **while** iter<$max_{iterations}$ **do**
10:     **for** each i **do**
11:         $v_i = \omega v_i + \varrho_p \tau_p (p_i - x_i) + \varrho_g \tau_g (g - x_i)$
12:         $x_i \leftarrow x_i + v_i$
13:         **if** $f(x_i) < f(p_i)$ **then**
14:             **update** position: $p_i \leftarrow x_i$
15:             **if** $f(p_i) < f(g)$ **then**
16:                 $g \leftarrow p_i$
17:             **end if**
18:         **end if**
19:     **end for**
20: **end while**

---

and Beyond. *IBM Journal of Research and Development*, 50(4/5):433–449, July-Sep 2006.

[2] S. Borkar, T. Karnik, S. Narendra, J. Tschanz, A. Keshavarzi, and V. De. Parameter Variations and Impact on Circuits and Microarchitecture. In *Proceedings of the 40th Annual Design Automation Conference*, DAC '03, pages 338–342, 2003.

[3] K. Bowman, S. Duvall, and J. Meindl. Impact of Die-to-Die and Within-Die Parameter Fluctuations on the Maximum Clock Frequency Distribution for Gigascale Integration. *IEEE Journal of Solid-State Circuits*, 37(2):183–190, February 2002.

[4] H.-S. Jeon, D.-H. You, and I.-C. Park. Fast frequency acquisition all-digital PLL using PVT calibration. In *Proceedings of the IEEE International Symposium on Circuits and Systems*, pages 2625–2628, 2008.

[5] C.-C. Kuo, M.-J. Lee, C.-N. Liu, and C.-J. Huang. Fast Statistical Analysis of Process Variation Effects Using Accurate PLL Behavioral Models. *IEEE Transactions on Circuits and Systems I: Regular Papers*, 56(6):1160–1172, June 2009.

[6] J. Lin, B. Haroun, T. Foo, J.-S. Wang, B. Helmick, S. Randall, T. Mayhugh, C. Barr, and J. Kirkpatric. A pvt tolerant 0.18mhz to 600mhz self-calibrated digital pll in 90nm cmos process. In *Proceedings of the IEEE International Solid-State Circuits Conference*, pages 488–541, 2004.

[7] D. Miyashita, H. Ishikuro, S. Kousai, H. Kobayashi, H. Majima, K. Agawa, and M. Hamada. A phase noise minimization of cmos vcos over wide tuning range and large pvt variations. In *Proceedings of the IEEE Custom Integrated Circuits Conference*, pages 583–586, 2005.

[8] J. Park, K. Choi, and D. Allstot. Parasitic-aware Design and Optimization of a Fully Integrated CMOS Wideband Amplifier. In *Asia and South Pacific Design Automation Conference, 2003. Proceedings of the ASP-DAC 2003.*, pages 904–907, Jan. 2003.

[9] R. Wu, J.-C. Wang, K.-W. Xia, and R.-X. Yang. Optimal Design on CMOS Operational Amplifier with QPSO Algorithm. In *International Conference on Wavelet Analysis and Pattern Recognition, 2008. ICWAPR '08.*, volume 2, pages 821 –825, Aug 2008.

[10] Y. Yang, L. Yang, and Z. Gao. A PVT Tolerant sub-mA PLL in 65nm CMOS Process. In *Proceedings of the 15th IEEE International Conference on Electronics, Circuits and Systems*, pages 998–1001, 2008.

A Denial-of-Service Resilient Wireless NoC Architecture

Amlan Ganguly
Department of Computer Engineering
Rochester Institute of Technology
Rochester, NY, USA
+1-585-475-4082
amlan.ganguly@rit.edu

Mohsin Yusuf Ahmed
Department of Computer Engineering
Rochester Institute of Technology
Rochester, NY, USA
+1-585-475-7700
mya6569@rit.edu

Anuroop Vidapalapati
Department of Electrical and
Microelectronic Engineering
Rochester Institute of Technology
Rochester, NY, USA
acv1076@rit.edu

ABSTRACT
Wireless Network-on-Chip (NoC) architectures have emerged as an enabling solution to design scalable NoC fabrics for massive many-core chips. However, such massive levels of integration of Intellectual Property (IP) cores make the chips vulnerable to malicious intrusions from untrustworthy processes or vendors. Hence, resilience to various types of hardware security threats is imperative in future many-core chips. In this paper we develop a design methodology to increase the resilience of a wireless NoC to Denial-of-Service (DoS) attacks. We demonstrate that the proposed architecture can sustain higher data transfer rates at lower energy dissipation with the spread of DoS attacks compared to conventional mesh based NoCs.

Categories and Subject Descriptors
B.8.0 [**Performance and Reliability**]: General

General Terms
Performance, Design, Security.

Keywords
Network-on-Chip, Wireless interconnect, Denial-of-Service.

1. INTRODUCTION
To enhance the performance of conventional metal interconnect-based Network-on-Chip (NoC) [1] on-chip wireless interconnects have emerged as a radically different technology. Low power, high bandwidth wireless links when deployed strategically in a NoC is capable of improving latency and power dissipation in data transfer [2]. While wireless NoC architectures make it possible to integrate an increasing number of cores on the same die within reasonable energy and delay budgets the many-core paradigm itself presents significant challenges pertaining to hardware trust and security [3]. As more parts of the IC design and manufacturing process are outsourced to reduce costs, the integration of many-core chips becomes vulnerable to intrusions by malicious circuitry. Untrustworthy vendors or fabrication processes can insert such malicious components into the chip. Such malicious circuitry is commonly known as hardware trojans. Typically these trojans when triggered can start malicious behavior in a normal chip. The effect of hardware trojans can manifest in several ways such as Denial-of-Service (DoS), data corruption, eavesdropping or leakage of sensitive information and hijacking or altering the configurations of the cores [4]. Such

Permission to make digital or hard copies of all or part of this work for personal or classroom use is granted without fee provided that copies are not made or distributed for profit or commercial advantage and that copies bear this notice and the full citation on the first page. To copy otherwise, or republish, to post on servers or to redistribute to lists, requires prior specific permission and/or a fee.
GLSVLSI'12, May 3–4, 2012, Salt Lake City, Utah, USA.
Copyright 2012 ACM 978-1-4503-1244-8/12/05...$10.00.

many-core chips deployed in sensitive applications in military, banking, space research and communication sectors can prove catastrophic. In the presence of such vulnerabilities in many-core chips the NoC fabric must be designed in a way to effectively eliminate or minimize its effects. DoS is one of the most common effects of hardware trojans. In this case an infected core can either inject excessive garbage traffic into the network or corrupt the router in the NoC associated with the malicious core. The affected router then fails to route data that passes through that switch. This can result in congested or non-functional NoC switches. In case of wireless NoCs the effect of DoS can be even more severe if switches associated with wireless interconnects (WiSs) are infected. This is because the on-chip wireless bandwidth is a limited resource which when optimally deployed results in gains in performance. In case the wireless bandwidth is excessively used for useless data transfer, traffic may be starving for access to the wireless medium for transmission. Thus the projected gains of using this emerging interconnect technology can be completely nullified by DoS attacks on the WiSs in the WiNoC.

In this work we study the effect of DoS attacks on the wireless NoC fabrics and compare that with the effect on a conventional wireline mesh NoC. Counter measures such as quarantining the affected parts in a network in case of security breaches often have significant latencies. In this work we propose a design methodology for wireless NoC architectures that can inherently sustain a better power-performance characteristics with the spread of a DoS attack across the NoC. This will help other complimentary counter measures in arresting the attack effectively before the whole NoC is infected.

2. RELATED WORK
In recent years significant effort has been dedicated to protect processors from security vulnerabilities [5, 6]. A bus-based SoC was designed to have enhanced security features against DoS attacks in [7]. With the advent of the many-core paradigm and the increasing importance of the on-chip communication fabric, research for enhancing the NoC with additional security features has gained momentum. In [4] a security measure for NoC fabrics was proposed by separating regions of the chip as well as data into secure and non-secure parts. The authors of [8] propose a NoC security protocol by incorporating counter measures through various approaches such as a single secure master IP that can manage the whole NoC, secure network interfaces, multiple virtual channels or verification of access rights. An encryption based NoC security protocol was proposed in [9]. A secure memory access protocol was proposed in [10]. In [11] a monitoring system inspired from biological systems was proposed to enhance security in a traditional wireline NoC. In [12] the authors designed a nature-inspired inherently fault-tolerant wireless NoC architecture. This work however, did not address the issue of targeted attacks in wireless NoCs. Although

bandwidth reduction attacks can have more pronounced effects on wireless NoCs no significant effort has been made to make them secure. In this work we propose a design methodology for wireless NoCs, which will be resilient to DoS or bandwidth reduction, attacks.

3. The DoS Attack Model

One of the most common manifestations of security attacks in a NoC fabric is DoS targeted at bandwidth reduction. DoS attacks in a WiNoC can have severe bandwidth reduction effects. Here we consider a situation when a hardware trojan in a core can trigger excessive injection of garbage traffic into the network. These packets will occupy all the virtual channels of the output ports in the switch associated with the core. Hence, this causes congestion in the switch and prevents traffic movement outwards from that switch. We will refer to this switch as the primary victim. With time, incoming traffic towards this particular switch will also be blocked, as there will be no available virtual channel in the output ports. This will cause congestion in the output ports of the switches, which are immediate neighbors of the primary victim. We will refer to these affected switches as the second level victims. This in turn will cause congestion in the third level victim switches and so on. Thus such a DoS attack will propagate from the primary victim to multiple levels of neighboring switches. In figure 1 we show this pattern of spreading of the DoS attack in a mesh based NoC starting with a switch at the centre. At each stage a new level of switches are infected starting from a single victim.

4. DoS Resilient Wireless NoC Architecture

Many naturally occurring large irregular networks like microbial colonies, cortical interconnections, the internet or social groups are found to display a striking resilience to faults either in the form of random failures or targeted attacks. Theoretical studies in complex networks reveal that certain types of network connectivity are inherently resilient to attacks [13]. A particular class of complex networks called small-world networks is characterized by many short-distance links between neighboring nodes as well as a few relatively long-distance direct shortcuts. Due to these shortcuts, small-world networks have very low average distance between nodes making them quite attractive for designing scalable NoC fabrics [14]. Such networks have a fairly homogeneous structure where all nodes have almost the same number of links. Hence, in such networks all the nodes contribute almost equally to the interconnectedness of the network. Due to this homogeneity of the network both targeted attacks and random failures marginally affect the overall connectivity of this network. The non-deterministic structure of the network coupled with its homogeneity would make it difficult for an attacker to isolate a particularly vulnerable node in the NoC. On the contrary, regularly designed NoC architectures such as mesh or tree have obvious vulnerabilities such as the central switch or the root of the

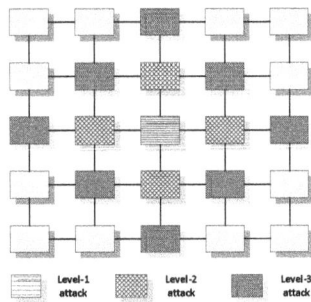

Figure 1. Spread of DoS attack across multiple levels.

tree respectively. Consequently, adopting novel architectures inspired by such natural complex networks in conjunction with the emerging interconnection technologies will enable design of high-performance many-core chips resilient to DoS attacks. Next, we outline the method of designing a DoS secure wireless NoC (SecWiNoC) based on the small-world topology. After that we describe an optimization step to further make the WiNoC resilient to spreading of the DoS over multiple levels of victims.

4.1 Small-world SecWiNoC Topology

In the proposed SecWiNoC topology, each core is connected to a NoC switch and the switches are interconnected using wireline and wireless links. The topology of the SecWiNoC is a small-world network where the links between switches are established following an inverse power law distribution as shown in (1).

$$P(i,j) = \frac{l_{ij}^{-\alpha} f_{ij}^{\beta}}{\sum_{\forall i} \sum_{\forall j} l_{ij}^{-\alpha} f_{ij}^{\beta}} \qquad (1)$$

Where, the probability of establishing a link, between two switches, i and j, $P(i,j)$, separated by an Euclidean distance of l_{ij} is proportional to the distance raised to a finite power [15]. The distance is obtained by considering a tile-based floorplan of the cores on the die. The frequency of traffic interaction between the cores, f_{ij}, is also factored into (1) so that more frequently communicating cores have a higher probability of having a direct link. This frequency is expressed as the percentage of traffic generated from i that is addressed to j and is based on the particular application mapped to the overall NoC. This optimizes the network architecture for non-uniform traffic scenarios. The parameters, α and β govern the nature of connectivity and the significance of the traffic pattern on the topology respectively. The value of α was chosen to be 1.8 to establish a small-world connectivity. β is chosen to be 1 to account for traffic interactions while establishing the links. To establish the network connectivity each pair of switch in the NoC is selected and a link is established between them with the probability given in (1). The network setup is repeated until a fully connected network is formed.

4.2 Optimization for DoS Resilience

Following the initial network setup, an optimization by means of a Simulated Annealing (SA) heuristics is performed in order to minimize the effect of spreading of the DoS attack. To perform SA, we establish a metric, which reflects the effect of spreading of DoS over the NoC. The average distance in number of hops between the switches in the SecWiNoC, μ indicates the interconnectedness of the NoC. The metric to be optimized should therefore be the decrease in μ as the DoS spreads over the network via multiple levels of victim nodes. Consequently, the optimization metric, ρ, for the SA algorithm is given in (2).

$$\rho = \frac{|\Delta\mu|}{l} \qquad (2)$$

Where, $\Delta\mu$ is the decrease in μ with l levels of switches being affected by the attack. The goal of the SA optimization is to minimize the metric ρ. In each iteration of the SA process, a new candidate network is created by randomly rewiring a link in the current network between a different source and destination pair, which are not already directly connected. The optimization metric is then computed for the candidate as the change in μ for 3 levels of DoS spread. In this work we have used Cauchy scheduling, where the temperature varies inversely with the number of iterations. The SA iterations were terminated when the moving average of the metric for the last 10 iterations was close enough to zero. The result of the SA optimization is the optimized SecWiNoC (OSecWiNoC) topology with minimal change in μ with spread in the DoS attack.

4.3 Adopted Physical Layer Design for the Interconnects

The basic architecture of the SecWiNoC is a hybrid network consisting of both wireline and wireless interconnects. As in [2] the wireless interconnects can be realized with miniature on-chip antennas based on carbon nanotube (CNT) structures. Antenna characteristics of CNTs in the THz frequency range have been investigated both theoretically and experimentally [16]. CNT antennas operating in the THz/optical frequency range can support very high data rates. As noted in [2] 24 different frequency channels can be created by external laser sources exciting the antennas enabling Frequency Division Multiplexing (FDM). At each WiS the carrier is modulated by Mach-Zehnder Modulator/Demodulators performing On-Off Keying (OOK) at 10Gbps. As only 24 wireless links can be created using the CNT antennas. The rest of the interconnects in the OsecWiNoC are realized with wireline links. If the wireline links are long enough to take more than one clock cycle for transmission of a flit they are pipelined by insertion of FIFO buffers such that between any two stages it is possible to transfer an entire flit in one clock cycle.

4.4 Adopted Data Routing Mechanism

We adopt a wormhole routing policy for the proposed NoC in both wired and wireless links. The small-world topology is essentially an irregular network. The adopted routing policy should not introduce substantial computational overheads and hence be distributed in nature. In addition it should be deadlock and livelock free as well. In order to achieve this we have adopted the Tree-based Routing Architecture for Irregular Networks (TRAIN) algorithm [17]. A Minimum Spanning Tree (MST) of the network is created with a randomly selcted node as the root, and data routed along the MST. The other links not included in the MST are then re-introduced as shortcuts. An allowed route never uses a link in the up direction along the tree after it has been in the down path once. The shortcuts are used for packet transmission only if they provide a shorter path to the destination than the route along the MST while flits are routed upwards along a branch. Hence, channel dependency cycles are prohibited and deadlock freedom is achieved. Livelock is avoided because each packet has a fixed path from its source to destination.

5. Experimental Results

In this section we present experimental results to show that the SecWiNoC is able to sustain higher data transfer rate compared to a traditional mesh architecture in presence of DoS attacks spreading over the NoC. We consider the DoS model as explained in section 3. We characterize the NoC architectures using a cycle accurate simulator which models the progress of data flits accurately per clock cycle accounting for flits that reach the destination as well as those that are dropped. One hundred thousand iterations were performed to reach stable results in each experiment, eliminating the effect of transients in the first few thousand cycles. A system size of 64 cores is considered for all our experiments to represent current trends in many-core designs. The width of all wired links is considered to be same as the flit size, which is 32 in this paper. The particular NoC switch architecture has three functional stages, namely, input arbitration, routing/switch traversal, and output arbitration. The input and output ports have four virtual channels per port, each having a

buffer depth of 2 flits. Each packet consists of 64 flits. The network switches are synthesized from a RTL level design using 65nm standard cell libraries from CMP (http://cmp.imag.fr), using Synopsys. The NoC switches are driven with a clock of frequency 2.5 GHz. The delays and energy dissipation on the wired links were obtained through Cadence simulations taking into account the specific lengths of each link based on the established connections in the 20mmx20mm die following the logical connectivity of the small-world topology. The energy dissipations on the wireless links were obtained through analytical and experimental findings in [16]. As outlined in [12] the energy dissipation on the longest wireless link considering all the overheads of various components is 0.33pJ/bit.

It has already been shown in [12] that the peak achievable throughput of a wireless NoC with CNT antennas is much higher than that of a conventional wireline mesh. Here, we show how the peak throughput of the proposed architectures varies as the DoS attack spreads across multiple levels of victims. Throughput is measured as the number of flits successfully routed to a core per cycle on an average. Figure 2 shows the decrease in throughput for a conventional mesh based NoC with 64 cores with spread of the DoS attack for uniform spatial traffic. As the hardware trojan can be inserted in any core affecting the switch attached to that core first, the origin of the DoS attack can also be at any of the switches in the mesh. We show 2 representative cases. First when the attack starts at a switch near the centre of the NoC and in the second case when the attack begins at a switch in one of the corners of the mesh. It can be observed that the sustainable peak throughput of a mesh falls rapidly with the spread of the attack when the central switch is the origin. This is because the central switch is responsible for a much larger volume of data transfer compared to the switches in the corners as it is in the path of a higher number of source-destination pairs. As the impact of the DoS spreads to the next level to the neighbors of the central switch another four switches are affected causing a large part of the NoC to be nonfunctional at the center of the chip. Thus it cannot sustain any measurable throughput in this case. Although a switch at a corner is responsible for less traffic the spread of the attack on second level switches affects the mesh severely and hence the throughput is drastically reduced. As can be seen, the spread of DoS to only second level of switches in both cases paralyzes the entire mesh NoC. The number of links (degree) of each switch can be different in the small-world topology. In figure 2 we also show results of case studies with DoS attacks originating in switches with both lowest and highest degrees for the SecWiNoC as well the OSecWiNoC. The degradation in throughput is more for the DoS originating in the switch with highest degree in both architectures. However, even in the worst case, the throughput is zero only after the attack spreads to 4 levels. We can see that the degradation in throughput for the optimized architecture is slower with the spread of the DoS attack

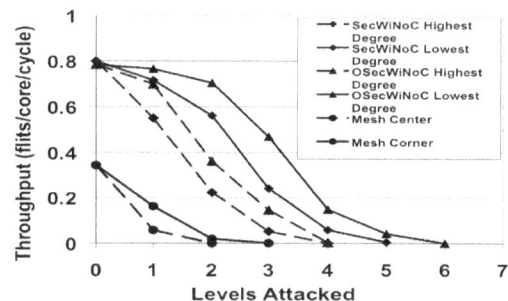

Figure 2. Decrease in throughput with spread of DoS

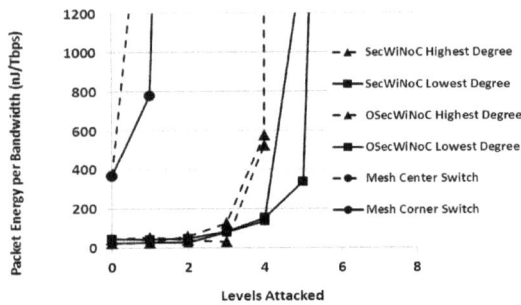

Figure 3. Packet Energy per Bandwidth of various architectures considered for best and worst case attacks.

for both cases. This is because the topology is optimized such that there is minimal change in average distance between cores even with the spread of the attack.

Figure 3 shows the packet energy per bandwidth of a mesh and SecWiNoC architectures as the DoS spreads from the originating location to its neighbors and so on. Packet energy per bandwidth is the average energy dissipation in transmission of a packet between source and destination per unit peak bandwidth. This represents the energy delay product and enables a fair comparison between NoCs with different data transfer rates. It can be observed that the proposed architectures have much lower packet energy per bandwidth due to the short average path lengths of the small-world topology compared to mesh as well as the low power wireless links. However, the packet energy increases drastically when the network cannot sustain any data transfer, as it takes infinitely long for data packets to get routed. This happens when the DoS spreads to the second level neighbors in a mesh NoC. However, the SecWiNoC resists the change in packet energy per bandwidth even when the attack spreads over more levels compared to a mesh. In the best case the optimized SecWiNoC architecture is able to achieve data transfer with reasonable energy dissipation and delay even when attack spreads to 5 levels.

In figure 4 we show the variation in throughput in an OSecWiNoC for attacks originating in a switch with highest degree for different application based traffic patterns. We study the resilience to DoS spread for hotspot, transpose, FFT and Matrix Multiplication traffic patterns. The performance degradation with spread of DoS attacks with non-uniform traffic patterns is slower than with uniform traffic. This is because certain paths are responsible for more traffic than others. Unless those paths are targeted, the DoS affects the NoC less severely.

In order to achieve this resilience to the spread of DoS attacks no additional hardware is used in the proposed SecWiNoC architecture. As the number of links in the entire NoC is same as that in a mesh the total number of ports of all switches combined is also same, making the overall switch area quite comparable to a

mesh. Based on the estimates in [2] the additional area due to the wireless transceivers and CNT antennas is 0.085mm^2.

6. CONCLUSIONS

Hardware trojans in future NoC based many-core systems can result in catastrophic failures. In this work we have proposed a design methodology for wireless NoC architectures such that the impact of a DoS attack originated by hardware trojan can be minimized. We have demonstrated that a small-world network based Wireless NoC architecture can sustain higher data throughput for a more severe spread of the attack compared to a conventional mesh architecture. In our future work we intend to develop a reactive routing mechanism to reroute the blocked data due to a DoS attack and compare the performance of the proposed methodology with other wireless NoC architectures.

7. REFERENCES

[1] L. Benini and G. D. Micheli, "Networks on Chips: A New SoC Paradigm," IEEE Computer, Vol. 35, Issue 1, January 2002, pp. 70-78.

[2] A. Ganguly, et al., "Scalable Hybrid Wireless Network-on-Chip Architectures for Multi-Core Systems", IEEE Transactions on Computers, 2010, vol. 60, issue 10, pp. 1485-1502.

[3] "Can we Trust Chips of the Future?", IEEE Design and Test of Computers, Vol. 28, 2011, pp. 96-103.

[4] S. Evain and J. Diguet, "From NoC Security Analysis to Design Solutions," Proc. IEEE Workshop Signal Processing Systems Design and Implementation, 2005, pp. 166-171.

[5] J. Coburn, et al., "SECA: Security-Enhanced Communication Architecture", Proc. of International Conference on Compilers, Architectures and Synthesis for Embedded Systems 2005, pp. 78-89.

[6] G. E. Suh, et al., "Design and Implementation of the AEGIS Single-Chip Secure Processor," Proc. of Annual International Symposium on Computer Architecture, 2005, pp. 25-26.

[7] L. Kim and J. D. Villasenor, "A System-On-Chip Bus Architecture for Thwarting Integrated Circuit Trojan Horses", IEEE Transactions on VLSI, Vol. 19, No. 10, 2011, pp. 1921-1926.

[8] J.P. Diguet, et al., "NoC-Centric Security of Reconfigurable SoC," Proc.of International Symposium on Networks-on-Chip, 2007, pp. 223-232.

[9] C.H. Gebotys and R.J. Gebotys, "A Framework for Security on NoC Technologies," Proc. of Annual Symposium on VLSI, 2003, pp. 113-117.

[10] L. Fiorin, et al., "Secure Memory Accesses on Networks-on-Chip," IEEE Transactions On Computers, Vol. 57, no. 9, 2008, pp. 1216-1229.

[11] A. Mandal, et al., "A Bio-inspired Framework for Secure System on Chip," Proc. of Workshop on SoC Architecture, Accelerators and Workload, 2010.

[12] A. Ganguly et al., "Complex Network Inspired Fault-Tolerant NoC Architectures with Wireless Links", Proc of International Symposium on Network-on-Chip, 2011, pp. 169-176.

[13] R. Albert, et al., "Error and Attack Tolerance of Complex Networks", Nature, Vol. 406, July 2000, pp. 378-382.

[14] U. Ogras and R. Marculescu, "It's a Small World After All: NoC Performance Optimization Via Long-Range Link Insertion", IEEE Transactions on VLSI, Vol. 14, No. 7, 2006, pp. 693-706.

[15] D. J. Watts and S. H. Strogatz, "Collective dynamics of 'small-world' networks," Nature 393, 440–442, 1998.

[16] K. Kempa, et al., "Carbon Nanotubes as Optical Antennae," Advanced Materials, vol. 19, 2007, pp. 421-426.

[17] H. Chi and C. Tang, "A Deadlock-Free Routing for Interconnection Networks with Irregular Topologies", Proc. Of International Conference on Parallel and Distributed Systems, 1997, pp. 88-95.

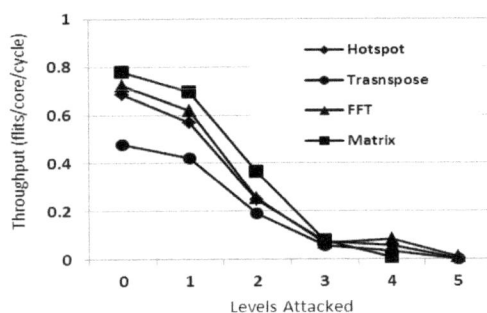

Figure 4. Throughput degradation with non-uniform traffic.

Sustainable Multi-Core Architecture with on-chip Wireless Links

Jacob Murray, Partha Pande, Behrooz Shirazi
School of Electrical Engineering and Computer Science
Washington State University
{jmurray, pande, shirazi}@eecs.wsu.edu

John Klingner
School of Computer Science
Cornell College
JKlingner12@cornellcollege.edu

ABSTRACT

Current commercial systems on chip (SoC) designs integrate an increasingly large number of pre-designed cores and their number is predicted to increase significantly in the near future. Specifically, molecular-scale computing will allow single or even multiple order-of-magnitude improvements in device densities. In the design of high-performance massive multi-core chips, power and temperature have become dominant constraints. Increased power consumption can raise chip temperature, which in turn can decrease chip reliability and performance and increase cooling costs. The new, ensuing possibilities in terms of single chip integration call for new paradigms, architectures, and infrastructures for high bandwidth and low-power interconnects. In this paper we demonstrate how small-world Network-on-Chip (NoC) architectures with long-range wireless links enable design of energy and thermally efficient sustainable multi-core platforms.

Categories and Subject Descriptors

C.5.4 [**Computer System Implementation**]: VLSI Systems

Keywords

Multi-Core, Small-World, Wireless NoC, Energy, Temperature.

1. INTRODUCTION

Modern multi-core chips are all pervasive in several domains ranging from climate forecasting and astronomical data analysis, to consumer electronics, smart phones, and biological applications. Design technologies in the era of massive integration present unprecedented advantages and challenges, the former being related to very high device densities, while the latter to soaring power dissipation issues. However, given the current trends in terms of power and performance figures, it is difficult to design future thousand-core platforms using existing methodologies and tools. Trying to build from existing techniques, traditionally applied to a relatively low number of processors, towards such large-scale systems is not appropriate. We need to look into new and far-reaching design methodologies to break the energy efficiency wall in massively integrated single-chip computing platforms [1]. NoC has emerged as an enabling methodology to integrate many embedded cores on a single die. The existing method of implementing a NoC with planar metal interconnects is deficient due to high latency and significant power consumption arising out of multi-hop links used in data exchanges. Increased power consumption will give rise to higher temperature, which in turn can decrease chip reliability and

performance and increase cooling cost [2]. This limitation of conventional NoCs can be addressed by drawing inspiration from the interconnection mechanism of natural complex networks. One possible innovative and novel approach is to replace multi-hop wired paths in a NoC by high-bandwidth single-hop long-range wireless links [3] [4]. The on-chip wireless links facilitate the design of a small-world NoC by enabling one-hop data transfers between distant nodes. In addition to reducing interconnect delay, eliminating multi-hop long distance wired communication reduces the energy dissipation as well. Consequently, this will enable lowering of temperature hotspots in specific regions of the chip. In this work our aim is to show how a small-world NoC architecture with long-range wireless links not only improves the latency and energy dissipation characteristics of a multi-core chip, but also helps to reduce the temperature.

2. RELATED WORK

The limitations and design challenges associated with existing NoC architectures are elaborated in [5], which highlights interactions among various open research problems of the NoC paradigm. NoCs have been shown to perform better by insertion of long range wired links following principles of small-world graphs [6]. Despite significant performance gains, the above schemes still require laying out long wires across the chip and hence performance improvements beyond a certain limit cannot be achieved [7].

Designs of small-world based hierarchical wireless NoC architectures were introduced and elaborated in [3] [4]. Recently, the design of a wireless NoC based on CMOS ultra wideband (UWB) technology was proposed [8].

Most of the existing works related to the design of wireless NoC demonstrate its advantages in terms of latency and energy dissipation. However, none of those addresses the correlation between the energy dissipation and temperature profile of a wireless NoC. In this work, we aim to bridge that gap by quantifying how a small-world based wireless NoC architecture reduces the heat dissipation and hence improves the thermal profile of a multi-core chip in addition to improving the latency and energy dissipation characteristics.

3. WIRELESS MESH ARCHITECTURE

Traditionally, a mesh is the most popular NoC architecture due to its simplicity and the regularity of grid structure [5]. However, one of the most important limitations of this architecture is the multi-hop communications between far apart cores, which gives rise to significant latency and energy overheads. To alleviate these shortcomings, long-range and single-hop wireless links are inserted as shortcuts on top of a mesh. It is shown that insertion of long-range wireless shortcuts in a conventional wired NoC has the potential for bringing significant improvements in performance and energy dissipation [3] [4]. Inserting the long-range links in a conventional wireline mesh reduces the average hop count, and increases the overall connectivity of the NoC. Furthermore, in this wireless mesh (WiMesh), by careful placement of wireless links depending on distance and traffic patterns we enable savings in latency, energy, and heat dissipation.

4. WIRELESS LINK PLACEMENT

We propose to use THz wireless links designed with carbon nanotube (CNT) antennas. It is already shown that this type of wireless link enables design of an energy efficient wireless NoC architecture [3]. By using multiband laser sources to excite CNT antennas, different frequency channels can be assigned to pairs of communicating source and destination nodes. This will require using antenna elements tuned to different frequencies for each pair, thus creating a form of frequency division multiplexing (FDM). High directional gains of these antennas aid in creating directed channels between source and destination pairs. In [9], it is shown that 24 continuous wave laser sources of different frequencies can be created. Thus, these 24 different frequencies can be assigned to multiple wireless links in the NoC in such a way that a single frequency channel is used only once to avoid signal interference on the same frequencies.

The placement of the links is dependent upon three main parameters, the number of cores, the number of long range links to be placed, and the traffic distribution. The aim of the wireless link placement is to minimize the hop count of the network. As discussed in [3], we optimize the average hop count weighted by the probability of traffic interactions among the cores. In this way equal importance is attached to both inter-core distance and frequency of communication. A single hop in this work is defined as the path length between a source and destination pair that can be traversed in one clock cycle. It is shown in [3] that for placement of wireless links in a NoC, the Simulated Annealing (SA)-based methodology converges to the optimal configuration much faster than the exhaustive search technique. Hence, we adopt a SA based optimization technique for placement of the wireless links in this work to get maximum benefits of using the wireless shortcuts. We also need to ensure that introduction of the wireless shortcuts does not give rise to deadlocks in data exchange. The routing adopted here is a combination of dimension order (X-Y) routing for the nodes without wireless links and South-East routing for the nodes with wireless shortcuts. This routing algorithm has been proven to be deadlock free in [6]. Figure 1 shows a sample link placement of a 16-core WiMesh network with FFT based traffic.

To benchmark the effectiveness of this WiMesh architecture, we evaluate its latency, energy dissipation and thermal profiles by using several SPLASH-2 benchmarks [10].

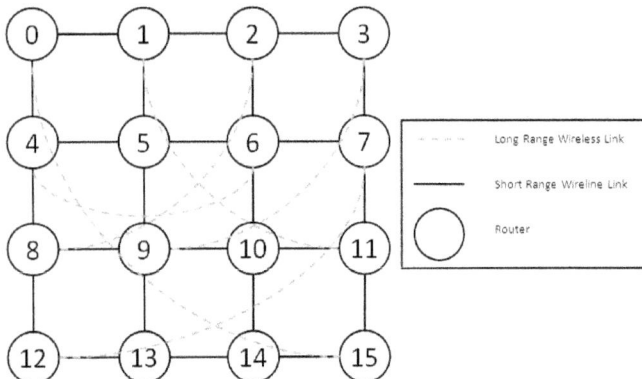

Figure 1. Sample Wireless Link Placement, 16-Core FFT

5. PERFORMANCE EVALUATION

In this section we characterize the performance of the proposed WiMesh architecture with long-range wireless links through detailed full system simulations. We use the GEM5 [11] platform, with Ruby and Garnet for carrying out detailed performance evaluations. The NoC routers are synthesized from a RTL level design using 65nm standard cell libraries from CMP (http://cmp.imag.fr), using Synopsys™ Design Vision. The NoC routers are driven with a clock of frequency 2.5 GHz. Three SPLASH-2 benchmarks, FFT, RADIX, and LU are considered to study the latency, energy dissipation, and thermal characteristics of the proposed architecture.

We consider a system of 64 alpha cores running Linux. We adopt wormhole routing for data exchange among the cores. The cores are arranged on a 20mm x 20mm die. The width of all wired links is considered to be the same as the flit size, which is 32 in this paper. The latency of the network was obtained through the Garnet output. The energy dissipations of network routers were obtained from the synthesized netlist by running Synopsys™ Prime Power. The energy dissipation on the wired links was obtained through HSPICE simulations taking into account the specific lengths of each link based on the established connections in the 20mm x 20mm die. Each wireless link can sustain a data rate of 10Gbps [3]. The energy dissipation of each wireless link was found to be 0.33pJ/bit, which is shown to be significantly less than metal wires [3].

The next subsection describes the performance metrics that we use to evaluate the characteristics of the proposed WiMesh architecture. This is followed by the detailed analysis of its performance compared to the conventional wireline mesh.

5.1 Performance Metrics

To characterize the performance of the WiMesh architecture we consider three parameters: latency, energy dissipation, and thermal profile.

Latency values were obtained through Garnet output. We consider average network latency per flit as the relevant parameter. It is the time (in clock cycles) that elapses from between the occurrence of a flit injection into the network at the source node and the occurrence of that flit reception at the destination node. As a measurement of energy, we determine the link and router energy per cycle for all links and routers. The total system energy is then a summation of the energy per cycle of every link and router.

Thermal effects have been known to have significant implications on device reliability and interconnects [2], and thus it is important to examine the thermal profile caused by the network, especially since the interconnect structure of a NoC can consume close to 50% of the overall power budget. As temperature is closely related to the energy dissipation of the IC, the thermal profile depends on the energy dissipation of the NoC, which is quantified in subsection 5.3. Lower energy corresponds to lower heat. To quantify the effects of the WiMesh architecture on heat, its thermal profile is evaluated. The temperature profile of the WiMesh is obtained by using the HotSpot tool [12]. In addition to the power profile of the NoC, the floor plan of the network is an input to the HotSpot simulation environment. With a complete floor plan and power profile of the NoC, the HotSpot tool is used to generate the thermal profile. HotSpot considers the floor plan and applies the power profile to it, and with this information, it calculates the volume power density. From this, the thermal profile is generated.

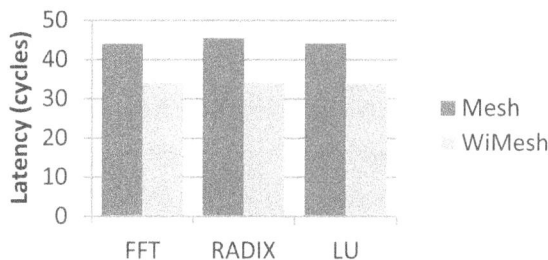

Figure 2. Latency of WiMesh vs. Traditional Mesh

We show that the proposed architecture regardless of the benchmark provides improvement in latency and energy dissipation. In the WiMesh, due to insertion of long-range wireless links, we reduce the communication density of the hotspot regions in the traditional mesh-based NoC. Consequently, we are able to achieve a better thermal profile.

5.2 Latency Characteristics

In this subsection, we evaluate the performance of the proposed architecture in terms of latency. Figure 2 shows the latency of the WiMesh versus the traditional mesh. As seen in Figure 2, the reduction in latency of FFT, RADIX, and LU are 22.57%, 24.88%, and 23.08%, respectively. The latency savings of the WiMesh over the conventional wired mesh was similar for all of the benchmarks considered. This happens due to the wireless link placement that takes into account interactions among the cores depending on the traffic pattern in addition to the physical distance between them. The latency savings seen is a result of the inherent properties of the small-world architecture. By taking a

wireless long-range link, we have effectively bypassed multiple hops that would have needed to be followed with a traditional mesh-based design.

5.3 Energy Dissipation

In this subsection we evaluate the energy dissipation characteristics of the WiMesh and compare that with the conventional mesh. The total energy of all routers and links in the NoC was determined. The energy savings can be seen in Figure 3. The total energy/cycle for the mesh running the FFT

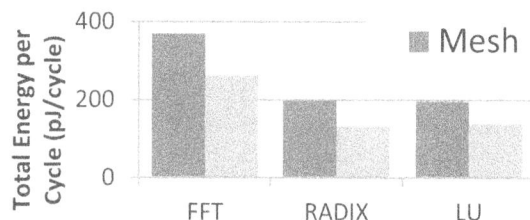

Figure 3. Energy/Cycle of WiMesh vs. Traditional Mesh

benchmark is 370pJ/cycle, whereas the WiMesh running the same benchmark dissipates 262pJ/cycle. This accounts to a 29.07% savings in energy. For RADIX traffic, the mesh dissipates 198pJ/cycle while the WiMesh dissipates 131pJ/cycle giving rise to an energy savings of 33.58%. Finally, the LU traffic running on mesh and WiMesh NoCs dissipate 196pJ/cycle and 138pJ/cycle respectively, which corresponds to a 29.54% savings in energy. From Figure 3 it can be seen that by appropriately placing the long-range wireless links according to the traffic

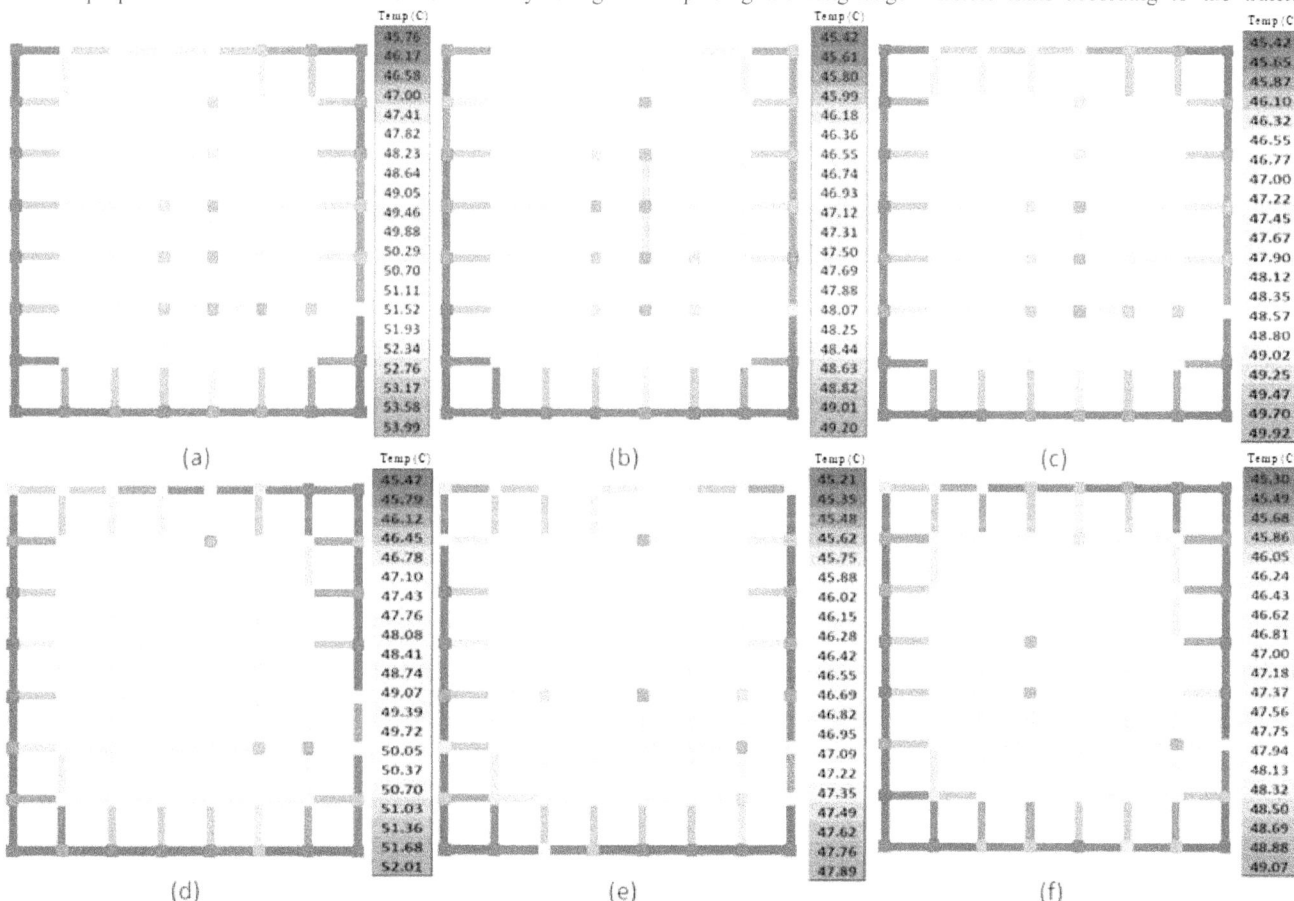

Figure 4. Temperature Profiles (a) FFT Mesh (b) RADIX Mesh (c) LU Mesh (d) FFT WiMesh (e) RADIX WiMesh (f) LU WiMesh

pattern, we can obtain an energy savings of between 29% and 34% for the different SPLASH-2 benchmarks. This energy savings is as a result of better connectivity of the small-world based architecture compared to the regular mesh topology. In addition, the low power and high bandwidth long-range wireless links also contribute to energy savings. We have found that the wireless links handle a significant amount of traffic in the WiMesh. The wireless links handled 33%, 33%, and 35% of all traffic for the FFT, RADIX, and LU benchmarks respectively. This is a significant amount of traffic considering there are only 24 long-range wireless links whereas there are 112 wired links. As a high amount of traffic is carried by the low power wireless links, the savings in energy dissipation is significant. This eventually helps in improving the thermal profile of the system.

5.4 Thermal Profile

In this subsection we evaluate the thermal profile of the WiMesh and compare that with the conventional mesh. As explained in 5.1, we use HotSpot to determine the thermal profile. Figure 4 depicts the temperature profiles of the WiMesh and the conventional wireline mesh in presence of FFT, RADIX and LU traffic patterns. As evident, the temperature range varies from one benchmark to another. This temperature difference can be explained by considering the average communication density, which is the average number of flits traversing through a router or link per cycle. Figure 5 shows the comparison between the communication densities of the benchmarks within their respective hotspot regions. The mesh architecture with FFT traffic has a larger average communication density compared to the other two. Hence, it has the highest temperature.

The important point to note here is that the WiMesh has decreased the temperature for all the benchmarks. It also alleviates the hotspot problem area forming in the central region of each mesh by dispersing it. As explained in section 5.3, with almost one third of the flits traversing on wireless links, we have reduced the average communication density of the hotspot region. To explain this phenomenon more clearly, we consider the average communication density of the routers and links that are within the hotspot region of interest. The average communication density was then determined for the WiMesh architecture over the same area of the traditional mesh hotspot for each benchmark. As can be seen, there is a large reduction in communication density within our problem region. To be more exact, the average link communication densities within the hotspot region of interest were reduced by 83.3%, 79.4%, and 75.4% over the FFT, RADIX, and LU traffics in the traditional mesh topology. The long-range links overwhelmingly take a lot of pressure off of the hotspot region by reducing the communication density by a significant amount. Similarly, savings of average router communication densities were 56.4%, 61.3%, and 49.7% over the FFT, RADIX, and LU traffics in the mesh topology respectively.

6. CONCLUSIONS AND FUTURE WORK

In this paper, we have demonstrated how a small-world wireless NoC improves both the energy and thermal profile of a multi-core chip. By adopting a small-world interconnection infrastructure inspired by natural complex networks, where long distance communications will be predominantly achieved through high performance specialized single-hop wireless links, communications can be made significantly more energy efficient. The low power and long-range wireless links carry a significant percentage of the overall traffic, and hence we are able to reduce the temperature hotspot regions in the system.

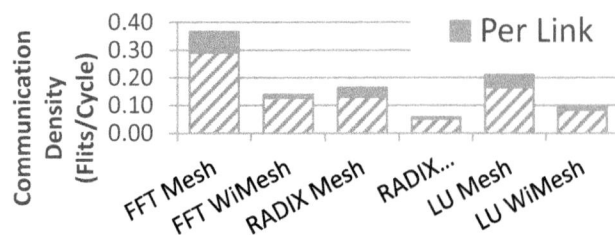

Figure 5. Average Communication Density of Hotspot Region

7. ACKNOWLEDGMENTS

This work was supported in part by the US National Science Foundation (NSF) CAREER grant (CCF-0845504).

8. REFERENCES

[1] Pande, P. et al. Sustainability through Massively Integrated Computing: Are We Ready to Break the Energy Efficiency Wall for Single-Chip Platforms? *Proceedings of IEEE Design, Automation and Test in Europe (DATE)*, (Grenoble, FR, 2011), 1-6.

[2] Shang, L., Peh, L.-S., Kumar, A. and Jha, N.K. Temperature-aware on-chip networks. *IEEE Micro: Micro's Top Picks from Computer Architecture Conferences*, Jan.-Feb. 2006

[3] Ganguly, A. et al. Scalable Hybrid Wireless Network-on-Chip Architectures for Multi-Core Systems. *IEEE Transactions on Computers*, 60 (10). 1485-1502.

[4] Deb, S. et al. Enhancing Performance of Network-on-Chip Architectures with Millimeter-Wave Wireless Interconnects. *Proceedings of IEEE International Conference on ASAP*, (Rennes, Fr, 2010), 73-80.

[5] Marculescu, R. et al. Outstanding Research Problems in NoC Design: System, Microarchitecture, and Circuit Perspectives. *IEEE Transaction on Computer-Aided Design of Integrated Circuits and Systems*, 28(1). 3-21.

[6] Ogras, U.Y. and Marculescu, R. It's a Small World After All: NoC Performance Optimization Via Long-Range Link Insertion. *IEEE Transactions on Very Large Scale Integration (VLSI) Systems*, 14(7). 693-706.

[7] Chang, K. et al. Performance Evaluation and Design Trade-Offs for Wireless Network-on-Chip Architectures. *ACM Journal on Emerging Technologies in Computing Systems*

[8] Zhao, D. and Wang, Y., SD-MAC: Design and Synthesis of a Hardware-Efficient Collision-Free QoS-Aware MAC Protocol for Wireless Network-on-Chip. *IEEE Transactions on Computers*, 57(9). 1230-1245

[9] Lee, B.G. et al. Ultrahigh-Bandwidth Silicon Photonic Nanowire Waveguides for On-Chip Networks. *IEEE Photonics Technology Letters*, 20(6). 398-400.

[10] Woo, S.C. et al. The SPLASH-2 Programs: Characterization and Methodological Considerations. *Proceedings of Annual International Symposium on Computer Architecture*, (Santa Margherita Ligure, IT, 1995), 24-36.

[11] Binkert, N. et al. The GEM5 Simulator. *ACM SIGARCH Computer Architecture News*, 39(2). 1-7.

[12] Skadron, K. et al. Temperature-Aware Microarchitecture. *Proceedings of the International Symposium on Computer Architecture*, (2003), 2-13.

SRAM Leakage in CMOS, FinFET and CNTFET Technologies

Zhe Zhang, Michael A. Turi and José G. Delgado-Frias
School of Electrical Engineering and Computer Science
Washington State University, Pullman, WA 99164-2752, USA

{zzhang, mturi, jdelgado}@eecs.wsu.edu

ABSTRACT

An in-depth study of the static power consumption in 6T and 8T SRAM cell designs based on 32nm CMOS, FinFET and CNTFET technologies is presented. In addition to the inverter leakage currents, memory cells that are not active when write or read operations occur draw current from/to the bus drivers increasing the total standby power consumption. The FinFET schemes yield substantially lower write (1023.5 pA) and read (522.5 pA) leakage currents in 8T cells, which are 10.4% and 4.4% of the amount in CMOS 8T cells. A CNTFET 6T cell consumes 1.9% and 2.8% of the leakage current drawn by a CMOS 6T cell for write and read.

Categories and Subject Descriptors

C.5.4 VLSI System: SRAM Design

General Terms

Design

Keywords

Carbon Nanotube FET; FinFET; 8T SRAM cell; leakage current.

1. INTRODUCTION

For the past four decades CMOS scaling has offered improved performance from one technology node to the next. However, as device scaling moves beyond the 32nm node, significant technology challenges will be faced. Currently two of the main challenges are: the considerable increase of standby power dissipation and the increasing variability in device characteristics which in turn affects circuit and system reliability. The aforementioned challenges will become more prominent as CMOS scaling approaches atomic and quantum-mechanical physics boundaries [1]. Efforts to extend silicon scaling through innovations in materials and device structure continue. Double-gate FinFETs are able to continue CMOS scaling [2-5]. One of the most important features of FinFETs is that the front and back gates may be made independent and biased to control the current and the device threshold voltage. This ability to control threshold voltage variations offers a means to manage the challenge of standby power dissipation. FinFET is considered a promising technology that can impact the immediate future due to its high-performance, low leakage power, reduced susceptibility to process variations, and ease of manufacture using current processes [6]. These features make FinFETs a strong candidate to bridge the technology gap between mainstream bulk CMOS and non-Silicon devices. Its low leakage power makes FinFETs a promising option for memory sub-systems. ITRS has recommended carbon-based nanoelectronics as the *Beyond CMOS* technology for accelerated development [2]. Semiconducting carbon nanotubes (CNTs) have an energy bandgap range of 0.5-0.65eV with a typical 1.4nm diameter [7]. CNTs have superior transport properties, low voltage bias and improved current density, and are less sensitive to many process parameter variations compared to conventional CMOS [8].

Memory modules are widely used in most digital systems. Leakage power is very important in memories. Memory access requires only one or very few memory rows at a given time; the great majority of memory cells draw only leakage power. The application of FinFET and CNTFET technologies to memories can save significant power. Kim et. al. [9] show CNTFET-based memory cells to outperform MOSFET and FinFET cells in power, delay and static noise margins.

This paper presents a leakage current study in 6T and 8T SRAM cells in three technologies: CMOS, FinFET and CNTFET. The SRAM cells and their designs are presented in Section 2. The cells' leakage currents in the three technologies are described in Section 3. Concluding remarks are presented in Section 4.

2. SRAM CELLS

Static random access memory (SRAM) is a major component of many digital systems. Fast memory access times and design for density have been two of the most important target design criteria for many years. However, with device scaling to achieve even faster designs, power supply voltages and device threshold voltages have scaled as well, leading to degradation of standby power.

2.1 Six- and Eight-Transistor SRAM Cells

The six-transistor (6T) static memory cell is widely accepted as the standard memory cell. It is designed to achieve fast read times with the inclusion of sense amplifiers. The standard 6T cell requires that a logic value and its inverse be placed on the bit lines during a write operation. The word line (WL) is raised to logic 1 and the logic levels on the bit lines passed into the cross-coupled inverter pair. Reading from the memory cell entails pre-charging the bit lines and then asserting logic 1 on the word line. The complexity of this cell is in arriving at the appropriate device sizes for proper functionality. Device sizing for a CMOS-based cell is driven primarily by area and functional operation constraints; sizing must be carefully performed to maintain a stored value, enable the cell to push a stored value to the bit lines with reasonable speed for a read operation, and to correctly transfer and overwrite new logic values into the cell for a write operation. An eight-transistor (8T) memory cell is shown in Figure 1. This 8T SRAM cell has a similar structure as a 6T cell with two additional transistors that decouple read and write operations. The read operation is performed by setting the read-word line to logic 1; the additional transistors (T7 and T8) discharge the RBit line that has been precharged before the Read-line is set. The 8T basic cell provides a more orthogonal design; read and write operations are performed by different transistors.

2.2 CMOS, FinFET and CNTFET Cells

Table 1 presents the performance of 6T and 8T SRAM cells in CMOS, FinFET, and CNTFET technologies. Performance is measured using a 16X16 SRAM array, representing a lookup table. All cell simulations in CMOS, FinFET, and CNTFET technologies are set up similarly. In addition to the loading from the other words on the lines, 4X-sized inverters are added to simulate the input capacitance of a pipeline stage's flip-flop. The tri-state inverters of the write driver and precharge transistors are 3X-sized. The write delay is the propagation delay from 50% of the write signal (with Bit/WBit and NBit/WNBit lines already set with the write value) to when the cell's internal Bit and NBit have crossed 50% of V_{DD}. The read delay is the propagation delay from 50% of the read signal to when the Bit (or NBit) line (for a 6T-cell) or the RBit line (for an 8T-cell) discharges from V_{DD} to 50% of V_{DD}. Energy is the total energy consumed in "write-0-read-0-write-1-read-1" process.

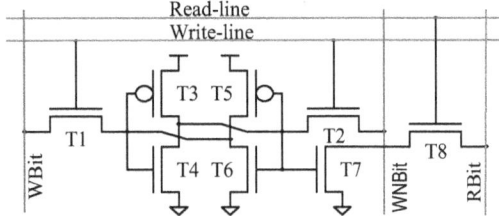

Figure 1. 8T SRAM Cell.

Table 1. CMOS, FinFET and CNTFET SRAM Performance

Parameter	CMOS		FinFET 6T		FinFET 8T		CNTFET	
	6T	8T	SG6	LP6	SGMS	LP_INV	6T	8T
Write d.(ps)	8.4	5.5	3.3	4.0	2.4	1.6	2.7	1.7
Read d. (ps)	28.4	15.9	4.5	10.6	2.3	4.4	18.6	9.3
Energy (fJ)	4.5	3.7	5.0	2.1	2.9	0.5	1.6	1.3

CMOS SRAM Cells. Berkeley 32nm PTM HP model [10] and HSPICE are used to perform simulations of the CMOS 6T and 8T cells. For CMOS 6T cell, the transistor widths for T1 (T2), T3 (T5), and T4 (T6) are 120nm, 80nm and 160nm, respectively. For CMOS 8T cell, the transistor widths are 80nm, 40nm, 40nm and 140nm for T1 (T2), T3 (T5), T4 (T6), and T7 (T8), respectively.

FinFET SRAM Cells. FinFETs are modeled using University of Florida's Spice3-UFDG (Linux ver. 3.7); a physics-based model calibrated to follow predicted results from Synopsys MEDICI and measured results from Motorola-fabricated symmetrical double-gate FinFETs [11]. Table 2 shows values of FinFET parameters used in this research. The L_G, T_{ox}, and R_{SD} values are from the "High-performance Logic Technology Requirements" in the 2007 ITRS Process Integration, Devices, and Structures report [2].

Table 2. FinFET device parameters

Parameter	Value	
N-Channel Surface Orientation	<110>	
Gate length (L_G)	30 nm	
Gate to source/drain underlap (L_{SD})	12 nm	
Fin height (H_{fin})	75 nm	
Fin thickness (T_{SI})	15 nm	
Oxide thickness (T_{ox})	1.2 nm	
Gate thickness (T_G)	20 nm	
Gate work function (Φ_G)	4.4 eV (n-type)	4.8 eV (p-type)
Low-Field Mobility for Thick T_{SI} (μ_0)	565 cm^2/(V-s) (n-type)	250 cm^2/(V-s) (p-type)
Fin body doping (N_{Body})	10^{15} cm^{-3}	
Source/drain doping (N_{DS})	10^{20} cm^{-3}	
Source/drain resistance (R_{SD})	170 Ω-μm	
Supply voltage (V_{DD})	1 V	

We examine the leakage currents of two 6T (*Shorted-Gate, SG6* and *Low-Power, LP6*) and two 8T (*Shorted-Gate Maximum-Swing, SGMS* and *Low-Power Inverters, LP_INV*) SRAM schemes. These were selected as the best performing schemes of those we examined in [12]. Scheme configurations are shown in Table 3. The *SGMS* scheme has a higher *Read-line* swing from -0.2V to 1.2V.

Table 3. FinFET SRAM scheme configuration summary[a]

Scheme	T1	T2	T3	T4	T5	T6	T7	T8
SG6	t / 1	t / 1	t / 1	t / 1	t / 1	t / 1		
LP6	- / 1	- / 1	+ / 1	- / 1	+ / 1	- / 1		
SGMS	t / 2	t / 2	t / 1	t / 1	t / 1	t / 1	t / 2	t / 2
LP_INV	t / 1	t / 1	+ / 1	- / 1	+ / 1	- / 1	t / 1	t / 1

a. **Table Key:** (configuration symbol) / (fin count)
Configuration Symbols [*FinFET in SG config.*:**t** = back gate (BG) tied to front gate / *FinFET in LP config.*:**-** = -0.2V BG bias; **+** =1.2V BG bias]

CNTFET SRAM Cells. Simulations have been performed using HSPICE and CNTFET models by Stanford University's Nanoelectronics Group [13]. Technology parameters and their values are: Gate length (L_G): 32 nm, Pitch: 20 nm, Work f. (Φ_{Metal}, Φ_{CNT}): 4.5 eV, Source/drain doping: 1.00%, Chirality of CNT: (19, 0), Oxide thickness (T_{ox}): 4.0 nm, Mobility (μ_0): $10^3 \sim 10^4$ cm^2/(V-s), and Supply voltage (V_{DD}): 0.9 V. The CNT numbers for 6T cell transistors T1 (T2), T3 (T5), and T4 (T6) are 3, 2 and 3, respectively; in CNTFET 8Tcells, the numbers of CNTs are 6, 1, 1 and 12 for T1 (T2), T3 (T5), T4 (T6), and T7 (T8), respectively. Data from Table 1 indicates that 8T CNTFET cell's maximum delay and energy are 50.1% and 18.7% smaller than the 6T CNTFET cell, respectively.

3. SRAM LEAKAGE CURRENTS

There are five major leakage currents in the 8T SRAM as shown in Figure 2. As mentioned earlier the 6T SRAM cell has the same structure, but with just six transistors (T1 to T6). The most commonly reported are the inverters' leakage currents. These currents flow through transistors T3 and T4 for inverter 1 (in the figure, T4 is off but subthreshold current flows through the transistor). Inverter 2 formed by transistors T5 and T6 has T5 off. These two leakage currents are usually referred as static current (or static power). In this paper these two currents are called I_{inv1} and I_{inv2}. In addition to the leakage of the memory cell inverters, leakage is present when a write or read operation occurs to a different SRAM cell in the column. Memory cells that are not selected for a write or read operation draw current from (or to) the drivers. Figure 2 shows a SRAM cell with a "0" stored (inverter 1 and 2 output 1 and 0, respectively). If a "1" is being written (Bit and NBit are 1 and 0, respectively) there is a leakage current from the driver of the Bit line through transistor T1; this leakage current is called I_{T1}. There is another leakage current from the cell to the driver of the NBit line through transistor T2; this leakage current is called I_{T2}.

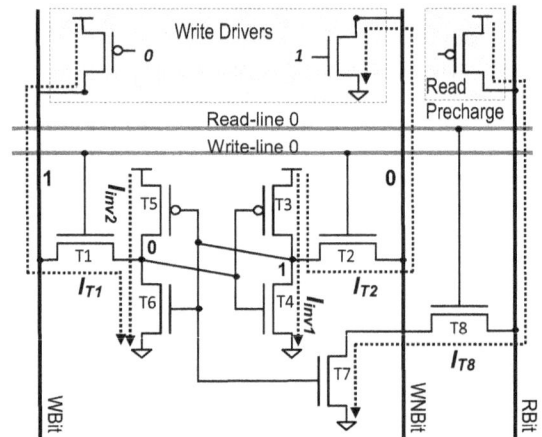

Figure 2. SRAM cell leakage currents under a write and read operations by other cells in the column.

When a read occurs in a 6T SRAM array, at precharge time both Bit and NBit lines (which are represented as WBit and WNBit in the figure) are set to "1." Using the example of Figure 2, current I_{T1} is the same as when there is a write while there is a much smaller current I_{T2}. The 8T SRAM cell has an independent read port. There is a leakage current from the read precharge circuitry through transistor T8. This leakage current is called I_{T8}. If a "0" is stored in the SRAM cell transistor, T7 is 'on'; Vds across T8 is approximately Vdd. On the other hand, if a "1" is stored in the cell, transistor T7 is 'off'; thus, there are two transistors in series that are 'off' (T7 and T8), presenting a larger resistance to current I_{T8}.

3.1 SRAM Leakage-CMOS Technology

In this section we report the leakage currents under all possible conditions of internal inverters and Bit lines. The following notation is used to indicate the status of Bit and NBit lines as well as Inv2 and Inv1 outputs. **Bit [Inv2 output Inv1 output] NBit**
The condition expressed as 1 [0 1] 0 is shown in Figure 2.

Table 4 shows the leakage currents under different write conditions for both 8T and 6T SRAM cells. As expected the static leakage currents (I_{inv1} and I_{inv2}) depend only on the stored value in the SRAM, i.e. the inverter's status. When the inverter's p-type transistor is off the static leakage current for that inverter is around 1691.7 pA and 1379 pA for the 8T and 6T cells, respectively. When inverter's n-type transistor is off the static leakage current for that inverter is around 4293.8 pA and 7491 pA (8T and 6T). For the write conditions, leakage currents I_{T1} and I_{T2} have a larger leakage current when there is drain source voltage (Vds) of about Vdd, the leakage current through either transistor T1 or T2 is between 1911 and 1894 pA for the 8T and between 5535 and 5487 pA for the 6T cell. On the other hand, when Vds=0V, the leakage current through either transistor T1 or T2 is between 0 and 17 pA for 8T and between 0 and 46 pA for 6T. The total leakage current for the write conditions varies from 6002.5 pA (when Vds for T1 and T2 is zero) to 9791 pA (when Vds=Vdd) for the 8T cell. The total leakage current for 6T under similar conditions varies from 8914 pA to 19894 pA. The minimum leakage current is mainly due to the static leakage current. The maximum leakage is when both leakage currents through transistors T1 and T2 are present. The 6T SRAM cell needs strong inverter's n-type transistors (T6 and T4); this in turn contributes to have a larger write leakage current which is between 1.48 and 2 times larger than the 8T.

Table 4. Write leakage (pA) for 8T CMOS cell vs 6T cell

Leakage current	8T write condition		6T write condition		6T/8T comparison	
	0 [0 1] 1	1 [0 1] 0	0 [0 1] 1	1 [0 1] 0	0[0..	1[0..
I_{inv1}	4293.8	4293.8	7489.1	7491.5	1.74	1.74
I_{inv2}	1691.7	1692.1	1378.1	1379.5	0.81	0.82
I_{T1}	0.0	1911.0	0.0	5535.1	1.00	2.90
I_{T2}	17.0	1894.0	46.5	5487.8	2.74	2.90
I_{TOTAL}	6002.5	9790.9	8913.7	19893.9	1.48	2.03

The 8T SRAM cell's read port needs to be considered when under two conditions: i) when a 0 is stored in the cell (condition [0 1]) which in turn sets transistor T7 'on;' ii) when 1 is stored that sets T7 'off.' Table 5 shows the leakage current through transistor T8, I_{T8}. It can be observed that the leakage current is significantly lower when 1 is stored in the cell. For the 6T cell the read conditions are similar to the write conditions. At precharge time all cells in the column experience the larger leakage current on either the Bit or NBit side, the other side has a low leakage current. This explains why the two cases are similar for the 6T as shown in Table 5. The 6T leakage is larger by 1.21X to 2X.

Table 5. Read leakage (pA) for CMOS SRAM 8T and 6T cells

Leakage current	8T		6T		6T/8T comp.	
	[0 1]	[1 0]	[0 1]	[1 0]	[0 1]	[1 0]
I_{T8}	5969.0	1242.4	*5582.3	*5580.7	0.94	44.93
I_{inv1}	4293.8	1692.1	7489.1	1378.1	1.74	0.81
I_{inv2}	1691.7	4293.8	1378.0	7489.1	0.81	1.74
I_{TOTAL}	11953.9	7228.3	14449.4	14447.9	1.21	2.00

*This leakage current is the sum of I_{T1} and I_{T2} for the 6T cell.

3.2 SRAM Leakage -FinFET Technology

Table 6 presents the leakage currents in the FinFET 6T SRAM schemes. There is a significant difference in the leakage currents I_{INV1} and I_{INV2} for each scheme. The n-type FinFETs have similar leakage for the I_{T1} and I_{T2} currents when the Bit and/or NBit lines hold values opposite to that of the cell. Its reverse-biased FinFETs allow the LP6 scheme to substantially reduce leakage compared to the SG6 scheme. A 96% reduction in most leakage currents is seen for the LP6 scheme. A 35% larger I_{T2} current is seen when the opposite value of the cell is set on the Bit/NBit lines.

Table 6. Write/read leakage (pA) for 6T FinFET SRAMs

Leak. Cur.	SG6			LP6		
	Write Cond.		Read C.	Write Cond.		Read C.
	0[0 1]1	1[0 1]0	1[0 1]1	0[0 1]1	1[0 1]0	1[0 1]1
I_{inv1}	501.4	501.4	501.4	17.6	17.6	17.6
I_{inv2}	90.6	90.6	90.6	3.6	3.6	3.6
I_{T1}	0.0	501.3	501.3	0.0	17.6	17.6
I_{T2}	0.0	501.3	0.0	0.0	678.8	0.0
I_{TOTAL}	592.0	1594.6	1093.3	21.2	717.6	38.9

The FinFET 8T schemes' leakage for the write paths and read path is presented in Tables 7 and 8. There are similarities between the leakage currents of the write paths for the 6T and 8T schemes. Both 8T schemes use SG FinFET configuration for transistors T1 and T2. Thus, LP_INV has the same I_{T1} and I_{T2} currents as SG6. SGMS has twice the amount of SG6's I_{T1} and I_{T2} leakage since its T1 and T2 FinFETs have two fins per device while SG6 is minimum sized. SGMS has SG configured T3-T6, thus it has the same I_{T1} and I_{T2} currents as SG6 while LP_INV has LP configured T3-T6, thus has equal I_{T1} and I_{T2} leakage to LP6.

Table 7. Write leakage (pA) for 8T FinFET cells vs 6T cells

Leak.	SGMS				LP_INV			
	8T		6T/8T cmp		8T		6T/8T cmp	
	0[0 1]1	1[0 1]0	0[0..	1[0..	0[0 1]1	1[0 1]0	0[0..	1[0..
I_{inv1}	501.4	501.4	1.00	1.00	17.6	17.6	1.00	1.00
I_{inv2}	90.6	90.6	1.00	1.00	3.6	3.6	1.00	1.00
I_{T1}	0.0	1002.6	1.00	0.50	0.0	501.3	1.00	0.04
I_{T2}	0.0	1002.7	1.00	0.50	0.0	501.3	1.00	1.35
I_{TOT}	592.0	2597.3	1.00	0.61	21.2	1023.9	1.00	0.70

For 6T/8T comparisons: SGMS vs. SG6; LP_INV vs. LP6

The 8T SRAM schemes see leakage reduction in the read path compared to the 6T schemes. Both 8T schemes use SG FinFETs on the read path for read speed, however, SGMS reverse-biases T7 and T8 when the read signal is inactive. This further reduces leakage current and is possible since it does not interfere with write operations due to the orthogonal read and write operations. When a "0" is stored in the cell and RBit=1, LP_INV has similar read path leakage as SG6 since two n-type FinFETs (one on and one off) form the path from the read bus to ground. The read path of SGMS has one n-type FinFET on, but with a back-gate reverse-biased, and one n-type with its front- and back-gate reverse-biased and sees substantial leakage reduction. When a "1" is stored in the cell and RBit=1, the 8T schemes have both n-type FinFETs 'off' and the *LP_INV* scheme has 61% reduced leakage.

Table 8. Read leakage (pA) for 8T FinFET cells vs 6T cells

Sch.	Leak.	(Precharge) RBit=1		6T/8T comp. LP6		6T/8T comp. SG6	
		[0 1]	[1 0]	[0 1]	[1 0]	[0 1]	[1 0]
SGMS	I_{T8}	0.6	1.0	29.3	17.6	835.5	501.3
	I_{TOT}	592.6	593.0	0.07	0.07	1.84	1.84
LP_INV	I_{T8}	501.3	193.3	0.04	0.09	1.00	2.59
	I_{TOT}	522.5	214.5	0.07	0.18	2.09	5.10

I_{TOT} for (Precharge) RBit=1 data of 8T cells equals $I_{T8} + I_{inv1} + I_{inv2}$
For 6T/8T I_{T8} comparisons, $I_{T1} + I_{T2}$ is used for 6T cells.

3.3 SRAM Leakage -CNTFET Technology

CNTFET has much less standby power than CMOS. When Vds of T1 (or T2) in both 8T and 6T is close to Vdd, a current of 135 pA (or 0.82 pA) is consumed. This is because in the CNTFET Model the drain and source are not interchangeable. Current density is much smaller when source voltage is higher than drain potential. T1 and T2 have their source connected to the internal node of the SRAM cell and their drain connected to the bit lines. In Table 9, leakage currents in the two inverters of the 8T cell are only 139.7 pA, 47.7% less than the 265.0 pA in 6T cell. The 8T cell inverters are of minimum size. When the data on the bitlines and the data stored in the cell are complementary, I_{T1} leakage is 134.8 pA, a 31.9% savings versus the 392.5 pA in 6T. Read leakage currents (I_{T8} and $I_{T1}+I_{T2}$ for 8T and 6T) are shown in Table 10. When a '0' is stored, I_{T8} is 252.8 pA, 1.76X higher than a 6T cell. The 8T cell is designed to have a fast read; thus, T7 and T8 are large. When a '1' is stored the 8T cell has lower leakage than the 6T cell.

Table 9. Write leakage (pA) for 8T CNTFET cells vs 6T cells

Leakage current	8T write condition		6T write condition		6T/8T comparison	
	0 [0 1] 1	1 [0 1] 0	0 [0 1] 1	1 [0 1] 0	0[0..	1[0..
I_{inv1}	75.48	75.53	119.3	119.3	1.58	1.58
I_{inv2}	56.17	56.18	137.5	137.5	2.45	2.45
I_{T1}	0.0	134.8	0.0	134.9	1.00	1.03
I_{T2}	8.1	0.815	8.1	0.843	1.00	1.03
I_{TOTAL}	139.7	267.3	265.0	392.5	1.89	1.47

Table 10. Read leakage (pA) for 8T CNTFET cells vs 6T cells

Leakage current	8T		6T		6T/8T comp.	
	[0 1]	[1 0]	[0 1]	[1 0]	[0 1]	[1 0]
I_{T8}	252.8	108.7	*143.0	*142.7	0.57	1.32
I_{inv1}	75.48	75.53	119.3	119.3	1.58	1.58
I_{inv2}	56.17	56.18	137.5	137.5	2.45	2.45
I_{TOTAL}	384.5	240.4	399.8	399.5	1.04	1.66

*This leakage current is the sum of I_{T1} and I_{T2} for the 6T cell.

4. CONCLUDING REMARKS

In this paper we have studied leakage currents in 32nm Bulk CMOS, FinFET, and CNTFET based SRAM cells. In large memory arrays, static power becomes a dominant factor in power consumption. It has a multiplicative effect. Table 11 shows leakage currents for the 8T and 6T SRAM cells. There are three major leakage currents: i) static leakage (st) that corresponds to the internal inverters leakage currents ($I_{inv1}+I_{inv2}$); ii) write leakage (wr) that represents the currents when a write occurs under the condition 1[01]0; and iii) read leakage (rd) current. The table includes comparisons between CMOS and the other two

technologies. For the 6T cells, FinFET *LP6* has the smallest static leakage current, 418X smaller than CMOS current, and the smallest read leakage current, 317X smaller than CMOS current. CNTFET cell has the smallest write leakage current; 60X smaller than the CMOS current. For the 8T cells, FinFET *LP_INV* has the smallest static leakage current; 282X smaller than the CMOS current. CNTFET cell has the smallest write leakage current; 28X smaller than the CMOS current. FinFET *SGMS* has the smallest read leakage current; 9948X smaller than the CMOS current.

Table 11. 8T and 6T SRAM leakage current comparison (pA)

Cell	Tech.	Leakage currents			CMOS comparison (CMOS/Tech)		
		st	wr	rd	st	wr	rd
		$I_{inv1}+I_{inv2}$	$I_{T1}+I_{T2}$	[*]			
8T	CMOS	5985.5	3805.0	5969.0	1.0	1.0	1.0
	FinFET SGMS	592	2005.3	0.6	10.1	1.9	9948
	LP_INV	21.2	1002.6	501.3	282	3.8	11.9
	CNTFET	131.6	135.6	252.8	45.5	28.1	23.6
6T	CMOS	8867.2	11023	5582.3	1.0	1.0	1.0
	FinFET SG6	592.0	1002.6	501.3	15.0	11.0	11.1
	LP6	21.2	696.4	17.6	418	15.8	317
	CNTFET	256.8	135.7	143.0	34.5	81.2	39.0

[*] Read leakage is I_{T8} and $I_{T1}+I_{T2}$ for the 8T and 6T cells, respectively.

5. ACKNOWLEDGMENTS

This research was supported in part by the Boeing Centennial Endowed Chair, School of EECS, Washington State University.

6. REFERENCES

[1] T-C Chen, "Overcoming Research Challenges for CMOS Scaling: Industry Directions," *Int. Conf. on Solid-State and IC Technology,* pp. 4-7, Oct. 2006.

[2] Int. Technology Roadmap for Semiconductors, www.itrs.net.

[3] T. Cakici, K. Kim, and K. Roy, "FinFET Based SRAM Design for Low Standby Power Applications," *8th Int. Symp. on Quality Electronic Design,* pp. 127-132, Mar. 2007.

[4] Y. B. Kim et al., "New SRAM Cell Design for Low Power and High Reliability Using 32nm Independent Gate FinFET Technology" *IEEE Int. Workshop on Design and Test of Nano Devices, Circuits and Systems,* pp. 25-28, Sept. 2008.

[5] T. King. "FinFETs for nanoscale CMOS digital integrated circuits," *Int. Conf. Computer-Aided Design,* pp. 207-210, Nov 2005.

[6] D. J. Frank et al., "Device scaling limits of Si MOSFETs and their application dependencies," *Proc. of the IEEE,* vol. 89, no. 3, pp. 259-288, Mar. 2001.

[7] P. H-S. Wong, "Field Effect Transistors-from Silicon MOSFETs to Carbon Nanotube FETs," *23rd Int. Conf. on Microelectronics,* pp. 103-107, May 2002.

[8] B.C. Paul et al., "Impact of a Process Variation on Nanowire and Nanotube Device Performance," *IEEE Trans. on Electron Devices,* vol. 54, no. 9, pp. 2369–76, Sept. 2007.

[9] Y. B. Kim et al., "A Low Power 8T SRAM Cell Design Technique for CNTFET," *IEEE Int. SoC Design Conf. (ISOCC),* pp. 176-179, Nov. 2008.

[10] Berkeley Predictive Tech. Model: www.eas.asu.edu/~ptm/.

[11] J. G. Fossum et al., "Recent upgrades and applications of UFDG," *2006 NSTI Nanotech. Conf. (Workshop on Compact Modeling),* pp. 674-679, May 2006.

[12] M. A. Turi and J. G. Delgado-Frias, "Performance-power tradeoffs of 8T FinFET SRAM cells," *54th IEEE Int. Midwest Symp. Circuits Syst.,* Aug. 2011.

[13] CNTFET Models. http://nano.stanford.edu/models.php

A Novel Hybrid FIFO Asynchronous Clock Domain Crossing Interfacing Method

Zaid Al-bayati, O. Ait Mohamed
ECE Department,
Concordia University,
Montréal, QC, Canada

{z_albaya,
ait}@ece.concordia.ca

Syed Rafay Hasan
ECE Department,
Tennessee Tech. University,
Cookeville, TN

shasan@tntech.edu

Yvon Savaria
EE Department,
École Polytechnique de Montréal,
Montréal, QC, Canada

yvon.savaria@polymtl.ca

ABSTRACT

Multi-clock domain circuits with Clock Domain Crossing (CDC) interfaces are emerging as an alternative to circuits with a global clock. CDC interfaces are susceptible to metastability, hence their design is very challenging. This paper presents a hybrid FIFO-asynchronous method for constructing robust CDC interfaces. The proposed design can handle arbitrary clock frequency ratios between the sender and receiver with random phase shifts. The proposed design avoids latency due to synchronizers with the asynchronous protocol modifications. Circuit simulation results confirm the operation and robustness of the design at maximum workloads, and arbitrary frequency ratios, over a temperature range of -50 to 50 degrees Celsius. The interface offers a maximum throughput of 606 million transfers per second without pausing the clock.

Categories and Subject Descriptors

B.4.3 [**Input/Output and Data Communications**]: Interconnections (Subsystems) – *asynchronous/synchronous operation*

General Terms

Performance, Design.

Keywords

GALS, clock domain crossing, CDC, pausable clocking, FIFO.

1. INTRODUCTION

Interfacing modules in multiple clock domains (MCDs) within a system-on-chip (SoC) has become very challenging in modern deep sub-micron technologies. The ITRS2009 road map [4] states that by 2015, about 25% of long interconnects in a SoC will comprise asynchronous handshaking. Researchers have explored several different asynchronous interfaces for MCDs. Pausable clocking is one such technique [1], which avoids synchronization failures, but requires separate clock generation circuits in each domain. One of the first such practical circuits is proposed in [5].

It has been observed that pausable clocking has some drawbacks. A known problem with pausable clocks is clock over-run [3]. This phenomenon occurs because in pausable clocking, each domain

should have its own clock source. This leads to clock edge discrepancy between the source and the registers, which may lead to malfunctioning of the system due to inability of instantaneous pausing of the clock at terminal registers. Furthermore, restarting the clock leads to period by period duration mismatches [7] known as jitter, caused by the dynamics of the clock source. The pausing of the clock halts the complete system, and the system loses performance. A misconception is to believe that pausing necessarily reduces energy consumption. If the clock is stopped, the system will normally take longer to finish its job. However, pausable clocking is an attractive option for interfacing modules. In pausable interfaces, there is no fear of metastability, thus it avoids latency due to two flip-flop synchronizers. Moreover, frequency ratios between communicating modules can be arbitrary and modules may have arbitrary phase discrepancy.

On the other hand, FIFO based GALS interfaces [2] use a circular queue to transfer data across clock domains. This technique requires synchronization of the full (empty) signals, which informs the sender (receiver) that the FIFO is full (empty). This synchronization is usually performed using multiple synchronizing flip-flops. FIFO based interfaces have their own pros and cons. They tend to have higher throughput in bulk transfers and they do not pause the clock. However, they introduce synchronizing schemes that incur latency and possibility of propagating metastability into the design.

In this paper, we propose a new hybrid interfacing methodology using the bundled data protocol and customized FIFOs. The use of FIFOs avoids performance penalty due to pausing and clock over-run issues as it decouples the locally synchronous modules from clock domain crossing interfaces. Moreover, bundled data protocol is used in a pseudo-deterministic way, similar to pausable clocking interfaces [5], hence, making the interface highly robust to metastability. The proposed technique is capable of handling dynamic phase variations, and rational clock frequency ratios between the communicating modules. Unlike conventional pausable clocking, this interface avoids completely pausing the locally synchronous modules. Viability of this design is demonstrated using extensive simulation results.

The rest of this paper is organized as follows: Section 2 describes conventional pausable clocking. Section 3 describes in detail our proposed interface. Section 4 shows the obtained results. Section 5 provides a discussion on the benefits of the proposed design over state-of-the-art designs, and Section 6 presents conclusions.

2. PRELIMINARIES

Because our design is closely related to pausable clocking, some preliminary information regarding CDC interfacing techniques is required to better understand our work. Pausable clocking stops the sender and receiver clocks during data transfers. This method requires controllable clock generation units, usually implemented as ring oscillators with mutual exclusion (ME) elements. Classically, bundled data protocol is used for interfacing utilizing D and P port control asynchronous state machines [5] (see Fig. 1), which use the following protocol:

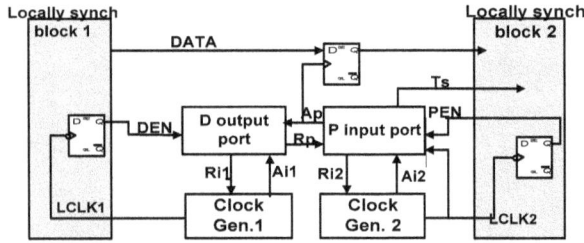

Figure 1. Conventional pausable clocking [5]

1) When the sender puts data on the data line, it enables the D port through a transition on DEN.

2) The D port sends a clock pausing request to the clock generation unit through Ri_1, consequently $LCLK_1$ in Fig. 1 pauses. Then, the D port receives the Ai signal and consequently the Rp signal is asserted, informing the receiver of a data transfer.

3) If the receiver can accept data (a transition on PEN indicates whether the receiver is ready to accept data or not), the P port raises the Ri_2 signal, which stops the receiver clock ($LCLK_2$).

4) Upon receiving Ai_2 assertion, the P port asserts Ap. This positive edge of Ap is used to latch the data in the data path.

5) Once data is latched, the protocol terminates with the release of the receiver clock followed by the release of the sender clock.

3. PROPOSED INTERFACE

The protocol is implemented in hardware as shown in Fig. 2. The middle blocks in Fig. 2, the D and P ports are asynchronous machines, borrowed from the conventional pausable clocking methodology in [5]. In addition to the ports, the interface circuit contains four additional blocks: a synchronous FIFO, two special circuits that are called protocol-pausers, one at each end, and an Ai generator block. Broadly, transfer requests from the sender block accumulate at the FIFO. This FIFO blocks the sender from sending more data if it sees the possibility of an overflow. The D and P ports are responsible for performing the asynchronous handshake between the two domains. These blocks are controlled by synchronous logic. The two pausers form the interface between synchronous and asynchronous logic. These two blocks pause the control signals and release them when they cannot cause timing violations at the sender or receiver, which is the most challenging task in this design. We will first discuss the transfer protocol followed by the hardware implementation of the various blocks.

3.1 The Protocol

Our proposed design is based on the bundled-data asynchronous

handshaking protocol. To achieve a level of determinism (and to avoid conventional synchronizers), a few modifications are done in the protocol. These modifications lead to a unique signal sequence described in the following:

1) When the sender requests to send data, the FIFO toggles the DEN signal activating the D port in Fig. 2.

2) The D port raises the Ri_1 signal, which in turn generates the Ai_1 signal using Ai_1 generator. The D port, upon receiving the Ai_1 signal, sends a request to send data to the receiver using the Rp signal.

3) If the receiver can accept data (a transition on PEN indicates whether the receiver is ready to accept data or not), the P port raises the Ri_2 signal. The protocol–pauser–R holds the processing of the request until the next positive edge of the receiver clock. With the assertion of $LCLK_2$, Ai_2 is also raised.

4) Upon receiving Ai_2, the P port asserts Ap. This positive edge of Ap is used to latch the data into the data flip-flops with a small deterministic delay with respect to the receiver clock. Therefore, data becomes available to the receiver side without violating its timing constraints.

5) Following the RTZ signaling, the handshake signals Rp and Ap are negated. The protocol-pauser-S generates a sender-safe ACK signal once the Ri_1 signal is negated.

Our design does not require pausing of the clock contrary to conventional pausable clocking. This has been achieved through the use of protocol pausers, which pause the transfers managed by the protocol rather than pausing the locally synchronous blocks. The signal transition graph (STG) representing the transfer protocol is shown in Fig. 3. The superscript ^{T} is used to indicate a signal transition that is either positive or negative.

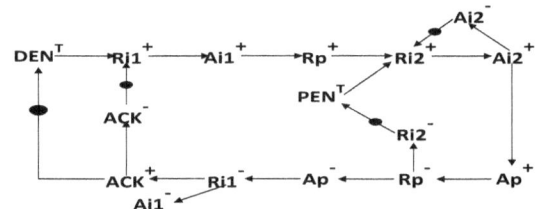

Figure 3. Signal Transition Graph of the proposed design

3.2 FIFO Implementation

The main goal of this FIFO is to decouple the terminating modules from the pseudo pausable interface. The FIFO keeps checking the overall state of the system, whenever the data input rate becomes faster than the data consumption rate, this FIFO asserts the hold signal. The Mealy state machine for a FIFO queue of size two is shown in Fig. 4.

More stages can be added in the FIFO with little modifications to the state machine. The state machine operates on a delayed version of the sender clock to allow transfer requests from the T

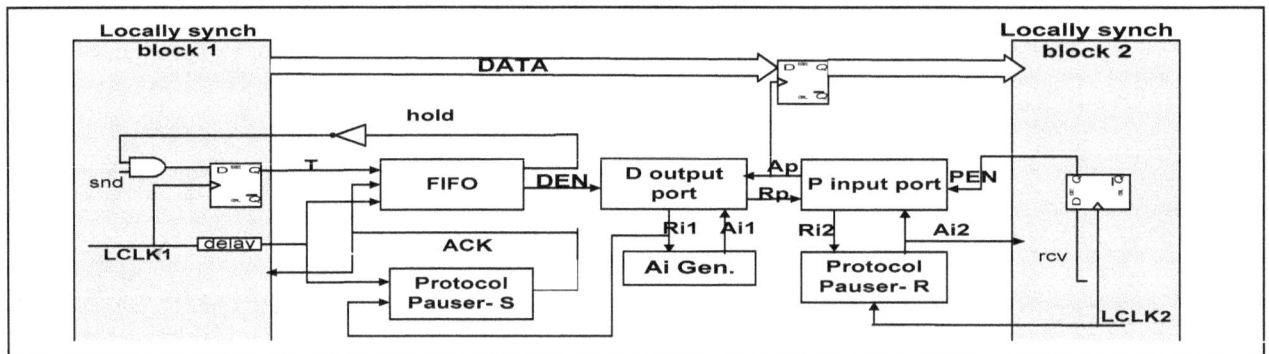

Figure 2. The proposed design

input to be latched during the same clock cycle in which they are produced by the sender. Upon receiving a request at its T (transfer) input, the FIFO issues the request through a transition on its DEN output which tells the D port to start a handshake cycle. DEN is kept at the same level until the FIFO receives an ACK. If a new request arrives at the T input during the handshake, it is accumulated in the queue and DEN is kept stable until an ACK arrives. The protocol-pauser-S makes sure that the ACK signal doesn't violate the timing constraints of the FIFO.

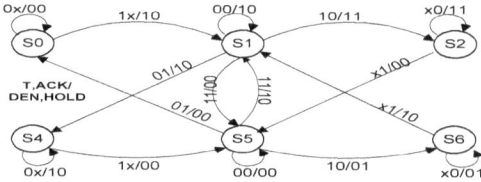

Figure 4. FSM of the two stage FIFO

3.3 Protocol-Pauser-S

The protocol-pauser-S is responsible for generating the ACK signal so that it is synchronous to the FIFO. The inputs of this block are the Ri_1 signal and the clock. The Ri_1 signal is generated by the asynchronous D port. Positive transition in Ri_1 is generated by the D port in response to a transition on DEN. Therefore, it is deterministic with respect to the sender clock, while negative transitions on Ri_1 are generated in response to a negative transition in the Ap signal, which is non-deterministic with respect to the sender clock. Hence, only one of the transitions in Ri_1 requires phase correction, which significantly simplifies the design. The circuit diagram for the Pauser-S is shown in Fig. 5.

The circuit consists of two Muller C-elements both having Ri_1 as one input. The other inputs to the two C-elements are the delayed version of the clock and its complement. The outputs of the C-elements are connected to a NOR gate. These gates act as a phase corrector for one transition of the Ri_1 signal.

The 3-gate combination immediately propagates the deterministic positive transition of the Ri_1 signal to node X. For negative transitions, this circuit blocks them unless they do not cause a timing violation in both C-elements concurrently. Timing window of such an occurrence is so small that it is filtered out using the NOR-gate delay. We demonstrated this metastability filtering behavior with our simulations in Section 5. As a second level precaution, a worst case delay requirement can be obtained using corner-analysis and this can be introduced at the output.

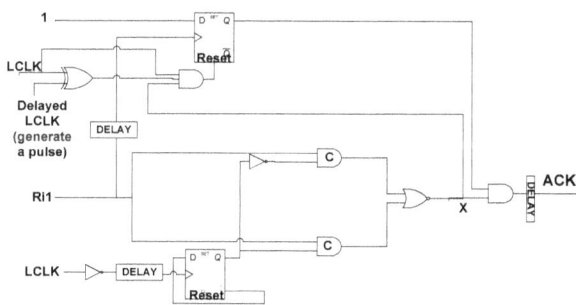

Figure 5. The protocol-pauser-S circuit diagram

The timing behavior of the pauser-S is shown in Fig. 6. The first event in the figure is the reset in the upper DFF. This occurs at the beginning of the sender clock cycle if Ri_1 is at logic '0'. In this case, the NOR gate output at node X is logic '1'. Therefore, the positive clock edge creates a reset pulse at the top flip-flop in Fig. 5. This brings the ACK signal to logic '0' (if it is initially high).

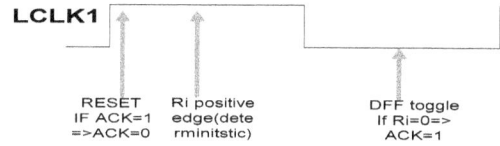

Figure 6. The behavior of the protocol-pauser-S

When the Ri_1 signal rises, node X becomes logic '0'. Ri_1 also acts as a clock for the upper flip-flop in Fig. 5, latching a '1' into the upper flip-flop. The delay on the Ri_1 line (connected to the upper flip-flop) makes sure that these two events occur in order. As long as Ri_1 remains high, node X will remain low and ACK will remain low. When the negative transition of Ri_1 occurs, node X continues to remain low since at least one of the outputs of C-elements remains high until the lower DFF toggles. The first toggle in the lower DFF that occurs after the negative transition of Ri_1 generates the ACK signal. Hence positive transitions in ACK become deterministic with respect to the sender clock.

The input to the lower DFF's clock is a delayed inverted version of the clock. This delayed version of the clock determines the position of the leftmost arrow in Fig. 6. If the negative transition of Ri_1 occurs after the positive transition in the inverted delayed version of the sender clock, the transition in Ri_1 is deferred to the next clock cycle. If Ri_1 and the lower DFF's toggle occur approximately at the same time, the simulation results in Section 5 demonstrate that the 2 parallel C-elements followed by a NOR and a delay produced by a gated AND followed by an inertial delay turn out to be an effective metastability filter.

3.4 Protocol-Pauser-R

The pauser-R operates in a similar manner to the pauser-S. It generates Ai_2 such that it does not violate receiver's constraints. Its design is similar to Fig. 5 with small modifications.

4. SIMULATION RESULTS

Electrical simulations of the proposed design were performed using the TSMC 90nm CMOS technology. As our protocol ensures that DFFs are only clocked when their input is stable, true single phase clocking (TSPC) DFFs [6] were used. Fig. 7 shows a proof of concept simulation of our implementation.

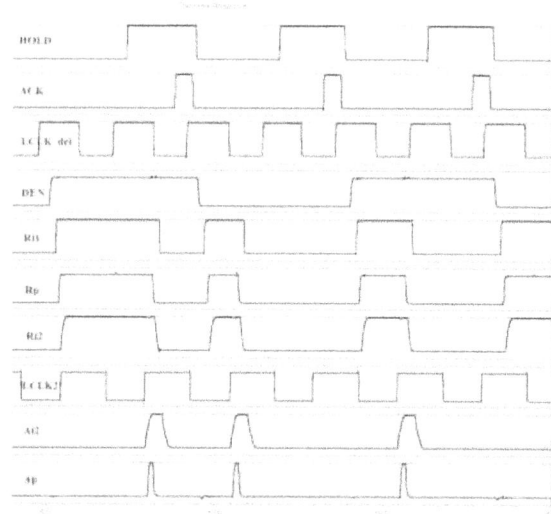

Figure 7. SPICE simulation

The figure shows that if the Ri_2 signal asserts closer to the receiver clock ($LCLK_2$) edge (i.e. after the toggle of the lower DFF as mentioned in section 3.3) then the Ai_2 signal asserts to logic '1' after one clock cycle. The second positive transition of Ri_2 in Fig. 7 shows that this transition occurs before the toggle of the DFFs, consequently asserting Ai_2 in the same cycle.

Extensive simulation results show that the design can operate robustly for several extreme variations, such as different workloads, frequency ratios (integer and non-integer), and extreme temperatures. We achieved operating frequencies of more than 1.5 GHz. Table 1 shows the performance of the design under these variations. The frequencies of the sender and receiver in this experiment are 1.51 and 2 GHz. Simulations have shown that the design can achieve a max throughput of 606 M items/s.

Table 1. Temperature and workload variations

	Max. workload			Once in 5 cycles	Once in 10 cycles
	-50°C	25°C	50°C		
Power (mW)	38.77	39.98	40.51	35.33	32.64
throughput (Mitems/s)	468	426	404	298	149

5. COMPARISON WITH OTHER STATE-OF-THE-ART TECHNIQUES

Comparing with the conventional pausable method [5], the new design does not pause the whole clock domain as is the case with pausable clocking. The overall throughput of the system considerably increases because the clocks of the sender and receiver continue to run. The proposed technique also relieves the requirement of using a ring oscillator avoiding pausable clocking problems discussed in Section 1. This comes at the price of an increase in area and power consumption compared to the original design. However, the interface is only needed when data is transferred between the blocks. If power consumption is very critical for the application, simple disabling logic can be used to turn off the entire block when data is not being transferred.

The comparison with FIFO based interfaces is summarized in Table 2. T_{snd} and T_{rec} refer to the clock periods of the sender and receiver respectively. The comparison depends on the workload and frequencies of the two domains. For the proposed interface, the pausers pause the protocol twice in each transfer cycle. This pausing makes the design flexible in terms of the frequency discrepancies between the communicating modules. The design functions correctly even if one module runs significantly faster than the other. On the other hand, FIFO based interfaces might fail if the receiver's frequency is more than 3 times the sender's frequency [2]. The cost of avoiding frequency restrictions is a decrease in the throughput of bulk transfers. However, many communicating modules do not send data every cycle. It is of interest that when there is one transfer request every 5 cycles at 1.51 GHz, a net transfer rate of 298 Mitems/s is obtained, and the interface imposes only a 1.3% throughput degradation.

A very important performance metric in such designs is latency. Table 2 reports latencies as functions of T_{snd} and T_{rec}. Replacing with values for T_{snd} and T_{rec} shows that for most cases, the proposed design has a lower latency than a FIFO design. For example, when operating at 1 GHz on the send and receive sides ($T_{snd} = T_{rec} = 1$ns), our design decreases latency of the interface by 39%. Our design has a lower latency because it does not require 2FF synchronizers. In summary, the FIFO design can provide higher maximum throughput, while the proposed design provides higher flexibility with different clock speeds, and lower latency.

The proposed design transforms mutually asynchronous signals into pseudo-deterministic ones. This enhances the robustness of

Table 2. Comparison with FIFO based technique

	Proposed design	FIFO based design
Min. latency (ns)	1.14	$0.5\, T_{snd} + 2.5\, T_{rec}$
Max. latency (ns)	$1.14 + T_{rec}$	$0.5\, T_{snd} + 3\, T_{rec}$
Min. throughput (items/s)	$\dfrac{1}{1.65ns + T_{rec} + T_{snd}}$	$\dfrac{1}{Max\{T_{rec}, T_{snd}\}}$
Max. throughput (items/s)	$\dfrac{1}{1.65ns}$	

the design to metastability. This is further illustrated in Fig. 8. The figure shows parametric analysis of the pauser-S block at 0.01 ps resolution. At one point, the ACK signal is generated at the clock edge immediately after (top waveform). When Ri goes down 0.01 ps later, ACK is paused until the next cycle (middle waveform) without getting into metastability.

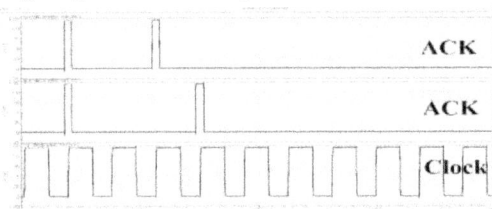

Figure 8. Parametric analysis at 0.01ps resolution

6. CONCLUSION

In this work we have proposed a new GALS interface that does not pause the clocks of the communicating systems and that requires no external synchronizers. The design introduces two blocks named protocol-pausers to avoid violating the timing constraints of the FIFO and the locally synchronous blocks. The blocks were implemented with TSMC 90nm CMOS technology and comprehensive simulations were performed. The proposed design was shown to be robust against workload and temperature variations and it was shown to work for arbitrary frequency ratios. Its system throughput is higher than pausable clocking interfaces and it offers lower latency compared to FIFO based designs.

7. REFERENCES

[1] Chapiro, D. 1984. Globally-Asynchronous Locally-Synchronous Systems. Ph.D.dissertion. Stanford University.

[2] Chelcea, T., and Nowick, S. M. 2004. Robust Interfaces for Mixed-Timing Systems. *IEEE Transactions on Very Large Scale Integration (VLSI) Systems*, Vol 12, Issue 8, 857-873.

[3] Dobkin, R., Ginosar, R., and Sotiriou, C.P. 2004. Data Synchronization Issues in GALS SoCs. in *10th IEEE International Symposium on Asynchronous Circuits and Systems*, April 2004, 170-179.

[4] ITRS 2009, available online at http://www.itrs.net/Links/2009ITRS/Home2009.htm.

[5] Muttersbach, J., Villiger, T., and Fichtner, W. 2000. Practical Design of Globally-Asynchronous Locally-Synchronous Systems. In *Proc. Int. Symp. on Advanced Research in Asynchronous Circuits and Systems* (ASYNC'00), 52-59.

[6] Rabaey, J. M., Chandrakasan, A.,and Nikolic, D. 2003. *Digital Integrated Circuits*, 2nd ed. Prentice Hall.

[7] Teehan, P., Greenstreet, M., and Lemieux, G. 2007. A Survey and Taxonomy of GALS Design Styles. In *IEEE Design & Test of Computers*, Vol. 24, Issue 5, 418-428.

Density-Reduction-Oriented Layer Assignment for Rectangle Escape Routing

Jin-Tai Yan and Jun-Min Chung
Department of Computer Science and Information
Engineering, Chung-Hua University,
Hsinchu, Taiwan, R. O. C.

Zhi-Wei Chen
College of Engineering,
Chung-Hua University,
Hsinchu, Taiwan, R. O. C.

ABSTRACT

Given a set of n buses in a pin array, the layer assignment(LA) for rectangle escape routing can be divided into five different problems: LA-1, opposite LA-2, corner LA-2, LA-3 and LA-4 problems for rectangle escape routing. Based on the optimality of a left-edge algorithm for interval packing, the LA-1 problem can be transformed into an interval packing problem and optimally solved in O($nlogn$) time. Furthermore, based on the definition of an exact low-bound and the concept of the density reduction, the opposite LA-2 problem can be optimally solved by using density-reduction-oriented layer assignment in O($nlogn$) time. Finally, by using the optimal results in the LA-1 and opposite LA-2 problems, the corner LA-2, LA-3 and LA-4 problems can be solved by using two-phase density-reduction-oriented layer assignment in O($nlogn$) time. Compared with Ma's approximation algorithm[6] for the LA-4 problem, the experimental results show that our proposed algorithm obtains the same optimal result but reduces 91.6% of CPU time for eight tested examples on the average.

Categories and Subject Descriptors
B.7.2 [**Integrated Circuits**]: Design Aids – *Placement and routing*

General Terms: Algorithms, Design

Keywords: PCB design, Escape routing, Layer assignment

1. INTRODUCTION

For bus-oriented escape routing, the layer assignment problem can be divided into layer assignment between two adjacent components[1-4] and layer assignment inside a component[5-6]. In layer assignment between two adjacent components, firstly, Kong et al. [1] proposed an optimal algorithm for bus sequencing. Given a set of buses, by using the proposed optimal algorithm for bus sequencing, a maximum subset of the unassigned buses is repeatedly assign onto a new layer in the heuristic algorithm. However, the proposed algorithm cannot minimize the number of assigned layers for bus-oriented escape routing. Recently, by finding the maximum matching in a bipartite matching problem, Yan et al.[3] propose an O($n^{2.38}$) optimal algorithm to minimize the number of assigned layers for bus-oriented escape routing.

However, the time complexity in the optimal algorithm is too high for bus-oriented escape routing. To reduce the time complexity, based on the optimality of interval packing, Yan et al.[4] further propose an O(n^2) optimal algorithm to minimize the number of assigned layers for bus-oriented escape routing. In layer assignment inside a component, firstly, H. Kong et al.[5] propose optimal algorithms for finding disjoint boundary rectangles on 2, 3 and 4 available boundaries inside a larger rectangle. By iteratively using the proposed optimal algorithm for a given set of buses inside a component, the layer assignment problem inside a component can be solved for PCB designs. However, the proposed algorithm cannot minimize the number of assigned layers for bus-oriented escape routing. Recently, Q. Ma et al.[6] formulate the rectangle escape problem(REP), show that the REP on 4 available boundaries is NP-complete, transform the REP into an integer linear program(ILP) problem and propose an approximation algorithm to solve the REP. However, the proposed approximation algorithm takes more CPU time for rectangle escape routing.

In this paper, given a set of n buses in a pin array, the layer assignment(LA) for rectangle escape routing can be divided into five different problems: LA-1, opposite LA-2, corner LA-2, LA-3 and LA-4 problems for rectangle escape routing. Based on the optimality of a left-edge algorithm for interval packing, the LA-1 problem can be transformed into an interval packing problem and optimally solved in O($nlogn$) time. Furthermore, based on the definition of an exact low-bound and the concept of the density reduction, the opposite LA-2 problem can be optimally solved by using density-reduction-oriented layer assignment in O($nlogn$) time. Finally, by using the optimal results in the LA-1 and opposite LA-2 problems, the corner LA-2, LA-3 and LA-4 problems can be solved by using two-phase density-reduction-oriented layer assignment in O($nlogn$) time. Compared with Ma's approximation algorithm[6] for the LA-4 problem, the experimental results show that our proposed algorithm obtains the same optimal result but reduces 91.6% of CPU time for eight tested examples on the average.

2. PROBLEM FORMULATION

In a PCB design, it is assumed that a pin array is located under a circuit component for IO connections. For rectangle escape routing, all the IO pins under a circuit component can be grouped into some buses and the IO pins in a bus must be escaped to one available boundary in the component with minimal detours. For any bus in the component, the minimum rectangular region which covers all the pins to be escaped to one available boundary in a single layer can be represented as a *pin rectangle*. For rectangle escape routing, the escape region of the bus to any available boundary can be represented as a *projection rectangle* in the mapping direction. For any bus, the projection of its pin rectangle to the left, right, top or bottom boundary can be defined as a left, right, top or bottom *projection interval*. For all the pins of a bus, B_i, in a pin array, the pin rectangle, R_i, and the four projection rectangles, R^l_i, R^r_i, R^t_i and R^b_i, of the bus and the left, right, top and bottom projection intervals, I^l_i, I^r_i, I^t_i and I^b_i, along four available boundaries can be obtained and shown in Figure 1.

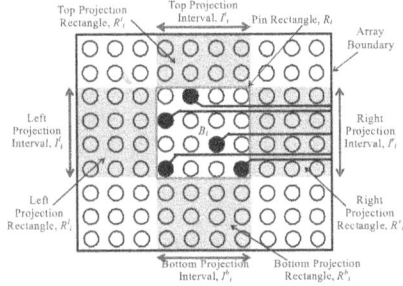

Figure 1 Projection rectangles and intervals for all the pins of a bus

Due to the high density of IO connections in a PCB design, all the buses in a pin array may not be assigned on the same layer. If their pin rectangles, R_i and R_j, overlap each other for any pair of buses, B_i and B_j, the two buses cannot be escaped on the same layer and the two buses must be assigned on two different layers for rectangle escape routing. Hence, the overlapping condition of the two pin rectangles, R_i and R_j, can be defined as a *compulsory conflict* between the buses, B_i and B_j, for rectangle escape routing. On the other hand, if their pin rectangles, R_i and R_j, do not overlap, the geometrical relation between the rectangles, R_i and R_j, can be divided into the *opposite relation* and the *diagonal relation* between the rectangles, R_i and R_j. If the assigned projection rectangles of two opposite or diagonal buses overlap, the two buses cannot be escaped on the same layer and the two buses must be assigned on two different layers for rectangle escape routing. In contrast, if the assigned projection rectangles of two opposite or diagonal buses do not overlap, the two buses may be routed on the same layer for rectangle escape routing. Hence, the overlapping condition of the projection rectangles, R_i and R_j, can be defined as a *conditional conflict* between the two opposite or diagonal uses, B_i and B_j, for rectangle escape routing.

If two buses are able to be assigned on the same layer, the two buses can be defined to be compatible for rectangle escape routing. If any pair of two buses on the same layer is compatible, all the buses on the layer can be defined to be compatible for rectangle escape routing. For the layer assignment of all the buses in rectangle escape routing, if the assigned buses in any layer are compatible for rectangle escape routing, the layer assignment for rectangle escape routing can be further defined as a *valid layer assignment* for rectangle escape routing. Given a set of n buses, $\{B_1, B_2,...,B_n\}$, represented as their pin rectangles, $\{R_1, R_2,...,R_n\}$, in a pin array, and a set of available boundaries, the layer assignment(LA) for rectangle escape routing is to assign one available projection rectangle, R^l_i, R^r_i, R^t_i or R^b_i, onto any bus, B_i, such that the number of the assigned layers is minimized to obtain a valid layer assignment for all the buses in $\{B_1, B_2,...,B_n\}$ and map all the buses in $\{B_1, B_2,...,B_n\}$, onto the assigned layers for rectangle escape routing. If r is the number of available boundaries in a LA problem, $1 \leq r \leq 4$, the LA problem on the r available boundaries can be defined as a LA-r problem.

As illustrated in Figure 2(a), it is assumed that the number of available boundaries in the LA problem is 4. Given a set of 11 buses, $\{A, B, C, D, E, F, G, H, I, J, K\}$, in a pin array for rectangle escape routing, the number of the assigned layers in the LA-4 problem can be minimized to be 2 and all the buses in the pin array can be assigned on two layers for rectangle escape routing. In the layer assignment of the 11 buses as illustrated in Figure 2(b), 7 buses,

A, B, E, F, G, I and K, can be assigned on the first layer and the other 4 buses, C, D, H and J, can be assigned on the second layer.

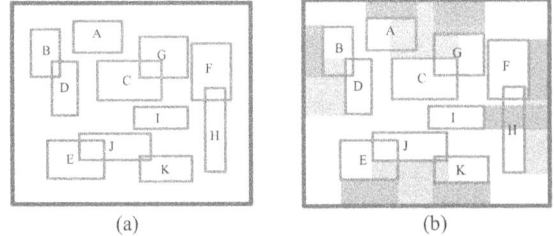

| (a) | (b) |

Figure 2 Layer assignment in the LA-4 problem

3. LAYER ASSIGNMENT FOR RECTANGLE RSCAPE PROBLEM

Given a set of n buses, $\{B_1, B_2,...,B_n\}$, in a pin array, the LA problem for rectangle escape routing can be divided into five different problems: *LA-1, opposite LA-2, corner LA-2, LA-3* and *LA-4 problems* according to the number and locations of the available boundaries in a pin array for rectangle escape routing.

3.1 Optimal Layer Assignment for LA-1 Problem

For the LA-1 problem, any bus, B_i, $1 \leq i \leq n$, only has one projection rectangle, R^l_i, R^r_i, R^t_i or R^b_i, to the available boundary for rectangle escape routing. Hence, any bus, B_i, $1 \leq i \leq n$, in the LA-1 problem can be simply represented as an interval, I_i, for its left(or top) projection interval, I^l_i(or I^t_i). It is known that the conditional conflict between two opposite buses, B_i and B_j, to the available boundary becomes a compulsory conflict and there is no conflict between two diagonal buses, B_i and B_j, in the problem. Hence, the overlapping condition between two represented intervals, I_i and I_j, becomes the only assignment constraint in the LA-1 problem.

For the represented intervals, I_1, I_2,... and I_n, of the buses, B_1, B_2,... and B_n, in the LA-1 problem, the endpoints, $p_{1,1}$, $p_{1,2}$, $p_{2,1}$, $p_{2,2}$,..., $p_{n,1}$ and $p_{n,2}$, of the n represented intervals can be found and the minimum range covering the n represented intervals can be defined as the *available range*, I^R, of the n represented intervals. According to the locations of the $2n$ endpoints, $p_{1,1}$, $p_{1,2}$, $p_{2,1}$, $p_{2,2}$,..., $p_{n,1}$ and $p_{n,2}$, the available range, I^R, can be further partitioned into some sub-ranges, I^r_1, I^r_2,... and I^r_m. In general, the *local density, d_i*, of the represented intervals, I_1, I_2,... and I_n, inside any sub-range, I^r_i, $1 \leq i \leq m$, can be defined as the number of the represented intervals overlapping the sub-range, I^r_i, and the *maximum density, d_{max}*, of the represented intervals, I_1, I_2,... and I_n, can be defined as the maximal local density, $Max\{d_1, d_2,...d_m\}$.

Since the overlapping condition between two represented intervals is the only assignment constraint in the LA-1 problem, the LA-1 problem on the available layers can be transformed into an interval packing problem on the available tracks and the interval packing problem can be optimally solved by using a left-edge algorithm[7]. According to the optimality of the left-edge algorithm, it is clear that the LA-1 problem on the available layers can be optimally solved by using a greedy left-edge algorithm in $O(n\log n)$ time, where n is the number of the buses inside a pin array. Clearly, the minimized number of the available layers is equal to the maximum density of all the represented intervals. Refer to the 11 buses in Figure 2(a) for rectangle escape routing, it is assumed that the only available boundary in the LA-1 problem is the top boundary in a pin array. The local densities of the 11 represented intervals inside the 17 partitioned sub-ranges can be computed and the maximum density of the 11 intervals can be obtained as 5. By using a greedy left-edge algorithm, the number of the assigned layers can be minimized to be 5 and all the buses in the pin array can be assigned on the 5 layers for rectangle escape routing.

3.2 Optimal Layer Assignment for Opposite LA-2 Problem

For the opposite LA-2 problem, any bus, B_i, $1 \leq i \leq n$, has two projection rectangles, R^l_i and R^r_i (or R^t_i and R^b_i), to the two opposite boundaries for rectangle escape routing. Hence, any bus, B_i, $1 \leq i \leq n$, in the LA-2 problem can be simply represented by using an interval, I_i, as its left (or top) projection interval, I^l_i (or I^t_i). As the represented intervals, I_i and I_j, of two opposite buses, B_i and B_j, in the LA-2 problem overlap, two different conditions must be considered. If the two buses, B_i and B_j, are escaped to two individual boundaries in the correct direction as illustrated in Figure 3(a), the two buses, B_i and B_j, can be assigned on the same layer. In contrast, if the two buses, B_i and B_j, are escaped to two individual boundaries in the reverse direction as illustrated in Figure 3(b), the two buses, B_i and B_j, must be assigned on two different layers. Additionally, it is known that there is no conflict between two diagonal buses, B_i and B_j, in the LA-2 problem. Hence, the overlapping condition between two represented intervals, I_i and I_j, and the escape directions of two opposite buses, B_i and B_j, are the assignment constraints in the LA-2 problem.

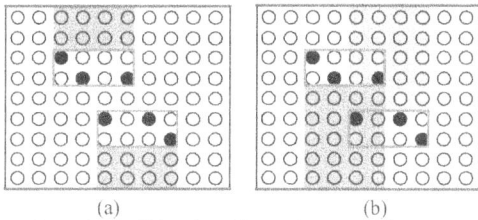

(a)   (b)

Figure 3 Conditional conflicts of two opposite buses

It is known that the maximum density of all the represented intervals is an exact low-bound for the number of the assigned layers in the LA-1 problem. Furthermore, the problem can be optimally solved by using a left-edge algorithm and the concept of layer-by-layer *density reduction* to assign the corresponding buses onto the available boundary. In the LA-2 problem, the *local interval density*, d^r_i, of all the represented intervals, I_1, I_2,... and I_n, inside any sub-range, I^r_i, $1 \leq i \leq m$, can be defined as the number of the represented intervals overlapping the sub-range, I^r_i. Additionally, the *local rectangle density*, d^R_i, of the represented rectangles, R_1, R_2,... and R_n, inside any sub-range, I^r_i, $1 \leq i \leq m$, can be defined as the maximal number of the represented rectangles overlapping each other inside any sub-range, I^r_i. Based on the conditional conflict between two opposite buses, the number of the assigned layers for the opposite boundaries inside the sub-range, I^r_i, can be restricted by a low-bound, $\lceil d^l_i / 2 \rceil$. On the other hand, the number of the assigned layers for the overlapped buses inside the sub-range, I^r_i, can be restricted by a low-bound, d^R_i. As a result, the *local density*, d_i, inside the sub-range, I^r_i, can be defined as the maximal value of two lower bounds, $Max\{\lceil d^l_i / 2 \rceil, d^R_i\}$, for a near exact low-bound and the *maximum density*, d_{max}, of the buses, B_1, B_2,... and B_n, in the LA-2 problem can be defined as the maximal local density, $Max\{d_1, d_2,...d_m\}$.

To minimize the number of the assigned layers in the opposite LA-2 problem, any conditional conflict between two opposite

buses must be avoided. Furthermore, the number of the assigned layers can be minimized by using the concept of layer-by-layer *density reduction* to assign the corresponding buses onto the available boundaries. For density-reduction-oriented layer assignment in the LA-2 problem, the reduction rule is that the maximum density of the represented intervals for the remaining buses must be as small as possible after assigning a feasible set of buses on a single layer. If the reduction rule is always obeyed in each layer assignment, the number of the assigned layers can be minimized. For the LA-2 problem, if the local density inside any sub-range is equal to the maximum density, the sub-range can be defined as a *reduction-oriented sub-range* and any bus inside a reduction-oriented sub-range can be defined as a *reduction-oriented bus*. Furthermore, a maximal set of the reduction-oriented buses overlapping each other inside the same sub-range can be defined as a *reduction-oriented group* and the *overlapping factor* of any bus in a reduction-oriented group can be defined as the number of the reduction-oriented buses in the group.

For density-reduction-oriented layer assignment, the bus assignment on a single layer can be divided into two steps: *Selection of critical buses on a single layer* and *Selection of compatible buses for single-layer routing*. In the selection of critical buses on a single layer, firstly, the *overlapping factor* of any bus inside all the reduction-oriented sub-ranges can be computed. Inside any reduction-oriented sub-range, one bus with the largest overlapping factor can be selected as a critical bus and the other bus which does not overlap the selected critical bus and has the largest overlapping factor can be further selected as the other critical bus. Clearly, the maximum density of the represented intervals for the remaining buses in most of the single-layer assignments can be reduced by 1. After selecting the critical buses inside all the reduction-oriented sub-ranges, the represented intervals of the remaining buses are sorted in an increasing order according to their left coordinates of all the intervals. By using a modified left-edge algorithm on two tracks, based on the locations of the selected critical buses, some feasible buses can be selected as *compatible buses* in the selection of compatible buses for single-layer routing. As a result, all the critical and compatible buses can be escaped to two opposite boundaries for single-layer routing. After completing layer-by-layer density-reduction-oriented layer assignment, all the buses inside a pin array in an opposite LA-2 problem can be optimally assigned in O($n\log n$) time.

Refer to the 11 buses in Figure 2(a) for rectangle escape routing, it is assumed that the available boundaries in the LA-2 problem is the top and bottom boundaries in a pin array. Firstly, the local densities of the 11 represented intervals inside the 17 sub-ranges can be shown in Figure 4(a). In density-reduction-oriented layer assignment, two reduction-oriented buses, G and K, can be selected from the original buses as two critical buses and the 4 compatible buses, A, B, E and F, can be selected for the two critical buses, G and K. As a result, 6 buses, A, B, E, F, G and K, can be assigned on the first layer. Furthermore, two reduction-oriented buses, C and J, can be selected from the remaining buses as two critical buses and the 2 compatible buses, D and H, can be selected for the two critical buses, C and J. As a result, 4 buses, C, D, H and J, can be further assigned on the second layer. Finally, the bus, I, can be selected from the remaining buses as a critical bus and the bus, I, can be assigned on the third layer as illustrated in Figure 4(b).

3.3 Efficient Layer Assignment for Corner LA-2, LA-3 and LA-4 Problems

Based on the concept of the density-reduction-oriented layer assignment, the LA-1 and opposite LA-2 problems can be optimally solved. Due to the difficulty in the definition of an exact low-bound in the corner LA-2, LA-3 and LA4 problems, the three problems cannot be optimally solved by using density-reduction-oriented layer assignment. By using the optimal results in the LA-1 and opposite LA-2 problems, an efficient two-phase density-reduction-oriented layer assignment is proposed to assign all the buses on the available layers in the three problems. For two-phase density-reduction-oriented layer assignment, firstly, the layer assignment to one boundary in the LA-1 problem can be applied as the first phase for the corner LA-2 problem and the layer assignment to two opposite boundaries in the opposite LA-2 problem can be applied as the first phase for the LA-3 and LA-4 problems. Furthermore, the layer assignment to the other boundary in the LA-1 problem can be applied as the second phase for the corner LA-2 and LA-3 problems and the layer assignment to the other two boundaries in the opposite LA-2 problem can be applied as the second phase for the LA-4 problem. Additionally, the escape directions of some assigned buses can be modified to shorten the escape length in the second phase. Based on the time complexities in the LA-1 and opposite LA-2 problems, all the buses inside a pin array in the three problems can be assigned in O($n\log n$) time.

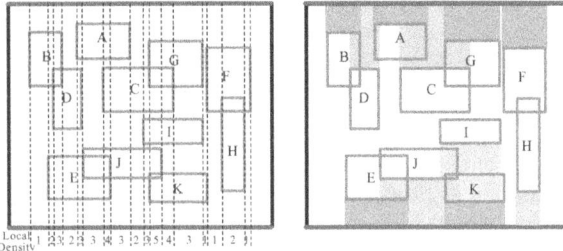

Figure 4 Layer assignment in the opposite LA-2 problem

Refer to the 11 buses, A, B, C, D, E, F, G, H, I, J and K in Figure 2(a) for rectangle escape routing, the available boundaries in the corner LA-2 problem is the top and right boundaries in a pin array. In density-reduction-oriented layer assignment for the corner LA-2 problem, the 4 buses, A, B, F and G, to the top boundary and the 2 buses, I and J, to the right boundary can be assigned on the first layer. Similarly, the 3 buses, C, D and H, to the top boundary and the bus, H, to the right boundary can be assigned on the second layer. Finally, the 2 buses, E and K, to the top boundary and the bus, K, to the right boundary can be assigned on the third layer as illustrated in Figure 5(a). In density-reduction-oriented layer assignment for the LA-3 problem, the 5 buses, A, B, E, G and K, to the top and bottom boundaries and the 2 buses, F and I, to the right boundary can be also assigned on the first layer. Finally, the 3 buses, C, D and J, to the top and bottom boundaries and the bus, H, to the right boundary can be assigned on the second layer as illustrated in Figure 5(b). In density-reduction-oriented layer assignment for the LA-4 problem, the 4 buses, A, E, G and K, to the top and bottom boundaries and the 3 buses, B, F and I, to the left and right boundaries can be assigned on the first layer. Finally, the 2 buses, C, and J, to the top and bottom boundaries and the 2 buses, D and H, to the left and right boundaries can be assigned on the second layer as illustrated in Figure 2(b).

4. EXPERIMENTAL RESULTS

For the layer assignment in rectangle escape routing, our proposed algorithms have been implemented by using standard C++ language and run on a Intel Core2 Quad Q9450 2.66GHz machine with 4GB memory. The first tested example is from the case, Ex2, in [6] and the other seven tested examples are obtained

according to the random generation of the left and right projection intervals for all the buses in a pin rectangle. In Table I, "#Bus" denotes the number of the buses in a pin array and "#Layer" denotes the number of the assigned layers for the given buses in a pin array. Furthermore, we also implement Ma's approximation algorithm[6] for the comparison of the final assignment result and CPU time for eight tested examples. In this algorithm, the ILP formulation has been solved by the open source linear solver lp_solve[8]. Compared with Ma's approximation algorithm[6] for the maximum density in rectangle escape routing with 4 available boundaries, the experimental results show that our proposed algorithm obtains the same optimal result but reduces 91.6% of CPU time for eight tested examples on the average.

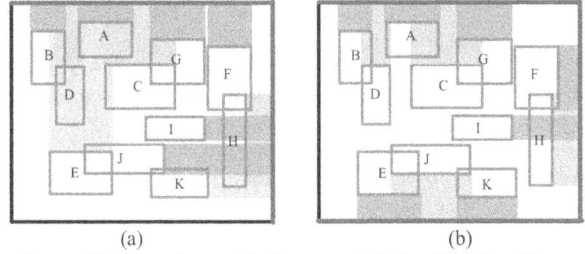

| (a) | (b) |

Figure 5 Layer assignment in the corner LA-2 and LA-3 problems

TABLE I EXPERIMENTAL RESULTS FOR LAYER ASSIGNMENT IN RECTANGLE ESCAPE ROUTING

	#Bus	Ma's Algorithm[6]		Our Algorithm	
		#Layer	CPU Time(s)	#Layer	CPU Time(s)
Ex01	20	2	0.06(100%)	2	0.02(33.3%)
Ex02	40	3	0.43(100%)	3	0.06(14.0%)
Ex03	80	4	1.78(100%)	4	0.16(9.0%)
Ex04	120	4	3.54(100%)	4	0.32(9.0%)
Ex05	160	5	5.33(100%)	5	0.49(9.2%)
Ex06	200	6	8.52(100%)	6	0.76(8.9%)
Ex07	240	7	13.27(100%)	7	1.13(8.5%)
Ex08	280	9	20.85(100%)	9	1.60(7.7%)
Total CPU Time			53.78(100%)		4.54(8.4%)

5. CONCLUSIONS

Based on the concept of the density-reduction-oriented layer assignment, the LA-1 and opposite LA-2 problems can be optimally solved. By using the optimal results in the LA-1 and opposite LA-2 problems, the corner LA-2, LA-3 and LA-4 problems can be further solved by using two-phase density-reduction-oriented layer assignment.

6. REFERENCES

[1] H. Kong, T. Yan, D. F. Wong and M. M. Ozdal, "Optimal bus sequencing for escape routing in dense PCBs," *International Conference on Computer-Aided Design*, pp.390–395, 2007.

[2] H. Kong, T. Yan, and M. D. F. Wong. Automatic bus planner for dense PCBs. *Design Automation Conference*, pp.326–331, 2009.

[3] T. Yan, H. Kong and D. F. Wong, "Optimal layer assignment for escape routing of buses," *International Conference on Computer-Aided Design*, pp.245–248, 2009.

[4] J. T. Yan and Z. W. Chen, "New optimal layer assignment for bus-oriented escape routing" *ACM Great Lakes Symposium on VLSI*, pp.205-210, 2011.

[5] H. Kong, Q. Ma, T. Yan and D. F. Wong, "An optimal algorithm for finding disjoint rectangles and its application to PCB routing," *Design Automation Conference*, pp.212–217, 2010.

[6] Q. Ma, H. Kong, D. F. Wong and F. Y. Young, "A provably good approximation algorithm for rectangle escape problem with application to PCB routing," *Asia and South Pacific Design Automation Conference*, pp.843–848, 2011.

[7] A. Hashimoto and J. Stevens, "Wire routing by optimizing channel assignment within large apertures," *the 8th Design Automation Workshop*, pp.155-169, 1971.

[8] lp_solve: an open source linear programming solver. [Online]. Available: http://sourceforge.net/projects/lpsolve

NBTI Effects on Tree-Like Clock Distribution Networks

Wei Liu, Sandeep Miryala, Valerio Tenace,
Andrea Calimera, Enrico Macii, Massimo Poncino
Politecnico di Torino, 10129, Torino, ITALY

ABSTRACT

Negative Bias Temperature Instability (NBTI) is considered one of the most critical device reliability concerns in nanometer CMOS technologies, because it causes devices to exhibit a temporal drift of performance over time.

In this work, we analyze the effects of this aging mechanism on tree-based Clock Distribution Networks (CDNs). Aging on clock tree can in fact impact the skew, causing a time-dependent failure of the circuit, if the aging conditions in different portions of the clock tree are non-uniform, like it happens in gated-clock tree in which one portion of the clock tree is selectively disabled to save power.

Characterization results collected through an in-house aging simulation framework provide valuable insights on the potential effects of various design parameters such as sizing and fanout of clock buffers and height of clock-trees.

Categories and Subject Descriptors

B.8.2 [**Performance and Reliability**]: Performance Analysis and Design Aids

General Terms

Reliability

Keywords

NBTI, clock tree, skew, clock buffer, reliability

1. INTRODUCTION AND BACKGROUND

In nanometer CMOS technologies, performance degradation induced by Negative Bias Temperature Instability (NBTI) has become the main threat to reliability of CMOS circuits and systems[2]. NBTI manifests as a time-dependent, non-linear increase of the threshold voltage (V_{th}) of PMOS devices, which, in turn, traduces to a progressive decrease in the transistor ON-current. There are two phases that characterize the appearance of NBTI[1]. Phase 1 is called the *stress phase*; it occurs in correspondence of a logic '0' at the gate terminal of the pMOS transistor ($V_{gs} = -V_{dd}$). During this phase Si-H bonds at silicon-oxide interface are broken, thus creating the positive interface traps that gets accumulated over time with "H" (Hydrogen) diffusing towards the gate. Accumulated charges at the silicon interface induce a progressive increase of the V_{th}. Phase 2 is called the *recovery phase*; it occurs in correspondence of a logic '1' ($V_{gs} = 0$). During this phase the H atoms diffuses back and anneals the broken Si-H bond. As a result the number of interface traps reduces causing a recovery of V_{th}. However, since not all the traps generated during the stress phase are recovered, the V_{th} recoveries only partially.

Beyond showing a value-dependent nature, NBTI-induced degradation is affected by several other parameters, such as the amount of applied stress, the operating temperature of active devices, the transition time of input signals, and the load capacitance. CMOS gates working under different electrical and operating conditions can thereby show substantial performance mismatch during their lifetime[4].

This issue is of paramount importance when considering the design of Clock Distribution Networks (CDNs), whose main function is keeping synchronization among latches and registers scattered across the chip area. The difference between the earliest and the latest arrival time from the clock source to any sink is known as clock *skew*. Skew must be bounded by a small value to guarantee correct sampling of data in registers.

The synthesis of a bounded-skew clock tree is a widely studied problem , however, existing methods can only guarantee that skew constraints are met at time zero, i.e., at the beginning of the circuit operations. Unfortunately, due to device aging like the one due to NBTI, clock buffers might be subject to different operating conditions thus exhibiting uneven aging profiles. This results in time-dependent phase shift between paths, and consequently, clock skew that increases as circuit operations progress.

Recent works [7, 9] have shown that NBTI is also very critical for gated clock-trees, in particular, those trees in which the clock gating conditions are not applied just at the leaf level (i.e., at the sinks, as done by most commercial clock synthesis tools), but are propagated upwards the clock tree so that a single cell can serve as gating element for an entire sub-tree[8]. In these situations, it is clear that gated portions of the CDN will age only when not gated. We can view thus the CDN as a set of regions, each with its activation function; all buffers of a region will then age proportionally to the probability of each activation function. This problem has been studied in [7, 9], which proposed solutions based on customized gating elements.

In this paper, we carry out an explorative analysis of other circuit parameters that affect skew variation due to NBTI-induced aging. More specifically, we show that the topological structure of the clock tree has significant impact in determining the delay degradation over time, in particular *the interrelation of clock tree height and buffer fanout*. Different design options are in fact available at design time: larger buffers may drive larger fanout reducing the height of the tree, while using a larger number of smaller buffers increases the degree of the tree (i.e., the fanout). Depending on such design variables clock paths may experience different fall/rise transition times and signal slew rates, which translates in a different sensitivity to NBTI.

Using an in-house fully automated NBTI-aware analysis tool, we show that clock tree structures with the same characteristics, i.e., same skew at time zero, have paths whose propagation delay have a different speed of aging, and, most important, that such difference in the aging speed amplifies over time in case of uneven activity of the branches, thereby causing NBTI-induced skew.

2. SIMULATION FRAMEWORK

The framework we have implemented for clock delay analysis relies on a SPICE-level simulator and, following a standard, commercial physical design flow, operates after clock tree synthesis. This implies that the RTL design went through logic synthesis, row-based layout placement and clock network synthesis and routing. At this point it is possible to extract accurately the interconnect parasitics, that are annotated on a SPEF file. For a given clock tree, the tool gathers the physical design and parasitic data, as well as the SPICE models provided by the library provider to create a SPICE-level netlist of the clock distribution network. Subsequently, this netlist is simulated using Synopsys HSPICE and the tool generates a final report that shows path delay degradation as well as clock skew variation results. Notice that aging effects due to NBTI are taken into account by inserting a stress factor into the models of buffer cells.

The clock analysis tool is based on customized and automated Tcl scripts that analyze and parse the input data and generate the SPICE-compliant netlist of the clock tree. This is modeled as a buffered distributed RC network linking the clock signal source to the registers. As buffers we used the aforementioned SPICE models of the technology library. For interconnects we used the detailed parasitic data contained in the SPEF file. Registers (i.e., the sinks of the clock tree) are modeled as load capacitances, whose values are taken again from the datasheet of the library. The clock signal is modeled by an ideal supply voltage that drives the root of the tree. In order to emulate the effects of NBTI on the clock network, the threshold voltage V_{th} on every PMOS transistors contained in each buffer is increased by a value which is function of the static zero probability and the age (number of years) of the circuit. These V_{th} variations have been obtained through a delay characterization using Synopsys MOSRA models.

3. CHARACTERIZATION RESULTS

3.1 Effects of NBTI on Clock Buffers

In this section we quantify the aging effects induced by NBTI on clock buffers. To this purpose we set up a tool chain that

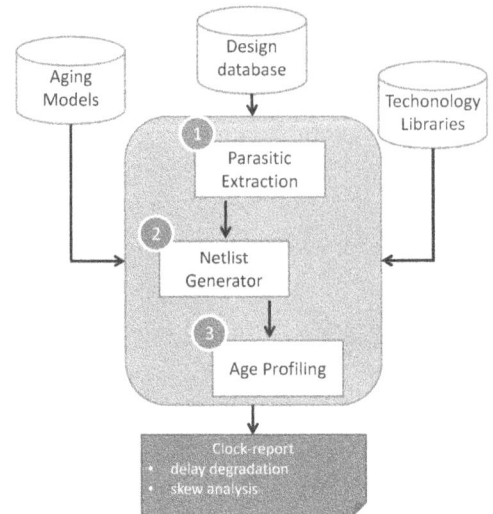

Figure 1: Overview of the Simulation Framework.

performs delay characterization of standard CMOS gate cells [3, 6, 5].

The effect of NBTI on the performance of clock tree buffers is shown in Figure 2 and Figure 3. In both figures the time reference is kept at 6 years and duty cycle of input pulse train is 0.5 (the typical value for a standard, ungated clock signal). In Figure 3 three different sized buffers (14X, 48X, and 62X) are plotted for increasing load capacitances. The NBTI induced performance degradation is shown as the percentage increase over the original value. It can be seen from the figure that the delay degradation increases significantly with increasing fanout: while a unit-fanout buffer exhibits a 10% delay degradation, this value approximately doubles for a fanout of 50, which is typical in clock trees. Dependence on buffer size is less marked, and becomes evident only at larger fanout values. As intuition would suggest, larger buffers exhibit a smaller degradation, even if the difference between the 48X and the 62X is negligible. The maximum values of load capacitance for these buffers are dictated by the technology library.

Figure 3, conversely, considers the impact of different input slew rate. The plot shows results for the minimum sized buffer (14X), for different fanout values. The plot shows that very different behavior exists depending on the fanout. For larger loads (FO4) delay is monotonically decreasing as input gets slower; on the contrary, for lighter loads the curves have a minimum, which moves towards smaller slew values (i.e., faster inputs). Furthermore, as fanout gets smaller, the degradation grows drastically for slower inputs, which have therefore to be avoided.

3.2 Effect of NBTI on Clock Tree

In order to test the methodology, we designed a set of synthetic clock trees; the choice of manually designed tree is motivated by the fact that in this way we have the opportunity of accurately controlling the topology and the distribution of the buffers in the tree, so that we can clearly expose the effects discussed in the previous section. All clock trees have zero skew (at time zero), and have been processed using the characterization data of a 45 nm industrial standard cell library. NBTI analysis is performed assuming after 6 years of operation.

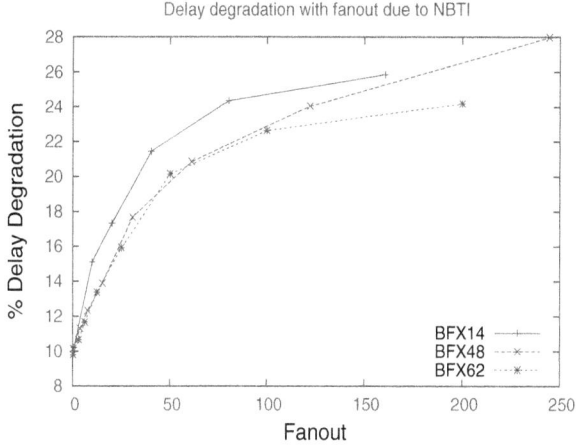

Figure 2: Delay variation Vs. Fanout after 6 years.

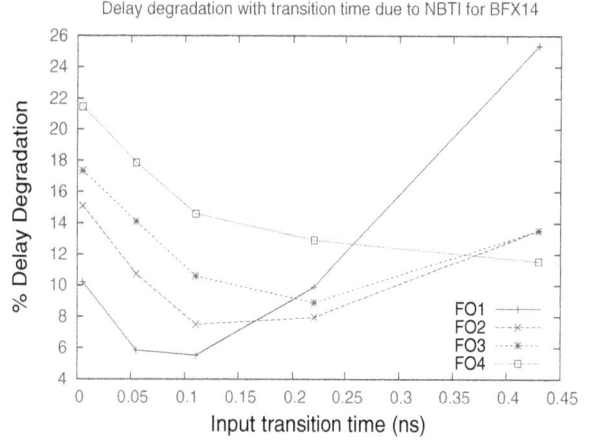

Figure 3: Delay variation Vs. Input Transition Time after 6 years.

The reference clock tree structures are illustrated in Figure 4 and Figure 5, where the clock source and some intermediate nodes are driven by buffers; leaves represent clock sinks. In *Tree1*, each clock buffer has a fanout of 4 while in *Tree2* each buffer has a fanout of 8. The total number of clock sinks in both trees are 64 and the number of buffering levels in the two trees differ by 1. Wiring between all tree nodes are assumed to have an equal length of 100 μm. Due to the symmetry of the structures, within each tree all sinks have the same delay from source thus both *Tree1* and *Tree2* have zero skew by construction.

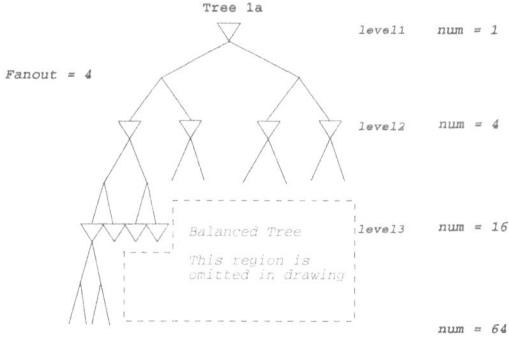

Figure 4: Clock tree *Tree1* with 3 level of buffers.

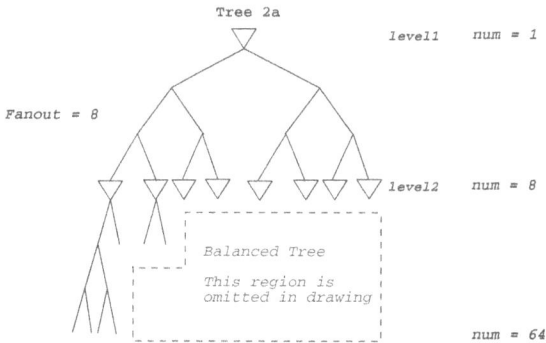

Figure 5: Clock tree *Tree2* with 2 level of buffers.

Starting from the reference clock trees, a number of transformations are applied to the clock tree structure with the purpose of studying the impact of buffer sizing and fanout on NBTI degradation. The operations performed include:

a) Original clock tree, all buffers have Small (14X) size.

b) Level 2 buffers changed to Medium (48X) size.

c) Level 2 buffers removed in *Tree1*.

d) All buffers changed to Medium (48X) size.

Since changes are applied to all buffers in the same level in each tree, the impact is on the propagation delay and not on the skew. The degradation analysis results are summarized in Table 1 where clock trees are named by appending the reference tree name with the operation performed. Columns $D0_{rise}$ and $D0_{fall}$ show the rise and fall delay without modeling NBTI aging effect. Columns $D6_{rise}$ and $D6_{fall}$ show the delay with NBTI aging effect taken into account. The percentage of changes (amount of delay degradation) are shown in the 4th and 7th columns.

Table 1: Rise and fall delays after 6 years.

	$D0_{rise}$ (ps)	$D6_{rise}$ (ps)	ΔD_{rise} %	$D0_{fall}$ (ps)	$D6_{fall}$ (ps)	ΔD_{fall} %
Tree1a	233	271	16.2	235	259	10.2
Tree1b	218	249	14.1	218	239	9.6
Tree1c	322	382	18.5	323	354	9.6
Tree1d	140	162	16.2	141	157	11.3
Tree2a	245	283	15.2	250	268	7.2
Tree2b	241	285	17.9	242	268	10.7
Tree2d	130	148	13.6	129	143	10.9

We can observe how propagation delay exhibits a significant degradation (about 16% for rise delay, and about 10% for fall delay) over the 6-year time horizon.

For *Tree2*, as there are only two levels of buffers, we only performed transformations b and d to the clock tree. Due to the larger fanout in *Tree2* compared to *Tree1*, only increasing the size of level 2 buffers does not significantly reduce its delay while the degradation even slightly increased in *Tree2b*. However, when the source buffer is also changed to Middle size as in *Tree2d*, the delay reduced by almost 50% and the degradation is decreased to 13.6%.

From these results, we can see that increasing the size of buffers can reduce the path delay and in most cases also

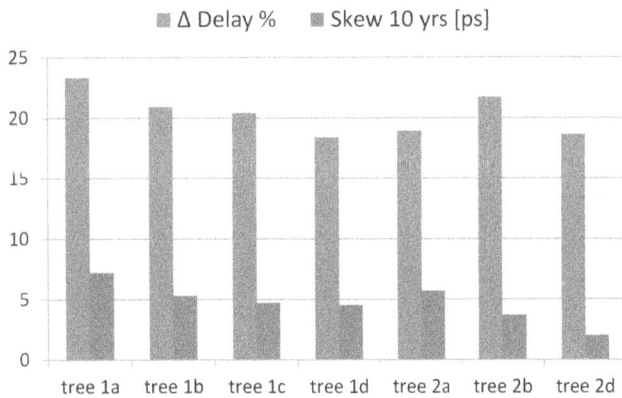

Figure 6: NBTI-induced skew variation due to uneven activity.

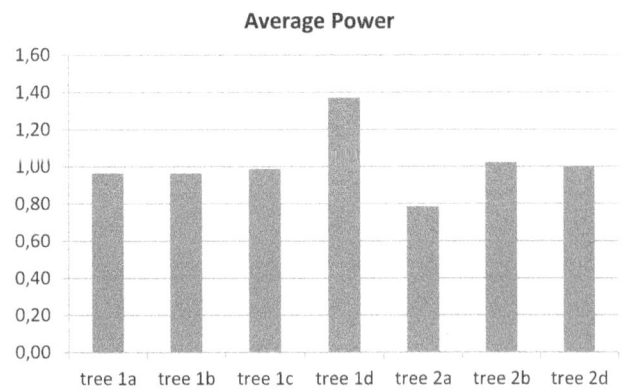

Figure 7: Power consumption comparison.

reduce the amount of NBTI-induced degradation. In addition, in the comparison of *Tree1c* and *Tree2a*, we can observe that for a given level of buffering, it is better to balance the fanout on each level. In fact, *Tree1c* has a larger capacitance and more than 3% of degradation than *Tree2a*.

It is therefore easy to understand that, depending on the topological structure of the tree (i.e., fanout and size of buffers), different paths may exhibit different aging rates. Such difference amplifies under uneven activity cases such as clock-gating, and may cause significant skew deviation.

Data on propagation delay are relevant because they expose the worst-case degradation on a single path; in terms of skew, this does not necessarily translate into an equivalent skew degradation because skew is a differential quantity. Therefore, degradation of root-to-sink delay can be interpreted as a loose upper bound for the clock skew.

In order to generate some skew, we have artificially introduced different amounts of activity in different branches of the clock tree. In particular, we have activated the rightmost and leftmost branches of the tree 10% and 90% of the time, respectively. Figure 6 reports delay variation and skew. Each group of bars shows the percentage of delay degradation on root-to-sink paths (left bars) and worst case clock skew (right bars); values refer to 10-year simulation. It is evident that configurations which are more sensitive to NBTI effects, i.e. those with larger delay variations, are also more prone to skew variation, e.g., *Tree 1a*. However, configurations that are less sensitive may not show the smallest skew; *Tree 1d*, for instance, has the smaller ΔDelay, but a skew variation that is 2x larger than *Tree 2d*. As a rule of thumb we can say that reducing buffer size is not convenient in terms of NBTI because it makes path delay more sensible, thus more susceptible to skew variation (*Tree 1a*); while using larger buffers reduces the NBTI effects only if the fanout is reduced (*Tree 1d* shows better performance than *Tree 1b*). Needless to say, modifying the topological structure of the tree playing with number of levels and buffers size may seriously affect the power consumption. Figure 7 shows the average power for the structures under test. *Tree1d*, that allows longer lifetime, shows larger power consumption, while *Tree2a*, seems to have the better aging-power trade-off. Needless to say, designers may privilege design choices that minimize their cost function.

4. CONCLUSIONS

In this paper, we proposed an NBTI induced aging analysis flow for clock trees. With the proposed tool, we investigated the impact of buffer size and fanout on the delay degradation using several tree structures. We show that increasing the buffer size and balance the fanout of each level can reduce the degradation.

5. REFERENCES

[1] M. Alam, H. Kufluoglu, D. Varghese, and S. Mahapatra. A comprehensive model for pmos nbti degradation: Recent progress. *Microelectronics Reliability*, 47(6):853 – 862, 2007.

[2] S. Borkar. Designing reliable systems from unreliable components: The challenges of transistor variability and degradation. *IEEE Micro*, 25(6):10–16, 2005.

[3] A. Calimera, E. Macii, and M. Poncino. Nbti-aware power gating for concurrent leakage and aging optimization. In *ISLPED-09: International Symposium on Low power Electronics and Design*, pages 127–132, 2009.

[4] A. Calimera, E. Macii, and M. Poncino. Nbti-aware sleep transistor design for reliable power-gating. In *Proceedings of the Great Lakes symposium on VLSI*, pages 333–338, 2009.

[5] A. Calimera, E. Macii, and M. Poncino. Analysis of nbti-induced snm degradation in power-gated sram cells. In *Proceedings of the International Symposium on Circuits and Systems*, pages 785–788, 2010.

[6] A. Calimera, E. Macii, and M. Poncino. Nbti-aware clustered power gating. *ACM Transactions on Design Automation of Electronic Systems (TODAES)*, 16(1):3, 2010.

[7] A. Chakraborty, G. Ganesan, A. Rajaram, and D. Pan. Analysis and optimization of nbti induced clock skew in gated clock trees. In *DATE-09: IEEE Design, Automation Test in Europe*, pages 296 –299, 2009.

[8] M. Donno, A. Ivaldi, L. Benini, and E. Macii. Clock-tree power optimization based on rtl clock-gating. In *DAC-03: IEEE Design Automation Conference*, pages 622–627, 2003.

[9] S.-H. Huang, C.-M. Chang, W.-P. Tu, and S.-B. Pan. Critical-pmos-aware clock tree design methodology for anti-aging zero skew clock gating. In *Proceedings of the Asia and South Pacific Design Automation Conference*, pages 480 –485, 2010.

A Framework for High-Level Synthesis of Heterogeneous MP-SoC

Youenn Corre, Jean-Philippe Diguet, Dominique Heller, and Loïc Lagadec

Lab-STICC, University of South Brittany, Rue Saint Maudé, 56100 Lorient, FRANCE

{youenn.corre, jean-philippe.diguet, dominique.heller}@univ-ubs.fr, loic.lagadec@univ-brest.fr

ABSTRACT

In this paper we propose an ESL synthesis framework which, from the C code of an application and a description of a generic architecture, automatically explores and generates a complete synthesizable version of a H-MPSoC architecture along with the adapted code application. We developed a Design Space Exploration (DSE) algorithm that merges hardware specialization, data-parallelism exploration, processor instantiation and task mapping according to user performance and cost constraints. We also inserted HLS in the DSE loop and get fast exploration of hardware acceleration. A new ESL framework is presented, it combines our contributions with some legacy tools issued from our and another team. We validated our framework with a case study of an MJPEG decoder.

Categories and Subject Descriptors: C.0 [**Computer Systems Organization**]: General –*System Architectures*; J.6 [**Computer Applications**]: Computer Aided Engineering –*Computer-aided design (CAD)* **Keywords:** Heterogeneous MP-SoC, High-level Synthesis, Design Space Exploration, FPGA, Embedded Systems

1. INTRODUCTION

FPGA allow for hardware specialization and so can improve performances with low clock frequencies compared to other solutions such as General Purpose Processors (GPP) or Graphics Processing Units (GPU). However the use of FPGA is not widely spread in industrial solutions. This is mainly due to the costly investment necessary to master FPGA design, which is still a complex and bug-prone process. Indeed, studies have shown that developing for FPGA is a very time-consuming task compared to other solutions [1]. The lack of standard for FPGA, such as x86 for GPP, is another reason since there is no guarantee for backward and onward compatibility of the designs. That is why tools are needed that ease the design by automating as most as possible the design choices and their implementation. These tools can rely on models of FPGA and architecture models to generate the code accordingly to the chosen target and thus resolve the lack of standard. Our goal is thus to provide a tool that simplifies the design and programming of H-MPSoC, especially in the domain of embedded systems and data-stream applications. This is done by automating the exploration and implementation of the design. The exploration is based on a scalable algorithm which allows to control the number of candidates at each stage of the DSE flow. Solutions are generated by means of a synthesizable specification of the hardware platform and C code that implement API for communication and synchronization between heterogeneous processors. The fact that the code is generated ensures that it is correct-by-construction and thus greatly reduces the test phase. So by automating those steps we provide a fast and easy methodology for H-MPSoC design, accessible to engineers with little knowledge about hardware design. Along with the overview of the whole framework, this paper focuses on the proposition of a fast and scalable DSE algorithm including High Level Synthesis (HLS) of specialized hardware accelerators. Other contributions of our framework that are not described here for the sake of brevity include the exploration of hardware specialization

GLSVLSI'12, May 3–4, 2012, Salt Lake City, Utah, USA.
Copyright 2012 ACM 978-1-4503-1244-8/12/05 ...$10.00.

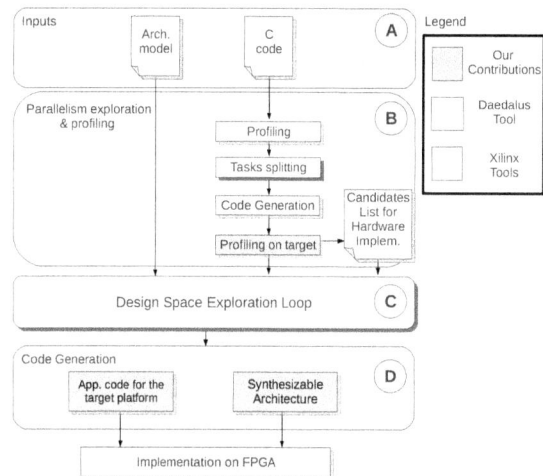

Figure 1: Flow of our framework.

and the exploration of data-parallelism through task duplication. This paper is organized as follow: in section 2 our approach is compared to existing work. We then make a brief overview of the flow of our tool in section 3. Section 4 describes the main contribution of this paper: the DSE algorithm. In section 5, we present a case study and we conclude in section 6.

2. RELATED WORK

Since our own framework is partly based on the Daedalus framework [2], we have a lot in common with it. We borrowed its performance simulation tool as well as its automatic transformation of the application into Kahn Process Network (KPN) [3]. This implies also that we use the same formalism as input. However we have added several improvements, which are the automated exploration, based on HLS, of the possible hardware accelerators, the exploration of the data parallelism and a faster DSE algorithm. SystemCoDesigner [4] is an ESL synthesis framework developed at the University of Erlangen-Nuremberg. It also introduces HLS with Forte Cynthesizer to generate hardware IP as RTL descriptions that feed a library of components. However we do not use the same MoC and language: while SystemCoDesigner uses a subset of SystemC (SysteMoC [5]), we focus on Data-flow applications and start from a KPN model simply specified in C, where a task represents the level of granularity of possible HW accelerators. We think that this is a much more simple approach, closer to programmer usual backgrounds. In [6], a H-MPSoC DSE approach based on a genetic algorithm is proposed. This tool provides scenario-based performances estimations in order to take into account both the intra-application and inter-applications dynamic behavior of the evaluated system. While this is an advantage over our framework, unlike us however it does not provide an exploration of the HW accelerators nor does it take into account data-parallelism.

3. FRAMEWORK FLOW

We first decomposed our design flow into steps and checked if available tools with standard inputs could deal with some of these steps so that we can focus on unsolved parts of the problem. As a consequence, parts of our flow reuse some existing tools that were best fitting our goal, as shown in Fig. 1 and 2. We made the following reuses from the Daedalus framework: (i) input formalism (YML) for architecture and KPN specification, such an XML specification can be produced with Model-Driven Engineering (MDE) tools based on a standard Y-chart MARTE specification [7], (ii) application code transformations compliant with the KPN MoC and (iii) performances simulation tool, Sesame [8]. We also use other tools such as GAUT [9] for the high-level synthesis of hardware IP, as well as Xilinx tools for the final implementation. However, those tools can be quite easily replaced by equivalent solutions as they are not part of the core of the framework flow. For example, the synthesis tool could be Catapult-C. Our framework has two main inputs (part A of Fig 1): a description of a generic architecture model and the C code of an application. The architecture model is used as a basis for the DSE. It describes the model of communication (shared memory, bus, etc.), the type of available processors, etc. The code of the application has first to be split into tasks (Part B of Fig.1); this stage is automatically performed by the Daedalus tool KPNGen [10] which generates KPN. The input has thus to be Daedalus compliant, i.e. the application is a Static Affine Nested Loop Program (SANLP) [11]. Each task becomes one process of the KPN and for each of them the designer must specify if they can be transformed into a hardware accelerator or if they can be duplicated. The newly-split application is then automatically adapted to perform an accurate profiling on the target softcore. Each task of the application is transformed into a POSIX thread communicating with the other tasks through POSIX message queues. We use a hardware timer, that counts the number of cycles required by each task to complete a software execution. The measures start after the read step and stop before the write step. It means that we separate processing and data transfer parts, in other words data are assumed to be available in the processor cache, memory buffer updates are estimated separately. The Sesame tool we use for performances evaluation needs the number of cycles taken for one iteration of the tasks as it estimates the time taken for the communication between tasks only according to the underlying architecture. These real measurements are used to order the tasks for their selection for hardware acceleration. The profiling also gives us for each task, the accurate number of cycles per execution. These are later used in the DSE for performance evaluation. Then the DSE process starts (Part C of Fig.1). The flow of the DSE is detailed in section 4. Basically, it explores different design dimensions such as the type and number of processors, mapping, scheduling, etc. in order to find solutions that fit the designer's objectives. Once a satisfying solution has been found, a synthesizable version of the hardware platform is generated along with the corresponding software code of the application (Part D of Fig.1). The generated code includes the insertion of the calls to communication and synchronization API. Commercial tools suite such as Xilinx ISE are used to implement the generated code onto the target FPGA.

4. DESIGN SPACE EXPLORATION

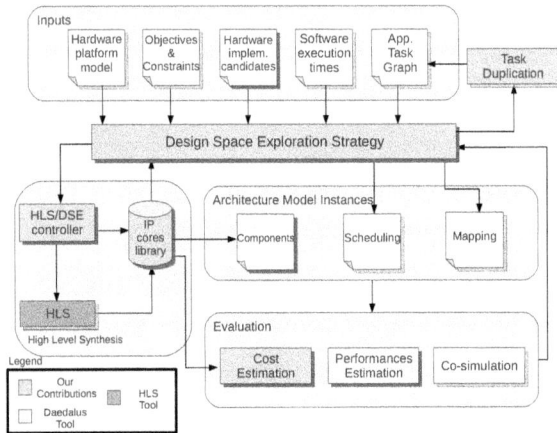

Figure 2: DSE flow.

Architecture model. Our tool targets scalable H-MPSoC architectures with distributed local memories where message passing is the paradigm of inter-task communications. The global interconnect implements point-to-point links that could be replaced by a network-on-chip in future work. The whole picture is described in Fig.3. The number of processors varies and each processor can be enhanced with one or more accelerators dedicated to a single task. An accelerator interface can be either a FIFO or

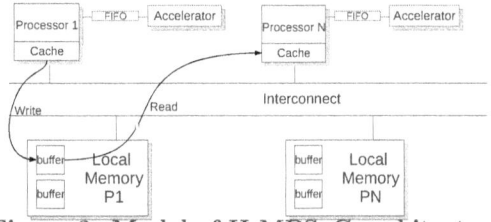

Figure 3: Model of H-MPSoC architecture

a Ping-pong memory connected to the processor local bus. Interprocessor communications are based on FIFO channels compliant with the KPN model. FIFOs are implemented as buffers in local shared memories with a dual port access. The communication cost depends on the size and amount of transferred data. Another important point is that message passing implementation is based on a zero-copy mechanism. It means that the output buffer of a Task i is implemented in the local memory of the Proc. j the task is running on. This output buffer is also the input buffer of the task that reads these data, therefore there is no copy from output to input buffers.

General principles. The DSE algorithm first starts by generating exhaustively all possible architectures with the available processors within the minimal and maximal bounds set by the designer. Then it explores ways to improve performances by adding a hardware accelerators. Then mappings and scheduling are explored and the performances are evaluated by Sesame. Along those steps, design space prunings are performed to reduce the DSE time. If no satisfying solution is found, data-parallelism is explored by means of task duplication. The final result is a selection of designs satisfying the constraints, they are proposed to the designer who makes the final choice.

Definitions. Hereafter we introduce a set of definitions that are necessary to understand the DSE algorithm presented in Algo.1: *AvailableProc* is the set of available types of processors (e.g. MicroBlaze, PowerPC, etc.). *PerfObjective*, *CostObjective* are the performance and cost objectives respectively. *minProc, maxProc* are the minimum and maximum numbers of possible processors in the design respectively. *SortedAccel* is the set of currently available accelerators sorted by increasing cost.*Solutions* is the set of selected architecture solutions. *Architecture* is a representation of an architecture model with a set of HW components, cost and performances evaluations. *NeedAcceleration* is the set of generated architectures, which do not meet the performance objective and consequently need hardware acceleration. *CandidateTasksToHW* is the set of application tasks identified as good candidates to hardware implementation. In case of task duplication, a unique implementation is considered and all tasks instances are grouped (*GroupedTask*). Tasks are sorted by decreasing execution time which came from the earlier profiling step. N is the size of a limited number of promising solutions all along the DSE stages. This parameter allows to balance the algorithm search-space/speed tradeoff. For instance $N = 1$, corresponds to a greedy algorithm while $N = \infty$ means an exhaustive exploration. T_{min} and T_{max} are intermediate performance metrics. T_{min} corresponds to an ideal case where the best performances are obtainable for a given architecture, it means that data are available in processor caches and an ideal distribution of computing time over processors according to Ahmdhal's law restrictions: $T_{min1} = \frac{ComputingTime + NWrite*WriteCost}{NumberOfProc}$ where: $NumberOfProc$ is the number of processors in the current architecture. $ComputingTime$ is the total time given by the Sesame tool for the monoprocessor architecture. It is based on the execution time of each task provided by the on-target profiling step. $ComputingTime$ is thus the sum of the computing times of each task i. If a task i is accelerated with a HW accelerator, then its computing time is updated and is equal to the division of the number of cycles taken for a software execution by the speedup obtained with the accelerator. $NWrite$ is the total write accesses count to the memory. $WriteCost$ is the cycle number for a write operation. $T_{min} = max\left(max_{i \in I}(ComputingTime(i)), T_{min1}\right)$ where I is the set of non duplicate and non accelerated tasks. T_{max} is the worst computation time, which is similar to T_{min}

except that there is no cache and consequently all data are read from the memory. This increases significantly the communication time: $T_{max} = \frac{ComputingTime}{NumberOfProc} + NWrite * WriteCost + NRead * CacheMissCost$, where: $NRead$ is the total number of read accesses. $CacheMissCost$ is the penalty for cache miss as a number of cycles. These values have been estimated once for the target architecture model with ad hoc profilings. α_0 parameter. T_{max} can be overestimated due to the no cache assumption and sequential data transfers but also on a possible underestimation introduced by the ideal balancing of cycles over available processors. These assumptions impact the accuracy of the upper bound, so a tuning variable α_0 is introduced to adapt the performance objective to the upper bound estimation. Its default value is set to 1. For a given architecture α is equal to $\frac{T_{max}}{PerfObjective}$.

Algorithm 1 Algorithm of the DSE

1: $S0$ = generateAllSWArch($AvailableProc, minProc, maxProc$)
2: **for all** Architecture A in $S0$ **do**
3:    evaluatePerformancesAndCost(A)
4:    **if** A.cost > $costObjective$ **then**
5:      discard(A)
6:    **else if** A.α < α_0 **then**
7:      $Solutions$.add(A)
8:    **else**
9:      $needAccelerator$.add(A)
10:    **end if**
11: **end for**
12: **for all** GroupedTasks T in $CandidateTasksToHW$ **do**
13:    $sortedAccel$ = launchHWExploration(T) //Call to HLS Estimator
14:    **for all** Architecture A in $needAccelerator$ **do**
15:      **for all** Accelerator acc in $sortedAccel$ **do**
16:        A2 = A.add(acc)
17:        // (Pruning 1) Test of matching between Task Instance / Accelerator Number
18:        evaluatePerformancesAndCost(A2)
19:        **if** A2.cost > $costObjective$ **then**
20:          discard(A2)
21:        **else if** A2.α < α_0 **then**
22:          $Solutions$.add(A2)
23:        **else**
24:          $needAccelerator$.add(A2)
25:        **end if**
26:      **end for**
27:    **end for**
28:    keepNBestSolutions($needAccelerator$)
29: **end for**
30: keepNBestSolutions($Solutions$)
31: **for all** Architecture A in $Solutions$ **do**
32:    generatedArchitectures = ExploreMappingAndScheduling(A)
33:    // (Pruning 2) 4 rules checking
34:    **for all** Architecture A2 in generatedArchitectures **do**
35:      evaluatePerformancesWithSesame(A2)
36:      evaluateCost(A2)
37:    **end for**
38:    $Solutions$.add(keepNBestSolutions(generatedArchitectures))
39:    //(Pruning 3) Selection of Pareto optimal solutions
40: **end for**
41: keepNBestSolutions($Solutions$)
42: **return** $Solutions$

DSE algorithm. The main DSE core algorithm is divided into two phases. First, a greedy algorithm explores ordered dimensions (architectures, task, accelerators) until the maximum number of solutions N is reached or that no candidate remains to be tested. Secondly, task mapping and scheduling are performed on all selected solutions. The algorithm relies on three loop nests described in the algorithm 1. The first step (line 1) is the generation of a set of architectures, which corresponds to the exhaustive possible architectures according to $Proc$ boundaries $minProc$ and $maxProc$. The set of N best solutions is initially empty. Then in the first loop nest (lines 2-11), all architectures are evaluated in terms of cost and performances (T_{min} and T_{max}). If the architecture cost is greater than the cost constraints, then the architecture is discarded and is no longer explored. If the evaluated architecture meets the performance objective (i.e. T_{max} or α is lower than the performance objective) then the set of results is added with the new valid solution. In all other cases, the algorithm will explore the possibility of adding hardware accelerators.

The second loop nest (lines 12-30) explores hardware task migration for all candidate tasks. These tasks have been tagged as accelerable by the designer based on profiling and on his knowledge of the application. The candidate tasks are sorted in a decreasing execution time order to favor first the highest possible speedup. The exploration starts with a synthesis of a set of accelerators performed by means of HLS. The resulting accelerators are also sorted in a decreasing order according to latency and integrated - in that order - in each evaluated architecture. Performances and cost are computed for each accelerator and the exploration stops as soon as the accelerator cost exceeds the objective. Then, the solutions set is updated and the algorithm moves on to the next candidate group of tasks if the number of solutions is lower than N. Note that a first pruning operation is introduced in this stage *(pruning 1)*. It consists in removing cases where the number of duplications of accelerated tasks is larger than the number of available associated hardware modules. When the second loop nest is achieved, a maximum of N best solutions is known but the execution time only relies on estimates. The last loop nest (lines 31-40) performs mapping and scheduling exploration steps based on the current N-solutions set. Once achieved, again the N best solutions are selected and sorted out.

Optimizations. Sesame considers all possible mappings and all available scheduling and we have observed that this strategy led to the greediest part of the DSE. So, contrary to Daedalus method that generates exhaustively all mappings, we generate only mappings that are significant and aggressively prune the exploration space. Thus we only generate mappings that respect the following rules. *R1:* If an architecture contains at least one accelerator, then the accelerated tasks must be mapped on the processor with the right accelerator. *R2:* If a task is duplicated then each duplicate task must be executed on a different processor in order to ensure that the parallelism is exploited. *R3:* Mappings must use all of the available processing units of the architecture, i.e. mappings where there is no task mapped on one or more processors are not considered. *R4:* Avoid symmetric mappings on identical processors i and j where i and j can be equivalently swapped. These rules for generating mappings significantly reduce the number of iterations compared to the exhaustive method. This has a huge impact as most of the DSE global time is spent on the simulation of performances. Thus by reducing the number of cases to simulate we significantly improve the global DSE time as illustrated in the results section. Another optimization we have made is that we have defined a library of accelerators where all candidates produced during the HLS exploration can be stored and classified with the relevant characteristics (function ID, performance, cost, communication model, target and so on). Indeed, standard functions can be frequently called in multiple projects and consequently re-explored in different DSE. So, if the exploration is called again in the same context, then the framework will avoid useless HLS loops. This library can also be updated with real synthesis results based on final implementations.

Post-processing steps. If no satisfying design is found at the end of the DSE loop, new potential data-parallelism can be explored. This is done at the application level by duplicating some of the tasks. Based on the tasks metadata given by the designer and their execution times, our tool automatically selects the tasks to duplicate and generates the code for the duplication according to KPN formalism. The code for the dispatch of data between the duplicate tasks as well as for its collection at their outputs is added. Since the performance estimation tool Sesame does not require the functional behavior of the task, dummy tasks are added in order to act as duplicate tasks during simulation. Once a satisfying solution has been found, the corresponding hardware and software code for the implementation are generated.

5. RESULTS

Architecture Model. We set the following assumptions on the architecture model we used. The final architecture must have a 100 MHz clock frequency, inter-processors communications are implemented with software FIFO channels in processor local memories and processor-coprocessor communications are implemented with hardware FIFO channels. We set the cost constraints to 17280 slices, 64 DSP blocks and 148 BRAM blocks, which cor-

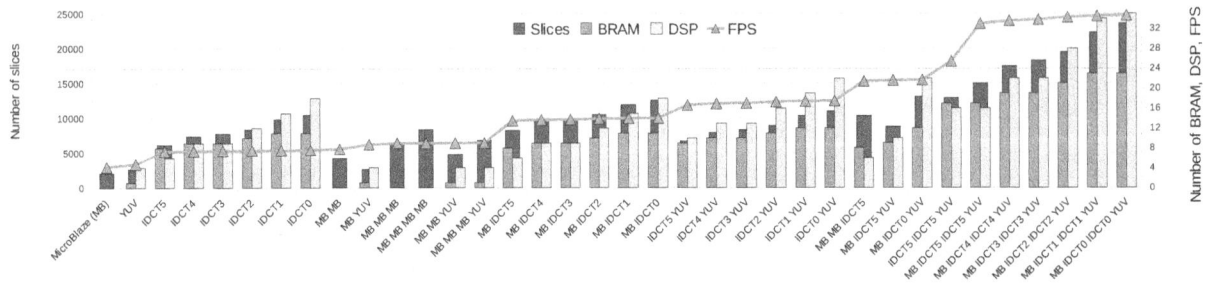

Figure 4: Selection of results of the DSE algorithm sorted by increasing performances.

respond to the resources available in the Xilinx XUPV5 FPGA board we used.

Application. In order to validate our framework, we considered an MJPEG video decoder application [12]. The goal is to have a minimal decoding speed of 24 frames per second (FPS). The application is split into five tasks: decoding, quantification, IDCT, YUV and display. The split application code is then automatically transformed in order to carry out the profiling step on the target, a MicroBlaze (MB). Each task is generated as a POSIX thread that can run on Xilkernel [13], the Xilinx' configurable OS.

On target profiling. The profiling results show that IDCT and YUV would be good candidates to hardware acceleration, so the designer marks them as "accelerable" for the DSE tool. Decoding is not considered in this case even though its computing time is superior to YUV, because the HLS tool used in this case study is not efficient for control-intensive task such as decoding. YUV is made of simple arithmetic operations and is already available as a fully synthesized accelerator in the library. On the other hand, IDCT is a more complex function which provides a large design space to be explored.

DSE. Once this step is achieved, the designer can launch the automatic DSE. In this example, we set the α variable to 1, its default value, and we set the size of the solution set N to 30. So the tool first generates a series of architectures with and without accelerators, and selects them by using the fast estimators. The hardware accelerator exploration resulted in 6 IDCT with different latencies (IDCT0...5 in Fig.4).

Task duplication. With the MJPEG splits in 5 tasks, the exploration did not produce any solution satisfying the 24 FPS objective. So the designer used the tool to increase the parallelism of the application by duplicating two tasks: the decoding and the IDCT. These two tasks have been chosen since they were the most time consuming ones in the application. A 9 tasks version has also been tested without providing significant improvements.

Exploration time. As an example, let's detail the exploration time for the 7 tasks version with up to four processors. If HLS-based estimations and Sesame simulations are not included (i.e. accelerator estimations are already available in library and no scheduling and mapping are performed) then the exploration time is about 3 seconds. If HLS-based estimation is active, then the time is about 5 minutes. As mentioned in section 4, a pruning function based on four rules has been added to the initial Sesame mapping process. This optimization leads to a total of 3000 mappings to be scheduled with Sesame. Based on a *first come, first served* policy, this scheduling step takes around 40 minutes. Without mapping pruning, 200 000 possibilities would be exhaustively generated and would require days to achieve the mapping/scheduling stage. Finally, the total time taken for the case study exploration based on our heuristic is about 45 minutes, which is really good for a complete H-MPSoC synthesis.

Results. A selection of the results of the exploration is shown in Fig.4 for versions of the MJPEG with 5 and 7 tasks, the 9 task is not presented since no improvement was observed. For each presented architecture we have selected the mapping providing the best result. We can see that only the solutions where IDCT and decoding tasks has been duplicated, can meet the minimal 24 FPS objective. Our tool managed to find a satisfying solution in 45 minutes. These results show that we can explore and then produce complete H-MPSoC solutions that respect user re-

quirements in a very reasonable amount of time. Moreover, the designer was involved in the DSE management for important and high level decisions but relieved from tedious and low level error-prone tasks.

6. CONCLUSION & FUTURE WORK

In this paper we have presented an ESL framework for H-MPSoC automated design space exploration and synthesis that includes an exploration of hardware accelerators and data-parallelism. We have shown that we could explore and generate a large number of architectures in a short amount of time and thus offering a choice to the designer between performance and cost. Our solution is compliant with upcoming H-MPSoC architectures and based on the standard KPN model of computation with C specification. The architecture generation is demonstrated on FPGA and based on a well-adopted MB softcore. Over 85% of the exploration time is spent in the evaluation of the mapping and scheduling with Sesame. Under going developments are the automatic generation of the .*mhs* and .*mss* Xilinx project files according to architecture models choices.

7. REFERENCES

[1] K. Benkrid. Reconfigurable Computing in the Multi-Core Era. In *International Workshop on Highly Efficient Accelerators and Reconfigurable Technologies (HEART) 2010*, 2010.

[2] M. Thompson et al. A framework for rapid system-level exploration, synthesis, and programming of multimedia MP-SoCs. In *5th conference on Hardware/software codesign and system synthesis*.

[3] G. Kahn. The semantics of a simple language for parallel programming. *Information processing*, 74:471–475, 1974.

[4] J. Keinert et al. SystemCoDesigner—an automatic ESL synthesis approach by design space exploration and behavioral synthesis for streaming applications. *ACM Transactions on Design Automation of Electronic Systems*, 14(1):1–23, 2009.

[5] J. Falk et al. Efficient representation and simulation of model-based designs in SystemC. In *Proceedings of the International Forum on Specification & Design Languages (FDL'06)*, pages 129–134.

[6] P. Van Stralen and A. Pimentel. Scenario-based design space exploration of mpsocs. In *Computer Design (ICCD), 2010 IEEE International Conference on*, pages 305–312, 2010.

[7] J. Vidal et al. A co-design approach for embedded system modeling and code generation with uml and marte. In *Design, Automation & Test in Europe Conference & Exhibition, 2009. DATE'09*. IEEE, 2009.

[8] A.D. Pimentel et al. A systematic approach to exploring embedded system architectures at multiple abstraction levels. *IEEE Transactions on Computers*, pages 99–112, 2006.

[9] P. Coussy et al. *GAUT: A High-Level Synthesis Tool for DSP applications*. Springer, 2008.

[10] S. Verdoolaege, H. Nikolov, and T. Stefanov. Pn: a tool for improved derivation of process networks. *EURASIP Journal on Embedded Systems*, 2007(1):19–19, 2007.

[11] H. Nikolov et al. Systematic and automated multiprocessor system design, programming, and implementation. *Computer-Aided Design of Integrated Circuits and Systems, IEEE Trans.*, 27(3):542–555, 2008.

[12] I. Augé et al. Platform-based design from parallel C specifications. *Computer-Aided Design of Integrated Circuits and Systems, IEEE Transactions on*, 24(12):1811–1826, 2005.

[13] Xilinx, OS and Libraries Document Collection (UG 643). http://www.xilinx.com/support/documentation/sw_manuals/xilinx12_3/oslib_rm.pdf.

Memory-based Computing for Performance and Energy Improvement in Multicore Architectures

Kamran Rahmani
University of Florida
kamran@cise.ufl.com

Prabhat Mishra
University of Florida
prabhat@cise.ufl.edu

Swarup Bhunia
Case Western Reserve Univ.
sxb21@case.edu

ABSTRACT

Memory-based computing (MBC) is promising for improving performance and energy efficiency in both data- and compute-intensive applications. In this paper, we propose a novel reconfigurable MBC framework for multicore architectures where each core uses caches for computation using Look Up Tables (LUTs). Experimental results demonstrate that on-demand memory-based computing in each core can significantly improve performance (up to 4.7X, 3.3X on average) as well as reduce energy consumption (up to 4.7X, 2X on average) in multicore architectures.

Categories and Subject Descriptors

C.1 [**PROCESSOR ARCHITECTURES**]: Adaptable architectures

General Terms

Design, Performance

Keywords

Multicore systems, memory-based computing, acceleration, energy optimization

1. INTRODUCTION

There are two common reconfigurable computing categories, FPGA [1], and CGRA [2]. FPGA suffers from poor technological scalability of performance and CGRA suffers from lack of flexibility to map diverse applications. In addition, both technologies fail to improve performance and energy efficiency in case of data-intensive applications. MBC is a promising alternative to address the above challenges. For ease of comparison, different features of all approaches are summarized in Table 1.

A promising solution to the challenges above is to use a reconfigurable memory-based computing (RMBC) framework. A RMBC framework uses dense 2-D memory arrays

Table 1: Comparison of reconfigurable frameworks

Feature	FPGA	CGRA	RMBC
Flexibility (compute-intensive)	High	Moderate	High
Flexibility (data-intensive)	Poor	Poor	High
Energy Efficiency	Poor	Moderate	High
Resource Utilization	Moderate	Moderate	High
Reconfiguration Latency	Moderate	Low	Moderate
Technology Scalability	Poor	Moderate	High

for computation by configuring them as LUTs to hold the configuration for the mapped application. RMBC addresses above challenges by making a trade-off between the flexibility and granularity. It is beneficial due to following reasons: i) it is amenable to efficient functional decomposition of complex operations (e.g. square root, trigonometric functions etc.) with fewer inputs (one operand often being constant coefficient) which are easier to implement as lookup table (LUT) than logic block; ii) they typically have larger than 1-bit pathwidth [3] which is amenable for sparse interconnect structure; and iii) most applications have high temporal locality in data that can be exploited to reduce computation overhead. In addition, RMBC leverages on traditional on-die cache memory which leads to minimum changes in hardware architecture. Necessary circuit and architecture level modifications however need to be incorporated into the conventional memory array before it can be used as a reconfigurable framework. This logic-memory duality benefits memory-intensive applications in addition to compute-intensive ones. In this paper we propose an architecture that uses RMBC in multicore architectures where each core has a tightly-coupled RMBC unit.

Paul and Bhunia proposed a novel MBC platform in [4]. This framework is however, primarily optimized for fine-grained combinatorial logic. In this paper, we optimize the same for coarse-grained regular data path, common to algorithmic tasks. The only work on MBC in a multicore architecture is done by Hajimiri et al. [5]. They proposed an architecture for improving reliability in multicore architectures using MBC. However, this improvement in reliability degrades performance. In this paper, our focus is on using MBC as an accelerator in a multicore architecture.

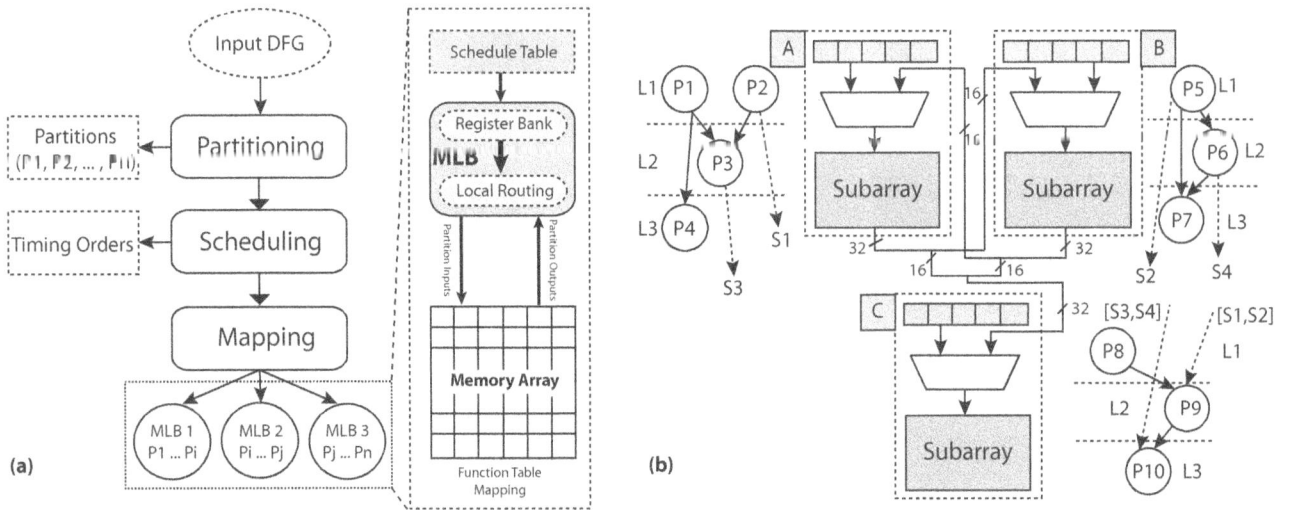

Figure 1: a) Functional block diagram for memory based computing b) Synchronization across multiple MLBs

2. AN OVERVIEW OF MBC

Figure 1(a) shows the functional block diagram of MBC model proposed in [4, 6]. The target application is partitioned into multi-input multi-output partitions, which are then mapped to multiple computing elements, each computing unit being referred to as a Memory Logic Block (MLB). Inside each MLB, the partitions are evaluated over multiple clock cycles in a topological manner. Intermediate partition outputs are stored in the register bank. LUTs are mapped to a dense 2-D memory array which is otherwise referred to as the function table. Schedule Table stores the microcode which determines the sequence of operations inside each computing element. A fast local routing network is used to select the intermediate partition outputs from the register bank. Figure 1(b) shows the communication and synchronization among multiple MLBs through a time-multiplexed usage of the local and global interconnect in the MBC model. As shown in Figure 1(b), S1 and S2 are outputs of MLB A and B at the end of cycle 1, while S3 and S4 are outputs at the end of cycle 2. Signals at the end of each cycle are transmitted over the same local/global channel to MLB C. Furthermore, intra-MLB cycle time is determined by the time taken to read the operands from the intermediate registers and the time to read the LUTs mapped to the subarray inside each MLB. Inter-MLB cycle time is determined by the LUT read time inside one MLB and its output to reach the inputs of LUTs in other MLBs.

3. MBC IN MULTICORE SYSTEMS

This section presents our RMBC framework for multicore architectures. The framework which normally acts as a computing resource can on-demand be transformed into data memory. Using memory for computation gives the dual benefit that the same can be used as normal storage for memory-intensive applications. Consequently, by customizing itself to different application requirements, it can provide speedup for a wide range of compute and data-intensive applications. RMBC framework operates as a Reconfigurable Functional Unit (RFU) in a conventional RISC pipeline. Consequently, it is important to consider the modifications of both

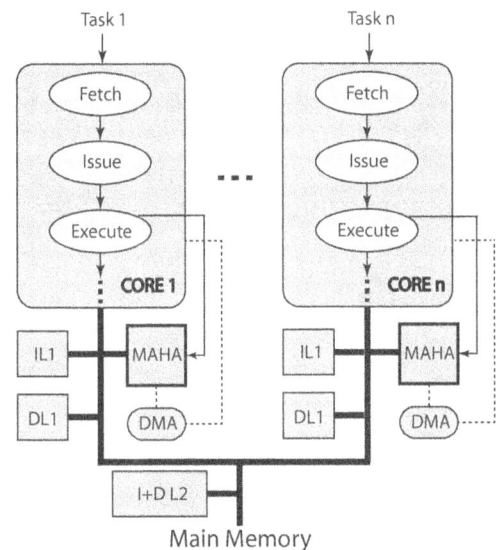

Figure 2: RMBC in a Multicore Architecture

hardware as well as software in the system. This modifications should enable the multicore architecture to on-demand execute a set of applications in the RMBC units. Figure 2 shows the proposed integration scheme of RMBC into a multicore architecture. In this architecture, each core has an independent RMBC unit that can be on-demand reconfigured in order to fulfill the application's requirements. Required modifications are discussed in the following sections.

3.0.1 Hardware Architecture

Constructing an RFU around the normal cache architecture imparts the exceptional advantage that the RMBC architecture can be interfaced with the main memory as well as with the processor register file for transfer of working dataset and configuration data. Since on-chip cache is already provisioned with these interfaces, the design overhead involved is minimal. However, for data-parallel applications for which large volume of data need to be loaded onto the RFU, large number of cycles will be wasted by the processor mediating between the cache and the main memory. The proposed

288

framework therefore utilizes a DMA interface between the RFU and the main memory similar to the scheme proposed in [7]. A data buffer is used to overlap computation in the RFU with loading/unloading of data from the main memory through the DMA controller. It is to be noted here that the *context memory* as described in [7] is virtually distributed as schedule table inside individual MLBs of the RMBC framework. This opens up the opportunity that sets of MLBs can operate and be reconfigured independently. Thus normal function in one set of MLBs proceed with the reconfiguration in others, effectively hiding the reconfiguration latency. Following are the sequence of actions taken by each core to initiate computation in the RMBC framework.

Write-back dirty blocks: Since the storage which serves as an RFU on-demand is otherwise part of the on-chip cache, dirty lines present in this storage need to be written back to the main memory before any RFU operation can proceed. This latency can be minimized by periodic cleaning of the dirty blocks.

Load reconfiguration data: RMBC exploits the large data bandwidth of memory interface to load the reconfiguration data into the MLBs in groups of 128 bits. In the worst case, when the entire 16KB memory inside an MLB requires reconfiguration, the total number of cycles required to configure a single MLB is only 1024. Additional 8 cycles are required to configure the schedule table. Programming the PI architecture however cannot be performed in parallel and requires 128 cycles to program the external connection for the 32-bit input to each MLB. The configuration data is brought into the RMBC framework from the main memory using the DMA controller. The core initiates this transfer by providing the starting address of reconfiguration data to the DMA controller.

Load working data: For loading the working data set, the DMA controller is supplied with the starting address by the core. Data from the main memory is first moved onto the data buffer which is then transferred over to selected MLBs. Unlike Morphosys, inputs for the application mapped to the RMBC framework also arrive at selected MLBs from the processor register file via normal load/store instructions. This helps to speed-up crypto applications and other combinatorial functions where data-parallelism is not inherent. In order to hide the latency of data arrival, the DMA controller is programmed to overlap data pre-fetch from the main memory with normal RFU operation.

3.0.2 Software Architecture

Integration of an RFU into a dynamic pipeline requires extension of the instruction set to incorporate the new RFU opcodes. As illustrated in Figure 3, such an extension is achieved for an Portable Instruction Set Architecture (PISA) through identification of the application and modification of the software toolflow to handle the RFU opcodes at different stages of compilation, and linking. The first step in the methodology is the identification of the hotspot or computational kernels where the code spends major portion of the execution time. The assembly instructions corresponding to these hotspots are determined from the disassembled binary. The subsequent step identifies the section of the C-code corresponding to the assembly instructions. A new library containing pseudo-assembler instructions is defined for the RFU functions using GCC inline assembler macro feature. This library is then used to compile the C-code.

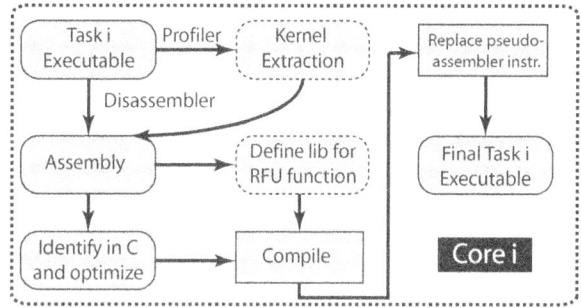

Figure 3: Software architecture for automatic translation of application kernel to RFU instructions

A post-processing on the compiled executable replaces the pseudo-assembler instruction with the correct machine opcode. This approach avoids modifying the C-compiler to take into account new opcodes corresponding to the RFU instructions. The correct machine opcode used to replace the pseudo-assembler instruction utilizes the unused opcode space of the host processor to define new RFU instructions.

4. EXPERIMENTS

4.1 Experimental Setup

To check the effectiveness of our proposed scheme, we used a widely used multicore simulator, M5 [8]. We enhanced M5 to make the required modifications in processor cores to support on-demand computation transfer to MBC. We configured the simulated system with a two-core, three-core, and four-core processor each of which runs at 500MHz. The DerivO3CPU model [8] in M5 is used which represents a detailed model of an out-of-order SMT-capable CPU which stalls during cache accesses and memory response handling. In addition, for MBC part we implemented a mapper using C that gets the critical-section of the application (kernel) DFG as input and provides the detailed MBC mapping information including required number of cycles and energy consumption per vector. The effectiveness of our proposed framework was investigated and validated for two different scenarios: i) improving the performance, and ii) improving memory sub-system energy consumption. For the second part, we applied the same energy model used in [9], which calculates both dynamic and static energy consumption, memory latency, CPU stall energy, and main memory fetch energy. We updated the dynamic energy consumption for each cache configuration using CACTI 6.0. We developed a Perl script that gets M5 simulation output as input and provides total consumed energy in memory hierarchy including all private instruction L1 and data L1 (for each core) and shared L2 caches in the multicore architecture.

In order to investigate our proposed framework, we formed different task sets to be executed together in a multicore environment. The tasks are independent and each task is assigned to a core. These tasks are collected from both compute- and memory-intensive applications. In the 2-core framework, the tasks are organized as *set1 (sha, atr), set2 (aes, ci), set3 (me, dwt),* and *set4 (census, dct).* Similarly, for 3-core scenario they are organized as *set5 (aes, sha, me),* and *set6 (ci, atr, dwt).* Likewise, we formed *set7 (aes, sha, me, atr)* for 4-core scenario. We simulated these scenarios and investigated the effect of using RMBC on performance as well as memory sub-system energy consumption.

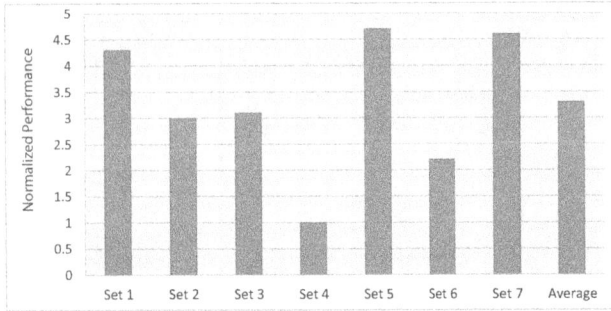

Figure 4: Performance of the multicore+RMBC normalized to the multicore only scenario

4.2 Performance Analysis

The metric that we used for performance comparison is the average of normalized performance improvements (cycle number) in all cores. Figure 4 shows the performance improvement in an RMBC-enabled multicore system normalized to the multicore system without RMBC. In most cases we have drastic improvements in performance. However as can be seen, this improvement depends on applications and their nature. For example in set1 we have 4.3X improvement in performance while it is 1.02X for set4. This variation shows that although RMBC is beneficial in both computation and memory intensive applications, some applications are not a good-fit when they are mapped to RMBC. Nonetheless, as it can be observed in most applications RMBC is a promising framework to improve the performance in multicore systems. In addition, it can be observed that generally the performance improvement increases as the number of cores increases. This is because of sharing resources, doing computation in RMBC in the cores releases some shared recourses (L2 and shared bus). In other words, while one or more cores are doing computation in RMBC, it is not using shared resources. This can make more resources available for other cores that are using it (doing computation in CPU); more resources lead to more performance for these cores. In summary, for our application sets the performance improvement is up to 4.7X for set5 and is 3.3X on average.

4.3 Energy Consumption

In order to investigate the effect of using RMBC on energy consumption, we compared the total energy consumption of memory sub-system in our proposed multicore framework. This metric comprises of total energy consumption in all caches plus RMBC units for RMBC-enabled system normalized to energy consumption in all caches for traditional multicore scenario. Figure 5 illustrates this metric for the same application sets discussed before. Like performance improvement, in most cases we have drastic reduction in energy consumption. As expected, in most cases there is a strong relation between improvement in performance and reduction in energy consumption. For example set1 that has 4X improvement in performance has 4X reduction in energy. As it can be observed, enabling RMBC in multicore systems is a promising approach to reduce energy consumption. For our application sets the energy reduction is up to 4.7X for set5 and is 2X on average.

5. CONCLUSIONS

We have presented RMBC, an embedded memory-based computing framework, which can serve as a reconfigurable

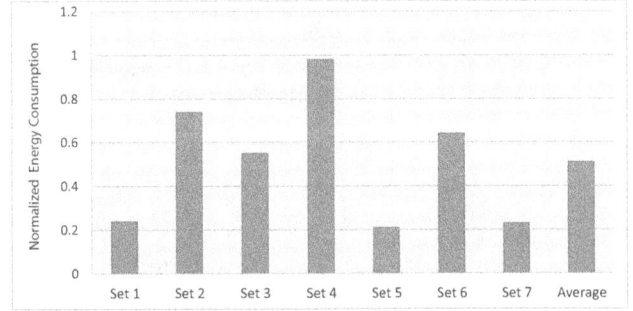

Figure 5: Energy consumption in multicore+RMBC normalized to the multicore only scenario

functional unit in multicore systems to provide hardware acceleration. The framework leverages on the high integration density offered by modern embedded memory and drastically reduces the requirements of programmable interconnects which impose major bottlenecks in terms of performance and technology scalability. Another important benefit of using a memory array is that it can be dynamically configured into a custom logic or memory block, unlike CGRA or FPGA. Such malleable nature of RMBC can be extremely useful for compute-intensive as well as data-intensive applications, which benefits from improved memory latency and/or bandwidth. Experimental results demonstrated that the proposed framework achieves significant improvement in performance (up to 4.7X, 3.3X on average) and energy consumption reduction (up to 4.7X, 2X on average).

6. ACKNOWLEDGMENTS

This work was partially supported by NSF grants ECCS-1002237, CCF-0903430, and CNS-0915376. We would like to thank Dr. Somnath Paul for his comments and insights on memory-based computing in embedded systems.

7. REFERENCES

[1] S. Hauck et al. The chimaera reconfigurable functional unit. *IEE Trans. on VLSI*, 2004.

[2] Y. Park et al. Cgra express: accelerating execution using dynamic operation fusion. In *CASES*, 2009.

[3] S. Majzoub et al. Reconfigurable platform evaluation through application mapping and performance analysis. In *IEEE SPIT*, 2006.

[4] S. Paul and S. Bhunia. A scalable memory-based reconfigurable computing framework for nanoscale crossbar. *IEEE Trans. on Nanotechnology*, 2010.

[5] H. Hajimiri et al. Reliability improvement in multicore architectures through computing in embedded memory. In *MWSCAS*, 2011.

[6] S. Paul, et al. Nanoscale reconfigurable computing using non-volatile 2-d sttram array. In *IEEE-NANO*, 2009.

[7] H. Singh et al. Morphosys: an integrated reconfigurable system for data-parallel and computation-intensive applications. *IEEE Trans. on Computers*, 2000.

[8] N. Binkert et al. The m5 simulator: Modeling networked systems. *IEEE Micro*, 2006.

[9] C. Zhang et al. A highly configurable cache architecture for embedded systems. In *Computer Architecture*, 2003.

Share Memory Aware Scheduler: Balancing Performance and Fairness

Xi Li[1, 2], Gangyong Jia[1, 2], Yun Chen[1, 2], Zongwei Zhu[1,2], Xuehai Zhou[1,2]

[1] Department of Computer Science and Technology, University of Science and Technology of China (USTC)
Hefei, 230027, China
[2] Suzhou Institute for Advanced Study, USTC, Suzhou, China
{llxx, xhzhou}@ustc.edu.cn; {gangyong, chenywx, zzw1988}@mail.ustc.edu.cn

Abstract

Optimizing system performance through scheduling has received a lot of attention. However, none of the existing approaches can balance the system performance improvement and the fair share of CPU time among threads. We present in this paper a share memory aware scheduler (SMAS). The key idea is to adopt thread group scheduling which partitions threads based on memory address space to reduce switching overhead and to give each thread a fair chance to occupy CPU time. There are three main contributions: 1) SMAS does well in balancing system performance and fairness among all threads; 2) to our knowledge, this is the first attempt to use share memory aware scheduler for system performance improvement; 3) we implement SMAS both in testbed and simulator for evaluation. The testbed results on a 2-core processor show that our proposed scheduler can improve performance of different performance parameters with neglected overhead in fairness, which reduced 0.128% in cache miss rate, 2.62% in run time, 13.15% in DTBL misses, 31.68% in ITLB misses and 46.15% in ITLB flushes maximum. Furthermore, our extensive simulation results for 4 and 8 cores demonstrate that SMAS is highly scalable.

Categories and Subject Descriptors: D.4.1 scheduling

General Terms: Management, Performance, Design.

Keywords: *Memory address space; scheduling; thread group; fairness; performance; share memory*

1. INTRODUCTION

More and more emerging applications are believed to be highly parallel. For example, the recent single-chip cloud computer (SCC) [1] from Intel integrates 48 IA-32 cores to run highly parallel workloads such as Recognition, Mining and Synthesis (RMS) benchmarks [2]. And many existing applications are highly parallel, such as web service application, file system application and so on.

Co-scheduling on CMP is a widely accepted scheme to improve performance [3][4]. Their main idea is to reduce the competition caused by hardware (mainly last level cache) sharing. These methods are effective for single-threaded scenario. On the other hand, the default scheduler employed in Linux 2.6.23, Completely Fair Scheduler (CFS) [8], can guarantee fair CPU time among all

This research was supported by the National Science Foundation of China under grants No.60873221 and Jiangsu production-teaching-research joint innovation project No.BY2009128.

threads according to each thread's static priority. CFS adopts cache affinity strategy to optimize performance, but it does not exploit the relationship among threads to improve performance.

Threads in the multi-threaded applications have some relationship, such as synchronizing in a point, communicating with each other, ordering in time and so on. Researchers have taken into account the workload of parallel threads which a group of parallel threads is generated at the fork-point of a parallel region and is synchronized at the join-point of the parallel region [8]. This synchronization point is called barrier. Same as this work, most existing performance researches [9][10] don't either involve multi-threaded characteristic or take advantaging of the relationship among multi-threaded threads.

We propose share memory aware scheduler to both improve performance and guarantee fairness in multi-threaded scenario. We prior choose the thread which is in the same memory address space with *current* thread when scheduling next thread, but we also give almost the same chance to occupy the CPU among all threads. Our SMAS features a three-phase design. The first phase is the scheduling within a core. When scheduling next thread, choose the thread with the same memory address space as that of the *current* one. After all threads of the same memory address space have been scheduled, we choose the next thread according to the CFS. The second phase is the scheduling among cores sharing last level cache. If some threads of the same memory address space with the current thread have not been scheduled, choose those, otherwise choose according to the *current* thread of other cores which sharing last level cache with the core. The third phase is the scheduling among cores not sharing last level cache. In this phase, we do not adopt any special handling. Scheduler among Cores without sharing last level cache is still independent. As a result, our scheduler can be scalable to many-cores.

We implement SMAS based on CFS in testbed and simulators to evaluate our scheduler overriding the default CFS. Experiment results show SMAS outperforms CFS in system performance with neglect overhead in fairness. In terms of five performance parameters, cache miss rate, run time, DTLB misses, ITLB misses and ITLB flushes, SMAS reduces 0.128%, 2.62%, 13.15%, 31.68% and 46.15% respectively.

To our knowledge, we are the first team to propose a share memory aware scheduler which can both improve the system performance and guarantee the fairness with neglect overhead in multi-threaded scenario.

2. DESIGN

2.1 Overview

The proposed scheduler is based on share memory and gives priority to the threads in the same memory address space. It differs with the default CFS in following two phases:

1. **Finding Thread Group**: Using thread group id of the *task_struct* in Linux kernel to find all the threads of the same thread group. It is critical to know whether threads belong to the same thread group.

2. **Finding Next Running Thread**: Choosing the next running thread based on the *current* thread. We choose a thread from the same thread group as the next running thread. When threre is no thread in the same thread group, we choose the next running thread using the default CFS strategy.

We apply these phases in an iterative process. Every scheduling time we first find threads in the same thread group, and then choose the next running thread among them or using default CFS strategy. We now elaborate these two phases.

2.2 Finding Thread Group
Our scheduler schedules threads according to thread group and gives higher priority to threads in the same thread group of the current running thread. So finding threads in the same thread group with the current running thread is essential.

In the Linux kernel, struct *task_struct* is used to represent each thread. Each thread in the kernel has a unique *task_struct*. Domain *tgid* in the *task_struct* is used to mark which thread group the thread is belonging to. All threads with the same *tgid* are in the same memory address space according recently thread-process model. Domain *thread_group* in the *task_struct* links the threads with the same *tgid*. Traveling the *thread_group* list can find all threads sharing the same memory address space. Figure 1(a) shows the default CFS scheduling queue organized into an *rb-tree*. Where each thread is marked with two numbers, the first one indicates the thread group it belongs to, and the second one shows the thread id in the thread group. For example, *thread12* represents a thread in group 1 with thread id is 2. Figure 1(b) demonstrates that through *thread_group* list, all threads in the same thread group are listed by the circular list on one core. All threads belonging thread group 1 are listed by the white circular list and all thread belonging thread group 2 are listed by the slash circular list. Figure 1(c) demonstrates that all threads in the same thread group are listed among two cores. By default, each core will have a unique scheduling queue.

In order to maintain the fairness of the CFS, all threads in the same thread group can only run one time when this group is currently running. After all threads in this group have run one time, we will choose another thread group. We add a domain called *access* to the struct *task_struct*. Domain *access* is used to mark whether the thread in the thread group has run. After the thread run on the CPU, domain *access* of the thread is marked as 1. When all threads in a thread group have run, all threads' domain *access* are marked 1, the scheduler will choose another thread group and reset the domain *access* of previous group to 0.

2.3 Finding Next Running Thread

2.3.1 On One Core
Scheduler needs to find the next running thread. In the default CFS, each time the scheduler chooses the *leftmost* thread in the *rb-tree*. In our scheduler, we will choose threads in the same thread group with the *current* thread. We first find the next running thread which is the *current* thread pointed in the list of the thread group. If the next thread's *access* is marked 1 which means the thread has run in this round, we skip it. If all the threads in the thread group have run in this round, sets all threads' *access* of the thread group to 0. We choose the next running thread according default method which chooses the *leftmost* thread. Each time when the *current* thread is

scheduled, the current thread's access is marked to 1. For example, if thread group 1 in Figure 1(b) is running, after *thread12* is scheduled, we need to choose the next running thread. We will first find *thread13* which is pointed by the *thread12* in the thread group list. If the *thread13*'s *access* is 0, the *thread13* will run on the core and the *thread12*'s access is marked to 1. Otherwise, if the *thread13*'s *access* is marked with 1, find the *thread13*'s next thread, *thread11*. If all threads' *access* of the thread group, *thread11*, *thread12* and *thread13*, are marked with 1. We choose the *leftmost* thread, *thread21*, to run next.

Figure 2 shows the process of finding the next thread in one core. We apply this process in an iterative process to find the next running thread on the single core circumstance.

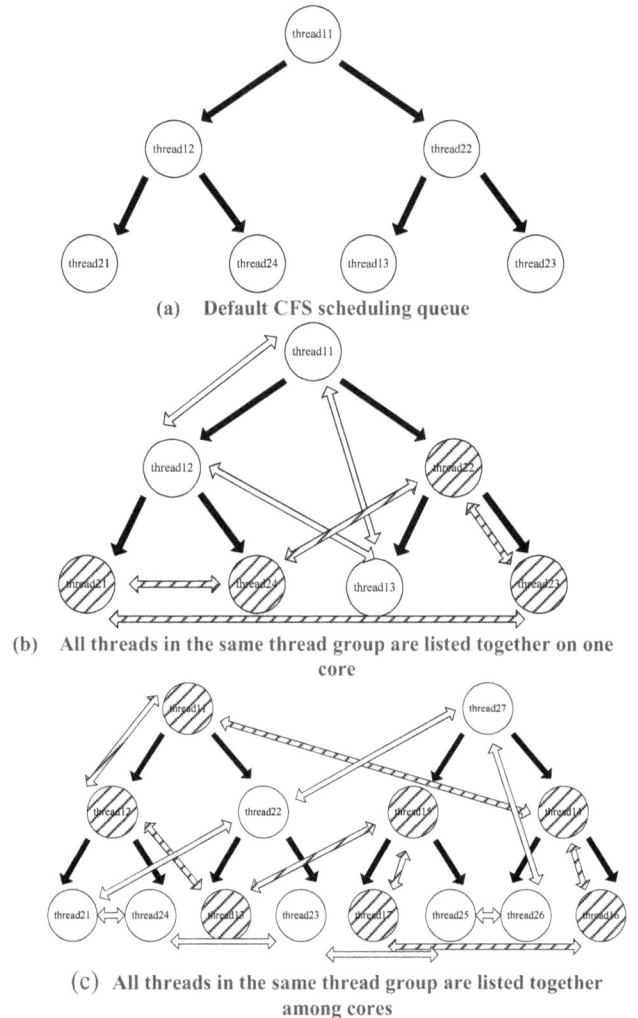

(a)   Default CFS scheduling queue

(b)   All threads in the same thread group are listed together on one core

(c)   All threads in the same thread group are listed together among cores

Figure 1 threads are listed according thread group

2.3.2 Among Sharing Last Level Cache's Multi Cores
To improve the performance of a multi-core system, when there are cores sharing the last level cache.

We run in parallel threads from the same thread group. When there exist threads' *access* are 0 in the same thread group with the *current* thread, we choose those threads to run next. This is the same as in the single core circumstance. When all threads' *access* of the current thread group being all marked with 1, the process becomes different.

We introduce a sharing variable *cur_tgid* for each core. This variable is used to record the thread group id of the running thread. Each variable *cur_tgid* can be read by all threads of all cores, but it can only be written by the threads of the core containing it. *cur_tgid* is equal to the current thread's *tgid*. Each *cur_tgid* shows which thread group the core is running. For example, if there are four cores sharing the last level cache, there will be four *cur_tgid* variables, *cur_tgid0*, *cur_tgid1*, *cur_tgid2* and *cur_tgid3* for core 0, core 1, core 2 and core 3 respectively.

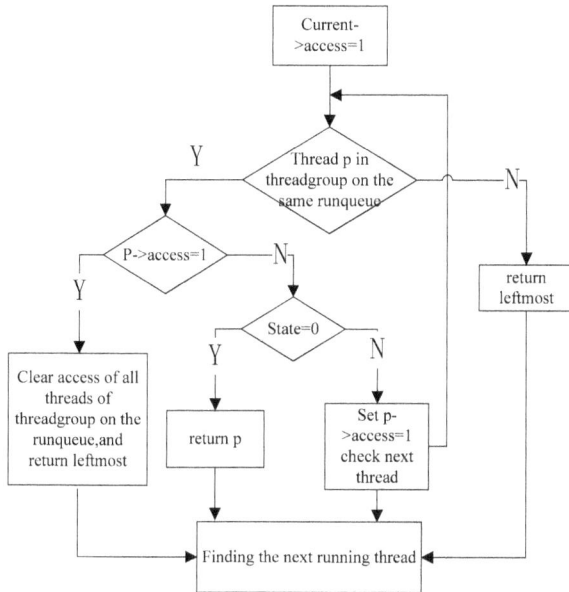

Figure 2 the design process of finding the next thread in one core

When all the threads in the same thread group on core 0 are marked with 1, choose the *leftmost* thread in the *rb_tree*. When threads in the same thread group on core 1 are all marked with 1, we don't choose the *leftmost* thread immediately. Instead, we first read the *cur_tgid0* variable and check whether there are some threads of the *cur_tgid0* thread group on core 1. If there is, we choose a thread in the *cur_tgid0* thread group; otherwise, we choose the *leftmost* thread. When threads in the same thread group on core 2 are all marked with 1, firstly, read the *cur_tgid0* variable. Check whether there are some threads of the *cur_tgid0* thread group on core 2. If existing, choose thread in the *cur_tgid0* thread group. Else read the *cur_tgid1* variable. If existing, choose thread in the *cur_tgid1* thread group. Else choose the *leftmost* thread. On core 3, similarly phases will be used.

We believe that if cores are not sharing last level cache scheduling queues will be independent. So our scheduler will be scalable to many-core.

3. EXPERIMENTS AND ANALYSIS

3.1 Testbed And Simulation Environment

Our testbed is a dual-core Intel E8200 processor. The operating system is Ubuntu 11.04 with Linux kernel 2.6.35. In order to evaluate our scheduler, we simultaneously run different combinations of selected from sysbench [7] and SPEC2006. In Table 1, the *number-appname* notation is the number of threads of the application with the name of *appname* for sysbench; for SPEC2006 workload, it is the number of copies of the application with the name of *appname*.

To stress test our scheduler in more cores system with a large-scale configuration, we conduct simulations using M5 [6] simulator. Its main parameters are listed in Table 2. Table 3 deploys each cache interconnecting of the simulated architecture.

We use performance monitoring unit to record performance parameters. We mainly record cache misses, running time, DTLB misses, ITLB misses and ITLB flushes as the performance parameters.

Table1 workload mixes

Mixes	sysbench, SPEC2006
mix1	12-sysbench cpu, 8-omnetpp
mix2	12-sysbench memory, 8-omnetpp
mix3	6-sysbench cpu, 6-sysbench memory, 8-omnetpp
mix4	24-sysbench cpu, 16-omnetpp
mix5	24-sysbench memory, 16-omnetpp
mix6	12-sysbench cpu, 12-sysbench memory, 16-omnetpp
mix7	48-sysbench cpu, 32-omnetpp
mix8	48-sysbench memory, 32-omnetpp
mix9	24-sysbench cpu, 24-sysbench memory, 32-omnetpp

Table 2 base simulation parameters

Processor Core	In-order, stalls on cache misses
L1 D Cache Size	16KB 2-way 64-byte line
L1 I Cache Size	16KB 2-way 64-byte line
L1 Access Time	1 cycle
L2 Cache Sizes	512KB 8-way
L2 Access Time	10 cycles Normal, 12 cycles Drowsy
Memory Access Time	200 cycles

Table 3 cache interconnect of each architecture

Architecture 1	Four cores, each core has private L1 data cache, L1 instruction cache, two cores share the L2 cache
Architecture 2	Four cores, each core has private L1 data cache, L1 instruction cache, four cores share the unique L2 cache
Architecture 3	Eight cores, each core has private L1 data cache, L1 instruction cache, four cores share the L2 cache

3.2 Performance Improving

The Table 4 has shows the performance parameters of the testbed both using CFS and SMAS. Scheduling threads in the same memory address space can reduce contending cache because of sharing the cache among threads.

Table 4 performance parameters of the testbed

	Mix1		Mix2		Mix3	
	CFS	SMAS	CFS	SMAS	CFS	SMAS
L2 cache miss rate	0.425%	0.262%	0.094%	0.009%	0.263%	0.127%
Run time(s)	47.4231	46.9052	97.1588	94.618	142.2743	140.0956
DTLB misses	32215	27978	26992646	26948703	18985561	18643707
ITLB misses	7334	7107	17895	12253	16880	11532
ITBL flushes	46	25	66	41	130	70

Figure 3(a) shows the real running sequence using CFS on one core. The slashed threads are multi-threaded threads. Without slashed threads are single-threaded threads. They are different with each other with the different number, for example, *thread11* is a multi-threaded thread and *thread12* is another multi-threaded thread, these two threads are in the same memory address space thread group, *thread2* is a single-threaded thread. In Figure 3(a) threads without slashed are more scattered among slashed threads than in Figure 3(b) which using SMAS. In Figure 3(b), slashed threads are always demonstrating in a regular way which six-slashed threads are arraying together and between two six-slashed threads are inserted by a few threads without slashed threads. Threads in the same memory address space thread group are less chance to be disturbed by other threads, so less TLB flushes and less L2 cache misses are occurring. Because of less TLB flushes and less L2 cache misses, the run time is less.

a) The real running sequence using the CFS on one core

b) The real running sequence using the SMAS on one core

Figure 3 the real running sequence on one core

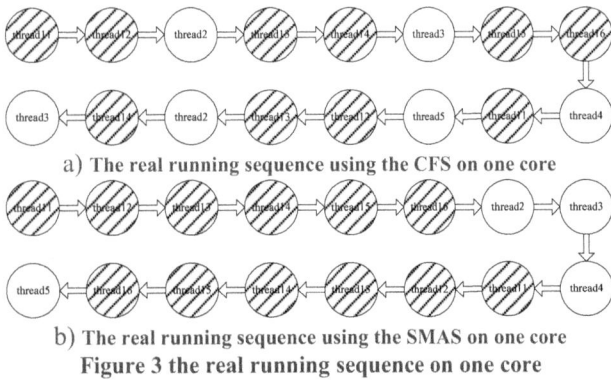

3.3 the Fairness of the SMAS

Comparing the Figure 3(a) and Figure 3(b) we can find the running sequence is almost the same if we suppose six-slashed threads are a entirety, we only moving threads without slashed to the entirety front in Figure 3(b) if the threads are in front of *thread11* in Figure 3(a) or moving threads to the entirety back if the threads are in back of *thread11*, the only difference is the running time which the threads without slashed will delay running. But the delay is predictable, after finishing the entirety the threads without slashed will run.

In this paper, we use each thread's occupying CPU time rate as the fairness parameter. Figure 4 shows the each application's occupying CPU time rate of the two scheduling. From this figure, we can find our SMAS is giving a little more CPU time to the multi-threaded applications. 6-sysbench-cpu occupies 62.30% of the CPU time in CFS, SMAS gives 67.20% of the CPU time, which gives 4.9% more CPU time in SMAS of 6-threaded application. For single-threaded application, SMAS gives 1.12%. The difference is just neglect. This illustrates SMAS is almost fairness among threads with neglect overhead. How to give completely fairness to all threads is our next step to research.

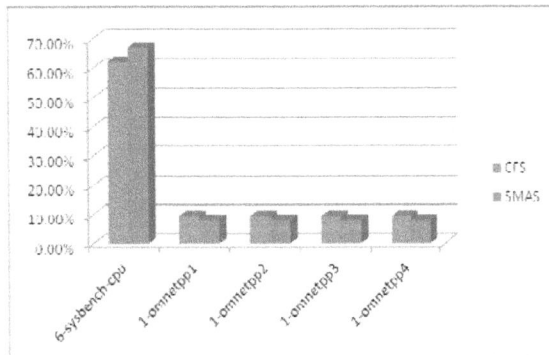

Figure 4 each application's occupying CPU time rate of the two scheduling

3.4 Scalable of SMAS

Table 5 shows the improvement in all five performance parameters of SMAS to CFS, which shows architecture1, architecture2 and architecture3's performance parameters in four cores and eight cores with different architecture. From these performance parameters, we can find our SMAS are almost getting the same improvement in architecuture1 and testbed, and in architecture2 and architecture3 are almost the same. There are two cores sharing the last level cache in architecture1, which is the same with the testbed. Four cores are sharing the last level cache which is the same in architecture2 and architecture3. But the total number of cores is

different in testbed and architecture1, architecture2 and architecure3. From these data, we can prove SMAS can be scalable to many cores under recently multi-core, many-core architecture which have limited cores sharing last level cache. We also find four cores sharing last level cache are only weakly better than two cores sharing last level cache, this may be related to the problem which is competion the sharing memory among threads. Next we will take into consider the reference problem.

Table 5 the performance improvement of the SMAS

	L2 cache miss rate	Run time	DTLB misses	ITLB misses	ITLB flushes
Architecture 1	0.133%	2.67%	13.25%	31.79%	46.31%
Architecture 2	0.182%	2.85%	17.7%	35.47%	48.18%
Architecture 3	0.179%	2.81%	16.9%	33.83%	46.85%

4. CONCLUSION

In this paper, we propose a share memory aware scheduler (SMAS), which takes advantage of the sharing memory to reduce the frequency of cache replacement to improve system performance with little degradation in the schedule's fairness. Most existing performance improvement studies are either not considering multi-threaded scenario or not taking advantage of the relationship among multi-threaded threads. This is the first attempt to utilize the relationship of sharing memory to improve system performance and guarantee fairness. Experimental results show that the proposed scheduler can improve performance over the default CFS scheduler with neglected overhead in fairness. Extensive simulation also shows that our scheduler is scalable to multi-cores systems. In the future, we will explore whether SMAS can be effective in reducing system's power and energy consumption.

5. ACKNOWLEDGMENT

We gratefully acknowledge Gang Qu at UMD for revising. We acknowledge the support of our industrial sponsor, China Software Center, Sony (China) Limited.

6. REFERENCES

[1] Intel. A 48-core ia-32 message passing processor with dvfs in 45nm cmos. In *to appeared in ISSCC*, 2010.

[2] Intel. Computer intensive, highly parallel applications and uses. 2005.

[3] D. Tam, R. Azimi, and M. Stumm. Thread Cluster: Sharing-Aware Scheduling on SMP-CMP-SMT Multiprocessors. In *Proceedings of the 2nd ACM European Conference on Computer Systems (EuroSys '07)*, 2007.

[4] Sergey Zhuravlev, Sergey Blagodurov and Alexandra Fedorova. Addressing Shared Resource Contention in Multicore Processors via Scheduling. *ASPLOS*, 2010.

[5] Qiong Cai, Jos´e Gonz´alez, Ryan Rakvic, Grigorios Magklis, Pedro Chaparro, and Antonio Gonz´alez. Meeting points: using thread criticality to adapt multicore hardware to parallel regions. *In PACT '08: Proceedings of the 17th international conference on Parallel architectures and compilation techniques*, pages 240–249, New York, NY, USA, 2008. ACM.

[6] N. Binkert, E. Hallnor, and S. Reinhardt. Network-Oriented Full-System Simulation using M5. In *Sixth Workshop on Computer Architecture Evaluation using Commercial Workloads*, 2003.

[7] Kopytov, A. SysBench: a system performance benchmark. http://sysbench.sourceforge.net/index.html. 2004.

[8] C. S. Pabla. Completely fair scheduler. *Linux J.*, 2009(184): 4, 2009.

[9] Josep Torrellas, A. Tucker, and A. Gupta. Evaluating the performance of Cache-Affinity Scheduling in Shared-Memory Multiprocessors. *Journal Of Parallel and Distributed Computing*. 1995.

[10] Vahid Kazempour, Alexandra Fedorova, Pouya Alagheband. Performance Implications of Cache Affinity on Multicore Processors. In Proceedings of Euro-Par. 2008.

Alleviating NBTI-induced Failure in Off-chip Output Drivers

Bhavitavya Bhadviya, Ayan Mandal, and Sunil P. Khatri
Texas A&M University, College Station, TX 77843, USA

ABSTRACT

Negative Bias Temperature Instability (NBTI) causes the threshold voltage of PMOS devices to degrade with time, resulting in a reduced lifetime of a CMOS IC. In this paper, we present an approach to mitigate the degradation due to NBTI for off-chip output drivers. Our approach is based on forcibly inducing relaxation in the individual fingers of the output driver (which is typically implemented in a multi-fingered fashion). The individual fingers are relaxed in a round-robin manner, such that at any given time, k out of n fingers of the driver are being relaxed. Our results show that the proposed approach significantly extends the lifetime of the output driver.

Categories and Subject Descriptors: B.8.1 [Hardware]: Reliability, Testing, and Fault-Tolerance

Keywords: NBTI

1. INTRODUCTION

In recent CMOS technologies, NBTI has become an important problem [1]. In this paper we present a novel approach to mitigate the effect of NBTI induced timing failures for off-chip output drivers. We refer to the proposed driver as the *NBTI-RR* driver. Our approach is based on relaxing individual PMOS fingers of the multi-fingered output driver, in a round-robin manner. In this way, at any given time, k out of n fingers of the output driver are being relaxed, and are recovering from their degraded threshold voltage (which was caused by the previous stress cycle). The other $n-k$ fingers are used to provide the output driver functionality. We provide the analytical formulas and methodology to derive the NBTI-induced threshold voltage degradation under stress-relaxation cycles induced by a round-robin operation of the fingers of the driver. Under the assumption that both the traditional and the NBTI-RR drivers have a total of n identical fingers implemented in 65nm technology, we demonstrate that the traditional output driver fails in 1.2 years when stressed 50% of the time, while the lifetime of the NBTI-RR driver is over 20 years. The control overhead incurred in implementing the round-robin scheduling policy is small (a minimal 0.71%). In the sequel, we refer to a *traditional output driver* as the state-of-the-art driver that is commonly used in industry (shown in Figure 1 a)). In this work, we define an NBTI induced failure to be the condition in which the propagation delay of the output driver degrades by 5% from its value at epoch (which corresponds to the beginning of the life of the CMOS IC). All plots, tables and figures in this paper are derived for a 65nm PTM [2] fabrication process.

2. PREVIOUS WORK

Various circuit level approaches have been proposed to mitigate the effects of NBTI for PMOS transistors like up-sizing the PMOS devices in an a-priori manner [3], adaptive body biasing technique [4], including on-chip sensors for flip-flops [5], flipping of the state of memory cells [6]. In contrast to these approaches, our proposed technique does not involve addition of sensors, compensators or any process sensitive analog circuitry, nor does it involve the addition of an extra body biasing voltage supply, or a-priori upsizing of the devices in the design. Also, the context of our approach is that of off-chip output drivers (or any circuit using a sufficiently large number of device fingers), unlike the papers mentioned above.

The work of [7] applies a round-robin based relaxation to *functional units* in a processor. Whenever the instruction load allows, functional units are relaxed. Similarly, the work of [8] applies a round-robin based relaxation schedule in the context of a multi-core system. However, these approaches would result in significant NBTI-based delay degradation under conditions of heavy load (when all functional units or cores are occupied). In contrast to these papers, our approach i) mitigates NBTI degradation regardless of system load, ii) applies a round-robin based relaxation schedule for output drivers which are implemented using multi-fingered devices and iii) does not require the existence of redundant blocks (such as multiple functional units or cores) before it can be employed.

In [9], the authors suggest a means to reduce the aging of the PMOS header devices in an MTCMOS based circuit. One of the three approaches they present is based on sequential relaxing the fingers of the PMOS sleep transistor. Although there is conceptual similarity of this technique to our approach, there are several differences. The authors of [9] devote a short paragraph to the approach, with no details about the control circuit used to achieve sequential relaxation. Also, the aging issues of this circuit are not discussed. Additionally, the analytic approach to determine the threshold voltage for the sequentially relaxed devices is not discussed. In contrast, the context of our work is completely different (off-chip output drivers versus MTCMOS based digital designs). In addition, we present results for the aging characteristics of the round-robin relaxed output drivers along with details about the control circuitry, including the aging characteristics of the control circuit. Also, analytical expressions for the V_{th} degradation of a PMOS device under a round-robin stress-relaxation regime are are provided in this paper.

Several analytical models for NBTI degradation exist in the literature [1, 10]. In [10], the authors have proposed a closed-form long term degradation model for both dynamic and static stress. We extend the long term degradation model, and come up with a more efficient technique to reduce the detrimental effects of NBTI for output drivers.

3. APPROACH

3.1 NBTI Background

Negative bias (stress) on a PMOS gate results in the degradation of its threshold voltage (an increase in $|V_{th}|$). Relaxation, on the other hand results in a partial recovery of the threshold voltage of the device (a decrease in $|V_{th}|$). Suppose a PMOS device is subjected to a stress condition (i.e. $V_{gs} = $ -VDD) starting at time t_0. In this case, the degradation in the threshold voltage of this device at time $t > t_0$ is given [10] by:

$$\Delta V_{th}(t) = (K_v \sqrt{(t - t_0)} + \sqrt[2n]{\Delta V_{th}(t_0)})^{2n} \quad (1)$$

Suppose that the stress condition ends at time t_1. At this point, the PMOS device undergoes a relaxation condition (i.e. $V_{gs} = 0$). The degraded threshold voltage due to stress now recovers by a quantity shown below [10]:

$$\Delta V_{th}(t) = V_{th}(t_1) \cdot (1 - \frac{2\xi_1 t_e + \sqrt{\xi_2 C(t - t_1)}}{(1 + \delta)t_{ox} + \sqrt{Ct}}) \quad (2)$$

In the above equations, the values of $K_v = (\frac{qt_{ox}}{\varepsilon_{ox}})^3 K^2 C_{ox}(V_{gs} - V_{th})\sqrt{C}e^{\frac{2E_{ox}}{E_0}}$. Note that K_v depends on V_{th}. In the above equations, the parameters n, ξ_1, ξ_2, t_e, C, δ, t_{ox}, K, ε_{ox}, C_{ox}, E_{ox} and E_0 are constants and are obtained from [10].

Equations 1 and 2 are referred to as *static* stress and relaxation equations respectively, since they refer to a condition where the device is continuously stressed or relaxed respectively.

In order to compute the long term effect of a sequence of m stress/relaxation cycles, it may be infeasible to evaluate each of Equations 1 and 2 m times. In the situation where the stress duration is $\alpha \cdot T_{clk}$ and the relaxation duration is $(1 - \alpha) \cdot T_{clk}$, where T_{clk} is the system clock period, the authors of [10] presented a tight upper bound of the asymptotic degradation of threshold voltage after m cycles, as follows:

$$\Delta V_{ths,m+1} = (\frac{\sqrt{K_v^2 \alpha T_{clk}}}{1 - \sqrt[2n]{\beta_m}})^{2n} \quad (3)$$

where $\beta_m = 1 - \frac{2\xi_1 t_e + \sqrt{\xi_2 C(1 - \alpha)T_{clk}}}{(1 + \delta)t_{ox} + \sqrt{mCT_{clk}}}$

The above scenario, in which the NBTI-induced threshold voltage degradation is accumulated as a consequence of stress and relaxation occurring in every clock cycle (with a stress duty cycle ratio of α) is referred to as *dynamic NBTI*. According to [10], Equation 3 is accurate for clock frequencies above a few hundred Hz, with approximation errors within 0.1%.

3.2 NBTI-RR and Off-chip Output Drivers

Consider a typical tri-stateable off-chip output driver shown in Figure 1 a). This driver behaves like a buffer which is tristated when *enable* is 0. When *enable* is 1, the circuit behaves as a non-inverting output driver. The size of the final inverter stage is typically about $1000\times$ [11] that of a minimum sized inverter in the process technology, due to the large load (typically 25pF to 50pF [12]) that the output driver is rated to drive. As a consequence, the final stage of the output driver is typically implemented as a multi-fingered device (with typically 10s of fingers connected in parallel) for layout efficiency. Figure 1 a) shows the final stage of the output driver to consist of n fingers. In general, the outputs of the NAND and NOR gates are buffered to drive the large load presented by the final stage of the driver. This buffer chain is omitted in Figure 1 a) for conciseness, but is modeled in our experiments and simulations.

The proposed NBTI-RR off-chip output driver is shown in Figure 1 b). This NBTI-RR driver also consists of n fingers in the final stage of the driver (just as in the case of the traditional driver). The key difference is that only $n - k$ of these fingers are used to drive the output node at any given time. The remaining k fingers are allowed to relax, in a round-robin manner. The control logic (which consists of a circular shift register with k contiguous bits whose values are driven to 1 when *enable* is 1) is shown as well. When *enable* = 0, the outputs of the R-R control circuit are driven such that the driver is tristated.

The NBTI-RR driver turns off (and thus relaxes) k fingers. Each of these fingers is turned off for kT seconds continuously. After this, each finger is operated (and thus it is stressed) for $(n - k)T$ seconds continuously. This stress/relaxation sequence is repeated indefinitely. Note that T can be much larger than T_{clk}, the clock period of the IC.

During the first stress phase which lasts $(n - k)T$ seconds, a finger incurs a threshold voltage degradation due to dynamic NBTI, as governed by Equation 3. We assume $\alpha = 0.5$ in our simulations for this stress condition, which means that the drivers drive a VDD output 50% of the time.

In the following relaxation phase which lasts kT seconds, the finger recovers the threshold voltage degradation incurred during the preceding stress phase, as governed by Equation 2.

Note that, from Equation 2, the threshold voltage recovery that is achieved after a relaxation phase that lasts for τ seconds is a fixed fraction of the threshold voltage degradation incurred as a consequence of the preceding stress phase. We define the recovery factor (RF) as the ratio of the stress-induced threshold voltage degradation V_{TS} that is recovered by a relaxation phase that lasts τ seconds. Let the recovered threshold voltage magnitude be $V_{TR}(\tau)$, after a relaxation phase that lasts τ seconds. This can be computed from the static recovery expression in Equation 2. Then, $RF(\tau) = \frac{V_{TR}(\tau)}{V_{TS}}$. Note that for the initial recovery is significant, but as τ increases, the incremental recovery diminishes significantly.

The key parameters we need to choose for the NBTI driver are the values of k and T. After fixing these parameters, the durations of the stress and relaxation phases for any finger are determined. The considerations in choosing k and T are discussed next:

- A large value of T means that the threshold voltage degradation during the stress phase is higher, since the duration of the stress phase is $(n - k)T$. Also, a larger fraction of the threshold voltage degradation during the stress phase is recovered during the recovery phase, since the value of T is large. However, based on Equation 2, threshold voltage recovery saturates for very large values of T. As a consequence, it is strictly harmful to select values of T greater than 1000s. Since a large fraction of the threshold voltage recovery occurs early in the relaxation phase, smaller values of T are typically better. We explore various T values in our experiments.

- A large value of k means that the relaxation phase for any finger is extended, resulting in a more complete threshold voltage recovery. However, for the same reasons presented in the previous bullet, very long relaxation phases present diminishing returns in terms of threshold voltage recovery. Further, we observe that the NBTI-RR driver employs $n - k$ drivers at any instant of time, while the traditional driver employs n drivers at all times. Therefore, if k is too large, then the nominal performance of the NBTI-RR driver will be significantly worse than the traditional driver – an unacceptable situation. In practice, we define the lifetime of the output driver to be the time at which the propagation delay of the driver degrades by $P\%$ compared to its propagation delay at epoch. Therefore, we select k as the maximum value such that the propagation delay of the NBTI-RR driver is no worse than $P\%$ more than that of the traditional driver at epoch. This gives us a means to determine the value of k.

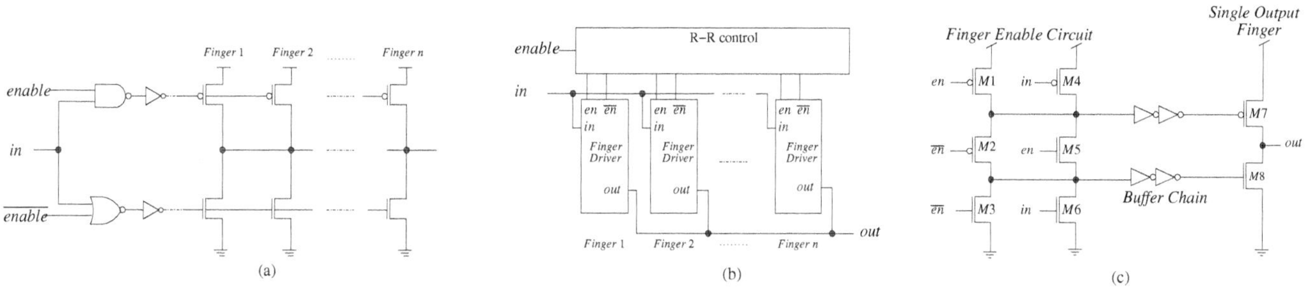

Figure 1: Off-chip Output Drivers a) Traditional Output Driver b) NBTI-RR Output Driver c) NBTI-RR Finger Driver Circuit

For a fair comparison, the delay of both drivers at time t are compared to that of the traditional driver's delay at epoch, when comparing driver lifetime.

3.3 Control Circuit Implementation

The control circuit of the NBTI-RR driver is described next. The NBTI-RR driver has n identical PMOS and NMOS finger driver circuits The NBTI-RR control circuit consists of an n bit shift register. Each flip-flop of the shift register has an asynchronous set and reset capability. The *reset* of each flip-flops is connected to the $\overline{enable}$ signal. Hence, when *enable* = 0, each flip-flop resets, and therefore all fingers are disabled. The *set* signal of n-k out of n flip-flops is connected to a positive edge detecting circuit of the *enable* signal Hence n-k out of n fingers are enabled at a positive edge of the *enable* signal. The circular shift register performs the round-robin operation, and disables k out of n fingers at each cycle of $DCLK$. $DCLK$ has a period T and is derived by dividing the system clock by T/T_{clk}. The device sizes of the control circuit are tuned such that the delay at epoch of the NBTI-RR driver (with all fingers turned on) and traditional driver are identical. The area overheads presented in the experiments are computed based on these sizes.

The finger driver circuit is shown in Figure 1 c). In this circuit, if *en* is 1, M2 and M5 are on, while M1 and M3 are off. Hence the gates of M7 and M8 are driven by $\overline{in}$, and the finger driver behaves as a buffer. On the other hand, when *en* is 0, M1 and M3 are on, and M2 and M5 are off. The gate of M7 is therefore driven to 1, and the gate of M8 is driven to 0, thereby tristating the output.

4. EXPERIMENTAL RESULTS

To validate the NBTI-RR approach, we performed several off-chip output driver simulations. All our simulations were performed in a 65nm PTM [2] process technology, with $t_{ox} = 12\,A^\circ$, $VDD = 1.1V$ and $|V_{th}| = 300mV$ for PMOS and NMOS devices at epoch. The values of all NBTI-related constants were obtained from [10]. Our NBTI induced threshold voltage computations exactly matched the results in [10]. We assume that the output driver operates at 1 GHz. In other words, a new value is driven out by the driver after T_{clk}=1ns. Each driver drives out a VDD value in every alternate clock cycle.

The traditional and NBTI-RR off-chip drivers we simulated are shown in Figure 1. The drivers were implemented with $n - 40$ fingers, and the final drive stage of each driver consisted of an inverter which was 2000× of a minimum sized driver, with $L = 1.5 \cdot l_{min}$ for improved ESD protection. Each finger was therefore sized at 50× the size of a minimum sized inverter. For both the NBTI-RR and traditional output drivers, identical sizing was employed for the finger driver devices. The NBTI-RR control circuit was sized so that the delay at epoch of the traditional driver and that of the NBTI-RR driver (with all fingers turned on) matched. Also, the output

load was assumed to be a lumped capacitor of value 25pF for both drivers.

We next discuss how we obtained the value of k. Consider Figure 2 (a). This figure plots the ratio of the increased propagation delay of the traditional off-chip output driver to the delay of the same driver at epoch (expressed as a percentage), versus the degradation in V_{th}. Based on this figure, we observe that when the threshold voltage degrades by 18.6 mV, the increase in propagation delay is 5% (recall that when the propagation delay increases by 5%, we claim that NBTI-induced failure has occurred). Therefore, the time taken for the threshold voltage of the PMOS driver of the traditional output driver to degrade by 18.6 mV is the lifetime of this driver. Using Equation 3, this lifetime is determined to be 1.2 years. The propagation delay (at epoch) of the NBTI-RR driver with $k = 1$ is worse than that of the traditional driver at epoch, since this NBTI-RR driver uses only 39 fingers instead of 40. We experimentally determined this increased propagation delay to be 2.7%. From Figure 2 (a), we can say that the NBTI-RR driver with $k = 1$ is equivalent to a traditional driver whose threshold voltage has already degraded by 10 mV at epoch. This in effect means that the NBTI-RR driver can tolerate an additional 8.6 mV of theshold voltage degradation before it fails. A similar exercise for the NBTI-RR driver with $k = 2$ showed that its increased propagation delay at epoch (compared to the traditional driver at epoch) was 5.38%, as illustrated in Figure 2 (a). This is because the NBTI-RR driver with $k = 2$ uses 38 fingers at all times (unlike the traditional driver which utilizes 40 fingers). Since this delay is greater than 5%, the NBTI-RR driver with $k = 2$ is not a valid option.

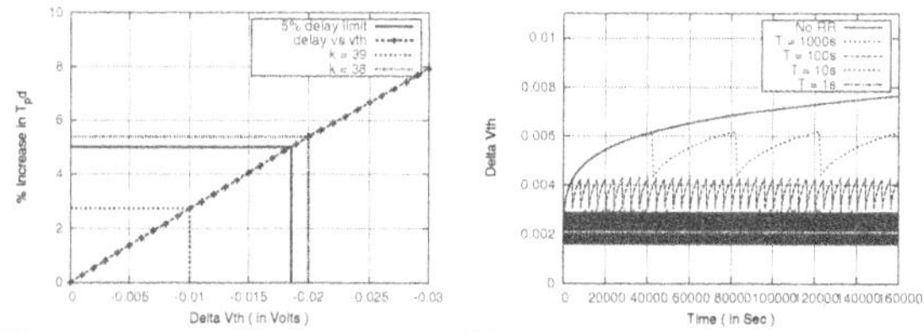

(a) Propagation Delay vs V_{th} Degradation

(b) V_{th} Degradation as a Function of T

Figure 2: Threshold Voltage Degradation and its Effect

Based on the above discussion, we utilize $k = 1$ for our NBTI-RR driver. We now discuss the methodology utilized to compute the threshold voltage for any finger of the NBTI-RR driver as a function of time. Given a value of T, we know that our NBTI-RR driver will be stressed for $39T$ seconds, and relax for T seconds. The methodology to compute the degraded threshold voltage after each stress and relaxation event is discussed next.

- For the first stress event, we compute the threshold voltage degradation (denoted as V_S^1) by utilizing Equation 3 for $39T$ seconds of stress (at a clock rate of 1 GHz).

- Now for the first relaxation event, we compute the threshold voltage after recovery (denoted as V_R^1) as $V_R^1 = V_S^1(1 - RF(T))$.

- In general, to compute the threshold voltage degradation of the jth stress event V_S^j, we first compute the time t^{j-1} (from epoch) when the threshold voltage degradation using dynamic NBTI reaches a value V_R^{j-1}. This is done by rewriting Equation 3 such that time is expressed as a function of threshold voltage degradation. Now we compute V_S^j using Equation 3, for a time duration corresponding to $39T + t^{j-1}$. This is reasonable since the NBTI-induced threshold voltage degradation V_R^{j-1} is identical to the threshold voltage degradation at time t^{j-1}. Hence, from a physical point of view, the number of interface traps present at time t^{j-1} is identical to that when the threshold voltage degradation is V_R^{j-1}. Hence the stress profile from t^{j-1} onwards should be identical to the stress profile if the device is stressed when its threshold voltage degradation is V_R^{j-1}.

- Next we compute the threshold voltage after recovery for the second relaxation event (denoted as V_R^j) as

$$V_R^j = V_S^j(1 - RF(T)).$$

Note that the difference between the values of V_R^j (V_S^j) and V_R^{j-1} (V_S^{j-1}) drops below 0.1 mV for values of $j > 4$.

Figure 2 (b) illustrates the values of V_R^j and V_S^j reached for the NBTI-RR driver, for varying T. The T values we used were 1000s, 100s, 10s and 1s. Note that for all these values of T, the maximum value of V_S^j reached is 6.1 mV. Since this value is less than the additional threshold voltage degradation of 8.6mV needed to cause a $P = 5\%$ degradation in the propagation delay of the NBTI-RR driver (compared to the delay of the regular driver at epoch), we can claim that these four values of T result in a off-chip output driver whose delay never degrades beyond the $P = 5\%$ limit over the duration shown in Figure 2 (b) (which corresponds to about 44 hours). Note that since a majority of the threshold voltage recovery (during relaxation) occurs very soon after the relaxation interval begins, using smaller values of T results in a lower maximum value of V_S^j, and hence yields an NBTI-RR driver with the smallest propagation delay degradation (compared to the traditional driver's propagation delay at epoch).

Driver	epoch	1yr	1.2yr	2yr	5yr	10yr	20yr
Traditional	0	4.8	(5.0)	(5.9)	(6.8)	(7.5)	(8.2)
RR(T=1s)	2.7	3.2	3.2	3.2	3.4	3.5	3.7
RR(T=10s)	2.7	3.3	3.3	3.4	3.7	3.9	4.2
RR(T=100s)	2.7	3.4	3.4	3.6	3.8	4.1	4.8
RR(T=1000s)	2.7	3.6	3.6	3.9	4.4	4.9	(>5)

Table 1: Delay Degradation (%) as a function of T

The percentage increase (with respect to the delay of the traditional driver at epoch) of the propagation delay of the traditional driver and the NBTI-RR driver for $T = 1000s$, 100s, 10s and 1s are shown in Table 1. Note that the delay of the control circuit is accounted for in this table. We observe that the lifetime of the traditional driver is only 1.2 years under 50% stress, while the lifetime achieved for the NBTI-RR driver is more than 20 years.

We estimated the active area overhead of the NBTI-RR approach, incurred due to the implentation of the control logic, to be 0.71% of the area of the traditional driver. Note that off-chip output drivers are extremely large inverters since they drive off-chip loads between 25pF and 50pF. Typical output drivers are ∼1500× to 2500× the size of a minimum sized inverter (2000× the size of a minimum inverter in our experiments), hence occupying a significant chip area. As a result, the overhead of the NBTI-RR control logic

is extremely small. By adding 0.71% more finger to the traditional driver, we only extend its lifetime from 1.2 years to 1.2172 years, whereas the delay of a driver implemented with our approach (with 0.71% overhead) never degrades beyond 5% of its epoch delay (for T=1s) even after 20 years.

The traditional driver consumes a total power of 18.72 mW including 2.16 mW consumed by its control circuit. In comparison, the NBTI-RR driver consumes a total power of 19.13 mW in which 2.71 mW is consumed by its control circuit. As the NBTI-RR driver uses a n bit shift register for its control circuit, it consumes on average 2.19% more power than traditional driver.

Driver	epoch	1yr	1.2yr	2yr	5yr	10yr	20yr
Traditional	0	1.2	1.2	1.6	2.1	3.1	4.5
NBTI-RR	0	0.9	0.9	1.1	1.9	2.4	3.3

Table 2: Delay degradation (%) of control circuit

The delay degradation of the control circuit with respect to its own delay at epoch is given in Table 2. It is observed that the NBTI-RR control circuit suffers less NBTI degradation than the control circuit of the traditional driver, since it uses smaller PMOS devices (n copies of the finger enable circuit, each driving a single finger as shown in Figure 1 c)), whereas the traditional driver control circuit has a bigger PMOS (the finger enable circuit drives all the n fingers as shown in Figure 1 a)). Hence the traditional control circuit suffers more NBTI degradation. Note that the overall degradation of the driver (reported in Table 1 include the effect of degradation of the driver circuits). Also, from Table 2, it is evident that the control circuitry degrades at a slower rate than the driver itself, and hence is not a reliability determining circuit.

5. REFERENCES

[1] M.A.Alam, H. Kufuloglu, D.Varghese, and S. Mahapatra, "A comprehensive model of PMOS NBTI degradation : Recent progress," *Microelectronics Reliability*, vol. 47, no. 6, pp. 853–862, 2007.

[2] "PTM website." http://www.eas.asu.edu/~ptm.

[3] B. Paul, K. Kang, H. Kufluoglu, M. A. Alam, and K. Roy. "Negative bias temperature instability: estimation and design for improved reliability of nanoscale circuits," *Transaction on Computer-Aided Design of Integrated Circuits and Systems*, vol. 26, pp. 743–751, April 2007.

[4] Z. Qi, Stan, and R. Mircea, "NBTI resilient circuits using adaptive body biasing," in *GLSVLSI '08: Proceedings of the 18th ACM Great Lakes symposium on VLSI*, pp. 285–290, ACM, 2008.

[5] M. Agarwal, B. C. Paul, M. Zhang, and S. Mitra, "Circuit failure prediction and its application to transistor aging," in *VTS '07: Proceedings of the 25th IEEE VLSI Test Symmposium*, pp. 277–286, IEEE Computer Society, 2007.

[6] S. V. Kumar, C. H. Kim, and S. S. Sapatnekar, "Impact of NBTI on SRAM read stability and design for reliability," in *ISQED '06: Proceedings of the 7th International Symposium on Quality Electronic Design*, pp. 210–218, IEEE Computer Society, 2006.

[7] T. Siddiqua and S. Gurumurthi, "A multi-level approach to reduce the impact of NBTI on processor functional units," in *GLSVLSI '10: Proceedings of the 20th symposium on Great lakes symposium on VLSI*, pp. 67–72, ACM, 2010.

[8] J. Sun, A. Kodi, A. Louri, and J. Wang, "NBTI aware workload balancing in multi-core systems," in *Proceedings of the 10th International Symposium on Quality of Electronic Design (ISQED 09)*, (Washington, DC, USA), pp. 833–838, 2009.

[9] A. Calimera, E. Macii, and M. Poncino, "NBTI-aware sleep transistor design for reliable power-gating," in *Proceedings of the 19th ACM Great Lakes symposium on VLSI*, (New York, NY, USA), pp. 333–338, 2009.

[10] S. Bhardwaj, W. Wang, R. Vattikonda, Y. Cao, and S. Vrudhula, "Scalable model for predicting the effect of negative bias temperature instability for reliable design," *IET Circuits, Devices and Systems*, vol. 2, no. 4, pp. 361–371, 2008.

[11] "DDR2 SDRAM device operation and timing diagram." Hynix Corporation. Technical Data Sheet.

[12] A. Aziz, "Package, power and I/O." http://users.ece.utexas.edu/ãdnan/vlsi-05-backup/lec20Packaging.ppt. Course notes. "Introduction to CMOS VLSI Design".

Mitigating Electromigration of Power Supply Networks Using Bidirectional Current Stress

Jing Xie, Vijaykrishnan Narayanan, Yuan Xie
The Pennsylvania State University, University Park, PA, USA
{jingxie, vijay, yuanxie}@cse.psu.edu

Abstract

Electromigration (EM) is one of the major reliability issues for
IC designs. The EM effect is observed as the shape change of metal
wires under uni-directional high density current. Such metal wire
distortions could result in open-circuit failures or short-circuit failures for the interconnects in integrated circuits. The current density
on power supply network is usually the highest one among all the
on-chip interconnects, and the current direction on power rails sel-
dom changes. Consequently, the power supply network is the most
EM-vulnerable component on a chip. We propose a novel solution based on the electromigration AC healing effect to extend the
lifetime of power supply networks. This solution uses simple control logics to apply balanced amount of current in both directions
of power rails. Therefore, power wires can perform self-healing
during function mode. This technique can be easily integrated into
different package plans with small area and performance overhead.
The post layout simulation shows 3X-10X increase of the mean
time to failure (MTTF) for the power rails.[1]

Categories and Subject Descriptors: B.8.1 [Performance and Re-
liability]:Reliability, Testing, and Fault-Tolerance

Keywords: Electromigration, power supply network, reliability.

1 Introduction

Electromigration (EM) is one of the key reliability concerns
in modern VLSI circuit designs. Electromigration occurs when a
surge of current going through metal wires. The drift of metal atoms
along with the flow of electrons causes a depletion of the metal upstream and a deposition of metal downstream along the current flow
direction. The upstream thinning increases the wire resistance and
ultimately results in open-circuit failures; while the downstream deposition may cause short-circuit failure to the nearby metal. Consequently, EM effect slows down the circuit through time, and in
the worst case can lead to the eventual loss of one or more connections and an intermittent failure of the entire circuit. As technology
scales, EM is aggravated with the ever-decreasing wire width and
rising temperature [1] [2].

Power supply network is one of the most vulnerable interconnects among all the on-chip wires, due to two reasons. First, the

[1]This work is supported in part by NSF grants 0643902,
0916887 and 0903432

current flow direction on power supply network does not change
as often as that in regular signal interconnects; Second, the current
density on power networks is usually significantly higher than that
of signal wires. Since high current density and uni-directional current flow are the two major contributors for the EM effect, mitigating the EM damage on power supply networks is one of the critical
reliability concerns for IC designers.

Electromigration problem has been well recognized and many
methods have been proposed to mitigate the EM effects in interconnects [3, 4, 5, 6, 2]. For example, Abella *et al.* proposed a method
to switch he power/ground supply wires by off-chip and on-chip
switches [3]. Lienig and Jerke [4] summarized a number of useful
design rules for preventing EM hazard. Xuan proposed an approach
by increasing the most vulnerable wire width [5]. Dasgupta and
Karri proposed a technique to mitigate EM by minimizing the maximum switching activity [6]. Other approaches include using copper
instead of aluminum, and covering bottom and sides of copper lines
with Tantalum liner [2]. Majority of the proposed solutions usually
result in large area/performance overhead, and usually become less
effective with increased on-chip temperature.

In this paper, we propose a circuit-level current compensation
method to make the metal wires "repair" themselves against the EM
effect. We also present an efficient algorithm for our EM-aware design, so that it can be integrated into the standard-cell place and
route flow. Compared to prior work, the reliability improvement
from this work does not diminish as temperature increases. To the
best of our knowledge, the proposed work represents the first design methodology for self-healing EM for power supply network
design.

2 Motivation and Background

The physical principle of EM is the motion of ions under the
influence of electric field [7]. This motion changes the shape of thin
metal wires under high current, and result in open-circuit failures or
short-circuit failures. EM-aware optimization is an important part
of high reliability circuit design.

The EM effect can be modeled by *Black Equation* [8] as follows
(MTTF is used to characterize the severity of EM):

$$MTTF_{EM} \propto J^{-n} \times e^{\frac{E_{aEM}}{kT}} \quad (1)$$

While technology scaling improves circuit performance, it dete-
riorates the EM effect. Smaller feature size leads to higher current
density. Suppose the scaling factor is z, technology scaling will
make EM z^2 times worse. Metal 1 does not scale as large as before,
but the supply voltage is almost constant these days. The real EM
problem can be more severe than z^2.

2.1 EM Effects on Power Supply Network

Different parts of a power network have their own EM severe-
ness. The power grid uses very wide high layer metals for whole
chip power delivery. The standard cell power rails convey current to
all transistors in each small block. The current density on these rails

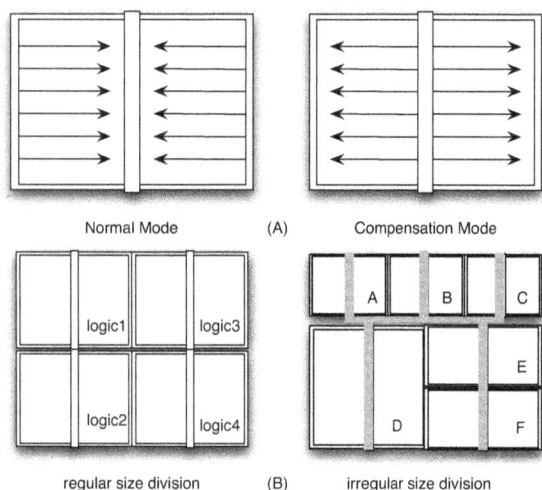

Figure 1. (A) A vertical Power/Ground strip (compensation strip) is added in the middle of the layout with two working modes (`normal mode`: power is supplied to the block from the P/G ring with the compensation strip in high-impedance state; `compensation mode`: the PAD supplies the compensation strip, with the regular P/G ring in high-impedance state); (B) chip layout divided into regular or irregular sizes with power grid.

are significantly higher than on power grids, because they usually use minimum width metal-1 layer.

The EM time to failure is found to increase with line width for long wires [9], which is usually the case for power rails and grids. On the other hand, if the metal length is under 10 *um*, narrow wires EM time to failure was observed to be long [10]. The power supply wires inside the standard cells meet this length requirement and are safe. After considering all the on-chip wires, the standard cell power rail has the highest risk of the EM failure.

2.2 Healing Effect

EM happens when long durations of uni-directional current applied. AC stress can provide healing effect in metal wires [11]. The experimental results of the time-to-failure under AC stress was discussed by Tao et al. [12]. Their result showed that uni-directional current will increase the resistance of metal wires. If opposite directional current is applied on wires, some but not all of the damage can be healed. The healing effect depends on the AC frequency. Given $|\bar{J}|^m = J_+ - J_-$, where J_+ and J_- are the current densities in opposite directions, the EM MTTF of a wire can be expressed as $\gamma(1-\eta)|\bar{J}|^m$ [13]. In the AC mode, η changes with frequency. The AC MTTF is high, when the frequency exceeds a threshold.

3 Electromigration Enhancement Design

Since EM influences the standard cell power rail most, we aim at reducing EM on power rails with AC stress self-healing. We change the topology of power networks to produce balanced bidirectional current on power rails.

3.1 Design Mechanism

An IC chip may have a complex power grid structure, but they can be divided recursively to the power ring structure. Consequently, our baseline design is a structure with a power ring and an array of standard cell rails.

Our mechanism is to apply a vertical power/ground (P/G) strip in the middle of the layout, which uses a different metal layer from the P/G ring as shown in Figure 1. This additional strip is a compensation power strip, which has similar width with the P/G ring.

This strip is connected to each standard cell power rail but is disconnected with the P/G ring.

There are two operation modes in this design: the normal mode and the compensation mode. The current flow directions on power rails are shown in Figure 1A. Both modes are driven by the same set of PADs to prevent PAD number increase. In the normal mode, power is supplied to the block from the P/G ring. The transistors connecting PADs and the compensation strip are off, thus the strip is in high-impedance state. In the compensation mode, the PAD supplies the compensation strip, and the P/G ring is in high-impedance state. If a block is too big to meet the IR drop requirement, it can be divided into regular or irregular sub-blocks with their compensation strips connected together as illustrated in Figure 1B. The sub-blocks switch into the normal or the compensation mode simultaneously.

3.2 Design Consideration

In the circuit implementation, several facts should be considered. The package plan, the switching performance overhead, and control gating overhead are essential to ensure the design to fit in all situations with minimum performance overhead.

3.2.1 Package plan influence on power grids

Two widely-used chip package methods are wire-bonding and flip-chip. For wire-bonding method, all the input signals including power supply sources are from the four edges of the chip. For flip-chip method, the minimum PAD pitch requirement is about 20 PADs/mm [14]. If half of the PADs are used for power supply, the distance between two power PADs is 200 *um*. Similar power grid spacing is designed to ensure reasonable IR drop for both package plans. Thus our proposed compensation grids should comply with these spacing constraint.

3.2.2 Switching performance overhead

Under the power grid spacing requirement, we investigated the power supply switching of the most power hungry circuit type - inverter chain. The P/G supply is at the two ends of the inverter chain. Signals *ctrl*1 and *ctrl*2 determine the on and off of power gating transistors. Signal integrity of output nodes during *ctrl* switching is a major concern.

The healing effect requires an AC frequency above 20 kHz for copper [12]. It is safe to use a 100 kHz switching frequency for the *ctrl* signals. The circuit frequency is around 1 GHz, then *ctrl* switches every 10k cycles.

We use an example circuits of a 128-inverter chain with 260 *um* power rail length under 130nm technology to evaluate the performance overhead. Simulation results show that non-overlap *ctrl*1 and *ctrl*2 can result in 10% latency overhead for rising edge and 4% latency overhead for falling edge. Having both *ctrl*1 and *ctrl*2 on for one additional cycle eliminates performance degradation at switching with 0.1% overlap.

3.2.3 Sizing the power-gating transistor

The size of the gating transistor determines the maximum current that can pass through it. However, larger transistors consume more chip area. For the 128-inverter chain, the gate size for *control* switching transistors should be above 3 *um* to achieve minimum performance impact and above 2 *um* to make the circuit functional under 130nm technology node.

3.3 Optimize MTTF with current balancing

The proposed mechanism is based on the principle of applying bidirectional current, but fully balanced AC stress at all nodes is not practical. Even if the current is balanced, EM still cannot be fully healed. Therefore, AC plus DC model is applied to estimate the best EM MTTF [15] under an unbalanced situation.

The healing effectiveness γ is described as:

$$\gamma = 1 - 2\left(\frac{f_0}{f}\right)^{1/n} \qquad (2)$$

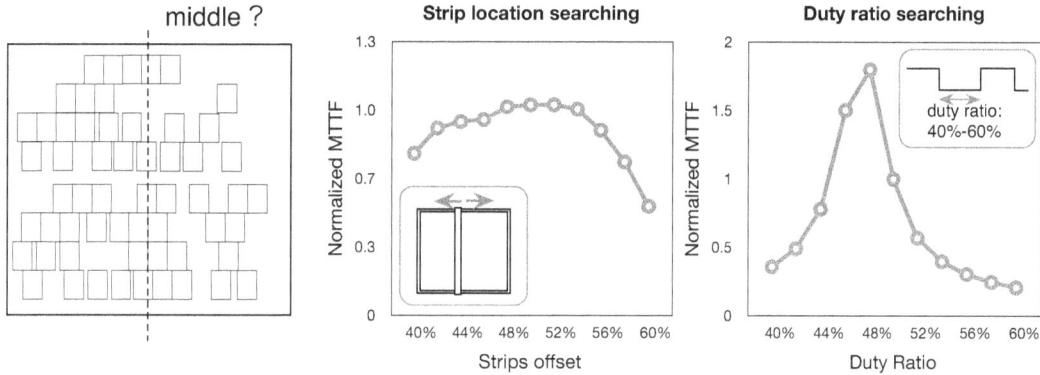

Figure 2. A single diagram of inbalanced placement. Approach 1 and Approach 2 simulation results.

where f_0 can be described as:

$$\frac{1}{2f_0} = \frac{A}{J_{DC}^{-n}} e^{E_a/kT} = MTTF_{DC} \qquad (3)$$

The higher the frequency is, the closer γ will approach to one. The current duty ratio r modifies the overall AC MTTF as:

$$MTTF_{AC} = \frac{A}{rJ_+ - \gamma(1-r)J_-^{-n}} e^{E_a/KT} \qquad (4)$$

J_+ and J_- stands for the current density in opposite directions.

We fully understand that the different input pattern will affect the current, but the most severe EM parts keep the same current direction all the time. Moreover, it is impossible to perform layout level simulaiton on architecture level benchmark/application. Consequently, the inbalanced placement of the standard cells becomes the main concern for the inbalanced bi-directional current, during the chip design stage. Based on this fact, we propose two approaches to optimize the EM MTTF.

●**Approach 1:** Change the compensation power strip locations, while keeping the duty ratio r of *ctrl1/ctrl2* signal at 50%. This method provides better MTTF and keeps the control logic simple. However, there are many blocks within a chip, such that a large number of compensation strip locations are required to be determined. Changing the strip locations to find the optimal solution will lead to repeated re-place and re-route, which increases the total design time significantly.

●**Approach 2:** Change the duty ratio r of AC stress, while fixing the compensation power strip in the middle of the power ring. The MTTF of the whole chip is a continuous function of r. However, this function is not derivable because it is a piecewise function constructed by choosing the worst single nodes' MTTF(r). Thus, EDA tools cannot derive the duty ratio for best MTTF. Sweeping r should be a time efficient algorithm. The step size of sweeping depends on the preciseness requirement of the MTTF optimization. We suggest sweeping no more than 16 points from 40% to 60% of r for reasonable design time and simple control logic.

We use an example benchmark circuit to evaluate the effectiveness of these two approaches(The simulations are based on a $554 \times 554\ um^2$ MUL unit using the $130nm$ GLOBALFOUNDRIES technology at 25^oC). The MTTF results are shown in Figure 2. It shows that the best MTTF of approach 1 have a similar value to the MTTF when placing the compensation strip in the middle of the chip. Meanwhile, the maximum MTTF in approach 2 is about two times the 50% duty cycle design. Consequently, we can conclude that the optimization for MTTF should place the compensation strip in the middle of the layout and sweep the duty cycle ratio for best MTTF.

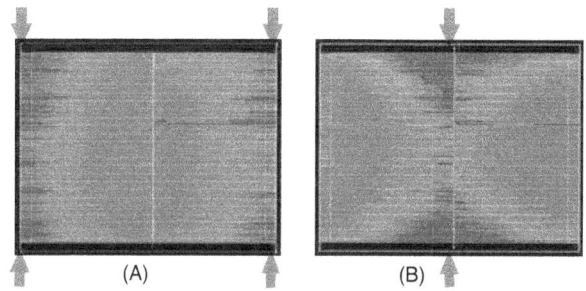

Figure 3. EM damage in an EM compensation design. Red parts are the most severe parts. (A) EM damage map in the normal mode; (B) EM damage map in the compensation mode.

4 Results

Three different units from OpenSparcT1 [16] are used to verify the proposed EM healing method. These units are Floating Point front-end Unit (FFU), Multiplex Unit (MUL) and Stream Processing Unit (SPU). They are chosen because they exhibit different functionality and have reasonable sizes.

The technology libraries used in this paper are the 130nm GLOBALFOUNDRIES process with 1.5 V supply voltage and the 45nm NCSU FreePDK process with 1.1 V supply voltage. The simulations were performed with the ambient temperature of 25^oC, and the on-chip temperature of 55^oC.

4.1 Experiment and Data Analysis

Our experiments and comparison are based on four sets of setup.
● The normal mode: the chip is driven by power ring.
● The compensation mode: the compensation strip drives the chip.
● The coarse bidirectional mode: half of the time the chip is driven by the power ring and another half by the compensation strip.
● The balanced bidirectional mode: the ratio of time driven by the power ring and the compensation strip is modified to balance the current in each directions.

Uni-directional current (DC) MTTF can be calculated from Black equation. For an AC stress with different forward and backward current density, its MTTF is related to the DC MTTF as: (use M to stand for MTTF. n=1.1 [1])

$$(M_{AC})^{-\frac{1}{n}} = \frac{1}{2}(M_{DC,+})^{-\frac{1}{n}} - \frac{\gamma}{2}(M_{DC,-})^{-\frac{1}{n}} \qquad (5)$$

In the balanced bidirectional method with optimized duty ratio

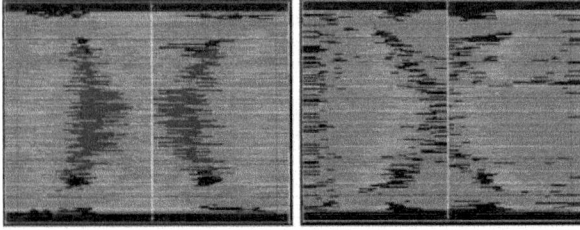

Figure 4. EM healing results. (left) The coarse bidirectional mode result; (right) the balance bidirectional mode result.

Table 1. Area ($um \times um$) and the Overhead

tech		spu	ffu	mul
45nm	area	980×980	837×837	418×418
	overhead	4%	4.70%	5.50%
130nm	area	1310×1310	1100×1100	554×554
	overhead	3%	3.60%	4.50%

r:

$$M_{AC} = \frac{1}{(r(M_{DC,+})^{-\frac{1}{n}} - (1-r)\gamma(M_{DC,-})^{-\frac{1}{n}})^n} \quad (6)$$

The *control* signal has 0.1% overlap to prevent performance degradation, which is small and treated as no overlap during calculation.

We use IR drop plot to determine the current directions. The locations that have the most severe EM issue do not change their current directions by input patterns. Figure 3A, B show the most severe EM locations in red. The goal is to improve EM MTTF at these locations. An example of coarse and balance bidirectional mode EM healing results for the FFU block is shown in Figure 4. The previous lowest MTTF points are healed.

When the feature size shrinks, finer power grid division is used for the same design to maintain reasonable IR drop. We compared designs under two processes. For example, MUL keeps the single power ring structure under 130nm process. A division into two blocks is used for 45nm process to meet IR drop requirements. Two compensation strips are applied and driven together. The area overhead for all experiment cases are no more than 5.5%(Table 1). Power grids will consume more area for smaller chips, but this trend is observed in all chip designs and is not an artifact of our mechanism.

We compared the MTTF of these 3 functional blocks under four experiment modes. The MTTF for different designs under the same technology and temperature are shown in Figure 5. It can be observed that for the normal mode (base-line design), the MTTF of these 3 designs are very close even though their physical design (floorplanning/placement/routing) are quite different. However, with the adding of compensation strip and bidirectional AC stress, MTTF can be improved dramatically. Comparing the two schemes applied self-healing, the balanced mode can achieve better improvement than the coarse mode. This improvement variation is related to power density and placement.

The trend for temperature and technology scaling also follows the theoretical analysis, as shown in Figure 5. For these two technologies comparison (130nm versus 45nm), the EM difference is about 10 times for all designs. The 45nm process is three generations smaller than the 130nm process ($z = 3$). This result is close to 9 times EM MTTF scaling assumption (z^2). A 30 degrees rising in temperature decreases the MTTF by ten times.

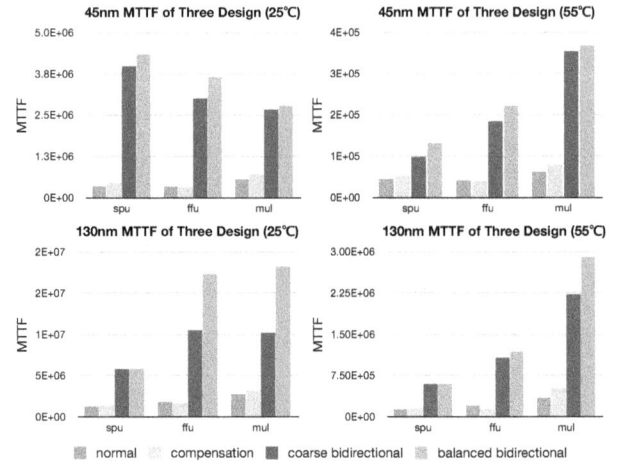

Figure 5. EM enhancement result for different design/technology node/temperature. The MTTF is based on hours.

5 Conclusion

Electromigration (EM) on the power supply network is one of the major reliability issues for IC designs. In this paper, we have proposed a novel solution based on the electromigration AC healing effect with compensation strip insertion. The proposed method uses simple control logics to apply balanced amount of current in both directions of power rails and therefore mitigate the EM effects. The post layout simulation on multiple designs with two technologies nodes (130nm and 45nm) shows 3X-10X increase of the mean time to failure (MTTF) with a small (3%-5.5%) area overhead.

6 References

[1] J. Srinivasan, S. Adve, P. Bose, and J. Rivers, "The case for lifetime reliability-aware microprocessors," in *ISCA*, 2004, p. 276.

[2] C.-K. Hu, R. Rosenberg *et al.*, "Scaling effect on electromigration in on-chip cu wiring," in *IITC*, 1999, p. 267.

[3] J. Abella, X. Vera *et al.*, "Refueling: Preventing wire degradation due to electromigration," *Micro, IEEE*, vol. 28, no. 6, p. 37, 2008.

[4] J. Lienig and G. Jerke, "Electromigration-aware physical design of integrated circuits," in *Intl. Conf. on VLSI Design*, 2005, p. 77.

[5] X.Xuan, "Analysis and design of reliable mixed-signal cmos cicuits," *Ph.D thesis, Georgia Inst. of Technology*, 2004.

[6] A. Dasgupta and R. Karri, "Electromigration reliability enhancement via bus activity distribution," in *DAC*, 1996. p. 353.

[7] J. W. Morris, C. U. Kim, and S. H. Kang. "The metallurgical control of electromigration failure in narrow conducting lines," *J. Optim. Theory Appl.*, vol. 48, no. 5, p. 43, 1996.

[8] J. Black, "Electromigration failure modes in aluminum metallization for semiconductor devices," *Proc. of the IEEE*, vol. 57, no. 9, p. 1587, 1969.

[9] L. Ting, J. May *et al.*, "Ac electromigration characterization and modeling of multilayered interconnects," in *IRPS*, 1993, p. 311.

[10] Y.-L. Cheng, W.-Y. Chang, and Y.-L. Wang, "Line-width dependency on electromigration performance for long and short copper interconnects," *JVST B*, vol. 28, no. 5, p. 973, 2010.

[11] Y.-L. Cheng, B.-J. Wei, and Y.-L. Wang, "Electromigration characteristics of copper dual damascene interconnects - line length and via number dependence," in *IPFA*, 2009, p. 723.

[12] J. Tao, J. Chen *et al.*, "Modeling and characterization of electromigration failures under bidirectional current stress," *IEEE Trans. Electron Devices*, vol. 43, no. 5, p. 800, 1996.

[13] J. Tao, N. Cheung, and C. Hu, "Metal electromigration damage healing under bidirectional current stress," *IEEE Electron Device Lett.*, vol. 14, no. 12, p. 554, 1993.

[14] H. P. Yeoh, M.-J. Lii, B. Sankman, and H. Azimi, "Flip chip pin grid array (fc-pga) packaging technology," in *EPTC*, 2000, p. 33.

[15] J. Tao, B.-K. Liew *et al.*, "Electromigration under time-varying current stress," *Microelectronics Reliability*, vol. 38, no. 3, p. 295, 1998.

[16] oracle, *http://www.opensparc.net/*.

Multiplexed Switch Box Architecture in Three-dimensional FPGAs to Reduce Silicon Area and Improve TSV Usage

Marzieh Morshedzadeh
Shahid Beheshti University, G. C.
morshedzadeh@sbu.ac.ir

Ali Jahanian
Shahid Beheshti University, G. C.
jahanian@sbu.ac.ir

ABSTRACT

In this paper, we propose a multiplexed 3D-switch box architecture that decreases the number of TSVs required for routing with a slight overhead in total wirelength. Our experimental results show that the presented architecture reduces the number of routing TSVs by about 48% in cost of less than 2% wirelength overhead.

Categories and Subject Descriptors

B.7.2 [**Integrated Circuits**]: Design Aides—*Placement and Routing*

General Terms

Algorithms, Design

Keywords

Multiplexing, Switch box, Three-dimensional FPGA

1. INTRODUCTION

Moore predicted that number of transistors in integrated circuits doubled each 1.5 to 2 years and this forecast has remained true for more than 40 years in the semiconductor field. However, this law is not always accurate and may be violated by approaching the technological limits in the fabrication process or even design productivity gap [1].

As an important and applicable instance of integrated circuit devices, FPGAs suffer from these limitations seriously. As complexity of FPGA designs becoming higher, routing congestion and also routability of them will be more critical. In modern FPGAs more than 70% of die area and power consumption relates to routing resources [2]. On the other hand, long wire segments in a large FPGA may lead to congestion and also performance degradation. Therefore, interconnect resource management in FPGA is very crucial. Three-dimensional integration technology has been proposed to overcome this problem by shortening the long wires into

short vertical vias. In 3D-circuits, large circuits are partitioned into some smaller pieces and stacked respectively on top of each other. The adjacent pieces connect to each other by Through Silicon Vias (TSVs) such that logic blocks spread between layers. This architecture can reduce the wire length and also routing congestion on the existence of a good placement and routing algorithm [3].

However, this technology suffers from few shortcomings. One of these drawbacks is that the number of TSVs that is used between layers is limited and therefore, TSV assignment should be performed correctly. TSVs have two important applications in 3D-chips; they not only can be used for vertical vias (routing TSV) but also can be utilized for cooling the chips (Thermal TSVs). On the other hand, the pitch of TSVs is greater than metal vias and they occupy considerable area of dies more than normal metal wires. Therefore, reducing the number of routing TSVs is very critical in quality of 3D-ICs because silicon area is saved and also more TSVs can be used for thermal management.

Some contributions are reported to reduce the FPGA TSVs in 3D-FPGAs, in recent years. Ababei et. al introduced a timing-driven partitioning and a simulated-annealing-based placement tool for 3D-FPGAs in [4]. Firstly, the tool divides circuit into a number of desired planes using min-cut hMetis algorithm [5] while trying to minimize the connections between layers at the same time. This leads to a minimization of the number of TSVs. Next, for each individual plane, place and route performed with a consideration of total wirelength and critical delay [4].

In [2], the authors addressed the problem of decrementing the area as well as the delay for 3D-FPGAs in different scenarios made up of different settings of number of layers, wafer bonding strategy and number of TSVs. They designed six different types of 3D-switch boxes that improved the number of used TSVs.

Lin et al. in [6] analyzed the performance benefit of monolithically stacked 3D-FPGAs. They separate logic blocks from programmable memory and interconnects and stacked them on different layers, resulting in improving logic density by about 3.2x and decreasing critical path delay as well as dynamic power consumption by about 1.7x than the baseline 2D-FPGA.

In the aforementioned researches, authors attempted to reduce TSV's by changing the structure of 3D-switch box. However, no one has provided a new solution instead of the 3D-switchbox. Therefore, we proposed a new method to reduce TSVs by using multiplexers. We analyzed the distribution of TSVs in many of 3D-FPGA designs and our

GLSVLSI'12, May 3–4, 2012, Salt Lake City, Utah, USA.
Copyright 2012 ACM 978-1-4503-1244-8/12/05 ...$10.00.

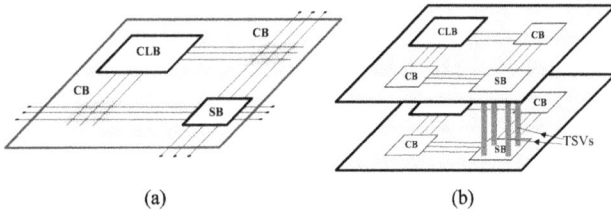

Figure 1: (a) 2D-FPGA tile (b) 3D-FPGA tile

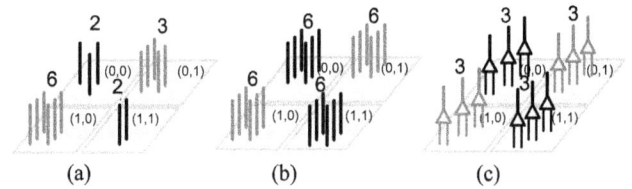

Figure 2: Various distributions for TSVs

analysis show that some TSVs can be multiplexed without significant reduction in performance. In this paper, we propose a new 3D-switching architecture and then, the FPGA routing algorithm is revised to multiplex non-critical TSVs reducing the number of TSVs.

This paper is outlined as follows. In Section 2, 3D-FPGA technology is reviewed. Sections 3 and 4 describe the proposed architecture and routing algorithm, respectively. In Section 4, experimental results are reported and analyzed and finally, Section 5 concludes the paper.

2. THREE-DIMENSIONAL FPGAS

Current two-dimensional FPGAs are normally designed based on the island-style architecture which consists of a matrix of similar tiles each of which comprising of one switch block (SB), one or two connection blocks (CB) and one configurable logic block (CLB) as shown in Figure 1-a.

Each CLB made up of lookup tables and flip-flops. They connect to the routing channels via connection blocks. Channels are also connected to other channels through switch boxes [7].

Traditionally, the structure of 3D-FPGAs inherits 2D-FPGAs' architecture by stacking more than one 2D-FPGA layer (aka Tiers) on top of each other. In 3D-FPGA architecture, a new type of via is used to connect metal layers in various 3D-planes. This type of via which goes through substrates, called as Through-Silicon Via (TSV). The thick vertical wires in Figure 1-b are TSVs. TSVs have two applications in 3D-ICs; firstly, they are used as vertical Vias and secondly can be utilized for sinking the heat of the internal tiers to out of the chip. These two kinds of TSVs are known as routing and thermal TSVs, respectively.

Switch box is one of the most important components of FPGA. In a 2D-FPGA, the switch boxes connect to the connection boxes as well as to the other switch boxes just in x and y direction. An important factor in a switch box structure is its flexibility (F_s). Flexibility of a switch box is the number of outgoing tracks connecting to an incoming track. Recent researches show that $F_s = 3$ resulting in a good tradeoff between routability and switch overhead in a 2D-FPGA. It is worth noting that in 2D-switchbox a single track need 6 switches for full connectivity.

In a 3D-kind FPGA, the structure of 2D-SBs changes a little for connecting the current layer to upper and lower layers independently. The only difference between 2D-SB and 3D-kind is that the tracks can go up and down through the TSVs. Therefore, number of switches is increased and also (F_s) becomes 5 for a 3D-SB. It means that for a complete connectivity in a single track 3D-switch box, 15 switches are needed. Every four 2D tracks connects to one upward and one downward TSV for maintaining the value of $F_s = 5$ in

3D-switch box. These tracks can only connect to each other in x and y directions. In other words, each TSV is connected to 4 two-dimensional tracks [4].

3. PROPOSED ARCHITECTURE

As mentioned before, TSV is an important element for connecting tiers together in 3D technology. However, the number of used TSVs is important because Actual pitch of a TSV in modern technologies is about $20\mu m$ to $40\mu m$. Therefore, reducing the number of TSVs may decrease the silicon area of the FPGA [2].

The main contribution of this paper is reducing the number of required routing TSVs by using a multiplexed architecture. Figure 2 represent a simplified example of the contribution. Figure 2-a shows a simple example of the TSV requirement in various regions of a design. In this figure, each grid shows a switch box of a plane in a 3D-FPGA. Shown tier requires 12 TSVs totally. This means that 3 TSV (on average) is required in each switch box.

By using an FPGA with maximum number of required TSV in each grid cell, all of the vertical wires can be routed without any detour (Figure 2-b). However, many of TSVs will be remain unused (i.e. 12 TSVs in this example). We suggested a new architecture in which all average number of TSVs is used in a multiplexed structure (Figure 2-c). In this structure, all potential TSVs are available but an average number of them can be used in a design. By this technique, the number of TSVs is reduced significantly without considerable routability degradation at cost of some auxiliary multiplexers.

The feasibility of this technique depends on the distribution of TSV requirement in real circuits. We analyzed many three-dimensional FPGA circuits to investigate the distribution of TSVs and also estimated the average number of required TSVs in switch box. In other means, these analyses represent the potential of 3D-FPGAs to save the number of routing TSVs using the multiplexed switch boxes.

We placed and routed some of largest MCNC circuits using TPR [4]. Then, mapped a grid network on each design and finally, calculated the parameters on this grid network in each design circuit. Table 1 shows the enumerated factors. In this table, columns *MaxTSV* and *AvgTSV* indicate the maximum and average number of TSVs in each region of benchmarks and column *Unused %)* represents the percentage of unused TSVs when FPGA is fabricated with Maximum number of required TSVs.

As can be seen in Table 1, more than 64% of TSVs can be saved by multiplexing. As mentioned before, a 3D-switch box architecture is proposed in this paper in which TSVs are multiplexed to use average number of TSVs rather than maximum of them. Then, a routing algorithm is presented to route vertical connections on multiplexed routing archi-

Table 1: Distribution of TSV requirement

Benchmark	MaxTSV	AvgTSV	Unused (%)
exp5	5	2.2	56.00
apex4	4.8	2	58.33
diffeq	3.4	1	70.59
dsip	2.4	1	58.33
misex3	4.4	2	54.55
alu4	4.2	1.8	57.14
des	3.2	1	68.75
seq	4.6	2.2	52.17
s298	3.2	1	68.75
apex2	5.2	1.8	65.38
frisk	3.6	1	72.22
elliptic	3.2	1	68.75
spla	4.6	1.4	69.57
ex1010	3.8	1.2	68.42
pdc	5	1.8	64.00
clma	4	1	75.00
Average			64.25%

Multiplexed TSV Routing Algorithm	
Step 1	Run TPR to place and route the design.
Step 2	Calculate TSV distribution in routed circuit. Create list of excess and demand switches.
Step 3	Route excess TSVs to demand switches.
Step 3-1	Augment the extra capacity to the demand nodes using a minimum cost flow algorithm.
Step 3-2	Route from excess switchbox to selected TSV on minimum cost path.
Step 3-3	Reroute from demand switchbox to target switchbox in underneath layer.
Step 4	Estimate the improvements and overheads.

Figure 4: Proposed algorithm for multiplexed 3D-Switch box architecture

tecture evaluating the benefits and overheads.

Figure (3-a) Shows two TSVs in a normal 3D-switch box with flexibility $F_s = 5$ and Figure (3-b) shows the same switch box after multiplexing. In this figure, thin solid lines are 2D connections, dashed lines show auxiliary switches for multiplexing the TSVs and thick solid lines are the TSVs. It is worth noting that all of the switches are fabricated on the wafer and output of multiplexers are connected to the TSVs.

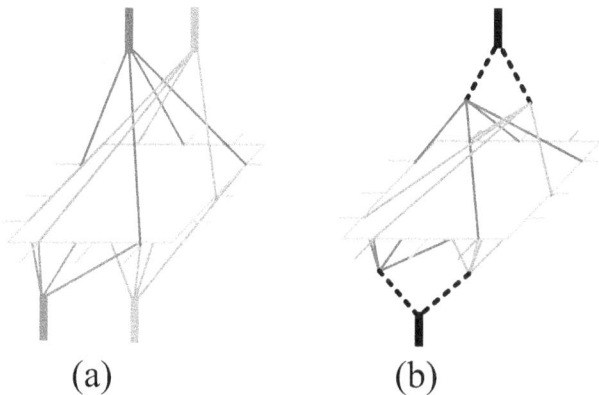

(a)   (b)

Figure 3: Proposed multiplexed 3D-Switch box

As shown in Figure 3-b, in the proposed architecture, each two TSVs are multiplexed as a TSV. Moreover, two auxiliary pass transistor are required for multiplexing each two TSVs. This is because of in our experience the maximum number of used TSVs was 6 before multiplexing and became 3 after multiplexing. Thus, it seems logical to use 2 to 1 multiplexer in each switch box. Therefore, in the way of each track to the desired TSV just a pass transistor added compared to normal state. Placing a pass transistor does not add a large impact on performance.

4. PROPOSED ROUTING ALGORITHM

In the proposed switch box architecture, each two TSVs

can be connected to one TSV through a multiplexer but just one of them can be selected in each multiplexer. Therefore, when two inputs of a multiplexer should be connected to TSV, one of them cannot be routed. In this case, it should be routed to a neighboring switch box that has free capacity and connected to a TSV via it.

One of the main challenges of the proposed architecture is the wirelength overhead due to detoured wires. We calculated and analyzed this overhead carefully and its results are reported in Section 5 of the article.The detailed description of the algorithm is as described in this sub-section.

Step 1: At first, design is placed and routed using a canonical three-dimensional FPGA tool. We used TPR as FPGA physical design engine.

Step 2: at this step, maximum number of required TSVs for switch boxes is calculated based on the routed circuit resulted by TPR. Accordingly, the average number of required TSVs per layer is calculated. Then, the number of TSVs crossing through each switch box is computed individually. At the end of this stage, capacity and overflow of the switch boxes are estimated and finally, list of excess and demand switch boxes are generated. Note that the switch boxes whose number of TSVs is lower than the average of the corresponding layer, are demand switch box and switch boxes that their number of TSVs are more than the average are considered as excess switch boxes.

Step 3: at this stage the excess switches are sorted based on the number of extra TSVs in a descending order. After that, switches are selected from the sorted list and find a nearest demand switch box using a minimum cost flow algorithm [8]. After selecting the demand switch, shortest path from excess switch box to demand switch box is found in the routing graph by using Dijkstra algorithm.

After routing the demand switches to free multiplexers all the wires pass through multiplexed TSVs to destination planes. However, detoured tracks in destination layer (after passing the corresponding TSV) should be routed to their destination terminal. Therefore, routing the wires in destination plane to exact final terminal point should be perfromed. we used Dijkstra algorithm to determine the shortest path for each track regarding the free capacity of FPGA channels of the new layer of 3D-FPGA in this paper.

Step 4: at this step, benefits and overheads of the design after multiplexing is evaluated.

5. EXPERIMENTAL RESULTS

We implemented the proposed algorithm using C++ on a 2.8GHz Dual-core Intel processor with 4 gigabytes of memory. We revised TPR place and route engine [4] which is a partitioning-based placement and routing tool for three dimensional FPGAs. We selected 16 biggest circuits of MCNC benchmarks and parameters of the FPGA is adjusted (in .arch file of TPR tool) based on 32nm technology node.

5.1 TSV Usage Improvement

As mentioned before, the main contribution of this paper is reducing the number of routing TSVs by the proposed multiplexed switch box architecture. Table 2 represents the TSV usage improvement of attempted benchmarks after applying the proposed architecture. In this table, column Dim shows the dimension of design and columns TPS and TNT represent the number of TSVs per switch box and total number of TSVs in each design, respectively. Finally, column TUI shows total TSV usage improvement (percentage of TSV reduction), respectively.

Table 2: TSV usage improvment

BM	Dim	Before MUX		After MUX		TUI
		TPS	TNT	TPS	TNT	(%)
exp5	15	5	1125	3	675	40.00
apex4	16*16	5	1280	3	768	40.00
diffeq	17*17	4	1156	2	578	50.00
dsip	17*17	3	867	2	578	33.33
misex3	17*17	5	1445	3	867	40.00
alu4	17*17	4	1156	2	578	50.00
des	18*18	4	1296	2	648	50.00
seq	18*18	4	1296	2	648	50.00
s298	19*19	4	1444	2	722	50.00
apex2	19*19	5	1805	3	1083	40.00
frisk	26*26	4	2704	2	1352	50.00
elliptic	27*27	4	2916	2	1458	50.00
spla	29*29	4	3364	2	1682	50.00
pdc	30*30	5	4500	3	2700	40.00
ex1010	29*29	4	3364	2	1682	50.00
clma	43*43	4	7396	2	3698	50.00
Average						45.83%

As can be seen in Table 2, total number of TSVs is reduced considerably (more than 45%) which reduces the silicon area and improves the TSV usage significantly.

5.2 Total Wirelength Overhead

In the proposed architecture, some TSVs should be detoured and transferred to other neighboring switch boxes to reach a free multiplexer that may resulted in wirelength overhead. We evaluated this overhead as the percentage of auxiliary FPGA tracks that should be used to route excess TSVs. Table 3 shows the total wirelength overhead in the proposed architecture. In this table, column TNT shows total number of TSVs in each design and columns NMT and NDT represent the number of multiplexed TSVs and detoured TSVs in each circuit, respectively. The next column (TO) shows total number of extra tracks (e.g. track overhead) that are required to route the detoured TSVs to closest free multiplexer. Column ($TOPT$) represents track

overhead per TSV and finally, column PTO shows the percentage of track overhead for each attempted benchmark.

Table 3: Total wirelengeth overhead

BM	TNT	NMT	NDT	TO	TOPT	PTO (%)
exp5	330.2	281.8	48.4	172.4	3.56	1.89
apex4	354.8	294.2	60.6	338	5.58	3.45
diffeq	134.6	102.4	32.2	84.4	2.62	1.06
dsip	78.4	64.4	14	32.8	2.34	0.36
misex3	381.4	337.8	43.6	158.4	3.63	1.85
alu4	322.4	277	45.4	116	2.56	1.35
des	140.6	114.4	26.2	58.8	2.24	0.48
seq	480.4	417.6	62.8	181.6	2.89	1.47
s298	175.4	142.2	33.2	80.8	2.43	0.71
apex2	447.8	371	76.8	236.8	3.08	1.73
fisk	487.8	356	131.8	736.4	5.59	3.09
elliptic	359.8	289	70.8	178.8	2.53	0.70
spla	716.6	552.8	163.8	1323.6	8.08	3.46
ex1010	570.2	451	119.2	559.6	4.69	1.61
pdc	1130.2	935	195.2	1558.4	7.98	3.22
clma	889	695.2	193.8	548.4	2.83	0.76
Average					3.92	1.70%

As can be seen in Table 3, the average of track overhead per TSVs is less than 4 tracks and total wirelength (e.g. track) overhead is less than 2%.

6. CONCLUSION

In this paper, we proposed a multiplexed switch box corresponding with a routing algorithm for reducing the number of routing TSVs three-dimensional FPGAs. The proposed method reduced the number of TSVs by about 48% at cost of less than 2% overhead of wirelength.

7. REFERENCES

[1] Interconnect Roadmap,[online] Available on: http://public.itrs.net, 2010.

[2] A. Gayasen et. al, "Designing a 3-D FPGA: switch box architecture and thermal issues,"In *IEEE Transactions on VLSI Systems*, pp.882-893, 2008.

[3] K. Siozios et. al, "Exploring alternative 3D-FPGA architectures: design methodology and CAD tool support,"In *Proceedings of the International Conference on FPGA*, pp. 652-655, 2007.

[4] C. Ababei et. al, "Placement and routing in 3D-integrated circuits,"In *IEEE Design and Test of Computers*, pp. 520-531, 2005.

[5] G. Karypis et al., "Multi-Level hypergraph partitioning: applications in VLSI design,"In *Proceedings of ACM/IEEE DAC*, pp. 526-529, 1997.

[6] M. Lin and A.E. Gamal, "Performance benefits of monolithically stacked 3D-FPGA,"In *IEEE Transactions on CAD*, pp. 216-229, 2007.

[7] C. Dong, C. Chilstedt and D. Chen, "Variation aware routing for three-dimensional FPGAS,"In International Symposium on VLSI, pp. 298-303, 2009.

[8] W.J. Cook, W.H. Cunningham, W.R. Pulleyblank, and A. Schrijver, Combinatorial optimization, Third Edition, Prentice Hall Pub., 2009.

A Scalable Threshold Logic Synthesis Method Using ZBDDs

Ashok kumar Palaniswamy
ECE Dept., Southern Illinois University,
Carbondale
Carbondale, IL-62901
ashok@engr.siu.edu

Spyros Tragoudas
ECE Dept., Southern Illinois University,
Carbondale
Carbondale, IL-62901
spyros@engr.siu.edu

ABSTRACT

A scalable synthesis method for large input threshold logic circuits using *Zero Suppressed Binary Decision Diagrams* is introduced. Existing synthesis methods require that a large input function must be initially decomposed using small input functions and this impacts the *synthesis cost*. The presented approach in this paper does not consider such restrictions. It is experimentally shown that the proposed method can synthesize the primary outputs of existing benchmarks without consulting the net-list, and the *synthesis cost* is significantly reduced over the existing methods.

Categories and Subject Descriptors: J.6.1 [COMPUTER-AIDED ENGINEERING]: Computer-aided design(CAD)

General Terms: Algorithms.

Keywords: Threshold Logic Gate, Threshold Networks.

1. INTRODUCTION

In this paper, we introduce an implicit scalable implementation of the existing non scalable synthesis methods for threshold logic. *Threshold Logic Gate(TLG)* is a special type of gate, which implements complex function with less number of gates. A *TLG* consists of n input variables and a threshold value T. Each input variable x_i is associated with a weight value w_i. The output of *TLG* is 1 only if $\sum_{i=1}^{n} w_i x_i \geq T$, otherwise it is 0. An n input function that can be implemented as a *TLG* is called as a *Threshold Function(TF)*. Nano devices are considered as the potential alternate for CMOS devices, where further scaling is not possible in near future [1]. Although *TLGs* were introduced in 1960s [2], they gained importance due to the recent developments in switching devices. *TLGs* are implementable with nano devices such as RTDs, MOBILE and neuMOS [3].

An n input function implemented by joining more than one *TLG* together is called a *Threshold Network(TN)*. Many synthesis techniques have been proposed for implementing circuits with *TLGs* [4–8]. Most of them are presented assuming an already synthesized net-list using CMOS gates

to represent large input function as a cluster of small input functions. This restricted synthesis approach has an impact on the *synthesis cost* (the total number of *TLGs* required to implement a function). The scalability of [4–6] is limited as it relies on *Linear Programming(LP)* for *TF* identification.

The methods in [7] [8] show how to substitute *LP* for *TF* identification by elegant algorithmic approaches. These methods operate on a CMOS circuit net-list where each node in the circuit is a small input function. Then each function is decomposed into a *TN*. The *TF* identification procedure in [7] is referred to as *Decomposition Method(DM)* in this paper. Two different approaches were proposed in [8]. The first one, referred to as the *MLFT*, is an improvement of the *DM*. The *MLFT* differs from the *DM* only in the aspect of decomposing a function into two sub-functions, which results in different *TN*. The *DM* and *MLFT* require the input function in *Complete Sum(CS)* [9] and they also require an appropriate *Variable Ordering(VO)*. Large input functions are binate functions, where the above tasks are more complex than for unate functions [2]. The second method of [8] uses *Binary Decision Diagram(BDD)* [10] to represent a function. It does not always find a *TF* unless it operates on the *VO* of *MLFT*, which unfortunately requires *CS* and impacts its scalability.

A recent method in [11] shows how to optimize the current mode framework of [12] so that *TLGs* with more than 50 inputs have much smaller delay than a network of CMOS gates. This motivated us to investigate methods that are able to identify whether a large input function can be implemented using a less number of *TLGs*, where each *TLG* may have more than 8 inputs which is the limitation of the methods in [7] [8]. The proposed method reduces the *synthesis cost* of *DM* and *MLFT* by working directly on the functionality of the *Primary Output(PO)* of the benchmark circuits without any dependence on input net-list synthesized for CMOS. This is illustrated by the following example.

Example 1: Consider a 5 input function $f = x_5(x_4+x_3+x_2) + x_4(x_2x_1+x_3(x_2+x_1)) + x_3x_2x_1$, which is synthesized in the net-list as $f = f_1 + f_2$, where $f_1 = x_4(x_2x_1+x_3(x_2+x_1))$ and $f_2 = x_5(x_4+x_3+x_2) + x_3x_2x_1$. By working on the net-list for f, *DM* and *MLFT* identify f_1 as a *TF*. However they do not identify f_2 as *TF* and decomposed it into a cluster of 2 *TFs* [7] [8]. Namely, $f_2 = f_3 + x_5(x_4+x_3+x_2)$, where $f_3 = x_3x_2x_1$ is a *TF*. At this point, f_1 and f_2 fail to satisfy *Function Containment* [7] to form f as a *TF*. Hence f is restructured as cluster of 3 *TLGs* such that $f = f_2 + x_4(x_2x_1 + x_3(x_2 + x_1))$. In contrast, the proposed method works directly on the functionality of f and identified it as a *TF*. This shows that the *synthesis cost* of both *DM* and *MLFT* are heavily influenced by the net-list representation.

This paper implements *DM* and *MLFT* implicitly using *Zero Suppressed Binary Decision Diagrams(ZBDDs)* [10] to overcome the *CS* and *VO*-related scalability problems, in order to decompose a function into *TN* on as-needed basis, and reduce the *TLG* count in the *TN*. This is referred to as the *Scalable Decomposition Method* (*SDM*). This paper is structured as follows. Section 2 presents the *SDM*. Section 3 presents experimental results. Section 4 concludes.

2. SCALABLE DECOMPOSITION METHOD

The *Complete Sum* (*CS*) is a reduced form of disjunction or union of all cubes which contains only the prime cubes and all the prime cubes. In *CS*, no cube includes any other cube and the consensus of any two cubes either does not exist or is contained in some other cube [9]. Consider an n input function f represented in *CS*, which is the union of $f_k (n \geq k \geq 1)$, where f_k is the union of all cubes of f having exactly k literals. The total number of occurrences of an input variable x_i in f_k is referred to as $f_k.x_i$. Two variables x_i and x_j of the f are said to have the same variable ordering(*VO*), if $f_k.x_i = f_k.x_j$ in all $f_k (n \geq k \geq 1)$ of f and this is denoted by $x_i \approx x_j$. If $f_k.x_i > f_k.x_j$ in f_k with the smallest value of k, then x_i is said to have higher order than x_j, and this denoted by $x_i > x_j$ [2].

The proposed *SDM* is an implicit scalable implementation of *DM* and *MLFT*. Since *DM* and *MLFT* insist on using *CS*, first it is shown how to store the *CS* of large input functions and how to obtain *VO* in a scalable manner using *CS* of function f by finding $f_k.x_i(n \geq k, i \geq 1)$ values and also the f is split into many sets of cubes $f_k(n \geq k \geq 1)$(*Section 2.1*). Then the decomposition tree is formed, where each node of the decomposition tree is a *ZBDD* and then the weight assignment for nodes in tree is done(*Section 2.2*).

2.1 Find VO and Splitting the Function

The input to *SDM* is an n input function represented in a *BDD*. Finding the *CS* of the given function amounts to the finding all the prime cubes of the given function [9]. The *CS* of n input function is generated and stored implicitly in a *ZBDD* as a set of cubes by using the approach presented in [13]. The *CS* of binate function contains both x_i and $\overline{x}_i$, and we stored them as distinct literals.

In *DM* and *MLFT*, the decomposition tree formation involves finding the divisor of the function at each node. This requires the *VO* of the node function f, which in turn requires the $f_k.x_i(n \geq k, i \geq 1)$ values of f. Additionally, *DM* decomposes the function f so that the right sub-function contains only the cubes in the f_k(least value on k) which contain the divisor. Algorithm 1 is an non enumerative *ZBDD* traversal algorithm, which calculates the $f_k.x_i(n \geq k, i \geq 1)$ values of an n input function f, and also splits the *ZBDD* of f into n *ZBDDs*, where each *ZBDD* is a $f_k(n \geq k \geq 1)$. The Algorithm 1 operation is explained with the example below.

Example 2: Consider the *CS* of an 6 input function $Y = \{x_6x_5, x_6x_4x_3, x_6x_4x_2, x_6x_4x_1, x_6x_3x_2, x_6x_3x_1, x_5x_4x_3, x_5x_4x_2x_1, x_5x_3x_2x_1\}$ represented as a set of cubes in a *ZBDD*. Figure 1 shows the operation of *ClassifyCubes* for the function Y. The *ZBDD* nodes are labeled in lexicographic order that reflects how they are visited.

Consider a node with the label A in the *ZBDD*. Let $A.f$ denotes the function at this node. Let field A_k be a *ZBDD*

Algorithm 1: *ClassifyCubes*

Input: P : *ZBDD* node.
Output: $\vec{P}$: Set of *ZBDDs* with occurrences value.
1 **if** $P.f = 1$ or $P.f = 0$ **then return** $\vec{P} = \phi$;
2 **else if** P *is Already Visited* **then return** $\vec{P}$;
3 **else**
4     $\vec{E} = $ ClassifyCubes($P.f_{\overline{x}_i}$);
5     $\vec{T} = $ ClassifyCubes($P.f_{x_i}$);
6     $P_1 = E_1$; $P_1.\vec{x} = E_1.\vec{x}$;
7   **if** $T.f = 1$ **then** $P_1 = P_1 \cup x_i$; $P_1.x_i = 1$;
8   **for** $k \leftarrow 2$ **to** n **do**
9        $P_k = E_k \cup (T_{k-1} * x_i)$;
10   **for** $j \leftarrow 1$ **to** n **do**
11   **if** $x_j = x_i$ **then** $P_k.x_j = |T_{k-1}|$;
12   **else** $P_k.x_j = E_k.x_j + T_{k-1}.x_j$;
13   **return** $\vec{P}$;

---

which contains all the cubes with exactly k number of literals of $A.f$. Clearly, $A.f = \cup_{k=1}^n A_k$. The total number of occurrences of an input variable x_i in a A_k is represented as $A_k.x_i$. For simplicity in notation, let $\vec{A} = A_k(n \geq k \geq 1)$ and the total number of occurrences of all input variables $x_i(n \geq i \geq 1)$ in a A_k be denoted as $A_k.\vec{x}$. For the function Y, the variable representing node A is x_6, for nodes I and B is x_5, for nodes J and C is x_4, for nodes M, K, H, D is x_3, for nodes L and F is x_2, and for node G is x_1. $A.f_{x_6}$ and $A.f_{\overline{x}_6}$ are the *if* and *else* nodes of A. All non-empty $P_k(n \geq k \geq 1)$ of each *ZBDD* node P of the function Y are shown in Figure 1.

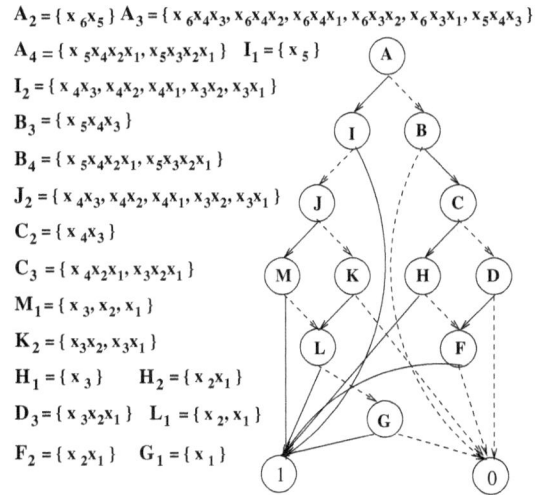

$A_2 = \{ x_6x_5 \}$ $A_3 = \{ x_6x_4x_3, x_6x_4x_2, x_6x_4x_1, x_6x_3x_2, x_6x_3x_1, x_5x_4x_3 \}$

$A_4 = \{ x_5x_4x_2x_1, x_5x_3x_2x_1 \}$ $I_1 = \{ x_5 \}$

$I_2 = \{ x_4x_3, x_4x_2, x_4x_1, x_3x_2, x_3x_1 \}$

$B_3 = \{ x_5x_4x_3 \}$

$B_4 = \{ x_5x_4x_2x_1, x_5x_3x_2x_1 \}$

$J_2 = \{ x_4x_3, x_4x_2, x_4x_1, x_3x_2, x_3x_1 \}$

$C_2 = \{ x_4x_3 \}$

$C_3 = \{ x_4x_2x_1, x_3x_2x_1 \}$

$M_1 = \{ x_3, x_2, x_1 \}$

$K_2 = \{ x_3x_2, x_3x_1 \}$

$H_1 = \{ x_3 \}$ $H_2 = \{ x_2x_1 \}$

$D_3 = \{ x_3x_2x_1 \}$ $L_1 = \{ x_2, x_1 \}$

$F_2 = \{ x_2x_1 \}$ $G_1 = \{ x_1 \}$

Figure 1: *ClassifyCubes* Example

The node traversal starts at the topmost pointer node(A) and traverses on *else* path, until it reaches the *Constant* 1 or 0 node. For node G, the *if* and *else* nodes are 1 and 0 respectively. Hence line 7 gives G_1 with $G_1.\vec{x} = (0,0,0,0,0,1)$. Similarly, we get F_2 with $F_2.\vec{x} = (0,0,0,0,1,1)$ and D_3 with $D_3.\vec{x} = (0,0,0,1,1,1)$. All others entries of type P_k for the nodes G, F and D are *empty*. For node H, the *else* path node is F, which is already visited(*line* 2) and gets E_2(*line* 4) and its *if* path is *1*. Hence node H gets H_1 with $H_1 . \overrightarrow{x} = (0, 0, 0, 1, 0, 0)$ and H_2 with $H_2 . \overrightarrow{x} = (0, 0, 0, 0, 1, 1)$. Then node C gets C_2 with $C_2 . \overrightarrow{x} = (0, 0, 1, 1, 0, 0)$ and C_3 with $C_3 . \overrightarrow{x} = (0, 0, 1, 1, 2, 2)$ (*line* 6 − 14).

For node B, the *if* and *else* path nodes are C and *0*. Hence it gets B_3 with $B_3 . \overrightarrow{x} = (0, 1, 1, 1, 0, 0)$ and B_4 with $B_4 . \overrightarrow{x} = (0, 2, 1, 1, 2, 2)$. That way all *ZBDD* nodes in the *else* path of node A are visited and their cube set fields are updated. Now *if* path of node A is taken for traversal and all the *ZBDD* nodes are visited and updated as explained above. Finally, node A has A_2 with $A_2 . \overrightarrow{x} = (1, 1, 0, 0, 0, 0)$ and A_3 with $A_3 . \overrightarrow{x} = (5, 1, 4, 4, 2, 2)$ and A_4 with $A_4 . \overrightarrow{x} = (0, 2, 1, 1, 2, 2)$. Thus the given function Y in a *ZBDD* is split and stored in $\overrightarrow{A}$ of top most *ZBDD* node A.

Since the calculation of number of occurrences of all input variables starts from *Constant 1* node and updated in bottom to top fashion, the value of $|T_{k-1}|$(*line* 11) can be calculated without storing the *ZBDD* in T_{k-1}. Hence *ClassifyCubes* can be used only with $P_k . \overrightarrow{x}$ without storing the *ZBDD* in the P_k field to determine the *VO*.

2.2 The ZBDD- based Decomposition Tree

In *MLFT*, both right and left sub-node functions are Shannon co-factors of the parent function f. However in *DM*, each right sub-node function is always the subset of f_k(least k value). In order to avoid finding the lowest f_k for each node function f of the decomposition tree, the initial function is split using *ClassifyCubes*(by storing the *ZBDD* in P_k) before forming decomposition tree of *DM*.

Algorithm 2 forms the decomposition tree of *DM* using *ZBDDs*. Only the non-empty f_ks of the function f by *ClassifyCubes* is the input to the *ZddTreeForm*. *ClassifyCubes* splits the function Y in example 2 into three cubes set(Y_2, Y_3, Y_4) along with total number of occurrences. The operation of *ZddTreeForm* for the function Y is explained using the Figure 2. The nodes in the decomposition tree are labeled in lexicographic order that reflects how they are formed. For each node the cubes inside the *square brackets* refers to the initial condition of the node. The cubes inside the *parenthesis* were updated after the decomposition of function using available cubes sets.

Algorithm 2: *ZddTreeForm*

Input: $\overrightarrow{S}$: Sets of *ZBDDs*, n : No of input variables.
Output: $\overrightarrow{T}$: Cluster of *ZBDD* nodes, i: No of nodes.
1 **for** $k \leftarrow 1$ **to** n **do**
2   **if** $T_j = 0$ **then** $T_j = S_k$; $T_j . \overrightarrow{x} = S_k . \overrightarrow{x}$;
3   **else** $T_j . \overrightarrow{x} =$ ClassifyCubes(T_j) ;
4     $div =$ ChoseDivisor($T_j . \overrightarrow{x}$) ;
       `//`$T_r(T_l)$ `right(left) sub-node of` T_j
5   ChildForm($\overrightarrow{T}$. j, i, div) ;
6     $[\overrightarrow{T}, i] =$ RecChildForm($\overrightarrow{T}$, r, i) ;
7     $j = l$;
8 **return** $[\overrightarrow{T}, i]$;

In Figure 2, initially node A is *0*, which is shown as [0] in node A. Hence the Y_2 is removed from Y and stored in node A(*line* 2). Procedure *ChoseDivisor* chose the input variable, which has highest value in the $Y_2 . \overrightarrow{x}$ as the divisor(*line* 4). The *VO* for the node A using $Y_2 . \overrightarrow{x}$ and $Y_3 . \overrightarrow{x}$ is $x_6 > x_5$.

Hence Procedure *ChildForm* forms the left and right sub-nodes of a parent node by Shannon decomposition, which gives nodes B and C as right and left sub-nodes for node A(*line* 5) with x_6 as the divisor. Now the node A is updated by performing union operation with remaining cube sets of Y, which is shown as (Y_3, Y_4) in node A. The value i is the count of total nodes in the decomposition tree. Now node B is taken by *RecChildForm* to form all the sub-nodes(*line* 6).

RecChildForm recursively decomposes the newly formed sub-nodes using *ChildForm* until the sub-node function is a *basic*. The nodes with the functionality of *Or*, *And*, *0*, and *1* are referred to as *basic* nodes. Here node B is a *basic* and the decomposition is moved to the next reference node C(*line* 7). Node C is initially *0*, hence Y_3 is removed from Y and added to the node C(*line* 2). The *VO* for node C using $Y_3 . \overrightarrow{x}$ is $x_6 > x_4 \approx x_3 > x_2 \approx x_1 > x_5$. Hence *ChildForm* gives nodes D and E as right and left sub-nodes with x_6 as the divisor. Then the node C is updated by the union of Y_4. Node D is not a *basic* and goes on recursive decomposition to form all its sub-nodes by *RecChildForm*, which form the nodes F, G, H, and I.

Then the next node E is taken for decomposition, which is not empty. Hence *ClassifyCubes* is used to find *VO* of node E(*line* 3). For node E, *ChildForm* forms nodes J and K as its sub-nodes with x_5 as divisor. Now node E is updated by union of Y_4. Since node J is a *basic*, the decomposition is moved to the node K. Node K is *0* initially, hence Y_4(last available cube set of Y) is added to it. Node K forms nodes L and M as its sub-nodes by *ChildForm*. Node L forms nodes N, O, P, and Q as its sub-nodes by *RecChildForm*. Thus the decomposition tree of function Y is formed by *ZddTreeForm*.

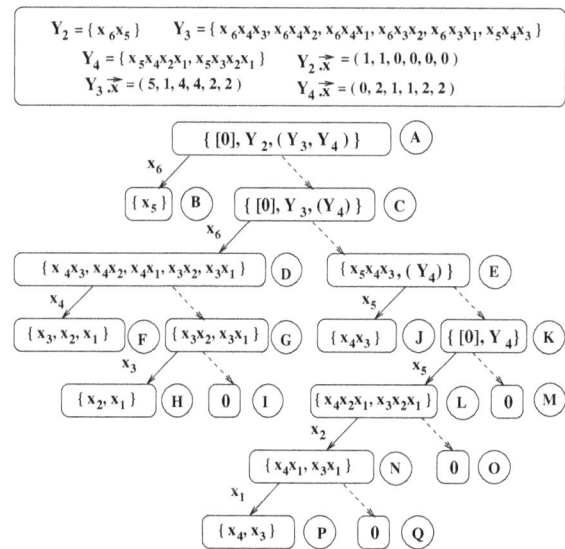

Figure 2: *ZddTreeForm* **Example**

Except the nodes A, C, and K, all other nodes use *ClassifyCubes* to find *VO*. In *ZddTreeForm*, *ClassifyCubes* is used without storing the *ZBDD* in P_k field. In *DM*, each right sub-node is the function of same size cubes and all its sub-nodes are subsets of that cubes set. This shows the advantage of using *ClassifyCubes* to form sets of cubes in the initial stage of the *SDM* for *DM*. However this is not needed in the *MLFT*, where the *decomposition tree* is formed by using only *ClassifyCubes* and *RecChildForm*.

Each node of the decomposition tree formed by *SDM* is a *ZBDD*. Since weight assignment on the decomposition tree by *DM* is bottom to top approach. Checking the weight assignment conditions of *DM* [7] and weight assignment on these nodes using *ZBDDs* are simple and straight forward. The weight assignment in the *MLFT* is same as the *DM* with additional optimization conditions [8].

3. EXPERIMENTAL RESULTS

The proposed method is implemented in the C language. Experiments were performed on a 750MHz SunBlade 1000 workstation with 1 Gb RAM on the ISCAS-85 combinational circuits [14]. In order to compare the efficiency of *SDM*, we implemented the *DM* in [7] for which is more challenging than *MLFT*. Experimental results listed for this method, assume an initial decomposition of clusters with up to 8 inputs. We have experimentally observed that we cannot store the *CS* and find $f_k.x_i(n \geq k, i \geq 1)$ values of functions that have more than 8 inputs. This emphasis the importance and necessity of *SDM* to handle large input functions.

In *SDM*, each primary output(*PO*) of the circuit is represented as a single n input function and the results obtained by them are presented. Let n denotes the *number of literals* of function, i.e, x_i and $\overline{x}_i$ are stored as distinct variables. Due to the memory constraints, only the functions(*POs* of Benchmark circuits) with up to 78 *literals* are listed in this experiment. The listed functions have up to 50 input variables because they are unate on some input variables. This is clearly a significant boost in the *TLG* synthesis scalability.

Table 1 elaborates on the efficiency of the *SDM* and compares it with *DM* and the time performance of *SDM*. The *PO* functions of all benchmark circuits are categorized into 5 groups based on the *number of literals* as shown in column 1. Column 2 shows the *Total number of PO functions* in each PO functions group. The average number of *TLGs* required per *PO* function by *DM* and *SDM* are shown in column 3 and 4 respectively. The percent decrease(P_D) in the average number of *TLGs* by the *SDM* in comparison with *DM* is also shown in column 5. On average, *SDM* results in P_D as 20 and also can be as big as 40. It also indicates the difference in *TLG* count between *DM* and *SDM* increases as the number of inputs of the function increases. These results clearly show the benefits of implementing *DM* implicitly.

No of Lits	PO	DM	SDM	P_D	T_{SDM}	% T_{VO}	% T_{ZT}
$3-10$	70	2	2	0.00	0.046	21.6	75.4
$11-25$	27	28	17	39.3	1.163	28.7	66.4
$26-40$	17	97	66	31.9	7.537	22.9	64.5
$41-60$	42	306	229	25.2	27.741	22.7	66.1
$61-78$	51	572	464	18.9	39.186	27.7	57.6

Table 1: Efficiency and Time Performance of SDM

In order to demonstrate the time performance of *SDM*, the average execution time taken to form the *TN* for each PO functions group is calculated, which given as the ratio of *Total Execution Time to form TN for all PO functions* to the *Total number of PO functions*. Column 6 shows the *average total execution time(T_{SDM})* taken to form the *TN*

of a *PO* function for each group. This shows the quickness of the *SDM* on large input functions. Column 6 and 7 shows the *percentage of execution time* taken for find *VO and Split*(T_{VO}) and *ZddTreeForm*(T_{ZT}) separately. This shows that more than 88% of the total execution time of *SDM* is taken for the decomposition tree formation($T_{VO} + T_{ZT}$). It also shows that approximately 25% of the execution time is utilized only for T_{VO}. The time for weight assignment is always less than T_{VO} and linear to the number of nodes.

4. CONCLUSION

In this paper, an implicit scalable implementation for existing methods to implement network of *TLGs* has been introduced. It is shown experimentally that percent decrease in *synthesis cost* is up to 40 and on average around 20 when working directly on the *PO* functionality instead of the existing methods that rely on pre-synthesized net-list. The proposed method can handle up to functions of up to 50 inputs without decomposition, where as existing methods can only handle up to 8 input functions.

5. ACKNOWLEDGMENTS

This research has been supported in part by the NSF under Grant No. 0702628 and the NSF I/UCRC for Embedded Systems at SIUC under Grant No. 0856039. Any opinions, findings, and conclusions or recommendations expressed in this material are those of the author(s) and do not necessarily reflect the views of the National Science Foundation.

6. REFERENCES

[1] International Technology Roadmap for Semiconductors,(2009) http://www.itrs.net.

[2] S. Muroga, "Threshold logic and its applications," John Wiley and Sons Inc. NY, USA, 1971.

[3] V. Beiu, J. M. Quintana, M. J. Avedilo, and R. Andonie, "Differential implementations of threshold logic gates," In Proc. International Symposium on SCS, vol. 2, pp. 489-492, Jul 2003.

[4] R. Zhang, P. Gupta, L. Zhong, and N. K. Jha, "Threshold network synthesis and optimization and its application to nanotechnologies," IEEE Transactions on CAD of IC and Systems, vol. 24, no. 1, pp. 107-118, Jan 2005.

[5] J. L. Subirats, J. M. Jerez, and L. Franco, "A new decomposition algorithm for threshold synthesis and generalization of boolean functions," IEEE Transactions on Circuits & Systems I, vol. 55, no. 10, pp. 3188-3196, Nov 2008.

[6] M. K. Goparaju, A. K. Palaniswamy, and S. Tragoudas, "A fault tolerance aware synthesis methodology for threshold logic gate networks," In Proc. IEEE Symposium on Defect and Fault Tolerance of VLSI Systems, pp. 176-183, Oct 2008.

[7] T. Gowda and S. Vrudhula, "Decomposition based approach for synthesis of multi-level threshold logic circuits," In Proc. ASP-Design Automation Conference, pp. 125-130, Mar 2008.

[8] T. Gowda, S. Vrudhula, N. Kulkarni, and K. Berezowski, "Identification of threshold functions and synthesis of threshold networks," IEEE Transactions on Computer-Aided Design of IC and Systems, vol. 30, no. 5, pp. 665-677, May 2011.

[9] Z. Kohavi, "Switching and finite automata theory," McGraw Hill, 1990.

[10] K. S. Brace, R. L. Rudell, and R. E. Bryant, "Efficient Implementation of a BDD Package," In Proc. 27th IEEE Design Automation Conference, pp. 40-45, 1990.

[11] C. Dara, S. Tragoudas, and T. Haniotakis, "Delay analysis for current mode threshold gates," Personal Communication with authors(unpublished manuscript).

[12] S. Bobba and I. N. Hajj, "Current mode threshold logic gates," In Proc. Intl Conference Computer Design, pp. 235-240, 2000.

[13] O. Coudert, J. C. Madre, H. Fraisse, and H. Touati, "Implicit prime cover computation: An overview," In Proc. Synthesis And SImulation Meet, 1993.

[14] Mcnc Benchmark Circuits, http://www.cbl.ncsu.edu/16080 /benchmarks/ISCAS85.

A Memristor-based TCAM (Ternary Content Addressable Memory) Cell: Design and Evaluation

Pilin Junsangsri and Fabrizio Lombardi
Department of Electrical and Computer Engineering
Northeastern University
Boston, MA 02115 USA

(1) 617-373-4854

junsangsri.p@husky.neu.edu, lombardi@ece.neu.edu

ABSTRACT

This paper presents a Ternary Content Addressable Memory (TCAM) cell that employs memristors as storage element. The TCAM cell requires two memristors in series to perform the traditional memory operations (read and write) as well as the search and matching operations for TCAM; this memory cell is analyzed with respect to different features (such as transistor sizing and voltage threshold) of the memristors to process fast and efficiently the ternary data. A comprehensive simulation based assessment of this cell is pursued by HSPICE.

Categories and Subject Descriptors

B.7.1 [**Integrated Circuits**]: Types and Design Styles – *Advanced Technologies, Memory Technologies, VLSI (Very Large Scale Integration).*

General Terms

Design

Keywords

Memory Cell, Modeling, Memristor, Ternary Content Addressable Memory, TCAM

1. INTRODUCTION

Ternary Content Addressable Memories (TCAMs) are widely employed for realizing networking circuits for address classification and packet filtering [1]. High-performance network routers require fast and high capacity TCAMs for improved look-up performance of routing tables. As network routers utilize network processors and memories (TCAMs and SRAMs), system performance strongly depends on searching throughput and storage capacity. However, the fabrication and design of TCAM chips with a large storage capacity have encountered substantial problems in CMOS. A significant problem is related to the design itself; ternary logic requires additional supply voltages if implemented by CMOS circuitry. Without the use of additional power rails, the size of a CMOS ternary memory cell is increased as two binary cells (i.e. at least 12 transistors) and additional transistors for the comparison are required. While a TCAM cell with a reduced transistor count

has been reported [2], stability problems are encountered if this cell is used in large storage chips. Also, a TCAM cell consumes significant power due to its operation. Power consumption may degrade chip reliability and affect packaging costs due to heat dissipation hardware. Although many approaches addressing power consumption have been reported [3-5], they result in circuit techniques that have substantial area overhead or deficiencies in noise immunity. Therefore, the high speed of a TCAM comes at a cost of increased silicon area and power consumption, two design parameters that designers strive to reduce. As TCAM applications grow (with an ever increasing demand on large storage), the power problem is further exacerbated. A reduction in power consumption without sacrificing speed or area is one of the main threads of recent research in large capacity memory design for TCAMs [6].

This paper proposes a new design for a ternary CAM (TCAM) that utilizes both MOSFETs and memristors. The TCAM cell is analyzed with respect to different features (such as transistor sizing and voltage threshold) of the memristors to process fast and efficiently the ternary data; a comprehensive simulation based assessment of this cell is pursued by HSPICE.

2. REVIEW OF MEMRISTOR

In circuit theory, the memristor (or memory resistor) is the 4th fundamental element that utilizes for its operation the relationship between flux and electric charge. This element was postulated by Chua in 1971 [8] based on the concept of symmetry with other circuit elements, such as the resistor, inductor and capacitor. However, it remained of theoretical interest for more than 30 years till HP Labs provided its physical implementation [9] based on a nano-scale thin film of titanium dioxide for its fabrication. The relationship between the flux and the electric charge of a memristor is given by [7]

$$d\varphi = M * dq \qquad (1)$$

where M is the memristance or memristor value (in Ω), φ is the flux through the magnetic field, and q is the electric charge. A memristor operates as a *variable resistor* whose value depends on the current or voltage across it, i.e. if there is a positive voltage across the memristor, its memristance will reduce to a small value (given by R_{ON}); if there is a negative voltage across the memristor, its memristance will increase up to a high value (given by R_{OFF}). Hereafter, the memristor is considered as a switching resistance device; as applicable to the HP Labs implementation [9], the rate of change for the memristance is usually linear provided its value is not close to the extreme values (R_{ON} and R_{OFF}). If the memristance value is close to the extreme values (R_{ON} or R_{OFF}), non-linearity is likely to occur for its rate of change. To model the characteristics of a memristor, different HSPICE models have been proposed in the technical literature [10-12]. In this paper, the memristor model of [10] is used, because it has been shown to closely resemble the HP Labs memristor parameters and operation [8].

3. PROPOSED TCAM CELL

A Content Addressable Memory (CAM) is a fully associative memory which can be classified into two types, binary CAM and ternary CAM (TCAM). A Binary CAM is primarily used as instruction or data cache, while a ternary CAM is mostly used for specific application tasks such as longest prefix matching task in network search engines. In TCAM arrays, the search operations are performed by comparing in parallel the input (searched) data against the entire list of entries stored in memory. The memory cells in a TCAM can store three states (i.e. '1', '0', '2'), while cells in a binary CAM store only two states (i.e. '0' and '1'). The additional state '2' is also referred to as the "mask" or "don't care" state; it is used for matching to either a '0' or '1' in the input search data process. Hence, a binary CAM is suitable for applications that require an exact match between the input data and the stored data; a TCAM is used for applications that allow both exact and partial matches.

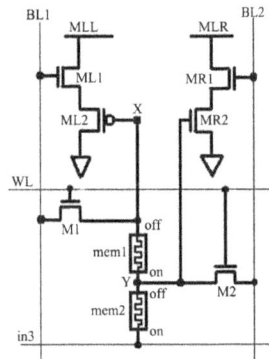

Figure 1. Proposed TCAM design using memristors

Due to its non-volatile characteristic, the memristor can be used as storage device. A Ternary Content Addressable Memory (TCAM) cell is proposed as shown in Figure 1. The three states of TCAM are defined using 2 memristors as follows.

For state '0', both memristors must be fully biased to the R_{OFF} state. The memristance range between R_{ON} and R_{OFF} is assumed to be large (as experimentally found in [9]), so if both memristors are in the R_{OFF} state, then total resistance of the memory cell is $2R_{OFF}$.

For state '1', both memristors must be in the R_{ON} state. So, the total resistance of the memory cell is $2R_{ON}$ i.e. a very low value compared with the resistance in state '0'.

For state '2' (i.e. the don't care state), one memristor must be in the R_{ON} state, while the other memristor must be in the R_{OFF} state. Therefore, the total resistance of the TCAM cell is $R_{OFF} + R_{ON}$.

A. Write Operation

The proposed TCAM cell has two memristors connected in series. The write operation consists of two distinct halves. In Figure 1 the write line (WL) is high during the write operation, data is provided through bit line 1 (BL1), bit line 2 (BL2), and input 3 line (in3) as follows.

For writing a '0', both memristors have to be in the R_{OFF} state. Then, WL must be enabled (ON or high), while BL1 and in3 are

low and high, respectively. During the first half of the write operation, BL2 is low and therefore mem2 is in the R_{OFF} state. In the second half of the write operation, BL2 is high for mem1 to be in the R_{OFF} state. Hence at completion of this process, both memristors are in the R_{OFF} state.

For writing a '1', both memristors must be in the R_{ON} state. This is similar to the write '0' operation; so, WL is high. BL1 is also high, while in3 is low. BL2 is low during the first half of the write operation, so that mem1 is in the R_{ON} state. During the second half of the write operation, BL2 is high such that mem2 is in the R_{ON} state also. Hence, both memristors are in the R_{ON} state.

For writing a '2', one memristor must be in the R_{ON} state, while the other memristor must be in the R_{OFF} state. So the write line is high, while BL1, BL2, in3 are low, high and low respectively. Therefore, mem1 is in the R_{OFF} state and mem2 is in the R_{ON} state.

B. Search Operation

The search operation in a TCAM cell checks whether there is a match between the searched (provided as input) and stored data. Two match lines (MLL and MLR) are used (Figure 1); these two lines are shown as separate lines to better understand the operations of the proposed TCAM cell and the discharge process (in practice these two lines can be combined into a single line). The search operation starts by precharging the voltage on MLL and MLR to high. Then, the searched data is input through BL1 and BL2. An input is provided at in3 to compare the data stored in the TCAM cell with the searched data. If the data stored in the TCAM cell is equal (matched) to the searched data, the match line is discharged. Else (no match), its voltage is kept unaltered.

Search '0': MLs must be precharged to V_{DD} prior to starting the search operation. For the search '0' operation, BL1 and BL2 are high and low respectively, i.e. ML1 is ON and MR1 is OFF. Then, the data input is placed through in3 to check the state of the TCAM cell. If the state of the TCAM cell is matched with the searched data (i.e. the TCAM state is '0' or '2'), MLL will be discharged. However, if the TCAM cell state is '1' (no match with the searched '0'), MLL remains the same and MLR is not affected by the search '0' operation.

The search operation can be better understood by considering Figure 1; each memristor is fully biased to its required state (R_{ON} or R_{OFF}). When a memristor is in the R_{ON} state, the voltage drop across it has a very low value (especially when compared with the R_{OFF} state). So, consider in Figure 1 the scenario when mem1 is in the R_{OFF} state, and mem2 is in the R_{ON} state. During the search operation, a high voltage (V_{DD}) is applied from in3 to both mem1 and mem2. Since mem2 is in the R_{ON} state, the voltage drop across mem2 has a low value and the voltage at node Y (V_Y) is approximately equal to V_{DD}. However, mem1 has a very high value (19kΩ in this case), so the voltage at node X (V_X) will slightly increase. As the search time is fast, this increase is not significant. By assuming that the search '0' operation is performed, the following cases can be distinguished. (1) If the TCAM cell is in state '0', both memristors must be in the R_{OFF} state, V_X and V_Y are very low (i.e. ML2 is ON and MR2 is OFF). For the search '0' operation, ML1 is ON and MR1 is OFF, a direct path exists from MLL to GND and MLL is discharged. (2) For state '2', mem1 must be in the R_{OFF} state and mem2 must be in the R_{ON} state, so V_X and V_Y are low and high respectively. Then, ML2 and MR2 are ON, a direct path from V_{DD} to GND exists via ML1 and ML2; MLL is discharged. (3) For state '1', both memristors must be in the R_{ON} state; so during the search operation, V_X and V_Y are high. Then, ML2 and MR2 are OFF and ON respectively. As MR1 is

OFF, there is no direct path from V_{DD} to GND; MLL and MLR retain their values, as result of the no-match.

Search '1': For the search '1' operation, BL1 and BL2 are low and high respectively. So, ML1 is OFF and MR1 is ON. As mentioned previously, if the TCAM state is matched with the searched data, ML is discharged, else no change will occur. (A) If the TCAM cell is in state '0', both memristors are in the R_{OFF} state; therefore, V_X and V_Y are very low (ML2 is ON and MR2 is OFF), then there is no direct path from V_{DD} to GND (i.e. no match is found). (B) If the TCAM cell is in state '1' or '2', mem2 is in the R_{ON} state and V_Y is high. As MR2 is ON, there is a direct path from MLR to GND, thus causing MLR to discharge.

Search '2': For the search '2' operation, the result is always a match because it is the "don't care" state. So, both BL1 and BL2 are high and ML1 and MR1 are ON. A direct path exists from MLL and MLR to GND, thus always resulting in a match.

4. SIMULATION RESULTS

In this section, the performance evaluation of the TCAM cell of Figure 1 is presented using HSPICE at 32nm CMOS technology. The model of [10] is employed for the memristor with a memristance range of 100-19kΩ.

A. Write Time

The write time is the time for the memristor to be in the desired state. By setting the voltage across the memristor to a constant value (equal to 0.9 V), it has been found that the time for fully biasing a single memristor to its state is approximately 200 ns. To fully charge both memristors, the write time can be estimated as follows (under the assumption that the voltage drop across M1 or M2 is given by 0.45V).

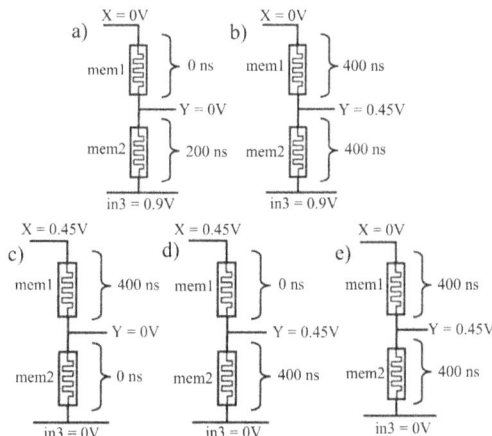

Figure 2. a) First Step of Write '0' Operation b) Second Step of Write '0' Operation c) First Step of Write '1' d) Second Step of Write '1' e) Write '2' Operation

Write '0' Operation: For the write '0' operation, both memristors must be in the R_{OFF} state. The voltages of the TCAM cell of Figure 1 when the write '0' operation is performed are shown in Figures 2(a) and 2(b). There are two steps for writing a '0': the first step is used to bias mem2 to R_{OFF}, and the second step is used to bias mem1 to R_{OFF}. Based on the Write '0' operation of Figures 2(a) and 2(b), the time required for both memristors to be in the correct states is given by 600 ns (200 ns for the first step and 400 ns for the second step). However if the voltage drops across a transistor is reduced (V_Y is increased to 0.72V), the time for the write '0' operation is faster. The first step of the write '0' operation takes 200ns, while the second step takes 280ns for writing to mem2 and mem1; so, the total time of the write '0' operation is nearly 480ns.

Write '1' Operation: Similarly to the write '0' operation, there are also two steps for the write '1' operation. The first step is used to bias mem1 to R_{ON}, while the second step is used to bias mem2 to R_{ON}. Due to the voltage drops across M1 and M2 in Figure 1, when BL1 and BL2 are both high, V_X and V_Y drop to 0.45V. The time for the write '1' operation is 800ns (400ns for the first step and 400ns for the second step). To reduce the time for the write '1' operation, the voltage drop across M1 and M2 must be reduced, i.e. V_X and V_Y are both equal to 0.72V, then the time is 280ns for each step, the total write time of this process is 560ns.

Write '2' Operation: For the write '2' operation, mem1 must be in the R_{OFF} state, while mem2 must be in the R_{ON} state. Since there is a voltage drop across M2, V_Y is equal to 0.45V. The time for the write 2 operation is 400ns. However if the voltage drop across M2 decreased, V_Y is increased. If V_Y is equal to 0.72V, the write time of this operation is 280ns.

From the above discussion, if the voltage drops across M1 and M2 are both 0.45V, the write time (T_W) of the proposed TCAM cell is at most 800ns. However, as the voltage drop is equal to 0.18V (0.9-0.72), then the write time is 560 ns.

B. Threshold Voltage Selection

Consider the threshold voltage of ML2, as related to V_X in Figure 1. Simulation has shown that when the TCAM cell is in state '1', both memristors must be in the R_{ON} state, i.e. V_X is nearly equal to V_{DD}. When the TCAM cell is in state '0' or '2' (the total memristance of these states is very high), V_X is just slightly higher than 0V. For selecting the threshold voltage of ML2, consider V_X during a search operation. V_X slightly increases when mem1 is in the R_{OFF} state (the TCAM cell is in state '0' or '2'). It is equal to 0.899V if mem1 is in the R_{ON} state (i.e. the TCAM cell is in state '1'). In the proposed design, the threshold voltage of ML2 is set to 0.735V because during the search operation the increase of V_X from 0 to 0.735V is sufficiently large to allow the match line to discharge. For the threshold voltage of MR2, MLR is discharged if '1' or '2' is searched in the TCAM cell. If the data in the TCAM cell is '1' or '2', mem2 is in the R_{ON} state; then during the search operation, V_Y is approximately equal to V_{DD} (0.899V in this case). So, the threshold voltage of MR2 can be selected to be any value lower than the supply voltage; hereafter the threshold voltage of MR2 is selected to be 0.735V to allow the match line to discharge as ML2. If the search '1' or '2' operation occurs, MLR is then discharged.

C. Search Operation

The search operation for the TCAM cell of Figure 1 is simulated. By using HSPICE, the simulation results show that if the data in the TCAM cell is matched with the input (searched) data, then correctly a match line (MLL or MLR) will be discharged, else, the match lines will be keeping the values unchanged.

Consider next the search time; the search time depends on the discharging rate of the match lines. However, as mentioned previously, during the search operation, if the TCAM cell is in state '0' or '2', V_X will increase from 0V to V_{DD} at a rate dependent on the memristance range. Consider state '2' of the TCAM cell during a search '0' operation. In the search '0' operation, BL1 is high and BL2 is low; also, ML1 is ON. To discharge MLL as outcome of the match operation, ML2 must be ON also. In Figure 1, ML2 (PMOS) is ON only when V_X is less than the threshold voltage. So, the match result of this search operation is accurate if and only if the time required for V_X to increase from 0 to the threshold voltage of ML2 is less than the time for the match line to be discharged to GND. By using the setting as mentioned previously, the simulation results show that the search time of the proposed TCAM cell is at most 8ns (i.e. less than 12ns required for reaching the threshold level of a memristor [7]). So during the search operation, memristors keep their states.

D. Transistor Sizing

The design of the proposed TCAM cell has been evaluated using different feature sizes. The simulation results are shown in Table 1 for different technology scaling. At the same supply voltage, the write time at 32nm is less than at 45 and 65nm; moreover when the supply voltage is increased, the write time of the proposed TCAM cell decreases. Since the voltage across a memristor is increased, the rate of change in memristance is also faster. Therefore, the write time at a higher supply voltage is faster.

Table 1. Comparison between transistor size and write time for a memristor range of $100 - 19k\Omega$

Technology	V_{DD} (V)	Writ time (ns)	Search time (ns)
32nm	1	270	6.08
	1.1	210	6.28
45nm	1	300	6.70
	1.1	230	7.20
65nm	1	330	6.81
	1.1	270	7.30

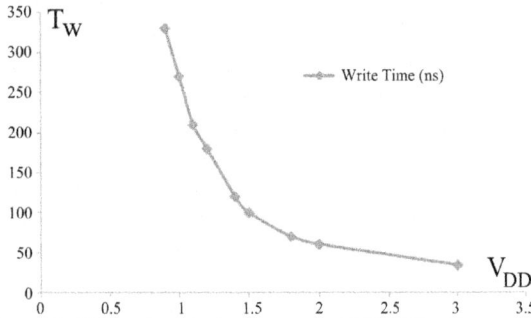

Figure 3. Write time (ns) of proposed TCAM cell vs supply voltage at 32nm technology

As for the search time, the simulation results in Table 1 show that the search time at 32nm is less than at 45 and 65nm; however at the same scaling, the simulation results show that the search time is slower at a higher supply voltage. This occurs because the match lines will take longer to completely discharge. However as shown in Table 1, the search time is less than the time for reaching the threshold level (12ns [7]), so a state change cannot occur.

The simulation results in Table 1 and Figure 3 show that the write time of the proposed TCAM cell at 32nm is faster. Moreover at a lower supply voltage, the rate of change of the memristors is very low if the supply voltage is reduced at low scaling (the supply voltage of 32nm is less than for 45nm and 65nm).

E. CAM Comparison

In this section, the TCAM cell (Figure 1) is operated as CAM by fixing mem1 to R_{ON} and varying mem2 (as dependent on the stored value, i.e. state '0' or '1'). To operate as a CAM, the voltage at BL1 in Figure 1 is fixed to V_{DD}, while the voltages at BL2 and in3 are dependent on the store data i.e. BL2 is at V_{DD} and in3 is at GND to store the data as R_{ON} (State '1'), vice versa for State '0'. The CAM cell operates using the same search operation as described previously for TCAM. Table 2 shows the comparison of the proposed CAM design with the MCAM of [7]. At the same supply voltage (0.9 V), the simulation results of Table 2 show that the proposed CAM by employing the TCAM design is better than

the MCAM of [7] in terms of number of supply voltages, number of transistors and write and search times. The write time of [7] requires the supply voltage rail of $\frac{V_{DD}}{2}$ to write data into the memory cell. The proposed cell requires V_{DD} only, so the write time of the proposed cell is faster than [7]. As the number of transistor in [7] is higher, also the cell of [7] will have a higher power dissipation than the proposed cell.

Table 2. Comparison between CAM of [7] and proposed CAM design at same V_{DD} (0.9V)

Metric	CAM [7]	Proposed CAM
Supply Voltage	$V_{DD}, \frac{V_{DD}}{2}, 0$	$V_{DD}, 0$
# of Transistors	7	6
Write Time (ns)	300	150
Search Time (ns)	12	8

5. REFERENCES

[1] H. Noda, K. Inoue, M. Kuroiwa, "A cost-efficient high-performance dynamic TCAM with pipelined hierarchical searching and shift redundancy architecture," *IEEE J. Solid-State Circuits*. vol. 40, no. 1, pp. 245-253, Jan. 2005.

[2] I. Arsovski, T. Chandler, and A. Sheikholeslami, "A ternary content-addressable memory (TCAM) based on 4T static storage and including a current-race sensing scheme," *IEEE J. Solid-State Circuits*, vol. 38, no. 1, pp. 155–158, Jan. 2003.

[3] I. Arsovski and A. Sheikholeslami, "A mismatch-dependent power allocation technique for match-line sensing in content-addressable memories," *IEEE J. Solid-State Circuits*, vol. 38, no. 11, pp. 1958–1966, Nov. 2003.

[4] K. Pagiamtzis and A. Sheikholeslami, "Pipelined match-lines and hierarchical search-lines for low-power content-addressable memories," in *Proc. IEEE Custom Integrated Circuits Conf.*, Sep. 2003, pp. 383–386.

[5] S. Choi, K. Sohn, M. W. Lee, S. Kim, H. M. Choi, D. Kim, U. R. Cho, H. G. Byun, Y. S. Shin, and H. J. Yoo, "A 0.7 fJ/bit/search, 2.2 ns search-time, hybrid type TCAM architecture," *IEEE Int. Solid-State Circuits Conf. Dig. Tech. Papers*, pp. 498–499, Feb. 2004

[6] K. Pagiamtzis, A. Sheikholeslami, "Content-addressable memory (CAM) circuits and architectures: a tutorial and survey," *IEEE J. Solid-State Circuits*, vol 41, no. 3, pp.712-727, Mar. 2006

[7] K. Eshraghian, K.R. Cho, O. Kavehei, S.K Kang, D, Abbott, S.M. Steve Kang, "Memristor MOS Content Addressable Memory (MCAM): Hybrid Architecture for Future High Performance Search Engines" *IEEE Transactions on VLSI Systems*, vol. 19, no. 8, pp. 1407-1417, 2011.

[8] L.O. Chua "Memristor – the missing circuit element" *IEEE Transactions on Circuit Theory*. Vol. CT-18 No.5 pp.507-519, Sep 1971

[9] D.B. Strukov, G. S. Snider, D.R. Stewart, R.S. Williams, "The missing memristor found", *Nature*, vol. 453, pp. 80-83, May 2008

[10] D. Batas and H. Fiedler, "A memristor spice implementation and a new approach for magnetic flux controlled memristor modeling," *Nanotechnology, IEEE Transactions on*, vol. PP, no. 99, pp. 1 –1, 2009

[11] Z. Biolek, D. Biolek, V. Biolova, "SPICE model of memristor with nonlinear dopant drift," *Radioengineering*, vol. 18, no. 2, pt. 2, pp. 210-214, 2009.

[12] A. Rak, G. Cserey, "Macromodeling of the Memristor in SPICE," *IEEE Trans. Computer-aided design of integrated circuits and systems*, vol. 29, no. 4, pp. 632-636, Apr 2010.

Extending Symmetric Variable-Pair Transitivities Using State-Space Transformations

Peter M. Maurer
Dept of Computer Science
Baylor University
Waco, TX 76798, USA
1-254-710-7305

Peter_Maurer@Baylor.edu

ABSTRACT

Detecting the symmetries of a Boolean function can lead to simpler implementations both at the hardware and software level. Large clusters of mutually symmetric variables are more advantageous than small clusters. One way to extend the symmetry of a function is to detect abstract two-cofactor relations in addition to ordinary symmetric relations. Unfortunately, ordinary symmetries are simply transitive but more complex types of relations are not. This paper shows how to convert the more complex relations into ordinary symmetries, allowing them to be used to form large clusters of symmetric variables, larger than would be possible using ordinary symmetries.

Categories and Subject Descriptors

D.7.2 [Integrated Circuits]: Design Aids: Simulation and Verification.

General Terms

Algorithms, Theory, Verification.

Keywords

Symmetric Boolean Functions, Logic simulation, Symmetry Detection

1. INTRODUCTION

Symmetric Boolean functions have been exploited in a number of different areas of design automation. [1-6] Our main interest is in the area of simulation. If a function is totally symmetric, or has large clusters of mutually symmetric variables, it can be simulated more efficiently than a non-symmetric function. In general, the larger the clusters of mutually symmetric variables, the more efficient the simulation.

One area of current research is to enumerate the various types of relations that can exist between the cofactors of a function, and associate symmetry types with these relations. [7-9] Other avenues of research use group-theoretic definitions of symmetry, and include symmetries that cannot be defined in terms of variable pairs, [10-12] however, this avenue of research is beyond the scope of the current paper. This is also true for other approaches that are not cofactor based. (See [13, 14] for example.)

Except for ordinary symmetries which are simply transitive, the transitive relations between various types of symmetric variable pairs are fairly complex, This makes it difficult or impossible to use these symmetry relations to form large clusters of symmetric variables. In this paper we will show how to extend the simple transitivity of ordinary symmetry to more general types of symmetry, creating large clusters of variables that cannot be created otherwise. Each of these clusters will be collapsed into a single clustered variable which can be used in place of simple variables for further symmetry detection, and for simulation. The use of clustered variables not only speeds up symmetry detection but also enables us to detect symmetric variable pairs that *do not manifest themselves as cofactor relations*. We are able to extend transitivity to all known types of symmetric variable pairs, and we can extend our algorithm to handle new types of cofactor relations that, to the best of our knowledge, have never been used for symmetry detection.

The output of our symmetry detection algorithm is a totally or partially symmetric function f and a collection of corrective functions that must be applied to the inputs and outputs of f. In many cases these corrective functions are simple, but in some cases they are not. In our practical implementations of this algorithm, we have confined ourselves to the cases with simple corrective functions, but in this paper we will include some of the details of the more complex cases.

2. THE COFACTOR RELATIONS

The cofactor of a Boolean function f is found by setting one or more inputs to a constant value, either 0 or 1. In this paper, we designate cofactors using subscripts consisting of 0, 1 and x. Suppose that f is a four-input function, and we have set the first two variables to 1. This will produce the cofactor f_{11xx}, which is a function of two variables. The ones and zeros designate the variables that have been replaced with constant values, and the x's represent the remaining variables. When there is no opportunity for confusion, we will drop the x's from the subscript and designate cofactors such as f_{11xx} as f_{11}.

In most cases, symmetry detection is done using relations between two-variable cofactors. There are six such relations, which listed in Figure 1 with their associated symmetry types. These relations are called the *classical relations*.

As shown in [15] the classical relations can be expressed in another form using the exclusive or function, which we designate as $\oplus$. The ordinary relation, $f_{01} = f_{10}$ becomes $f_{01} \oplus f_{10} = \overline{0}$, and the multi-phase relation becomes $f_{00} \oplus f_{11} = \overline{0}$, where $\overline{0}$ represents the zero function.

A slightly more complex relation can be obtained by replacing the zero function with the constant-one function as in $f_{01} \oplus f_{10} = \overline{1}$. This gives us six new relations that are normally associated with six types of anti-symmetry. (The term anti-symmetry was introduced in [16]. Anti-symmetry is also known as skew symmetry [17] and negative symmetry [18].) These six new relations are given in Figure 2 along with their associated symmetries.

Type	Condition
Ordinary	$f_{01} = f_{10}$
Multi-Phase	$f_{00} = f_{11}$
Single-Variable	$f_{10} = f_{11}$
Single Variable	$f_{01} = f_{11}$
Multi-Phase Single Variable	$f_{00} = f_{01}$
Multi-Phase Single Variable	$f_{00} = f_{10}$

Figure 1. The Classical Relations.

Cofactor relations can be expanded in several ways, only a few of which have been investigated. The constant functions $\overline{0}$ and $\overline{1}$ can be replaced with other functions to create relations of the form $f_{01} \oplus f_{10} = g$ and the XOR operation can be replaced with another two-input function such as AND and OR to give relations of the form $f_{01}f_{10} = \overline{0}$ and $f_{10} + f_{01} = g$. It is possible to use cofactors in more than two variables to detect higher-order symmetries between three or more variables. These types of cofactor relations have not been extensively studied.

Relations between three or four two-variable cofactors give rise to the Kronecker symmetries. [9]. These relations use the XOR function, and are listed in Figure 3. Three and four cofactor relations can be extended in the same way as two-cofactor relations.

Type	Condition
Anti	$f_{01} \oplus f_{10} = \overline{1}$
Anti Multi-Phase	$f_{00} \oplus f_{11} = \overline{1}$
Anti Single-Variable	$f_{10} \oplus f_{11} = \overline{1}$
Anti Single Variable	$f_{01} \oplus f_{11} = \overline{1}$
Anti Multi-Phase Single Variable	$f_{00} \oplus f_{01} = \overline{1}$
Anti Multi-Phase Single Variable	$f_{00} \oplus f_{10} = \overline{1}$

Figure 2. The Anti-Relations.

3. THE STATE SPACE

In its most elementary form, the state-space of a Boolean function is an n-dimensional hypercube containing either Boolean values or various cofactors of the function. Each dimension of the hypercube corresponds to a variable. In the extreme case, there will be one dimension for each input variable and the nodes will contain Boolean values. However, when symmetries are detected, two dimensions can be collapsed into a single dimension. The collapsed dimension corresponds to a clustered variable, and contains three or more nodes. We call this structure a hyperlinear structure to emphasize its non-cubical nature. This structure has been used in the algorithms of [19] and [20], but has been limited to the relations of Figure 1.

During symmetry detection new cofactors are computed and new symmetry tests are performed. The hyperlinear structure expands when new cofactors are computed, and collapses when symmetric variable pairs are detected. If the process runs to completion, the value of each node will be reduced to a single Boolean value. However, symmetry detection can be terminated at any point using either time or space constraints. This enables the detection process to be run in "anytime" fashion as in [18]. The process is adaptable to many existing algorithms, particularly those that already use cofactor relations for symmetry detection. It is especially well suited to techniques such as that detailed in [21], which use the cofactors that already exist in a ROBDD.

Type	Condition
K_0	$f_{01} \oplus f_{10} \oplus f_{11} = \overline{0}$
K_1	$f_{00} \oplus f_{10} \oplus f_{11} = \overline{0}$
K_2	$f_{00} \oplus f_{01} \oplus f_{11} = \overline{0}$
K_3	$f_{00} \oplus f_{01} \oplus f_{10} = \overline{0}$
K_4	$f_{00} \oplus f_{01} \oplus f_{10} \oplus f_{11} = \overline{0}$
Anti-K_0	$f_{01} \oplus f_{10} \oplus f_{11} = \overline{1}$
Anti-K_1	$f_{00} \oplus f_{10} \oplus f_{11} = \overline{1}$
Anti-K_2	$f_{00} \oplus f_{01} \oplus f_{11} = \overline{1}$
Anti-K_3	$f_{00} \oplus f_{01} \oplus f_{10} = \overline{1}$
Anti-K_4	$f_{00} \oplus f_{01} \oplus f_{10} \oplus f_{11} = \overline{1}$

Figure 3. The Kronecker Relations.

When a symmetry other than an ordinary symmetry is detected, a state-space transformation is applied to convert the symmetry to an ordinary symmetry. Corrective functions are then added to the input and output of the function to compensate for the transformation. *Any* of the relations discussed in Section 2 can be converted to ordinary symmetries, but in some cases the corrective functions are too complex for these relations to be useful.

The technique described here is based on the technique presented in [20]. However the algorithm of [20] is limited to the classical symmetries.

4. SYMMETRY DETECTION

The detection algorithm begins by computing four cofactors of the function and placing them in a two-dimensional hypercube. Figure 4 illustrates how this would be done for the function $abc + abd + acd + bcd$ and its four cofactors, 0, cd, cd, and $c + d$. The physical structure is a single-dimensional array which is indexed using a private indexing function to make it appear multi-dimensional. The use of a private indexing function permits the number of dimensions in the logical structure to change during the symmetry detection process. Furthermore, the logical structure can be indexed in unconventional ways to perform certain state-space transformations.

It is obvious from Figure 4 that there is a an ordinary symmetry between variables a and b. This causes the logical structure to collapse into a single dimension as in Figure 5. The vertices of Figure 5 are indexed by the number of ones in the input variables a and b.

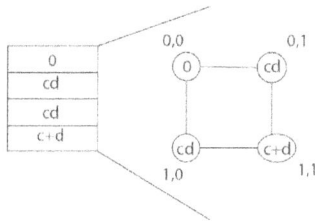

Figure 4. The Array and Logical Structure.

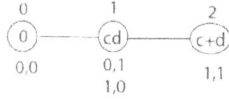

Figure 5. A Collapsed Logical Structure.

Collapsing reduces the size and the number of dimensions, and it permits symmetric variable pairs to be combined into a single composite variable. Composite variables can be treated more or less like simple variables, thus simplifying the detection process.

Testing for the other five classical relations can be done by selecting various pairs of nodes from the structure of Figure 5. However, rather than doing this, we modify the private indexing function and test for ordinary symmetry. For multi-phase symmetry, we modify the private indexing function to index one dimension in reverse order and test for ordinary symmetry. Figure 6 shows how reversing the second dimension of Figure 4 repositions the nodes. (The comparison is always between the node labeled "0,1" and the node labeled "1,0", wherever these indices may appear.) Reversing the dimension is the same as negating the associated input variable. We call this the *phase* transformation.

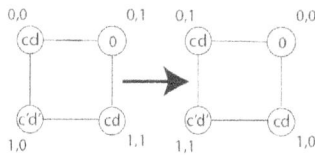

Figure 6. Reverse Indexing the Second Dimension.

When a multi-phase symmetry is detected, the structure is collapsed using the phase transformation. This makes the transformation permanent for all future operations. To correct for the phase transformation it is necessary to place an inverter on the input representing the reversed dimension.

To detect the single-variable symmetries it is necessary to reverse the indexing of odd-numbered rows or the odd-numbered columns. (Rows and columns are numbered starting with zero.) We call this type of transformation the *conjugate* transformation. It is equivalent to transforming the inputs with a matrix is over the field GF2. (In GF2 there are two values 0, and 1 with AND replacing multiplication and XOR replacing addition. See [20] for the theoretical development.) Figure 7 gives a matrix and shows the effect of multiplying the inputs by the given matrix. This results in testing for a single-variable symmetry when comparing the nodes labeled "0,1" and "1,0". Strictly speaking, the type of symmetry detected using the conjugate transformation is *conjugate symmetry*, not single-variable symmetry. Every single-variable symmetry is also a conjugate symmetry, but there are conjugate symmetries that do not manifest themselves as single-variable symmetries. (As in the function $ab' + ac + b'c$, which has an ordinary symmetry between a and b and a conjugate symmetry between the composite pair ab and the variable c.)

The conjugate transformation can detect these types of symmetries as well as single-variable symmetries.

$$\begin{pmatrix} 1 & 1 \\ 0 & 1 \end{pmatrix}$$

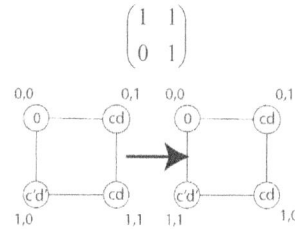

Figure 7. Reverse Indexing the Odd Rows.

Again, the conjugate transformation is used during the collapse of the logical structure, making the transformation permanent. The matrix used during the detection is retained and used to construct the corrector function. If more than one conjugate symmetry is detected, the matrices are multiplied together in the order they are first used. To correct for conjugate transformations it is necessary to add one or more XOR gates to the inputs of the function. The final matrix is used to determine the inputs of these XOR gates. Each column of the matrix corresponds to an input of the transformed function, and each row corresponds to one of the original inputs. The outputs of these gates replace the original inputs. The columns of the transformation matrix determine the inputs of the XOR gates.

There are two types of single-variable symmetries. Reversing the indexing of the odd rows gives one type, reversing the odd columns gives the other type. The multi-phase single-variable symmetries can be tested by combining a phase transformation with a conjugate transformation.

Detecting anti symmetries requires a transformation of the node contents, not just a transformation of indices. Correcting these sorts of transformations requires an output corrective function rather than an input corrective function. Figure 8 shows the *anti* transformation required to detect an anti-symmetry. Odd numbered rows of the hyperlinear structure are inverted. In Figure 8 it is not technically necessary to complement node 1,1, but doing so simplifies the corrector function. Like the previous transformations, anti transformations are temporary until a symmetry is detected, at which time the transformation becomes permanent.

To correct the transformation of Figure 8, the output of the function must be XORed with variable corresponding to the vertical dimension. The anti transformation can be combined with the phase and conjugate transformations to detect all conditions listed in Figure 2.

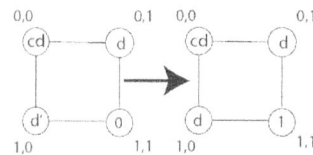

Figure 8. Anti-Symmetry Detection.

Although our symmetry detection algorithms do not check for relations of the form $f_{01} \oplus f_{10} = g$ we could easily adapt them to do so. Figure 9 shows the state-space transformation for detecting this type of relation. The function g is XORed with the odd rows of the logical structure. Once a symmetry is detected, the logical structure is collapsed and the transformation becomes permanent. The corrective function for $f_{01} \oplus f_{10} = g$ relations is shown in

Figure 10, where F is the result of the state-space transformation and a is the input variable corresponding to the vertical dimension. The corrector function requires the computation of the function g so it should be easy to compute, or a function that is already computed elsewhere in the circuit.

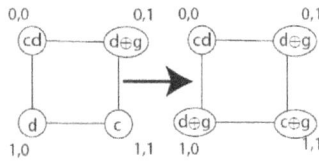

Figure 9. Detection for Relation $f_{01} \oplus f_{10} = g$.

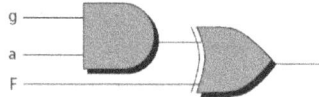

Figure 10. Output Correction for $f_{01} \oplus f_{10} = g$ Relations.

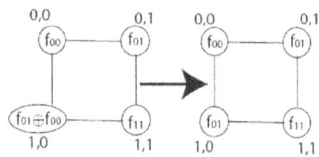

Figure 11. Structure for $f_{01} \oplus f_{00} \oplus f_{10} = \overline{0}$ Relations.

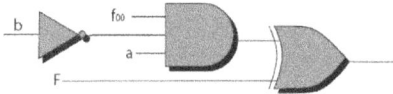

Figure 12. Corrector for $f_{01} \oplus f_{00} \oplus f_{10} = \overline{0}$ Relations.

For relations of the form $f_{01} \oplus f_{00} \oplus f_{10} = \overline{0}$ the situation is more complex because the output corrector function must include at least one cofactor of the function. The two-dimensional structure and its state-space transformation, the *K transformation*, are shown in Figure 11. The corrector function is given in Figure 12. This function could be simplified by XORing node 1,1 with f_{00}, but this trick does not scale up to the more complex structures described in the next section. Combining the phase, conjugate, and anti transformations with the K transformation permits all relations given in Figure 3 to be tested.

5. COMPLEX LOGICAL STRUCTURES

As the number of variables represented in the logical structure increases, so does the complexity of the structure. In some algorithms, the search for symmetric variable pairs is done from the top down by starting over with the original function f and computing new cofactors. Instead of doing this, our algorithm proceeds by computing the cofactors of the cofactors already in the structure. Figure 12 shows how the structure of Figure 4 can be expanded to three dimensions if no symmetry is detected between a and b.

In Figure 13, two comparisons are necessary to detect the symmetry between variables a and b. Node 0,1,0 must be compared to node 1,0,0, and node 0,1,1 must be compared to node 1,0,1. There are now three pairs of variables that must be compared, (a,b), (a,c) and (b,c). Dimensions must be examined two at a time, and all planes in that pair of dimensions must be examined.

When a symmetric pair is detected in a three-dimensional cube, it is collapsed as shown in Figure 14. In all further operations the two variables a and b will be treated as a composite variable taking the values 0, 1 and 2, corresponding to the number of ones in the two inputs.

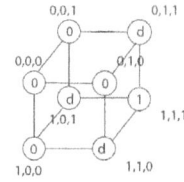

Figure 13. Three Dimensions.

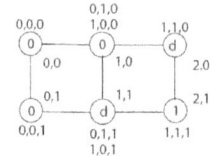

Figure 14. Collapsed 3-Dimensional Structure.

Since all detected symmetries are converted to ordinary symmetries a complex analysis of symmetry transitivities is not needed. The primary problem is detecting symmetries with respect to composite variables. When comparing a composite variable to another variable, there will be more than one anti-diagonal, as in Figure 14. To check for ordinary symmetry, it is necessary that all vertices along each diagonal have the same value. Node 0,1 will be compared to node 1,0 and node 1,1 will be compared to node 2,0. It is permissible for different diagonals to have different values.

When testing for skew and conjugate symmetries, reverse-indexing the dimension corresponding to a composite variable is equivalent to negating each individual variable in the composite. Thus reverse-indexing can be used to detect multi-phase and conjugate symmetries with respect to composite variables.

For anti-symmetries and more general relations, we need the following theorem which clarifies the situation with respect to composite variables.

Theorem 1. Given a Boolean function f, if the relation $f_{01} \oplus f_{10} = \overline{0}/\overline{1}$ is true for variables a and b and if there is an ordinary symmetry between variables b and c then the relation is also true for variables a and c.

Proof. Let f_{000} designate the cofactors of f with respect to the variables a b and c in that order. By assumption, $f_{01x} \oplus f_{10x} = \overline{0}/\overline{1}$ which implies that $f_{010} \oplus f_{100} = \overline{0}/\overline{1}$ and $f_{011} \oplus f_{101} = \overline{0}/\overline{1}$. Because there is an ordinary symmetry between b and c, $f_{x01} = f_{x10}$, which implies that $f_{001} = f_{010}$ and $f_{101} = f_{110}$. Substituting these into the previous two equations, we get $f_{001} \oplus f_{100} = \overline{0}/\overline{1}$ and $f_{011} \oplus f_{110} = \overline{0}/\overline{1}$ which implies that $f_{0x1} \oplus f_{1x0} = \overline{0}/\overline{1}$, the desired condition.

Theorem 1 says that if a condition exists between any element of a composite variable c and another variable v, then that condition exists between v and every other member of c. Theorem 1 also applies to relations of the form $f_{01} \oplus f_{10} = g$ if g is symmetric or if the variables b and c are adjacent. Theorem 2 gives us a similar result for the Kronecker relations.

Theorem 2. Given a Boolean function f, if any of the relations

$$f_{01} \oplus f_{10} \oplus f_{11} = \overline{0}/\overline{1}, \qquad\qquad f_{01} \oplus f_{10} \oplus f_{00} = \overline{0}/\overline{1}$$

$$f_{01} \oplus f_{11} \oplus f_{00} = \overline{0}/\overline{1}, \qquad f_{11} \oplus f_{10} \oplus f_{00} = \overline{0}/\overline{1} \qquad \text{or}$$

$$f_{00} \oplus f_{01} \oplus f_{10} \oplus f_{11} = \overline{0}/\overline{1} \text{ exists between variables } a \text{ and } b,$$

and if an ordinary symmetry exists between variables b and c, then the same relation also exists for variables a and c.

The proof of Theorem 2 is essentially identical to that for Theorem 1.

Theorems 1 and 2 are helpful in determining how to test for non-classical symmetries with composite variables. In Figure 14 we are comparing two composite variables, each of which consists of three mutually symmetric variables. Five diagonals must be compared.

If there is an anti-symmetry between any two variables from the respective composites, Theorem 1 states that there must be a total of nine anti-symmetries, one for each pair of variables. If we break out each of these nine symmetries in the manner of the Theorem 1 proof, we see that the relation $f_x \oplus f_y = \overline{1}$ must exist between every pair of consecutive nodes along each diagonal. Another way of expressing the relation $f_x \oplus f_y = \overline{1}$ is $f_x = f'_y$, which implies that a cofactor and its complement must alternate along each diagonal. General relations of the form $f_{01} \oplus f_{10} = g$ are more complicated because the comparison involves cofactors of g, and if g is not symmetric, the comparisons along a diagonal may not be uniform.

For Kronecker symmetries, applying Theorem 2 shows that the relation must exist for every "square" of nodes along a diagonal. For example, for the relation $f_{01} \oplus f_{10} \oplus f_{00} = \overline{0}$ to be true between the composite variables of Figure 14, it is necessary that $node_{01} \oplus node_{10} \oplus node_{00} = \overline{0}$, $node_{20} \oplus node_{21} \oplus node_{10} = \overline{0}$, $node_{21} \oplus node_{02} \oplus node_{01} = \overline{0}$, etc. There are nine comparisons in all.

In our comparison algorithm, we can detect Kronecker symmetries between composite variables, but we make no use of these symmetries in practical applications because the output-corrective functions are too complex. Not only do they involve cofactors of the function, but they are non-uniform. The corrective function for Figure 14 would require the computation of nine cofactors, one for each of the variable pairs.

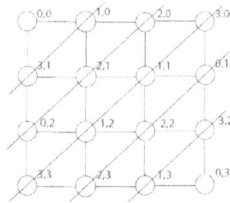

Figure 14. Comparing Composite variables.

For anti-symmetries, the corrective function is simple. Any time an anti-symmetry is detected, the odd rows of the hyperlinear structure are inverted. The variables corresponding to these rows are added to an inversion list. Regardless of the number of anti-symmetries, a single XOR gate is used on the output of the

function. The inputs to this gate are the output of the function and the variables on the inversion list.

6. BDDs FROM STATE SPACES

At the end of the symmetry detection process we are left with the state space of a symmetric function and a list of elementary corrective functions. The state-space is converted to a BDD one dimension at a time. The first dimension of Figure 14 would be converted as shown in Figure 15. The leaf-nodes in this diagram are the columns of the state-space of Figure 14. Duplicate leaves are combined, nodes pointing to duplicate leaves are removed and the conversion proceeds in a recursive manner at the leaf nodes. This process is repeated until the leaves contain elementary Boolean values.

7. EXPERIMENTAL DATA

To determine the effectiveness of our symmetry detection algorithms, we ran our algorithms on the ISCAS85 benchmarks. [22] Each circuit was first partitioned into a collection of fanout-free networks, each one of which represents a single-output function. Because some of these fanout-free networks are extremely large, they were further partitioned into circuits with no more than eight inputs. Our partitioning algorithm is capable of using any number of inputs as a limit, but we find that a limit of eight tends to maximize the number of symmetries for most circuits. (This issue is discussed in greater depth in [20]). Figure 16 shows the total number of symmetries of each type detected by our algorithms. The K3 column contains the results for the positive three-cofactor relations, and K4 gives the results for the positive four-cofactor relation.

Although Figure 16 shows a number of Kronecker symmetries for some circuits, these results are misleading because of a phenomenon we call *symmetry masking*. Sometimes detecting one type of symmetry will prevent other types of symmetry from being detected. (Although sometimes the reverse is true, such as for c7552.) Figure 17 shows the number of classical symmetries detected when detection of Kronecker symmetries is suppressed. In several cases, the Kronecker symmetries are being detected at the cost of detecting classical symmetries (see c5315, for example). Because the corrective functions for classical symmetries are far simpler than those for Kronecker symmetries, it is clearly undesirable to detect Kronecker symmetries for these circuits.

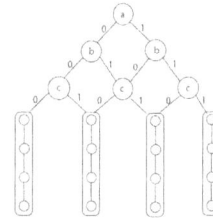

Figure 15. Converting to BDDs.

This phenomenon is far less pronounced for anti-symmetries, as Figure 18 shows. In most cases the anti-symmetries are detected with little or no suppression of classical symmetries. This is especially true for c6288, for which the number of detected symmetries almost doubles, without suppressing any classical symmetries.

8. CONCLUSION

We have presented a symmetry detection technique that permits the natural transitivity of classical symmetries to be extended to virtually any cofactor relation. For some types of relations, namely the multi-phase, single-variable and anti symmetries, the cost of this transitivity is negligible. For others, especially the Kronecker relations, the cost is higher.

Circuit	Classical	Anti	K3	K4
c432	103	0	0	0
c499	158	0	0	24
c880	144	20	19	8
c1355	142	104	0	0
c1908	146	0	3	0
c2670	356	34	0	1
c3540	461	7	1	15
c5315	588	32	12	82
c6288	480	464	0	0
c7552	959	89	32	0

Figure 16. Results for all symmetry types.

Circuit	Classical
c432	103
c499	174
c880	152
c1355	142
c1908	146
c2670	358
c3540	464
c5315	669
c6288	480
c7552	953

Figure 17. Classical Symmetries Alone.

However, our experimental data suggests that detecting Kronecker symmetries may not provide a significant advantage, even if simple corrective functions could be found. One area that would bear further investigation is relations of the type $f_{01} \oplus f_{10} = g$ which have corrective functions that are not significantly more complex than those for anti-symmetry.

Circuit	Classical	Anti
c432	103	0
c499	174	0
c880	152	19
c1355	142	104
c1908	146	0
c2670	357	34
c3540	463	5
c5315	664	32
c6288	480	464
c7552	951	99

Figure 18. Classical and Anti-Symmetries.

9. REFERENCES

1. C. E. Shannon, "The synthesis of two-terminal switching circuits," *Bell System Technical Journal*, Vol.28, *No.1*, pp. 59-98, 1949.
2. C. R. Edwards and S. L. Hurst, "A digital synthesis procedure under function symmetries and mapping methods," *IEEE Transactions on Computers*, Vol.27, *No.11*, pp. 985-997, 1978.
3. D. Moller, P. Molitor, R. Drechsler and J. W. G. U. Frankfurt, "Symmetry based variable ordering for ROBDDs," *IFIP Workshop on Logic and Architecture Synthesis*, pp. 47-53, 1994.
4. C. Scholl, D. Moller, P. Molitor and R. Drechsler, "BDD minimization using symmetries," *IEEE Transactions on Computer-Aided Design of Integrated Circuits and Systems*, Vol.18, *No.2*, pp. 81-100, 1999.
5. T. Sasao, "A new expansion of symmetric functions and their application to non-disjoint functional decompositions for LUT type FPGAs," *IEEE International Workshop on Logic Synthesis*, pp. 105-110, 2000.
6. V. N. Kravets and K. A. Sakallah, "Constructive library-aware synthesis using symmetries," *Design Automation and Test in Europe*, pp. 208-213, 2000.
7. C. C. Tsai and M. Marek-Sadowska, "Boolean functions classification via fixed polarity Reed-Muller forms," *IEEE Transactions on Computers*, Vol.46, *No.2*, pp. 173-186, 1997.
8. M. Chrzanowska-Jeske, "Generalized symmetric variables," *The 8th IEEE International Conference on Electronics, Circuits and Systems*, pp. 1147-1150, Vol. 3, 2001.
9. M. Chrzanowska-Jeske, A. Mishchenko and J. R. Burch, "Linear Cofactor Relationships in Boolean Functions," *IEEE Transactions on Computer-Aided Design of Integrated Circuits and Systems*, Vol.25, *No.6*, pp. 1011-1023, 2006.
10. P. M. Maurer, "An application of group theory to the analysis of symmetric gates," Department of Computer Science, Baylor University, Waco, TX 76798, http://hdl.handle.net/2104/5438, 2009.
11. V. N. Kravets and K. A. Sakallah, "Generalized symmetries in boolean functions," *IEEE International Conference on Computer Aided Design*, pp. 526-532, 2000.
12. J. Mohnke, P. Molitor and S. Malik, "Limits of using signatures for permutation independent Boolean comparison," *Formal Methods Syst. Des.*, Vol.21, *No.2*, pp. 167-191, 2002.
13. C. C. Tsai and M. Marek-Sadowska, "Detecting symmetric variables in boolean functions using generalized reed-muller forms," *IEEE International Symposium on Circuits and Systems*, pp. 287-290, Vol. 1, 1994.
14. S. L. Hurst, "Detection of symmetries in combinatorial functions by spectral means," *IEE Journal on Electronic Circuits and Systems*, Vol.1, *No.5*, pp. 173-180, 1977.
15. M. Chrzanowska-Jeske, "Generalized symmetric and generalized pseudo-symmetric functions," *Proceedings of the 6th IEEE International Conference on Electronics, Circuits and Systems*, pp. 343-346, Vol. 1, 1999.
16. J. Rice and J. Muzio, "Antisymmetries in the realization of boolean functions," *IEEE International Symposium on Circuits and Systems*, pp. 69-72, Vol. 4, 2002.
17. C. C. Tsai and M. Marek-Sadowska, "Generalized Reed-Muller forms as a tool to detect symmetries," *IEEE Transactions on Computers*, Vol.45, *No.1*, pp. 33-40, 1996.
18. N. Kettle and A. King, "An anytime algorithm for generalized symmetry detection in ROBDDs," *IEEE Transactions on Computer-Aided Design of Integrated Circuits and Systems*, Vol.27, *No.4*, pp. 764-777, 2008.
19. P. M. Maurer. Efficient event-driven simulation by exploiting the output observability of gate clusters. *Computer-Aided Design of Integrated Circuits and Systems, IEEE Transactions on* Vol.22, *No.11*, pp. 1471-1486.
20. P. M. Maurer, "Conjugate Symmetry," *Formal Methods Syst. Des.*, Vol.38, *No.3*, pp. 263-288, 2011.
21. D. Moller, J. Mohnke and M. Weber, "Detection of symmetry of boolean functions represented by ROBDDs," *IEEE International Conference on Computer-Aided Design*, pp. 680-684, 1993.
22. F. Brglez, P. Pownall and R. Hum. Accelerated ATPG and fault grading via testability analysis. Presented at Proceedings of IEEE Int. Symposium on Circuits and Systems.

Crosslink Insertion for Variation-Driven Clock Network Construction

Fuqiang Qian, Haitong Tian, Evangeline Young

Department of Computer Science and Engineering
The Chinese University of Hong Kong
{fqqian, httian, fyyoung}@cse.cuhk.edu.hk

ABSTRACT

Link based non-tree clock network is an effective and economic way to reduce clock skew caused by variations. However, it is still an open topic where links should be inserted in order to achieve largest skew reduction with smaller extra resources. We propose a new method using linear program to solve this problem in this paper. In our approach, clock skew in a non-tree clock network is computed using the delay model in [13] and the information is used to select the node pairs for link insertion. Tradeoff between crosslink length and skew reduction effect is explored. Based on the analysis, we propose a new algorithm to insert crosslinks into a clock network. We compare our work with the method in [1] and a recent work [4] which inserts links between internal nodes of a tree. Experiments show that our method can reduce skew under variations effectively.

Categories and Subject Descriptors

B.7.2 [**INTEGRATED CIRCUITS**]: Design Aids---Placement and routing

General Terms

Algorithm, Design

Keywords

Clock, Skew, Non-tree, Crosslink

1. INTRODUCTION

Clock skew limits the performance of a synchronous digital system. On the other hand, the effect of process variation is becoming more significant. Clock skew caused by variations is now one of the most critical problems that the design of large and high performance system is facing today.

Non-tree clock network is a promising way to address the skew variation problem. Clock mesh, because of its inherent redundancy, is more tolerant to process variation and is able to provide lower skew variability compared with traditional clock tree. However, clock mesh has the disadvantage of resulting in

much more power dissipation. A clock distribution network that combines the advantages of tree and mesh is described in [5]. The design is used on several microprocessor chips, achieving very low skew. For analysis and optimization, Venkataraman *et al.* [6] managed to reduce a clock mesh by retaining only the critical edges. Their strategy offers designers a possibility of trade-off between power and tolerance to process variation. In paper [7], a sliding window based scheme is used to analyze the latency in a clock mesh. A comprehensive and automated framework for planning, synthesizing and optimization of clock mesh networks is proposed in [8].

Compared to the mesh structure, clock network constructed by inserting crosslinks will consume much less power. Rajaram *et al.* [1] propose a framework to construct such non-tree network and present an analysis on the effect of link insertion on skew variability. Lam *et al.* [9] present a statistical based non-tree clock network construction technique. They use statistical timing analysis to obtain the distribution of skew in a clock tree network. The works in [10, 11] explore the challenges in buffered clock tree link insertion and construct a buffered clock network with crosslinks. In [12], wire sizing is performed to improve skew variability in a non-tree topology with links which are generated using the minimum weight matching-based method in [1]. Recently, a link insertion scheme that inserts crosslinks at higher level internal nodes instead of sink nodes in a clock tree is proposed [4]. Their work reduces skew variability and total crosslink length.

Although there are previous works on non-tree clock network construction with crosslinks, some important questions are still unanswered. Most existing works on link insertion attempt to reduce skew variability while use the minimum wirelength. However, the tradeoff between the length of a link and its ability to reduce clock skew is not analytically studied. In this work, efforts are made towards solving these problems. We use the load redistribution and tree decomposition technique [13] to obtain the delay and skew values in a non-tree clock network. We then formulate the crosslink insertion problem as a sequence of linear programs, with an objective to find a pair of nodes to insert a crosslink such that the skew variability can be reduced the most while the wirelength increase due to the link insertion is constrained. By applying this technique recursively, we can add a user-defined number of crosslinks and the clock skew will be reduced progressively. Simulation results show that our method can lead to significant skew reduction under variations. The power consumption is also reduced comparing with previous works. The major contributions of this paper can be summarized as follows:

• Our work computes signal delay and clock skew in a non-tree clock network analytically when inserting links. The link insertion process thus produces more effective result.

• We formulate the problem of finding a node pair for link insertion in a non-tree clock network as a sequence of linear optimization problems, with an objective function to consider the tradeoff between clock skew reduction and link capacitance.

• We devise a method to speed up the linear program so that even for a clock network with several thousand sinks, the problem can be solved in a reasonable amount of time.

The paper is organized as following. Techniques for computing signal delay and clock skew in non-tree clock networks will be discussed in Section 2. In Section 3, our method to construct a non-tree clock network with crosslinks will be presented. The experimental results will be reported in Section 4. Finally, concluding remarks will be made in Section 5.

2. SIGNAL DELAY AND CLCOK SKEW IN NON-TREE CLOCK NETWORKS

In this section, the delay calculation technique in general RC network will be reviewed. This method can be used to evaluate node delays and clock skew in a non-tree network with crosslinks. Some analysis of the load redistribution effect of a link will be provided.

Calculation of the delays in a non-tree network is non-trivial, because there are loops between the nodes, which makes possible that one node is driving and loading another node at the same time. The relationships between nodes are not explicit in a non-tree clock network. Several techniques have been suggested to model the delay in general RC network [3, 13, 14].

Based on the Elmore delay model, a technique called tree decomposition [13] is proposed to calculate the delays of all the nodes in a general RC network with resistance loops. Suppose that N is an arbitrary node in a general RC network and N has k neighboring nodes, denoted by M_i where $i = 1, ..., k$. Node N is connected to M_i through edge E_i with a resistance of value R_i. Denote the load capacitance of N as C. The idea is to partition and distribute C into N's k neighboring edges. Let C_i be the equivalent load distributed to edge E_i, note that C_i can be negative. According to the calculation of the Elmore delay, if the delay of node M_i is T_i, the delay of node N will be $T = T_i + R_iC_i$. Then there is a set of constraints as follows.

$$\Sigma_{i=1,...,k} C_i = C \qquad (1)$$

$$T = T_i + R_iC_i \qquad i = 1, ..., k \quad (2)$$

Applying the above idea, a general RC network can be decomposed into a tree. Since the delays in a tree can be determined efficiently, we are able to calculate the delays and clock skews in the original RC network.

Consider the effect of inserting a link between two nodes N_i and N_j with delays D_i and D_j respectively in a tree as shown in Fig.1(a). Assume that the load capacitances of node N_i and N_j are C_i and C_j respectively before link insertion. After a link is inserted between node N_i and N_j, node N_j becomes one of the neighboring nodes of N_i. The loading capacitance of node N_i will be redistributed as in Fig.2(b). In this example, the load capacitance of node N_i is redistributed. Note that we can also split N_j instead of N_i. Suppose $C_{i,1}$ is the capacitance remaining at node N_i and $C_{i,2}$ is the load redistributed from node N_i to node N_j through the inserted link. Let D_i' and D_j' be the new delays at node N_i and N_j respectively, which can be calculated after this decomposing step.

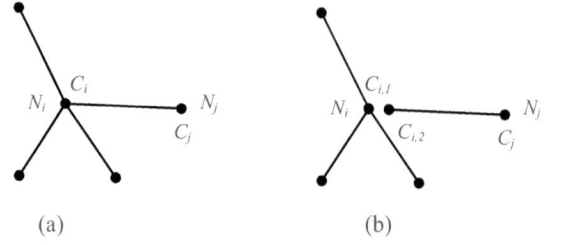

Fig.1. Load Redistribution through a Link

According the set of constraints (1) and (2), we have the following equality constraints for this particular example:

$$C_{i,1} + C_{i,2} = C_i \qquad (3)$$

$$D_i' = D_j' + D_{link} \qquad (4)$$

The delay difference between node N_i and node N_j becomes $D_{link} = R_{link}C_{i,2}$. Note that if $C_{i,2}$ is positive, it indicates that node N_j is actually driving node N_i after this link insertion step and node N_j load part of node N_i's capacitance, which is $C_{i,2}$. If $C_{i,2}$ is negative, it indicates that node N_i is driving node N_j instead. This load redistribution is also applicable for node pairs with non-zero skew.

The Elmore delay of node N_i in a tree T is

$$\sum_k R_{i,k} C_k \qquad (5)$$

where $k \in$ all nodes in T, $R_{i,k}$ is the shared path resistance between node N_i and node N_k and C_k is the capacitance of node N_k. For a given tree, the resistances of edges and capacitances of the nodes are known, so the delays at nodes N_i and N_j can be written as follows with constants k, where $i = 1, 2, 3, 4, 5, 6$.

$$D_i = k_1C_i + k_2C_j + k_3 \qquad (6)$$

$$D_j = k_4C_i + k_5C_j + k_6 \qquad (7)$$

The skew between node N_i and N_j before link insertion is

$$q_{ij} = D_i - D_j \qquad (8)$$

After a link is inserted between N_i and N_j, the delays become

$$D_i' = k_1(C_i - C_{i,2}) + k_2(C_j + C_{i,2}) + k_3 \qquad (9)$$

$$D_j' = k_4(C_i - C_{i,2}) + k_5(C_j + C_{i,2}) + k_6 \qquad (10)$$

The skew becomes $q_{ij}' = D_i' - D_j'$, given by

$$q_{ij}' = q_{ij} - (k_1 - k_4)C_{i,2} + (k_2 - k_5)C_{i,2} \qquad (11)$$

which equals $R_{link}C_{i,2}$, so we have

$$C_{i,2} = q_{ij}/(k_1 - k_2 - k_4 + k_5 + R_{link}) \qquad (12)$$

The skew after link insertion is $q_{ij}' = R_{link}C_{i,2} = R_{link} q_{ij}/(k_1 - k_2 - k_4 + k_5 + R_{link})$, the skew is thus scaled by $R_{link}/(k_1 - k_2 - k_4 + k_5 + R_{link})$. Therefore, the skew reduction effect of a link is affected by the resistance of the link R_{link}. If a link resistance is large, its skew reduction effect will be small. Although the above analysis is based on a structure that is a tree, the same argument can be applied even after several links are inserted since a non-tree network can be represented by a tree structure by applying the above load redistribution method.

3. LINK INSERTION FOR NON-TREE CLOCK NETWORK

Based on the techniques discussed in section 2, we are able to calculate signal delays in a clock network. In our approach, the sink load capacitances are modeled as variables because they are affected by various variations. Each sink node's delay will change according to the values of the sink capacitances. We use the worst case skew value as a metric to evaluate the skew. A worst case skew is defined as the maximum skew that might appear in the clock network with variations under consideration.

Adding links between the node pairs which are most susceptible to have the worst case clock skew will be beneficial. Therefore, one important step is to identify the node pairs which will have large skew under variations. A shared path length l_{ij} is the length of the path shared between the path from the root s to sink node S_i and the path from s to S_j. It is obvious that if two nodes share a long common path from the source in a tree topology, the delay difference will be small. If the shared path length l_{ij} is small or even zero, the delay difference between S_i and S_j can be high. Those node pairs with small shared path length are topologically far away. In a clock tree, adding a link to those topologically far away node pairs will be beneficial because they are likely to have large clock skew. For example, in Fig. 2, adding a link between node b and node c will generally be better than adding a link between node b and node d, because node pair (b, d) is less likely to have large skew.

When the clock network is no longer a tree structure after some crosslinks are inserted, the relationship between nodes will not be explicit. It is hard to tell if a node pair is topologically far away or not. In Fig. 2, if we consider node pair (a, b) and node pair (d, c), node pair (a, b) are topologically closer if no links are inserted, since the lowest common ancestor (LCA) of (a, b) is at level 1 while the LCA of (d, c) is at the root (level 0). However, if a link exists between node b and node c, it will be difficult to tell which node pair are topologically closer thus less likely to exhibit worse skew. In this case, we will use the non-tree delay calculation technique (Section 2) to obtain the worst case skew of each sink node pair in order to decide which pairs will likely have large skew.

In our approach, we will insert the first link between the two subtrees of the root. We will find the node pair with the smallest physical distance and thus the smallest resistance. Starting from the second link, we will formulate a sequence of linear programs to find the most beneficial node pair for inserting a link. Fig. 3 shows an overall flow of our algorithm to insert links. In the following section we will discuss in more details of the sequence of linear programs.

Algorithm: Add Links
Input: A clock distribution network
Output: m crosslinks
1. Insert the first link between node pair (a, b) where a and b are in the left and right subtree of the root and are closest to each other physically
2. $k \leftarrow 1$
3. While $k < m$
4.   For each sink node pair (u, v)
5.   Construct an LP to find the largest $f_{u,v}$ value
       * $f_{u,v}$ is an objective function described below
6.   Return the node pair with the largest $f_{u,v}$ value for adding a link
7. $k \leftarrow k + 1$
8. End

Fig.3. Link Insertion Overview

3.1 Linear Programs for Selecting Node Pairs

Given a non-tree clock network, we will first make use of the techniques discussed in Section 2 to decompose the network into a tree structure and redistribute the load capacitances. An example is illustrated in Fig. 4(a). In Fig. 4(a), a link k is added between sink nodes S_1 and S_2, with load capacitance C_1 and C_2 respectively. First we add half of the link capacitance C_{link} to endpoints S_1 and S_2 of the link. When the tree is decomposed, we pick one of the link endpoints to decompose. In this example, we pick node S_1 to be viewed as being split into two nodes as in Fig. 4(b). The new node S_3 is connected to S_2 through the added link. The new loading capacitance at node S_1 is $C_1 + C_{link}/2 - c_k$ where c_k is the capacitance distributed to node S_2. Suppose that there are m links in the clock distribution network, similar decomposition can be done for all the links. We can use m variables $c_1, c_2, ..., c_m$ to describe the load redistribution from one endpoint of a link to the other. Finally, we obtain a tree structure with some way of load distribution. The Elmore delays in the clock network can then be found.

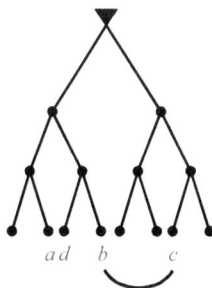

Fig. 2. Clock Network with Links

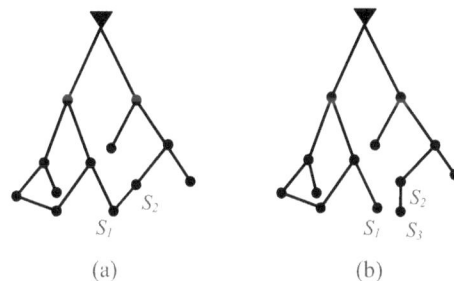

(a)   (b)

Fig. 4. Decomposition of Clock Network with Links

Once the network is decomposed into trees and the load capacitances are redistributed, according to equation (5), we can write the Elmore delay of each sink node as a linear function of the sink load capacitances. For a buffered clock network, the delay can also be expressed as a linear function of the sink capacitances. Let n be the total number of sinks and m be the number of links, the delay of a sink node S_u is

$$t_u = f_u(C_1, C_2, ..., C_n, c_1, c_2, ..., c_m) \quad (13)$$

The skew between sink node S_u and S_v is

$$q_{u,v} = |t_u - t_v| \quad (14)$$

Then we can formulate the worst case skew between a pair of nodes under variations. However, adding a link between the pair with the worst skew might involve very long route and a significant increase in wirelength. If the length of the link is large, the resistance of the link will be high. Firstly, this will make the clock network not practical. Secondly, a large wirelength will lower the skew reduction effect of a link as discussed in Section 2. In Fig. 2, if we do not consider the resistance of the link, adding a link between node b and node c will be better than adding a link between node a and node b, since node a and node b share some parts of their paths from the source and the delay difference will be smaller. However, if the distance between node b and node c is larger than the distance between node a and node b, it will be difficult to say which link is better. A tradeoff between the length of a link and its clock skew reduction effect is needed. Therefore, when selecting node pairs, we consider both clock skew and link length to find the node pairs which are most beneficial for link insertion. We thus define an objective function to strike a balance between the worst case skew and the link resistance $r_{u,v}$ as follows:

$$f_{u,v} = q_{u,v}/r_{u,v} \quad (15)$$

This objective function is established empirically. From the experimental results, we find that this objective function can achieve low clock skew effectively. We can find the maximum of $f_{u,v}$ subject to changes in the load capacitance values and a set of linear equality constraints as described above.

The sink load capacitances have upper and lower bounds. For each sink node S_i

$$l_i < C_i < u_i \quad (16)$$

We also have a set of linear equality constraints to describe the delay equalities due to the links. Let $S_{left(k)}$ and $S_{right(k)}$ be the two endpoints of link k. W.l.o.g, let $S_{left(k)}$ be the node being split and $S_{left(k)}$ becomes $S_{left(k)}$ and $S_{mid(k)}$ after splitting. Assume that the delays at $S_{left(k)}$ and $S_{right(k)}$ are t_{k1} and t_{k2} respectively, and the delay at $S_{mid(k)}$ is t_{k3}. D^k_{link} is the delay on the link connecting $S_{mid(k)}$ and $S_{right(k)}$. Assume that the link resistance is R^k_{link}, the equality constraints due to the kth link can be written as:

$$t_{k1} = t_{k3} = t_{k2} + D^k_{link} \quad (17)$$

$$D^k_{link} = R^k_{link}c_k \quad (18)$$

To summarize, the optimization problem for maximizing the objective function $f_{u,v}$ between node S_u and S_v is:

Maximize: $\qquad f_{u,v} = |t_u - t_v|/r_{u,v}$

Constraints:

$$t_u = f_u(C_1, C_2, ..., C_n, c_1, c_2, ..., c_m) \quad (19)$$

$$t_v = f_v(C_1, C_2, ..., C_n, c_1, c_2, ..., c_m) \quad (20)$$

$$l_i < C_i < u_i \qquad i = 1,...,n$$

$$t_{k1} = t_{k2} + D^k_{link} \qquad k = 1,...,m \quad (21)$$

In our implementation, instead of calling a linear solver, we will first use Gaussian elimination to solve the k equality constraints (21), so the k variables c_i can be expressed by C_1, $C_2, ..., C_n$. Equation (19) and (20) become

$$t_u = f_u'(C_1, C_2, ..., C_n) \quad (22)$$

$$t_v = f_v'(C_1, C_2, ..., C_n) \quad (23)$$

Then the maximum objective value subject to changes in the sink load capacitances can be computed directly from the upper and lower capacitance bounds.

The maximum $f_{u,v}$ (called $p_{u,v}$) for a pair of node S_u and S_v can be obtained by solving the above linear program. Finally, we can find a node pair with the largest $p_{u,v}$. A crosslink is added between this node pair to reduce the clock skew while constraining the wirelength of the link. By applying the technique recursively, we can add a user-defined number of crosslinks to construct a link based clock network.

3.2 Reducing the Number of Optimizations

Although the optimization problem for a pair of nodes can be solved quickly as formulated above, we need to run it for all pairs of nodes to get the node pair with p_{max}. The running time will increase quickly when the number of nodes increases to thousands. Therefore, we devised a method to speed up the process when the number of sinks is large. In order to reduce the running time, instead of trying all pairs of sinks, we will choose some node pairs which are more likely to have the p_{max}. Actually there are some pairs that we do not need to consider, e.g., nodes that are topologically close to each other. This is because these node pairs share a long mutual sub-path from the source and the delay differences between them will not be big.

Starting from the sink node level, we will travel up to find a set of internal nodes I such that the set S of subtrees rooted at these internal nodes should cover all the sink nodes. The detailed procedure to find the internal nodes is shown in Fig. 5. Compared with the total number of sinks, the number of these internal nodes will be smaller. Since the sink nodes within a subtree in S will likely to have similar delays, we will not consider their skews. We will pick one node from each subtree in S to find their skew. In our implementation, we will also consider the number of sinks in each subtree in S and limit them to be less than or equal to 16. In this way, even benchmarks with a large number of sinks can be solved in a reasonable amount of time.

```
Algorithm: Find Nodes
Input: A clock network
Output: Internal node set I
1. I ← Φ
For each sink node S_i
2. If S_i is contained in a subtree rooted at a node in I
3.    Continue
4. Else
5.    k ← 1, r ← 1, I_i ← S_i
6.    While r < lower bound in the sink number of a subtree
7.       k ← k + 1, I_i ← parent of I_i, temp ← I_i
8.       r ← number of sinks contained in the subtree rooted at I_i
9.    End
10.   If r > upper bound in the sink number of a subtree
11.      I_i ← temp
12.   Add I_i to I
```

Fig. 5. Find a Set of Internal Nodes I

4. EXPERIMENTS

Our algorithm to select node pairs for non-tree clock network construction with crosslinks is implemented in C. The experiments are performed on a 3.2 GHz Intel CPU Linux machine with 4GB memory. To verify the effectiveness of our method, we will compare our link insertion scheme with the work in [1] and the work in [4]. These works did not consider non-tree delay when inserting links. We implemented the minimum weight matching-based link insertion method which is the best method in [1]. We start with the same zero skew clock tree which is obtained from the bounded skew tree method [2]. The benchmarks and the bounded skew tree code are downloaded from the GSRC bookshelf (http://vlsicad.eecs.umich.edu/BK/). For comparison with the work in [4], experiments are performed on the ISPD 2010 contest benchmark. In our implementation and simulations, the per unit wire resistance is 10^{-4} Ω/nm and the per unit wire capacitance is 2×10^{-4} fF/nm. *Ngspice* is used to simulate our clock distribution network. Process variations are accounted in the simulations for VDD variations and sink capacitance variations. We allow 15% variations in the VDD and the sink capacitances and all the variations follow a normal distribution. For each clock network, 500 *spice* simulations are performed to obtain the worst case skew (WCS).

All the benchmark information and the experimental results of our work and the method in [1] are shown in Table 1. The benchmark sizes, worst case skews (WCS), and the wire capacitances of the clock trees are given. We list the number of links inserted, WCS results, link capacitance results of the Link-M method and our work. The Link-M method refers to the minimum weight matching based link insertion method in [1]. From the table, we observe that our method always outperforms the Link-M method in terms of clock skew reduction. Besides, for a same number of links, Link-M method costs 249% link capacitance comparing with ours on average. The reason is that the Link-M method does not consider non-tree delay when inserting links, which may result in ineffective crosslink insertion into the clock tree. By considering the balance between clock skew and link resistance, we can achieve more clock skew reduction, and the link capacitance can also be reduced. We list the CPU time in seconds in the table. The CPU time for the Link-M method is ignorable so it is not listed in the table. The technique discussed in Section 3.3 can be used to control the running time and we can see the running time is acceptable.

To compare our proposed method with the most recent work [4] which inserts links in the internal nodes of a tree and insert links while constructing the tree, we obtain the clock network results from the authors of [4]. The links added between the internal nodes are removed before our method is applied to generate crosslinks into the clock network. Table 2 compares our work with the method in [4]. LCS refers to the local clock skew constraint on the benchmark in ISPD 2010 High Performance Clock Network Synthesis Contest [15]. Our work uses less link resources to achieve similar LCS results. Note that the CPU time of our work in the table is for the link insertion phase only, while the tree construction time is not included. Please note that the two results are not directly comparable, because in [4], the links are inserted while the trees are being constructed, so their tree construction performs in such a way to favor the link insertion. In our case, we take their trees and insert the links in a post processing way, so the results are hard to be compared. However, we still want to display the comparisons to show that our method

can also handle the clock trees with buffer and can perform actually quite well comparing with [4] in which the link insertion and the tree construction are performed simultaneously.

5. CONCLUSION

In this paper, signal delay and clock skew in non-tree clock networks are discussed. Based on the analysis, a new method is proposed to select node pairs for link insertion in a clock network. Experimental results show that this method can be applied to insert links effectively.

6. REFERENCES

[1] A. Rajaram, J. Hu, and R. Mahapatra. Reducing clock skew variability via crosslinks. *IEEE Trans. Computer-Aided Design of Integrated Circuits and Systems*, vol. 25, no. 6, 2006, 1176–1182.

[2] J. Cong, A. B. Kahng, C.-K. Koh, and C.-W. A. Tsao. Bounded-skew clock and steiner routing under Elmore delay. In *Proceedings of the 1995 IEEE/ACM International conference on Computer-aided design*, 1995, 66–71.

[3] P. Chan and K. Karplus. Computing signal delay in general RC networks by tree/link partitioning. *IEEE Trans. Computer-Aided Design of Integrated Circuits and Systems*, vol. 9, no. 8, 1990, 898–902.

[4] T. Mittal and C.-K. Koh. Cross link insertion for improving tolerance to variations in clock network synthesis. In *Proceedings of the 2011 international symposium on Physical design*, 2011, 29–36.

[5] P. Restle, T. McNamara, D. Webber, P. Camporese, K. Eng, K. Jenkins, D. Allen, M. Rohn, M. Quaranta, D. Boerstler, *et al.*. A clock distribution network for microprocessors. *IEEE Journal of Solid-State Circuits*, vol. 36, no. 5, 2001, 792–799.

[6] G. Venkataraman, Z. Feng, J. Hu, and P. Li, Combinatorial algorithms for fast clock mesh optimization. In *Proceedings of the 2006 IEEE/ACM international conference on Computer-aided design*, 2006, 563–567.

[7] H. Chen, C. Yeh, G. Wilke, S. Reddy, H. Nguyen, W. Walker and R. Murgai. A sliding window scheme for accurate clock mesh analysis. In *Proceedings of the 2005 IEEE/ACM International conference on Computer-aided design*, 2005, 939–946.

[8] A. Rajaram and D. Z. Pan. MeshWorks: An efficient framework for planning, synthesis and optimization of clock mesh networks. In *Proceedings of the 2008 Asia and South Pacific Design Automation Conference*, 2008, 250–257.

[9] W. -C. D. Lam, J. Jain, C. -K. Koh, V. Balakrishnan, and Yiran Chen. Statistical based link insertion for robust clock network design. In *Proceedings of the 2005 IEEE/ACM International conference on Computer-aided design*, 2005, 588-891.

[10] G. Venkataraman, N. Jayakumar, J. Hu, P. Li, S. Khatri, A. Rajaram, P. McGuinness, and C. Albert. Practical techniques for minimizing skew and its variation in buffered clock networks. In *Proceedings of the 2005 IEEE/ACM International conference on Computer-aided design*, 2005, 592–596.

Table 1. Benchmark information, worst case skew and wire cap results with 15% variations in vdd and sink capacitances

Bench -mark	#Sinks	WCS (ps)	Wire Cap (fF)	#Links	Link-M WCS (ps) a (a/b)	Our WCS (ps) b	Link-M Link Cap (fF) c (c/d)	Our Link Cap (fF) d	CPU (s)
r1	267	0.364	264.1	4	0.314 (1.33)	0.236	9.7 (3.24)	2.9	0.81
				16	0.162 (1.41)	0.115	16.3 (1.23)	13.2	3.21
				22	0.122 (1.20)	0.102	60.0 (3.14)	19.1	4.52
r2	598	0.802	523.7	4	0.772 (1.68)	0.460	6.9 (4.93)	1.4	9.23
				20	0.286 (1.39)	0.206	21.0 (1.60)	13.1	46.47
				40	0.192 (1.42)	0.135	40.7 (1.46)	27.9	94.09
r3	862	1.250	678.2	6	0.470 (1.09)	0.433	10.2 (1.96)	5.2	2.74
				24	0.305 (1.08)	0.282	25.9 (1.07)	24.3	11.73
				48	0.229 (1.36)	0.168	101.6 (1.91)	53.2	27.21
r4	1903	2.054	1357.8	8	1.565 (1.55)	1.008	41.6 (8.67)	4.8	9.21
				40	0.448 (1.21)	0.370	57.3 (1.23)	46.5	64.34
				68	0.336 (1.03)	0.327	125.3 (1.42)	88.3	140.3
r5	3101	2.612	2005.5	8	1.451 (1.14)	1.276	43.2 (7.20)	6.0	36.17
				40	0.740 (1.14)	0.648	76.2 (1.67)	45.7	229.8
				72	0.589 (1.16)	0.506	95.9 (1.08)	89.0	510.5
s1423	74	0.048	21.30	4	0.042 (1.83)	0.023	1.33 (1.17)	1.14	0.03
				8	0.036 (2.57)	0.014	2.77 (1.07)	2.60	0.03
s5378	179	0.061	35.10	4	0.063 (1.75)	0.036	1.60 (2.86)	0.56	0.29
				12	0.025 (1.09)	0.023	5.94 (2.57)	2.31	0.89
s15850	597	0.136	89.65	4	0.111 (1.34)	0.083	0.56 (1.27)	0.44	10.62
				22	0.053 (1.39)	0.038	3.88 (1.56)	2.48	59.82

Table 2. Comparison of our method with the work in [4]

Bench -mark	#Sinks	Method	LCS (ps)	Ratio of Link Cap	CPU(s)
01	1107	[4]	7.88	1.007	1092
		Our work	10.67	1	52.4
02	2249	[4]	8.32	1.223	4314
		Our work	8.17	1	424
03	1200	[4]	6.34	1.073	383
		Our work	5.89	1	10.6
04	1845	[4]	7.42	1.001	934
		Our work	7.33	1	33.8
05	1016	[4]	5.90	1.199	278
		Our work	5.87	1	5.36
06	981	[4]	6.78	1.003	285
		Our work	8.98	1	6.69
07	1915	[4]	6.77	1.228	818
		Our work	6.05	1	49.8
08	1134	[4]	6.42	1.225	327
		Our work	8.59	1	8.34

[11] Anand Rajaram and David Z. Pan. Variation tolerant buffered clock network synthesis with cross links. In *Proceedings of the 2006 international symposium on Physical design*, 2006, 157–164.

[12] Z. Li, Y. Zhou, and W. Shi. Wire sizing for non-tree topology. *IEEE Trans. Computer-Aided Design of Integrated Circuits and Systems*, vol. 26, no. 5, 2007, 872–880.

[13] T. M. Lin and C. A. Mead. Signal delay in general RC networks. *IEEE Trans. Computer-Aided Design of Integrated Circuits and Systems*, vol. 3, no. 4, 1984, 331–349.

[14] P. K. Chan and M. D. F. Schlag. Bounds on signal delay in RC mesh networks. *IEEE Trans. Computer-Aided Design of Integrated Circuits and Systems*, vol. 8, no. 6, 1989, 581–589.

[15] http://archive.sigda.org/ispd/contests/10/ispd10cns.html.

WRIP: Logic Restructuring Techniques for Wirelength-Driven Incremental Placement

Xing Wei, Wai-Chung Tang,
Yu-Liang Wu
Department of Computer Science and
Engineering
The Chinese University of Hong Kong
{xwei, wctang, ylw}@cse.cuhk.edu.hk

Cliff Sze,
Charles Alpert
IBM Austin Research Center
11400 Burnet Road
Austin TX 78758
{csze, alpert}@us.ibm.com

ABSTRACT

This paper presents WRIP—a **W**irelength-driven **R**ewiring-based **I**ncremental **P**lacement which effectively reduces wirelength of the optimized placement of industrial large-scale standard cell designs. WRIP uses a powerful logic synthesis technique called logic rewiring which restructures the local circuits while preserving the logic functionality and reduces the wirelength under an accurate estimation of the half perimeter wirelength (HPWL) metric. We integrated WRIP into an industrial EDA tool and tested it upon several real designs with hundreds of thousands of movable objects. Tested on circuits which has been fully optimized by the state-of-the-art industrial placement tool, our experiments showed that on average WRIP reduces wirelength by 2.25% after placement and 2.45% after global routing in HPWL and Steiner WL model respectively. The runtime of WRIP is only about half an hour for the largest tested ASIC circuit. This is the first attempt to fully integrate powerful logic synthesis into industrial placement tools with real-life effectiveness and efficiency.

Categories and Subject Descriptors

B.6 [**LOGIC DESIGN**]: Design Aids; B.6.3 [**Automatic synthesis**]: Optimization

General Terms

Algorithms

Keywords

logic rewiring, HPWL, wirelength driven, incremental placement

1. INTRODUCTION

It is a well-known fact that wire delay dominated the total path delay as we entered the sub-micron era. This greatly increases the importance of the VLSI cell placement problem because gate locations directly determine the minimum achievable path delays. The classic wirelength-driven placement problem, which seeks to minimize total wirelength in the designs, has been the core engine of modern backend physical design tools. It is mainly because (1) total wirelength is usually a first-order estimation of circuit timing; (2) timing-driven placement tools often use net weights to minimize delay along critical paths and it is straight-forward to apply wirelength-driven core placement engine toward the objective of weighted wirelength minimization; (3) routing congestion has evolved as the new bottleneck of physical design flows, and congestion mitigation is often highly correlated with wirelength minimization.

The problem of wirelength driven placement has been well-researched for the last couple of decades. Due to the ISPD placement contests in 2005 and 2006, placement algorithms have been greatly improved since then. The basic problem formulation often seeks to minimize total wirelength, HPWL or other objectives under the constraint that cells do not overlap each other and density targets. In the literature, many placement algorithms have been proposed to handle wirelength minimization, such as simulated annealing [1, 2], partitioning/clustering [3, 4], and analytical placement [5, 6].

Besides the global placement framework, incremental placement and post-placement optimization algorithms are also proposed in [7, 8]. An incremental placer starts with an initial placement as its input, and focuses on further optimization for wirelength, timing, or other objectives. A timing-driven incremental placement is proposed in [7] that uses linear programming to meet delay constraints. In [8], a safe netlist transformations which do not create any illegal cell overlap during the incremental placement is proposed.

As designs are getting bigger with higher frequency targets and relatively less routing resource [9], physical synthesis has became very challenging for post-90nm technology. Traditional design flows of logic synthesis followed by placement and routing is no longer good enough for modern designs. In other words, we observed the necessity to make use of more powerful logic transforms during core placement steps in order to further improve the placement quality.

This paper aims at demonstrating a practical framework to integrate logic synthesis into the classical wirelength-driven placement problem. We discovered that logic rewiring is fast and very effective in our proposed incremental placement flow for wirelength minimization, which makes it the perfect logic synthesis engine for this purpose. More discussion on logic rewiring will be included in Section 2.

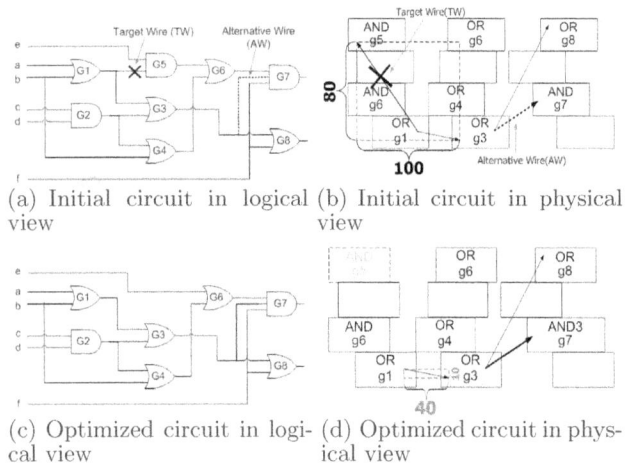

(a) Initial circuit in logical view (b) Initial circuit in physical view

(c) Optimized circuit in logical view (d) Optimized circuit in physical view

Figure 1: Reduce HPWL by restructuring circuit

An example of our incremental placement step is shown in Figure 1. Figure 1(a) is a sample circuit in logical view while Figure 1(b) is the corresponding placement of the gates. The half-perimeter wirelength (HPWL) of the net driven by $g1$ is 180. By logic rewiring techniques, we observed that if we add a logical connection from $g3$ to $g7$, the connection from $g1$ to $g5$ can be removed without changing the logical functionality of the whole circuit. As one can verify, the output logic function of $g7$ is exactly the same $((e(a+b)+b+cd)f$ verse $(e+b+cd)(a+b+cd)f)$. The circuit (logical and physical view) after rewiring is drawn in Figure 1(c) and (d). As shown in the figures, the HPWL of the net driven by $g1$ is reduced to 50 while the HPWL of net driven by $g3$ does not increase. After the rewiring, the total HPWL is reduced by 130.

In this paper, we present a **W**irelength-driven **R**ewiring-based **I**ncremental **P**lacement framework (WRIP), which uses a powerful logic restructuring technique called logic rewiring to further improve half perimeter wirelength (H-PWL) of initial placement solutions. In the prior arts, most global placement or incremental placement methods focus on placing cells at better locations to improve total wirelength or timing while little logic restructuring technique is involved to further optimize the placement. There exists little study on combining logical and physical synthesis [8, 10]. Furthermore, these approaches suffer from the runtime issue caused by the high complexity of logic restructuring algorithms which prevents the industrial placer from fully combining with these logical restructuring techniques. In [10], the algorithm takes more than 1000 seconds optimizing a circuit with about 7000 cells, which is not scalable for industrial large-scale standard cell designs. The contribution of this paper can be summarized as follows.

1. It is in fact the first work to practically integrate logic synthesis transforms into industrial placement tools with high effectiveness and efficiency for real VLSI designs.

2. A thorough analysis of HPWL estimation for incremental logic synthesis is presented and is utilized to guide the local circuit restructuring.

3. An efficient algorithm is proposed which can reduce HPWL of placements already well optimized by industrial tools.

The rest of this paper is organized as follows. Section 2 formally defines the problem as well as the related terminologies. Section 3 presents our wirelength-driven rewiring-based incremental placement framework. Section 4 reports the experimental results. Finally, we conclude our work in Section 5.

2. PRELIMINARIES AND PROBLEM DEFINITION

The circuit is formulated as a Boolean network in logical view which can be represented by a Directed Acyclic Graph (DAG) where each node represents a logic gate and a directed edge corresponds to a $\mathbf{wire}(i, j)$ where the gate i is an input of gate j. In this paper, we define i to be the source and j to be the sink of the wire (i, j). We also define i is the fanin of j and j is the fanout of i. With our notations, an n-pin net would be equivalent to a set of $n - 1$ wires which have the same source and thus have 1 fanin and $n - 1$ fanouts. We use O_i to denote a cone rooted on node i. O_i is a sub-graph consisting of i and some of its predecessors, such that for any node $w \in O_i$, there is a path from w to i and all fanouts of $w \in O_i$. The maximum cone of i is called the fanin cone of i. WRIP is able to transform the Boolean logic of the local circuit by creating new wires to replace certain specified wires.

Half-perimeter wirelength(HPWL) of a net is defined as the sum of width and height of the bounding box of the net. HPWL has been widely used by different placement algorithms, and is used mainly as the figure of merit in this paper. WRIP works very differently when compared to other traditional HPWL-driven placement algorithms which do not change the Boolean logic function of any gate. WRIP can be applied on fully optimized circuits and still achieve substantial HPWL reduction. WRIP uses logic rewiring to disconnect wires whose removal can greatly reduce HPWL and to create new wires which will not increase HPWL more than the reduction. The HPWL of the whole circuit can be improved by restructuring the local circuits iteratively. Figure 2 shows the HPWL change accompanied by circuit restructuring in different situations.

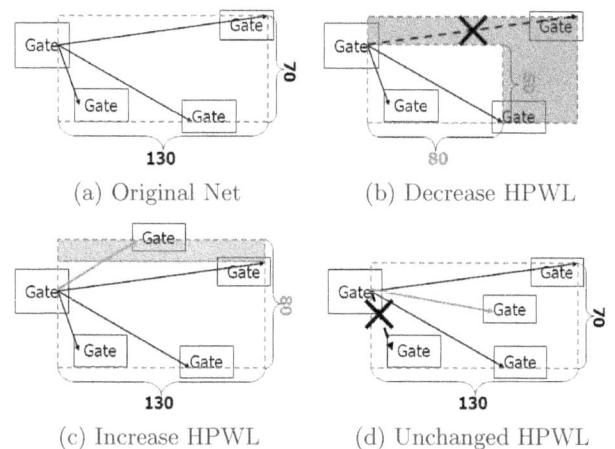

(a) Original Net (b) Decrease HPWL

(c) Increase HPWL (d) Unchanged HPWL

Figure 2: Impact on HPWL by logic rewiring

Figure 2(a) is the original net whose HPWL is 200. The HPWL of net may change if we disconnect a fanout/wire of the net or add a new fanout/wire into the net. The HPWL of the net may decrease to 130 if the wire is removed as shown in Figure 2(b). The HPWL may increase to 210 if a new wire is added as shown in Figure 2(c). Sometimes the HPWL will not change even if some wires are disconnected or added as shown in Figure 2(d). In this paper, we use a bounding box area to denote the HPWL, and the shaded/colored part of the bounding box is the change in HPWL.

Logic rewiring [11, 12] is a circuit restructuring technique referring to the process of replacing a certain wire (target wire, TW) by another wire (alternative wire, AW) without changing the functionality of the circuit. The addition and removal of wires are sometimes accompanied by changes in logic gates corresponding to the AW and TW.

An example of logic rewiring has been shown in Fig. 1(a) and (c). The target wire is $(g1, g5)$ and the alternate wire is $(g3, g7)$. Gate $g7$ is expanded into a 3-input AND gate (AND3) after addition of the alternative wire. Some theories and techniques in logic rewiring can be found in [11–13].

Given the above definitions, the wirelength-driven rewiring-based incremental placement problem can be formally defined as follows: *Given a logic netlist and its physical placement, the wirelength-driven rewiring-based incremental placement problem is to further reduce the total wirelength by (1) restructuring local circuit while preserving the functionality and (2) changing local placement while keeping the placement free of overlap.*

3. ALGORITHM OF WRIP

In this section, we will introduce how WRIP reduces HPWL based on the logic rewiring technique. As logic rewiring creates alternative wires to replace existing target wires, the basic idea is to use logic rewiring to iteratively remove each **troublesome** target wire by adding its alternative wire to yield a new placement with a less total HPWL. In this paper, we refer to a "troublesome" wire as a wire when being removed the total HPWL can be largely reduced.

We first introduce the overall framework of the algorithm. Then we will explain in detail the identification of target wires and how to find the best alternative wire for each TW.

3.1 The Overall Framework

We implemented a rewiring engine using one of the latest rewiring schemes [13]. By calling the engine with the specific target wire, we are given a set of AW choices with each being a feasible alternative wire to replace the TW. We also proposed a rewiring manager to keep track of the AW/TW sets and to restructure the circuit while preserving the logic functionality. The manager examines each wire to identify all "troublesome" wires and keep them as the set of target wires. For each target wire, we invoke the rewiring engine and the set of alternative wire candidates will be returned. The manager will then select the best AW to obtain the best HPWL reduction.

Our framework is integrated into an industrial placement tool and WRIP is performed at the end of the placement optimization phase. WRIP changes the connections of circuits and the related gates if needed. For example, we have to convert a NAND4 gate into a NAND3 if one input is disconnected. At the end of WRIP, legalization is invoked to

remove the overlaps of the placement. The overall flow of WRIP is shown in Figure 3.

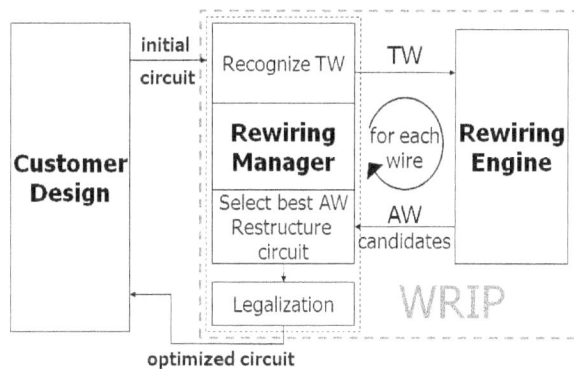

Figure 3: Framework of WRIP

3.2 Target Wire Recognition

In our algorithm, WRIP examines each target wire and calculates the HPWL reduction if it is removed. So, we temporarily remove the wire and estimate the change in HPWL. If the HPWL change is negative, WRIP confirms the wire removal and passes the target wire to the rewiring engine. The HPWL estimation is shown in Figure 4 and Figure 5.

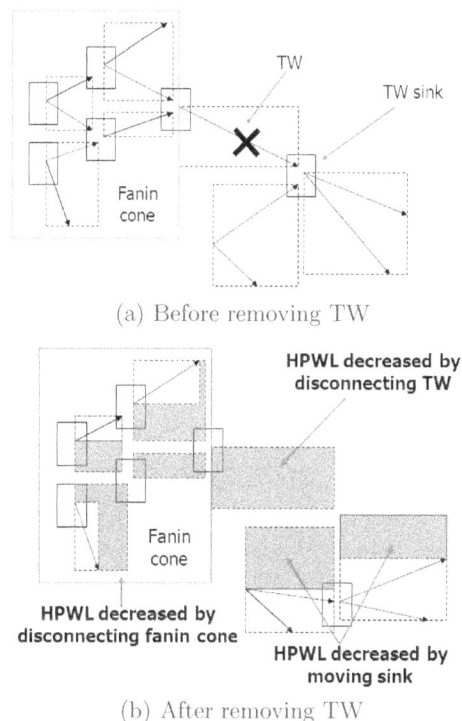

(a) Before removing TW

(b) After removing TW

Figure 4: Estimate HPWL of single fanout net

Figure 4 shows an example of estimating the HPWL reduction when a single fanout net is removed. The estimated HPWL reduction consists of three parts. An obvious one is the HPWL reduction due to disconnecting the target wire, which is shown by the center blue area in Figure 4(b). As the source gate has only one fanout, the gate will become redundant and the fanin cone of the source can also be removed without affecting the overall logic functionality. The

(a) Before removing TW   (b) After removing TW

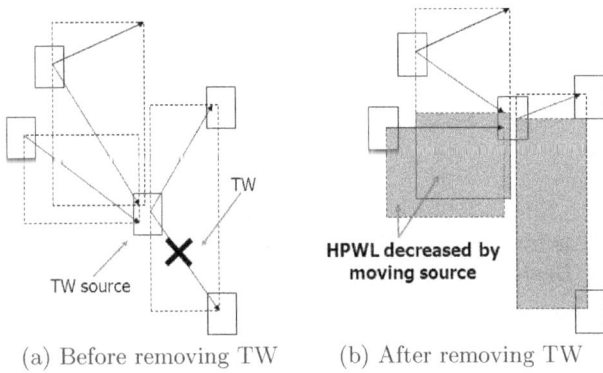

Figure 5: Estimate HPWL of multiple fanout Net

(a) Before adding AW   (b) After adding AW

Figure 7: HPWL increase due to AW addition

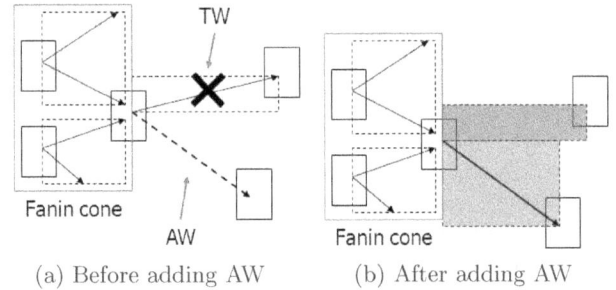

estimated HPWL reduction due to disconnecting the fanin cone of the source gate is shown by the left blue area in Figure 4(b). After the TW is removed, the sink gate of the TW can be relocated to a better location to reduce the HPWL of its fanin and fanout nets. It has been shown that the local HPWL is minimal if the gate is placed at the x-median and y-median of all its source pins of the fanin nets and sink pins of its fanout nets [14, 15]. Such HPWL reduction is shown by the right blue area in Figure 4(b).

Figure 5 shows the steps to estimate HPWL reduction when TW is in an n-pin nets where $n \geq 2$. The TW source gate cannot be removed after removing the TW so WRIP cannot simply disconnect fanin cone. In this case, the potential HPWL reduction of disconnecting the fanin cone will always be 0. But we can relocate TW source gate to achieve further HPWL reduction. The change in HPWL by moving TW source is shown by the left blue area in Figure 5(b).

We show the HPWL updating due to TW removals in Figure 6. The HPWL reductions by removing TWs and relocating TW sink gates are always realizable. When the target net has only one fanout, we can get an extra HPWL reduction by deleting the fanin cone of TW source gate. When the net has ≥ 2 fanouts, we can achieve another extra HPWL reduction by moving the TW source to the optimal region as described previously.

(a) Before adding AW   (b) After adding AW

Figure 8: Revise HPWL estimation
when TW&AW have the same source

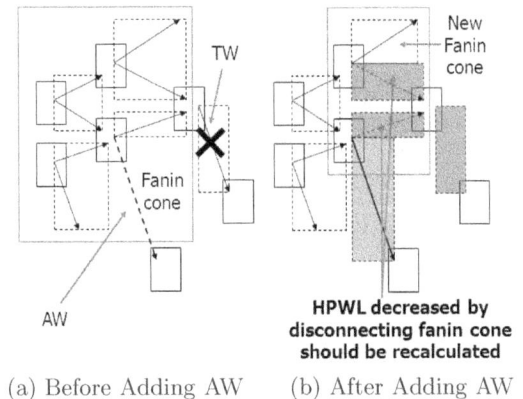

	HPWL Reduction			
	Disconnect TW	Move Sink	Disconnect fanin cone	Move Source
Condition	Always available	Always available	Target net has only 1 fanout	Target net has 2+ fanouts

Figure 6: HPWL reduction due to TW removal

3.3 Alternative Wire Selection

When we attempt to remove a target wire, the rewiring engine usually returns more than one AWs, each of which can replace the target wire. In this section, we will discuss how to select the best AW among candidates to gain most HPWL reduction. For each AW candidate, we temporarily add it into the netlist and estimate the HPWL increase. More importantly, we need to *recalculate* the HPWL reduction due to AW addition because TW and AW may share the

same source or sink gates. Figure 7, 8 and 9 show different situations for adding an AW.

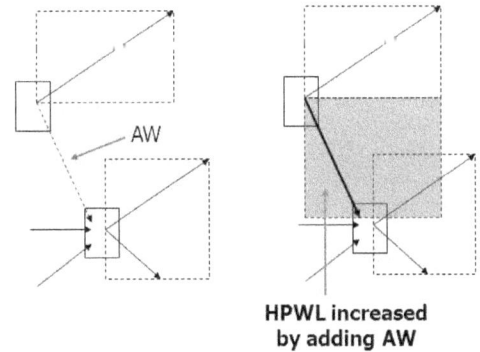

(a) Before Adding AW   (b) After Adding AW

Figure 9: Revise HPWL estimation
when AW is in the fanin cone of TW

Figure 7 shows the most common case when adding the AW into the circuit. The brown area shown in Figure 7(b) is the HPWL increase due to the AW addition.

Figure 8 shows a special case that TW and AW have the same source gate. In this case, as the source gate has only one fanout which is removed, the fanin cone of the source could have been disconnected before adding the AW and we could have HPWL reduction by disconnecting fanin cone. After adding the AW, the previously mentioned source gate cannot be removed because it has to drive the AW. The HP-WL reduction by disconnecting fanin cone has to be revised to 0 for none fanin is able to be disconnected which is shown in Figure 8(b).

Figure 9 shows another special situation that the source gate of AW belongs to the fanin cone of TW. In this case, the fanin cone will change after creating AW and the HPWL reduction by disconnecting fanin cone has to be recalculated. The estimation revision of HPWL reduction is shown in Figure 9(b).

We summarize the estimation of HPWL reduction considering both TW removal and AW addition in Figure 10. The estimation is composed of 5 parts. WRIP estimates the HPWL reduction by disconnecting TW, moving TW sink and source gates, deleting the fanin cone of the source gate before calling rewiring engine. Once an AW is identified, the HPWL reduction by disconnecting fanin cone will be revised and the HPWL increase by AW addition will be calculated. The estimation of total HPWL reduction is computed as:

$$
\begin{aligned}
\Delta TotalHPWL = {} & \Delta HPWL_{disconnectTW} \\
& + \Delta HPWL_{moveTWsink} \\
& + \Delta HPWL_{moveTWsource} \\
& + \Delta HPWL_{disconnectTWfanincone} \\
& + \Delta HPWL_{createAW} \qquad (1)
\end{aligned}
$$

TW&AW Pair	HPWL Reduction				
	Disconnect TW	Move TW sink	Disconnect TW fanin cone	Move TW source	Add AW
Different source	√	√	√	√	√
Same source	√	√	0	√	√
AW in TW fanin cone	√	√	recalculate	√	√

Figure 10: HPWL reduction considering TW and AW

The total flow of WRIP is shown in Algorithm 1. We traverse each wire and identify "troublesome" target wires if the TW removal can potentially reduce HPWL. We call rewiring engine with the target wire and it generates a list of alternative wire candidates. For each TW, WRIP estimates the HPWL reduction for each TW&AW pair and chooses the best pair which achieves the best HPWL reduction.

Table 1: Benchmark information

Circuit	HPWL	#Cell	#Net	#Wire
case1	17307504	98631	100315	316350
case2	50933622	318703	324102	846651
case3	55531283	292141	295995	823244
case4	52551332	295781	300312	826153
case5	48316683	322038	327605	865159
case6	156464726	394915	408776	1340470
total	381105150	1722209	1757105	5018027

4. EXPERIMENTAL RESULTS

We implement WRIP in C++ which includes both the rewiring engine [13] and the rewiring manager (Algorithm 1), and integrate it into an industrial EDA tool. All the

---

Algorithm 1 WRIP Algorithm

```
    for each target wire TW = (i, j) do
2:      temporarily remove TW and estimate HPWL
        reduction(ΔHPWL) by the following 4 lines
        estimate ΔHPWL by disconnecting TW
4:      estimate ΔHPWL by moving TW sink i
        estimate ΔHPWL by moving TW source j
6:      estimate ΔHPWL by disconnecting fanin cone of i
        calculate total HPWL reduction after removing TW
8:      if total HPWL reduction <= 0 then
            continue
10:     end if
        call rewiring engine and collect all AW candidates for TW
12:     bestHPWLReduction = 0
        bestAW = NULL
14:     for each AW candidate do
            temporarily add AW into the netlist
16:         revise ΔHPWL by disconnecting fanin cone
            estimate HPWL increased by adding AW
18:         calculate total HPWL reduction after creating new
            wire(ΔTotalHPWL)
            if bestHPWLReduction < ΔTotalHPWL then
20:             bestHPWLReduction = ΔTotalHPWL
                bestAW = currentCandidate
22:         end if
        end for
24:     if bestAW ≠ NULL then
            permanently remove TW and add bestAW into cir-
            cuit
26:     end if
    end for
28: legalize optimized circuit and minimize total cell move-
    ment [16]
```

Table 2: Real versus estimated HPWL reduction

Circuit	Real HPWL Reduction	Estimated HPWL Reduction	Difference	Difference Ratio
case1	263149	262519	630	0.24%
case2	494169	532313	38144	7.72%
case3	963262	1235655	272392	28.28%
case4	1138092	1519884	381792	33.55%
case5	889164	970413	81249	9.14%
case6	2910768	2959316	48548	1.64%
total	6658604	7480100	822755	12.36%

experiments are performed on a Linux workstation with Intel(R) Xeon(R) CPUs X7350 running at 2.93GHz. We tested WRIP on 6 real high-performance ASIC designs. Table 1 shows the basic information of benchmarks.

In Table 1, column 2 shows the HPWL at the end of an industrial EDA placement tool [17,18]; column 3 is the number of cells which includes sequential cells such as registers and latches, arithmetic cells such as adders and subtractors, and simple combinational cells such as NAND, MUX, AOI gates, etc; column 4 is the number of nets while column 5 is the number of wires, which implies that each circuit net has about 3 fanouts on average.

Table 2 demonstrates the accuracy of our estimation of HPWL reduction. We compare the estimated HPWL reduction with real HPWL reduction obtained after the diffusion-

Table 3: Effectiveness of our framework for placement and global routing

Circuit	placement				global routing		
	Initial HPWL	HPWL reduction	Reduction ratio	Runtime (s)	Initial Steiner WL	Steiner WL reduction	Reduction ratio
case1	17307504	337819	1.95%	130	20487793	509694	2.43%
case2	50933622	640299	1.26%	352	65201994	1123606	1.69%
case3	55531283	1210168	2.18%	443	67858981	1956429	2.80%
case4	52551332	1433974	2.73%	490	66365291	2437595	3.54%
case5	48316683	1137464	2.35%	463	58014446	1856664	3.10%
case6	156464726	3818605	2.44%	1858	199685147	4114180	2.02%
total	381105150	8578329	2.25%	3736	477613652	11998168	2.45%

based legalization. The results show merely 12.36% of difference comparing the estimated and real reduction on average. WRIP involves gate creation, deletion and movement, all of which may trigger legalization which would lead to a change in HPWL reduction. Compared to final HPWL reduction shown in column 2, we can see that the estimated HPWL reduction shown in column 3 is a little bit more optimistic for almost all circuits except case 1. Among our test cases, the cell density of designs 3 and 4 are much higher than other designs so that legalization has a bigger impact to the change of HPWL reduction, as shown in row 4 and 5.

The first 4 columns of table 3 shows the HPWL reduction of WRIP. Column 2 shows the initial HPWL and column 3 and 4 show the HPWL reduction obtained by WRIP. Column 5 shows the total runtime (for both rewiring engine and manager) for each design. The results show that WRIP can reduce total HPWL by 2.25% on average on top of a fully optimized industrial placement and the optimization can be finished in half an hour for the largest circuit.

We find that the HPWL reduction we gain can be well reflected in the subsequent physical design stage. We compare the wirelength between initial circuit and optimized circuit after performing global routing [17,18] respectively and use a more accurate Steiner wirelength model to calculate the total wirelength after global routing. The results are shown in the last three columns of table 3. We can see the Steiner wirelength can be reduced by 2.45% on average.

5. CONCLUSIONS

In this paper, we proposed WRIP, the wirelength-driven rewiring-based incremental placement which utilizes both logical synthesis and placement optimization to further improve total HPWL upon fully optimized designs. WRIP uses a powerful logic restructuring technique called logic rewiring and a thorough and accurate estimation technique for HPWL reduction was presented to guide the optimization effectively. We integrate WRIP into an industrial EDA tool and test it on real high-performance ASIC designs. Experiments showed that 2.25% total HPWL reduction and 2.45% total Steiner WL reduction are obtained for placement and global routing respectively, which is a useful step toward full integration between logic synthesis and physical design.

ACKNOWLEDGEMENT

This work has been in part supported by Hong Kong CUHK-DG2050501, RGC2150648 and ITS/261/09FP grants.

6. REFERENCES

[1] J.Vygen, "Timberwolf 3.2: A new standard cell placement and global routing package," in *Proc. Design Automation Conference*, pp. 432–439, 1986.

[2] M.Wang, X.Yang, and M.Sarrafzadeh, "Dragon2000: Standard-cell placement tool for large industry circuits," in *Proc. IEEE/ACM International Conference on Computer-Aided Design*, pp. 260–263, 2000.

[3] T.Chan, J.Cong, T.Kong, and J.Shinnerl, "Multilevel optimization for large-scale circuit placement," in *Proc. IEEE/ACM International Conference on Computer-Aided Design*, pp. 171–176, 2000.

[4] J. Z.Yan, C. Chu, and W.-K. Mak, "Safechoice: A novel approach to hypergraph clustering for wirelength-driven placement," *IEEE Trans. Comput.-Aided Design Integr. Circuits Syst.*, vol. 30, pp. 1020–1033, July 2011.

[5] N. Viswanathan and C. C.-N. Chu, "Fastplace: Efficient analytical placement using cell shifting, iterative local refinement and a hybrid net model," *IEEE Trans. Comput.-Aided Design Integr. Circuits Syst.*, vol. 24, pp. 722–733, May 2005.

[6] P.Spindler, U.Schlichtmann, and F.M.Johannes, "Kraftwerk2: a fast force-directed quadratic placement approach using an accurate net model," *IEEE Trans. Comput.-Aided Design Integr. Circuits Syst.*, vol. 27, pp. 1398–1411, Aug. 2008.

[7] C.Wonjoon and K.Bazargan, "Incremental placement for timing optimization," in *Proc. IEEE/ACM International Conference on Computer-Aided Design*, pp. 463–466, 2003.

[8] K. hui Chang, I. L. Markov, and V. Bertacco, "Saferesynth: A new technique for physical synthesis," *IEEE Trans. VLSI Syst.*, vol. 41, pp. 544–556, July 2008.

[9] A. C. J., Z. Li, M. M. D., N. Gi-Joon, R. J. A., and T. Gustavo, "What makes a design difficult to route," in *Proc. IEEE/ACM ISPD*, pp. 7–12, 2010.

[10] S. M.Plaza and I. L.Markov, "Optimizing nonmonotonic interconnect using functional simulation and logic restructuring," *IEEE Trans. Comput.-Aided Design Integr. Circuits Syst.*, vol. 27, pp. 2107–2119, December 2008.

[11] S.-C. Chang, L. Van Ginneken, and M. Marek-Sadowska, "Circuit optimization by rewiring," *IEEE Trans. Comput.*, vol. 48, pp. 962–970, Sep 1999.

[12] C. W. Chang and M. Marek-Sadowska, "Theory of wire addition and removal in combinational boolean networks," *Microelectronic Engineering*, vol. 84, no. 2, pp. 229–243, 2007.

[13] X. Yang, T.-K. Lam, and Y.-L. Wu, "ECR: A low complexity generalized error cancellation rewiring scheme," in *Proc. Design Automation Conference*, pp. 511–516, 2010.

[14] S. Goto, "An efficient algorithm for the two-dimensional placement problem in electrical circuit layout," *IEEE Trans. Comput.-Aided Design Integr. Circuits Syst.*, vol. 28, pp. 12–18, Jan. 1981.

[15] N. Viswanathan, M. Pan, and C. Chu, "Fastplace: An efficient multilevel force-directed placement algorithm," *Modern Circuit Placement: Best Practices and Results*, no. 8, pp. 193–226, 2007.

[16] R. Haoxing, P. D. Z., A. C. J., and V. Paul, "Diffusion-based placement migration," in *Proc. Design Automation Conference*, pp. 515–520, 2005.

[17] L. Trevillyan, D. Kung, R. Ruri, L. N. Reddy, and M. A. Kazda, "An integrated environment for technology closure of deep-submicron ic designs," in *IEEE Design and Test of Computers*, 2004.

[18] C. J. Alpert, S. Karandikar, Z. Li, G.-J. Nam, S. T. Quay, H. Ren, C. Sze, P. G. Villarrubia, and M. Yildiz, "Techniques for fast physical synthesis," in *Proceedings of the IEEE*, 2007.

STEP: A Unified Design Methodology for Secure Test and IP Core Protection

Pranav Yeolekar[1], Rishad A. Shafik[2], Jimson Mathew[3], Dhiraj K. Pradhan[4], Saraju P. Mohanty[5]

[1,2,3,4]Dept. of Computer Science, University of Bristol, Bristol, BS8 1UB, UK

[5]NanoSystem Design Laboratory (NSDL), University of North Texas, Denton, TX 76203, USA

csras@bristol.ac.uk[2], jimson@cs.bris.ac.uk[3], pradhan@cs.bris.ac.uk[4], saraju.mohanty@unt.edu[5]

ABSTRACT

Intellectual property (IP) core based embedded systems design is a pervasive practice in the semiconductor industry due to shorter time-to-market and tougher cost competitions. Protecting the design information in these IP cores and securing test from various attacks are two emerging challenges in today's embedded systems design. Recently reported techniques address these challenges considering secure test and IP core protection separately. However, for ensuring high security during IP core functionality and also during test, joint consideration of secure test and IP core protection is much needed. In this paper, we propose a novel and unified design methodology, called STEP (**S**ecure **TE**st and IP core **P**rotection), which addresses the joint objective of secure test and IP core protection. The aim of STEP design methodology is to achieve high security at low system cost using the same key integrated hardware during test and IP core functionality. We evaluate the effectiveness of STEP design methodology considering advanced encryption standard (AES) system as a case study. We show that proposed design methodology benefits from high security and test accuracy, requiring up to 9% higher area and 20% power overheads.

Categories and Subject Descriptors: B.7 HardwareIntegrated Circuits; K.6.5Computing MilieuxSecurity and Protection-*Physical Security*

Keywords: Security and protection; intellectual property core

1. INTRODUCTION

With continued technology scaling to unprecedented levels, embedded systems design is increasingly becoming complex. This is further exacerbated by the shorter time-to-market demands with design constraints related to power, performance and reliability. To address such design complexity, designers have resorted to highly modular, reusable and effective design approach using intellectual property (IP) cores. Although such design approach has proven to be effective to date, an emerging challenge for IP core based design approach is to securely protect the design information from design pirates or hackers. These design pirates or hackers intrude into the design information through various tampering or attack mechanisms, including reverse engineering techniques, power and timing analysis and fault injection or even through stealing fabrication masks, etc [1, 2, 3]. The design information can be then misused by them in the following

two ways. Firstly, the design information can be used to build counterfeit and competitive products, causing direct financial losses [4]. Secondly, the design information can be altered deliberately, inflicting damage of reputation and more financial losses. Hence, protection of IP core design information and functionality is one of the major concerns for semiconductor industry [5, 6].

Traditional IP core design methodology integrates design for test (DfT) features in the hardware design. The premise of the DfT features is that the original and also the added hardware can be validated against various defects or faults to ensure correct functionality [7]. Scan chain based testing is considered as a *de facto* standard of DfT due to its simplicity of design and high fault coverage [8]. It is implemented through insertion of a chain of flip-flops between logic blocks for providing with a mechanism to observe responses of these logic blocks using different test patterns. However, since these scan chains directly reveal the internal state of the logic blocks and their circuits, extracting design information from them becomes easier for design pirates or hackers through response analysis or side channel attack during testing [9]. Hence, secure testing is a critical requirement for DfT [10, 11].

Over the years, researchers have proposed various techniques and methodologies to address IP core protection and secure test. For example, an IP core protection approach using locking of combinational logic circuits was proposed by Roy *et al* [12]. Their protection approach uses separate locking key for every single chip and enables a licensing technique allowing only approved users to be able to unlock the device. Chakraborty *et al* [13] proposed another protection approach using hardware obfuscation technique at netlist level. In this approach, every chip requires activation by a specific input sequence. When activation does not occur, response of the hardware changes randomly. Among others, IP core protection techniques using watermarking were proposed by Castillo *et al* [14] and Kahng *et al* [4]. The watermarking is incorporated by hosting the bits of a digital signature during design specification using combinational logic within the original design. To secure the design from various attack mechanisms during scan chain based testing, a number of different other techniques have been shown. For example, scan chain scrambling technique by Hely *et al* [15], scan chain randomization technique by Lee *et al* [9] and scan chain replacement approach with de Bruijin graph based shift register chains by Fujiwara *et al* [16] were proposed. The main idea in these works is to make side channel based attack difficult by dividing scan chains into sub-chains and making information in the scan chains unpredictable to the attacker. Another secure DfT approach using flipped scan chains was shown by Sengar *et al* [17]. In their approach, inverters are inserted randomly in the scan chains for protection.

The above works address IP protection and secure test separately [11, 12, 13, 17]. However, such consideration do not automatically complement security and protection during test and also during normal IP core functionality. For example, with an IP core protection technique alone, it is still exposed to security threats during testing as it is possible to reverse engineer the bitstream through side channel attacks [7]. Similarly, with a secure DfT alone, it is possible to carry out a response analysis during normal operation to extract design information [14]. To provide with security and protection at all times, it is important that secure DfT and IP core protection are considered as a joint objective, which is the main of this paper. However, system design with such joint objective is confronted with conflicting design requirements with the system cost. This is because, design for IP core protection introduces extra hardware resources. Due to addition of these hardware resources, either fault coverage obtained during testing will need to be compromised, or more scan chains and test patterns would be needed to ensure required fault and test coverage. Moreover, to ensure security during testing, further hardware resources and test patterns would be required, causing high system overhead. As a result of such design requirements with possible overheads involved, design for secure test and IP core protection is highly challenging [7].

In this paper, we propose a *novel* and *unified* STEP (**S**ecure **TE**st and IP **P**rotection) design methodology to address secure test and IP core protection as a joint objective. The aim is to use the same hardware resource for secure test and IP protection to achieve low system cost. We show that STEP is simple to implement and features high security and test accuracy at low system overheads. To the best of our knowledge, this is the first paper that addresses unified methodology with such joint design objectives. The rest of this paper is organized as follows. Section 2 presents the proposed unified design methodology, STEP, for secure test and intellectual property (IP) core protection, while Section 3 details the secure test and IP core protection architectures generated through STEP using an advanced encryption standard (AES) design as a case study. Section 4 presents the comparative system costs and security analysis of the secure AES systems generated using the STEP design methodology with the insecure AES systems designed using traditional methodology. Section 5 concludes the paper.

2. PROPOSED DESIGN METHODOLOGY

In this work, we propose a unified design methodology, STEP, for secure test and IP core protection. The STEP design methodology removes the need to add dedicated hardware separately for security and protection in the system and hence gives low area and power overheads (Section 4 details results of different systems overheads and security analysis). Figure 1 shows the STEP design methodology, highlighting the four major design phases. The first two design phases deal with traditional design methodology based on functional design and design for test (DfT) using scan chains. The other two phases integrate security features into scan chain based test and also incorporate IP core protection. The detailed descriptions of STEP design phases follow.

2.1 Phase I: *Functional Design*

This phase includes the design specification at register-transfer level (RTL). This is followed by simulation to validate functionality of the RTL design. Once validated, the design is then synthesized, which generates netlist of the design. Using this netlist, power and area analysis are carried out. To validate the functionality of the post synthesis design, gate-level simulation is also carried out.

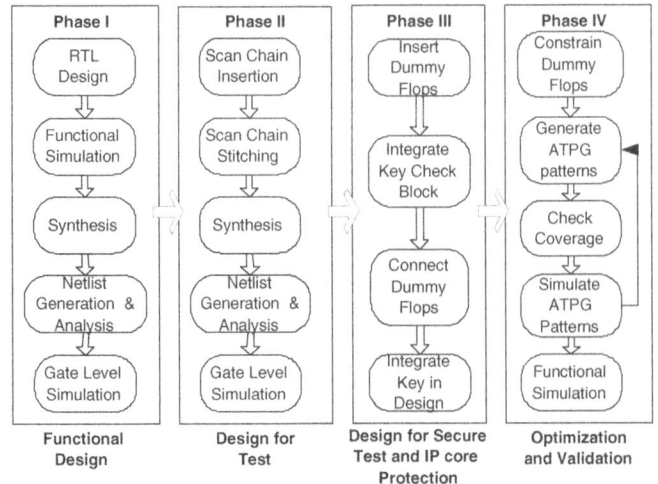

Figure 1: Proposed unified design methodology, STEP, for secure test and IP core protection

2.2 Phase II: *Design for Test*

This phase implements the design for test (DfT) strategy through insertion of scan chains and scan chain stitching in the netlist generated in Phase I. This is done through replacing the original flip-flops by the scan flip-flops and stitched together to form the scan chain. This is then followed by synthesis to generate the new netlist with DfT features. Using this netlist, area and power analyses are carried out to determine the overheads caused by introducing DfT in this design phase. Finally, gate level simulation is carried out to validate the functional behavior of the design (Figure 1).

2.3 Phase III: *Design for Secure Test and IP Core Protection*

This is the most crucial part of the design methodology as hardware changes are made to introduce security in the design. These hardware changes include insertion of dummy flip-flops in the design to form a shift register and integrating key checking hardware block to the design at the netlist level. Due to insertion of dummy flip-flops, the scan chains are broken in the design in Phase II (Figure 1). As a result, the complexity of determining secret information through scan-based side channel attacks increases substantially, making the scan chains secure. To provide further security and protection during test and also during normal operation in the IP core, random key generation and comparison hardware is integrated in the system. The random key generation is carried out through a pseudo-random bit-sequence (PRBS) generator. During testing operation, this PRBS key generator receives seed from scan chain data, while during normal operation the PRBS key generator receives predefined seed for generating a random sequence of numbers. Such key generation makes it very hard for a design hacker to extract bitstreams through reverse engineering. Further details of the key based mechanism for secure test and IP core protection are presented in Section 3.2 using a case study of an AES system.

2.4 Phase IV: *Optimization and Validation*

In this final phase, design optimization and validation is carried out to minimize system cost in terms of area and power. First, the the number and placements of the dummy flip-flops are constrained to minimize the system cost. Then the test patterns for scan chains are generated through automatic test pattern generation (ATPG). With the given test patterns, the effectiveness of the secure test (Phase III) is found out and fault coverage is analyzed through the covered and uncovered faults. Pattern generation and fault coverage analysis is continued until desired coverage is obtained. When desirable coverage is achieved, simulations are carried out to validate the effectiveness and functionality of the system with integrated secure test and IP core protection.

The unified design methodology outlined above can be used to generate a system with integrated secure test and IP core protection architectures. The secure test and IP core protection architectures implemented on an AES system are shown next.

3. SECURE TEST AND IP CORE PROTECTION ARCHITECTURES

In this section, the proposed unified design methodology, STEP (Section 2), is employed for secure test and IP core protection of an advanced encryption standard (AES) benchmark system [18]. The AES system has been chosen as a case study, as also used in [10, 17], since it is widely used in various critical cryptographic applications in finance, banking, security, etc. In the following, secure test and IP core protection architectures of an AES system, generated by proposed STEP design methodology, are presented in details.

3.1 Secure Test Architecture

Figure 2 shows secure test architecture of an AES system generated using STEP design methodology (Figure 1). For demonstration purposes, only two scan chains are shown. As can be seen, to incorporate secure test in the test architecture, dummy flip flops are inserted randomly in the design (Phase III, Section 2). The addition of these dummy flip-flops into the scan chains increase the complexity of determining secret information through scan-based side channel attacks and thus make scan chain based testing secure.

To incorporate further security in the test architecture, key integrated security hardware block is introduced. This hardware block consists of a key checker and a pseudo-random bit-sequence generator (PRBS) (Figure 2). The key checker holds hard-coded secret key, which is only available to a licensed or approved user. The key checker checks this key against the key input from all dummy flops that is N bits wide. The PRBS generator feeds pseudo-random sequences on every clock cycle using the seed from the scan chain inputs.

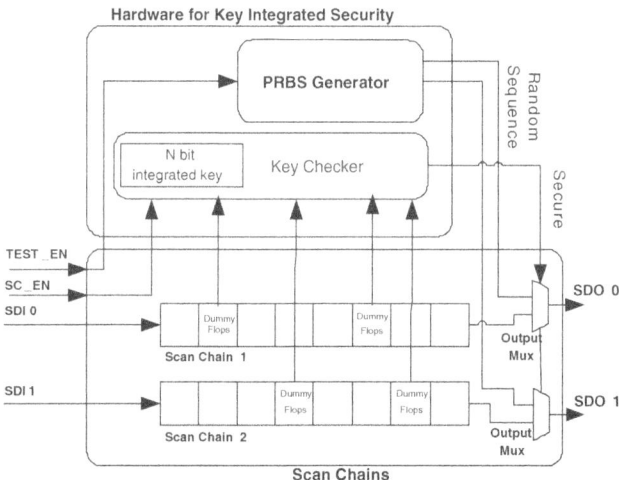

Figure 2: Secure test architecture generated by STEP design methodology for an AES system

With the added hardware resources, the operational sequence in the secure test architecture generated by STEP design methodology is given below:

(a) **Enable Testing Mode:** The secure testing is enabled by HIGH $TEST\_EN$ signal. This also enables key checking mechanism as SC_EN is set to LOW.

(b) **Scan Cycles:** When testing is enabled, the data is shifted into the scan chain through $SDI0$ and $SDI1$ and the response is checked at the output signals $SDO0$ and $SDO1$. During this time, LOW SC_EN acts as a select line for the scan multiplexers. The data shifting happens in $LOAD$ and $SHIFT$ cycles. During $LOAD$ cycle, the internal data from the combinational logic are loaded into the scan chains and during $SHIFT$ cycle, these data are shifted out.

(c) **Key Checking:** To enable these shifted data at the output multiplexer, the key checker must check the hard-coded key in it with the N bits key stored in N dummy flip-flops during every $LOAD$ cycle. When a key match takes place, the key checker generates output as LOW $Secure$ signal, which acts as the select line for enabling the shifted scan data at the output multiplexer (as $SDO0$ and $SDO1$). In case of mismatch, HIGH $Secure$ output signal is generated, which acts as a select line for the output of PRBS generator. The random sequence generated by PRBS is enabled at the output multiplexer. Hence, unapproved users without the secure key fail to see any meaningful sequence at the output multiplexer.

Using the above secure test mechanism with key integrated security hardware, it becomes extremely hard for a design pirate or hacker to extract the design information. This is because, the design hacker will need access to the following three information to successfully extract design information: (a) the size of the random key, N, (b) the position of dummy flip-flops, and (c) the seed used in PRBS key generator. However, the addition of such hardware resources also add to the system overheads and costs. Section 4 presents the resulting system costs and security analysis for the secure test architecture.

3.2 IP Core Protection Architecture

The basic principle of IP core protection in STEP design methodology is to use variable keys during operation, which is an effective technique for protection against unsolicited design attacks and intrusion [7]. Figure 3 shows the block diagram of an IP core architecture incorporating variable key protection. Due to unified design methodology, the same hardware is used for IP core protection during normal operation. However, the following operational changes are incorporated for variable key based protection:

- The dummy flip-flops now forms an N bits shift register.

- The PRBS generator is now used as an internal variable key generator using pre-defined seed.

- The key checker now checks for variable key sequence in every iteration instead of the hard-coded key that was used during secure test operation.

- The first scan chain input ($SDI0$) is now used as the input for the N bits shift register formed by the dummy registers.

With the above changes, the operating sequence of the IP protection architecture generated by STEP design methodology is as follows:

(a) **Enable Functional Mode:** When the $TEST\_EN$ pin is LOW, the chip enters into functional mode. During functional mode, SC_EN is is set to HIGH. This enables the logic data input at the output of the scan multiplexers.

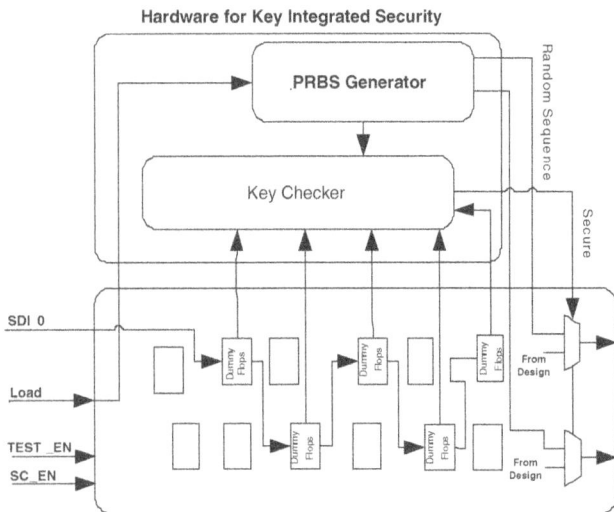

Figure 3: IP core protection architecture generated by STEP design methodology for an AES system

(b) **Variable Key Generation:** The PRBS generates a new key during every new iteration in the AES core with a given pre-defined seed. This forms a variable key generation scheme.

(c) **Key Checking:** The variable key from the PRBS is then compared within the key checker against the key stored in the N bits shift registers. These shift registers are formed through random inter-connection scheme among the dummy flip-flops within the scan chains (Figure 3). The key sequence is loaded into these shift registers through the scan input $SDI0$. When there is a key match, the key checker generates a HIGH *Secure* signal enabling the design logic data to be selected at the output. When there is no key match, the key checker generates a LOW *Secure* signal enabling the previously generated random sequence from PRBS to be selected at the output.

With the added key integrated security hardware through STEP design methodology, the AES system only works as expected for approved or licensed users, who have access to the given key sequence. Due to variable key integration in the IP core architecture, it provides with highly protected IP core against any security threats in terms of reverse engineering or even other response analyses techniques. This is because, for extracting actual design information, the hacker must decode the following three information: (a) the variable key sequence, (b) the inter-connections of dummy flip-flops used to form a shift register to shift and hold a variable key sequence, and (c) the seed used for PRBS. The following section presents the details of the resulting system costs due to addition of the key integrated security hardware for IP core protection, emphasizing the achievable security level through STEP design methodology.

4. RESULTS AND ANALYSIS

To evaluate the effectiveness of the proposed design methodology, three secure AES systems with varied complexity (i.e. number of S-boxes) are designed with the proposed STEP design methodology (Section 2). These secure AES systems are then compared with insecure systems of the same generated using traditional design flow (Phases I and II in STEP, Section 2). The comparative evaluations are carried out through the following comparisons: area, power, testability and security. The comparative analyses follow.

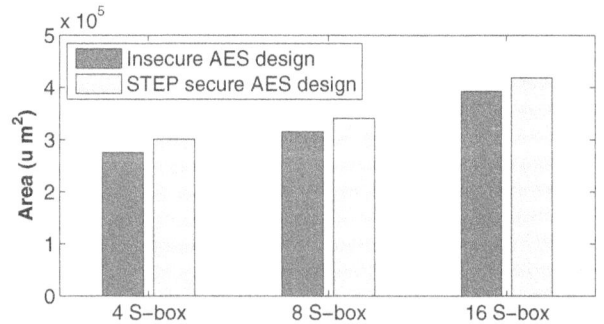

Figure 4: Area comparisons between secure AES systems using the proposed unified design methodology, STEP (Figure 1), and insecure AES systems

4.1 Area Comparisons

Figure 4 shows the comparative areas (in μm^2) of the three secure and insecure AES systems found through post-synthesis evaluations in Synopsys Design Compiler^{TM}. From Figure 4 two observations can be made. The first observation is related to the fact that with higher complexity of the AES systems, the resulting area of AES systems increases as expected. For example, as complexity increases from 4 S-box to 16 S-box, the area increases by about 39% and 42% for the secure AES (through STEP) and for traditional insecure AES, respectively. The second observation is that the secure AES systems designed using the STEP design methodology (Section 2) gives higher area (in μm^2) than the insecure AES systems. The higher area for secure AES is expected due to addition of key integrated security hardware in secure test and IP core protection architectures (Section 3). However, due to unified design methodology in STEP using the same hardware for both secure test and IP core protection, the area overhead is kept low. From Figure 4 it can be seen that up to 9% area overhead is caused for incorporating security in the 8 S-box AES system, when compared with that of 8 S-box insecure AES system.

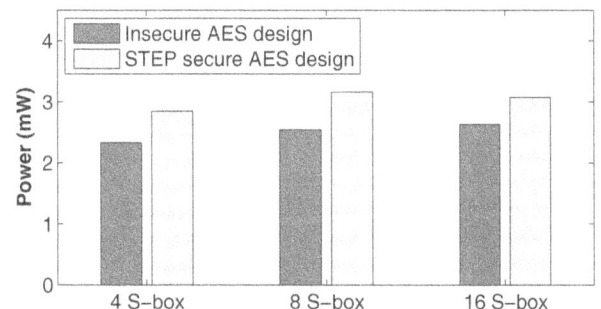

Figure 5: Power comparisons between secure AES systems using the proposed unified design methodology, STEP (Figure 1), and insecure AES systems

4.2 Power Comparisons

Figure 5 shows the comparative power consumptions (in mW) between three secure AES systems with proposed STEP design methodology (Section 2) and insecure AES systems using traditional design methodology (Phase I and Phase II, Figure 1). The power consumptions were evaluated using Synopsys Design Compiler^{TM}. As can be seen, with higher complexity of the AES, the power consumption increases. This is expected because with higher AES complexity (i.e. with higher S-box designs), the number of AES iterations and also the computations carried out over a given time increases [18]. For example, as AES complexity increases from 4 S-box to 16 S-box for the secure AES systems, the power consumption increases by about 13%. From Figure 5 it can also be seen that the power con-

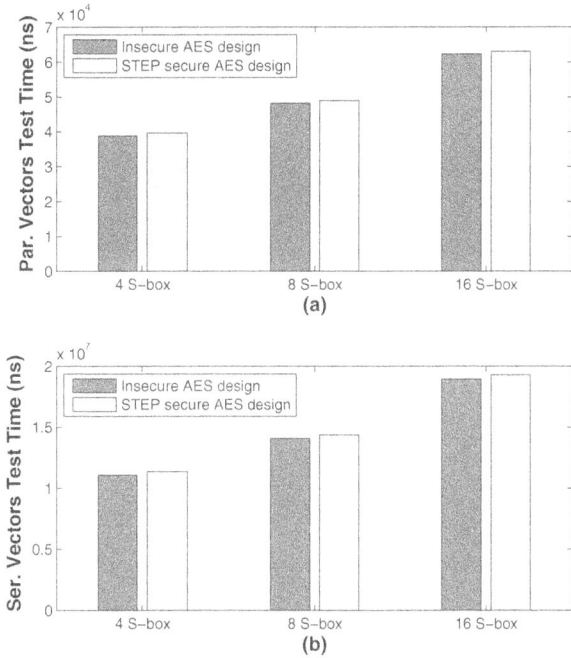

Figure 6: Comparative test times (in ns) for (a) parallel vectors, and (b) serial vectors using secure design methodology (Section 2) and insecure design methodology

sumption is higher for secure AES systems when compared with that of similar insecure AES systems. For example, the power consumption increases by up to 20% for the proposed secure 8 S-box AES system, when compared with the same of an insecure 8 S-box AES system. The higher power consumptions in secure design is due to addition of key integrated security hardware in unified design methodology in STEP (Section 3).

4.3 Test Time and Fault Coverage Analysis

Since the secure test and IP core protection architectures generated by proposed STEP design methodology integrates extra hardware (Section 3), it is important that the test capabilities are compared between the secure and insecure AES systems. To this end, Figures 6(a) and (b) show the comparative test times taken by different secure and insecure AES systems using the parallel and serial test vectors. These test vectors were input through Synopsys Tetra Max^{TM}. From these figures, two observation can be made. Firstly, as expected it can be seen that the test times are considerably lower for parallel vectors (Figure 6(a)) compared to the serial vectors (Figure 6(b)). This is because parallel test vectors significantly reduce the time required for the scan chain data to be loaded and shifted. Secondly, as can be seen, the secure AES systems generated using the proposed unified STEP design methodology takes more more test time for both parallel and serial test vectors. This is because secure AES systems use fixed (in testing mode) and variable (in functional mode) key based hardware to incorporate secure test and IP core protection (Section 3). The key generation, loading and checking mechanism within this integrated security hardware require extra test time (i.e. up to 2% extra delay for 4 S-box secure AES system) compared to the original test times in the insecure AES systems.

Test capabilities of the secure AES systems are further evaluated in terms of the required number of test patterns for achieving a specified fault coverage. Table 1 shows the number of inserted faults, corresponding fault coverage obtained and the number of test patterns used for testing in different secure AES

Table 1: Total faults injected, fault coverage and number of test patterns tested in different secure AES systems generated using the proposed STEP design methodology (Section 2)

AES System	no. of faults	Fault Coverage	Test Patterns
4 S-box	94316	99.04	775
8 S-box	116256	99.03	963
16 S-box	160412	99.02	1244

systems. Columns 1 and 2 show the AES designs and number of faults injected, while columns 3 and 4 show the corresponding fault coverage and the number of test patterns used. As can be seen, with increased design complexity, higher number of faults need to be investigated and tested for due to increased number of iterations and area of the AES (Section 4.1). For these given number of faults, 99% fault coverage can be effectively achieved using the secure test architecture (Section 3.1). However, this fault coverage is achieved using various numbers of test patterns (column 5, Table 1). As expected, as the design complexity increase, the number of test patterns used also increases. For example, from 8 S-box secure AES design to 16 S-box secure AES design the number of test patterns increase by about 29%.

To compare the test capabilities between secure and insecure AES systems, Figure 7 shows the comparative number of test patterns used by the secure and insecure AES designs for a given fault coverage (i.e. 99% fault coverage). These test patterns were generated using special testbenches in Synopsys Tetra Max^{TM} tool. As can be seen, the secure test in AES system gen-

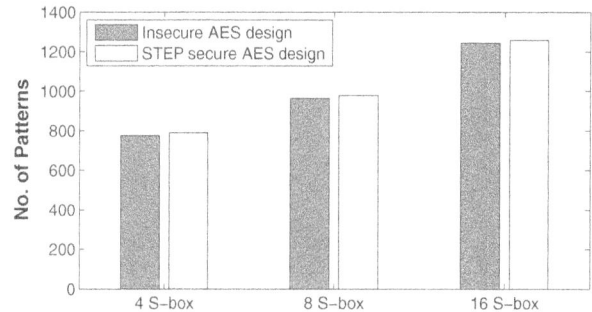

Figure 7: Comparative number of test patterns for similar test coverage between secure AES designs and insecure AES designs

erated using STEP design methodology uses up to 2% higher number of test patterns for achieving similar test coverage as that of the test in insecure AES system. The extra test patterns in secure test can be explained as follows. The extra key integrated security hardware used in secure test architecture (Section 3.1) requires more scan chains and hence more test patterns are needed to achieve similar fault coverage.

4.4 Security Analysis

The proposed STEP design methodology gives high security advantage at the cost of up to 9% area, 20% power and 2% test overheads (Sections 4.1, 4.2 and 4.3). To understand the effective security advantage in the system, in the following hacking scenarios of secure test and IP core protection are briefly explained.

4.4.1 Test Security Analysis

To successfully hack into the secure test architecture, a hacker must extract the following information (Section 3.1): (a) the size of the random key, N, (b) the positions of N dummy flip-flops within S total flip-flops within scan chain, and (c) the seed used in PRBS key generator, R. Assuming that the hacker stores his guessed random key and PRBS seed in an M bits number and that $M \geq N$, the number of combinations hacker has to try for guessing N (C_N) and R (C_R) correctly are

$$C_N = 2^M \quad , \quad C_R = 2^M \quad . \tag{1}$$

Also, to guess the correct position information of the dummy flip-flops the hacker will have to try another C_{ff-pos} combinations, given by

$$C_{ff-pos} = G \begin{pmatrix} S \\ N \end{pmatrix} \quad , \tag{2}$$

where S is the size of a scan chain with dummy flip-flops, N is the number of dummy flip-flops and G is the number of scan chains. Since for each N and R guess, the hacker will have to try to locate the dummy flip-flop positions, the total number of combinations the hacker would need to try for successfully breaking into the secure test system is given by number of combinations given in (1) and (2), i.e.

$$C_{test} = C_N \, C_R \, C_{ff-pos} = 2^{2M} \, G \begin{pmatrix} S \\ N \end{pmatrix} \quad , \tag{3}$$

which is extremely challenging.

4.4.2 IP Core Protection Analysis

For a successful attack in the IP core architecture, a hacker must extract the following information: (a) the sequence of k variable keys, (b) the protocol to shift in the key, i.e. a given interconnection of N connections out of total S scan chain flip-flops, and (c) the seed used for PRBS key generator, $\mathcal{R}$ (Section 3.2).

Considering k keys in the sequence, the number of combinations the hacker have to try for getting the correct sequence (C_{seq}) and the seed ($C_{\mathcal{R}}$) in an M bits number are given as

$$C_{seq} = 2^{kM} \quad , \quad C_{\mathcal{R}} = 2^M \quad . \tag{4}$$

For correctly guessing the interconnection scheme among N dummy flip-flops and also to identify their positions within G number of scan chains of length S each, the hacker will have to try C_{guess}^{ff-con} combinations, given by

$$C_{ff-con} = G \, N! \begin{pmatrix} S \\ N \end{pmatrix} \quad . \tag{5}$$

Since for each N and R guess, the hacker will have to try to locate the dummy flip-flop positions and connections at the same time, the total number of combinations the hacker would need to try for successfully breaking into the secure IP core proection is given by (4) and (5), i.e.

$$C_{IP} = C_{seq} C_{\mathcal{R}} C_{ff-con} = 2^{M(k+1)} \, G \, N! \begin{pmatrix} S \\ N \end{pmatrix} \quad , \tag{6}$$

which is again extremely challenging.

As can be seen from (3) and (6), STEP design methodology provides high security advantage over insecure design methodologies requiring the hacker to generate large number of combinations to extract the design information. As an example, considering N=32, G=8, S=132 and k=4 for an 8 S-box AES system, a total of $C_{test} = 6.7 \times 10^{50}$ and $C_{IP} = 1.4 \times 10^{115}$ combinations are required for breaking into secure test and IP core protection, respectively. This can be further made more challenging by increasing the number of combinations through the use of more and longer scan chains (i.e. higher G and S) with higher number of dummy flip-flops (i.e. higher N). However, this will impose higher system costs in terms of area, power and test times or accuracy.

5. CONCLUSIONS

We have presented a novel design methodology for secure test and IP core protection. We have shown that the proposed unified design methodology, STEP (**S**ecure **TE**st and IP core **P**rotection), is simple to implement and employs the same key integrated security hardware for providing with security and protection during test and IP core functionality (Sections 2 and 3). Due to such use of unified security hardware, the STEP design methodology benefits from high security at low system costs. To evaluate the effectiveness of the proposed design methodology, different AES systems were designed and compared with similar insecure systems as case studies. The comparisons showed that our methodology offers significantly high security requiring high order of magnitude combinations required by the hacker to break into the security and protection of an 8 S-box AES system. This security advantage is achieved at the cost of 9% higher area, 20% higher power and 2% higher test times overheads (Section 4) without affecting the test capabilities.

6. REFERENCES

[1] L. Spitzner, "The Honeynet Project: trapping the hackers," IEEE Security & Privacy, vol.1, no.2, pp. 15- 23, 2003.

[2] E. Oswald, S. Mangard,ä"Counteracting Power Analysis Attacks by Masking," Chapter inäSecure Integrated Circuits and Systems. ISBN 978-0-387-71829-3, pp. 159 – 178. January 2010.

[3] D. Boneh, R. A. DeMillo, and R. J. Lipton. "On the importance of checking cryptographic protocols for faults," in Lecture Notes in Computer Science, 1233, pp. 37-Ű51, 1997.

[4] A. B. Kahng, J. Lach, W. H. Mangione-Smith, S. Mantik, I. L. Markov, "Constraint-based watermarking techniques for design IP protection," IEEE TCAD, vol. 20, no. 10, pp. 1236–1252, Oct. 2001.

[5] Y. M. Alkabani and F. Koushanfar, "Active hardware metering for intellectual property protection and security" In Proceedings of 16th USENIX Security Symposium, pp. 291–306, Niels Provos (Ed.). Berkeley, CA, USA.

[6] D. C. Musker, "Protecting and exploiting intellectual property in electronics", in Proc. IBC Conf., 1998.

[7] M.A. Razzaq, V. Singh, and A. Singh, "SSTKR: Secure and testable scan design through test key randomization", in 20th IEEE Asian Test Symposium (ATS), New Delhi, India, Nov. 2011.

[8] S. Wang, W. Wei, "A Technique to Reduce Peak Current and Average Power Dissipation in Scan Designs by Limited Capture," in Proc. of Asia-Pacific Design Automation Conference, ASP-DAC, pp.810–816, 23-26 Jan. 2007.

[9] J. Lee, M. Tehranipoor, C. Patel, and J. Plusquellic, "Securing Designs against Scan-Based Side-Channel Attacks", Dependable and Secure Computing, IEEE Transactions on, vol.4, no.4, pp. 325-336, Oct.-Dec. 2007

[10] B. Yang; K. Wu; R. Karri, "Scan Based Side Channel Attack on Dedicated Hardware Implementations of Data Encryption Standard," in Proc. International Test Conference, pp 339-344, 2004.

[11] U. Chandran and D. Zhao, "SS-KTC: A High-Testability Low-Overhead Scan Architecture with Multi-Level Security Integration," in Proc. of IEEE VLSI Test Symposium, 2007.

[12] Jarrod A. Roy, Farinaz Koushanfar and Igor L. Marko, "EPIC: Ending Piracy of Integrated Circuits", in Proc. of DATE, 2008.

[13] R.S Chakraborty,S. Bhunia, "HARPOON: An Obfuscation-Based SoC Design Methodology for Hardware Protection", in IEEE Trans. Computer Aided Design of Integrated Circuits and Systems, vol. 28, no. 10, pp. 1493–1502, Oct. 2009.

[14] E. Castillo, U. Meyer-Baese, A. Garcia, L. Parilla, and A. Lloris, "IPP-HDL: Efficient intellectual property protection scheme for IP cores", IEEE Trans. of Very Large Scale Integration (VLSI) Systems, vol. 15, no. 5, pp. 578-590, May 2007.

[15] D. Hely, M.-L. Flottes, F. Bancel, B. Rouzeyre, and N. Berard. "Scan design and secure chip," in Proc. 10th IEEE International On-Line Testing Symposium, 2004.

[16] H. Fujiwara and M. E. J. Obien, "Secure and testable scan design using extended de Bruijn graph", in Proc. 15th Asia and South Pacific Design Automation Conference, 2010.

[17] G. Sengar, D. Mukhopadhyay, D. R. Chowdhury, "Secured Flipped Scan Chain Model for Crypto-Architecture," IEEE TCAD, vol. 26,no.11, Nov. 2007.

[18] National Institute of Standards and Technology, Advanced Encryption Standard (AES), Federal Information Processing Standards Publication 197, 2001.

Towards Systolic Hardware Acceleration for Local Complexity Analysis of Massive Genomic Data

Agathoklis Papadopoulos[1], Vasilis J. Promponas[2], Theocharis Theocharides[1]

[1]KIOS Research Center
ECE Department, University of Cyprus
123 Kyreneias Ave, Nicosia, Cyprus
{papadopoulos.agathoklis, ttheocharides}@ucy.ac.cy

[2]Bioinformatics Research Laboratory
Department of Biological Sciences, University of Cyprus
75 Kallipoleos Str., Nicosia, Cyprus
vprobon@ucy.ac.cy

ABSTRACT
Modern biological research has greatly benefited from genomics. Such research however requires extensive computational power, traditionally employed on large-scale cluster machines as well as multi-core systems. Recent research in reconfigurable architectures however suggests that FPGA-based acceleration of genomic algorithms greatly improves the performance and energy efficiency when compared to multi-core systems and clusters. In this work, we present an initial attempt for massive systolic acceleration of the popular CAST algorithm employed by biologists for complexity analysis of genomic data. CAST is used for detecting (and subsequently masking) low-complexity regions (LCRs) in protein sequences. We designed and implemented a high-performance hardware-accelerated version of CAST for which we built an FPGA prototype, and benchmarked its performance against serial and multithreaded versions of the CAST algorithm in software. The proposed architecture achieves remarkable speedup compared to both serial and multithreaded CAST implementations ranging from approx. 100x-9500x, depending on the dataset features, such as low-complexity content and sequence length distribution. Such performance may enable complex analyses of voluminous sequence datasets, and has the potential to interoperate with other hardware architectures for protein sequence analysis.

Categories and Subject Descriptors
B.7.1 [**Integrated Circuits**]: Types and Design Styles – *Algorithms implemented in hardware.*

General Terms
Algorithms, Performance, Design.

Keywords
Bioinformatics, Reconfigurable Parallel Architectures, Genomics, Hardware Acceleration

1. INTRODUCTION
Genomics – that is the study of the complete genomes of biological species (including human) – has undoubtedly revolutionized the way biological research is currently taking place. This is particularly the case with novel high-throughput sequencing technologies that enable whole genome DNA sequencing of simple unicellular organisms in a matter of days [1]. With the anticipation of the, so called, $1000 genome it is expected that genomic science will soon become a central part of clinical practice [2]. Importantly, the huge amounts of sequence data currently produced worldwide at an ever increasing pace require extensive downstream computational analysis. Typical computational pipelines for genomics have at their core a computationally intensive sequence comparison component; this is justified by the empirical observation that genes and proteins with similar sequences usually perform similar functions. Therefore, sequence similarity search serves for inferring functional and structural analogy for biological macromolecules [3, 4].

Such complex algorithms have traditionally been implemented on high-performance computing clusters and multiprocessor systems [5], and recently, with the emergence of multicore and manycore architectures, on clusters and workstations equipped with such CPUs. However, there has been a tremendous amount of effort in designing efficient systems that can take advantage of the inherent parallelism opportunities of such algorithms; one of the most powerful supercomputers in the world (IBM Blue Gene) has been designed with the objective of performing large scale protein folding simulations [6]. Furthermore, the specialized nature of bioinformatics algorithms limits the number of end-users for such platforms, thus keeping the costs of possible custom hardware solutions extremely high. As such, alternative technologies that can better balance the cost and performance constraints can be more efficient for targeting the bioinformatics research communities.

Given the cost-performance benefits of FPGAs, recent advances in computational biology suggest that reconfigurable hardware might be indeed a powerful alternative to cluster-based supercomputers [7]. Such architectures employ the flexibility of general-purpose computing that biologists can utilize without steep learning curves, as well as the hardware suitable for parallel implementations of the algorithms that general-purpose processors lack. Thus, FPGAs can balance out performance, manufacturing costs and hardware resource constraints; furthermore, they offer higher programmability than ASICs and are easier accessible to non-expert end-users. Among the most benefited algorithms when mapped on such architectures, include massive genomic data processing which typically involves sequence comparison.

Sequence comparison is usually performed by aligning sequences, i.e. by algorithmically identifying the optimal correspondence of individual positions (residues) between the compared sequences, given a scoring scheme. When a newly identified protein sequence is submitted for comparison against a database of already known proteins, the similarity score alone is not sufficient to pinpoint important biological relationships for functional inference. Thus, robust statistical computations [8] have been used to reliably identify those similarities that are unlikely to have arisen simply by chance in a haystack of unrelated hits. Such measures involve complex data processing, with unpredictable data flow behavior; data dependencies involved in such algorithms are therefore a significant drawback when employing traditional superscalar and multi-threaded or multicore CPU architectures. Furthermore, memory bandwidth and memory management is another drawback; even assumptions related to such applications (for example, the assumption that the compared sequences are random sequences generated by sampling from a single amino acid residue distribution, i.e. the distribution of the protein sequence database), for a significant fraction of known proteins, these assumptions do not hold [9, 10].

Several improvements have been suggested by the bioinformatics community for improving the computational behavior of such algorithms; however, the improvements themselves involve similarly complicated algorithms that too have significant computational overheads. Such algorithms include the SEG [10] and CAST [11] algorithms. Masking protein sequences with CAST has been empirically shown to result into similarity searches of superior specificity (low false positive rate) without sacrificing sensitivity [11]. However, a major obstacle against the wider use of the CAST algorithm is its relatively low computational performance due to its iterative nature[1]. With the exponential growth of sequence databases [12], even the fastest algorithms for sequence comparison fall short.

Along these lines, this paper presents an initial attempt in implementing massively parallel, systolic architectures that can be deployed on cost-effective FPGAs, to efficiently map sequence comparison bioinformatics algorithms, and in particular the CAST algorithm. The proposed architecture consists of a chain of processing elements (PEs) that executes the CAST algorithm in a pipelined manner and can be expanded vertically with minor modifications (introduction of vertical data flow) to operate in a fully systolic manner. This paper focuses on describing and evaluating the design and flow of operations of a single horizontal chain; the full architecture is part of ongoing work.

The paper is organized as follows. Section II presents a brief background on the CAST algorithm and relevant related work. Section III presents the proposed architecture, and an experimental prototype on an FPGA, along with the results, is presented in Section IV. Section V concludes the paper.

2. BACKGROUND – RELATED WORK
2.1 Background
Sequence alignment of biological macromolecules is a routine computational procedure practiced daily in most biological research laboratories throughout the world. Thus, a large set of tools have been developed over the years trying to provide improved solutions to the sequence comparison problem over the traditional dynamic programming methods. Rapid and sensitive algorithms, such as the BLAST heuristic algorithm [13], have become the mostly used tools for homology searches in sequence databases. The wide utilization of the BLAST suite of programs is clearly reflected by the fact that the two key methodological papers describing the methods [13,14] have been collectively cited more than 60000 times [15].

```
Algorithm CAST
Input: A protein sequence S
Output: The sequence masked for LCRs

residues   <- (A, C, D, .., Y)
hscore     <- (0, 0, 0, .., 0)
from       <- 0
to         <- 0
neutral    <- X

do
       for each res in residues
              (hscore[res], from, to) <- detectBias(S, res)

       hscoreMAX <- max(hscore[A], hscore[C], .., hscore[Y])
       lcrType <- residues[ argmax(hscore[A], hscore[C], .., hscore[Y]) ]

       if (hscoreMAX>=Threshold)
              S <- mask(S, lcrType, from, to)

while (hscoreMAX>=Threshold)

function detectBias(Sequence: S, Residue:r)
       (maxscore, from, to) <- alignSmithWaterman(S, poly-residue)
       return (maxscore, from, to)

function mask(Sequence: S, Residue:lcrType, Start:from, End:to)
       for pos in (from .. to)
              if (S[pos] equals lcrType)
                     S[pos] <- neutral
       return (S)
```

Figure 1: Pseudocode for the CAST algorithm. LCR detection is performed by iteratively comparing input S with degenerate homopolymers of the 20 naturaly occuring amino acids.

When high sequence similarity is detected between a query sequence of unknown function and an annotated database entry, reliable function prediction for the query can be obtained. However, LCRs may result in erroneous function predictions, as the high score observed is rather due to the bias effect and not due to genuine homology. Thus, masking LCRs can significantly improve the reliability of homology detection and the quality of function prediction.

The CAST algorithm [11] is an iterative method for identifying and masking LCRs in protein sequences and has shown better quality results over other proposed methods, such as SEG [10]. In principle, the CAST algorithm compares protein sequences against an artificial database consisting of 20 degenerate protein sequences of arbitrary length, each one being a homopolymer based on one of the 20 natural amino acid residue types. CAST identifies LCRs in a single linear pass for each amino acid type (Fig. 1) with only linear memory required. The most remarkable feature of CAST is that not only it detects LCRs, but also identifies the type of amino acid residue causing the bias. Thus, selective masking can be performed in a more subtle and specific manner compared to other masking methods (e.g. [10]).

The LCR detection step requires the use of a substitution matrix, where matching non-identical residue types may give positive scores as well. Thus, an LCR biased in one residue type may lead to high scores for biases of a similar but not identical type (e.g. arginine/lysine). Therefore, CAST iteratively performs LCR detection and masking steps to prevent unnecessary masking due to cross-dependencies between amino acid residue types. An empirically defined threshold value T for the similarity score serves as a LCR selection criterion. With the use of the BLOSUM62[2] substitution matrix, the optimal value $T = 40$ is used. In practice, a variant of BLOSUM62 serves as the default scoring matrix: the scores of each residue type against the neutral type 'X', are computed as the mean value of the amino acid substitution scores for the respective residue type.

---

1 It is worth mentioning that proteins with LCRs seem to play important roles in several natural biological processes (including human disease) [22]. However, in this work we mainly focus on the application of massive LCR detection in large protein datasets for masking, as a preprocessing step in sequence database search.

---

2 BLOSUM62 is one of the substitution matrixes commonly used for calculating scores between evolutionarily divergent protein sequences. While is possible to use any other substitution matrix, BLOSUM62 is the substitution matrix of choice for this work.

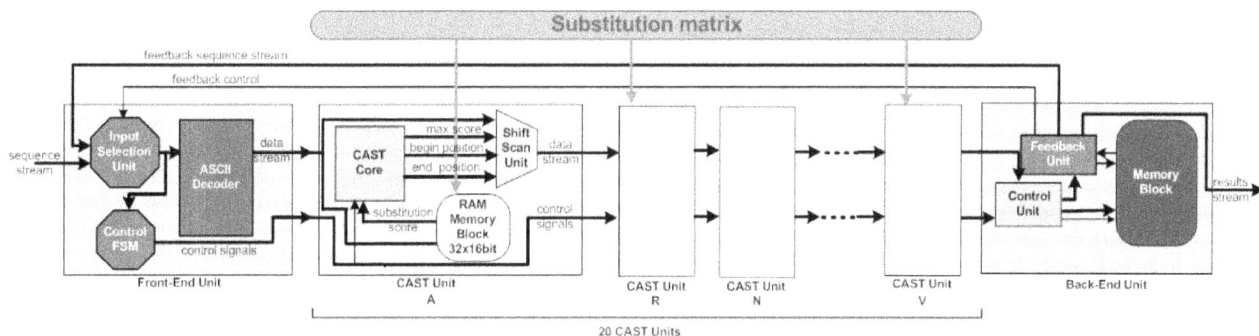

Figure 2: Proposed CAST Architecture

The algorithm shown in Fig.1 receives as input a protein sequence, and, searches for the LCR candidates (highest scoring segments) of each natural amino acid type. It then selects the maximum score between the searched segments, and if that score is less than the threshold T, it ends outputting the discovered LCRs; otherwise, it replaces each occurrence of the max scoring residue type in the highest scoring segment region with an 'X' (i.e. a neutral amino acid) and iterates through the updated sequence. Further details of the algorithm can be found in [11].

2.2 Related Work

The protein database has been growing exponentially over the years, and the execution time of existing tools implemented traditionally in software grows exponentially even on high-end computer systems [16]. Recently, however, application-specific reconfigurable hardware solutions, utilizing FPGAs have emerged as promising solutions. Several researchers proposed architectures for bioinformatics on FPGAs that exhibit performance improvements, such as those in [17 -20].

The reconfigurable architecture of the *BLAST* algorithm proposed in [17] showed remarkable speedups up to 1400x over the software implementation of *BLASTp*. Along the same lines, the FPGA-based implementation of *PSI-BLAST* proposed in [18] showed speedup over 20x compared to the existing software solutions. In a relevant sequence analysis problem, an FPGA accelerator of the *GOR* algorithm used for protein secondary structure prediction, showed speedup factors above 430x over the original *GOR* and more than 110x over the multi-threaded software version [19]. Recently, a Network-on-Chip-based hardware accelerator for biological sequence alignment reported speedup over three orders of magnitude over traditional CPU architectures [7]. Preprocessing tools implemented on hardware can further reduce execution times since major tools – such as *BLAST* – have already been implemented on reconfigurable fabric with astonishing results. A prefiltering approach for further improving performance of *BLAST* is already implemented on reconfigurable fabric [20].

To the best of our knowledge, the architecture proposed in this paper is the first attempt to implement a hardware architecture that can be used for identifying and masking LCRs in protein sequences.

3. PROPOSED ARCHITECTURE

The proposed architecture –shown in Fig.2 – consists of a series of interconnected PEs that take advantage of the natively parallel characteristics of the CAST algorithm. The system receives a stream of protein sequence represented in FASTA format[3], compares the stream with the twenty degenerate sequences in parallel, extracts current iteration's LCR, masks it and re-iterates the masked sequence until all LCRs are discovered. Finally, the system outputs each of the LCRs score and position in the sequence.

The CAST architecture consists of three different types of PEs: a front-end unit, a number of CAST processing units and a back-end unit. The input sequence under consideration is streamed into the system through an I/O controller and each unit propagates the necessary signals to the next one in a pipelined manner until the propagated stream reaches the back-end unit. The back-end unit then sends it to the external I/O controller or redirects the updated stream back for the next iteration.

3.1 Front-end Unit

The front-end unit (FEU) is responsible for receiving one ASCII symbol of the input sequence per cycle, and to generate the control signals needed by the CAST processing units. The received symbol is recoded to a lesser bit representation in order to reduce the hardware resources needed. The recoding is done by the *ASCII decoder unit*. The *Control Finite State Machine (FSM)* of the FEU interprets the received symbol and decides if it is the beginning or end of the sequence, generating the appropriate control signals which are propagated to the CAST units. The FEU also facilitates the logic necessary to control the number of iterations needed. Whenever the need for another iteration occurs, the unit re-feeds the updated sequence received from the back-end unit back to the architecture, holding any new pending sequence from the I/O controller.

3.2 CAST Unit

The CAST processing units perform the core of the computation. The architecture has twenty identical CAST units, each one responsible for one of the natural amino acids. They are interconnected in a row-wise pipelined manner. Each unit calculates the high scoring segment (HSS) – LCR candidates -- of the input sequence for their respective amino acid residue type. A CAST unit consists of a *local memory block* holding the subset of the substitution matrix referring to its respective amino acid residue, a *CAST core* performing the score calculations needed to identify the boundaries of the HSS, and a *shift/scan unit* for propagating the signals needed by the adjacent CAST unit.

The *local memory block* is initialized with the substitution values from a centralized memory block holding the complete matrix

---

3 FASTA format is using English alphabet letters to represent amino acids in proteins, represented in ASCII text. [23].

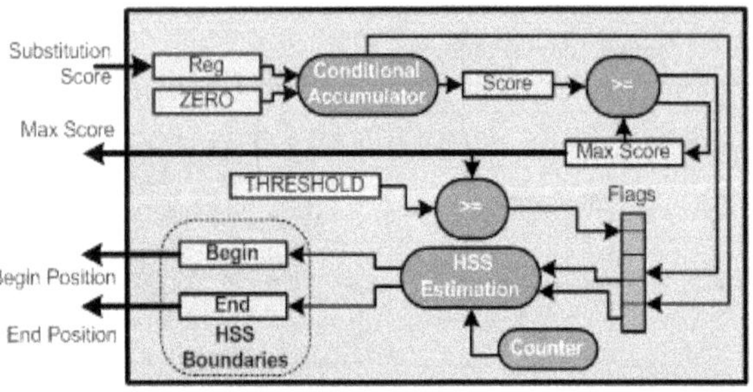

Figure 3: CAST core architecture

prior the arrival of the first input sequence. The substitution matrixes used in Bioinformatics are rather small in size (i.e. BLOSUM62 variant used, requires less than 500 bytes) and can be stored in a centralized multi-ported memory block. However, the number of accesses to the centralized memory block to fetch the substitution values needed in the calculations for all of CAST units each cycle is a potential bottleneck. Hence, the substitution values were stored in small local memory blocks, mapped on the FPGA fabric, so that they can be accessed asynchronously, as fast as possible. This approach eliminates time spent in memory accesses, as the data is fetched and processed in one cycle.

The *CAST core* – shown in Fig.3 – derives the score for the sequence streamed in and holds the statistics needed for identifying HSS and its score. HSS is identified by storing its beginning and ending position in the sequence. A counter measuring the current position of the symbol in the input sequence is used for that purpose. The score is updated every cycle by accumulating the substitution scores of each amino acid in the sequence. If the accumulated score falls below zero, the score is set to zero and remains at this value until a positive substitution score is arrived. This score value is then compared against the current max score. The max score is increased if needed, and it is also compared against the threshold set by the CAST algorithm. The various modules embedded within the *CAST core* are used for marking the HSS boundaries using a sliding window approach. Those boundaries and the max score are fed to the *Shift & Scan* unit.

The *Shift & Scan unit* is responsible for propagating the signals needed for the next stage of calculations. The actions of the unit are decided by the control signals propagated through the CAST Unit. The unit operates in two modes; *shift* and *scan*. When in *shift* mode, the unit propagates the stream without changes while the system is computing the HSS. When in *scan* mode, the unit propagates the highest value between the CAST Unit's max score and the propagated max score received from the left adjacent CAST unit. This method allows the back-end unit to receive the sequence's maximum HSS score as soon as possible. Whenever the feedback unit informs the corresponding CAST unit that its respective amino acid HSS is actually the current iteration's LCR, the *shift & scan* unit transmits the boundaries of the LCR, first the beginning and then the ending position.

3.3 Back-end Unit

The back-end unit (BEU) decides whether the sequence analysis is complete and propagates it to the external I/O controller or redirects the updated sequence back to the FEU for calculating the next LCR. This decision is taken by the control unit and is based on the existence of HSSs in the current iteration with higher value than the threshold parameter of the CAST algorithm. The symbol stream received from the last CAST unit while in calculation

mode, are stored in a local *Line Memory* of a size equal to the size of the maximum sequence length used in the software implementations of the algorithm. The memory size can be easily increased to accommodate larger protein sequences. *Line Memory* is implemented as a dual-port memory block. The stream representing the sequence is deemed necessary to be stored locally as the nature of the CAST algorithm requires a number of iterations dynamically decided at runtime. Upon receiving confirmation that all symbols of the sequence are processed, the control unit waits for the HSS with the maximum score to propagate through the architecture and mark it as current iteration's LCR. It then transmits the necessary signals to the feedback unit to start reading the sequence from the Line Memory and stream it either back to the FEU for another round of calculations, or to the system output. The sequence is dynamically updated by the feedback unit as the symbols corresponding to the highest scoring amino-acid are replaced with 'X' while passing through the feedback unit.

3.4 Overall system data flow

The local memory blocks belonging to CAST units are first initialized with their respective substitution matrix subsets. The input sequence is then streamed and processed one symbol per cycle. The FEU recognizes the beginning of the sequence, sets the appropriate control signals and starts propagating the symbols to the CAST units. When a new sequence is streamed in each CAST unit, the unit first resets and waits for the symbols to arrive. Next, the unit searches through its local memory block for the symbol's substitution value, and it then updates the necessary internal registers for the scores and HSS boundaries for each symbol propagated through.

The BEU stores the streamed sequence to its Line Memory block and waits for the end of sequence signal. When this happens, the BEU generates the appropriate signals for the CAST units and waits to receive the maximum scoring HSS. The maximum score is propagated through the architecture, and, when it reaches the BEU, is marked as current iteration's LCR. The CAST unit having that LCR then, propagates the region boundaries to BEU. If the current iteration's LCR score is higher than the CAST algorithm threshold, the BEU updates the sequence stored in its Line Memory and signals to the FEU to hold any new incoming sequences. The updated sequence is then fed back to the FEU for the next iteration. However, if the threshold has not been reached, the BEU sends the LCR boundaries and scores to the system output. In the latter case, the FEU is free to receive the next protein sequence.

4. EXPERIMENTAL PLATFORM AND RESULTS

4.1 Experimental Platform and Benchmarks

A prototype of the proposed hardware architecture was implemented on a Virtex-5 LX110T FPGA running on a 100MHz clock, and evaluated using actual protein databases [24] that were loaded to a DRAM memory using a Compact Flash card. Each local memory block of each CAST unit was mapped as distributed RAM, for more flexibility and speed. Mapping the memory blocks as distributed RAM on the available Virtex-5 LUTs, allows for better utilization of LUT resources from the Xilinx tools as well. The line memory of the BEU is implemented as a standard dual-port memory and is mapped on the BRAM resources of the FPGA.

Table 1: Protein Sequence Datasets used for Evaluation

Dataset Name	# of sequences	Average length	Details
Case.I	1	1695	Longest sequence of haem database
Case.II	250	311.6	Random sequences from haem database
Case.III	500	299.2	Random sequences from haem database
Case.IV	1243	292.3	Random sequences from haem database
Case.V	113	764	Sequences longer than 600 from haem database
Haem	1743	305.1	All sequences from haem database
p.falciparum	5491	755.9	All sequences From p.falciparum database

To compare the proposed hardware architecture against the CAST software implementations, we used a number of datasets stemming from two actual protein sequence databases. The first database belongs to *Haemophilus influenza* bacteria (*haem database*) and the second consists of protein sequences from the malaria parasite genome *Plasmodium falciparum* (p.falciparum database). The haem database was also partitioned in smaller sequence databases (labeled Case I to Case V) in order to have different sized datasets and datasets with different sequence lengths. Table 1 lists all the test databases used during the evaluation process.

4.2 CAST Performance Comparisons

The original CAST implementation presented in [8] could be an initial comparison reference; however, for obvious reasons, comparing a parallel hardware architecture with a sequential single-threaded software implementation is unfair. Thus, we developed a multi-threaded version of CAST. We used both versions for comparison purposes, with the sequential CAST (sCAST) optimized for uniprocessor systems and multi-threaded CAST (mCAST) optimized for multicore systems. The multi-threaded version has been developed for fully utilizing modern OS thread scheduling capabilities by using C++ threading libraries for both Linux and Windows operating systems. The multi-threaded software takes advantage of the parallel nature of the first step of the CAST algorithm where the input sequence(s) are compared against the twenty amino acid residue types in order to find the respective highest scoring segments (Fig. 1). The search for those segments can be done in parallel since there are no data dependencies. Thus, the main program can create up to twenty

Table 2: Single-Threaded VS Multi-threaded Software

	Intel Core2 Duo 8400@3GHz/4GBRAM Windows XP			Intel Core i7 720@2.66GHz/8GB RAM Windows Server 2008		
	sCAST (ms)	mCAST (ms)	X[a]	sCAST (ms)	mCAST (ms)	X[a]
Case.I	238	288	**1.23**	133	85,1	**1.56**
Case.II	552	1185	**0.46**	532	63,6	**8.36**
Case.III	1031	673	**1.53**	1090	365.6	**2.98**
Case.IV	541	671	**0.80**	549	376.1	**1.46**
Case.V	1502	1797	**0.83**	931	535.5	**1.74**
Haem	4.2s	4.8s	**0.85**	4.0s	2.4s	**1.66**
p.falciparum	960s	1166s	**0.82**	528s	313s	**1.69**

a. Speedup of mCAST over sCAST

threads for calculating the high scoring segments. The mCAST was compared against the sCAST presented in [8]. The execution times for sCAST and mCAST have been calculated while suppressing I/O system calls for fair comparison with the proposed hardware architecture; the software under these conditions focuses on reading inputs and computing the algorithm, similarly with the hardware approach.

The two software versions were executed on two different computer systems: an *Intel Core2 Duo E8400@3GHz/4GB RAM* running *Windows XP,* and an *Intel Core i7 720@2.66GHz/8GB RAM* running *Windows Server 2008* for all sequence databases listed in Table 1. The execution times, presented in Table 2, indicate that mCAST offers minimal performance improvements for the first system, as the *Intel Core2 Duo E8400* supports up to two threads in parallel. The remaining threads have to wait to be scheduled and the resulting continuous context switch rather burdens performance than helping. The second system supports mCAST much better, as its processor supports up to eight simultaneous threads. The speedup however does not reach the ideal 8x (except Case.II) because of the frequent memory accesses and the time used by the OS. Thus, the speedup for small-to-average sized sequences is not significant as the performance gain stemming from the parallel execution is of the same order as the time consumed by the OS for creating and scheduling the threads. The results also indicate that multithreading does not help for datasets having longer sequences. For Case.II and Case.III datasets, the speedup reaches above 2x, since the sequences in these datasets need more iterations – larger number of LCRs – and the time spent on memory accesses and by the OS for context switch, becomes less important. mCAST therefore, still does not achieve full potential for several datasets due to memory and control issues.

These results demonstrate how the parallel hardware architecture can potentially replace such systems. We compare the results of the proposed hardware architecture against the software implementations. For comparison purposes, we use the mCAST running on *Intel Core i7 720*. Moreover, we compared the proposed hardware results with the results of the SEG algorithm running on *Intel Core i7 720* for each of our test databases. Evaluating the proposed architecture over SEG execution times is important, as the SEG algorithm is used traditionally in the BLAST suite. Table 3 shows the comparisons between the execution time achieved by mCAST and SEG, against the execution time of a single instance of the hardware architecture.

The hardware architecture results outperform both mCAST and SEG software implementations. In fact, results are orders of magnitude better for all test sequences, when compared to mCAST. Furthermore, the proposed architecture outperforms SEG with a speedup above 50x for most test cases. The

Table 3: Software VS Proposed Hardware Architecture

	H/W (ms)	mCAST (ms)	H/W Speedup	SEG (ms)	H/W Speedup
Case.I	**0.01**	85	**8500**	1	**100**
Case.II	**0.57**	64	**112**	58	**102**
Case.III	**1.12**	366	**326**	30	**27**
Case.IV	**0.59**	376	**637**	29	**49**
Case.V	**0.63**	535	**850**	33	**52**
Haem	**4.08**	2370	**580**	209	**51**
p.falciparum	**33.46**	313302	**9363**	4357	**130**

Table 4: Synthesis Results for the Xilinx LX110T FPGA

FPGA Resources	Used / Available	Usage Percentage
Slice Registers	4259 out of 69120	6%
Number of Slice LUTs	7802 out of 69120	11%
Block Rams	1 out of 148	0.6%
DSP48Es	0 out of 64	0%
Frequency	145MHz	
I/O Bandwidth	8bits/cycle	

p.falciparum database, which has a large number of compositionally biased regions, demonstrates the capabilities of the hardware approach, as the proposed hardware architecture has a 9360x speedup over mCAST and 130x speedup over SEG. Table 3 indicates the limitations of LCR masking on general-purpose high-end processors, as they cannot efficiently execute the algorithms due to memory bandwidth issues and limitations in parallelism. On the other hand, a custom-build hardware system running on a cheap FPGA board can significantly accelerate performance. Moreover, it is anticipated that a fully systolic architecture will exhibit even higher performance gains.

4.3 Hardware Synthesis Results for Xilinx Virtex-5 FPGA

We lastly give the hardware synthesis results in Table 4, in order to give a complete picture of the required hardware resources needed for mapping the proposed architecture on a relatively cheap, off-the-shelf FPGA. The proposed architecture accelerates the execution time of CAST algorithm using just 10% of the LX110T FPGA. Such overheads are small enough to allow other FPGA-based bioinformatics algorithms such as BLAST to be combined with CAST on the same FPGA, further enhancing the performance of such systems.

5. CONCLUSION AND FUTURE WORK

This paper presented an architecture for accelerated masking of LCRs in protein sequences. The proposed architecture is based on the CAST algorithm, enabling very fast and selective masking in very large genomic datasets. We observe that for a cheap, off-the-shelf FPGA, the speedup is over two orders of magnitude with a very conservative hardware design approach. The performance results for the proposed algorithm outperform multi-threaded implementations of the CAST algorithm by 100x-1000x times. Even compared to the significantly faster SEG, the proposed architecture runs 20 to 130 times faster.

We plan on developing a fully systolic architecture for parallelizing the input sequences. The expected systolic array will take advantage of the inherent parallelism, to further increase performance gains for CAST algorithm. Issues, such as dynamic load balancing and I/O communication channels for using multiple sequences simultaneously, as well as power consumption will be addressed as well.

6. ACKNOWLEDGMENTS

This work was supported in part by the European Regional Development Fund and national funding sources from Greece and Cyprus, as part of the European Union's Cross Border Cooperation Programme 'Greece – Cyprus 2007-2013'.

7. REFERENCES

[1] M. L. Metzker, "Sequencing technologies - the next generation," *Nature Reviews Genetics*, vol. 11, no. 1, pp. 31-46, 2008.

[2] E. R. Mardis, "Anticipating the $1,000 genome," *Genome Biology*, vol. 7, no. 112, 2006.

[3] R. Nair and B. Rost, "Sequence conserved for subcellular localization," *Protein Science*, vol. 11, no. 12, pp. 2836-2847, 2002.

[4] A. J. Enright, S. van Dongen, and C. A. Ouzounis, "An efficient algorithm for large-scale detection of protein families," *Nucleic Acids Research*, vol. 30, no. 7, pp. 1575-1884, 2002.

[5] A. E. Darling, L. Carey, and W.-chun Feng, "The design, implementation, and evaluation of mpiBLAST," in *ClusterWorld*, 2003, pp. 13-15.

[6] A. Gara et al., "Overview of the Blue Gene/L system architecture," *IBM Journal of Research and Development*, vol. 49, no. 2, pp. 195-212, Mar. 2005.

[7] S. Sarkar, G. R. Kulkarni, P. P. Pande, and A. Kalyanaraman, "Network-on-Chip Hardware Accelerators for Biological Sequence Alignment," *IEEE Transactions on Computers*, vol. 59, no. 1, pp. 29-41, Jan. 2010.

[8] S. Karlin and S. F. Altschul, "Methods for assessing the statistical significance of molecular sequence features by using general scoring schemes," *Proceedings of the National Academy of Sciences of the United States of America*, vol. 87, no. 6, pp. 2264-2268, 1990.

[9] J. C. Wootton, "Sequences with 'unusual' amino acid compositions," *Current Opinion in Structural Biology*, vol. 4, pp. 413-421, 1994.

[10] J. C. Wootton and S. Federhen, "Statistics of local complexity in amino acid sequences and sequence databases," *Computers & Chemistry*, vol. 17, no. 2, pp. 149-163, Jun. 1993.

[11] V. J. Promponas et al., "CAST: an iterative algorithm for the complexity analysis of sequence tracts," *Bioinformatics*, vol. 16, no. 10, pp. 915-922, 2000.

[12] D. Benton et al., "GenBank," *Nucleic Acids Research*, vol. 36, p. D25-D30, 2008.

[13] S. F. Altschul et al., "Basic local alignment search tool," *Journal of Molecular Biology*, vol. 215, no. 3, pp. 403-410, 1990.

[14] S. F. Altschul et al., "Gapped BLAST and PSI-BLAST: A new generation of protein database search programs," *Nucleic Acids Research*, vol. 25, no. 17, pp. 3389-3402, 1997.

[15] "Scopus®." [Online]. Available: http://www.scopus.com. [Accessed: 09-Sep-2011].

[16] R. C. Edgar, "Search and clustering orders of magnitude faster than BLAST.," *Bioinformatics (Oxford, England)*, vol. 26, no. 19, pp. 2460-1, Oct-2010.

[17] E. Sotiriades and A. Dollas, "A General Reconfigurable Architecture for the BLAST Algorithm," *The Journal of VLSI Signal Processing Systems for Signal, Image, and Video Technology*, vol. 48, no. 3, pp. 189-208, Jun. 2007.

[18] S. Kasap, K. Benkrid, and Y. Liu, "A high performance fpga-based implementation of position specific iterated blast," in *Proceeding of the ACM/SIGDA international symposium on Field programmable gate arrays - FPGA '09*, 2009, p. 249.

[19] F. Xia, Y. Dou, G. Lei, and Y. Tan, "FPGA accelerator for protein secondary structure prediction based on the GOR algorithm.," *BMC bioinformatics*, vol. 12 Suppl 1, no. 1, p. S5, Jan. 2011.

[20] P. Afratis, E. Sotiriades, G. Chrysos, S. Fytraki, and D. Pnevmatikatos, "A rate-based prefiltering approach to blast acceleration," in *2008 International Conference on Field Programmable Logic and Applications*, 2008, pp. 631-634.

[21] Xilinx Tech Topics, "Virtex Series High-Performance Communications Channel," [Online]. Available: http://www.nalanda.nitc.ac.in/industry/appnotes/xilinx/documents/products/virtex/techtopic/selectlink.htm. [Accessed: 03-Sep-2011].

[22] W. Haerty and G. B. Golding, "Low-complexity sequences and single amino acid repeats: not just 'junk' peptide sequences," *Genome*, vol. 53, pp. 753-762, 2010.

[23] "IUPAC code table." [Online]. Available: http://www.dna.affrc.go.jp/misc/MPsrch/InfoIUPAC.html. [Accessed: 09-Sep-2011].

[24] "National Center for Biotechnology Information (NCBI) Databases." [Online]. Available: http://www.ncbi.nlm.nih.gov/ [Accessed: 09-Sep-2011].

A Dual-Phase Compression Mechanism for Hybrid DRAM/PCM Main Memory Architectures

Seungcheol Baek, Hyung Gyu Lee, Jongman Kim
School of Elec. and Comp. Eng.
Georgia Institute of Technology
{bsc11235, hyunggyu, jkim}@gatech.edu

Chrysostomos Nicopoulos
Department of Elec. and Comp. Eng.
University of Cyprus
nicopoulos@ucy.ac.cy

ABSTRACT

Phase-Change Memory (PCM) is emerging as a promising new memory technology, due to its inherent ability to scale deeply into the nanoscale regime. However, PCM is still marred by a duet of potentially show-stopping deficiencies: poor write performance and limited durability. These weaknesses have urged designers to develop various supporting architectural techniques to aid and complement the operation of the PCM, while mitigating its innate flaws. One promising such solution is the deployment of hybridized memory architectures that fuse DRAM and PCM, in order to combine the best attributes of each technology. In this paper, we introduce a Dual-Phase Compression (DPC) scheme specifically optimized for DRAM/PCM hybrid environments. Extensive simulations with traces from real applications running on a full-system simulator of a multicore system demonstrate 35.1% performance improvement and 29.3% energy reduction, on average, as compared to a baseline DRAM/PCM hybrid implementation.

Categories and Subject Descriptors

C.0.4 [**Computer Systems Organization**]: General—*System architecture*

1. INTRODUCTION

Over the last few years, several new memory technologies have been developed in an effort to address some of the shortcomings of DRAM technology. One of the most promising new actors is Phase-Change Memory (PCM), which is gaining a foothold in the research community, predominantly due to its attractive ability to scale very deeply into the low nanometer regime, low power consumption, non-volatility, and fast read performance [9]. However, there are some crippling limitations that prevent PCM from completely ousting DRAM from future systems: low write performance and limited long-term endurance. These drawbacks have led designers toward the adoption of various architectural techniques, which attempt to mitigate the deficiencies of PCM technology.

Notable examples of such schemes include the re-organization of memory buffers [14], Flip-N-Write [5], and redundant-bit write removal and row shifting [21]. A higher-level memory system architectural approach employs a *partial-write* technique that flushes only a specific part of a dirty cache line from the traditional cache

[14]. Similarly, a *segment-swapping* mechanism can be used to periodically swap memory segments of high and low write accesses in order to achieve a *wearleveling* effect (i.e., prolong the lifetime of PCM). Although these schemes reduce the actual number of cell operations within the PCM, they not only require extensive modifications to the PCM device itself, but they also involve *multi-layered modifications* in the memory sub-system and require a subsequent orchestration of activities at *multiple levels* of the hierarchy.

Another approach to circumventing the undesired characteristics of PCM is the adoption of *hybridized* memory structures, which utilize both DRAM and PCM memories, in order to combine the best of both worlds [4, 16, 7]. The conventional way to hybridize DRAM and PCM is by using the DRAM as an off-chip cache (i.e., an additional level in the memory hierarchy). Although this DRAM cache significantly reduces the number of PCM accesses, DRAM/PCM memory hybrids are still not enough – by themselves – to adequately address all issues pertaining to the use of PCM. Thus, substantial support from other layers in the memory sub-system is also pertinent.

The aforementioned observations constitute our primary motivation. The elemental driver of this work is two-fold: (a) consider the aspects of performance, energy consumption, and lifetime of the PCM simultaneously, and (b) deploy an architecture that does *not* require any support from, or any modification to, the various layers of the cache/memory hierarchy. This latter goal implies that the new design must be transparent to the operation of the remaining components of the memory sub-system; i.e., the proposed solution must be self-contained and self-supported.

Toward this end, we hereby propose a practical *Dual-Phase Compression technique (DPC)* optimized for hybrid DRAM/PCM main memory architectures. The main contributions of this work are:

- A compression mechanism tailored to hybrid DRAM/PCM main memory systems, whereby the compression process is divided into two distinct phases for better performance, energy consumption, *and* wearleveling. The first phase optimizes DRAM cache accesses by utilizing a simplified, low-latency *word-level* compression algorithm. This phase dramatically reduces the number of PCM accesses – thereby increasing the effective memory capacity – without adversely affecting access latencies. The second phase adopts a *bit-level* compression algorithm to further reduce the number of PCM accesses and to increase the lifetime of the PCM.

- A low-overhead wearleveling technique, called Compression-based Segment Rotation (CSR), which enhances the lifetime of the PCM by exploiting the remaining memory space after compression. This mechanism acts on the data after compression and intelligently balances wear-out within the PCM.

- The devised architecture can be cost-effectively implemented by slightly modifying the memory controller. It does *not* require any architectural support from other entity in the CMP.

Extensive simulations driven by traces extracted from *real application workloads* running on a full-system simulator demonstrate that the dual-phase mechanism yields 35.1% performance improvement and 29.3% energy reduction, on average, as compared to a baseline DRAM/PCM hybrid implementation. Additionally, we demonstrate that the co-operative use of our CSR scheme and a simple existing wearleveling technique (also implemented within the memory controller) ensure a lifetime for the PCM within a range of 4.42 to 40 years.

2. BACKGROUND & RELATED WORK

2.1 Fundamentals of PCM

PCM exploits the unique trait of calcogenide glass to switch between two states - crystalline (SET, low resistance) and amorphous (RESET, high resistance) through the passage of an electric current. The PCM's write operation commences with the melting of the cell under hot temperature, followed by fast quenching (RESET operation), or slow quenching (SET operation), depending on the write value. This heating-and-cooling process generally takes more than 100 ns (i.e., it is relatively slow compared to DRAM), and it also severely affects the endurance of PCM, limiting its lifetime to within 10^6 to 10^8 write cycles in current PCM technology. In addition, this power-hungry write operation prevents the exploitation of burst memory-write operations, which, in turn, decreases the bandwidth of write operations [12]. On the other hand, PCM's read performance is similar to that of DRAM. Actually, PCM's read latency is currently 2 times slower than DRAM, but PCM has been shown to support the DDR2 interface for read operations [11], which will soon bring its performance up to par with DRAM. Thus, reducing the number of PCM writes and the data size of write accesses are crucial goals when dealing with PCM.

2.2 Compression techniques in PCM

Compression techniques have generally been used to reduce the memory traffic and increase the effective capacity of memory systems. However, they require additional, non-negligible latency overhead during the compression/decompression processes. In addition, address remapping schemes, as well as somewhat complex compaction and/or reallocation processes are mandatory, because of the variable size of the compressed data. Hence, compression techniques have not been widely adopted in real products. However, despite said non-negligible overhead, compression techniques constitute an attractive solution within the context of PCM, because the reduced data size after compression can increase the performance (especially the write performance), energy consumption, and endurance simultaneously. Furthermore, if the remaining (saved) space is not used to store additional data, the overhead of address remapping and data compaction and/or reallocation can be eliminated. Fortunately, one can easily give up the capacity benefits of compression, because the main attraction of PCM is to provide more space than DRAM at the same cost. Hence, the remaining space (after compression) can instead be used to further enhance the lifetime of the PCM. We will demonstrate this effect later on in the paper.

Compression techniques have been investigated for PCM-based main memory systems in [20]. However, compression is only one of several alternatives explored in that paper, and it is only triggered when the other alternative schemes are not expected to work. Furthermore, the compression algorithm employed was not specifically optimized for the nuances of a PCM-based system. Consequently, the contribution of data compression was not found to be so important.

The authors of [13] indicate that conventional data compression techniques may not be so effective for PCM, because mismatches in the sizes of new and old compressed data segments may generate additional write traffic for compaction, and may require address

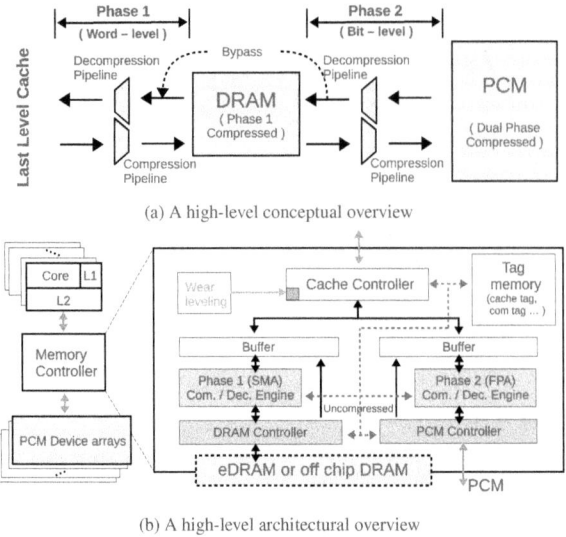

(a) A high-level conceptual overview

(b) A high-level architectural overview

Figure 1: A high-level conceptual / architectural overview of the proposed Dual-Phase Compression Mechanism.

space remapping. Therefore, the incorporation of *efficient* and *effective* data compression for PCM-based main memory systems necessitates optimizations at both the algorithmic and architectural levels; simply applying conventional compression techniques to PCM-based systems will not yield the desired effects.

Recently, encoding-based memory compression [18] has been introduced for reducing the number of PCM writes using Frequent Pattern Compression [19]. The encoding engine was embedded inside the PCM chip and required some modification of the PCM row-array structure. This implementation has some disadvantages, such as *resource duplication* (all individual memory chips must have the same encoding tables and logic, thus increasing the energy consumption of the PCM device), updating the Frequent Value table is not easy, and all the tag bits must be stored within the PCM device at a *fixed position* (meaning that the wearleveling mechanism must also consider the position of the tag bits). Moreover, the proposed scheme of [18] focuses only on enhancing the energy consumption and endurance of PCM.

On the other hand, our proposed scheme can be effectively implemented within the memory controller, and does *not* require any architectural (or other) support from the system, rendering its potential incorporation into future CMPs very straightforward. Most importantly, the technique proposed in this paper enhances the performance, energy consumption, and endurance simultaneously, by applying a fine-tuned compression technique specifically tailored to the characteristics of PCM technology.

3. A DUAL-PHASE COMPRESSION (DPC)

Our goal in this work is to exploit the benefits of compression techniques for PCM without the aforementioned limitations of compression. To this end, we distribute the compression process into two phases, as shown in Figure 1(a). The first phase of the compression process compresses the data at the **word-level** using a *Successive Matching Algorithm (SMA)* and aims to increase the effective DRAM cache capacity with very little delay overhead. The second phase operates on the compressed data of the first phase using a *Frequent Pattern Algorithm (FPA)* [2] at the **bit-level**, in order to achieve even more data compaction. The proposed compression mechanism also employs an efficient wearleveling technique to prolong the lifetime of the PCM, *without* resorting to any remapping activities.

Figure 1(b) presents the accompanying memory controller architecture, which consists of a DRAM cache controller, two compres-

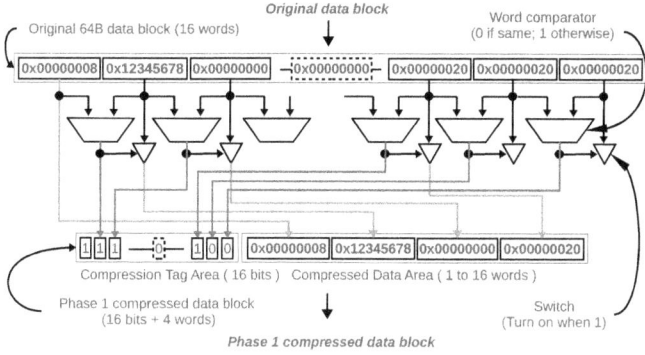

Figure 2: Phase 1 of the Dual-Phase Compression scheme employs a hardware-based *Successive Matching Algorithm (SMA)*.

sion engines, DRAM and PCM device controllers, a tag memory space, and (embedded) DRAMs. In DRAM/PCM hybrid memory implementations, the DRAM serves as a cache for the PCM (i.e., it forms an extra level in the memory hierarchy). Note that all our proposed components, including the tag memory, reside solely in the memory controller.

3.1 Phase 1 – Word-Level Compression with a Successive Matching Algorithm (SMA)

Figure 2 illustrates the SMA utilized in the first phase of the DPC architecture, which is very lightweight, simple, and extremely fast. In this phase, a 64B data block (assuming a last-level cache line containing 16 32-bit words) is compressed at the granularity of a single word (i.e., word-level compression). As the term "successive matching" suggests, compression is only performed between *adjacent* (consecutive) words in the 64B block. While compressing the original 64B data block, the i and $i-1$ words are compared for a possible match ($i = 0$ is the least significant word of the original 64B data block). The resulting compressed data block comprises two parts: the *Compression Tag Area* and the *Compressed Data Area*. The Compression Tag Area consists of 16 bits, with each bit corresponding to one word in the original 64B data block (remember, a 64B block holds 16 words). If two consecutive words are found to be identical, a '0' is written in the corresponding bit of the Compression Tag Area. Otherwise, a '1' is written. The first bit of the Compression Tag Area is always '1', since there is no $i-1$ word to compare with. If the i-th tag bit is '1', then the i-th word is written from the original data block in the Compressed Data Area (Figure 2). Otherwise, nothing is written.

Looking at the scenario presented in Figure 2, the first three tag bits are set to '1', because the first three words in the original data block are different. However, the 4th-13th tag bits are all set to '0', because the 3rd word in the original data block is repeated multiple times. Similarly, the last three words in the original data block are the same. Hence, the 14th tag bit is set to '1', while the 15th and 16th bits are set to '0', to indicate compression. The storage overhead of this scheme, in terms of extra bits, is only one bit for each word (i.e., 3% for a 32-bit data path system), which is the cost of the *Compression Tag Area*. If there are *at least two identical consecutive words* in the original 16-word data block, then the overhead is fully amortized and yields immediate compaction benefits.

In order to fully take advantage of the compression in Phase 1, the DRAM cache must be able to accept more compressed lines than conventional uncompressed lines. In other words, it must be able to accommodate cache lines of variable size without wasting available space due to placement restrictions. In order to achieve this, we exploit *decoupled, variable-segment caching* [1], but with our SMA-based compression. The proposed DRAM cache architecture is depicted in Figure 3. While each set is *physically* 4-way

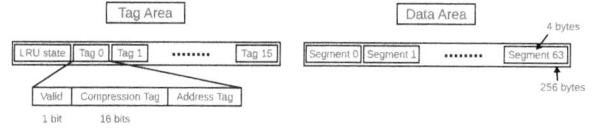

Figure 3: Illustration of a single set of the proposed DRAM cache structure, which is a similar implementation to a *decoupled variable-segment DRAM cache* [1].

set associative, we overlay a *logical* 16-way set associativity by using more tags and a segmented data area. More specifically, the data array of the cache is broken into 4B (single-word) segments, with 64 segments statically allocated to each cache set.

As previously described, the SMA compresses each 64B cache line into anything between 1 (fully compressed) to 16 (uncompressed) words. Hence, each set in the proposed DRAM cache structure of Figure 3 can, technically, hold up to 64 compressed cache lines (in the extreme case, where each cache line is compressed to a single 32-bit word). In order to support such high number of cache lines per set, the Tag Area size (see Figure 3) would have to increase dramatically. So, the critical question raised is whether real applications would benefit from such extreme capability (i.e., 64 compressed cache lines per set). Based on extensive simulations using traces from a full-system simulator, real applications hardly yield more than 10 compressed cache lines per set. Thus, we set the maximum number of cache lines that one set can hold to 16, which means that each set can potentially increase its effective capacity by up to four times (when storing 16 compressed cache lines).

Each tag in the Tag Area (Figure 3) consists of a valid bit, the 16-bit Compression Tag (from Figure 2), and the address tag. The overhead, as opposed to a conventional cache, is the extra 16-bit Compression Tag, which is required for all 16 tags of each set.

Data segments are stored contiguously in Address Tag order. The offset for the first data segment of cache line k (in a particular set) is

$$\text{segment\_offset}(k) = \sum_{i=0}^{k-1} \text{actual\_size}(i)$$

where $k \le 15$ and actual_size(i) is the size of ith cache line (whether compressed or not), in bytes, as given by

$$\text{actual\_size}(i) = \sum \text{num\_tag}(i) \times 4 \text{ bytes}$$

where num_tag(i) is the number of '1's in the valid compression tag(i), $0 \le i \le 15$. The 'segment_offset' and 'actual_size' parameters are used to target specific segments in the Data Area of the cache for address-tag matching.

To enable rapid access to a specific word in a cache line, it is imperative for the technique to be able to swiftly locate the requested word in the *compressed* cache line. This, in fact, is one of the potential drawbacks of compression schemes: how does one locate a specific item within the compressed data (a) without having to decompress, and (b) without having to search the entire compressed block? The proposed SMA mechanism avoids *both* of these complications by providing a *critical-word-first* capability: to access the i-th word from the original uncompressed data block, the controller simply counts the number of tag bits set to '1' in the 16-bit Compression Tag Area of the set (see Figure 2) from tag bit 0 to tag bit $i-1$. This number immediately tells the location of the requested i-th word in the compressed data block. This procedure pinpoints the required word *directly* with no need for decompression or searching. This is a valuable attribute that contributes to the enormous performance benefits that will be demonstrated in Section 5. Although our DRAM cache incurs 4% storage overhead, due to the presence of the Compression Tags, this overhead is

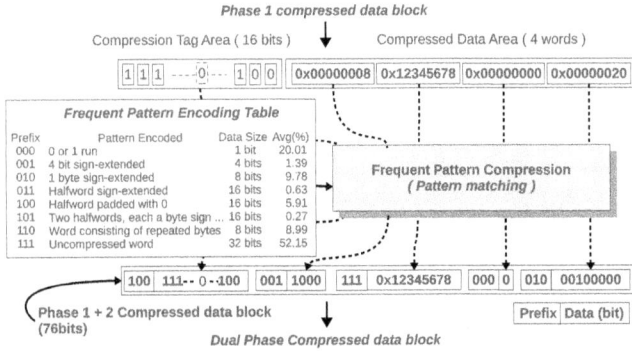

Figure 4: Phase 2 of the Dual-Phase Compression scheme employs a hardware-based *Frequent Pattern Algorithm (FPA)*.

fully amortized by the vast increases in the average *effective* DRAM cache capacities achieved.

We have also analyzed the hardware overhead of the control logic of the proposed mechanism. Its area cost in terms of gate count is only 0.81% (specifically, 394 gates) of a commercial DDR2 SDRAM memory controller, while it incurs only 1 additional memory clock per compression/decompression operation. This overhead is included in our evaluation/simulations later on.

3.2 Phase 2 – Bit-Level Compression with a Frequent Pattern Algorithm (FPA)

Phase 2 of our DPC architecture attempts to compact the compressed data of Phase 1 even more for storage into the PCM. In order to identify this additional redundancy that can be further compressed, the mechanism has to operate at a finer granularity than the word-level approach of Phase 1.

One of the most efficient and fastest bit-level compression algorithms is FPA [2]. FPA uses a Frequent Pattern Encoding table, as illustrated in Figure 4. The table includes eight bit patterns that tend to appear frequently in real application workloads [6]. These eight patterns are known a priori and are converted to 3-bit prefix values (2^3=8) and a corresponding data value in their compressed form (see Figure 4). For example, a so called '0'-run (a pattern of all consecutive zeros in the word) corresponds to the first pattern in the table of Figure 4, which compresses to a prefix of '000'. In the example of the same figure, the first word of the compressed data from Phase 1 is compressed to the '000' prefix and a 1-bit data value of '0', because the original word is a '0'-run word. As shown, after the FPC procedure of Phase 2, the compressed data block from Phase 1 is further compressed to 76 bits.

Despite the pre-compression of Phase 1, it turns out that the same eight frequent patterns identified in [2, 6] also appear in the compressed (at the word-level) cache lines of Phase 1, as shown in the third column of the table of Figure 4. This is because only word-level compression is performed in Phase 1 to minimize latency. Even though the *frequencies* of the eight patterns are smaller than those in uncompressed data, there are still *significant* numbers of occurrences of the frequent patterns of the FPA algorithm of [6]. So we can safely use the same frequent pattern encoding table as the one shown in Figure 4.

The hardware implementation cost of the second compression phase is extracted from [1, 6], and we also include this additional overhead (timing and energy) in our evaluations.

4. A COMPRESSION-BASED SEGMENT ROTATION (CSR) SCHEME

Compression techniques are quite useful not only for enhancing performance, but also for prolonging the lifetime of the PCM. In this section, we propose a Compression-based Segmented Rotation (CSR) mechanism, which is a type of intra-line wearleveling tech-

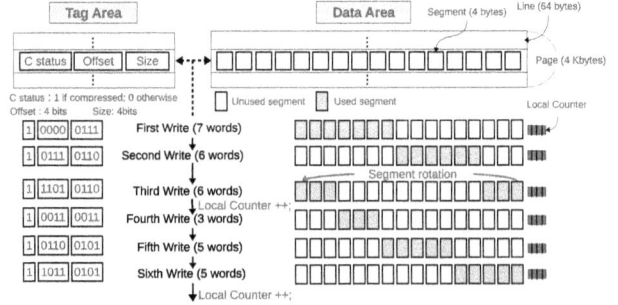

Figure 5: The fundamental operating principle of the proposed Compression-based Segment Rotation (CSR) process for local wearleveling within a line.

nique added to our proposed DPC scheme with minimal overhead. The minimum unit of flushed-out data from the DRAM cache is a cache line, i.e., 64B in this paper. However, after applying our DPC mechanism, the physical size of a cache line is reduced to less than half of the original, on average. In most general-purpose compression algorithms, the remaining space is used for storing other data. Instead, our CSR scheme uses this remaining space to enhance the lifetime of the PCM (in this case, address remapping is no longer necessary) under the assumption that the PCM device already provides enough memory space. This is not unrealistic, since PCM is an extremely scalable memory architecture offering massive storage density.

Figure 5 depicts the basic principle of the proposed CSR technique, which is only performed at the *line-level*. As shown in the figure, the first write operation only uses the first 7 segments of the available line space, because the *compressed* size of the flushed-out cache line happens to be 7 words long (i.e., 28 bytes). At the second write, the data is stored in the *following* segments, from the 8th to the 13th (occupying 5 segments in the line). If the data spills out of the last segment of the line, the remaining information wraps around to the first segments of the line while increasing the line counter value by one, as shown in the case of the third write in Figure 5. By doing this, six write requests to the same address (the example of Figure 5) only increase the write count of the line by two. For efficient implementation, we use 9 bits of additional space per line (i.e., a 1.8% space overhead in the case of 64B line size, but this will decrease if the line is larger than 64B) to indicate the compression status (C status - 1 bit) and the offset and size of valid compressed data. Since this 9-bit tag data is stored into on-chip memory, and is read and updated at the same time as the accessing of the tag area of the DRAM cache, there is no additional timing overhead in accessing this data.

Although this CSR scheme only focuses on wearleveling within a line, we have found that our mechanism can efficiently co-operate with a local-counter-based page swapping technique [13], which is also implemented in the memory controller together with the CSR scheme. In this case, the lifetime of the PCM device is extended to more than 5 years, even under memory-intensive applications, as illustrated in Figure 7 in Section 5.3. This is crucial, since no additional wearleveling technique is required (such as partial-writes [14], Flip-N-Write [5], and line-level wearleveling [16]).

5. EVALUATION

5.1 Simulation Framework

We have developed a trace-driven simulator for evaluating our proposed Dual-Phase Compression architecture. The traces have been extracted from the Simics/GEMS full system simulator [15] with timing information in CPU cycles. All the details of the simulation configuration are described in Table 1. We investigate a total of four memory system configurations: (1) DRAM-only, (2) PCM-

Number of CMP Cores	16
Processor Core Type	UltraSPARC-III+ (in-order), 2 GHz
L1 caches (Private)	I- and D-cache: 64 KB, 4-way 64 B block
L2 caches (Shared)	1 MB, 4-way 64 B block
DRAM memory (large)	DDR2 8 GB
DRAM memory (small)	DDR2 64 MB

Table 2: Memory access characteristics of the benchmark applications used (PARSEC benchmark suite [3])

Group #	Application	Description	R/W rate	MMA rate[1]
1	canneal	Simulated cache-aware annealing	1.83	12.6
	dedup	Compression with data deduplications	1.39	4.7
2	fluidanimate	Fluid dynamics for animation purposes	1.38	4.6
	x264	H.264 video encoding	1.78	2.3
3	vips	Image processing	1.32	1.6
	freqmine	Frequent itemset mining	1.55	1.5
4	ferret	Content similarity search server	1.52	0.9
	bodytrack	Body tracking of a person	1.71	0.8

[1] Main Memory Accesses per 1K CPU cycles

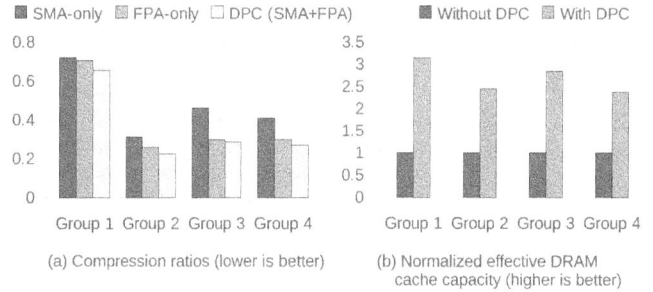

(a) Compression ratios (lower is better)

(b) Normalized effective DRAM cache capacity (higher is better)

Figure 6: Compression ratios (=compressed-size/original-size) and the effect of SMA in the DRAM cache of the DRAM/PCM hybrid memory architecture.

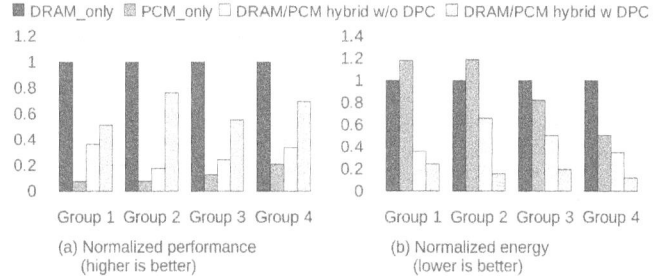

(a) Normalized performance (higher is better)

(b) Normalized energy (lower is better)

Figure 7: Performance and Energy simulation results. The Performance and Energy results are *normalized to the DRAM-only configuration*.

only, (3) regular DRAM/PCM hybrid with 64-entry write queue, and (4) DRAM/PCM hybrid with our Dual-Phase Compression (DPC) technique. 8 GB off-chip DRAM modules and 8 GB off-chip PCM modules are used for the DRAM-only and PCM-only memory configurations, respectively, while a 64 MB DRAM and 8 GB PCM modules are used for the DRAM/PCM hybrid memory configuration. We set the size of DRAM cache to 64 MB even though 256 MB DRAM cache is a sweet spot for most applications in the baseline DRAM cache, because the proposed DPC technique increases the effective capacity of the DRAM cache by 2 to 3 times, on the average. So we expect similar results to using DRAM cache modules of well over 128 MB. This 64 MB DRAM cache can be implemented as a small discrete component (memory module), or it can be embedded into the controller as embedded DRAM (eDRAM). The energy and performance models used for the PCM and DRAM devices in our evaluations are derived from state-of-the-art academic and industrial references [8, 12, 17, 10].

We selected 8 benchmarks from the PARSEC benchmark suite [3]; their memory-access characteristics are shown in Table 2. We categorized 8 benchmarks into four groups (Group 1 to 4), based on their memory access rates. Group 1 consists of the two most memory-intensive applications of Table 2, whereas Group 4 consists of the two least memory-intensive applications. Groups 2 and 3 lie in-between Groups 1 and 4, in terms of memory access rates.

5.2 Performance and Energy Improvements

Since the main objective of the paper is to apply an optimized dual-phase compression mechanism for DRAM/PCM hybrid main memory systems, we first evaluate the compression ability of the algorithms that we used in each phase. As shown in Figure 6(a), the proposed DPC always exhibits better compression ratio than each of the individual compression algorithms. This is attributed to the fact that DPC performs combined global and local compression.

The DPC technique can also increase the effective capacity of a 64 MB uncompressed DRAM cache to anywhere from 144 to 192 MB, as shown in Figure 6(b). This means that the performance of the proposed DRAM cache is always higher than a normal DRAM cache with the same physical size. Alternatively, one can have a smaller-sized DRAM cache without experiencing any performance degradation, due to the increased *effective* cache capacity resulting from the DPC mechanism.

Figure 7(a) shows performance comparisons between the four memory configurations. Each configuration is normalized to that of the DRAM-only configuration. Note that a higher bar indicates better performance. As expected, our proposed architecture

– DRAM/PCM hybrid with DPC – shows the second-best performance for all groups of applications. This is because the DPC mechanism (1) reduces the number of PCM writes (dirty cache misses) by increasing the effective DRAM cache size and hit ratio in Phase 1, and (2) it also significantly reduces the PCM write-block size of dirty misses. Remember that PCM cannot support a burst-write operation, due to its heavy power requirement. Therefore, reducing the data block size even by a few words can enhance the performance significantly. On average, our proposed DPC mechanism enhances the performance by 35.1%, as compared to the baseline hybrid configuration.

We also compare the energy consumption of each configuration, as shown in Figure 7(b). Similar to Figure 7(a), the energy consumption for each configuration is normalized to that of the DRAM-only. In this figure, a smaller bar indicates better energy efficiency. Since the DRAM-only configuration uses a large-size DRAM (8 GB), its energy consumption is substantially higher than the hybrid configurations. The PCM-only configuration also consumes a large amount of energy especially in memory intensive applications (Groups 1 and 2), because the number of energy-hungry write operations is never reduced compared with DRAM-only configuration. If we focus on the two DRAM/PCM hybrid configurations, the DPC-augmented one always consumes less energy than the other one. The reasons are similar to the ones described in our performance analysis: storing compressed data markedly reduces the number of energy-hungry PCM write operations. On average, the proposed DRAM/PCM hybrid with DPC enhances energy efficiency by 29.3%, as compared to the same hybrid configuration without the DPC technique.

5.3 Enhancing the lifetime of the PCM device

Figure 8 compares the PCM lifetime of the DRAM/PCM hybrid main memory architecture with and without our DPC scheme. We use 10^8 as the maximum number of write cycles that can be performed to any cell before the cell fails (worst-case analysis). To aid understanding, we divide the results into two graphs. Figure 8(a), whose y-axis is displayed in units of days, directly compares

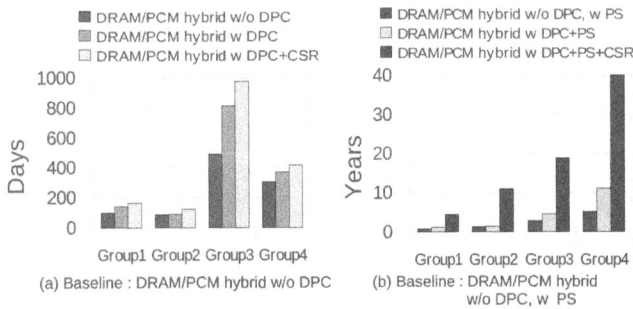

Figure 8: Assessment of the lifetime of the PCM device.

the effectiveness of the DPC technique. Without any wearleveling techniques, the lifetime of the baseline hybrid configuration ranges from 98 to 488 days. After applying the proposed DPC scheme (but without CSR), its lifetime can be prolonged by up to 1.66 times. Finally, applying DPC+CSR increases the lifetime by up to 2 times, as compared with the baseline hybrid configuration.

Even though the proposed DPC architecture can significantly increase the lifetime of the PCM, it is still not enough to ensure a reasonable lifetime for the PCM under memory-intensive applications (Groups 1 and 2). This is because the proposed CSR design only focuses on wearleveling within a line. Hence, we analyze the effect of CSR when this scheme is combined with an existing Page-level Swapping (PS) technique [13], which is also implemented in the memory controller. As shown in Figure 8(b), when our proposed wearleveling scheme is combined with the page-swapping technique (DRAM/PCM hybrid w DPC+PS+CSR), the lifetime of the PCM is extended to multiple years (the y-axis is displayed in units of years). The lifetime ranges from 4.42 to 40 years, while the combination of the DRAM/PCM hybrid with PS but without our proposed DPC mechanism ranges only from 0.7 to 5.1 years.

6. CONCLUSION

In this paper, we present a novel enhancement mechanism for DRAM/PCM hybrid memory architectures. The proposed Dual-Phase Compression (DPC) scheme dramatically reduces the high-cost PCM write accesses by increasing the effective capacity of the DRAM cache, and by accessing the data in compressed mode. In addition, the DPC mechanism is architected in such a way as to be a *self-contained* and *self-supported* solution, which does *not* require any support from – or any modification to – the various layers of the cache/memory hierarchy. Our extensive simulations using traces extracted from a full-system simulator running real applications demonstrate that the DPC mechanism achieves 35.1% performance improvement and 29.3% energy reduction, on average, as compared to a baseline DRAM/PCM hybrid implementation. Additionally, a Compression-based Segmented Rotation (CSR) wearleveling mechanism is introduced and combined with the DPC scheme. The combined design is shown to significantly prolong the lifetime of the DRAM/PCM hybrid by up to 2 times. If page-swapping and inter-line wearleveling are also employed, the resulting architecture exhibits endurance well beyond the typical useful lifetime of a computer system.

7. ACKNOWLEDGMENTS

This work is partially supported by KORUSTECH(KT)-2008-DC-AP-FS0-0003. It also falls under the Cyprus Research Promotion Foundation's Framework Programme for Research, Technological Development and Innovation 2009-10 (DESMI 2009-10), co-funded by the Republic of Cyprus and the European Regional Development Fund, and specifically under Grant ΤΠΕ/ΠΛΗΡΟ/06-09(ΒΙΕ)/09.

8. REFERENCES

[1] A. R. Alameldeen and D. A. Wood. Adaptive cache compression for high-performance processors. In *Proceedings of the ISCA 2004*.

[2] A. R. Alameldeen and D. A. Wood. Frequent pattern compression: A significance-based compression scheme for l2 caches. In *Technical Report 1500, Computer Sciences Department, University of Wisconsin-Madison*, April 2004.

[3] R. Bagrodia and *et al.* Parsec: A parallel simulation environment for complex systems. *Computer*, 31:77–85, October 1998.

[4] A. Bivens and *et al.* Architectural design for next generation heterogeneous memory systems. *IEEE International Memory Workshop*, pages 1 – 4, May 2010.

[5] S. Cho and H. Lee. Flip-N-Write: A simple deterministic technique to improve PRAM write performance, energy and endurance. In *Proceedings of the MICRO 2009*.

[6] R. Das and *et al.* Performance and power optimization through data compression in network-on-chip architectures. *Proceedings of the HPCA 2008*, pages 215 – 225.

[7] G. Dhiman, R. Ayoub, and T. Rosing. PDRAM: a hybrid PRAM and DRAM main memory system. In *Proceedings of the DAC 2009*, pages 664–469.

[8] A. P. Ferreira and *et al.* Using PCM in next-generation embedded space applications. *Proceedings of the RTAS 2010*.

[9] R. F. Freitas and W. W. Wilcke. Storage-class memory: The next storage system technology. *IBM Journal of Research and Development*, 52(4/5):439 – 447, 2008.

[10] HP Labs. CACTI: an integrated cache and memory access time, cycle time, area, leakage and dynamic power model. *http://www.hpl.hp.com/research/cacti/*.

[11] H. Jung and *et al.* A 58nm 1.8v 1Gb PRAM with 6.4MB/s program BW. In *Proceedings of IEEE ISSCC 2011*.

[12] S. Kang and *et al.* A 0.1/spl mu/m 1.8V 256Mb 66MHz synchronous burst PRAM. In *Proceedings of the ISSCC 2006*, pages 487 – 496.

[13] J. Kong and H. Zhou. Improving privacy and lifetime of PCM-based main memory. In *Proceedings of the DSN 2010*.

[14] B. C. Lee, E. Ipek, O. Mutlu, and D. Burger. Architecting phase change memory as a scalable DRAM alternative. In *Proceedings of the ISCA 2009*, pages 2–13.

[15] P. S. Magnusson and *et al.* Simics: A full system simulation platform. *Computer*, 35:50–58, February 2002.

[16] M. K. Qureshi, V. Srinivasan, and J. A. Rivers. Scalable high performance main memory system using phase-change memory technology. In *Proceedings of the ISCA 2009*.

[17] Samsung Electronics. DDR2 registered SDRAM module, M393T5160QZA. *Datasheet*.

[18] G. Sun, D. Niu, J. Ouyang, and Y. Xie. A frequent-value based pram memory architecture. In *Proceedings of the ASP-DAC 2011*, pages 211–216.

[19] J. Yang, Y. Zhang, and R. Gupta. Frequent value compression in data caches. In *Proceedings of the MICRO 2000*, pages 258–265.

[20] W. Zhang and T. Li. Characterizing and mitigating the impact of process variations on phase changed based memory systems. In *Proceedings of the MICRO 2009*.

[21] P. Zhou, B. Zhao, J. Yang, and Y. Zhang. A durable and energy efficient main memory using phase change memory technology. In *Proceedings of the ISCA 2009*, pages 14–23.

Verilog-AMS-PAM: Verilog-AMS integrated with Parasitic-Aware Metamodels for Ultra-Fast and Layout-Accurate Mixed-Signal Design Exploration

Geng Zheng[1], Saraju P. Mohanty[2], Elias Kougianos[3], Oleg Garitselov[4]
NanoSystem Design Laboratory (NSDL)[1,2,3,4]
Dept.of Computer Science and Engineering[1,2,4] and Dept. of Electrical Engineering Technology[3]
University of North Texas Denton, TX 76203.[1,2,3,4]
gengzheng@my.unt.edu[1], saraju.mohanty@unt.edu[2], eliask@unt.edu[3]

ABSTRACT

Current Verilog-AMS system level modeling does not capture the physical design (layout) information of the target design as it is meant to be fast behavioral simulation only. Thus, the results of behavioral simulation can be very inaccurate. In this paper a **paradigm shift of the current trend** is presented that integrates layout level information (with full parasitics) in Verilog-AMS through polynomial metamodels such that system-level simulation of a mixed-signal circuit/system is realistic and as accurate as the true parasitic netlist simulation. As a specific case study, a voltage-controlled oscillator (VCO) Verilog-AMS behavioral model and design flow are proposed to assist fast PLL design exploration. Based on a quadratic polynomial metamodel, the PLL simulation achieves approximately a $10\times$ speedup compared to the layout extracted, parasitic netlist. The simulations using this behavioral model attain high accuracy. The observed error for the simulated lock time and average power consumption are 0.7% and 3%, respectively. This behavioral metamodel approach bridges the gap between layout accurate but fast simulation and design space exploration.

Categories and Subject Descriptors

B.7.1 [**Integrated Circuits**]: Types and Design Styles—VLSI (very large scale integration)

Keywords

Polynomial and Nonpolynomial Metamodel, Mixed-Signal Design, Behavioral Simulation, Verilog-AMS Modeling, PLL.

1. INTRODUCTION

Parasitics greatly degrade the performance of nano-CMOS circuit designs. They cause significant mismatch between schematic and layout circuit simulations. To account for the parasitic effects and achieve design closure, numerous iterations at the layout stage are usually required. This process requires great amounts of time and effort. Layout-accurate verification is the major obstacle be-

cause the iteration time is mainly spent on layout modification and simulations. Behavioral models that are capable of representing circuit layout have the potential to dramatically shorten the design cycle [8, 7, 12]. Parasitic effects, however, are not discussed in most works due to the inherent inability of function-based behavioral models to account for them. Also, circuit models in these works are commonly implemented as Verilog-A modules rather than Verilog-AMS modules which are more flexible in terms of functionality. Modeling techniques such as model order reduction [14] and symbolic model generation [3] have been also proposed but they only work for relatively small circuits. It may be noted that **the terms macromodel and metamodel are often used interchangeably** in the literature. However, while macromodels are simplified models of a circuit and system that use the same simulator [2], metamodels are mathematical algorithms that can decouple the design and simulations to a pure behavioral tool such as MATLAB [4].

A metamodeling technique for nano-CMOS AMS circuits was proposed in [5]. The models built with this method accurately reflect parasitic effects. In the present work, an accurate VCO behavioral model is proposed based on this approach. This behavioral model is implemented using the Verilog-AMS language which enables fast simulations. Combining the metamodeling technique and Verilog-AMS simulation, the design verification process achieves a large speedup and maintains reasonably high accuracy. In fact, not only the proposed Verilog-AMS behavioral model can help the design space exploration and optimization, it can also assist the verification process of complex System-on-Chip (SoC) designs. A phase-locked loop (PLL) design with an LC-tank VCO using 180 nm CMOS process is used to demonstrate the modeling technique, design flow and implementation method. Among different PLL architectures, the charge-pump PLL (CPPLL) has been widely used in various system due to its simplicity and effectiveness.

The **novel contributions** of this paper are as follows:

1. An accurate and efficient quadratic polynomial metamodel for a 180 nm LC VCO design is developed.

2. A Verilog-AMS module is constructed to implement the VCO metamodel.

3. A parameterized netlist approach using the VCO layout netlist after full parasitic extraction is used to capture parasitic effects by the metamodel.

4. Metamodel-integrated PLL simulations are presented and the accuracy and speed of the proposed VCO behavioral Verilog-AMS model are discussed.

5. A metamodel-assisted PLL optimization flow is demonstrated.

The rest of this paper is organized as follows: Section 2 discusses previous works relevant to PLL behavioral modeling. Section 3 describes the metamodeling technique and the proposed Verilog-AMS VCO behavioral module. Section 4 presents the PLL simulation flow and methodology with the proposed VCO behavioral model. Section 5 demonstrates the PLL optimization with the assistance of the metamodel. Section 6 concludes this work and discusses future research.

2. RELATED PRIOR RESEARCH

Verilog-A behavioral modules of linear VCOs were used in [9] for PLL jitter characterization and in [15] for aiding a hierarchical CPPLL sizing method. No parasitic effects were included in this model. A characterization technique is developed in [11] to extract circuit parameters, including parasitic effects. The authors also adopted the linear VCO model which may be sufficient for performing verification on fixed designs, but is not useful for design exploration since the VCO linearity condition is not always valid. The VCO behavioral models developed in [1] and [6] use a table-lookup approach inside Verilog-A modules, which is not efficient for global design space exploration. An event-driven analog modeling approach was proposed in [13] which used the Verilog-AMS wreal data type to improve the model efficiency. However, it is not clear how the VCO gain and output frequency were modeled.

3. LAYOUT-ACCURATE POLYNOMIAL METAMODELING OF CPPLL

3.1 High-level Description of the CPPLL

A typical CPPLL consists of a phase/frequency detector (PFD), a charge-pump (CP), a loop filter (LF), and a VCO. If the PLL needs to perform frequency synthesis, a frequency divider (FD) will also be employed. The system level topology of a CPPLL is shown in Fig. 1. In this paper, we focus on developing a VCO behavioral model that can accurately mimic the VCO *physical* design. The model is constructed using the Verilog-AMS language to enable fast design exploration. The other parts of the PLL are modeled with hardware description languages or at schematic level in order to simulate the whole PLL system.

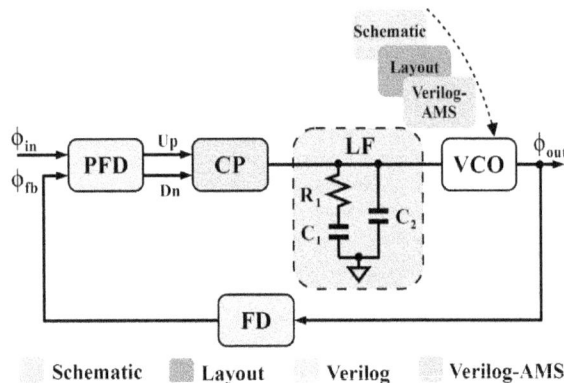

Figure 1: The CPPLL configuration in this paper.

3.2 CPPLL Verilog-AMS Behavioral Model

Mixed-signal systems such as CPPLLs can be simulated using mixed-signal simulators which have two kernels—an event-driven digital kernel and a continuous-time analog kernel [10]. Calling the

analog kernel in a mixed-signal simulation is generally far more computationally expensive than calling the digital kernel. Thus for fast design verification with given accuracy requirements, models that will cause unnecessary analog events should be avoided.

Fig. 1 illustrates the CPPLL configuration in this work. The LF consists of three simple passive components R_1, C_1, and C_2. Modeling it behaviorally does not improve the simulation efficiency noticeably. Therefore the SPICE model is used for the LF and it is implemented in a schematic view. The PFD and FD are pure digital circuits. The frequency of the FD output ϕ_{fb} is $1/N$ of that of the VCO output ϕ_{out}, where N is the FD division ratio. The PFD activates its output Up or Dn to vary the VCO output until ϕ_{fb} and ϕ_{in} are aligned and have the same frequency. They introduce non-idealities to the system via their signal delay, and the rise/fall time. These non-idealities can be easily described in the digital domain. Thus the behavior of these two blocks is implemented using the Verilog language. The CP has digital inputs and analog output so it is implemented as a Verilog-AMS module.

Three different views have been implemented for the VCO: (1) schematic, (2) layout with parasitics, and (3) Verilog-AMS. Fig. 2 shows the schematic and layout views of the LC VCO design. Both schematic and layout views use SPICE models for simulation. While the layout view includes the parasitic elements therefore takes longer to simulate, it results in accurate estimate of the real silicon. Table 1 lists the number of elements in the schematic view and parasitic extracted layout view. The parasitics consist of Resistance (R), Capacitance (C), and self (L) and mutual inductance (K).

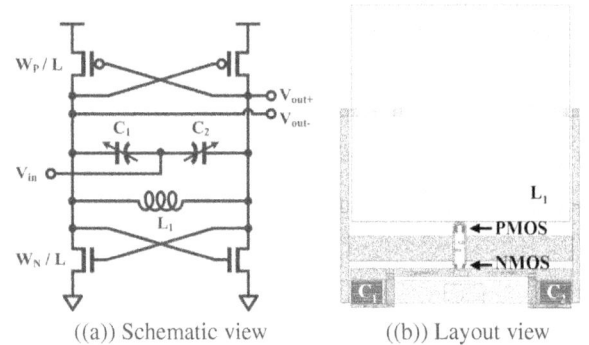

((a)) Schematic view   ((b)) Layout view

Figure 2: The LC VCO schematic and layout views. $L = 180$ nm; $W_P = 20$ μm; $W_N = 10$ μm.

Table 1: Element Counts for LC VCO Schematic and Layout.

Elements	Schematic	Layout
Transistor	4	4
Inductor	1	10
Capacitor	2	38
Resistor	0	560
Total	7	612

The Verilog-AMS view implements an accurate behavioral model. The modeling approach is detailed in Section 3.3.

3.3 VCO Polynomial Metamodeling

The VCO behavior is mainly determined by its voltage frequency transfer curve. A common way to model a VCO to assume that the

VCO is perfectly linear and model it with the following:

$$f_{osc} = f_0 + K_{VCO}V_C, \qquad (1)$$

where f_{osc} is the oscillation frequency, f_0 is the center frequency, K_{VCO} is the gain, and V_C is the control voltage at the VCO input. This linear model can be implemented by sampling two data points on the VCO transfer curve. When performing design exploration, however, the linearity is not guaranteed, which leads to invalid simulation results. Also, parasitic effects from layout extraction further degrade the accuracy of this modeling approach. To account for the non-linearity and layout parasitics, the metamodeling approach suggested in [4] is used.

We chose to implement polynomial metamodels because they have the following advantages: (1) they are simple closed form equations which are easy to implement; (2) their form is flexible so that one can quickly examine and compare the accuracy of polynomial models with different degree; (3) they have been widely used and their properties are well understood. The polynomial metamodel used in this paper is as follows:

$$f(\mathbf{x}) = \sum_{i=0}^{K-1} \beta_i x_1^{p_{1i}} x_2^{p_{2i}} x_3^{p_{3i}}, \qquad (2)$$

where x_1, x_2, and x_3 are three input variables corresponding to W_P, W_N, and V_C in this work, respectively. K is the number of basis functions this model has and β_i is the coefficient for the basis function. $f(\mathbf{x})$ is the output that approximates the true model. In order to construct the metamodel for a given VCO design, for each basis function the coefficient β_i and the power terms p_{1i}, p_{2i}, and p_{3i} for each input variable need to be obtained. This is done in three steps: first, a set of input variables $[x_1\ x_2\ x_3]$ is generated using the Latin Hypercube Sampling (LHS) technique; second, circuit simulations are performed and the outputs for each set of inputs are saved; third, with the inputs and outputs from previous steps, the coefficients and the power terms that lead to a model with good fit are computed. In order to incorporate the parasitic effects into the model without repeating the layout for each simulation, the netlist for the extracted layout view is parameterized for W_P and W_N.

In this work, we are interested in the VCO output frequency and its power consumption. Therefore two respective metamodels are built. They share the same power terms for the input variables, while the coefficients β_i in the two models are different. After these values are computed, they are written into a text file which will be read by the VCO Verilog-AMS module to implement the model. A quadratic polynomial metamodel with first order interaction has been implemented. Table 2 shows the layout of the text file storing the values for the power terms and the coefficients for this model obtained from 100 samples. In the table, $\beta_{i,f}$ and $\beta_{i,p}$ are the coefficients for the frequency and power consumption models, respectively. These values are read into the Verilog-AMS module during the initialization process.

Fig. 3 shows a portion of the VCO Verilog-AMS module. The part of the basis function related to the input variables W_P and W_N is constructed in the `initial` block. The remainder of the basis functions are constructed in the `always` block since the third variable V_C needs to be updated continuously during the simulation. The output signal of this module is implemented to be digital logic type to reduce the computation cost. As in the PFD and FD modules, the non-idealities associated with this output signal can be modeled in the digital domain.

This Verilog-AMS module can be easily reconfigured for metamodels with different degrees by changing the parameter K. In Fig. 4, the simulation results of the VCO transfer curves for the design in

Table 2: Layout of the text file storing the power terms and coefficients for the VCO quadratic polynomial metamodel

i	p_{1i}	p_{2i}	p_{3i}	$\beta_{i,f}$	$\beta_{i,p}$
0	0,	0,	0,	2.113e+009,	1.385e-005
1	1,	0,	0,	-3.214e+012,	44.459e+000
2	2,	0,	0,	3.456e+016,	-2.804e+005
3	0,	1,	0,	6.869e+012,	39.729e+000
4	1,	1,	0,	-1.021e+017,	2.911e+005
5	0,	2,	0,	-2.071e+017,	-1.080e+006
6	0,	0,	1,	3.513e+008,	-8.271e-004
7	1,	0,	1,	-2.565e+012,	-31.282e+000
8	0,	1,	1,	-5.331e+012,	-11.392e+000
9	0,	0,	2,	0.000e+000,	1.041e-003

```
'timescale 10ps / 1ps
'include "disciplines.vams"
module vco_metamodel (out, in);
... ...
parameter integer K;
initial
begin
    out = 0;   // Initialize vco digital output
    ... ...  // Declare ports and data types
    metaf = $fopen("metamodel.csv", "r");
    while (!$feof(metaf))
    begin
        readfile = $fscanf(metaf,
                "%e,%e,%e,%e,%e\n",
                p1, p2, p3, betaf, betap);
        bf[i] = pow(wp,p1) * pow(wn,p2) * betaf;
        bp[i] = pow(wp,p1) * pow(wn,p2) * betap;
        pv[i] = p3;
        i = i + 1;
    end
    $fclose(metaf);
    ... ...
end
always
begin
    vc = V(in);
    ... ...
    freq = 0;
    power = 0;
    for (i = 1; i <= K; i = i + 1)
    begin
        freq = freq + bf[i] * pow(vc, pv[i]);
        power = power + bp[i] * pow(vc, pv[i]);
    end
    ... ...
    #(0.5 / freq / 10p)
    out = ~out;
end
... ...
endmodule
```

Figure 3: Example of the VCO Verilog-AMS source code implementing the polynomial metamodel.

Fig. 2 are shown. The parasitics cause a large difference between the schematic and layout results both in the VCO center frequency and the gain. Metamodel 1 is the Verilog-AMS module with the quadratic model from 100 samples. Metamodel 2 is the module with a 5-th degree polynomial model from 500 samples. Metamodel 2 does not have significant improvements over Metamodel 1. Thus Metamodel 1 is used in the PLL simulations shown in Sections 4 and 5. Differences between the transfer curves of layout and metamodel Verilog-AMS views can still be observed, which means a better metamodel may be used to further improve the accuracy. However, as will be seen in Section 4, this polynomial metamodel is sufficient for system level PLL verification to simulate lock time and average power dissipation.

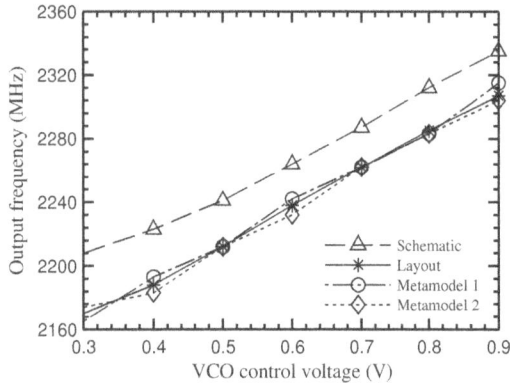

Figure 4: VCO transfer curves for threes different views.

4. VERILOG-AMS-PAM BASED SIMULATION OF PLL

In this section, we demonstrate PLL simulations with the VCO design shown in Fig. 2. The PLL configuration shown in Fig. 1 is used. The PFD and FD are in Verilog view. The CP is in Verilog-AMS view and the LF is in schematic view. The views for the aforementioned blocks were not changed throughout all simulation runs. The VCO view was changed from schematic, to layout with parasitics, and then to Verilog-AMS views. Two Verilog-AMS views have been implemented—one for the linear model and one for the quadratic metamodel proposed in Section 3.3. The results for using different VCO views are compared.

A 550 MHz input clock ϕ_{in} is assigned to the PLL input. The FD has a division ration of 4. Thus the desired frequency for the PLL output clock ϕ_{out} is 2200 MHz. Fig. 5 shows the ϕ_{out} frequencies from 500 ns transient simulations with different VCO views. Although the PLLs with the different VCO views are all able to lock to the same correct frequency, the one with the schematic view shows quite different locking behavior compared to the one with the layout views. This mismatch is due to the parasitic effects which greatly change the VCO characteristics. The one with the linear model shows improvements over the the schematic since the parasitics have been taken into account. However, it still has significant errors, for example, in the lock time. The PLL with the metamodel Verilog-AMS view offers the best approximation of the true model and accurately estimated the lock time. To further understand the behavior of the PLL with different VCO views, the critical analog signal V_C was inspected.

Fig. 6 compares the V_C waveform from the four simulations. Again, the metamodel Verilog-AMS view provides the best approximation of the layout view behavior. The PLL with the schematic VCO view can just barely lock to 2200 MHz since V_C is approach-

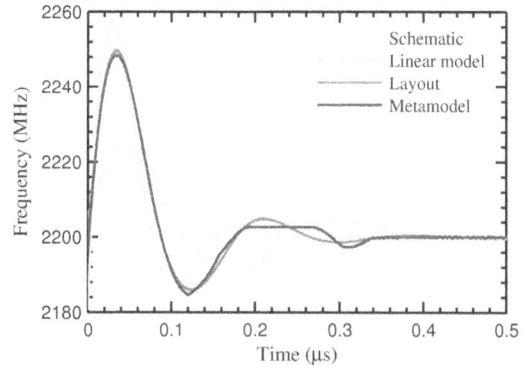

Figure 5: PLL output frequency from AMS simulation with three different VCO views.

ing the NMOS threshold. This shows that the center frequency and the gain of the schematic VCO are very different from the layout. These also confirm the VCO transfer curves plotted in Fig. 4.

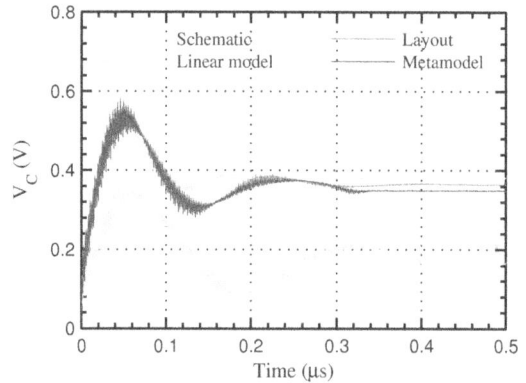

Figure 6: VCO control voltage waveforms from PLL simulations.

The Verilog-AMS metamodel also facilitates estimation of the power consumption. Fig. 7 shows the average VCO power consumption per fifty cycles in the four simulations. It once again confirms that the Verilog-AMS metamodel can better model the layout counterpart. Table 3 shows the PLL simulation results to compare the accuracy of the linear model and the proposed metamodel.

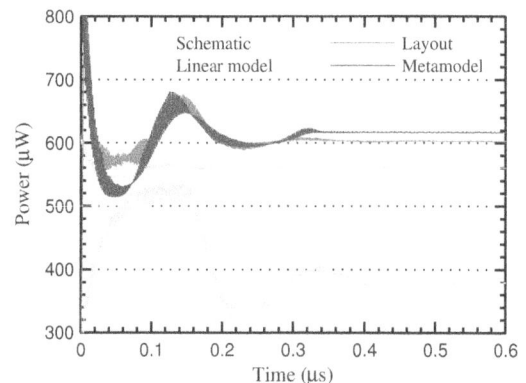

Figure 7: Average VCO power consumption per 50 cycles

In Table 3, the estimated PLL lock time is listed. The one from the simulation with the VCO layout view serves as the true model. The errors resulting from the other two models are computed. The metamodel achieves a very low error rate of 0.7 %, while the linear model causes a large error of 31.7 %. f_{Locked} is the PLL output frequency when it is locked. P_{Locked} is the average VCO power consumption when the PLL is locked. Again, the metamodel resulted in a good estimate of the true power dissipation. The V_C root-mean-square error (RMSE) of the models for the 500 ns simulations are also listed.

Table 3: Comparison of PLL Simulations with Different VCO Modules

	Layout	Linear Model	Metamodel
Lock time (ns)	335.4	229.1	332.9
Error %	0.0 %	31.7 %	0.7 %
f_{Locked} (MHz)	2199.99	2199.99	2199.99
Error %	0.0 %	0.0 %	0.0 %
P_{Locked} (μW)	602	560	620
Error %	0.0 %	7.0 %	3.0 %
V_C RMSE (mV)	0	33.508	10.889

Table 4 compares the runtimes for the PLL transient simulations with the layout, schematic, and Verilog-AMS metamodel VCO views. The Verilog-AMS metamodel achieves roughly a 10× speedup compared to the layout. Note that in practice the VCO design may contain more complex circuitry which leads to longer runtime for a simulation run. The runtime for simulation with the Verilog-AMS module will not be different. Thus the speedup will be more significant in that case.

Table 4: Comparison of The Speed of The PLL Simulations with Different VCO Modules

	Layout	Schematic	Metamodel
Runtime	80.5 s	40.3 s	8.7 s
Normalized speed	1×	∼ 2×	∼ 10×

5. PLL OPTIMIZATION USING METAMODELS

The metamodel and its Verilog-AMS implementation can also accelerate design optimization. In this section, we demonstrate how the metamodel assists PLL optimization. Modern low-power devices are been used in many applications. The wake up time for these device is crucial, which requires short lock time if PLLs are employed. The goal of this optimization is to find a design with minimized lock time and low power consumption. The transistor sizes W_P and W_N of the LC VCO are chosen as the design variables to be optimized. A simple optimization flow is developed to highlight the use of the metamodel and its Verilog-AMS implementation. In practice, more sophisticated flows can be used to handle problems of larger sizes. Table 5 summarizes the optimization flow.

In the first step, the ranges of the design variables W_P and W_N are defined to be 10–30 μm and 5–15 μm, respectively. Within each range, 31 values are evenly selected, which results in a total of 961 possible designs. The design space is then reduced by applying

Table 5: Summary of the optimization flow

Step #	Action		Design Space (total design counts)
1	Define design space	→	961
2	Shrink design space with tuning range constraint	→	320
3	Run AMS simulation to obtain design choices with minimized lock time	→	5
4	Select optimal design with low-power consideration	→	1
5	Verify the final design with layout simulation	→	Done

the tuning range constraint. We define the desired VCO frequency tuning range to be 2180–2300 MHz. A metamodel is used in this step to calculate the VCO tuning range for each design without performing circuit simulations. Only 320 designs are left after this step. Verilog-AMS simulations are then run to obtain the PLL lock time for these designs. The simulations only took 30 minutes to complete due to the use of the Verilog-AMS module. The top five designs with the minimum lock time are saved. These designs are listed in Table 6 along with their average power consumption when the PLL is locked.

Table 6: Comparison of the choices for optimal design

Choice #	W_P (μm)	W_N (μm)	Lock Time (ns)	P_{Locked} (μW)
1	23.2	5	328.6	504
2	21.4	5	328.7	486
3	21.4	5.3	330.4	494
4	22	5	330.4	492
5	22.6	5.3	330.4	506

Choice 2 from Table 6 is selected as the final design for its lowest power consumption. Fig. 8 shows the top five design candidates in the design space of 961 designs. Although the lock time can be further minimized the resulting designs would violate the tuning range requirements. Table 7 compares the original LC VCO design (baseline) shown in Fig. 2 and the optimal design. The optimization reduces both the lock time and the power consumption. Fig. 9 shows the PLL simulation with the VCO layout view of the optimal design relocks from 2180 MHz to 2300 MHz. This simulation finalizes the verification of the optimal design.

6. CONCLUSIONS

A Verilog-AMS behavioral model based on quadratic polynomial metamodeling for a 180 nm LC VCO has been proposed. With this behavioral model, fast and accurate PLL design verification and optimization have been demonstrated. The behavioral model can be further improved but is sufficient for lock time and average power estimation. Future research includes developing behavioral

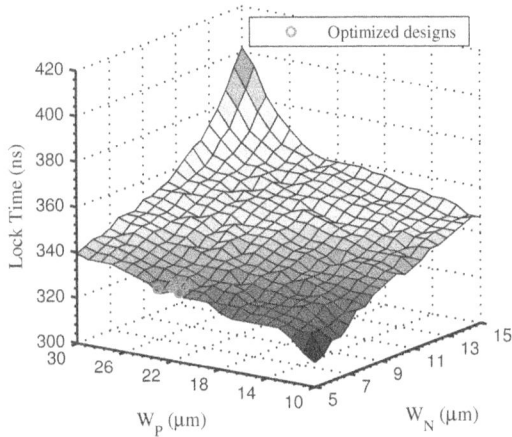

Figure 8: Sizing the transistors for lock time.

Table 7: Comparison of baseline and optimized designs

	Baseline	Optimal	Reduction
W_P / W_N ($\mu m / \mu m$)	20 / 10	21.4 / 5	–
Lock time (ns)	335.4	320.4	15.0
P_{Locked} (μW)	602	455	147
Tuning Range (MHz)	2170–2304	2173–2321	–

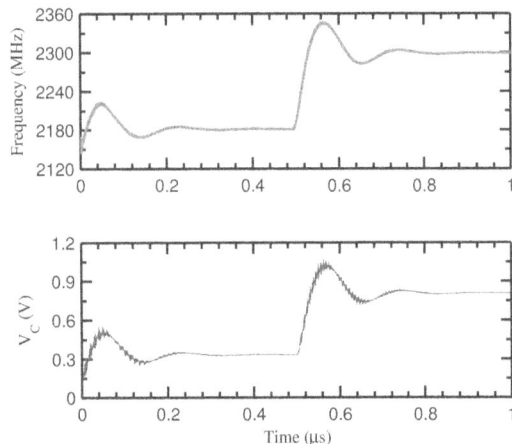

Figure 9: Simulation showing that the PLL first locks to 2180 MHz and then relocks to 2300 MHz.

models that incorporate parasitics for the rest of the PLL building blocks and studying different metamodeling methods.

7. REFERENCES

[1] S. Ali, L. Ke, R. Wilcock, and P. Wilson. Improved performance and variation modelling for hierarchical-based optimisation of analogue integrated circuits. In *Proc. DATE '09. Design, Automation & Test in Europe Conf. & Exhibition*, pages 712–717, 2009.

[2] S. Basu, B. Kommineni, and R. Vemuri. Variation-Aware Macromodeling and Synthesis of Analog Circuits Using Spline Center and Range Method and Dynamically Reduced Design Space. In *Proc. Int. Conf. VLSI Des., VLSID*, pages 433–438, 2009.

[3] C. Borchers. Symbolic behavioral model generation of nonlinear analog circuits. 45(10):1362–1371, 1998.

[4] O. Garitselov, S. P. Mohanty, and E. Kougianos. A Comparative Study of Metamodels for Fast and Accurate Simulation of Nano-CMOS Circuits. *IEEE Transactions on Semiconductor Manufacturing*, 2012.

[5] O. Garitselov, S. P. Mohanty, E. Kougianos, and P. Patra. Nano-cmos mixed-signal circuit metamodeling techniques: A comparative study. In *Proc. Int Electronic System Design (ISED) Symp*, pages 191–196, 2010.

[6] I. Harasymiv, M. Dietrich, and U. Knochel. Fast mixed-mode PLL simulation using behavioral baseband models of voltage-controlled oscillators and frequency dividers. In *Proc. XIth Int Symbolic and Numerical Methods, Modeling and Applications to Circuit Design (SM2ACD) Workshop*, pages 1–6, 2010.

[7] M. Hsieh and G. E. Sobelman. Modeling and verification of high-speed wired links with Verilog-AMS. In *Proc. IEEE Int. Symp. Circuits and Systems ISCAS 2006*, pages 2105–2108, 2006.

[8] C. Hung, W. Wuen, M. Chou, and K. A. Wen. A unified behavior model of low noise amplifier for system-level simulation. In *Proc. 9th European Conf. Wireless Technology*, pages 139–142, 2006.

[9] Q. Jing, T. Riad, and S. Chan. Characterizing PLL jitter from power supply fluctuation using mixed-signal simulations. In *Proc. 2nd Asia Symp. Quality Electronic Design (ASQED)*, pages 112–117, 2010.

[10] K. Kundert and O. Zinke. *The Designer's Guide to Verilog-AMS*. Kluwer Academic Publishers, Boston, 1st edition, May 2004.

[11] C. Kuo. An efficient approach to build accurate behavioral models of PLL designs. *IEICE Transactions on Fundamentals of Electronics, Communications and Computer Sciences*, E89-A(2):391–398, Feb. 2006.

[12] S. Pam, A. K. Bhattacharya, and S. Mukhopadhyay. An efficient method for bottom-up extraction of analog behavioral model parameters. In *Proc. 23rd Int. Conf. VLSI Design VLSID '10*, pages 363–368, 2010.

[13] Y. Wang, C. Van-Meersbergen, H.-W. Groh, and S. Heinen. Event driven analog modeling for the verification of PLL frequency synthesizers. In *Proc. IEEE Behavioral Modeling and Simulation Workshop BMAS 2009*, pages 25–30, 2009.

[14] J. Wood, D. E. Root, and N. B. Tufillaro. A behavioral modeling approach to nonlinear model-order reduction for RF/microwave ICs and systems. 52(9):2274–2284, 2004.

[15] J. Zou, D. Mueller, H. Graeb, U. Schlichtmann, E. Hennig, and R. Sommer. Fast automatic sizing of a charge pump phase-locked loop based on behavioral models. In *Proc. IEEE Int. Behavioral Modeling and Simulation Workshop BMAS 2005*, pages 100–105, 2005.

Efficient Folded VLSI Architectures for Linear Prediction Error Filters

Sayed Ahmad Salehi[*]
Isfahan University of Technology
Department of Electrical and Computer Engineering
Digital Signal Processing Research Lab
P.O.Box 84156-83111, Isfahan, Iran
sa.salehi@ec.iut.ac.ir,fattahi@cc.iut.ac.ir

Rasoul Amirfattahi[*]

Keshab K. Parhi[†]
[†]Electrical and Computer Engineering
University of Minnesota
200 Union Street SE
Minneapolis, MN 55455
parhi@umn.edu

ABSTRACT

In this paper we propose two efficient low-area, low-power folded VLSI architectures for linear prediction error filter. One of them is based on the split-Levinson-Durbin and requires half computational complexity of an architecture based on the Levinson-Durbin algorithm. The other one is based on the Schur algorithm. Using folding method, the number of multipliers and adders is minimized. In addition, by modifications in data scheduling, the number of required multiplexers are also decreased. Comparison with previous architectures demonstrates the efficiency of the proposed architectures with respect to hardware and computational complexity.

Categories and Subject Descriptors

B.7.1 [**Integrated Circuits**]: Types and Design Styles---*Algorithms implemented in hardware, VLSI (very large scale integration).*

General Terms

Algorithms, Design.

Keywords

Levinson-Durbin, Schur, Split-Levinson, VLSI, Low-power, Folding transformation

1. INTRODUCTION

Linear Prediction (LP) analysis [1,2] is a key tool in digital signal processing including speech coding, channel equalization, feature extraction, etc. With rapidly growing microelectronics technology, high-speed, cost effective VLSI devices are available for implementation of signal processing algorithms like LP filters. However, the objective of this paper is to design architectures for linear prediction for low-speed applications such as biomedical signal processing. Regarding the computational complexity of LP analysis, past VLSI design approaches have considered parallel architectures [3-6]. Due to the use of LP error filter in power limited applications such as mobile devices and body implantable chips, there is a demand for low-area low-power hardware implementation of LP filters.

A prediction-error filter can predict a future value (forward prediction) or a past value (backward prediction) of a signal

source modeled as a stationary discrete-time random process. A typical linear prediction-error filter is shown in Figure 1 [1]. The prediction error ($e_M(n)$) is the difference between the predicted value and actual value of next input sample. Using Wiener filter theory the predictor tap weights, a, can be determined to minimize the prediction error [1]. This includes calculating r, autocorrelation of input samples, and solving equation (1).

$$R.a = r \qquad (1)$$

Where R represents the autocorrelation matrix and is toeplitz. We assume R to be known, and the objective is to compute the prediction filter coefficients and prediction error $e_M(n)$.

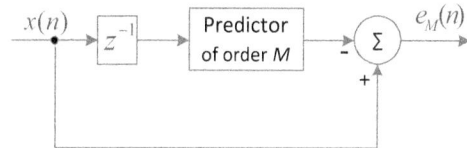

Figure 1: Prediction-error filter

The Levinson-Durbin algorithm (LDA) is the most popular method for solving (1) and computing the prediction coefficients of the system. The LDA is an iterative approach summarized as follows:

0.　$a_0(0) = 1, P_1 = r(0)$

$for\ m = 1, \dots, M$

1.　$k_m = -\frac{\sum_{i=0}^{m-1} r(m-i).a_{m-1}(i)}{P_m}$

2.　$a_m(i) = a_{m-1}(i) + k_m.a_{m-1}(m-i) \qquad i = 0,1,\dots,m$

3.　$P_m = P_{m-1}(1 - k_m^2)$

$end\ for.$

Here $r(i)$'s represent the autocorrelation coefficients of input ($x(n)$), M is the order of the predictor, k_m is the mth order reflection or partial correlation (PARCOR) coefficient, P_m is the mth order minimum square error for predictor, and $a_m(i)$ denotes the tap weights for predictor filter of order m.

Using LDA, $(\sum_{m=1}^{M} 2m) + 2M = M^2 + 3M$ multiplications are required to compute $a_M(i)$. Then convolving $x(n)$ and $a_M(i)$ produces error coefficients ($e_M(n)$). In prior work [5,7], the LDA recursion was reconstructed into a more hardware-friendly form. Here P_m in line 1 of LDA was replaced with equation (2). So P_m

no longer is a variable in the recursion and line 3 of LDA is left out of hardware consideration. In this case $(\sum_{m=1}^{M} 3m) = \frac{3}{2}M^2 + \frac{3}{2}M$ multiplications are required.

$$P_m = r(0) - \sum_{i=1}^{m-1} r(i).a_{m-1}(i) \qquad (2)$$

Direct implementation of LDA does not result in an efficient VLSI architecture, with respect to the computation complexity and required area. Therefore other versions of the algorithm can be used. The computational complexity of different modifications to LDA has been briefly described in [8]. An interesting development is exploiting some inherent redundancy of LDA to improve computational efficiency of the algorithm. This development has been introduced as split-LDA or immitance-domain Levinson algorithm [9,10]. In this paper, we propose a novel architecture based on split-LDA. Furthermore, we propose a minimum area VLSI architecture by clever mapping of the Schur algorithm to a folded architecture [11,12]. We will describe split-LDA and Schur algorithms in the next section. Their corresponding VLSI architectures are presented in Section 3. In Section 4 proposed architectures are compared with previous architectures and some conclusions are drawn in Section5.

2. BACKGROUND
2.1 The Split-LDA
The idea behind split-LDA is achieving symmetry by calculating the sum of forward and backward prediction coefficients [10]. When the autocorrelation vector **r** is real, the backward predictor tap weights, $c_m(i)$, are the same as forward predictor taps, $a_m(i)$, but in reverse order (i.e., $c_m(i) = a_m(m-i)$). So $s_m(i) = a_m(i) + c_m(i)$ is symmetric (i.e., $s_m(i) = s_m(m-i)$) and fully specified by half of its coefficients. One way of reducing complexity is to propagate $s_m(i)$ rather than $a_m(i)$. This converts the LDA to the following three-term recursion called split-LDA [9,10]:

$0.\ \tau_0 = r(0), \gamma_0 = 0,\ s_0 = [2], s_1 = \begin{bmatrix} 1 \\ 1 \end{bmatrix}$

$for\ m = 1,2, \dots, M$

1. $\tau_m = \sum_{i=0}^{m} r(i)s_m(i)$
2. $\alpha_m = \frac{\tau_m}{\tau_{m-1}}$
3. $\mathrm{k_m} = -1 + {\alpha_m}/{(1 - \mathrm{k_{m-1}})}$
4. $s_{m+1}(i) = s_m(i) + s_m(i-1) - \alpha_m s_{m-1}(i-1)$

$$i = 0,1, \dots, m$$

$end\ for$

5. $a_M(i) = a_M(i-1) + s_{M+1}(i) - (1 - k_M)s_M(i-1)$
$$i = 1,2, \dots, M$$

τ_m and α_m are notations used only for simplicity. Lines 1 and 4 can be folded due to the symmetry of s_m. Line 1 can be rewritten as:

$$\begin{cases} \tau_m = \sum_{i=0}^{\frac{m-1}{2}} [(r(i) + r(m-i)]s_m(i) & m\ is\ odd \\ \\ \tau_m = r\left(\frac{m}{2}\right)s_m\left(\frac{m}{2}\right) + \sum_{i=0}^{\frac{m}{2}-1} [(r(i) + r(m-i)]s_m(i) & m\ is\ even \end{cases}$$

and line 4 can be rewritten as

$$s_{m+1}(i) = s_m(i) + s_m(i-1) - \alpha_m s_{m-1}(i-1)$$
$$i = 0,1, \dots, \frac{m}{2}$$

Because lines 1 and 4 are folded in half, the total number of multiplications per recursion will be $2\left(\frac{m}{2}\right) = m$, as compared with $2m$ for classical LDA. This is how 50% reduction in computational complexity arises. But the number of additions and divisions is increased. This is because of lines 3 and 5 of the algorithm. Fortunately it is possible to calculate $e_M(n)$ without using $a_M(i)$ by the following equation:

$$e_{m+1}(n) = e_m(n) + e_m(n-1) - \alpha_m\,e_{m-1}(n-1)$$
$$m = 1, \dots, M-1\ \&\ \ n = 1, \dots, N \qquad (3)$$
$$e_0(n) = 2x(n)\ \ \&\ \ e_1(n) = x(n) + x(n-1).$$

Thus, only α_m is needed and lines 3 and 5 of split-LDA can be ignored for implementation of linear error prediction filter.

2.2 The Schur Algorithm
Although the Schur algorithm and LDA were developed independently, it is easy to show they are the same and the Schur algorithm can be obtained from LDA [13]. The Schur algorithm calculates reflection coefficients (k_m) without calculating tap weights of prediction filter (a_m) [13,8]. In the linear prediction context the forward and backward Schur variables can be written in the form:

$$g_m^+(i) = E[f_m(n)x(n-i)] = \sum_{k=0}^{m} a_m(k).r(i-k) \qquad (4)$$

$$g_m^-(i) = E[b_m(n)x(n-i)] = \sum_{k=0}^{m} c_m(k).r(i-k) \qquad (5)$$

Where $f_m(n)$ and $b_m(n)$, respectively, represent forward and backward prediction errors. The following identities are well-known orthogonality relations in LP theory:

$$g_m^+(i) = 0 \qquad for\ i = 1, \dots, m$$
$$g_m^-(i) = 0 \qquad for\ i = 0, \dots, m-1$$

Therefore, for Mth order predictor, the Schur variables are calculated only for $m \leq i \leq M$. A recursion for these variables describes the Schur algorithm as follows [13]:

$0.\quad \begin{cases} g_0^+(i) = r(i) & i = 1, \dots, M \\ g_0^-(i) = r(i) & i = 0,1, \dots, M-1 \end{cases}$

$for\ m = 1,2, \dots, M$

1. $\qquad k_m = \frac{g_{m-1}^+(m)}{g_{m-1}^-(m-1)}$

$\quad for\ i = m, \dots, M$

2. $\qquad g_m^+(i) = g_{m-1}^+(i) - k_m g_{m-1}^-(i-1)$
3. $\qquad g_m^-(i) = g_{m-1}^-(i-1) - k_m g_{m-1}^+(i)$

$\quad end\ for$
$end\ for$

After calculation of k_m it can be used to obtain $f_M(n)$ and $b_M(n)$ by (6).

$$f_m(n) = f_{m-1}(n) - k_m b_{m-1}(n-1) \qquad (6)$$
$$b_m(n) = b_{m-1}(n-1) - k_m f_{m-1}(n) \qquad \begin{array}{l} m = 1, \dots, M \\ n = 1, \dots, N \end{array}$$
$$f_0(n) = b_0(n) = x(n)$$

The required number of multiplications in this algorithm to calculate k_M, is $M(M+1)$. This is about twice the required multiplications for split-LDA. But this algorithm is attractive for hardware implementation as its symmetric properties can be used for designing a low area folded architecture. Such an architecture is described in Section 3.2.

3. PROPOSED VLSI ARCHITECTURES

3.1 The Split-LDA

Figure 2 shows mapping of split-LDA to a hardware architecture.

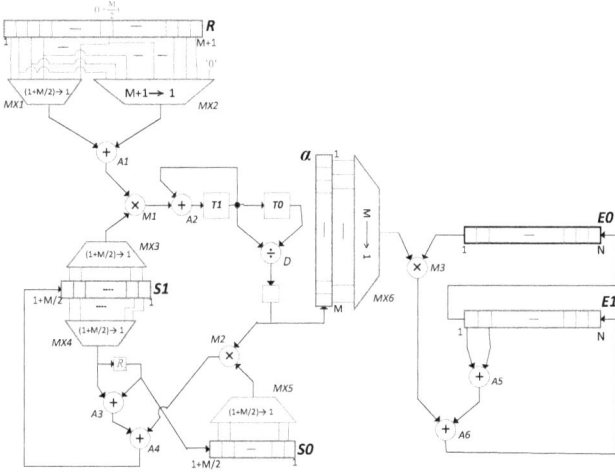

Figure 2: Mapping of split-LDA to hardware architecture

In this architecture for each vector and variable in the algorithm, a buffer or register is considered as in Table 1.

Table 1. Mapping of variables in split-LDA to buffers in Figure 2

variable (in split-LDA)	buffer (in Figure 2)	size	initial value
r	R	$1 + M$	$r(0:M)$
s_{m-1}	$S0$	$\frac{M}{2} + 1$	$2, 0, \ldots, 0$
s_m, s_{m+1}	$S1$	$\frac{M}{2} + 1$	$1, 1, 0, \ldots, 0$
τ_m	$T1$	1	0
τ_{m-1}	$T0$	1	$r(0)$
α_m	α	M	$0, \ldots, 0$
e_{m-1}	$E0$	N	$x(n)$
e_m, e_{m+1}	$E1$	N	$x(n) + x(n-1)$

N is the number of input samples desired to calculate error prediction for them and it is independent of M, the order of the filter. Notation $E1(i)$, for example, refers to ith register in buffer **E1**. To describe the circuit functionality we consider 5 stages based on the equations of the split-LDA in Section 2.1:

1. Initialization: At first initializations are applied based on the last column of Table 1.
2. τ_m calculation: In this stage based on line 1 of split-LDA the value of τ_m is computed by $A1$, $A2$ and $M1$ in Figure 2. Before beginning this step in each recursion, the old value of $T1$ moves to $T0$ and its current value is set to zero. At the end of mth recursion the calculated τ_m is stored in $T1$. One of the $MX2$'s input in Figure 2 is bonded to '0'. It is due to calculating $\left[r\left(\frac{m}{2}\right) + 0 \right] . s_m\left(\frac{m}{2}\right)$ when m is even.

3. α_m calculation: Dividing the final value of $T1$ by $T0$ produces the required α_m at each recursion. Since the division is computed in one clock cycle, this stage is active for only one cycle in every recursion.
4. s_m calculation: In this stage values of $S1$ and $S0$ are updated. Based on line 4 of split-LDA, $M2$, $A3$, and $A4$ are used to calculate s_{m+1} which is directly saved in $S1$. Also the old value of $S1$ is transferred to $S0$.
5. e_m calculation: Based on equation (3) this stage calculates $e_m(n)$ using $M3$, $A5$ and $A6$. This stage is similar to s_m calculation but here the length of the updating vector, e_{m+1}, in each recursion is fixed and equals N. Therefore it is possible to implement it without using multiplexer as shown in Figure 2. We can use $E0(1)$, $E1(1)$ and $E1(2)$, respectively, as $e_{m-1}(n-1)$, $e_m(n-1)$ and $e_m(n)$ to compute $e_{m+1}(n)$. Because of shifting to left in next clock cycle the same registers can be used again to compute $e_{m+1}(n+1)$. After the last recursion $E1$ contains e_M which is the final Mth order error prediction. This stage lasts N clock cycles and performs N multiplications and $2N$ additions.

s_m calculation and e_m calculation either can be carried out by different adders and multipliers or can be folded to the same adder and multiplier. The goal of this paper is to present low power, low area architecture for LP filter. Therefore, the folded architecture in Figure 3 is proposed. In this architecture, $A3$, $A4$ and $M2$ are used for both s_m calculation and e_m calculation.

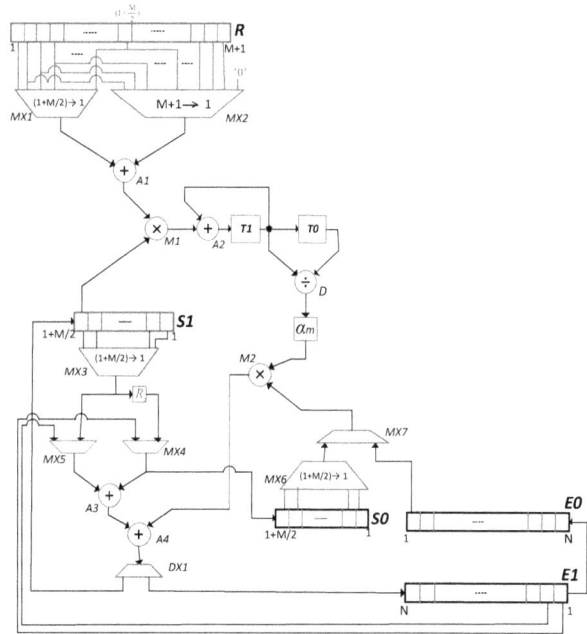

Figure 3: Proposed folded architecture based on split-LDA

The functionality of the circuit can be displayed by the flow chart in Figure 4. After initialization, τ_m calculation starts. For $m \geq 2$, s_m calculation also starts with τ_m calculation. Because s_m calculation and e_m calculation are carried out using the same hardware resource, they can't be done concurrently. Both of these two stages can start with τ_m calculation but it is better to start s_m calculation due to two reasons. First: for e_m calculation only α_m is required and calculated s_{m+1} is used to compute α_{m+1}; So starting e_m calculation immediately after α_m calculation decreases overall latency of the circuit by M clock cycles. Second: both τ_m calculation and s_m calculation have the same latency of

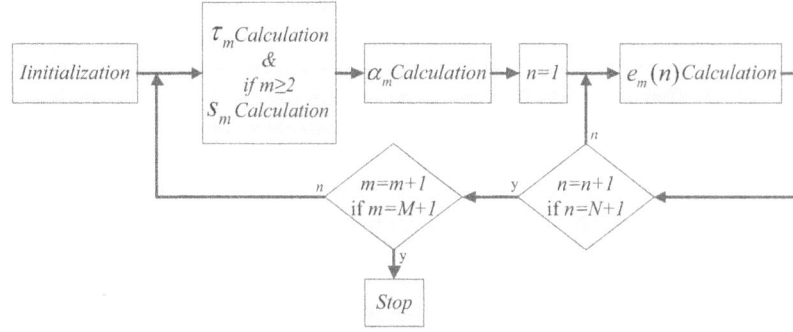

Figure 4: Flowchart for folded architecture based on split-LDA

$\left(\frac{m}{2}+1\right)$ clock cycles while the latency of e_m calculation is N clock cycles. Same latency allows to some simplification in control complexity and improvement in circuit such as removing multiplexer *MX3*.

After calculation of τ_m, it is used by α_m calculation stage to produce α_m. Then e_m calculation uses this α_m to produce $e_m(n)$. If it is not $e_M(n)$, next recursion starts by s_m calculation and τ_m calculation. Unlike the architecture in Figure 2, this circuit does not need any shift register to save $\alpha_1, \alpha_2, \dots, \alpha_M$. Here all the N values of $e_m(n)$ are calculated before starting new α_m calculation.

The total number of required clock cycles (L_t) to calculate $e_M(n)$ for N input samples is the sum of required clock cycles for M recursions:

$$L_t = \underbrace{\sum_{m=1}^{M}(\frac{m}{2}+1)}_{\tau_m \text{ and } s_m} + \underbrace{M}_{\alpha_m} + \underbrace{NM}_{e_m} = \frac{M^2}{4} + M\left(\frac{9}{4}+N\right).$$

3.2 The Schur Algorithm

Figure 5(a) shows a circuit based on lines 1, 2 and 3 of the Schur algorithm, described in Section 2.2 and Figure 5(b) shows implementation of (6).

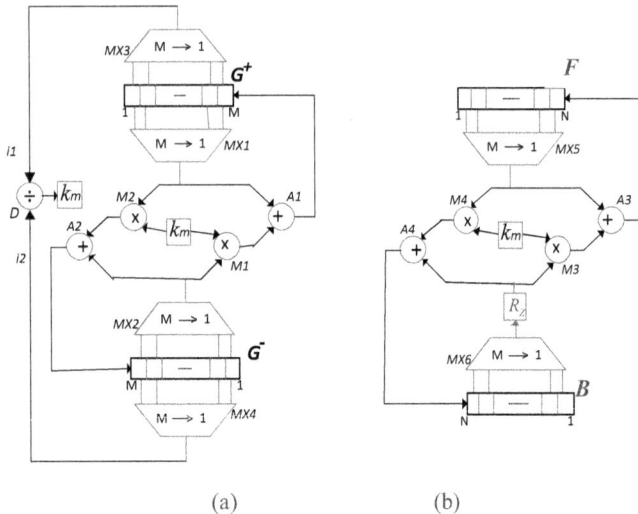

(a)   (b)

Figure 5: Mapping of Schur algorithm to hardware architecture

This architecture has a dedicated buffer for every vector in the Shur algorithm as shown in Table 2.

Table 2. Mapping of vectors in the Schur algorithm to buffers in Figure 5

variable	buffer	size	initial value
g_m^+	G^+	M	$r(1:M)$
g_m^-	G^-	M	$r(0:M-1)$
f_m	F	N	$x(n)$
b_m	B	N	$x(n)$

In mth recursion, for each clock cycle a register in G^+ which contains $g_{m-1}^+(i)$ $i = m, \dots, M$, is selected by *MX1*. At the same time a register in G^- which contains $g_{m-1}^-(i-1)$ $i = m, \dots, M$, is selected by *MX2*. The outputs of *MX1* and *MX2* are used by *M1* and *A1* to compute $g_m^+(i)$ and by *M2* and *A2* to compute $g_m^-(i)$. Table 3 shows values of G^+ and divider inputs after each recursion. Based on line 1 of the algorithm, at the end of mth recursion *MX3* selects the element of G^+ which contains $g_m^+(m+1)$. The output of *MX3* is used as D_{i1} (i_1 input of divider D) to calculate k_{m+1}. Positions of these data in G^+ are declared by black dot in Table 3.

Table 3. G^+ elements at the end of each recursion

m	G^+						D	
	1	2	3	4	...	M	$i1$	$i2$
0	$r(1)$ •	$r(2)$	$r(3)$	$r(4)$	...	$r(M)$	$r(1)$	$r(0)$
1	$g_1^+(1)$	$g_1^+(2)$ •	$g_1^+(3)$	$g_1^+(4)$	...	$g_1^+(M)$	$g_1^+(2)$	$g_1^-(1)$
2		$g_2^+(2)$	$g_2^+(3)$ •	$g_2^+(4)$	...	$g_2^+(M)$	$g_2^+(3)$	$g_2^-(2)$

If the circuit calculates $g_m^+(i)$ and $g_m^-(i)$ in reverse order of i (i.e., $i = M, \dots, m$) then Table 4 shows the content of G^+ at the end of each recursion.

Table 4. G^+ elements at the end of each recursion after rescheduling

m	G^+						D	
	1	2	3	...	$M-1$	M	$i1$	$i2$
0	$r(M)$	$r(M-1)$	...	...	$r(2)$	$r(1)$ •	$r(1)$	$r(0)$
1		$g_1^+(M-1)$	$g_1^+(m-2)$	...	$g_1^+(3)$	$g_1^+(2)$ •	$g_1^+(2)$	$g_1^-(1)$
2			$g_2^+(M-1)$	...	$g_2^+(4)$	$g_2^+(3)$ •	$g_2^+(3)$	$g_2^-(2)$

360

Table 4 also describes data selected for D_{i1}. In new scheduling all of the black dotted values are located in the same column. This means that the datum for D_{i1} is provided only from Mth register of G^+. So $MX3$ is not needed any more. By the same reason, $MX4$ can be deleted from the circuit and the value for D_{i2} is provided only from Mth register of G^-.

Table 5 shows signal values for circuit shown in Figure 5(b) at the end of each recursion. F and B are shift registers of length N (input data number). Equation (6) shows that in calculating $f_m(n)$ and $b_m(n)$, the value of $b_m(N)$ is not used. It is considered to be zero when stored in B because after next recursion it will be used as $b_{m-1}(-1) = 0$ for the following recursion. In Figure 5(b) R_z causes one sample time delay to implement $b_{m-1}(n-1)$ term in (6).

Table 5. k_m, F, B and R_z contents at the end of each recursion

m	k_m	F				B				R_z
		1	2	...	N	1	2	...	N	
0	k_1	$x(1)$	$x(2)$	...	$x(N)$	$x(1)$	$x(2)$	...	$x(N) \to 0$	0
1	k_2	$f_1(1)$	$f_1(2)$	...	$f_1(N)$	$b_1(1)$	$b_1(2)$	...	$b_1(N) \to 0$	0 $(x(N))$
2	k_3	$f_2(1)$	$f_2(2)$	...	$f_2(N)$	$b_2(1)$	$b_2(2)$	...	$b_2(N) \to 0$	0 $(b_1(N))$

By exploiting symmetry in Figures 5(a) and 5(b), it is possible to achieve a folded structure with the minimum number of multipliers and adders. This folded architecture is represented in Figure 6.

To describe how this architecture works, we divide its scheduling to 6 phases.

1. Initialization: A one time initialization sets the initial values in Table 2.

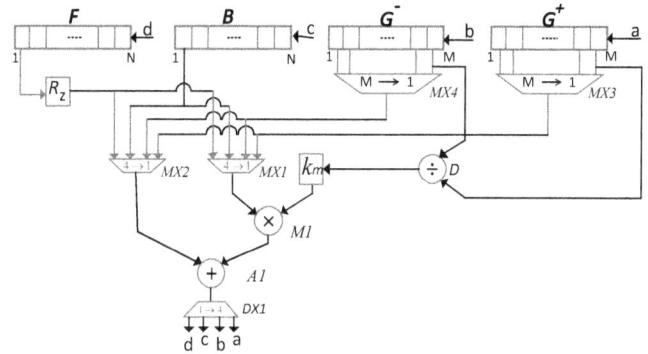

Figure 6: Folded architecture based on the Schur algorithm

2. k_m calculation: In this phase which is done at the beginning of each recursion, the output of D is latched to K_m register. (Line 1 of schur algorithm).

3. g_m^+ calculation: In this case multiplexer $MX1$ chooses the input from G^- to be multiplied by k_m and $MX2$ chooses the input from G^+ to be added to the output of multiplier $M1$. Also the output of adder A1 is routed to G^+ (Line 2 of schur algorithm).

4. g_m^- calculation: In this case multiplexer $MX1$ chooses the input from G^+ to be multiplied by k_m and $MX2$ chooses the input from G^- to be added to the output of the multiplier $M1$. Also the output of adder $A1$ is routed to G^- (Line 3 of schur algorithm).

5. f_m calculation: In this case multiplexers $MX1$ and $MX2$ choose inputs from B and F as inputs of $M1$ and $A1$, respectively. Also the adder's output is shifted to F.

6. b_m calculation: In this case multiplexers $MX1$ and $MX2$ choose inputs from F and B as inputs of $M1$ and $A1$, respectively. Also the adder's output is shifted to B.

The flow chart in Figure 7 shows the required sequence to compute $e_M(n)$ for N input samples. e_m calculation is completed before g_m calculation because it needs only k_m from current recursion. In other words equation (6) is computed between line 1 and 2 of the Schur algorithm.

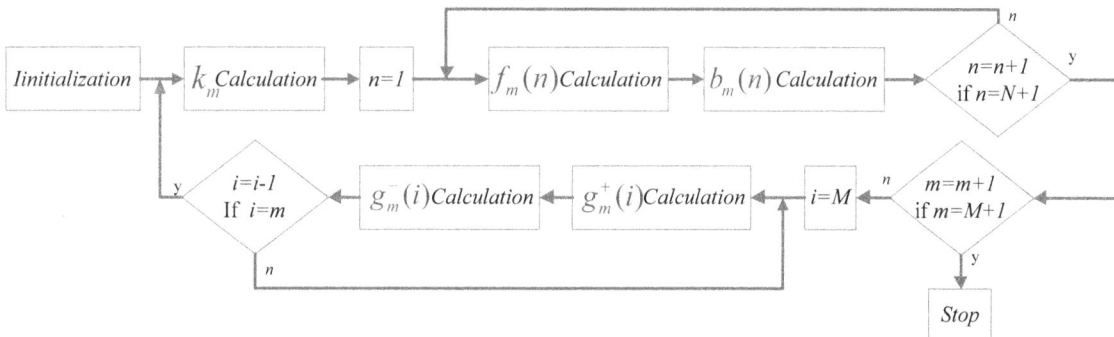

Figure 7: Flowchart for folded architecture based on the Schur algorithm

Table 6. Comparison of proposed architectures with previous architectures

Architecture	Multiplier	Adder	Multiplexer			Demultiplexer			Complexity Order	Latency	Critical Path
			$M \to 1$	$\frac{M}{2} \to 1$	$2 \to 1$	$1 \to 2$	$1 \to 4$	$1 \to M$			
Xu [5]	$3(M-1)+n_f$	$3(M-1)$	2	M *	-	-	-	-	$3\frac{M^2}{2}$	$3M+L_f$	T_m+4T_a
Fazlali [7]	$3+n_f$	5	7	-	1	-	-	1	$3\frac{M^2}{2}$	M^2+2M+L_f	T_m+2T_a
Proposed (Split-LD)	2	2	1 **	3 **	3	1	-	-	$\frac{M^2}{2}$	$\frac{M^2}{4}+M(\frac{9}{4}+N)$	T_m+2T_a
Proposed (Schur)	1	1	2	-	2	-	1	-	M^2	$M^2+M(N+2)$	T_m+T_a

In architecture [5] there are M multiplexers with 2 to M inputs. We consider the average.

** *For split-LDA multiplexers are $(1+M) \to 1$ and $(1+\frac{M}{2}) \to 1$.*

4. COMPARISON

This section compares some architectures for the LDA based error prediction filter including the two proposed architectures. Table 6 summarizes the comparison. [5] and [7] proposed architectures only for calculation of $a_M(i)$. So to calculate $e_M(n)$, they need additional resources and latency for filtering input samples with $a_M(i)$ tap weights. n_f in Table 6 represents additional multipliers and L_f represents additional latency. Other extra resources like adders and multiplexers are not considered. Exploiting existing symmetry in the Schur algorithm, and folding, the corresponding VLSI architecture requires only one multiplier and one adder. Also using shift register and applying some improvement in data flow can decrease the number of required multiplexers. The proposed architecture based on split-LDA has almost the same properties. Even though the number of adders and multipliers is not as small as the Schur architecture, its computation complexity is less than that of other architectures. This translates to saving in energy consumption. Also the latency of split-LDA is less than that of the Schur architecture and the architecture in [7]. Latency is less in split-LDA, because the denominator of the divider, τ_{m-1}, has been computed in previous recursion. There are some VLSI architectures with latency of order M but the number of multipliers in all of them is of order M [5,1]. Regarding the number of adders, multipliers, multiplexers and computation complexity, the proposed architectures have better properties than other architectures. The Schur based architecture has the same capability as [7] to implement dynamic order Linear Prediction filter because the value of minimum mean-square prediction error (P_m) is available in $g_m^-(m)$ and can be used to decide to either stop or continue the recursion. [14] and [15] show in detail that although the reflection coefficient's error due to finite precision implementation of the split-LDA is larger than the respective error in the Schur algorithm, both split-LDA and the Shur algorithm are stable for well-conditioned systems. This means if M is not very large and reflection coefficients are not absolutely close to 1, the residual error is very small.

5. CONCLUSION

Based on some developments on LDA, this paper has presented two folded architectures for implementation of linear prediction-error filter. One of them requires only one multiplier and one adder, and requires less multiplexers compared to known previous architectures. The other one has nearly half computational complexity than previous architectures. These results provide low-area, low-power architectures suitable for ultra-low power applications.

6. REFERENCES

[1] Haykin, S. 2002. *Adaptive Filter Theory,* Prentice Hall.

[2] Makhoul, J. 1975. Linear prediction: a tutorial review. *Proceedings of the IEEE*, 63(4):561–580.

[3] Kung, S.Y. and Hu,Y. H. 1983. A highly concurrent algorithm and pipelined architecture for solving toeplitz systems. *IEEE Trans. Acoust. Speech, Signal Proc.*, vol.-ASSP-31, pp. 66-76.

[4] Nudd, G. R. and Nash, J. G. 1985. *Application of Concurrent VLSI Systems to Two-Dimensional Signal Processing.* VLSI and Modern Signal Processing,Englewood Cliffs, NJ, Prentice-Hall.

[5] Xu, F. J., Ariyaeeinia A. and Sotudeh, R. 2005. "Migrate Levinson-Durbin Based Linear Prediction Coding Algorithm Into FPGAs," *Proc. Int. Conf. on ICECS*, pp. 1-4.

[6] Hwang, Y. T. and Han, J. C. 1999. A novel FPGA design of a high throughput rate adaptive prediction error filter. *AP-ASIC '99. IEEE Asia Pacific Conference on*, 202 – 205.

[7] Fazlali, B. and Eshghi, M. 2011. A pipeline design for implementation of LPC feature extraction system based on Levinson-Durbin algorithm. *(ICEE), 19th Iranian Conference on Electrical Engineering*, Page(s): 1 – 5.

[8] Nagarajan, S. and Sankar, R.1998. Efficient implementation of linear predictive coding algorithms. *Southeastcon '98. Proceedings. IEEE* , 69 – 72.

[9] Delsarte P. and Genin,Y. 1986. The split levinson algorithm. *IEEE Trans. Acoust. Speech, Signal Processing.*, vol.-ASSP-34, pp. 470-478.

[10] Bistritz, Y., Lev-Ari, H. and Kailath, T. 1989. Immittance-domain levinson algorithms. *Information Theory, IEEE Transactions on*, Vol 35 Issue 3. 675 – 682

[11] Parhi, K. K. 1999. *VLSI Digital Signal Processing Systems: Design and Implementation.* Hoboken, NJ: Wiley,

[12] Parhi, K. K., Wang, C. Y. and Brown, A. P. 1992. Synthesis of control circuits in folded pipelined DSP architectures. *IEEE Journal of Solid-State Circuits,* vol.27,no. 1, 29-43

[13] Orfanidis,S. J. 2007. *Optimum Signal Processing .* Macmillan ,2nd ed.

[14] Glaros, N. and Carayannis, G. 1994. Exact and first-order error analysis of the Shur and split Shur algorithms: theory and practice. *Signal Processing, IEEE Tran. on*, 1919 –1938

[15] Krishna, H., Yunbiao, W. 1993.The split levinson algorithm is weakly stable. *SIAM Journal of Numer. Anal.*, 1498-150

Synergistic Integration of Code Encryption and Compression in Embedded Systems

Kamran Rahmani, Hadi Hajimiri, Kartik Shrivastava, Prabhat Mishra
Department of Computer & Information Science & Engineering
University of Florida, Gainesville, Florida, US
{kamran, hadi, kshrivas, prabhat}@cise.ufl.com

ABSTRACT

Code encryption is a promising approach that encrypts the application binary to protect it from reverse engineering and tampering, and decrypts the instructions during runtime. A major challenge is to trade-off between the security level and runtime decryption overhead. In this paper, we explore a synergistic combination of various code compression algorithms with code encryption techniques to reduce this overhead. Since decryption overhead (time) is linearly dependent on code size, it is promising to employ compression to reduce code size, and thereby achieve the advantages of both compression and encryption. Experimental results demonstrate that our proposed scheme can employ efficient encryption techniques while significantly improve the performance up to 2.3X (1.5X on average) and reduce energy consumption up to 57% (26% on average), compared to using encryption alone.

Categories and Subject Descriptors

C.3 [**SPECIAL-PURPOSE AND APPLICATION-BASED SYSTEMS**]: Real-time and embedded systems

General Terms

Design, Performance, Security

Keywords

Embedded systems, code compression, code encryption, energy optimization, performance optimization

1. INTRODUCTION

Embedded systems are used everywhere - starting from everyday appliances to complex safety-critical systems. In many scenarios, it is becoming exceedingly essential to keep these devices authentic and confidential. Encryption is widely used as a reliable way of protecting critical data for storage and transmission. For example, it is useful to encrypt

network messages while sending them out through communication media and to decrypt them at the receiving end. Likewise, ciphering files while storing them on the hard disk protects them from being read in case the hardware itself is compromised. The process of encrypting binary code is different from that of other static data. Encrypting static data is mainly concerned with the complexity of the ciphering algorithm and the mode of operation. Binaries themselves can be encrypted as static data in the secondary storage and then decrypted while loading them in the main memory. For security reasons, it may be required to keep encrypted binary code in the main memory. This is required specially when the bus between main memory and secondary storage is not secure. In this situation, repeated fetching and decryption of blocks of code is going to produce an immense overhead which would render the execution extremely slow and infeasible in many scenarios.

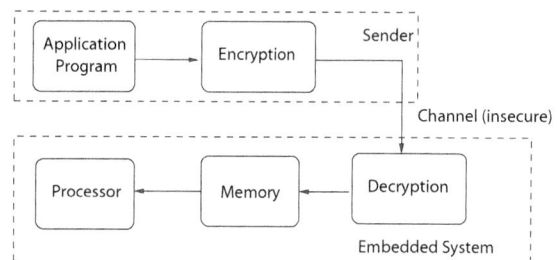

Figure 1: An Overview of Encryption Framework

Figure 1 shows the framework in which encryption and decryption are performed on application binaries in an embedded system. There are two stages of processing. During the offline stage, the binary code of the embedded system is first encrypted. Next, when the program is loaded, the code passes through the insecure channel between sender and memory and it is decrypted inside the embedded system. The encrypted code is stored in the primary memory and accessed by the processor. The decryption is done online (for each fetch). The time it takes to decrypt the code affects the performance of the system. Reducing the code size with code compression can reduce this decryption overhead significantly.

Figure 2 shows how code compression and associated decompression are performed in embedded systems. There are two stages similar to Figure 1. The first stage is offline, in which the code is compressed. The code is decompressed between the main memory and the processor to increase the effective memory size as well as to improve the performance.

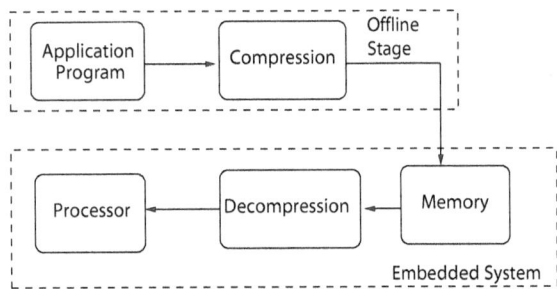

Figure 2: An Overview of Compression Framework

The decompression is done online (during runtime). Hence, it is critical to have a fast decompression hardware. Our scheme integrates code encryption with code compression, attempting to make code execution secure and efficient at the same time.

The rest of the paper is organized as follows. Section 2 provides an overview of existing compression and encryption techniques. Section 3 presents our approach of combining encryption and compression. Section 4 presents our experimental results. Finally Section 5 concludes the paper.

2. RELATED WORK

Code compression techniques were first developed for embedded systems by Wolfe and Channin [1]. Nam et al. [17] used dictionary based compression to compress VLIW instructions. Larin and Conte used Huffman based compression on embedded systems [16]. Tunstall coding was used by Xie et al. [19] to perform variable to fixed length compression. Usage of variable sized block was further exploited by Lin et al. [4], when they proposed LZW-based code compression of embedded processors. Seong et al. [18] proposed a bitmask-based compression (BMC) that remembers mismatches using bitmasks. Hajimiri et al. [6, 7] used code compression with cache reconfiguration. In this paper, we explore Huffman coding, dictionary-based compression and BMC with various encryption techniques.

Private key cryptography has been in use since the early 20th century in which both parties operating on the data had the same key to encrypt and decrypt. This type of shared key cryptography is of two types: block and stream. A block cipher operates on a block of data while a stream cipher works by combining the data with a stream of pseudorandom bits. Example of block ciphers include AES and DES. RC4 is an example of stream cipher. However, the problem of sharing the private key forced people to change to public key cryptography. RSA is an example of public key cryptography. In this paper, we explore AES, DES and RC4 with various compression techniques.

There are few efforts to combine both encryption and compression together. Johnson et al. [10] proposed a method to compress encrypted data using Low Density Parity Check codes (LDPC) and they have shown their performance on OTP encrypted data. However, their method is not suitable in embedded systems, since LDPC compression is NP hard. Also, they have used their algorithm only on OTP encrypted data, which is not considered a good encryption scheme. Ruan et al. [15] improved the Shannon-Fano-Elias technique of encrypting compressed data by improving the code length. However, the intensive decryption/decompression of these codes are not applicable in embedded systems.

Shaw et al. [5], developed a method to combine compression and encryption. The compression schemes used by them, which comprises of codebooks is lossy in nature. This may be suitable for data, but certainly not applicable for code, since it will lead to incorrect functionality. Cypress, developed by Lekatsas et al. [11] has integrated compression and encryption. They deal with both code and data sequences for multimedia embedded systems.

3. CODE ENCRYPTION & COMPRESSION

Runtime decryption of encrypted code significantly increases instruction fetch delay. The main challenge is how to combine encryption with compression to keep performance as high as possible while maintaining the security of the system. As the decryption time is proportional to the size of code, combining encryption with compression is promising to improve the overall performance. However, combining both encryption and compression may lead to a number of problems. The main problem is that both decryption and decompression are slow and hence may prevent the full utilization of the processor performance. In order to get the best possible processor utilization, the decompression unit should be such that the rate at which instructions are fetched is equal to the rate at which the instructions are decompressed. This section describes the challenges associated with integration of encryption and compression and presents mechanisms to address these challenges.

3.1 Encryption followed by Compression

There are two ways of combining encryption with compression. The first scenario is shown in Figure 3. In this combination, the code encryption is followed by compression. The problem is that most compression algorithms take advantage of the repeating patterns in the uncompressed data set. Encrypted data generally has high entropy and therefore, has less similarity in patterns. As a result, as our experimental results show, it is difficult to compress those data.

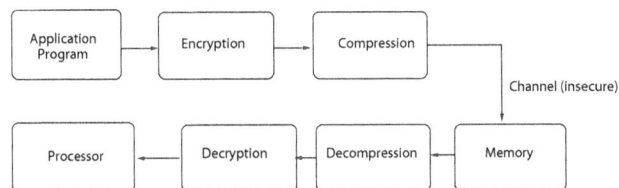

Figure 3: Encryption followed by Compression

3.2 Compression followed by Encryption

This is the most useful way of combining encryption and compression. In this combination the code compression is followed by encryption. It is beneficial to compress the unencrypted code by exploiting the regular pattern. Moreover, this compressed data can be easily encrypted and sent across the insecure channel to the receiving end. The decryption and the decompression units can do the rest of the work. This scenario is shown in Figure 4.

3.3 Placement of Cache

This section describes the challenges and opportunities associated with cache placement.

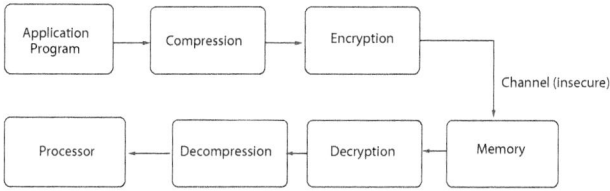

Figure 4: Compression followed by Encryption

3.3.1 PCDD Architecture

Figure 5 shows a configuration where both the decryptor and decompressor are put together between the cache and the main memory. Here the job of the decryptor and decompressor is to both decrypt and decompress a block of code from the memory and provide the cache with a block of regular code. We refer this architecture as Processor-Cache-Decompressor-Decryptor (**PCDD**) architecture.

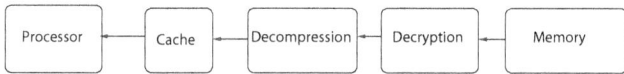

Figure 5: Processor-Cache-Decompressor-Decryptor (PCDD)

In this scheme the granularity of decompression is an instruction block of the cache. Larger block size would not necessarily mean a reduction in the total number of fetches from the cache. That would depend on the actual binary and the size of the basic blocks in the code. Now we would like to analyze the effect of encryption and compression on the overall system performance. Assuming a uniform compression ratio throughout the program code, the following equations present a basic mathematical model for execution of encrypted and compressed code. In PCDD architecture let,

C=Compression ratio[1] of the code
M_b=Cycles taken to fetch a cache block from memory
E_b=Cycles taken to decrypt an encrypted cache block
R_b=Cycles taken to decompress a compressed cache block
N_b=Number of blocks the cache fetches at runtime
T_{n1}= Total cycles to fetch blocks from memory
T_{e1}= Total cycles to fetch and decrypt the code
T_{ec1}= Total cycles to fetch, decrypt and decompress the code
Then,

$$T_{n1} = N_b.M_b \qquad (1)$$

$$T_{e1} = N_b.(M_b + E_b) \qquad (2)$$

$$T_{ec1} = C.N_b.(M_b + E_b + R_b) \qquad (3)$$

Note that T_{n1} equals to the total number of cycles that it will take to fetch instructions from the memory if the code is neither compressed nor encrypted. Similarly, T_{e1} equals to the total number of cycles that it will take if the code is only encrypted. Likewise, T_{ec1} gives the total number of cycles it will take if the code is both encrypted and compressed.

Now we would like to investigate how decompression and decryption affect the performance of the system. Equation

---

[1]Compression Ratio = Compressed Code Size / Original Code Size

4 gives the ratio of cycles of encrypted and compressed code over regular code (S_{N1}) and Equation 5 gives the ratio for encrypted and compressed code over encrypted only code (S_{E1}).

$$S_{N1} = \frac{T_{ec1}}{T_{n1}} = C\left(1 + \frac{E_b + R_b}{M_b}\right) \qquad (4)$$

$$S_{E1} = \frac{T_{ec1}}{T_{e1}} = C\left(1 + \frac{R_b}{M_b + E_b}\right) \qquad (5)$$

The goal is to make S_{N1} and S_{E1} as low as possible. The obvious way to do so is to have a lower compression ratio C and a low decompression latency R_b. We need to make a trade-off between these two factors. For example, Huffman coding gives a great compression but its decompression is slow. On the other hand, simple dictionary based compression gives low/moderate compression with faster decompression. Bitmask-based compression [18] provides a trade-off between these two aspects by providing good compression ratio with fast decompression.

3.3.2 PDCD Architecture

We can gain more benefit of code compression by storing compressed code in the cache. This can be done by putting the decompression unit between processor and cache. Figure 6 shows this architecture. In this scheme the encrypted code is fetched as blocks from the memory by the decryption unit, which are then decrypted and sent back to the cache. We refer this architecture as Processor-Decompressor-Cache-Decryptor (**PDCD**) architecture.

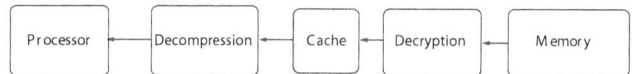

Figure 6: Processor-Decompressor-Cache-Decryptor (PDCD)

The advantage of PDCD over PCDD is that the compressed code is kept in the cache. Hence, the effective cache size and cache hits increase. However as the processor fetches instructions directly from the decompression unit, there is decompression latency for each instruction fetch. This approach requires fast decompression hardware that can decompress one instruction per cycle. The overhead of such fast decompression units can be minimized by pipelining the decompression in the system. We model this architecture as follows (remaining parameters are same as before).

R=Cycles taken to decompress a word of compressed text
N= Number of instruction word fetches from the processor at run time
T_{n2}= Total cycles to fetch a block from memory to the cache when executing regular code
T_{e2}= Total cycles to fetch and decrypt the code
T_{ec2}= Total cycles to fetch, decrypt and decompress the code

$$T_{n2} = N_b.M_b \qquad (6)$$

$$T_{e2} = N_b.(M_b + E_b) \qquad (7)$$

$$T_{ec2} = C.(N_b.(M_b + E_b) + N.R) \qquad (8)$$

$$S_{N2} = \frac{T_{ec2}}{T_{n2}} = C\left(1 + \frac{N_b.E_b + N.R}{N_b.M_b}\right) \qquad (9)$$

$$S_{E2} = \frac{T_{ec2}}{T_{e2}} = C\left(1 + \frac{N.R}{N_b(M_b + E_b)}\right) \qquad (10)$$

Like PCDD, the aim is to minimize S_{N2} and S_{E2}. Choosing compression algorithm should be a trade-off between compression ratio and decompression unit speed. On the other hand, encryption algorithm should be chosen such that S_{N2} is least. More secure algorithms need more cycles to decrypt. Hence, choice of an encryption algorithm leads to a trade-off between needed security and expected speed that depends on embedded system requirements. For example, AES will have larger decryption latency than DES. Hence S_{E2} would be larger for DES and S_{N2} would be larger for AES, i.e., execution of the encrypted and compressed code will be slower for AES compared to DES. *Interestingly, the effect of compression would be more significant for AES as compression will hide more latency.*

4. EXPERIMENTS

4.1 Experimental Setup

In order to explore different combination of code encryption and compression tradeoffs, we examined *cjpeg, djpeg, epic, adpcm* (*rawcaudio* and *rawdaudio*), *g.721* (*encode, decode*) benchmarks from the Mediabench [3] and *dijkstra, patricia, crc32* from Mibench [12] compiled for the Alpha target architecture. All applications were executed with the default input sets provided with the benchmarks suites. Since the space is limited, we present the result for five of these benchmarks. However, the result is consistent for the remaining benchmarks.

Code encryption and compression are performed offline. In order to extract the code (instruction) part from executable binaries we used ECOFF[2] header files provided in SimpleScalar toolset [9]. The text segment is extracted from the binary and compression is performed on it, giving compressed text segment as a result. Since the decompression unit must be able to start execution from any of the jump targets, branch targets should be aligned in the compressed code. In addition, the mapping of old addresses (in the original uncompressed code) to new addresses (in the compressed code) is kept in a jump table. This compressed text is then encrypted and a new binary file is created using the compressed-encrypted text, the dictionary, the jump-mapping table and the rest of the segments from the original file.

Three different code compression techniques including bitmask-based, dictionary-based and Huffman code compression were used. To attain the best achievable compression ratios, in compression algorithms, for each application we examined dictionaries of 1 KB, 2KB, 4KB, and 8 KB. In addition, for bitmask-based compression similar to Seong et al. [18] we tried three mask sets including one 2-bit sliding, 1-bit sliding and 2-bit fixed, and 1-bit sliding and 2-bit fixed masks. We found that dictionary size of 2KB is the best choice for this set of benchmarks. Also, we examined compression word sizes of 8 bits, 16 bits, and 32 bits. We found

out that 16 bits word size is the best choice for dictionary-based and Huffman compression algorithms. We used AES (128 bits block), DES (64 bits block), and RC4 encryption algorithms to examine the effect of compression on different classes of encryption methods (from strong-slow algorithm to weak-fast one).

To obtain cache hit and miss statistics, we modified the SimpleScalar toolset to be able to decrypt, decompress, and simulate encrypted-compressed applications based on PDCD architecture. Decompression unit can decompress the next instruction by one cycle (in pipelined mode) if it finds the entire needed bits in its buffer. Otherwise, it takes one cycle (or more cycles, if cache miss occurs) to fetch the needed bits into its buffer and one more cycle to decompress the next instruction. In decryptor, we used 18 [2], 11 [13], and 7 [14] cycles per byte latencies for AES, DES, and RC4 algorithms, respectively. Correctness of the compression and encryption algorithms was verified by comparing the outputs of encrypted-compressed applications with regular versions.

We applied the same energy model used in [20], which calculates both dynamic and static energy consumption, memory latency, CPU stall energy, and main memory fetch energy. The energy model includes decompression and decryption overhead energy. We used a single 1KB instruction direct cache with a line size of 16 bytes for all simulations. We refer it as *base cache*. We updated the dynamic energy consumption for this cache configuration using CACTI 4.2 [8].

4.2 Performance Improvement

Figure 7 shows the performance of applications in different combination of AES and compression algorithms normalized to the AES encryption only method. It confirms that code compression can improve performance in many scenarios while used with AES encryption algorithm. As we can see, performance improvement varies significantly and depends on the application binary. For instance, in the case of application *g721_enc*, applying compression would result in 1.1X, 1.2X, and 1.4X performance improvements for dictionary-based, Huffman, and bitmask-based algorithms, respectively. This improvement is up to 2.3X for bitmask-based algorithm in *cjpeg* application. On the other hand, in *rawdaudio* application we do not see any noticeable improvement. The improvement is significant if application code size and its behavior is such that it needs larger cache size than base cache (like *cjpeg*), and is negligible if application fits in the base cache effectively (like *rawdaudio*). The average performance improvements are 1.2X, 1.2X, and 1.5X for dictionary-based, Huffman, and bitmask-based algorithms, respectively. For ease of comparison, original numbers are shown in Table 1.

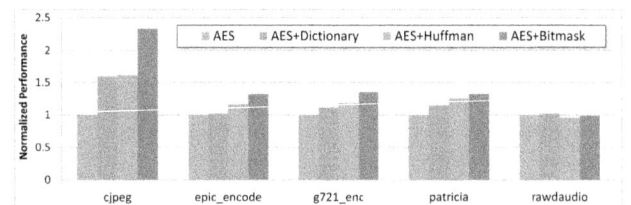

Figure 7: Performance of different compression algorithms with AES encryption (normalized to AES without compression)

---

[2]Extended Common Object File Format

Table 1: Instruction Per Cycle (IPC) for combination of AES and different compression algortihms

Benchmark	AES	+Dic	+Huffman	+BMC
cjpeg	0.063	0.100	0.101	0.146
epic_encode	0.262	0.268	0.304	0.346
g721_enc	0.029	0.033	0.034	0.040
patricia	0.015	0.018	0.019	0.020
rawdaudio	1.352	1.387	1.300	1.342

Table 2: Energy consumption (nanojoule) for combination of AES and different compression algortihms

Benchmark	AES	+Dic	+Huffman	+BMC
cjpeg	51468	32350	31908	22005
epic_encode	34443	33797	29793	25987
g721_enc	3710426	3327708	3142287	2756331
patricia	717480	625254	571496	542503
rawdaudio	796	779	824	800

Figure 8 illustrates the performance of applications for different combinations of DES and compression algorithms normalized to the DES encryption only method. As behavior of applications remains same for different encryptions, we see improvement pattern similar to AES case. For instance in the case of application *g721_enc*, similar to AES, applying compression would result in 1.1X, 1.2X, and 1.4X performance improvements for dictionary-based, Huffman, and bitmask-based algorithms, respectively. This improvement is up to 2.2X in *cjpeg* application. The average performance improvements are 1.2X, 1.2X, and 1.4X for dictionary-based, Huffman, and bitmask-based algorithms, respectively.

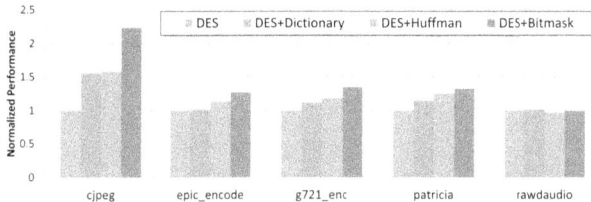

Figure 8: Performance of different compression algorithms with DES encryption (normalized to DES without compression)

Performance improvements for different combinations of RC4 and compression algorithms is shown in Figure 9. As we discussed earlier, performance improvement is less in the case of faster decryption unit. We see this happens in RC4 that is faster and less secure than AES. For instance, in *cjpeg* we have 2.1X improvement in performance for RC4 with bitmask-based compression that is 2.3X for corresponding AES case. As cache misses are same in both cases, the improvement is larger for longer decryption latency. The average performance improvements are 1.2X, 1.2X, and 1.4X for dictionary-based, Huffman, and bitmask-based algorithms, respectively.

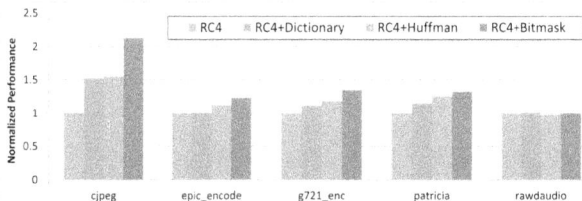

Figure 9: Performance of different compression algorithms with RC4 encryption (normalized to RC4 without compression)

On average we get most improvement in performance when we use bitmask-based compression with encryption algo-

rithms. For this set of benchmarks, application code size is reduced by 15%-25%, 30%-35%, and 30%-45% for dictionary-based, Huffman, and bitmask-based compression, respectively. Bitmask-based compression is the best choice in terms of compression for this set of applications. The reason is that because of large similarity in instructions (that lets us use masks) we can use large 32 bits words and reduce the code size even more than Huffman algorithm (with restricted dictionary size).

The decompression hardware for dictionary-based compression is simple but average improvement is small. Bitmask-based compression is the best choice to be combined with all the three encryption algorithms in terms of performance improvement. This can result in up to 2.3X (1.5X on average) improvement in performance. This improvement can satisfy real-time requirements in many embedded applications while keeping them safe by using encryption methods.

4.3 Energy Savings

Energy consumption in instruction cache subsystem for different combinations of AES and compression algorithms is shown in Figure 10. As compression reduces the miss ratio in cache, it reduces the power consumption of the system. For instance in the case of *patricia* application, we have reduction of energy by 13%, 20%, and 24% for AES combined with dictionary-based, Huffman and bit-mask based compression, respectively. Energy saving can be even more significant when cache size is a bottleneck in the application. For example, we can save up to 57% of the total energy in the *cjpeg* application by combining bitmask-based compression with AES encryption. The average energy savings are 13%, 17%, and 26% for dictionary-based, Huffman, and bitmask-based algorithms, respectively. For ease of comparison, original numbers are shown in Table 2.

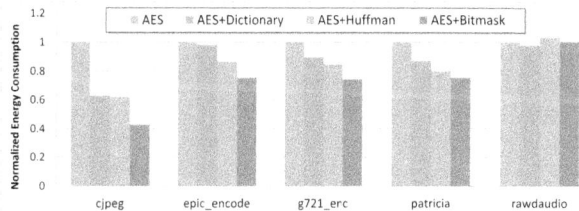

Figure 10: Energy consumption of different compression algorithms with AES encryption (normalized to AES without compression)

We have similar savings in other algorithms. Figure 11 illustrates energy consumption for different combinations of DES and compression algorithms. The average energy savings are 11%, 16%, and 26% for dictionary-based, Huffman, and bitmask-based algorithms, respectively.

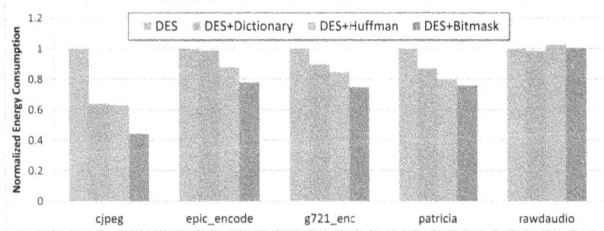

Figure 11: Energy consumption of different compression algorithms with DES encryption (normalized to DES without compression)

Energy consumption for different combinations of RC4 and compression algorithms is shown in Figure 12. Like performance improvement, because of less latency in RC4 compared to AES, we have less energy saving in RC4 compared to AES. For instance in *epic_encode* we have 20% energy saving for RC4 with bitmask-based compression whereas, it was 25% for corresponding AES case. On average, energy savings are 12%, 16%, and 25% for dictionary-based, Huffman, and bitmask-based algorithms, respectively.

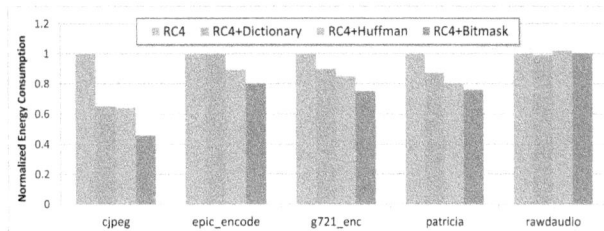

Figure 12: Energy consumption of different compression algorithms with RC4 encryption (normalized to RC4 without compression)

In summary, like performance improvement, integration of compression with encryption would be useful in terms of energy consumption when the application needs larger instruction cache. As simulation results show, bitmask-based compression is the best choice in terms of both performance improvement and energy saving. On average, by using this algorithm with AES encryption we can save 26% of total energy and improve the performance by 47%.

5. CONCLUSIONS

Encryption and compression are important for embedded systems. While the former provides code security and prevent tampering by third party, the latter is used to minimize the code size and thus reduce power and memory requirements as well as improve the overall performance. In this paper, we have demonstrated that it is useful to first compress the code and then encrypt it, employing a Processor-Decompressor-Cache-Decryptor architecture. Since code size is reduced due to compression, the decryptor has to operate on less amount of code, which makes it faster. Our experimental results demonstrated up to 2.3X (1.5X on average) improvement in performance and up to 57% (26% on average) energy saving by combining compression and encryption compared to employing encryption alone. This improvement can enable use of encryption in embedded systems.

6. ACKNOWLEDGMENTS

This work was partially supported by NSF grant CNS-0915376. We would like to thank Kanad Basu and Yogesh Sharma for their comments and suggestions.

7. REFERENCES

[1] A. Wolfe et al. Executing compressed programs on an embedded RISC architecture. In *Proc. of MICRO*, 1992.

[2] B. Schneier et al. Performance comparison of the aes submissions. In *Proc. of AES Candidate Conference*, 1999.

[3] C. Lee et al. Mediabench: A tool for evaluating and synthesizing multimedia and communication systems. In *Proc. of MICRO*, 1997.

[4] C. Lin et al. LZW-based code compression for VLIW embedded systems. In *Proc. of DATE*, 2004.

[5] C. Shaw et al. A pipeline architecture for encompression (encryption + compression) technology. In *Proc. of VLSI Design*, 2003.

[6] H. Hajimiri et al. Synergistic integration of dynamic cache reconfiguration and code compression in embedded systems. In *Proc. of IGCC*, 2011.

[7] H. Hajimiri et al. Compression-aware dynamic cache reconfiguration for embedded systems. *SUSCOM*, 2012.

[8] http://www.hpl.hp.com. *CACTI. HP Labs, CACTI 4.2.*

[9] http://www.simplescalar.com. *Simplescalar.*

[10] M. Johnson. On compressing encrypted data. *IEEE Transactions on Signal Processing*, 2004.

[11] H. Lekatsas et al. Cypress: compression and encryption of data and code for embedded multimedia systems. *IEEE Design & Test of Computers*, 2004.

[12] M.R. Guthaus et al. Brown. Mibench: A free, commercially representative embedded benchmark suite. In *Proc. of WWC*, 2001.

[13] A. Nadeem and M. Javed. A performance comparison of data encryption algorithms. In *Proc. of ICICT*, 2005.

[14] P. Prasithsangaree and P. Krishnamurthy. Analysis of energy consumption of rc4 and aes algorithms in wireless lans. In *IEEE GLOBECOM* , 2003.

[15] X. Ruan. Using improved shannon-fano-elias codes for data encryption. *IEEE International Symposium on Information Theory*, 2006.

[16] S. Larin and T. Conte. Compiler-driven cached code compression schemes for embedded ilp processors. In *Proc. of MICRO*, 1999.

[17] S. Nam et al. Improving dictionary-based code compression in VLIW architectures. *IEICE Trans. on Fundamentals*, 1999.

[18] S. Seong and P. Mishra. Bitmask-based code compression for embedded systems. *IEEE TCAD*, 2008.

[19] Y. Xie et al. Code compression for VLIW processors using variable-to-fixed coding. In *Proc. of ISSS*, 2002.

[20] C. Zhang et al. A highly configurable cache architecture for embedded systems. In *Proc. of Computer Architecture*, 2003.

Author Index

Abella, Jaume 245

Acquaviva, Andrea 71

Ahmed, Mohsin Yusuf 259

Ait Mohamed, Otmane 271

Al-bayati, Zaid 271

Allam, Osman 55

Alpert, Charles 327

Amirfattahi, Rasoul 357

Andrus, Curtis 147

Arafeh, Abdalrahman 21

Atienza, David 15

Ayinala, Manohar 63

Baek, Seungcheol 345

Bahar, R. Iris 39

Belzer, Benjamin 165

Ben Atitallah, Rabie 239

Berezowski, Krzysztof 67

Bhadviya, Bhavitavya 295

Bhunia, Swarup 287

Biernat, Janusz 67

Bombieri, Nicola 71

Bousquet, Laurent 87

Brenner, David 177

Briki, Aroua 153

Brisk, Philip 103

Burleson, Wayne 159

Calimera, Andrea 279

Chang, Kevin 165

Chavet, Cyrille 153

Chen, Huan 141

Chen, Jifeng 45

Chen, Pengpeng 123

Chen, Yun 291

Chen, Zhi-Wei 75, 275

Chung, Jun-Min 275

Corbetta, Simone 33

Corre, Youenn 283

Cosic, Miralem 165

Coussy, Philippe 153

Cremona, Fabio 39

Davis, Al 1

de Paula, Flavio M. 189

Deb, Sujay 165

Dekeyser, Jean-Luc 239

Delgado-Frias, José G. 267

Diguet, Jean-Philippe 283

Donato, Marco 39

Dong, Sheqin 171

Eeckhout, Lieven 55

Eyerman, Stijn 55

Firouzi, Farshad 201

Fornaciari, William 33

Friedman, Eby G. 129

Friedman, Joseph S. 209

Fu, Yuzhuo 99

Fummi, Franco 71

Ganguly, Amlan 165, 259

Garitselov, Oleg 255, 351

Goto, Satoshi 171

Grissom, Daniel 103

Gruber, Dominik 195

Guerra e Silva, Luis 135

Guthaus, Matthew R. 147

Hajimiri, Hadi 363

Han, Jie 221

Hasan, Syed Rafay 271

Hasan Babu, Hafiz Md. 215

Heittmann, Arne 227

Heller, Dominique 283

Heo, Deukhyoun 165

Hu, Alan J. 189

Huang, Libo 59

Huang, Ming-Chien 75

Inoue, Keisuke 79

Ismail, Yehea I. 209

Ivanov, André 189

Jabeur, Kotb 3

Jahanian, Ali 303

Jakushokas, Renatas 129

Jamal, Lafifa 215

Jeon, Heungjun 83

Jia, Gangyong 291

Jiao, Jiajia 99

Jin, Warren 39

Joshi, Ajay 9

Jung, In-Seok 251

Junsangsri, Pilin 311

Kaneko, Mineo 79, 215

Khatri, Sunil P. 295

Kiamehr, Saman 201

Kim, Jongman 345

Kim, Kyundong 91

Kim, Yong-Bin 83, 111, 251

Klingner, John 263

Kougianos, Elias 255, 351

Kudithipudi, Dhireesha 177

Lagadec, Loïc 283

Le Beux, Sébastien 3

Lee, Hyung Gyu 345

Li, Xi 291

Liu, Wei 279

Lombardi, Fabrizio 221, 311

Luo, Rong 123

Lysecky, Roman 27

Macii, Enrico 279

Mandal, Ayan 295

Maric, Bojan 245

Marques-Silva, Joao 141

Martin, Eric 153

Mathew, Jimson 333

Maurer, Peter M. 315

Mazumdar, Kaushik 51

Merkel, Cory 177

Miryala, Sandeep 279

Mishra, Prabhat 287, 363

Mitra, Sajib Kumar 215

Miwa, Shinobu 91, 233

Mohanty, Saraju P. 255, 333, 351

Morshedzadeh, Marzieh 303

Mundy, Joseph 39

Murray, Jacob 263

Nakamura, Hiroshi 91, 233

Narayanan, Vijaykrishnan 299

Niar, Smail 239

Nicopoulos, Chrysostomos 345

Nileshwar, Varadaraj Kamath 27

Noll, Tobias G. 227

O'Connor, Ian 3

Ostermann, Timm 195

Palaniswamy, Ashok Kumar 307

Pande, Partha P. 165, 263

Papadopoulos, Agathoklis 339

Parhi, Keshab 63, 357

Patronik, Piotr 67

Patterson, William 39

Piestrak, Stanislaw J. 67

Poncino, Massimo 279

Pradhan, Dhiraj K. 333

Promponas, Vasilis J. 339

Qian, Fuqiang 321

Rahmani, Kamran 287, 363

Rangaraju, Nikhil 209

Rethinagiri, Santhosh Kumar 239

Ruggiero, Martino 15

Sait, Sadiq M. 21

Salama, Khaled Nabil 207

Salehi, Sayed Ahmad 357

Savaria, Yvon 271

Sengupta, Dipanjan 189

Senn, Eric 239

Shafik, Rishad A. 333

Shirazi, Behrooz 263

Shrivastava, Aviral 67

Shrivastava, Kartik 363

Simeu, Emmanuel 87

Sridhar, Arvind 15

Stan, Mircea R. 51

Sze, Cliff 327

Tahoori, Mehdi B. 107, 201

Takeda, Seidai 91, 233

Tang, Wai-Chung 327

Taskin, Baris 117

Tehranipoor, Mohammad 45,
                                    95, 183

Tenace, Valerio 279

Teng, Ying 117

Tessier, Russell 159

Theocharides, Theocharis 339

Tian, Haitong 321

Tragoudas, Spyros 307

Turi, Michael A. 267

Tuzzio, Nicholas 95

Usami, Kimiyoshi 233

Valero, Mateo 245

Veneris, Andreas 189

Vidapalapati, Anuroop 259

Vincenzi, Alessandro 15

Vinco, Sara 71

Wang, Kan 171

Wang, Shuo 45, 183

Wang, Zhiying 59

Wei, Wei 221

Wei, Xing 327

Wessels, Bruce W. 209

Wu, Yu-Liang 327

Xiao, Kan 95

Xiao, Nong 59

Xie, Jing 299

Xie, Yuan 299

Yakymets, Nataliya 3

Yan, Jin-Tai 75, 275

Yang, Huazhong 123

Yang, Jing 111

Yeolekar, Pranav 333

Young, Evangeline 321

Zamani, Masoud 107

Zangeneh, Mahmoud 9

Zaslavsky, Alexander 39

Zhang, Xuehui 95

Zhang, Zhe 267

Zhao, Bo 123

Zhao, Jia 159

Zheng, Geng 255, 351

Zhou, Xuehai 291

Zhu, Zongwei 291